W9-DAV-875

Machine Elements in Mechanical Design

Robert L. Mott, P.E.
University of Dayton

Charles E. Merrill Publishing Company
A Bell & Howell Company
Columbus Toronto London Sydney

To my wife, Marge
our children, Lynné, Robert, Jr., and Stephen
and my Mother and Father

Published by
Charles E. Merrill Publishing Company
A Bell & Howell Company
Columbus, Ohio 43216

This book was set in Times Roman and Optima.
Production Coordination: Martha Morss
Cover Designer: Cathy Watterson
Cover Photo: Ken Whitmore, Joan Kramer and Associates

Library of Congress Card Number: 84-061746
International Standard Book Number: 0-675-20326-0

Printed in the United States of America

2 3 4 5 6 7 8—88 87 86

Contents

17 ELECTRIC MOTORS

18 POWER SCREWS, BALL SCREWS, AND FASTENERS

19 BOLTED CONNECTIONS, WELDED JOINTS, AND MACHINE FRAMES

20 DESIGN PROJECTS

APPENDIXES

MERRILL TITLES IN MECHANICAL AND CIVIL TECHNOLOGY

Bateson	*Introduction to Control System Technology,* 2nd edition, (8255-2)
Keyser	*Materials Science in Engineering,* 3rd edition, (8182-3)
Lamit and Lloyd	*Drafting for Electronics* (20200-0)
Lamit, Wahler and Higgins	*Workbook in Drafting for Electronics* (20417-8)
Mott	*Applied Fluid Mechanics,* 2nd edition, (8305-2)
Mott	*Machine Elements in Mechanical Design* (20326-0)
Rolle	*Introduction to Thermodynamics,* 2nd edition, (8268-4)

Preface

The objective of this book is to provide the concepts, procedures, data, and decision analysis techniques necessary to design machine elements commonly found in mechanical devices and systems. Students completing a course of study using this book should be able to execute original designs for machine elements and integrate the elements into a system composed of several elements.

This process requires a consideration of the performance requirements of an individual element and of the interfaces between elements as they work together to form a system. For example, a gear must be designed to transmit power at a given speed. The design must specify the number of teeth, pitch, tooth form, face width, pitch diameter, material, and method of heat treatment. But the gear design also affects, and is affected by, the mating gear, the shaft carrying the gear, and the environment in which it is to operate. Furthermore, the shaft must be supported by bearings, which must be contained in a housing. Thus, the designer should keep the complete system in mind while designing each individual element. This book will help the student approach design problems in this way.

This text is designed for those interested in practical mechanical design. The emphasis is on the use of readily available materials and processes and appropriate design approaches to achieve a safe, efficient design. It is assumed that the person using the book will be the designer, that is, the person responsible for determining the configuration of a machine or a part of a machine. Where practical, all design equations, data, and procedures needed to make design decisions are specified.

It is expected that students using this book will have had a good background in statics, strength of materials, college algebra, and trigonometry. Helpful, but not required, would be knowledge of kinematics, industrial mechanisms, dynamics, materials, and manufacturing processes.

Among the important features of this book are the following:

1. It is designed to be used at the undergraduate level in a first course in machine design.

2. The large list of topics allows the instructor some choice in the design of the course. The format is also appropriate for a two-course sequence and as a reference for mechanical design project courses.
3. Students should be able to extend their efforts into topics not covered in classroom instruction because explanations of principles are straightforward and include many example problems.
4. The practical presentation of the material leads to feasible design decisions and is useful to practicing designers.
5. The text advocates and demonstrates use of computer programs in cases requiring long, laborious solution procedures. The programs show a flow chart, the program code, and sample output. Using the interactive BASIC language allows the operator to make decisions at several points within the program while the computer performs all computations (see the chapters on spur gears, wormgearing, springs, columns, shafts, and shrink fits). These sections also stress the usefulness of the programmable calculator and other computer languages.
6. References to other books, standards, and technical papers assist the instructor in presenting alternate approaches or extending the depth of treatment.
7. In addition to the emphasis on original design of machine elements, much of the discussion covers commercially available machine elements and devices, since many design projects require an optimum combination of new, uniquely designed parts and purchased components.
8. For some topics the focus is on aiding the designer in selecting commercially available components, such as rolling contact bearings, flexible couplings, ball screws, electric motors, belt drives, chain drives, clutches, and brakes.
9. Computations and problem solutions use both the International System of Units (SI) and the U.S. Customary System (inch-pound-second), approximately equally. The basic reference for the usage of SI units is Standard E380, *Metric Practice Guide,* issued by the American Society for Testing and Materials (ASTM).
10. An extensive Appendix is included along with detailed tables in many chapters that permit the reader to make real design decisions, using only this text.

My appreciation is extended to all who provided help and encouragement to me in the preparation of this book. I thank the editorial and production staffs of the Charles E. Merrill Publishing Company, the several manufacturers who provided illustrations and data about their products, my colleagues at the University of Dayton and elsewhere in the profession, and my family. I also thank those who reviewed the manuscript: Raymond Neathery (Oklahoma State University), Joseph Shelley (Trenton State College, New Jersey), George Cavaliere (New York City Technical College), Tom Vanderloop (University of Wisconsin-Stout), and Rob Spanovich (St. Paul Technical Institute, Minneapolis). I especially thank my students, past, present, and future.

Notes Concerning Computer Programs in This Book

The design of machine elements frequently involves many steps and calculations. Also, it is often desirable to generate several trial designs before specifying one to approach an optimum design.

For this reason, the use of computer programs is advocated. The time-consuming and laborious calculations can be done in a matter of seconds by the computer, and calculation errors are reduced. Many of the decisions normally done by the designer can be performed by the computer, such as looking up a standard wire size for a spring in a table, calculating the allowable stress for a material, or identifying that a computed stress is higher than the allowable. The designer, however, must remain in control and be involved in the design process. For this reason, the programs should be interactive, using a language such as BASIC to allow decision making during the execution of the program. Thus, the designer does the thinking while the computer performs the computations rapidly and accurately.

Each program in the book is accompanied by a flowchart that shows the logic of the program in diagram form. In addition, the programs and their sample output are annotated to help you follow the steps and to relate the output to the statements in the program. Program lines that are *INPUT statements,* calling for the operator to type in data, are highlighted in the lefthand margin with a ►. The lines in the output which correspond to these input lines are highlighted in the same way and the line number of the program calling for the input data is identified in the righthand margin.

Many of the statements in the programs are *PRINT statements,* which help to break up the output into an attractive format and prompt the operator to input data to the program. These statements make the program easy to use, but they tend to increase the length of the program.

The programs are written in standard BASIC language that is available on most computer systems and most personal computers. There are several specific versions of the BASIC language, and any of these can be used. Any other language suitable for technical computation may also be used, such as FORTRAN or PASCAL. Students are encouraged to create their own versions and to translate the sample programs into another language.

The variable names used in the programs conform to the requirements of the standard, simple BASIC language; that is, they are a single letter or a letter followed by a single number. The correlation of the variable names with the terms used in the programs is given in the discussion accompanying each program.

Most of the programs include parts that allow the operator to make decisions on how to continue on the basis of results at several points in the program. Some of the programs use subroutines to perform a set of operations. Some include computations using equations that are the results of curve-fitting techniques that replace the task of looking for a value in a table or a graph.

1 Machine Elements in Mechanical Design

1-1 OVERVIEW

Design of machine elements is an integral part of the larger and more general field of mechanical design. In mechanical design, a designer or design engineer creates a device or system to satisfy a specific need. A device typically involves moving parts that transmit power and accomplish a specific pattern of motion. A mechanical system is composed of several mechanical devices.

The products of mechanical design are useful in many fields. Consumer products such as lawn mowers, chain saws, power tools, garage door openers, and household appliances are all mechanical devices (see Figures 1-1 and 1-2). Mechanical systems used in manufacturing industries include machine tools, automated assembly and packaging systems, and material handling devices such as conveyors, cranes, and transfer machines (see Figures 1-3, 1-4, and 1-5).

In construction, mechanical design principles are used to create reliable and efficient power shovels, tractors, backhoes, mobile cranes, front end loaders, earth movers, graders, dump trucks, road pavers, concrete mixers, and many other devices. Figure 1-6 shows a tractor with a front end loader attachment. In agriculture, applications include tractors, combines, corn pickers, hay balers, and grain conveyors (see Figures 1-7 and 1-8).

Transportation equipment (automobiles, buses, trucks, aircraft, and ships) includes hundreds of mechanical devices: suspension components, door and window hardware, and the drive train. Aircraft include retractable landing gear, flap and rudder actuators, and cargo handling devices. Ships may include elevators, winches, cranes, and rotating radar antennas. Figures 1-9 and 1-10 illustrate systems that use these devices.

Design Skills

Designers use a wide range of skills and knowledge in their daily work, including the following:

Drafting	Statics
Manufacturing processes	Dynamics
Strength of materials	Properties of materials
Kinematics	Mechanisms
Design of machine elements	Mechanical design

Designers also draw on knowledge of fluid mechanics, thermodynamics, heat transfer, fluid power, electrical controls, and chemical processing. This book will not discuss these specialized fields; its focus is on mechanical hardware.

1-2 INTEGRATION OF MACHINE ELEMENTS INTO MECHANICAL DESIGN

Mechanical design is the process of designing or selecting mechanical components and putting them together to accomplish a desired function. Of course, machine elements must be compatible, fit well together, and perform safely and efficiently. The designer must consider not only the performance of the element being designed at a given time but also the elements with which it must interface.

To illustrate how the design of machine elements must be integrated with a larger mechanical design, let us consider the design of a speed reducer for a small tractor. The

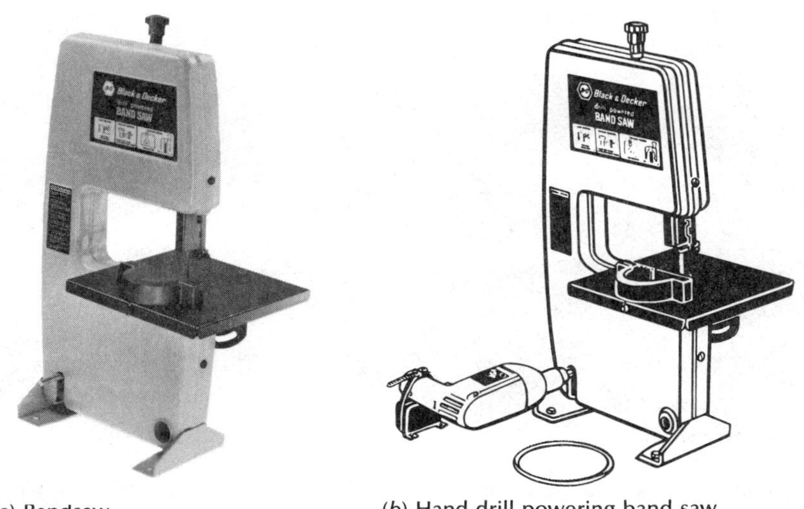

(*a*) Bandsaw (*b*) Hand drill powering band saw

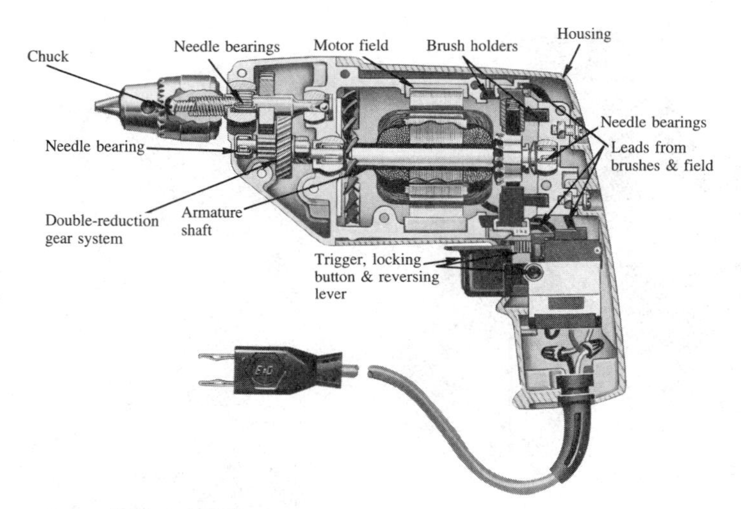

(*c*) Parts of a hand drill

Figure 1-1 Drill-Powered Bandsaw (Courtesy of Black & Decker (U.S.) Inc.)

Figure 1-2 Chain Saw (Copyright McCulloch Corporation, Los Angeles, Calif.)

(a) Power chain conveyor

(b) Drive system

(c) Conveyor chain

Figure 1-3 Chain Conveyor (Richards-Wilcox Manufacturing Company, Aurora, Ill.)

Figure 1-4 Automated Parts Handling Device (Pickomatic Systems, Inc., Sterling Heights, Mich.)

Figure 1-5 Industrial Crane (Air Technical Industries, Mentor, Ohio)

Figure 1-6 Tractor with a Front End Loader Attachment (J. I. Case, Racine, Wis.)

Figure 1-7 Tractor Pulling an Implement (J. I. Case, Racine, Wis.)

Figure 1-8 Cutaway of a Tractor (J. I. Case, Racine, Wisconsin)

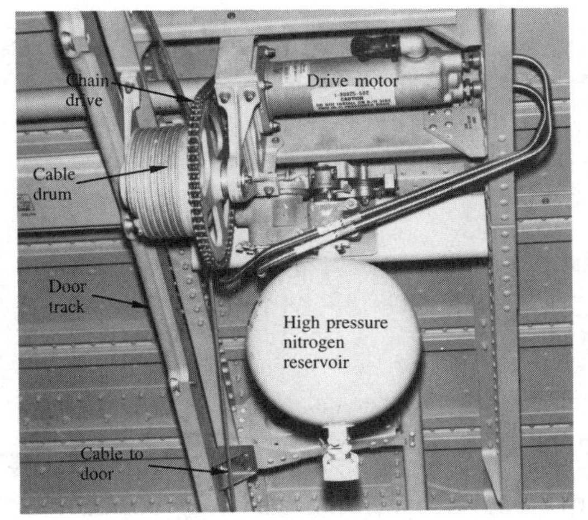

(a) Photograph of installed mechanism

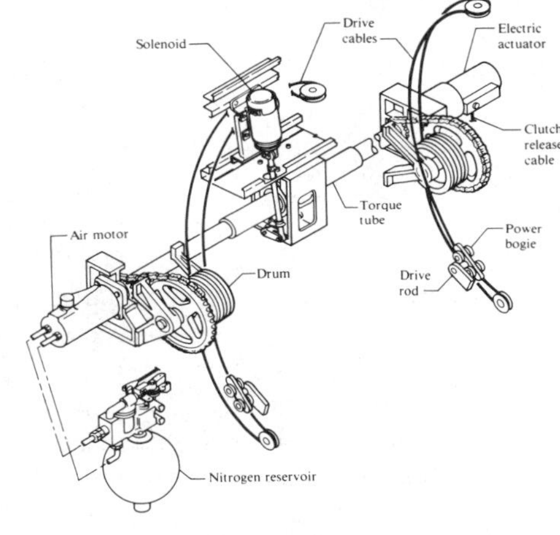

(b) Cabin door drive mechanism

Figure 1-9 Aircraft Door Drive Mechanism (McDonnell Douglas Corporation, Long Beach, Calif.)

shaft speed at the wheels should be 292 revolutions per minute (rpm), but the engine operates at 2 000 rpm while delivering 15 horsepower (hp). The speed reducer would take the output from the engine and reduce the speed to that required for the wheels.

To accomplish the speed reduction you decide to design a double reduction, spur–gear speed reducer. You specify four gears, three shafts, six bearings, and a housing to hold the individual elements in proper relation to each other, as shown in Figure 1-11. You should recognize that you have already made many design decisions by rendering such a sketch. First, you chose *spur gears* rather than helical gears, a worm and wormgear, or bevel gears. In fact, other types of speed reduction devices—belt drives, chain drives, or many others—could be appropriate.

The arrangement of the gears, the placement of the bearings so that they straddle the gears, and the general configuration of the housing are also design decisions. The design process cannot rationally proceed until these kinds of decisions are made. Notice that the sketch (Figure 1-11) is where *integration* of the elements into a whole design begins. When the overall design is conceptualized, the design of the individual machine elements

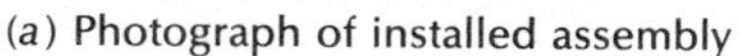

(*a*) Photograph of installed assembly

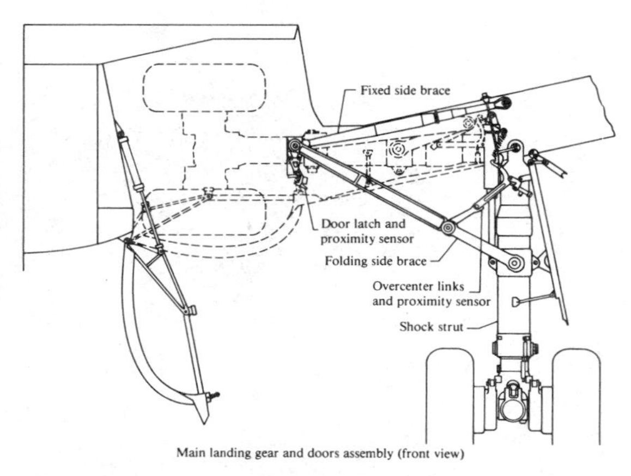

(*b*) Drawing showing landing gear in stowed and deployed positions

Figure 1-10 Aircraft Landing Gear Assembly (McDonnell Douglas Corporation, Long Beach, Calif.)

in the speed reducer can proceed. As each element is discussed, scan the relevant chapters of the book. In the chapters on gears, for example, you can see how some manufacturers have designed gear-type speed reducers.

For the gear pairs, you must specify the number of teeth in each gear, the pitch of the teeth, the pitch diameters, the face width, and the material and its heat treatment. These specifications depend on considerations of strength and wear of the gear teeth and the motion requirements (kinematics). You must also recognize that the gears must be mounted on shafts in a manner that ensures proper location of the gears, adequate torque transmitting capability from the gears to the shafts (as through keys), and safe shaft design.

Having designed the gear pairs, you consider the shaft design. The shaft is loaded in bending and torsion because of the forces acting at the gear teeth. Thus its design must consider strength and rigidity and permit the mounting of the gears and bearings. Shafts of varying diameters may be used to provide shoulders against which to seat the gears and bearings. There may be keyseats cut into the shaft. The input and output shafts will extend

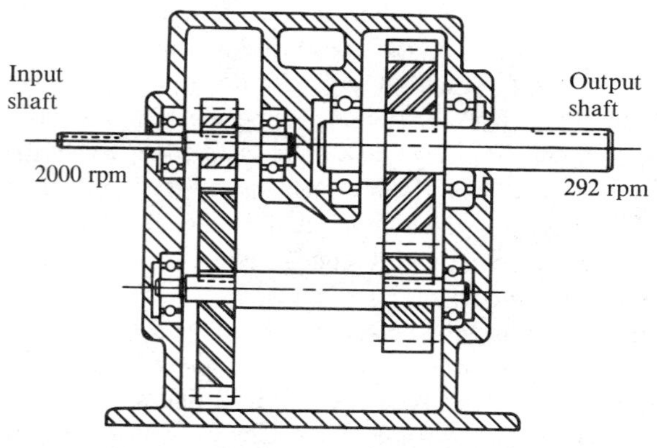

Figure 1-11 Conceptual Design for a Speed Reducer

beyond the housing to permit coupling with the driving motor and the driven machine. The type of coupling must be considered, as it can have a dramatic effect on the shaft stress analysis.

Design of the bearings is next. If rolling contact bearings are to be used, you will most probably select commercially available bearings from a manufacturer's catalog, rather than design a unique one. You must first determine the magnitude of the loads on each bearing from the shaft analysis and the gear designs. The rotational speed and reasonable design life of the bearings and their compatibility with the shaft on which they are to be mounted must also be considered. For example, on the basis of the shaft analysis you could specify the minimum allowable diameter at each bearing seat location to ensure safe stress levels. The bearing selected to support a particular part of the shaft, then, must have a bore (inside diameter) no smaller than the safe diameter of the shaft. Of course, the bearing should not be grossly larger than necessary. When a specific bearing is selected, the diameter of the shaft at the bearing seat location and allowable tolerances have to be specified, according to the bearing manufacturer's recommendations, to achieve proper operation and life expectancy of the bearing.

Now the keys and keyseats can be designed. The diameter of the shaft at the key determines the key's basic size (width and height). The torque that must be transmitted is used in strength calculations to specify key length and material. Once the working components are designed, the housing design can begin.

The housing design process must be both creative and practical. What provisions should be made to mount the bearings accurately and to transmit the bearing loads safely through the case to the structure on which the speed reducer is mounted? How will the various elements be assembled into the housing? How will the gears and bearings be lubricated? What housing material should be used? Should the housing be a casting, a weldment, or an assembly of machined parts?

The design process as outlined here implies that the design can progress in sequence: from the gears to the shafts, to the bearings, to the keys and couplings, and finally to the housing. It would be rare, however, to follow this logical path only once for a given design. Usually the designer must go back many times to adjust the design of certain components affected by changes in other components. This process, *iteration,* continues until an acceptable overall design is achieved. Frequently prototypes are developed and tested during iteration.

Because of the usual need for several iterations, and because many of the design procedures require long, complex calculations, computer programs or programmable calculators are often useful in performing the design analysis. *Interactive* programs allow you, the designer, to make design decisions during the design process. In this way, many trials can be made in a short time and the effects of changing various parameters can be investigated. The sample programs presented in this book are written in the BASIC language and are designed to be used interactively. You are encouraged to use the given programs, to write your own, to convert the programs into another language such as FORTRAN, or to implement them on a programmable calculator.

1-3 FUNCTIONS AND DESIGN REQUIREMENTS

The ultimate objective of mechanical design, of course, is to produce a useful device that is safe, efficient, economical, and practical to manufacture. When beginning the design of a machine or an individual machine element, it is important to define the functions of the device clearly and completely. Statements of functions tell what the device is supposed to do. They may be general, but they should always employ action phrases. For example, the following statements describe the functions of the speed reducer discussed in Section 1-2:

1. To receive power from the input source (such as an electric motor or an engine) through a rotating shaft.
2. To transmit the power through machine elements that reduce the rotational speed to a specified value.
3. To deliver the power at the lower speed to a driven machine or to the wheels of a vehicle.

After the functions are defined, a set of design requirements is prepared. In contrast to the broad and general function statements, design requirements must be detailed and specific, giving quantitative data wherever possible. Look at the following set of design requirements for the speed reducer:

1. The reducer must transmit 15.0 hp.
2. The input is from a two-cylinder gasoline engine with a rotational speed of 2 000 rpm.
3. The output delivers the power at a rotational speed in the range of 290 to 295 rpm.
4. A mechanical efficiency of greater than 95 percent is desirable.
5. The minimum output torque capacity of the reducer should be 3 050 pound-inches (lb · in).
6. The reducer output is connected to the drive shaft for the wheels of a farm tractor. Moderate shock will be encountered.
7. The input and output shafts must be in-line.
8. The reducer is to be bolted to a rigid steel frame of the tractor on a horizontal surface.
9. Small size is desirable. The reducer must be installed on a square surface 20 in × 20 in, with a maximum height of 24 in.
10. The tractor is expected to operate 8 hours (h) per day, 5 days per week, with a design life of 10 years.
11. The reducer must be protected from the weather and capable of operating anywhere in the United States at temperatures ranging from 0 to 130°F.
12. Flexible couplings will be used on the input and output shafts to prohibit axial and bending loads from being transmitted to the reducer.
13. The production quantity is 10 000 units per year.
14. A moderate cost is critical to successful marketing.

Careful preparation of function statements and design requirements will ensure that the design effort is focused on the desired results. Much time and money can be wasted on designs that, although technically sound, do not meet design requirements. Design requirements should include everything that is needed, but at the same time they should offer ample opportunity for innovation.

1-4 MACHINE DESIGN CRITERIA

Design approaches may have to be adjusted to be compatible with certain industries and markets. For example, devices to be incorporated in aircraft should have low weight, whereas a part for a machine in a manufacturing plant does not normally have weight restrictions. Such considerations can have dramatic effects on the design process.

In approaching a design, the designer should establish criteria that will guide the decision-making processes inherent in any project. Because each design problem has many alternative solutions, each can be evaluated in terms of the list of criteria. There may not be a single best design, but designers should work toward an *optimum* design. That is, the design should maximize benefits and minimize disadvantages.

The following are general criteria for machine design:

Safety

Performance (the degree to which the design meets or exceeds the design objectives)

Reliability (a high probability that the device will reach or exceed its design life)

Ease of manufacture

Ease of service or replacement of parts

Ease of operation

Low initial cost

Low operating and maintenance costs

Small size and low weight

Low noise and vibration; smooth operation

Use of readily available materials and purchased components

Judicious use of uniquely designed parts along with commercially available components

Appearance that is attractive and appropriate to the application

You or your coworkers can supply additional criteria and determine the relative importance of the criteria applied in each design. Certainly safety is of paramount concern, and the designer is legally liable if someone is injured because of a design error. You must take into account the reasonably foreseeable uses of the device and ensure the safety of those operating it or those who may be close by. Adequate performance is a primary criterion. The importance of the remaining criteria varies from design to design.

1-5 UNIT SYSTEMS

Computations in this book will be performed by using either the American Customary Unit System (inch-pound-second) or the International System (SI). Table 1-1 lists the typical units used in the study of machine design. *SI*, the abbreviation of Le Système International d'Unités, is the standard for metric units throughout the world. For convenience, the term *SI units* will be used instead of *metric units*.

Prefixes applied to the basic units indicate order of magnitude. Only those prefixes listed in Table 1-2, which differ by a factor of 1 000, should be used in technical calculations. The final result for a quantity should be reported as a number between 0.1 and 10 000, times some multiple of 1 000. Then, the unit with the appropriate prefix should be specified. Table 1-3 lists examples of proper SI notation.

Sometimes you have to convert a unit from one system to another. Appendix A-2 provides a table of conversion factors. Also, you should be familiar with the typical order of magnitude of the quantities encountered in machine design so you can judge the reasonableness of design calculations (see Table 1-4 for several examples).

Distinction must be made between the terms *force, mass,* and *weight*. *Mass* is the quantity of matter in a body. A *force* is a push or pull applied to a body that results in a change in the body's motion or in some deformation of the body. Clearly these are two

Table 1-1 Typical Units Used in Machine Design

Quantity	*U.S. Customary Unit*	*SI Unit*
Length or distance	inch (in) foot (ft)	meter (m) millimeter (mm)
Area	square inch (in^2)	square meter (m^2) or square millimeter (mm^2)
Force	pound (lb) Kip (K) (1 000 lb)	newton (N) ($1\ N = 1\ kg \cdot m/s^2$)
Mass	slug ($lb \cdot s^2/ft$)	kilogram (kg)
Time	second (s)	second (s)
Angle	degree (deg)	radian (rad) or degree (deg)
Temperature	degrees Fahrenheit (°F)	degrees Celsius (°C)
Torque or moment	pound-inch ($lb \cdot in$) or pound-foot ($lb \cdot ft$)	newton-meter ($N \cdot m$)
Energy or work	pound-inch ($lb \cdot in$)	joule (J) ($1\ J = 1\ N \cdot m$)
Power	horsepower (hp) ($1\ hp = 550\ lb \cdot ft/s$)	watt (W) or kilowatts (kW) ($1\ W = 1\ J/s = 1\ N \cdot m/s$)
Stress, pressure, modulus of elasticity	pounds per square inch (lb/in^2, or psi) Kips per square inch (K/in^2, or Ksi)	pascal (Pa) ($1\ Pa = 1\ N/m^2$) kilopascal (kPa) ($1\ kPa = 10^3\ Pa$) megapascal (MPa) ($1\ MPa = 10^6\ Pa$) gigapascal (GPa) ($1\ GPa = 10^9\ Pa$)
Section modulus	inches cubed (in^3)	meters cubed (m^3) or millimeters cubed (mm^3)
Moment of inertia	inches to the fourth power (in^4)	meters to the fourth power (m^4) or millimeters to the fourth power (mm^4)
Rotational speed	revolutions per min (rpm)	radians per second (rad/s)

Table 1-2 Prefixes Used with SI Units

Prefix	*SI Symbol*	*Factor*
micro-	μ	$10^{-6} = 0.000\,001$
milli-	m	$10^{-3} = 0.001$
kilo-	k	$10^3 = 1\,000$
mega-	M	$10^6 = 1\,000\,000$
giga-	G	$10^9 = 1\,000\,000\,000$

Table 1-3 Quantities Expressed in SI Units

Computed Result	*Reported Result*
0.001 65 m	1.65×10^{-3} m, or 1.65 mm
32 540 N	32.54×10^3 N, or 32.54 kN
1.583×10^5 W	158.3×10^3 W, or 158.3 kW; or $0.158\,3 \times 10^6$ W, or 0.158 3 MW
2.07×10^{11} Pa	207×10^9 Pa, or 207 GPa

different physical phenomena, but the distinction is not always understood. The units for force and mass used in this text are listed in Table 1-1.

The term *weight,* as used in this book, refers to the amount of *force* required to support a body against the influence of gravity. Thus, in response to "What is the weight of 75 kg of steel?" we would use the relationship between force and mass from physics:

$$F = ma \qquad \text{or} \qquad w = mg$$

Table 1-4 Typical Order of Magnitude for Commonly Encountered Quantities

Quantity	*U.S. Customary Unit*	*SI Unit*
Dimensions of a wood standard 2 × 4	1.50 in × 3.50 in	38 mm × 89 mm
Moment of inertia of a 2 × 4 (3.50 in side vertical)	5.36 in^4	2.23×10^6 mm^4, or 2.23×10^{-6} m^4
Section modulus of a 2 × 4 (3.50 in side vertical)	3.06 in^3	5.02×10^4 mm^3, or 5.02×10^{-5} m^3
Force required to lift 1.0 gal of gasoline	6.01 lb	26.7 N
Density of water	1.94 $slugs/ft^3$	1 000 kg/m^3, or 1.0 Mg/m^3
Compressed air pressure in a factory	100 psi	690 kPa
Yield point of AISI 1040 hot-rolled steel	42 000 psi, or 42 Ksi	290 MPa
Modulus of elasticity of steel	30 000 000 psi, or 30×10^6 psi	207 GPa

where F = force, m = mass, a = acceleration, w = weight, and g = the acceleration due to gravity. We will use

$$g = 32.2 \text{ ft/s}^2 \qquad \text{or} \qquad g = 9.81 \text{ m/s}^2$$

Then, to compute the weight:

$$w = mg = 75 \text{ kg}(9.81 \text{ m/s}^2)$$
$$w = 736 \text{ kg} \cdot \text{m/s}^2 = 736 \text{ N}$$

Remember that, as shown in Table 1-1, the newton (N) is equivalent to 1.0 kg · m/s^2. In fact, the newton is defined as the force required to give a mass of 1.0 kg an acceleration of 1.0 m/s^2. In our example, then, we would say that the 75-kg mass of steel has a weight of 736 N.

1-6 DESIGN CALCULATIONS

As you study this book and in your career as a designer, you will make many design calculations. It is important to record the calculations neatly, completely, and in an orderly fashion. You may have to explain to others how you approached the design, which data you used, and which assumptions and judgments you made. In some cases, someone else will actually check your work when you are not there to comment on it or to answer questions. Also, an accurate record of your design calculations is often useful if changes in design are likely. In all these situations, you are going to be asked to communicate your design to someone else in written and graphic form.

To prepare a careful design record, you will usually take the following steps:

1. Identify the machine element being designed and the nature of the design calculation.
2. Draw a sketch of the element, showing all features that affect performance or stress analysis.
3. Show in a sketch the forces acting on the element (the free-body diagram) and provide other drawings to clarify the actual physical situation.
4. Identify the kind of analysis to be performed, such as stress due to bending, deflection of a beam, buckling of a column, and so on.
5. List all given data and assumptions.

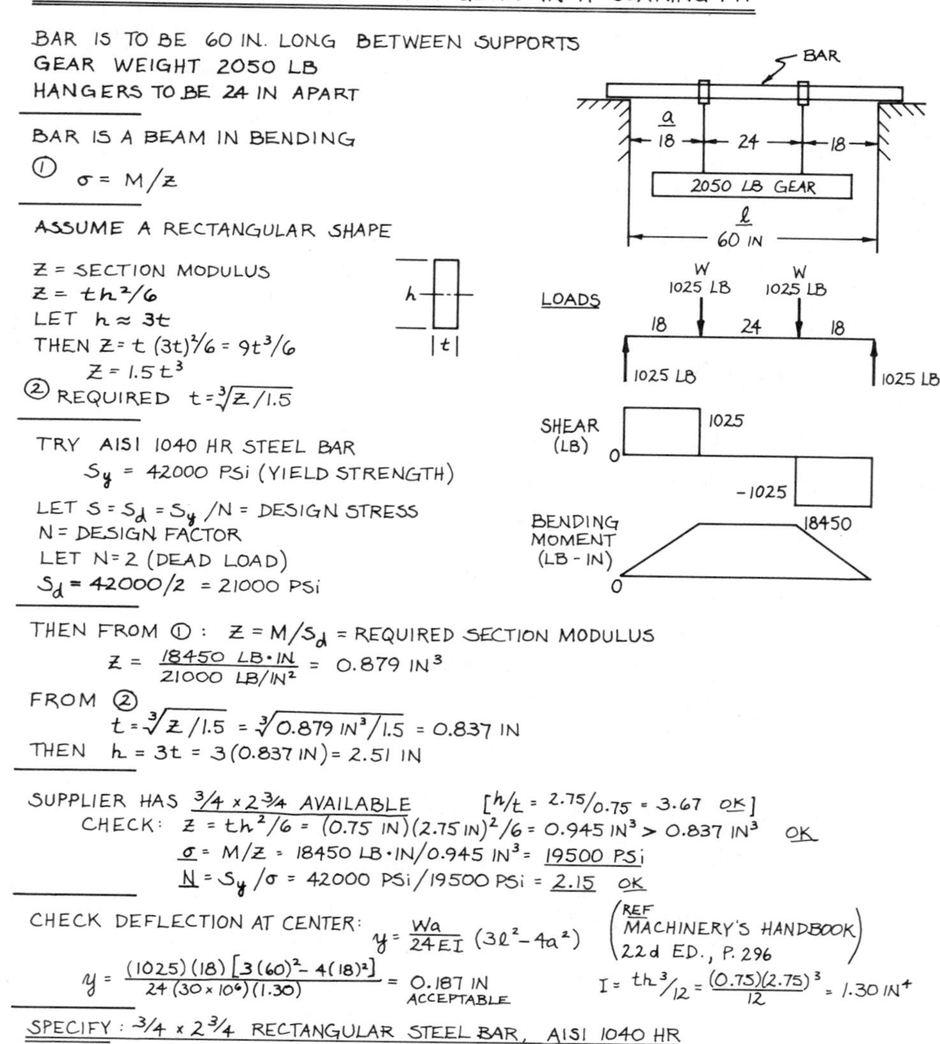

R. L. MOTT

DESIGN OF A BAR TO SUPPORT A GEAR IN A SOAKING PIT

BAR IS TO BE 60 IN. LONG BETWEEN SUPPORTS
GEAR WEIGHT 2050 LB
HANGERS TO BE 24 IN APART

BAR IS A BEAM IN BENDING
① $\sigma = M/Z$

ASSUME A RECTANGULAR SHAPE

Z = SECTION MODULUS
$Z = th^2/6$
LET $h \approx 3t$
THEN $Z = t\,(3t)^2/6 = 9t^3/6$
$Z = 1.5t^3$
② REQUIRED $t = \sqrt[3]{Z/1.5}$

TRY AISI 1040 HR STEEL BAR
S_y = 42000 PSI (YIELD STRENGTH)

LET $S = S_d = S_y/N$ = DESIGN STRESS
N = DESIGN FACTOR
LET N = 2 (DEAD LOAD)
$S_d = 42000/2 = 21000$ PSI

THEN FROM ①: $Z = M/S_d$ = REQUIRED SECTION MODULUS
$Z = \dfrac{18450 \text{ LB·IN}}{21000 \text{ LB/IN}^2} = 0.879 \text{ IN}^3$

FROM ②
$t = \sqrt[3]{Z/1.5} = \sqrt[3]{0.879 \text{ IN}^3/1.5} = 0.837$ IN
THEN $h = 3t = 3(0.837 \text{ IN}) = 2.51$ IN

SUPPLIER HAS 3/4 x 2 3/4 AVAILABLE [$h/t = 2.75/0.75 = 3.67$ OK]
CHECK: $Z = th^2/6 = (0.75 \text{ IN})(2.75 \text{ IN})^2/6 = 0.945 \text{ IN}^3 > 0.837 \text{ IN}^3$ OK
$\sigma = M/Z = 18450 \text{ LB·IN}/0.945 \text{ IN}^3 = 19500$ PSI
$N = S_y/\sigma = 42000 \text{ PSI}/19500 \text{ PSI} = 2.15$ OK

CHECK DEFLECTION AT CENTER: $y = \dfrac{Wa}{24EI}(3l^2 - 4a^2)$ (REF MACHINERY'S HANDBOOK 22d ED., P. 296)

$y = \dfrac{(1025)(18)[3(60)^2 - 4(18)^2]}{24(30 \times 10^6)(1.30)} = 0.187$ IN ACCEPTABLE

$I = th^3/12 = \dfrac{(0.75)(2.75)^3}{12} = 1.30 \text{ IN}^4$

SPECIFY: 3/4 x 2 3/4 RECTANGULAR STEEL BAR, AISI 1040 HR

6. Write the formulas to be used in symbol form and clearly indicate the values and units of the variables involved. If a formula is not well known to a potential reader of your work, give the source. The reader may want to refer to it to evaluate the appropriateness of the formula.
7. Solve each formula for the desired variable.
8. Insert data, check units, and perform computations.
9. Judge the reasonableness of the result.
10. If the result is not reasonable, change the design decisions and recompute. Perhaps a different geometry or material would be more appropriate.
11. When a reasonable, satisfactory result has been achieved, specify the final values for all important design parameters, using standard sizes, convenient dimensions, readily available materials, and so on.

PROBLEMS

For the devices described in items 1–14, write a set of functions and design requirements in a similar manner to those in Section 1-3. You or the instructor may add more specific information to the general descriptions given.

1. The hood latch for an automobile
2. A hydraulic jack used for car repair
3. A portable crane to be used in small garages and homes
4. A machine to crush soft drink or beer cans
5. An automatic transfer device for a production line
6. A device to raise a 55-gallon (gal) drum of bulk materials and dump the contents into a hopper
7. A paper feed device for a copier
8. A conveyor to elevate and load gravel into a truck
9. A crane to lift building materials from the ground to the top of a building during construction
10. A machine to insert toothpaste tubes into cartons
11. A machine to insert 24 cartons of toothpaste into a shipping container
12. A gripper for a robot to grasp a spare tire assembly and insert it into the trunk of an automobile on an assembly line
13. A table for positioning a weldment in relation to a robotic welder
14. A garage door opener

For the following problems, perform the indicated conversion of units. (Refer to Appendix A-2 for conversion factors.) Express the results with the appropriate prefix as illustrated in Tables 1-2 and 1-3.

15. Convert a shaft diameter of 1.75 in to mm.
16. Convert the length of a conveyor from 46 ft to meters.
17. Convert the torque developed by a motor of 12 550 lb · in to N · m.

Refer to Appendix A-16 for problems 18–21.

18. A wide flange steel beam shape, W12×14, has a cross-sectional area of 4.12 in^2. Convert the area to mm^2.
19. The W12×14 beam shape has a section modulus of 14.8 in^3. Convert it to mm^3.
20. The W12×14 beam shape has a moment of inertia of 88.0 in^4. Convert it to mm^4.
21. What standard steel equal leg angle would have a cross-sectional area closest to (but greater than) 750 mm^2?
22. An electric motor is rated at 7.5 hp. What is its rating in watts (W)?
23. A vendor lists the ultimate tensile strength of a steel to be 127 000 psi. Compute the strength in MPa.
24. Compute the weight of a steel shaft, 35.0 mm in diameter and 675 mm long. (See Appendix A-3 for the density of steel.)

2 Materials in Mechanical Design

2-1 OVERVIEW

Satisfactory performance of machine parts and systems depends greatly on the materials the designer selects. The designer must understand how materials behave, what properties of the material affect the performance of the parts, and how to interpret the large amount of data available on material properties. The designer's ability to communicate effectively with material suppliers, metallurgists, manufacturing process personnel, heat treatment personnel, plastics molders, machinists, and quality assurance specialists often has a strong influence on a design's success.

This chapter focuses on the use of material property data in design decisions rather than the metallurgy or chemistry of the materials. Part of the chapter serves as a glossary of terms related to material properties and processing for your reference as you use the rest of the book. Also, this chapter describes the appendixes; study these data tables as you read the text. Then, whenever materials are discussed in later chapters, you will be familiar with the types of data available.

Metals and plastics are the most common materials used in design. A discussion of metal properties is followed by specific sections on carbon and alloy steel, structural steel, stainless steel, cast iron, aluminum, titanium, bronze, and powdered metals. The characteristics of plastics are described along with descriptions of several types of plastics commonly used for mechanical and structural parts.

2-2 PROPERTIES OF METALS

The way the machine element or system is used determines which properties are important in a given design exercise. This section will cover a few of the many properties with which you should be familiar.

Tensile strength
Yield strength
Proportional limit
Elastic limit
Ductility
Percent elongation
Modulus of elasticity in tension
Modulus of elasticity in shear
Poisson's ratio
Shear strength
Impact strength
Hardness
Machinability
Density
Thermal conductivity
Electrical resistivity
Coefficient of thermal expansion

Many of these properties are determined from the standard tensile test during which the stress-strain diagram is produced. Figure 2-1 is a typical diagram for a plain carbon steel, sometimes referred to as a *mild steel*. The following properties can be determined directly from such a diagram or from the data obtained during the stress-strain test.

Tensile Strength, s_u

The peak of the stress-strain curve is considered the *ultimate tensile strength,* sometimes called the *ultimate strength* or, simply, the *tensile strength*. At this point during the test the highest *apparent stress* on a test bar of the material is measured. The curve appears to drop off after the peak, which may be interpreted as a lower stress in the test bar. However, notice that the instrumentation used to create the diagram is actually plotting

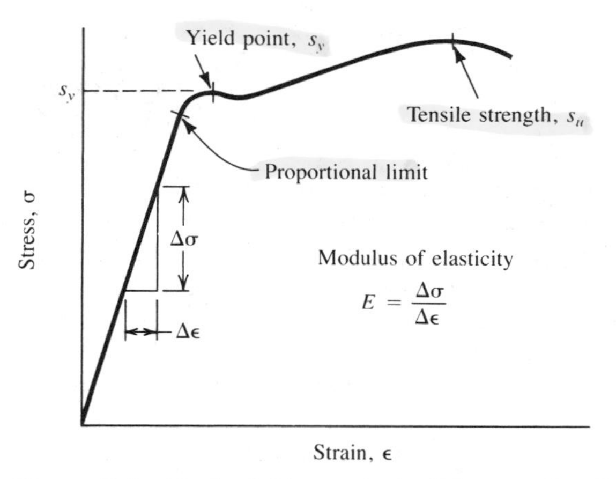

Figure 2-1 Typical Stress-strain Diagram for Steel

load versus deflection rather than *true stress versus strain*. The apparent stress is computed by dividing the load by the original cross-sectional area of the test bar. After the peak of the curve is reached, there is a pronounced decrease in the bar's diameter, referred to as *necking down*. Thus the load acts over a smaller area, and the *actual stress* continues to increase until failure. It is very difficult to follow the reduction in diameter during the necking down process, so it has become customary to use the peak of the curve as the tensile strength, although it is a more conservative value.

Yield Strength, s_y

That portion of the stress-strain diagram where there is a large increase in strain with little or no increase in stress is called the *yield strength*. This property indicates that the material has, in fact, yielded or elongated plastically, permanently, and to a large degree. If the point of yielding is quite noticeable, as it is in Figure 2-1, the property is called the *yield point* rather than the yield strength.

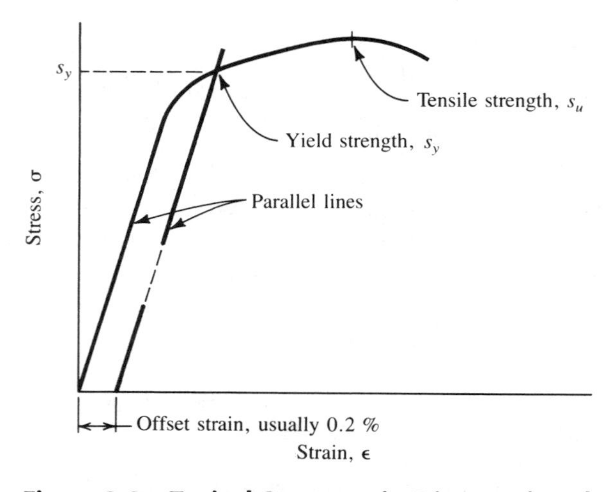

Figure 2-2 Typical Stress-strain Diagram for Aluminum and Other Metals Having No Yield Point

Figure 2-2 shows the stress-strain diagram form that is typical of a nonferrous metal such as aluminum or titanium or of certain high-strength steels. Notice that there is no pronounced yield point, but the material has actually yielded at or near the stress level indicated as s_y. That point is determined by the *offset method,* in which a line is drawn parallel to the straight-line portion of the curve and offset to the right by a set amount, usually 0.20 percent strain (0.002 in/in). The intersection of this line and the stress-strain curve defines the material's yield strength. In this book, the term *yield strength* will be used for S_y, regardless of whether the material exhibits a true yield point or the offset method is used.

Proportional Limit

That point on the stress-strain curve where it deviates from a straight line is called the *proportional limit.* That is, at or above that stress value, stress is no longer proportional to strain. Below the proportional limit, Hooke's law applies; stress is proportional to strain.

Elastic Limit

At some point, called the *elastic limit,* a material experiences some amount of plastic strain and thus will not return to its original shape after release of the load. Below that level, the material behaves completely elastically. The proportional limit and elastic limit lie quite close to the yield strength. Because they are difficult to determine, they are rarely reported.

Modulus of Elasticity in Tension (E)

For that part of the stress-strain diagram that is straight, stress is proportional to strain and the value of E is the constant of proportionality. That is,

$$E = \text{stress/strain} = \sigma/\epsilon$$

This is the slope of the straight-line portion of the diagram. The modulus of elasticity indicates the stiffness of the material, or its resistance to deformation.

Ductility and Percent Elongation

Ductility is the degree to which a material will deform before ultimate fracture. The opposite of ductility is brittleness. When ductile materials are used in machine members, impending failure is detected easily and sudden failure is unlikely. Also, ductile materials normally resist the repeated loads on machine elements better than brittle materials.

The usual measure of ductility is the percent elongation of the material after fracture in a standard tensile test. Figure 2-3 shows a typical standard tensile specimen before and after the test. Before the test, gage marks are placed on the bar, usually 2.00 in apart. Then, after the bar is broken, the two parts are fitted back together and the final length between the gage marks is measured. The percent elongation is the difference between the final length and the original length divided by the original length, converted to a percentage. That is,

$$\frac{L_f - L_o}{L_o} \times 100\% = \text{Percent elongation}$$

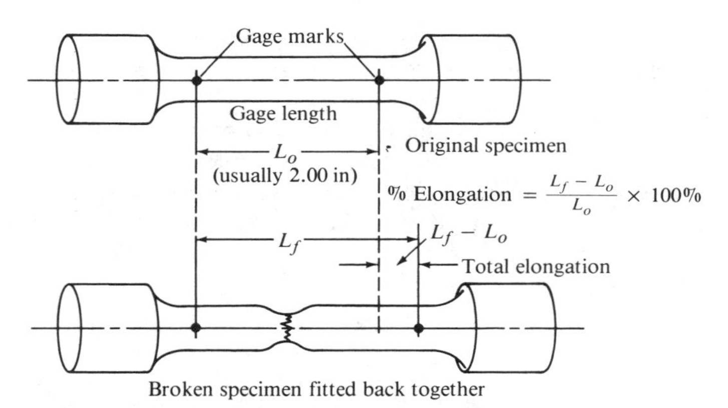

Figure 2-3 Measurement of Percent Elongation

The percent elongation is assumed to be based on a gage length of 2.00 in unless some other gage length is specifically indicated. Theoretically, a material is considered ductile if its percent elongation is greater than 5 percent. For practical reasons, it is advisable to use a material with a value of 12 percent or higher for machine members subject to repeated loads or shock or impact.

Percent reduction in area is another indication of ductility. This value is found by comparing the original value with the final area at the break for the tensile test specimen.

Shear Strength

Both the yield strength and the ultimate strength in shear are important properties of materials. Unfortunately, these values are seldom reported. We will use the following estimates:

$$s_{ys} = s_y/2 = 0.50s_y = \text{Yield strength in shear}$$

$$s_{us} = 0.75s_u = \text{Ultimate strength in shear}$$

Poisson's Ratio

When a material is subjected to a tensile strain, there is a simultaneous shortening of the cross-sectional dimensions perpendicular to the direction of the tensile strain. The ratio of the shortening strain to the tensile strain is called *Poisson's ratio,* usually denoted by ν, the Greek letter nu. (The Greek letter mu, μ, is sometimes used for this ratio.) Typical values for Poisson's ratio are 0.25 for cast iron, 0.27 for steel, and 0.33 for aluminum and titanium.

Modulus of Elasticity in Shear

The *modulus of elasticity in shear* is the ratio of shearing stress to shearing strain, denoted by G. This property indicates a material's stiffness under shear loading, that is, the resistance to shear deformation. The value of G is frequently reported, and there is a simple relationship between E, G, and Poisson's ratio:

$$G = \frac{E}{2(1 + \nu)}$$

Hardness

The resistance of a material to indentation by a penetrator is an indication of its *hardness*. Several types of devices, procedures, and penetrators measure hardness; the Brinell hardness tester and the Rockwell hardness tester are most frequently used for machine elements. For steels, the Brinell hardness tester employs a hardened steel ball 10 mm in diameter as the penetrator under a load of 3 000-kg force. The load causes a permanent indentation in the test material, and the diameter of the indentation is related to the Brinell hardness number, which is abbreviated BHN or HB. The actual quantity being measured is the load divided by the contact area of the indentation. For steels, the value of HB ranges from approximately 100 for an annealed low-carbon steel to over 700 for high-strength, high-alloy steels in the as-quenched condition. In the high ranges, above HB 500, the penetrator is sometimes made of tungsten carbide rather than steel. For softer metals, a 500-kg load is used.

The Rockwell hardness tester uses a hardened steel ball with a 1/16-in diameter under a load of 100-kg force for softer metals, and the resulting hardness is listed as Rockwell B, R_B, or HRB. For harder metals, such as heat-treated alloy steels, the Rockwell C scale is used. A load of 150-kg force is placed on a diamond penetrator (a *brale* penetrator) made in a sphero-conical shape. Rockwell C hardness is sometimes referred to as R_C or HRC. Many other Rockwell scales are used.

The Brinell and Rockwell methods are based on different parameters and lead to quite different numbers. However, since they both measure hardness, there is a correlation between them, as seen in Appendix A-2. It is also important to note that, especially for highly hardenable alloy steels, there is a nearly linear relationship between the Brinell hardness number and the tensile strength of the steel, according to the equation

$$0.50(\text{HB}) = \text{Approximate tensile strength (Ksi)}$$

This relationship is shown in Figure 2-4.

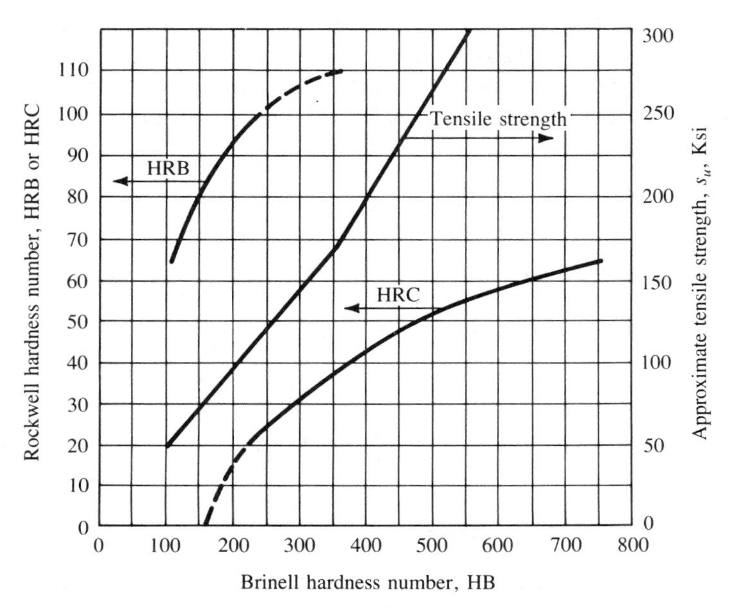

Figure 2-4 Hardness Conversions

To compare the hardness scales with the tensile strength, consider the following table.

MATERIAL AND CONDITION	HARDNESS			TENSILE STRENGTH	
	HB	*HRB*	*HRC*	*Ksi*	*MPa*
1020 Annealed	121	70	—	60	414
1040 Hot-Rolled	144	79	—	72	496
4140 Annealed	197	93	13	95	655
4140 OQT 1000	341	109	37	168	1160
4140 OQT 700	461	—	49	231	1590

There is some overlap between the HRB and HRC scales. Normally, HRB is used for the softer metals and ranges from approximately 60 to 100, and HRC is used for harder metals and ranges from 20 to 65. Using HRB numbers above 100 or HRC numbers below 20 is not recommended. Those shown in the preceding table are for comparison purposes only.

Hardness in a steel indicates wear resistance as well as strength. Wear resistance will be discussed in later chapters, particularly with regard to gear teeth.

Machinability

Machinability is related to the ease with which a material can be machined to a good surface finish with reasonable tool life. Production rates are directly affected by the machinability. It is difficult to define measurable properties related to machinability, so machinability is usually reported in comparative terms, relating the performance of a given material with some standard.

Impact Strength

Parts subjected to suddenly applied loads, shock, or impact need high values of impact strength. A measure of toughness, *impact strength* is measured by subjecting a notched specimen to a rapidly moving striker having a relatively high mass. The specimen absorbs energy from the striker as fracture occurs, and the amount of energy absorbed is reported as the impact strength. The two most widely used testing methods for impact strength are the Charpy and the Izod tests.

Physical Properties

Here we will discuss density, coefficient of thermal expansion, thermal conductivity, and electrical resistivity.

Density

Density is defined as the mass per unit volume of the material. Its usual units are kg/m^3 in the SI Unit system and lb/in^3 in the U.S. Customary Unit system, where the pound unit is taken to be pounds-mass.

Coefficient of Thermal Expansion

The *coefficient of thermal expansion* is a measure of the change in length of a material subjected to a change in temperature. It is defined by the relation

$$\alpha = \frac{\text{Change in length}}{L_o(\Delta T)} = \frac{\text{strain}}{(\Delta T)} = \frac{\epsilon}{(\Delta T)}$$

where L_o is the original length and ΔT is the change in temperature. Virtually all metals and plastics expand with increasing temperature. But different materials expand at different rates. For machines and structures containing parts of more than one material, the different rates can have a significant effect on the performance of the assembly and on the stresses produced.

Thermal Conductivity

Thermal conductivity is that property of a material that indicates its ability to transfer heat. Where machine elements operate in hot environments or significant internal heat is generated, the ability of the elements or of the machine's housing to transfer heat away can affect machine performance. For example, wormgear speed reducers typically generate frictional heat due to the rubbing contact between the worm and the wormgear teeth. If not adequately transferred, heat causes the lubricant to lose its effectiveness, allowing rapid gear tooth wear.

Electrical Resistivity

For machine elements that conduct electricity while carrying loads, the electrical resistivity of the material is as important as its strength. *Electrical resistivity* is a measure of the resistance offered by a given thickness of the material and is measured in ohm-centimeters (ohm · cm). *Electrical conductivity,* a measure of the capacity of the material to conduct electric current, is sometimes used instead of resistivity. It is often reported as a percentage of the conductivity of a reference material, usually the International Annealed Copper Standard.

2-3 CARBON AND ALLOY STEEL

Steel is possibly the most widely used material for machine elements because of its properties of high strength, high stiffness, durability, and relative ease of fabrication. Many types of steels are available, and classification is not precise. This section will discuss the methods used for designating steels and describe the most frequently used types.

The term *steel* refers to an alloy of iron, carbon, manganese, and one or more other significant elements. Carbon has a very strong effect on the strength, hardness, and ductility of any steel alloy. The other elements affect hardenability, toughness, corrosion resistance, machinability, and strength retention at high temperatures. The primary alloying elements present in the various alloy steels are sulfur, phosphorus, silicon, nickel, chromium, molybdenum, and vanadium.

Designation Systems

Three national organizations, the American Iron and Steel Institute (AISI), the Society of Automotive Engineers (SAE), and the American Society for Testing and Materials (ASTM), have developed uniform systems for designating steels used in machine and structural applications.

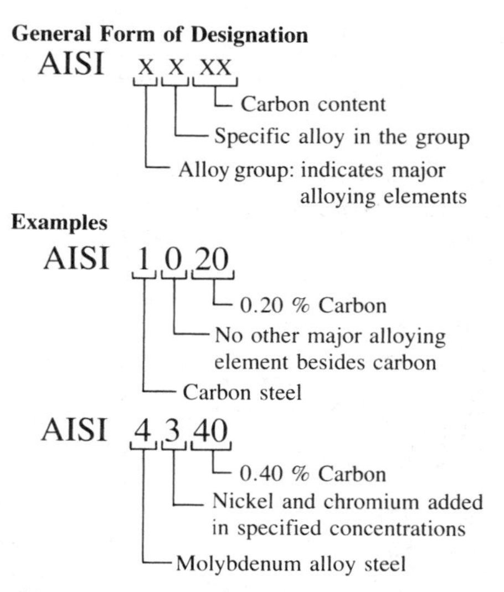

Figure 2-5 Steel Designation System

For most carbon and alloy steels, the AISI and SAE systems use four-digit designations in which the last two digits indicate the amount of carbon in the steel, and the first two digits indicate the specific alloy group, which identifies the primary alloying elements other than carbon. Figure 2-5 illustrates the system. Table 2-1 identifies the alloy groups.

Although most steel alloys contain less than 1.0 percent carbon, it is included in the designation because of its effect on the properties of steel. As Figure 2-5 illustrates, the last two digits indicate carbon content in hundredths of a percent. For example, when the last two digits are 20, the alloy includes approximately 0.20 percent carbon. Some variation is allowed. The carbon content in a steel with *20 points of carbon* ranges from 0.18 to 0.23 percent.

As carbon content increases, strength and hardness also increase under the same conditions of processing and heat treatment. Since ductility decreases with increasing carbon content, selecting a suitable steel involves some compromise between strength and ductility.

As a rough classification scheme, a *low-carbon steel* is one having fewer than 30 points of carbon (0.30 percent). These steels have relatively low strength but good formability. In machine element applications where high strength is not required, low-carbon steels are frequently specified. If wear is a potential problem, low-carbon steels can be carburized (discussed in Section 2-4) to increase the carbon content in the very outer surface of the part and to improve the combination of properties.

Medium-carbon steels contain 30 to 50 points of carbon (0.30–0.50 percent). Most machine elements having moderate to high strength requirements with fairly good ductility and moderate hardness requirements come from this group.

High-carbon steels have 50 to 95 points of carbon (0.50–0.95 percent). The high carbon content provides better wear properties suitable for applications requiring durable cutting edges and those where surfaces are subjected to constant abrasion. Tools, knives, chisels, and many agricultural implement components are among these uses.

A *bearing steel* nominally contains 1.0 percent carbon. Common grades are 50100, 51100, and 52100; the usual four-digit designation is replaced by five digits, indicating 100 points of carbon.

Symbols

d = pitch diameter of pinion (smaller) gear
D = " " " larger gear
N_p = number of teeth on pinion
N_G = " " " " gear
P_c = Circular Pitch
P_d = Diametral Pitch
D_o = Outside Diameter
a = addendum
b = dedendum
c = clearance
D_r = Root Diameter
C = center Distance

GEARING TERMS

MOTT

263 **Pitch Circle** - Remain tangent with other gears pitch circle

263 **Pitch Point** - Point of tangency of pitch diameters

264 **Circular Pitch** (Inches) - arc length : distance from a point of a gear at the pitch circle to a corresponding point on the next adjacent tooth. ey left tooth face to left tooth face

outdated

264 **Diametral Pitch** (TEETH/INCH) - the number of teeth per inch of pitch diameter.

267 Addendum - radial distance from pitch circle to the

Dedendum - radial distance from pitch circle to the bottom of the tooth space.

Clearance - radial distance from the top of a tooth to the bottom of the tooth spacing of the mating gear when the tooth is fully engaged.

Outside Diameter - diameter of the circle that encloses the outside of the gear teeth.

ROOT DIAMETER - diameter of the circle that contains the bottom of the tooth space.

Center Distance - distance from the center of the pinion to the center of the gear.

Table 2-1 Alloy Groups in the AISI Numbering System

10xx	Plain carbon steel: No significant alloying element except carbon and manganese; less than 1.0 percent manganese. Also called *nonresulfurized*.
11xx	Free cutting steel: Resulfurized. Sulfur content (typically 0.10 percent) improves machinability.
12xx	Free cutting steel: Resulfurized and rephosphorized. Presence of increased sulfur and phosphorus improves machinability and surface finish.
12Lxx	Free cutting steel: Lead added to 12xx steel further improves machinability.
13xx	Manganese steel: Nonresulfurized. Presence of approximately 1.75 percent manganese increases hardenability.
15xx	Carbon steel: Nonresulfurized; greater than 1.0 percent manganese.
23xx	Nickel steel: Nominally 3.5 percent nickel.
25xx	Nickel steel: Nominally 5.0 percent nickel.
31xx	Nickel-chromium steel: Nominally 1.25 percent Ni, 0.65 percent Cr.
33xx	Nickel-chromium steel: Nominally 3.5 percent Ni, 1.5 percent Cr.
40xx	Molybdenum steel: 0.25 percent Mo.
41xx	Chromium-molybdenum steel: 0.95 percent Cr, 0.2 percent Mo.
43xx	Nickel-chromium-molybdenum steel: 1.8 percent Ni; 0.5 percent or 0.8 percent Cr; 0.25 percent Mo.
44xx	Molybdenum steel: 0.5 percent Mo.
46xx	Nickel-molybdenum steel: 1.8 percent Ni, 0.25 percent Mo.
48xx	Nickel-molybdenum steel: 3.5 percent Ni, 0.25 percent Mo.
5xxx	Chromium steel: 0.4 percent Cr.
51xx	Chromium steel: Nominally 0.8 percent Cr.
51100	Chromium steel: Nominally 1.0 percent Cr; bearing steel, 1.0 percent C.
52100	Chromium steel: Nominally 1.45 percent Cr; bearing steel, 1.0 percent C.
61xx	Chromium-vanadium steel: 0.50–1.10 percent Cr, 0.15 percent V.
86xx	Nickel-chromium-molybdenum steel: 0.55 percent Ni, 0.5 percent Cr, 0.20 percent Mo.
87xx	Nickel-chromium-molybdenum steel: 0.55 percent Ni, 0.5 percent Cr, 0.25 percent Mo.
92xx	Silicon steel: 2.0 percent silicon.
93xx	Nickel-chromium-molybdenum steel: 3.25 percent Ni, 1.2 percent Cr, 0.12 percent Mo.

As indicated in Table 2-1, sulfur, phosphorus, and lead improve the machinability of steels and are added in significant amounts to the 11xx, 12xx, and 12Lxx grades. These grades are used for screw machine parts requiring high production rates although the resulting parts are not subjected to high stresses or wear conditions. In the other alloys, these elements are controlled to a very low level because of their adverse effects, such as increased brittleness.

Table 2-2 Uses of Some Steels

AISI Number	*Applications*
1015	Formed sheet metal parts; machined parts (may be carburized)
1030	General purpose bar-shaped parts, levers, links, keys
1045	Shafts, gears
1080	Agricultural equipment parts (plowshares, disks, rake teeth, mower teeth) subjected to abrasion; springs
1112	Screw machine parts
4140	Gears, shafts, forgings
4340	Gears, shafts, parts requiring good through-hardening
4640	Gears, shafts, cams
5150	Heavy-duty shafts, springs, gears
52100	Bearing races, balls, and rollers (bearing steel)
6150	Gears, forgings, shafts, springs
8650	Gears, shafts
9260	Springs

Nickel improves the toughness, hardenability, and corrosion resistance of steel and is included in most of the alloy steels. Chromium improves hardenability, wear and abrasion resistance, and strength at elevated temperatures. In high concentrations, chromium provides significant corrosion resistance, as discussed in the section on stainless steels. Molybdenum also improves hardenability and high-temperature strength.

Table 2-2 lists some common steels and their typical uses.

The steel selected for a particular application must be economical and provide optimum properties of strength, ductility, toughness, machinability, and formability. Frequently, metallurgists, manufacturing engineers, and heat treatment specialists are consulted.

2-4 CONDITIONS FOR STEELS AND HEAT TREATMENT

The final properties of steels are dramatically affected by the way they are produced. Some processes involve mechanical working, such as rolling to a particular shape or drawing through dies. In machine design, many bar-shaped parts, shafts, wire, and structural members are produced in these ways. But most machine parts, particularly those carrying heavy loads, are heat-treated to produce high strength with acceptable toughness and ductility.

Carbon steel bar and sheet forms are usually delivered in the *as-rolled condition;* that is, they are rolled at an elevated temperature that eases the rolling process. The rolling can also be done cold to improve strength and surface finish. Cold-drawn bar and wire have the highest strength of the worked forms, along with a very good surface finish. However, when a material is designated to be *as-rolled,* it should be assumed that it was hot-rolled.

Heat Treating

Heat treating is any process in which steel is subjected to elevated temperatures to modify its properties. Of the several processes available, those most used for machine steels are annealing, normalizing, through-hardening (quench and temper), and case hardening.

Annealing

Annealing is performed by heating the steel to approximately 1 500°F (815°C) and then slow-cooling it to room temperature. This treatment produces a soft, low-strength form of the material, free of significant internal stresses. Parts are frequently cold-formed or machined in the annealed condition. The process also sometimes follows welding, machining, or cold-forming to relieve residual stresses and thereby minimize subsequent distortion.

Normalizing

Normalizing is performed in a similar manner to annealing but at a higher temperature, above the transformation range where austenite is formed, approximately 1 600°F (870°C). The result is a uniform internal structure in the steel and somewhat higher strength than annealing produces. Machinability and toughness are usually improved over the as-rolled condition.

Through-hardening

Through-hardening is accomplished by heating the steel to above the transformation range where austenite forms and then rapidly cooling in a quenching medium. The rapid cooling causes the formation of martensite, the hard, strong form of steel. The degree to which martensite forms depends on the alloy's composition. An alloy containing a minimum of 80 percent of its structure in the martensite form over the entire cross section has *high hardenability*. This is an important property to look for when selecting a steel requiring high strength and hardness. The common quenching media are water, brine, and special mineral oils. The selection of a quenching medium depends on the rate at which cooling should proceed. Most machine steels use either oil or water quenching.

Tempering

Tempering is usually performed immediately after quenching and involves reheating the steel to a temperature of 400 to 1 300°F (200–700°C) and then slow-cooling back to room temperature. This process modifies the steel's properties: tensile strength and yield strength decrease with increasing tempering temperature, whereas ductility improves, as indicated by an increase in the percent elongation. Thus the designer can tailor the properties of the steel to meet specific requirements. Furthermore, the steel in its as-quenched condition has high internal stresses and is usually quite brittle. Machine parts should normally be tempered at 700°F (370°C) or higher after quenching.

To illustrate the effects of tempering on the properties of steels, several charts in Appendix A-4 show graphs of strength versus tempering temperature. Included in these charts are the tensile strength, yield point, percent elongation, percent reduction of area, and the hardness number HB, all plotted in relation to tempering temperature. Note the difference in the shape of the curves and the absolute values of the strength and hardness when comparing the plain carbon AISI 1040 steel with the alloy steel AISI 4340. Even though both have the same nominal carbon content, the alloy steel reaches a much higher strength and hardness. Note also the as-quenched hardness in the upper right part of the heading of the charts; it indicates the degree to which a given alloy can be hardened. When using the case hardening processes described next, the as-quenched hardness becomes very important.

Appendix A-3 lists the range of properties that can be expected for several grades of carbon and alloy steels. The alloys are listed with their AISI numbers and conditions. For the heat-treated conditions, the designation reads, for example, AISI 4340 OQT 1000, which indicates that the alloy was oil-quenched and tempered at 1 000°F. Expressing the properties at the 400°F and 1 300°F tempering temperatures indicates the end points of the possible range of properties that can be expected for that alloy. To specify a strength between these limits, you could refer to graphs such as those shown in Appendix A-4 or determine the required heat-treatment process from a specialist. For the purposes of material specification in this book, a rough interpolation between given values will be satisfactory.

Case Hardening

In many cases, the bulk of the part requires only moderate strength although the surface must have a very high hardness. In gear teeth, for example, high surface hardness is necessary to resist wear as the mating teeth come into contact several million times during the expected life of the gears. At each contact a high stress develops at the surface of the teeth. For applications such as this, *case hardening* is used; the surface (or *case*) of the part is given a high hardness to a depth of perhaps 0.010 to 0.040 in (0.25–1.00 mm), although the interior of the part (the *core*) is affected only slightly if at all. The advantage of surface hardening is that, as the surface receives the required wear-resisting hardness, the core of the part remains in a more ductile form, resistant to impact and fatigue. The processes used most often for case hardening are flame hardening, induction hardening, carburizing, nitriding, cyaniding, and carbo-nitriding.

Figure 2-6 shows a drawing of a typical case-hardened gear tooth section, clearly showing the hard case surrounding the softer, more ductile core. Case hardening is used in applications requiring high wear and abrasion resistance in normal service (gear teeth, crane wheels, wire rope sheaves, and heavy-duty shafts).

The processes of flame hardening and induction hardening involve the rapid heating of the surface of the part for a limited time so that a small controlled depth of the material reaches the transformation range. Upon immediate quenching, only that part above the transformation range would produce the high level of martensite required for high hardness.

Flame hardening uses a concentrated flame impinging on a localized area for a controlled amount of time to heat the part, followed by quenching in a bath or by a stream of water or oil. *Induction hardening* is a process in which the part is surrounded by a coil through which high-frequency electric current is passed. Because of the electrical conductivity of the steel, current is *induced* primarily near the surface of the part. The resistance of the material to the flow of current results in a heating effect. Controlling the electrical

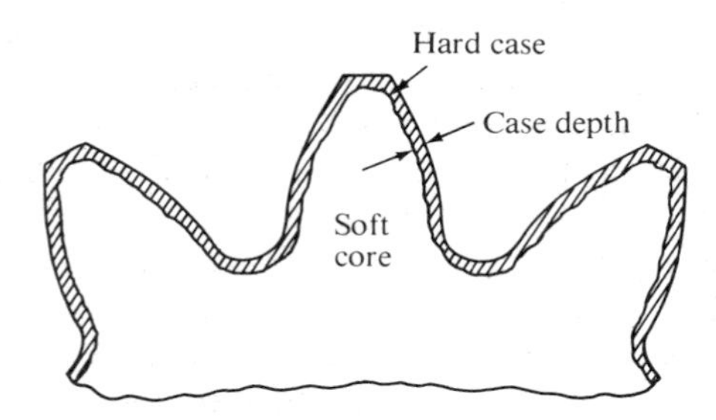

Figure 2-6 Typical Case Hardened Gear Tooth Section

power and frequency of the induction system, and the time of exposure, determines the depth to which the material reaches the transformation temperature. Rapid quenching after heating hardens the surface.

It should be noted that, for flame or induction hardening to be effective, the material must have a good hardenability. Usually the goal of case hardening is to produce a case hardness in the range of Rockwell C hardness HRC 55 to 60 (Brinell hardness approximately HB 550 to 650). Therefore, the material must be capable of being hardened to the desired level. Carbon and alloy steels with fewer than 30 points of carbon typically cannot meet this requirement. So the alloy steels with 40 points or more of carbon are the usual types given flame or induction hardening treatments.

The remaining case hardening processes, carburizing, nitriding, cyaniding, and carbo-nitriding, actually alter the composition of the surface of the material by exposing it to carbon-bearing gases, liquids, or solids at high temperatures that produce carbon and diffuse it into the surface of the part. The concentration and depth of penetration of carbon depend on the nature of the carbon-bearing material and the time of exposure. Nitriding and cyaniding typically result in very hard, thin cases that are good for general wear resistance. Where high load carrying capability in addition to wear resistance is required, as with gear teeth, carburizing is preferred because of the thicker case.

Several steels are produced as carburizing grades. Among these are 1015, 1020, 1022, 1117, 1118, 4118, 4320, 4620, 4820, and 8620. Appendix A-5 lists the expected properties of these carburized steels. Note when evaluating a material for use that the core properties determine its ability to withstand prevailing stresses, and the case hardness indicates its wear resistance. Carburizing, properly done, will virtually always produce a case hardness of from HRC 55 to 64 (Rockwell C hardness) or HB 550 to 700 (Brinell hardness).

Carburizing has several variations that allow the designer to tailor the properties to meet specific requirements. The exposure to the carbon atmosphere takes place at a temperature of approximately 1 700°F (920°C) and usually takes 8 h. Immediate quenching achieves the highest strength, although the case is somewhat brittle. Normally, a part is allowed to cool slowly after carburizing. It is then reheated to approximately 1 500°F (815°C) and then quenched. A tempering at the relatively low temperature of either 300 or 450°F (150 or 230°C) follows, to relieve stresses induced by quenching. As can be seen in Appendix A-5, the higher tempering temperature lowers the core strength and the case hardness by a small amount but, in general, improves the part's toughness. The process just described is *single quenching and tempering*.

When a part is quenched in oil and tempered at 450°F, for example, the condition is *case hardening by carburizing, SOQT 450*. Reheating after the first quench and quenching again further refines the case and core properties; this process is *case hardening by carburizing, DOQT 450*. These are the conditions listed in Appendix A-5.

2-5 STAINLESS STEELS

The term *stainless steel* characterizes the high level of corrosion resistance offered by alloys in this group. To be classified as a stainless steel, the alloy must have a chromium content of at least 10 percent. Most have 12 to 18 percent chromium.

The AISI designates most stainless steels by its 200, 300, and 400 series. Another designation system is the Unified Numbering System (UNS) developed by the SAE and the ASTM. Appendix A-6 lists the properties of several grades, giving both designations.

The three main groups of stainless steels are austenitic, ferritic, and martensitic. *Austenitic* stainless steels fall into the AISI 200 and 300 series. They are general purpose grades with moderate strength. Most are not heat treatable and their final properties are determined by the amount of working, the resulting temper referred to as ¼ hard, ½ hard, ¾ hard, and full hard. These alloys are nonmagnetic and typically used in food processing equipment.

Ferritic stainless steels belong to the AISI 400 series, designated as 405, 409, 430, 446, and so on. They are magnetic and perform well at elevated temperatures, from 1 300 to 1 900°F (700–1 040°C), depending on the alloy. They are non-heat-treatable but they can be cold-worked to improve properties. Typical applications include heat exchanger tubing, petroleum refining equipment, automotive trim, furnace parts, and chemical equipment.

Martensitic stainless steels are also members of the AISI 400 series, including 403, 410, 414, 416, 420, 431, and 440 types. They are magnetic, can be heat treated, and have higher strength than the 200 and 300 series, while retaining good toughness. Typical uses include turbine engine parts, cutlery, scissors, pump parts, valve parts, surgical instruments, aircraft fittings, and marine hardware.

There are many other grades of stainless steels, many of which are proprietary to particular manufacturers. A group used for high-strength applications in aerospace, marine, and vehicular applications is of the precipitation hardening type. They develop very high strengths with heat treatments at relatively low temperatures, from 900 to 1 150°F (480–620°C). This characteristic helps to minimize distortion during treatment. Some examples are 17-4PH, 15-5PH, 17-7PH, PH15-7Mo, and AMS362 stainless steels.

2-6 STRUCTURAL STEEL

Most structural steels are designated by ASTM numbers established by the American Society for Testing and Materials. The most common grade is ASTM A36, which has a minimum yield point of 36 000 psi (248 MPa) and is very ductile. It is basically a low-carbon, hot-rolled steel available in sheet, plate, bar, and structural shapes such as wide-flange beams, American Standard beams, channels, and angles. The geometric properties of some of each of these sections are listed in Appendix A-16.

Other standard structural steels are ASTM A283, A284, and A285, which have yield points in the range of 21 000 to 33 000 psi, produced mostly in sheet and plate form.

Many higher-strength grades of structural steel are available for use in construction, vehicular, and machine applications. They provide yield points in the range from 42 000 to 100 000 psi (290–700 MPa). Some of these grades, referred to as high-strength, low-alloy (HSLA) steels, are ASTM A242, A440, A441, A514, A572, A588, and A607.

Appendix A-7 lists the properties of several structural steels.

2-7 CAST IRON

Large gears, machine structures, brackets, linkage parts, and other important machine parts are made from cast iron. The several types and grades available span wide ranges of strength, ductility, machinability, and wear resistance that are attractive in many applications. The three most commonly used types of cast iron are gray iron, ductile iron, and malleable iron. Appendix A-8 lists the properties of several cast irons.

Gray iron is available in grades having tensile strengths ranging from 20 000 to 60 000 psi (138–414 MPa). Its ultimate compressive strength is much higher, three to five times as high as the tensile strength. One disadvantage of gray iron is that it is brittle and, therefore, should not be used in applications where impact loading is likely. But it has excellent wear resistance, is relatively easy to machine, has good vibration damping ability, and can be surface hardened. Applications include engine blocks, gears, brake parts, and machine bases. The gray irons are rated by the ASTM specification A48-76 in classes 20, 25, 30, 40, 50, and 60, where the number refers to the minimum tensile strength in Kips/in^2 (Ksi). For example, class 40 gray iron has a minimum tensile strength of 40 Ksi or 40 000 psi (276 MPa). Because it is brittle, gray iron does not exhibit the property of yield strength.

Ductile irons have higher strengths than the gray irons and, as the name implies, are more ductile. However, their ductility is still much lower than that of typical steels. A three-part grade designation is used for ductile iron in the ASTM A536-77 specification. The first number refers to the tensile strength in Ksi, the second is the yield strength in Ksi, and the third is the approximate percent elongation. For example, the grade 80-55-06 has a tensile strength of 80 Ksi (552 MPa), a yield strength of 55 Ksi (379 MPa), and a 6 percent elongation in 2.00 in. Higher-strength cast parts such as crankshafts and gears are made from ductile iron.

Malleable iron is a group of heat-treatable cast irons with moderate to high strength, high modulus of elasticity (stiffness), good machinability, and good wear resistance. The five-digit designation roughly indicates the yield strength and the expected percent elongation of the iron. For example, grade 40010 has a yield strength of 40 Ksi (276 MPa) and a 10 percent elongation. The strength properties listed in Appendix A-8 are for the non-heat-treated condition. Higher strengths would result from heat treating.

2-8 POWDERED METALS

Making parts with intricate shapes by powder metallurgy can sometimes eliminate the need for extensive machining. Metal powders are available in many formulations whose properties approach the wrought form of the metal. The processing involves preparing a preform by compacting the powder in a die under high pressure. Sintering at a high temperature to fuse the powder into a uniform mass is the next step. Repressing is sometimes done to improve properties or dimensional accuracy of the part. Typical parts made by the powder metallurgy (PM) process are gears, gear segments, cams, eccentrics, and various machine parts having oddly shaped holes or projections. Dimensional tolerances of 0.001 to 0.005 in (0.025–0.125 mm) are typical. One disadvantage of PM parts is that they are usually brittle and should not be used in applications where impact loading is expected. Another important application is in sintered bearings, which are made to a relatively low density with consequent high porosity. The bearing is impregnated with a lubricant that may be sufficient for the life of the part. This type of material is discussed further in Chapter 14.

Manufacturers of metal powders have many proprietary formulations and grades. However, the Metal Powder Industries Federation (MPIF) is promoting standardization of materials. Figure 2-7 shows photographs of some powder metal parts, and the table in Appendix A-9 lists properties of some metal powders.

2-9 TOOL STEELS AND CARBIDES

When cutting edges and/or high wear resistance are required of a machine part, one of several tool steels or carbides may be selected. Table 2-3 lists the general properties and applications for the most common tool steels.

Figure 2-7 Examples of Powder Metal Components (Imperial Clevite Inc., Powder Metal Products Division, Salem, Ind.)

2-10 ALUMINUM

Aluminum is widely used for structural and mechanical applications. Chief among its attractive properties are light weight, good corrosion resistance, relative ease of forming and machining, and pleasing appearance. Its density is approximately one-third that of steel. However, its strength is somewhat lower, also.

The standard designations for aluminum alloys listed by the Aluminum Association use a four-digit system. The first digit indicates the alloy type according to the major alloying element. Table 2-4 lists the commonly used alloy groups. The second digit, if it is other than zero, indicates modifications of another alloy or limits placed on impurities in the alloy. The presence of impurities is particularly important for electrical conductors. Within each group there are several specific alloys, indicated by the last two digits in the designation.

Table 2-5 lists several common alloys, along with the forms in which they are typically produced in some of their major applications. Of the fifty or more available alloys Table 2-5 lists several that span the range of typical applications. The table should aid you in selecting a suitable alloy for a particular application.

The mechanical properties of the aluminum alloys are highly dependent on their condition. For this reason, the specification of an alloy is incomplete without a reference to its *temper*. The following list describes the usual tempers given to aluminum alloys. Note that some alloys respond to heat treating, and others are processed by strain hardening. *Strain hardening* is controlled cold working of the alloy, in which increased working increases hardness and strength while reducing ductility. The following are the tempers available:

F—as fabricated: No special control of properties is provided. Actual limits are unknown. This temper should be accepted only when the part can be thoroughly tested prior to service.

O—annealed: A thermal treatment that results in the softest and lowest strength condition. Sometimes specified to obtain the most workable form of the alloy. The resulting part can be heat-treated for improved properties if it is made from alloys in the 2xxx, 4xxx, 6xxx, or 7xxx series. Also, the working itself may provide some improvement in properties similar to that produced by strain hardening for alloys in the 1xxx, 3xxx, and 5xxx series.

Table 2-3 Tool Steels and Carbides

	Group	Wear Resistance	Toughness	Applications
T	Tungsten base high-speed steel	Very high	Low	Tool bits, taps, drills, milling cutters, broaches, reamers, hobs
M	Molybdenum base high-speed steel	Very high	Low	Variety of cutting tools
H	Chromium base hot-work steel	Fair	Good	Dies for extrusion, forging, die casting, highly stressed aircraft parts
H	Tungsten base hot-work steel	Good	Good	Dies for extrusion, forging
H	Molybdenum base hot-work steel	High	Medium	Dies for extrusion, forging
D	High-carbon, high-chromium cold-work steel	Good	Poor	Forming dies, gauges, molds, thread-rolling dies, abrasion-resistant liners
A	Air hardening cold-work steel	Good	Fair	Slitters, intricate die shapes
O	Oil hardening cold-work steel	Good	Fair	Short-run dies, gauges
S	Shock-resistant tool steels	Fair	Excellent	Impact-loaded tools, chisels, hammers
P	Mold tool steel	Varies	High	Dies for low-temperature die casting, molds for plastics
L	Low-alloy tool steel	Medium	High	Clutch plates, feed fingers, chuck parts
F	Carbon-tungsten tool steel	Varies	Varies	Wire drawing dies, paper-cutting knives
W	Water hardening tool steel	Medium	Good	Dies for cold-heading, coining, cutlery
—	Carbides[1]	—	—	Inserts for cutting tools

[1]Includes columbium carbide, molybdenum carbide, niobium carbide, tantalum carbide, titanium carbide, tungsten carbide (WC or W_2C).

Sources: Penton/IPC. *Machine Design Magazine Materials Reference Issue* 56, no. 8(April 1984); Busche, Michael G. "Mechanical Properties and Tests/A to Z," *Materials Engineering* 65, no. 6(June 1967): 93–104.

H—strain hardened: A process of cold working under controlled conditions that produces improved, predictable properties for alloys in the 1xxx, 3xxx, and 5xxx groups. The greater the amount of cold work, the higher the strength and hardness, although the ductility is decreased. The H designation is followed by two or more digits (usually 12, 14, 16, or 18) that indicate progressively higher strength. However, several other designations are used.

T—heat treated: A series of controlled heating and cooling processes applied to alloys in the 2xxx, 4xxx, 6xxx, and 7xxx groups. The letter *T* is followed by one or more numbers to indicate specific processes. The more common designations for mechanical and structural products are T4 and T6.

Table 2-4 Aluminum Alloy Groups

Alloy Designations (by Major Alloying Element)	
1xxx 99.00 percent or greater aluminum content	
2xxx Copper	5xxx Magnesium
3xxx Manganese	6xxx Magnesium and silicon
4xxx Silicon	7xxx Zinc

Table 2-5 Common Aluminum Alloys and Their Uses

Alloy	*Applications*	*Forms*
1060	Chemical equipment and tanks	Sheet, plate, tube
1350	Electrical conductors	Sheet, plate, tube, rod, bar, wire, pipe, shapes
2014	Aircraft structures and vehicle frames	Sheet, plate, tube, rod, bar, wire, shapes, forgings
2024	Aircraft structures, wheels, machine parts	Sheet, plate, tube, rod, bar, wire, shapes, rivets
2219	Parts subjected to high temperatures (to 600°F)	Sheet, plate, tube, rod, bar, shapes, forgings
3003	Chemical equipment, tanks, cooking utensils, architectural parts	Sheet, plate, tube, rod, bar, wire, shapes, pipe, rivets, forgings
5052	Hydraulic tube, appliances, sheet metal fabrications	Sheet, plate, tube, rod, bar, wire, rivets
6061	Structures, vehicle frames and parts, marine uses	All forms
6063	Furniture, architectural hardware	Tube, pipe, extruded shapes
7001	High-strength structures	Tube, extruded shapes
7075	Aircraft and heavy-duty structures	All forms except pipe

Property data for aluminum alloys are included in Appendix A-10. Because these data are typical values, not guaranteed values, the supplier should be consulted for data at the time of purchase.

For mechanical design applications, alloy 6061 is one of the most versatile types. Note that it is available in virtually all forms, has good strength and corrosion resistance, and is heat-treatable to obtain a wide variety of properties. It has good weldability. In its softer forms, it is easily formed and worked. Then, if higher strength is required, it can be heat-treated after forming. However, it has low machinability.

2-11 TITANIUM

The applications of titanium include aerospace structures and components, chemical tanks and processing equipment, fluids-handling devices, and marine hardware. It has very good corrosion resistance and a high strength-to-weight ratio. Its stiffness and density are between those of steel and aluminum: a modulus of elasticity of approximately 16×10^6 psi (110 GPa) and a density of 0.160 lb/in^3. Typical yield strengths range from 25 Ksi to 175 Ksi (172–1 210 MPa). Disadvantages of titanium include relatively high cost and difficult machining.

The classification of titanium alloys usually falls into four types: commercially pure alpha titanium, alpha alloys, alpha-beta alloys, and beta alloys. Appendix A-11 shows the properties of some of these grades. The term *alpha* refers to the hexagonal close-packed metallurgical structure that forms at low temperatures, and *beta* refers to the high-temperature, body-centered cubic structure.

The grades of commercially pure titanium indicate the approximate expected yield strength of the material. For example, TI-50A has an expected yield strength of 50 000 psi (345 MPa). As a class, these alloys exhibit only moderate strength but good ductility.

One popular grade of alpha alloy is titanium alloyed with 0.20 percent palladium (Pd). Its properties are listed in the Appendix A-11 table for one heat-treat condition. Some alloys can improve high-temperature strength and weldability.

Generally speaking, the alpha-beta alloys and the beta alloys are stronger forms of titanium. They are heat-treatable for close control of their properties. Since several alloys are available, a designer can tailor the properties to meet special needs for formability, machinability, forgeability, corrosion resistance, high-temperature strength, weldability, and creep resistance, as well as basic room temperature strength and ductility.

2-12 BRONZE

Bronze is a class of alloys of copper with several different elements, one of which is usually tin. They are useful in gears, bearings, and other applications where good strength and high wear resistance are desirable.

Wrought bronze alloys are available in four types:

Phosphor bronze: Copper-tin-phosphorus alloy

Leaded phosphor bronze: Copper-tin-lead-phosphorus alloy

Aluminum bronze: Copper-aluminum alloy

Silicon bronze: Copper-silicon alloy

Cast bronze alloys have four main types:

Tin bronze: Copper-tin alloy

Leaded tin bronze: Copper-tin-lead alloy

Nickel tin bronze: Copper-tin-nickel alloy

Aluminum bronze: Copper-aluminum alloy

The cast alloy called *manganese bronze* is actually a high-strength form of brass because it contains zinc, the characteristic alloying element of the brass family. Manganese bronze contains copper, zinc, tin, and manganese.

In the UNS, copper alloys are designated by the letter *C*, followed by a five-digit number. Numbers from 10000 to 79900 refer to wrought alloys; 80000 to 99900 refer to casting alloys. See Appendix A-12 for typical properties.

2-13 PLASTICS

Plastics include a wide variety of materials formed of large molecules called *polymers*. The thousands of different plastics are created by combining different chemicals to form long molecular chains.

One method of classifying plastics is by using the terms *thermoplastic* and *thermosetting*. In general, the *thermoplastic* materials can be formed repeatedly by heating or molding because their basic chemical structure is unchanged from its initial linear form. *Thermosetting* plastics do undergo some change during forming and result in a structure in which the molecules are cross-linked and form a network of interconnected molecules. Some designers recommend the terms *linear* and *cross-linked* in place of the more familiar "thermoplastic" and "thermosetting".

Selecting a plastic for a particular application is a complex process and usually involves compromises among properties related to strength, weight, toughness, ease of processing, appearance, life, and other properties. Designers should consult suppliers and evaluate several candidate plastics before setting final specifications.

The properties of plastics related to design performance and safety are similar to those of metals. Data sheets usually report the tensile strength, modulus of elasticity in tension, hardness, impact strength (measured by the Izod test), and strength and modulus in flexure (bending). Other properties to be considered are density, electrical resistivity and dielectric constant, thermal expansion coefficient, water absorption, and appearance factors (color, clarity, weather resistance, and so on).

Listed next are several plastics used for load-carrying parts and therefore of interest to the designer of machine elements. This listing shows the main advantages and uses of a sample of the many plastics available; Appendix A-13 lists typical properties.

Thermoplastics

Nylon: Good strength, wear resistance, and toughness; wide range of possible properties depending on fillers and formulations. Used for structural parts, mechanical devices such as gears and bearings, and parts needing wear resistance.

Acrylonitrile-butadiene-styrene (ABS): Good impact resistance, rigidity, moderate strength. Used for housings, helmets, cases, appliance parts, pipe and pipe fittings.

Polycarbonate: Excellent toughness, impact resistance, and dimensional stability. Used for cams, gears, housings, electrical connectors, food processing products, helmets, and pump and meter parts.

Acrylic: Good weather resistance and impact resistance; can be made with excellent transparency, translucent, or opaque with color. Used for glazing, lenses, signs, and housings.

Polyvinyl chloride (PVC): Good strength, weather resistance, and rigidity. Used for pipe, electrical conduit, small housings, ductwork, and moldings.

Polyimide: Good strength and wear resistance, very good retention of properties at elevated temperatures up to 500°F. Used for bearings, seals, rotating vanes, and electrical parts.

Acetal: High strength, stiffness, hardness, and wear resistance; low friction; good weather resistance and chemical resistance. Used for gears, bushings, sprockets, conveyor parts, and plumbing products.

Polyurethane elastomer: A rubberlike material with exceptional toughness and abrasion resistance; good heat resistance and resistance to oils. Used for wheels, rollers, gears, sprockets, conveyor parts, and tubing.

Thermosets

Phenolic: High rigidity, good moldability and dimensional stability, very good electrical properties. Used for load-carrying parts in electrical equipment, switchgear, terminal strips, small housings, handles for appliances and cooking utensils, gears, and structural and mechanical parts. Alkyd, allyl, and amino thermosets have properties and uses similar to those of the phenolics.

Polyester: Known as *fiberglass* when reinforced with glass fibers. High strength and stiffness, good weather resistance. Used for housings, structural shapes, and panels.

REFERENCES

1. Adamas Carbide Corporation. *A Dictionary of Carbide Terms.* Kenilworth, N.J.: Adamas Carbide Corporation, n.d.
2. Aluminum Association. *Aluminum Standards and Data.* New York: Aluminum Association, 1982.
3. "The Basics of Carburizing." *American Machinist* 120(October 1976): 127–129.
4. Bethlehem Steel Corporation. *Modern Steels and Their Properties.* Bethlehem, Pa.: Bethlehem Steel Corporation, 1980.
5. Earle M. Jorgensen Company. *Steel and Aluminum Stock List and Reference Book.* Dayton: Earle M. Jorgensen Company, 1979.
6. Kazanas, H. C. *Properties and Uses of Ferrous and Nonferrous Metals.* Ann Arbor, Mich.: Prakken Publications, 1979.
7. Penton/IPC. *Machine Design Magazine Materials Reference Issue* 56, no. 8(April 1984).
8. Busche, Michael G. "Mechanical Properties and Tests/A to Z," *Materials Engineering* 65, no. 6 (June 1967): 93–104.

QUESTIONS

1. Define *ultimate tensile strength.*
2. Define *yield point.*
3. Define *yield strength* and tell how it is measured.
4. What types of materials would have a yield point?
5. What is the difference between proportional limit and elastic limit?
6. Define *Hooke's law.*
7. What property of a material is a measure of its stiffness?
8. What property of a material is a measure of its ductility?
9. If a material is reported to have a percent elongation in a 2.00-in gage length of 2 percent, is it ductile?
10. Define *Poisson's ratio.*
11. If a material has a tensile modulus of elasticity of 114 GPa and a Poisson's ratio of 0.33, what is its modulus of elasticity in shear?
12. A material is reported to have a Brinell hardness of 525. What is its approximate hardness on the Rockwell C scale?
13. A steel is reported to have a Brinell hardness of 450. What is its approximate tensile strength?

Questions 14-17: What is wrong with the following statements?

14. "After annealing, the steel bracket had a Brinell hardness of 750."
15. "The hardness of that steel shaft is HRB 120."
16. "The hardness of that bronze casting is HRC 12."
17. "Based on the fact that this aluminum plate has a hardness of HB 150, its approximate tensile strength is 75 Ksi."
18. Name two tests used to measure impact strength.
19. What are the principal constituents in steels?

20. What are the principal alloying elements in AISI 4340 steel?
21. How much carbon is in AISI 4340 steel?
22. What is the typical carbon content of a low-carbon steel? For a medium-carbon steel? For a high-carbon steel?
23. How much carbon does a bearing steel typically contain?
24. What is the main difference between AISI 1213 steel and AISI 12L13 steel?
25. Name four materials that are commonly used for shafts.
26. Name four materials that are typically used for gears.
27. Describe the properties desirable for the auger blades of a post hole digger and suggest a suitable material.
28. Appendix A-3 lists AISI 5160 OQT 1000. Describe the basic composition of this material, how it was processed, and its properties in relation to other steels listed in that table.
29. If a shovel blade is made from AISI 1040 steel, would you recommend flame hardening to give its edge a surface hardness of HRC 40? Explain.
30. Describe the differences between through-hardening and carburizing.
31. Describe the process of induction hardening.
32. Name ten steels used for carburizing. What is their approximate carbon content prior to carburizing?
33. What types of stainless steels are nonmagnetic?
34. What is the principal alloying element that gives a stainless steel corrosion resistance?
35. Of what material is a typical wide-flange beam made?
36. With regard to structural steels, what does the term *HSLA* mean? What strengths are available in HSLA steel?
37. Name three types of cast iron.
38. Describe the following cast iron materials according to type, tensile strength, yield strength, ductility, and stiffness:
    ASTM A48-76, Grade 30
    ASTM A536-77, Grade 100-70-03
    ASTM A47-77, Grade 35018
    ASTM A220-76, Grade 70003
39. Describe the process of making parts from powdered metals.
40. What properties are typical for parts made from heat-treated F-0005-S steel powder?
41. What are the typical properties of and uses for Group D tool steels?
42. What does the suffix *O* in aluminum 6061-O represent?
43. What does the suffix *H* in aluminum 3003-H14 represent?
44. What does the suffix *T* in aluminum 6061-T6 represent?
45. Name the aluminum alloy and condition that has the highest strength of those listed in Appendix A-10.
46. Which is one of the most versatile aluminum alloys for mechanical and structural uses?
47. Name three typical uses for titanium alloys.
48. What is the principal constituent of bronze?
49. Describe the bronze having the UNS designation C86200.
50. Name two typical uses for bronze in machine design.
51. Describe the difference between thermosetting plastics and thermoplastics.
52. Suggest a suitable plastic material for each of the following uses:
    a. Gears
    b. Football helmets
    c. Transparent shield
    d. Structural housing
    e. Pipe
    f. Wheels
    g. Electrical switchgear structural part

3 Stress Analysis

3-1 OVERVIEW

A designer is responsible for ensuring the safety of the components and systems he or she designs. Although many factors affect safety, analysis of machine members carrying loads must demonstrate that the levels of stress are safe. This implies, of course, that nothing actually breaks under reasonably foreseeable loads and operating conditions. Safety may also be compromised if components are permitted to deflect excessively even though nothing breaks.

Most of this book emphasizes developing special methods to analyze and design machine elements. But these methods are all based on the fundamentals of stress analysis. This chapter presents a brief review of these fundamentals. It is assumed that you have completed a course in strength of materials but that you may need some review.

In this book, every design approach will ensure that the stress level is below yield in ductile materials, automatically ensuring that the part will not break under a static load. For brittle materials, we will ensure that the stress levels are well below the ultimate tensile strength. Where deflection is critical to safety or performance of a part, it will also be analyzed.

Two other failure modes that apply to machine members are fatigue and wear. *Fatigue* is the response of a part subjected to repeated loads (see Chapter 5). Wear is discussed within the chapters devoted to the elements, such as gears, bearings, and chains, for which it is a major concern.

This chapter focuses on the basic types of stress analysis: direct tensile and compressive stresses, direct shear stress, torsional shear stress, vertical shear stress, and stress due to bending. The fundamental relationships for deflection due to these kinds of stresses are also presented. The last section reviews stress concentrations.

Combined stresses, including the use of Mohr's circle, are discussed in Chapter 4. Chapter 5 presents methods of designing for different modes of failure and treats design examples. Most of the examples given in this chapter are of the analysis type; that is, the stress resulting from a given manner of loading is computed to review the stress analysis.

3-2 DIRECT STRESSES: TENSION AND COMPRESSION

Stress can be defined as the internal resistance offered by a unit area of a material to an externally applied load. *Normal stresses* (σ) are either *tensile* (positive) or *compressive* (negative).

For a load-carrying member in which the external load is uniformly distributed across the cross-sectional area of the member, the magnitude of the stress can be calculated from the direct stress formula:

$$\sigma = \text{Force/area} = F/A \qquad (3\text{-}1)$$

The units for stress are always expressed as *force per unit area*. (See Table 1-1.) In the U.S. Customary Unit system, the units for stress are generally expressed in lb/in^2 or psi. For convenience, stress is sometimes expressed as Kips/in^2 (Ksi). In the SI system the unit for force is the newton (N). Thus the standard unit for stress is N/m^2, called the *pascal* (Pa). Since the pascal is a very small unit, the typical stress level in machine elements, particularly in metallic parts, is in the *megapascal* (10^6 pascal) range. It is also good to remember that

$$1.0\ \text{MPa} = 1.0\ \text{N/mm}^2$$

This relationship is useful since the cross-sectional area of typical machine parts is expressed in mm^2. For example, if a tensile force of 9 500 N is applied to a round bar having a diameter of 12 mm, the direct tensile stress is

$$\sigma = F/A$$
$$A = (\pi D^2)/4 = [(\pi)(12\text{ mm})^2]/4 = 113\text{ mm}^2$$
$$\sigma = (9\,500\text{ N})/(113\text{ mm}^2) = 84.0\text{ N/mm}^2 = 84.0\text{ MPa}$$

The conditions on the use of equation (3-1) are as follows:

1. The load-carrying member must be straight.
2. The line of action of the load must pass through the centroid of the cross section of the member.
3. The member must be of uniform cross section near where the stress is being computed.
4. The material must be homogeneous and isotropic.
5. In the case of compression members, the member must be short to prevent buckling. The conditions under which buckling is expected are discussed in Chapter 6.

Deformation under Direct Axial Loading

The following formula computes the stretch due to a direct axial tensile load or the shortening due to a direct axial compressive load:

$$\delta = FL/EA \tag{3-2}$$

In this equation, δ = the total deformation of the member carrying the axial load. Also:

F = Direct axial load
L = Original total length of the member
E = Modulus of elasticity of the material
A = Cross-sectional area of the member

Noting that $\sigma = F/A$, the deformation can also be computed from:

$$\delta = \sigma L/E \tag{3-3}$$

For example, if the tension rod analyzed for stress previously has a length of 3 600 mm and is made from steel having E = 207 GPa, its total elongation would be

$$\delta = \frac{\sigma L}{E} = \frac{(84.0 \times 10^6\text{ N/m}^2)(3\,600\text{ mm})}{(207 \times 10^9\text{ N/m}^2)} = 1.46\text{ mm}$$

3-3 DIRECT SHEAR STRESS

Direct shear stress occurs when the applied force tends to cut through the member as scissors or shears do or when a punch and die are used to punch a slug of material from a sheet. Another important example of direct shear in machine design is the tendency for a key to be sheared off at the section between the shaft and the hub of a machine element when transmitting torque. Figure 3-1 shows the action.

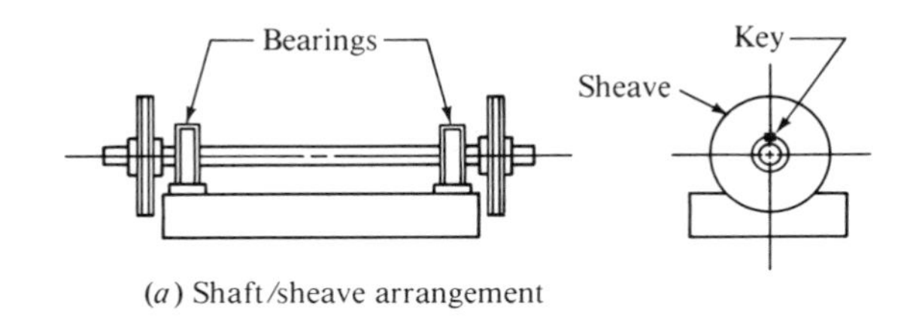

(a) Shaft/sheave arrangement

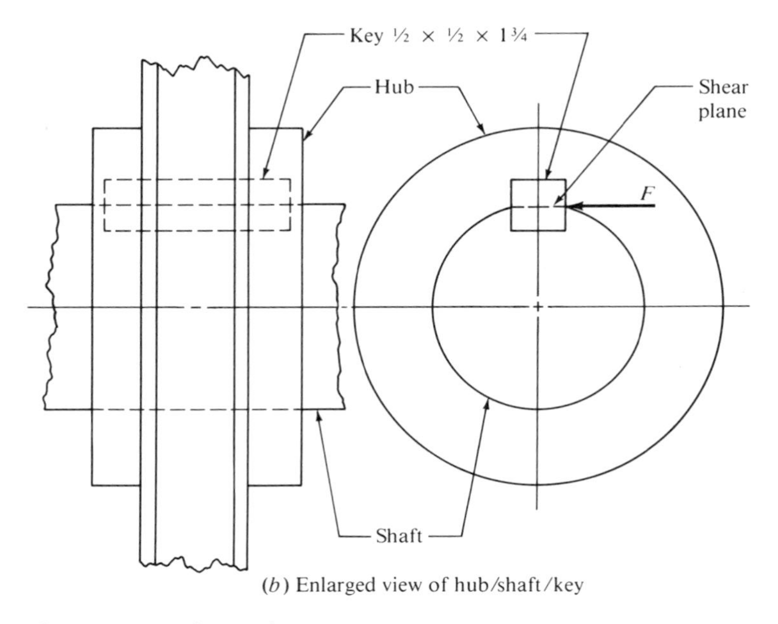

(b) Enlarged view of hub/shaft/key

Figure 3-1 Direct Shear on a Key

The method of computing direct shear stress is similar to that used for computing direct tensile stress because the applied force is assumed to be uniformly distributed across the cross section of the part that is resisting the force. But that kind of stress is *shear stress* rather than *normal stress*. The symbol used for shear stress is the Greek letter tau (τ). The formula for direct shear stress can thus be written:

$$\tau = \text{Shearing force/area in shear} = F/A_s \tag{3-4}$$

This is more properly called the *average shearing stress,* but we will make the simplifying assumption that the stress is uniformly distributed across the shear area.

For example, consider the key in Figure 3-1. Assume that it is square, 0.50 in on a side, and has a length of 1.75 in. Also assume that the force that acts at the interface between the shaft and the hub of the sheave is 12 500 lb. Then the direct shear stress on the key is

$$\tau = F/A_s$$

$$A_s = (0.50 \text{ in})(1.75 \text{ in}) = 0.875 \text{ in}^2$$

$$\tau = (12\,500 \text{ lb})/(0.875 \text{ in}^2) = 14\,300 \text{ lb/in}^2 = 14\,300 \text{ psi}$$

3-4 TORSIONAL SHEAR STRESS

When a *torque,* or twisting moment, is applied to a member, it tends to deform by twisting, causing a rotation of one part of the member relative to another. Such twisting causes a shear stress in the member. For a small element of the member the nature of the stress is the same as that experienced under direct shear stress. However, in *torsional shear,* the distribution of stress is *not* uniform.

The most frequent case of torsional shear in machine design is that of a round circular shaft transmitting power. The relationship between the power (P), rotational speed (n), and torque (T) in a shaft is described by the equation

$$T = P/n \tag{3-5}$$

In Si units, power is expressed in the unit of *watt* (N · m/s) and the rotational speed is in radians per second (rad/s). For example, let's compute the amount of torque in a shaft transmitting 750 W of power while rotating at 183 rad/s. (Note: This is equivalent to the output of a 1.0-hp, 4-pole electric motor, operating at its rated speed of 1 750 rpm. See Chapter 17.)

$$T = (750 \text{ N} \cdot \text{m/s})/(183 \text{ rad/s})$$

$$T = 4.10 \text{ N} \cdot \text{m/rad} = 4.10 \text{ N} \cdot \text{m}$$

In such calculations, the unit of N · m /rad is dimensionally correct, and some advocate its use. Most, however, consider the radian to be dimensionless, resulting in torque in N · m or other familiar units of force times distance.

In the U.S. Customary Unit system, power is typically expressed as *horsepower,* equal to 550 ft · lb/s. The typical unit for rotational speed is rpm, or revolutions per minute. But the most convenient unit for torque is the pound-inch (lb · in). Considering all these quantities and making the necessary conversions of units, the following formula is used to compute the torque in a shaft carrying a certain power P (in hp) while rotating at a speed of n rpm.

$$T = (63\,000P)/n \tag{3-6}$$

The resulting torque will be in pound-inches. You should verify the value of the constant, 63 000.

For example, let's compute the torque on a shaft transmitting 1.0 hp while rotating at 1 750 rpm. Note that these are approximately the same conditions for which the torque was computed in the example using SI units.

$$T = [63\,000(1.0)]/1\,750 = 36.0 \text{ lb-in}$$

Torsional Shear Stress Formula

When subjected to a torque, the outer surface of a solid round shaft experiences the greatest shearing strain and therefore the largest torsional shear stress. The value of the maximum torsional shear stress is found from

$$\tau_{max} = Tc/J \tag{3-7}$$

where c is the radius of the shaft and J is the polar moment of inertia. See Appendix A-1 for formulas for J.

Example. Compute the maximum torsional shear stress in a shaft having a diameter of 10 mm when it carries a torque of 4.10 N · m.

$$J = \pi D^4/32 = [(\pi)(10 \text{ mm})^4]/32 = 982 \text{ mm}^4$$

$$c = D/2 = 5.0 \text{ mm}$$

$$\tau = \frac{(4.10 \text{ N} \cdot \text{m})(5.0 \text{ mm})}{982 \text{ mm}^4} \frac{10^3 \text{ mm}}{\text{m}} = 20.9 \text{ N/mm}^2 = 20.9 \text{ MPa}$$

If it is desired to compute the torsional shear stress at some point inside the shaft, the more general formula is used:

$$\tau = Ty/J \tag{3-8}$$

where y is the radial distance from the center of the shaft to the point of interest. Figure 3-2 shows graphically that this equation is based on the linear variation of the torsional shear stress from zero at the center of the shaft to the maximum value at the outer surface.

Equations (3-7) and (3-8) apply also to hollow shafts (Figure 3-3 shows the distribution of shear stress). Again note that the maximum shear stress occurs at the outer surface. Also note that the entire cross section carries a relatively high stress level. Because of this, the hollow shaft is more efficient. Notice that the material near the center of the solid shaft is not highly stressed.

For design, it is convenient to define the *polar section modulus*, Z_p, where

$$Z_p = J/c \tag{3-9}$$

Then the equation for the maximum torsional shear stress is

$$\tau_{max} = T/Z_p \tag{3-10}$$

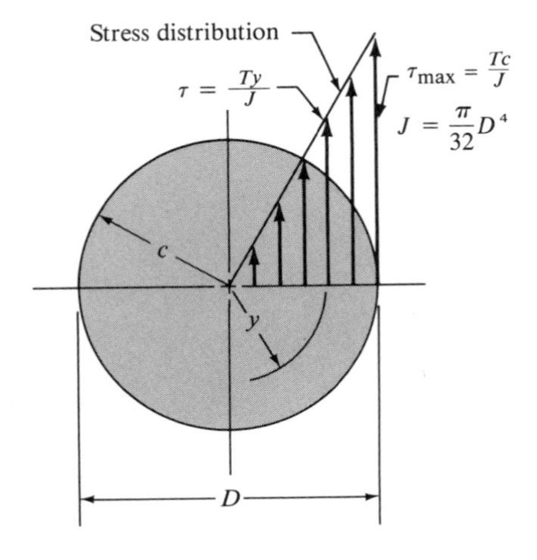

Figure 3-2 Stress Distribution in Solid Shaft

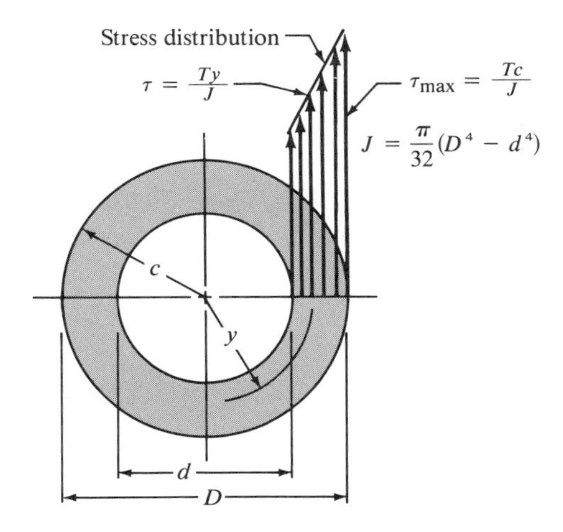

Figure 3-3 Stress Distribution in Hollow Shaft

Formulas for the polar section modulus are also in Appendix A-1.

Torsional Deformation

When a shaft is subjected to a torque, it undergoes a twisting in which one cross section is rotated relative to other cross sections in the shaft. The angle of twist is computed from

$$\theta = TL/GJ \tag{3-11}$$

In this formula:

θ = Angle of twist (radians)

L = Length of the shaft over which the angle of twist is being computed

G = Modulus of elasticity of the shaft material in *shear*

Example. Compute the angle of twist of a 10-mm diameter shaft carrying 4.10 N · m of torque if it is 250 mm long and made of steel with $G = 80$ GPa.

For consistency, let $T = 4.10 \times 10^3$ N · mm and $G = 80 \times 10^3$ N/mm². Then

$$\theta = \frac{(4.10 \times 10^3 \text{ N} \cdot \text{mm})(250 \text{ mm})}{(80 \times 10^3 \text{ N/mm}^2)(982 \text{ mm}^4)} = 0.013 \text{ rad}$$

Using π rad = 180 degrees, this is equivalent to 0.75 degrees.

Torsion in Members Having Noncircular Cross Sections

The behavior of members having noncircular cross sections when subjected to torsion is radically different from that for members having circular cross sections. However, the factors of most use in machine design are the maximum stress and the total angle of twist for such members. The formulas for these factors can be expressed in similar forms to the formulas used for members of circular cross section (solid and hollow round shafts).

The following two formulas can be used.

$$\tau_{max} = T/Q \tag{3-12}$$

$$\theta = TL/GK \tag{3-13}$$

Note that equations (3-12) and (3-13) are similar to equations (3-10) and (3-11), with the substitution of Q for Z_p and K for J. Refer to Figure 3-4 for the methods of determining the values for K and Q for several types of cross sections useful in machine design. These

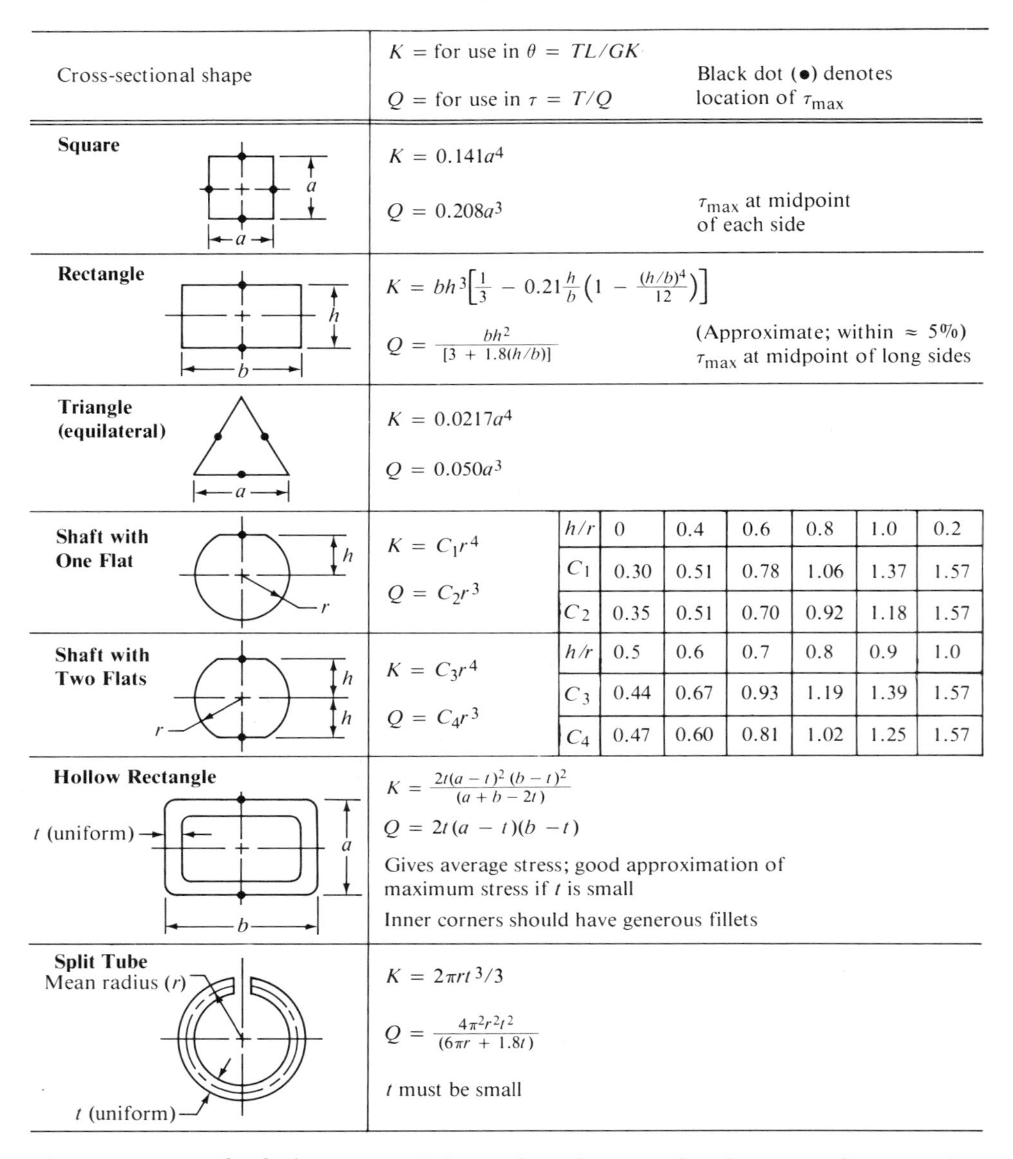

Cross-sectional shape	K = for use in $\theta = TL/GK$; Q = for use in $\tau = T/Q$	Black dot (•) denotes location of τ_{max}
Square	$K = 0.141a^4$; $Q = 0.208a^3$	τ_{max} at midpoint of each side
Rectangle	$K = bh^3\left[\frac{1}{3} - 0.21\frac{h}{b}\left(1 - \frac{(h/b)^4}{12}\right)\right]$; $Q = \frac{bh^2}{[3 + 1.8(h/b)]}$	(Approximate; within ≈ 5%) τ_{max} at midpoint of long sides
Triangle (equilateral)	$K = 0.0217a^4$; $Q = 0.050a^3$	
Shaft with One Flat	$K = C_1r^4$; $Q = C_2r^3$	See table below
Shaft with Two Flats	$K = C_3r^4$; $Q = C_4r^3$	See table below
Hollow Rectangle (t (uniform))	$K = \frac{2t(a-t)^2(b-t)^2}{(a+b-2t)}$; $Q = 2t(a-t)(b-t)$	Gives average stress; good approximation of maximum stress if t is small. Inner corners should have generous fillets
Split Tube Mean radius (r), t (uniform)	$K = 2\pi rt^3/3$; $Q = \frac{4\pi^2r^2t^2}{(6\pi r + 1.8t)}$	t must be small

h/r	0	0.4	0.6	0.8	1.0	0.2
C_1	0.30	0.51	0.78	1.06	1.37	1.57
C_2	0.35	0.51	0.70	0.92	1.18	1.57

h/r	0.5	0.6	0.7	0.8	0.9	1.0
C_3	0.44	0.67	0.93	1.19	1.39	1.57
C_4	0.47	0.60	0.81	1.02	1.25	1.57

Figure 3-4 Methods for Determining Values for *K* and *P* for Several Types of Cross Sections

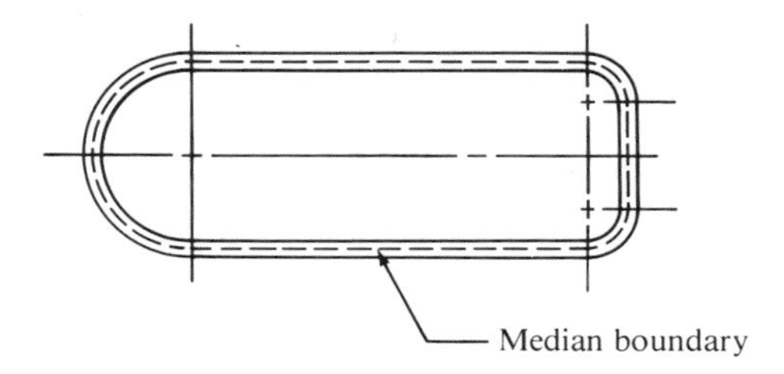

Figure 3-5 Closed, Thin-walled Tube with a Constant Wall Thickness

values are appropriate only if the ends of the member are free to deform. If either end is fixed, as by welding to a solid structure, the resulting stress and angular twist are quite different. See the References listed at the end of this chapter.

Torsion in Closed, Thin-walled Tubes

A general approach for closed, thin-walled tubes of virtually any shape uses equations (3-12) and (3-13) with special methods of evaluating K and Q. Figure 3-5 shows such a tube having a constant wall thickness. The values of K and Q are

$$K = 4A^2t/U \quad (3\text{-}14)$$

$$Q = 2tA \quad (3\text{-}15)$$

where

A = Area enclosed by the median boundary (indicated by the dashed line in Figure 3-5)

t = Wall thickness (must be uniform and thin)

U = Length of the median boundary

The shear stress computed by this approach is the *average stress* in the tube wall. However, if the wall thickness t is small (thin-wall), the stress is nearly uniform throughout the wall, and this approach will yield a close approximation of the maximum stress. For the analysis of tubular sections having nonuniform wall thickness, refer to the references (2, 4, 7).

To design a member to resist torsion only, or torsion and bending combined, it is advisable to select hollow tubes, either round or rectangular, or some other closed shape. They possess good efficiency both in bending and torsion.

The term *open tube* refers to a shape that appears to be a tube but is not completely closed. For example, some tubing is made by starting with a thin, flat strip of steel that is roll-formed into the desired shape (round, square, rectangular, and so on) and then welded along its entire length. Figure 3-6 shows an example for a thin-walled hollow tube. For the closed tube,

$$J = \pi(D^4 - d^4)/32$$

$$J = \pi(3.500^4 - 3.188^4)/32 = 4.592 \text{ in}^4$$

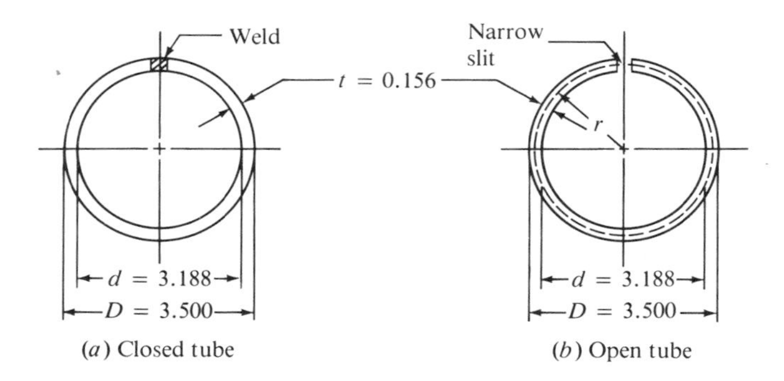

Figure 3-6 Comparison of Closed and Open Tubes

For the open tube before the slit is welded, from Figure 3-4,

$$K = 2\pi r t^3/3$$

$$K = [(2)(\pi)(1.672)(0.156)^3]/3 = 0.0133\ \text{in}^4$$

Note that the angle of twist is inversely proportional to J for the closed tube (equation [3-11]) or to K for the open tube (equation [3-12]). Therefore, the ratio of the stiffness of the closed tube to that of the open (slit) tube is

$$\text{Ratio} = J/K = 4.592/0.0133 = 345$$

Thus, for a given applied torque, the slit tube would twist 345 times as much as the closed tube.

3-5 VERTICAL SHEARING STRESS

A beam carrying loads transverse to its axis will experience shearing forces, denoted by V. In the analysis of beams, it is usual to compute the variation in shearing force across the entire length of the beam and to draw the *shearing force diagram*. Then the resulting vertical shearing stress can be computed from

$$\tau = VQ/It \tag{3-16}$$

In this formula, I is the rectangular moment of inertia of the cross section of the beam and t is the thickness of the section at the place where the shearing stress is to be computed. For most section shapes, the maximum vertical shearing stress occurs at the neutral axis. Specifically, if the thickness is not less at a place away from the neutral axis, then it is assured that the maximum vertical shearing stress occurs at the neutral axis.

The term Q in equation (3-16) is called the *statical moment* of the section: the moment of the area above where the shearing stress is to be calculated. That is,

$$Q = A_p y \tag{3-17}$$

where A_p is that part of the area of the section above the place where the stress is to be computed and y is the distance from the neutral axis of the section to the centroid of the area A_p. Figure 3-7 shows three examples of how Q is computed in typical beam cross sections. In each, the maximum vertical shearing stress occurs at the neutral axis.

An example of the use of equation 3-16 can be shown by referring to part (d) of Figure 3-7, which shows a beam carrying two concentrated loads. The shearing force diagram indicates that the maximum shearing force of 1 000 lb occurs between points *A* and *B*. For the rectangular cross section as shown in Figure 3-7(a)

$$I = th^3/12 = [(2)(8)^3]/12 = 85.3 \text{ in}^4$$
$$Q = A_p y = t(h/2)(h/4) = (2)(8/2)(8/4) = 16.0 \text{ in}^3$$

Then the maximum shearing stress is

$$\tau = \frac{(1\,000 \text{ lb})(16.0 \text{ in}^3)}{(85.3 \text{ in}^4)(2.0 \text{ in})} = 93.8 \text{ lb/in}^2 = 93.8 \text{ psi}$$

Part (d) of Figure 3-7 shows this value and the parabolic shape of the shearing stress distribution across the cross section of the rectangular beam.

It should be noted that the vertical shearing stress is equal to the *horizontal shearing stress* because any element of material subjected to a shear stress on one face must have a shear stress on the adjacent face of the same magnitude for the element to be in equilibrium. Figure 3-8 shows this phenomenon.

In most beams, the magnitude of the vertical shearing stress is quite small compared to the bending stress (see the following section). For this reason it is frequently not computed at all. Those cases where it is of importance include the following:

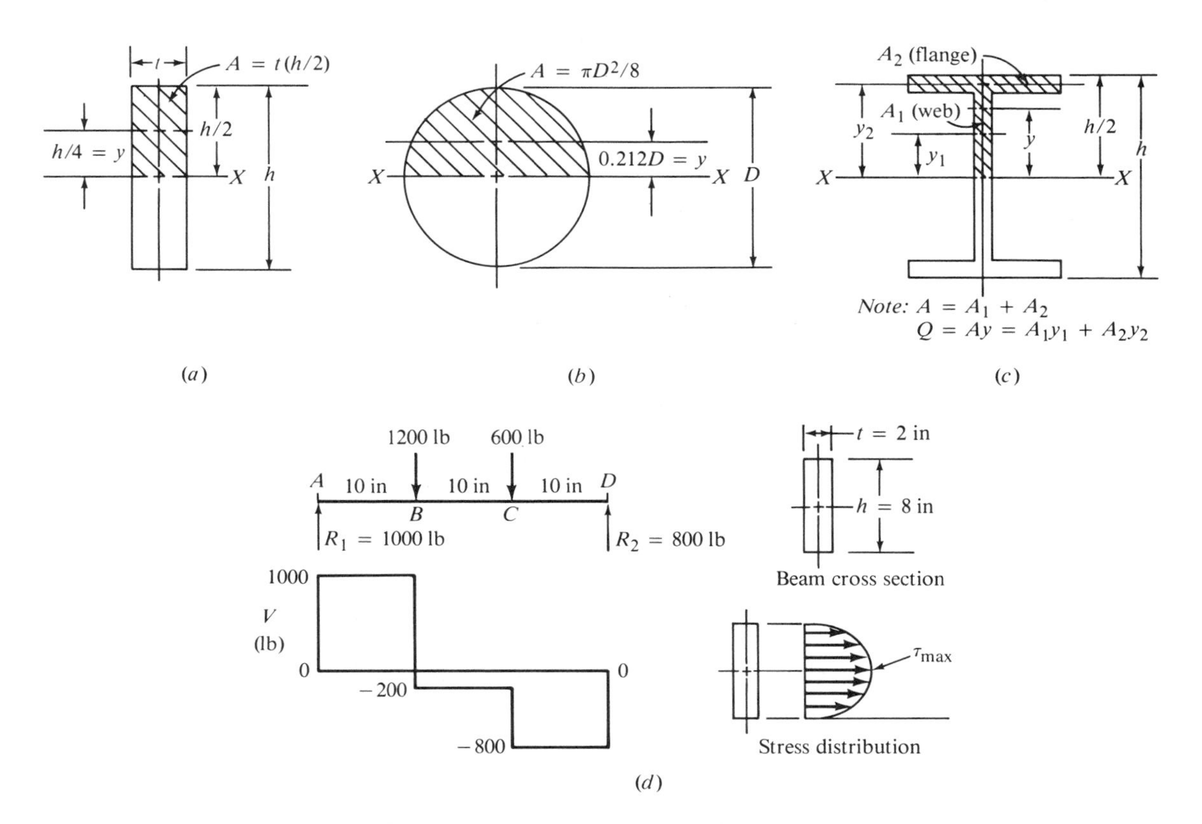

Figure 3-7 Statical Moment Q and Vertical Shear Stress for Beams

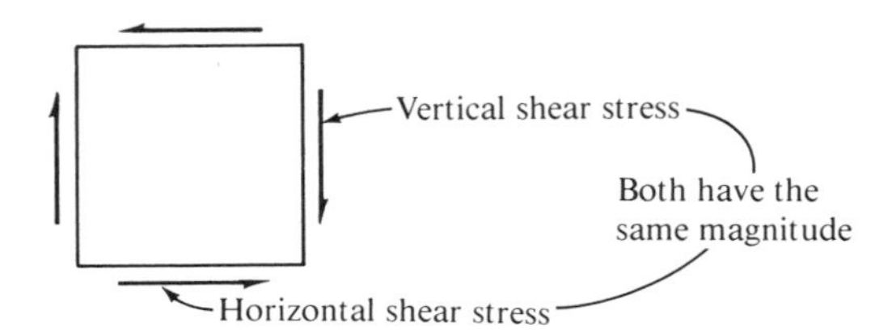

Figure 3-8 Shear Stresses on an Element

When the material of the beam has a relatively low shear strength (such as wood)

When the bending moment is zero or small (and thus the bending stress is small), for example at the ends of simply supported beams and for short beams

When the thickness of the section carrying the shear force is small, as in sections made from rolled sheet, some extruded shapes, and rolled structural shapes such as wide-flange beams

Special Shear Stress Formulas

Equation (3-18) can be cumbersome because of the need to evaluate the statical moment, Q. There are several commonly used cross sections that have special, easy-to-use formulas for the maximum vertical shearing stress.

Rectangle

$$\tau = 3V/2A \qquad \text{(exact)} \tag{3-18}$$

where A is the total cross-sectional area of the beam.

Circle

$$\tau = 4V/3A \qquad \text{(exact)} \tag{3-19}$$

I-shape

$$\tau = V/th \qquad \text{(approximate: about 15 percent low)} \tag{3-20}$$

where t is the web thickness and h is the height of the web (for example, a wide-flange beam).

Thin-walled Tube

$$\tau = 2V/A \qquad \text{(approximate: a little high)} \tag{3-21}$$

In all of these cases, the maximum shear stress occurs at the neutral axis.

3-6 STRESS DUE TO BENDING

A *beam* is a member that carries loads transverse to its axis. Such loads produce bending moments in the beam, which result in the development of bending stresses. Bending stresses are *normal stresses,* that is, either tensile or compressive. The maximum bending stress in a beam cross section will occur in the part farthest from the neutral axis of the section. At that point, the *flexure formula* gives the stress:

$$\sigma = Mc/I \tag{3-22}$$

where M is the magnitude of the bending moment at the section; I is the moment of inertia of the cross section with respect to its neutral axis; and c is the distance from the neutral axis to the outermost fiber of the beam cross section. The magnitude of the bending stress varies linearly within the cross section from a value of zero at the neutral axis, to the maximum tensile stress on one side of the neutral axis, and to the maximum compressive stress on the other side. Figure 3-9 shows a typical stress distribution in a beam cross section.

The flexure formula was developed subject to the following conditions:

1. The beam must be in pure bending. Shear stresses must be zero or negligible. No axial loads are present.
2. The beam must not twist or be subjected to a torsional load.
3. The material of the beam must obey Hooke's law.
4. The modulus of elasticity of the material must be the same in both tension and compression.
5. The beam is initially straight and has a constant cross section.
6. Any plane cross section of the beam remains plane during bending.
7. No part of the beam shape fails because of local buckling or wrinkling.

If condition 1 is not strictly met, the analysis can be continued by using the method of combined stresses presented in Chapter 4. In most practical beams, which are long relative to their height, shear stresses are sufficiently small as to be negligible. Furthermore, the maximum bending stress occurs at the outermost fibers of the beam section,

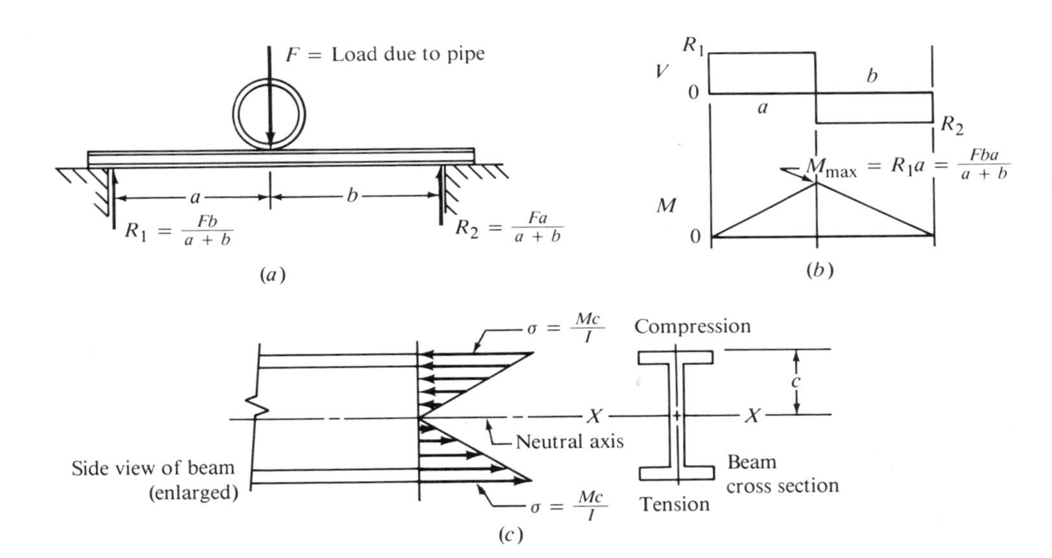

Figure 3-9 Typical Bending Stress Distribution in a Beam Cross Section

where the shear stress is in fact zero. A beam with varying cross section, which would violate condition 5, can be analyzed by using stress concentration factors discussed later in this chapter.

For design, it is convenient to define the term *section modulus*, Z, where $Z = I/c$. The flexure formula then becomes

$$\sigma = M/Z \tag{3-23}$$

Since I and c are geometrical properties of the cross section of the beam, Z is, also. Then, in design, it is usual to define a design stress, σ_d; with the bending moment known,

$$Z = M/\sigma_d \tag{3-24}$$

This results in the required value of the section modulus; from it the required dimensions of the beam cross section can be determined.

Example. For the beam shown in Figure 3-9, the load F due to the pipe is 12 000 lb. The distances are $a = 4$ ft and $b = 6$ ft. Determine the required section modulus for the beam to limit the stress due to bending to 21 600 psi, the recommended design stress for structural steel in static bending.

Using the relations given in the figure,

$$M_{max} = \frac{Fba}{a + b} = \frac{(12\,000 \text{ lb})(6 \text{ ft})(4 \text{ ft})}{(6 \text{ ft} + 4 \text{ ft})} = 28\,800 \text{ lb} \cdot \text{ft}$$

$$Z = \frac{M}{\sigma_d} = \frac{28\,800 \text{ lb} \cdot \text{ft}}{21\,600 \text{ lb/in}^2} \frac{12 \text{ in}}{\text{ft}} = 16.0 \text{ in}^3$$

We could select a beam section having at least this value for the section modulus. An American Standard beam shape, S8X23, has $Z = 16.2$ in^3, as found in Appendix A-16.

The beam section must be loaded in a way that ensures symmetrical bending; that is, there is no tendency for the section to twist under the load. Figure 3-10 shows several shapes that are typically used for beams having a vertical axis of symmetry. If the line of action of the loads on such sections passes through the axis of symmetry, then there is no tendency for the section to twist, and the flexure formula applies.

When there is no vertical axis of symmetry, as with the sections shown in Figure 3-11, care must be exercised in placement of the loads. If the line of action of the loads were as shown as F_1 in the figure, the beam would twist and bend, so the flexure formula would not give accurate results for the stress in the section. For such sections, the load must be placed in line with the *flexural center,* sometimes called the *shear center.* Figure 3-11 shows the approximate location of the flexural center for these shapes (indicated by the symbol Q). Applying the load in line with Q, as shown with the forces labeled F_2, would result in pure bending. A table of formulas for the location of the flexural center is available for the reader (7).

3-7 BEAM DEFLECTIONS

The bending loads applied to a beam cause it to deflect in a direction perpendicular to its axis. A beam that was originally straight will deform to a slightly curved shape. In most cases, the critical factor is either the maximum deflection of the beam or its deflection at specific locations.

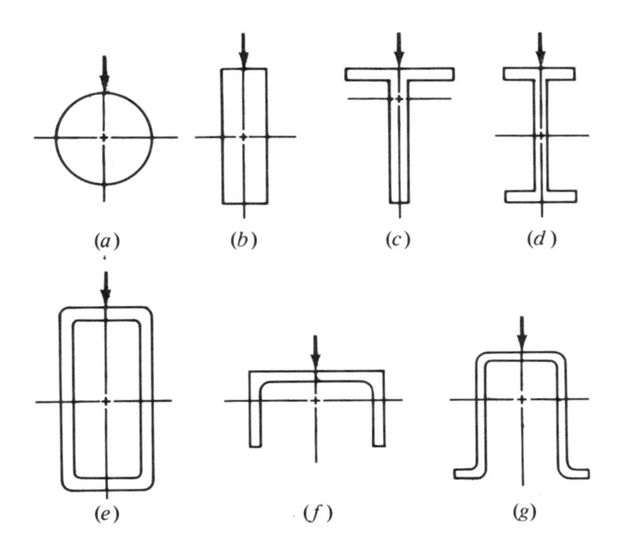

Figure 3-10 Symmetrical Sections. Load applied through axis of symmetry.

Consider the double-reduction speed reducer shown in Figure 3-12. The four gears (A, B, C, D) are mounted on three shafts, each of which is supported by two bearings. The action of the gears in transmitting power creates a set of forces that in turn act on the shafts to cause bending. One component of the total force on the gear teeth acts in a direction that tends to separate the two gears. Thus, gear A is forced upward while gear B is forced downward. For good gear performance, the net deflection of one gear to the other should not exceed 0.005 in (0.13 mm) for medium-sized industrial gearing.

To evaluate the design, there are many methods of computing shaft deflections. We will review briefly those using deflection formulas, superposition, and a general analytical approach.

A set of formulas for computing the deflection of beams at any point or at selected points is useful in many practical problems. Appendix A-14 includes several cases.

For many additional cases, superposition is useful if the actual loading can be divided into parts that can be computed by available formulas. The deflection for each loading is computed separately, then the individual deflections are summed at the points of interest.

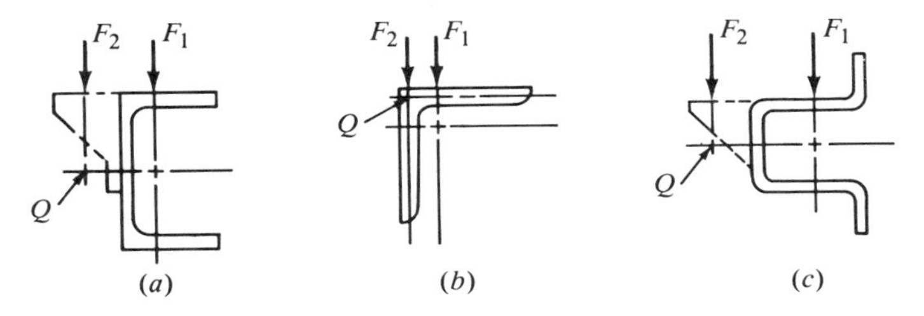

Figure 3-11 Nonsymmetrical Sections. Load applied as at F_1 would cause twisting; loads applied as at F_2 through flexural center Q would cause pure bending.

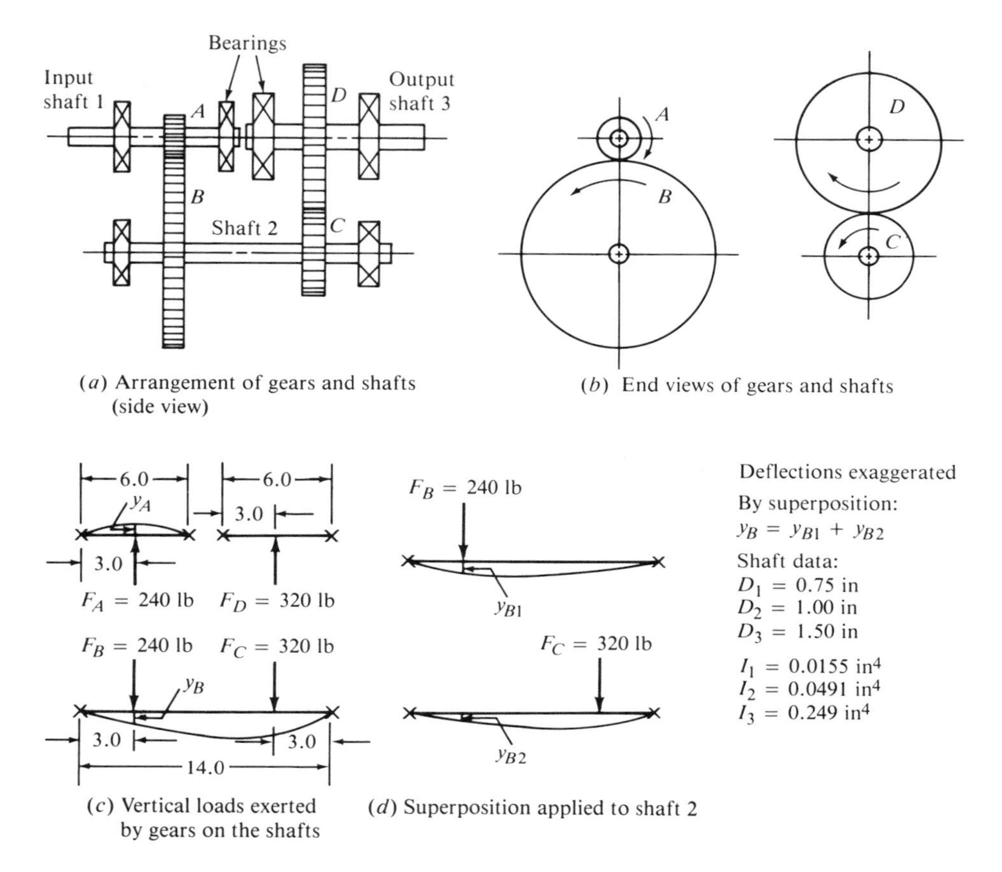

Figure 3-12 Shaft Deflection Analysis for a Double-reduction Speed Reducer

Example. For the two gears, A and B, in Figure 3-12, compute the relative motion between them in the plane of the paper that is due to the forces shown in part (c). These *separating forces,* or *normal forces,* are discussed in Chapters 11 and 12. It is customary to consider the loads at the gears and the reactions at the bearings to be concentrated.

Shaft 1 is a simply supported beam with a single concentrated load at its center (see case 2 in Appendix A-14):

$$y_A = \frac{F_A L_1^3}{48EI} = \frac{(240)(6.0)^3}{48(30 \times 10^6)(0.015\,5)} = 0.002\,3 \text{ in}$$

Shaft 2 is a simply supported beam carrying two nonsymmetrical loads. None of the given beam deflection cases applies directly. Superposition allows us to consider each load, F_B and F_C, separately, as indicated in part (d) of Figure 3-12, and then adding them together.

First we can compute the deflection at B due only to the 240-lb force at B, calling it y_{B1}:

$$y_{B1} = -\frac{F_B a^2 b^2}{3EI_2 L_2} = -\frac{(240)(3.0)^2(11.0)^2}{3(30 \times 10^6)(0.049\,1)(14)} = -0.004\,2 \text{ in}$$

Next we can compute the deflection at B due only to the 320-lb force at C, calling it y_{B2}.

$$y_{B2} = -\frac{F_C av}{6EI_2 L_2}(L_2^2 - v^2 - a^2)$$

$$y_{B2} = -\frac{(320)(3.0)(11.0)}{6(30 \times 10^6)(0.0491)(14)}[(14)^2 - (11.0)^2 - (3.0)^2]$$

$$y_{B2} = -0.0056 \text{ in}$$

Then the total deflection at gear B is

$$y_B = y_{B1} + y_{B2} = -0.0042 - 0.0056 = -0.0098 \text{ in}$$

Because shaft 1 deflects upward and shaft 2 deflects downward, the total *relative deflection* is the sum of y_A and y_B:

$$y_{\text{total}} = y_A + y_B = 0.0023 + 0.0098 = 0.0121 \text{ in}$$

This is a very large deflection for this application.

Equations for Deflected Beam Shape

The general principles relating the deflection of a beam to the loading on the beam and its manner of support are presented here. The result will be a set of relationships among the load, vertical shearing force, bending moment, slope of the deflected beam shape, and the actual deflection curve for the beam. Figure 3-13 shows diagrams for these five factors, with θ as the slope and y indicating deflection of the beam from its initial straight position. The product of modulus of elasticity and the moment of inertia, EI, for the beam is a measure of its stiffness or resistance to bending deflection. It is convenient to combine EI with the slope and deflection values to maintain a proper relationship, as discussed next.

One fundamental concept for beams in bending is

$$\frac{M}{EI} = \frac{d^2y}{dx^2}$$

where M is the bending moment, x is the position on the beam measured along its length, and y is the deflection. Thus, if it is desired to create an equation of the form $y = f(x)$ (that is, y as a function of x), it would be related to the other factors as follows:

$$y = f(x)$$

$$\theta = dy/dx$$

$$\frac{M}{EI} = \frac{d^2y}{dx^2}$$

$$\frac{V}{EI} = \frac{d^3y}{dx^3}$$

$$\frac{w}{EI} = \frac{d^4y}{dx^4}$$

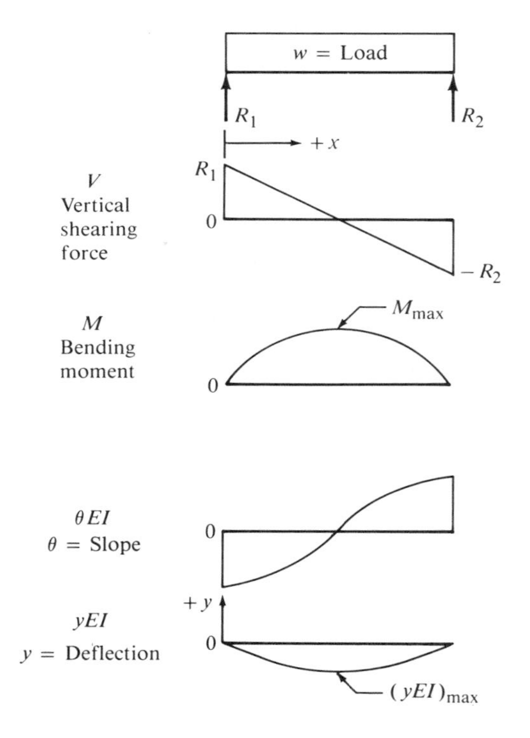

Figure 3-13 Relationships of Load, Vertical Shearing Force, Bending Moment, Slope of Deflected Beam Shape, and Actual Deflection Curve of a Beam

where w is a general term for the load distribution on the beam. The last two equations follow from the observation that there is a derivative (slope) relationship between shear and bending moment and between load and shear.

In practice, the fundamental equations above are used in reverse. That is, the load distribution as a function of x is known and the equations for the other factors are derived by successive integrations. The results are

$$w = f(x)$$

$$V = \int w\,dx + V_0$$

$$M = \int V\,dx + M_0$$

where V_0 and M_0 are constants of integration evaluated from the boundary conditions. In many cases, the load, shear, and bending moment diagrams can be drawn in the conventional manner, and the equations for shear or bending moment can be created directly by the principles of analytic geometry. With M as a function of x, the slope and deflection relations can be found:

$$\theta EI = \int M\,dx + C_1$$

$$yEI = \int \theta EI\,dx + C_2$$

The constants of integration must be evaluated from boundary conditions.

3-8 BEAMS WITH CONCENTRATED BENDING MOMENTS

Figures 3-9 and 3-13 show beams loaded only with concentrated forces or distributed loads. For such loading in any combination, the moment diagram is continuous. That is, there are no points of abrupt change in the value of the bending moment. Many machine elements such as cranks, levers, and brackets carry loads whose line of action is offset from the centroidal axis of the beam in such a way that a concentrated moment is exerted on the beam. Consider the examples shown in Figures 3-14, 3-15, and 3-16.

The bell crank in Figure 3-14 is a part of a linkage in which the 80-lb horizontal force is transferred to F_2 vertically. The crank pivots about O. To design the part of the crank from A to O, we can draw a free body diagram of only that part. Then at O we show the moment equal to 80 lb times 1.50 in, along with the vertical reaction at the pivot equal to F_2. From statics we can show that $F_2 = 60$ lb. The shear and moment diagrams can then be drawn in the conventional manner, the result looking similar to a cantilever. The moment of 120 lb · in is an example of a concentrated bending moment.

Figure 3-15 represents a print head. The force F presses the character to be printed against the ribbon, which imprints the character on the paper backed up by the fixed platen. The free body diagram and the shear and moment diagrams are also shown. Note that the free body diagram shows only the horizontal part of the print head being designed. Thus the concentrated bending moment of F times h acts at its right end. Also note that the moment arm h is the distance from the line of action of the force to the *neutral axis of the horizontal beam*.

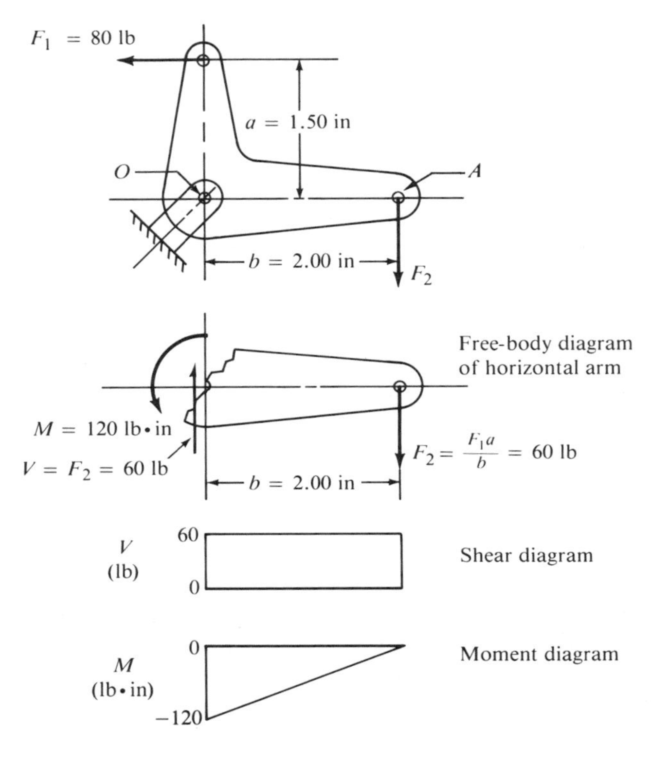

Figure 3-14 Bending Moment in a Bell Crank

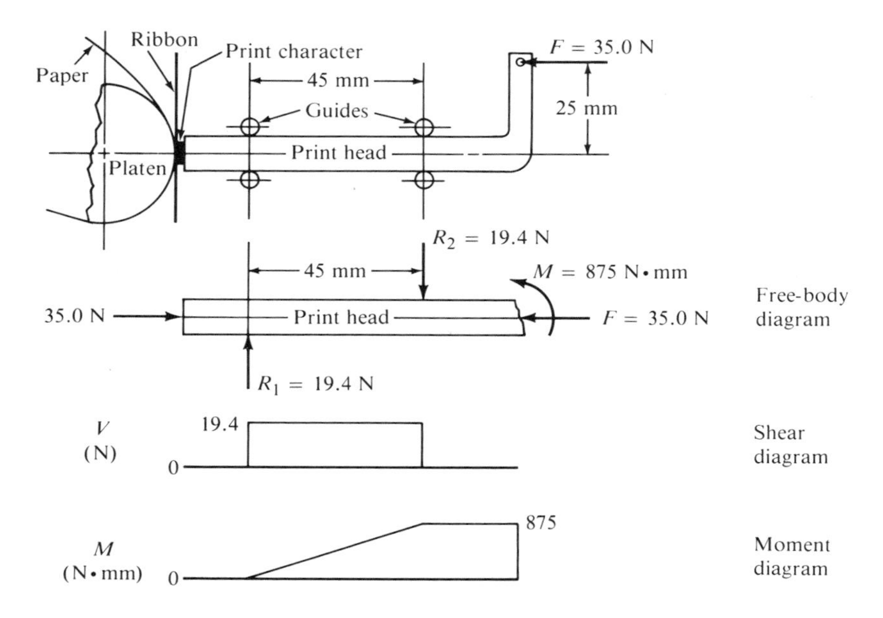

Figure 3-15 Bending Moment on a Print Head

Figure 3-16 is a crank in which the downward force exerts a torque on the rod equal to 60 lb times 5.0 in, or 300 lb · in. This torque is resisted at the left end of the rod. But because the force acts at the end of the crank handle, it also exerts a concentrated bending moment on the rod at the point of attachment B, equal to 60 lb times 3.0 in, or 180 lb · in. The free body diagram can be drawn by considering only the rod itself and only the factors affecting bending. At B we show the moment of 180 lb · in and also the downward force of 60 lb. These are the effects of the crank on the rod. The principles of statics are the basis for computing the reactions at the supports. The shear diagram follows the classic rules. The moment diagram clearly illustrates the effect of the concentrated bending moment, since the abrupt change in moment occurs at B.

Note that when the concentrated bending moment acts on the beam in a *counterclockwise* direction, the moment diagram *drops*; when a *clockwise* concentrated moment acts, the moment diagram *rises*.

3-9 STRESS CONCENTRATIONS

The formulas reviewed earlier for computing simple stresses due to direct tensile and compressive forces, bending moments, and torsional moments are applicable under certain conditions. One condition is that the geometry of the member is uniform throughout the section of interest.

In many typical machine design situations, inherent geometric discontinuities are necessary for the parts to perform their desired functions. For example, shafts carrying gears, chain sprockets, or belt sheaves usually have several diameters that create a series of shoulders that seat the power transmission members and support bearings. Grooves in the shaft allow the installation of retaining rings. Keyseats milled into the shaft enable keys to drive the elements. Similarly, tension members in linkages may be designed with retaining ring grooves, radial holes for pins, screw threads, or reduced sections.

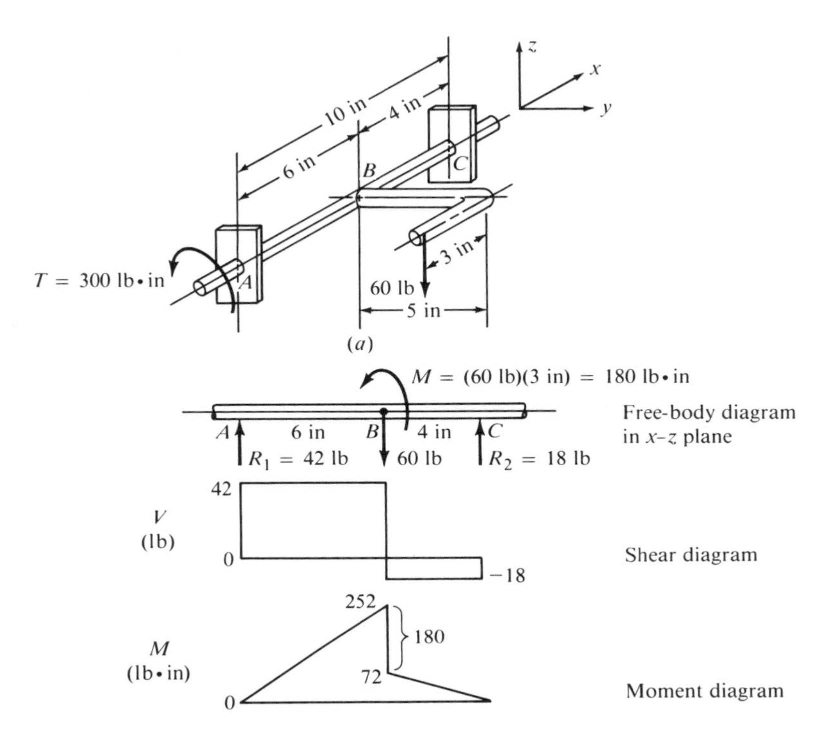

Figure 3-16 Bending Moment on a Shaft Carrying a Crank

Any of these geometric discontinuities will cause the actual maximum stress in the part to be higher than the simple formulas predict. Defining *stress concentration factors* as the factors by which the actual maximum stress exceeds the nominal stress predicted from the simple equations allows the designer to analyze these situations. The symbol for these factors is K_t. In general, the K_t factors are used as follows:

$$\sigma_{\text{max}} = K_t\,\sigma_{\text{nom}} \qquad \text{or} \qquad \tau_{\text{max}} = K_t\,\tau_{\text{nom}}$$

depending on the kind of stress produced for the particular loading. The value of K_t depends on the shape of the discontinuity, the specific geometry, and the type of stress. Appendix A-15 includes several charts for stress concentration factors. Note that the charts indicate the method of computing the nominal stress. Usually the nominal stress is computed by using the net section in the vicinity of the discontinuity. For example, for a flat plate with a hole in it subjected to a tensile force, the nominal stress is computed as the force divided by the minimum cross-sectional area through the location of the hole.

But there are other cases in which the gross area is used in calculating the nominal stress. Keyseats, for example, are analyzed by applying the stress concentration factor to the computed stress in the full-diameter portion of the shaft.

Figure 3-17 shows an experimental device that demonstrates the phenomenon of stress concentrations.

Example. Compute the maximum stress in a round bar subjected to an axial tensile force of 9 800 N. The geometry is shown in Figure 3-18.

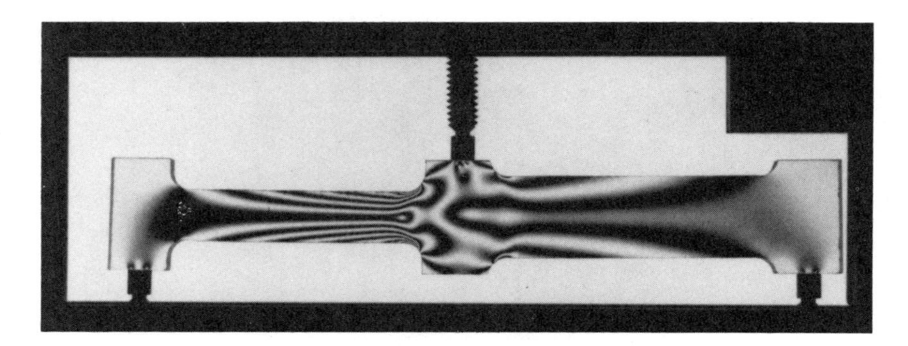

Figure 3-17 Illustration of Stress Concentrations (Source: Measurements Group, Inc., Raleigh, North Carolina, U.S.A.)

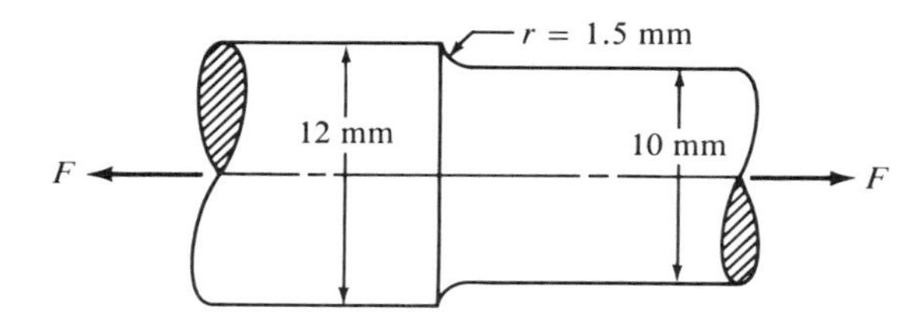

Figure 3-18 Stepped Round Bar Subjected to Axial Tensile Force

Solution. Appendix A-15 indicates that the nominal stress is computed for the smaller of the two diameters of the bar. The stress concentration factor depends on the ratio of the two diameters and the ratio of the fillet radius to the smaller diameter.

$$D/d = 12 \text{ mm}/10 \text{ mm} = 1.20$$
$$r/d = 1.5 \text{ mm}/10 \text{ mm} = 0.15$$

From these values we can find $K_t = 1.60$. The stress is

$$A = \pi d^2/4 = [(\pi)(10 \text{ mm})^2]/4 = 78.5 \text{ mm}^2$$
$$\sigma = \frac{K_t F}{A} = \frac{1.60(9\,800 \text{ N})}{78.5 \text{ mm}^2} = 199.6 \text{ N/mm}^2 = 199.6 \text{ MPa}$$

The following are guidelines on the use of stress concentration factors.

1. The worst case occurs for those areas in tension.
2. Always use stress concentration factors in analyzing members under fatigue loading because fatigue cracks usually initiate near points of high local tensile stress.
3. Stress concentrations can be ignored for static loading of ductile materials because, if the local maximum stress exceeds the yield strength of the material, the load is redistributed. The resulting member is actually stronger after the local yielding occurs.
4. The stress concentration factors in Appendix A-15 are theoretical values based only on the geometry of the member and the manner of loading.

5. Use stress concentration factors when analyzing brittle materials under either static or fatigue loading. Because the material does not yield, the stress redistribution described in item 3 cannot occur.
6. Some materials are less sensitive to stress concentrations than others. One measurement of this phenomenon is *notch sensitivity, q,* defined in terms of the theoretical stress concentration factor and a strength reduction factor, K_f:

$$q = \frac{K_f - 1}{K_t - 1}$$

From this, the strength reduction factor is

$$K_f = 1 + q(K_t - 1)$$

High-strength, highly heat-treated steels and other high-strength alloys have a value of q near unity (1.0). For such materials, $K_f = K_t$. For lower-strength, more ductile materials, the value of q ranges from approximately 0.5 to 0.9. Thus, ignoring the notch sensitivity produces a conservatively safe result.
7. Even scratches, nicks, corrosion, excessive surface roughness, and plating can cause stress concentrations. Chapter 5 discusses the care essential to manufacturing, handling, and assembling components subjected to fatigue loading.

REFERENCES

1. Blake, Alexander. *Practical Stress Analysis in Engineering Design*. New York: Marcel Dekker, 1982.
2. Boresi, A. P., O. M. Sidebottom, F. B. Seely, and J. O. Smith. *Advanced Mechanics of Materials,* 3d ed. New York: John Wiley & Sons, 1978.
3. Mott, R. L. *Applied Strength of Materials*. Englewood Cliffs, N.J.: Prentice-Hall, 1980.
4. Muvdi, B. B., and J. W. McNabb. *Engineering Mechanics of Materials*. New York: Macmillan, 1980.
5. Neathery, R. F. *Applied Strength of Materials*. New York: John Wiley & Sons, 1982.
6. Peterson, R. E. *Stress Concentration Factors*. New York: John Wiley & Sons, 1974.
7. Roark, R. J., and W. C. Young. *Formulas for Stress and Strain,* 5th ed. New York: McGraw-Hill Book Company, 1975.

PROBLEMS

1. A tensile member in a machine structure is subjected to a steady load of 4.50 kN. It has a length of 750 mm and is made from a steel tube having an outside diameter of 18 mm and an inside diameter of 12 mm. Compute the tensile stress in the tube and the axial deformation.
2. A tensile load of 5.00 kN is applied to a square bar, 12 mm on a side and having a length of 1.65 m. Compute the stress and axial deformation in the bar if it is made from (a) AISI 1020 hot-rolled steel, (b) AISI 8650 OQT 1000 steel, (c) ductile iron A536-77 (60-40-18), (d) aluminum 6061-T6, (e) titanium Ti-6Al-4V, (f) rigid PVC plastic, and (g) phenolic plastic.
3. An aluminum rod is made in the form of a hollow square tube, 2.25 in outside, with a wall thickness of 0.120 in. Its length is 16.0 in. What axial compressive force would cause the tube to shorten by 0.004 in? Compute the resulting compressive stress in the aluminum.
4. Compute the stress in the middle portion of the rod *AC* in Figure 3-19 if the vertical force on the boom is 2 500 lb. The rod is rectangular, 1.50 in by 3.50 in.
5. Compute the forces in the two angled rods in Figure 3-20 for an applied force, $F = 1\,500$ lb, if the angle θ is 45 degrees.

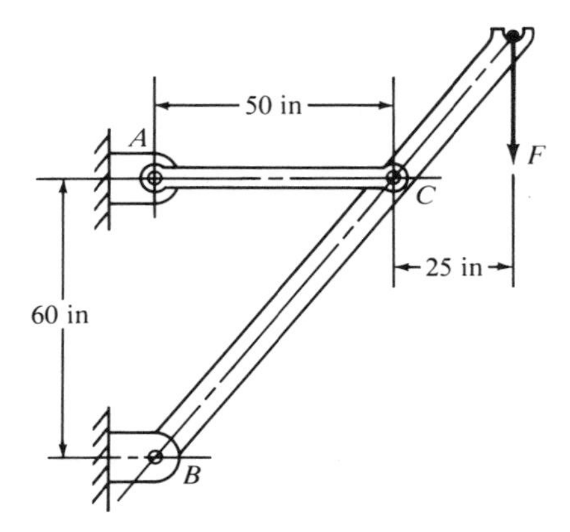

Figure 3-19 (Problem 4)

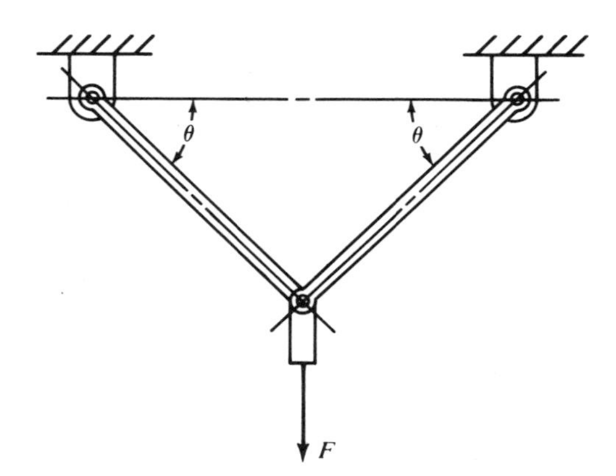

Figure 3-20 (Problem 5)

6. If the rods from problem 5 are circular, determine their required diameter if the load is static and the allowable stress is 18 000 psi.
7. Repeat problems 5 and 6 if the angle θ is 15 degrees.
8. Compute the torsional shear stress in a circular shaft with a diameter of 50 mm that is subjected to a torque of 800 N · m.
9. If the shaft of problem 8 is 850 mm long and made of steel, compute the angle of twist of one end in relation to the other.
10. Compute the torsional shear stress due to a torque of 88.0 lb · in in a circular shaft having a 0.40-in diameter.
11. Compute the torsional shear stress in a solid circular shaft having a diameter of 1.25 in that is transmitting 110 hp at a speed of 560 rpm.
12. Compute the torsional shear stress in a hollow shaft with an outside diameter of 40 mm and an inside diameter of 30 mm when transmitting 28 kilowatts (kW) of power at a speed of 45 rad/s.
13. Compute the angle of twist for the hollow shaft of problem 12 over a length of 400 mm. The shaft is steel.
14. A square steel bar, 25 mm on a side and 650 mm long, is subjected to a torque of 230 N · m. Compute the shear stress and the angle of twist for the bar.
15. A 3.00-in diameter steel bar has a flat milled on one side as shown in Figure 3-21. If the shaft is 44.0 in long and carries a torque of 10 600 lb · in, compute the stress and angle of twist.
16. A commercial steel supplier lists rectangular steel tubing having outside dimensions of 4.00 by 2.00 in and a wall thickness of 0.109 in. Compute the maximum torque that can be applied to such a tube if the shear stress is to be limited to 6 000 psi. For this torque, compute the angle of twist of the tube over a length of 6.5 ft.
17. A beam is simply supported and carries the load shown in Figure 3-22. Specify suitable dimensions for the beam if it is steel and the stress is limited to 18 000 psi, for the following shapes:
    a. Square
    b. Rectangle with height three times the width
    c. Rectangle with height one-third the width
    d. Solid circular section

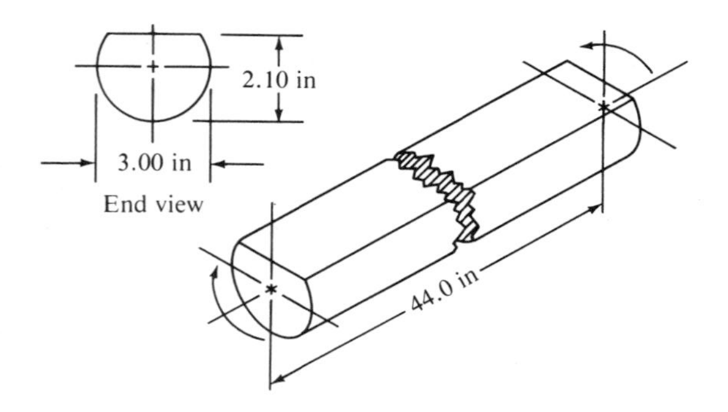

Figure 3-21 (Problem 15)

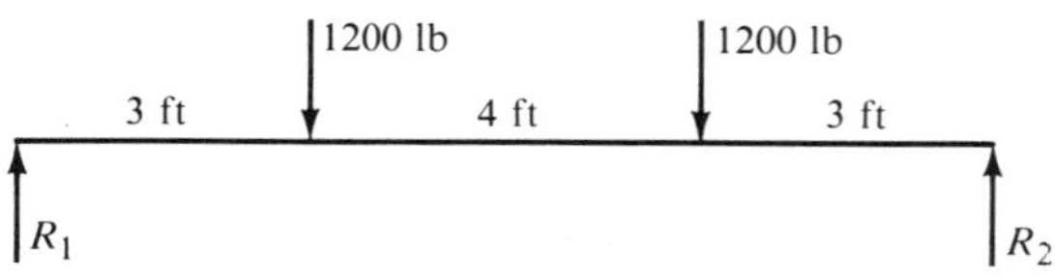

Figure 3-22 (Problem 17)

e. American Standard beam section
f. American Standard channel with the legs down
g. Standard structural tubing

18. For each beam of problem 17, compute its weight if the steel weighs 0.283 lb/in^3.
19. For each beam of problem 17, compute the maximum deflection and the deflection at the loads.
20. For the beam loading of Figure 3-23, draw the complete shearing force and bending moment diagrams and determine the bending moments at points A, B, and C.
21. For the beam loading of Figure 3-23, design the beam, choosing a shape that will be reasonably efficient and will limit the stress to 100 MPa.
22. Figure 3-24 shows a beam made from a TS4 structural tubing with a wall thickness of 0.237 in. Compute the deflection at points A and B for two cases: (a) the simple cantilever and (b) the supported cantilever.
23. Select an aluminum I-beam shape to carry the load shown in Figure 3-25 with a maximum stress of 12 000 psi. Then compute the deflection at each load.
24. Figure 3-26 represents a wood joist for a platform, carrying a uniformly distributed load of 120 lb/ft and two concentrated loads appled by some machinery. Compute the maximum stress due to bending in the joist and the maximum vertical shear stress.

For problems 25 through 36 draw the free body diagram of only the horizontal beam portion of the given figures. Then draw the complete shear and bending moment diagrams. Where used, the symbol X indicates a simple support capable of exerting a reaction force in any direction but having no moment resistance. For

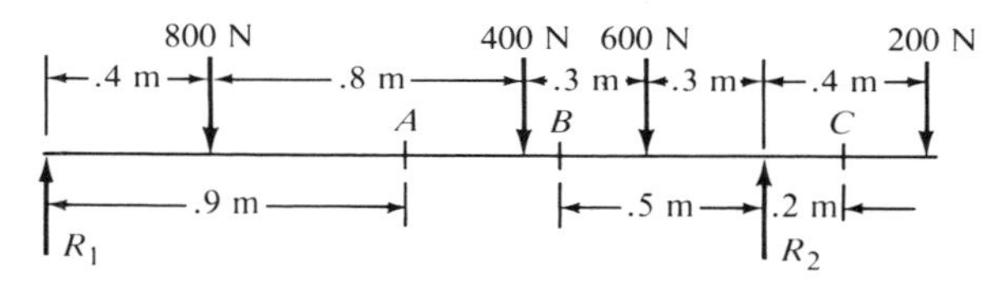

Figure 3-23 (Problems 20 and 21)

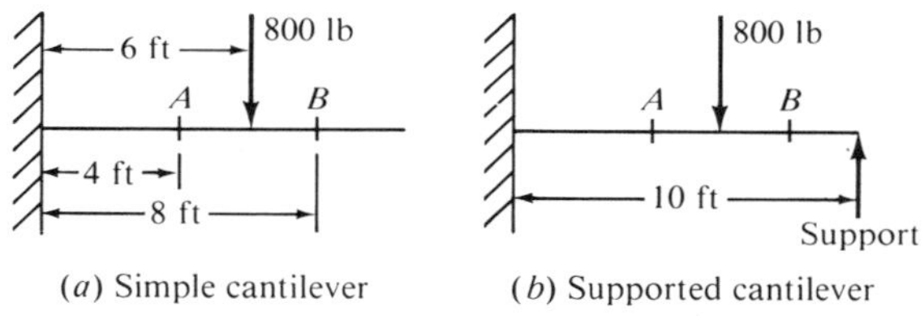

Figure 3-24 (Problem 22)

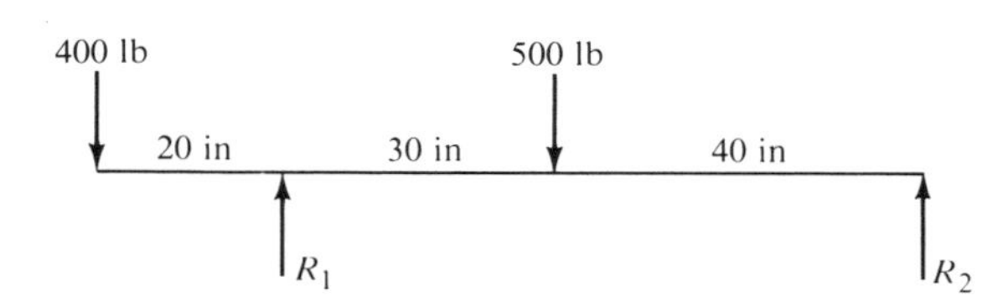

Figure 3-25 (Problem 23)

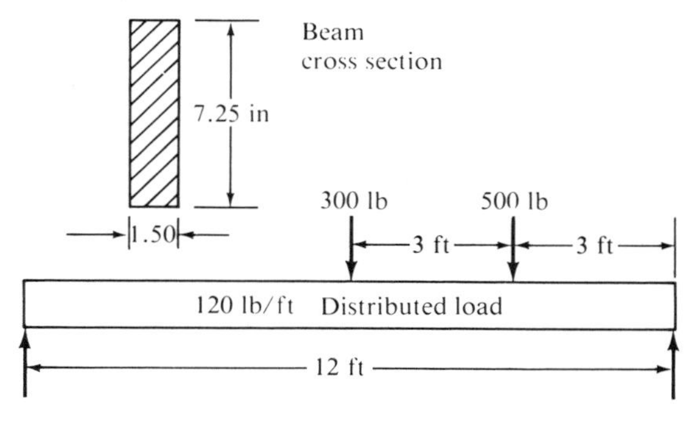

Figure 3-26 (Problem 24)

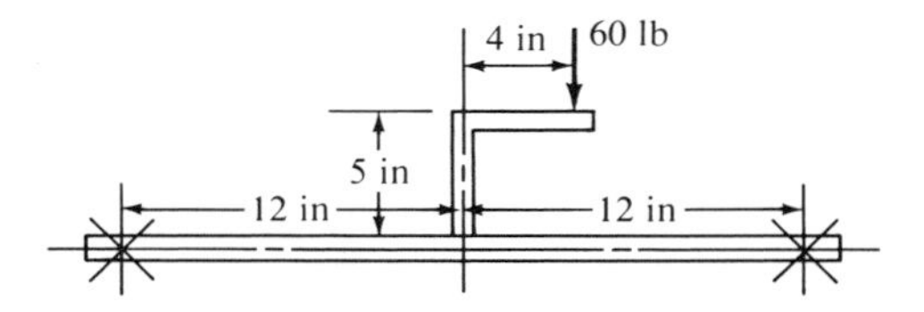

Figure 3-27 (Problem 25)

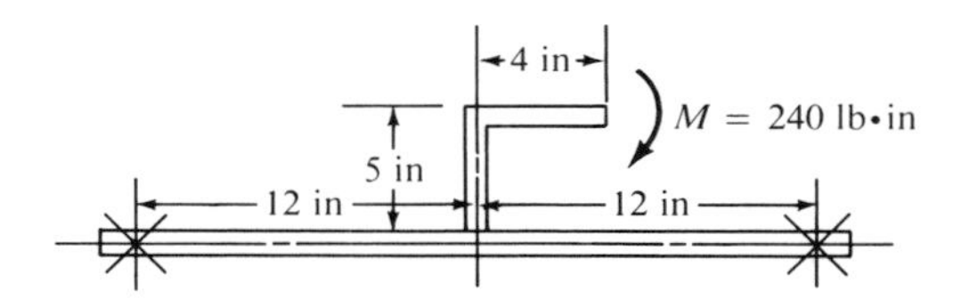

Figure 3-28 (Problem 26)

beams having unbalanced axial loads, you may specify which support offers the reaction.

25. Use Figure 3-27.
26. Use Figure 3-28.
27. Use Figure 3-29.
28. Use Figure 3-30.
29. Use Figure 3-31.
30. Use Figure 3-32.
31. Use Figure 3-33.
32. Use Figure 3-34.
33. Use Figure 3-35.
34. Use Figure 3-36.
35. Use Figure 3-37.
36. Use Figure 3-38.
37. Figure 3-39 shows a valve stem from an engine subjected to an axial tensile load applied by the valve spring. For a force of 1.25 kN, compute the maximum stress at the fillet under the shoulder.

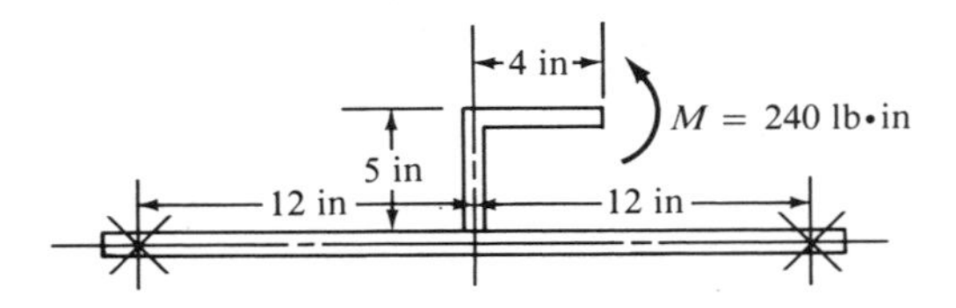

Figure 3-29 (Problem 27)

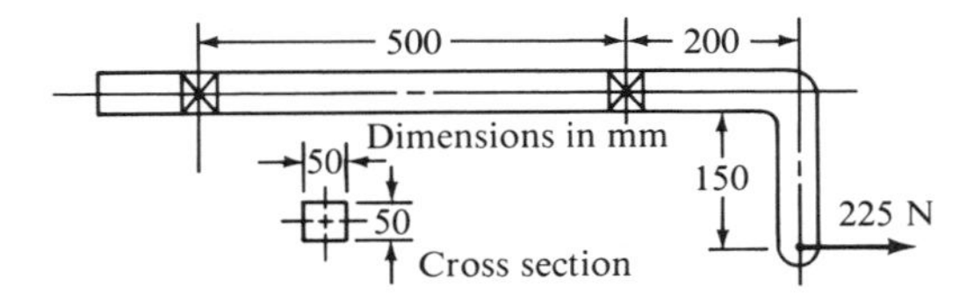

Figure 3-30 (Problem 28)

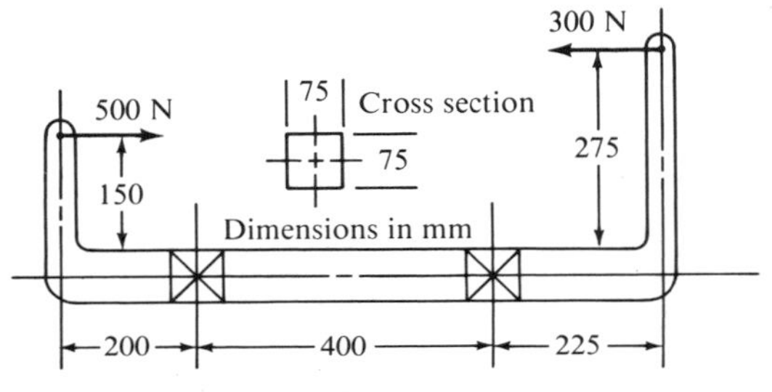

Figure 3-31 (Problem 29)

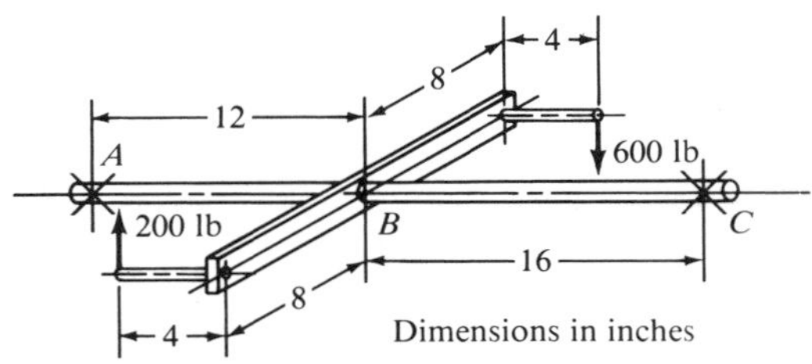

Figure 3-32 (Problem 30)

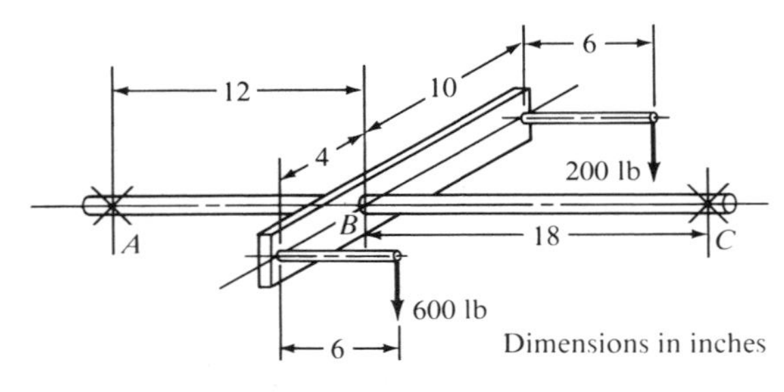

Figure 3-33 (Problem 31)

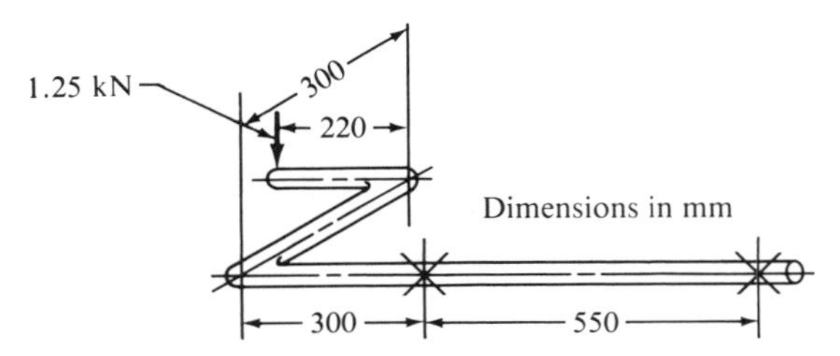

Figure 3-34 (Problem 32)

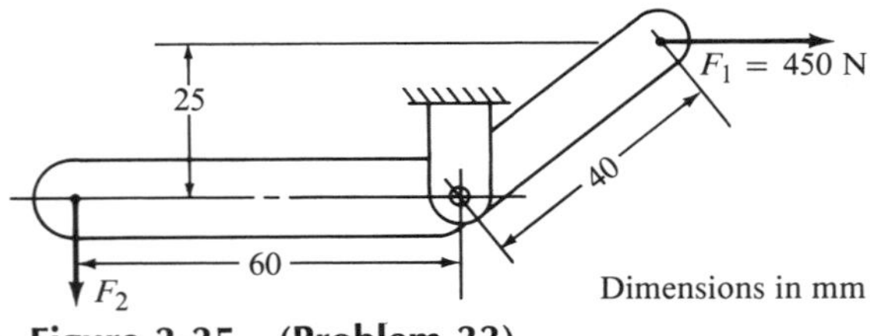

Figure 3-35 (Problem 33)

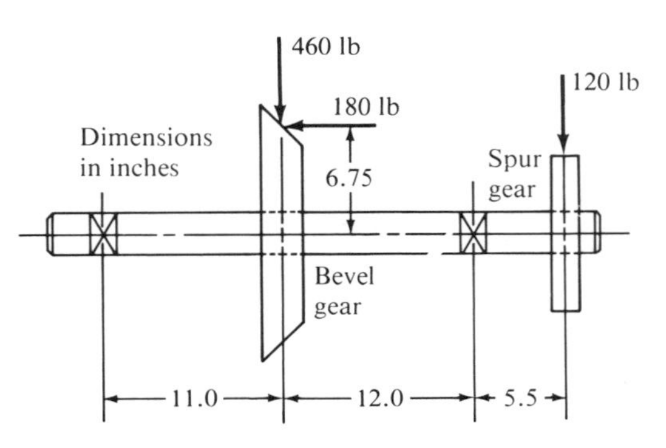

Figure 3-36 (Problem 34)

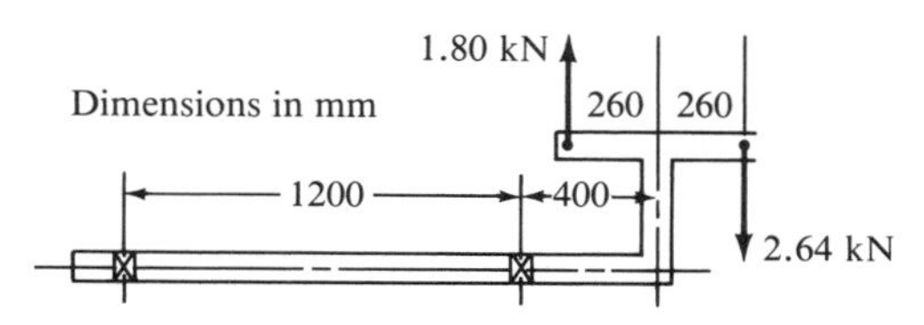

Figure 3-37 (Problem 35)

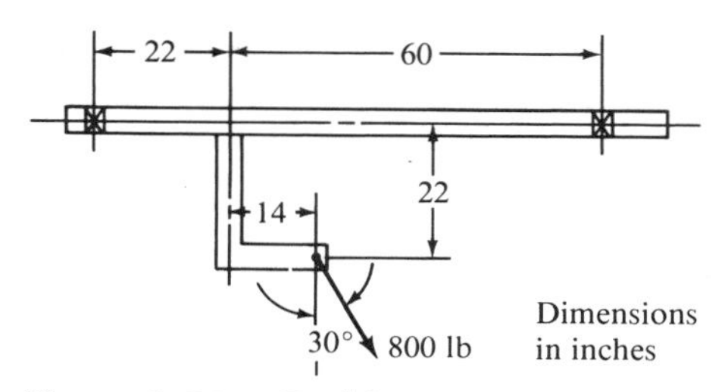

Figure 3-38 (Problem 36)

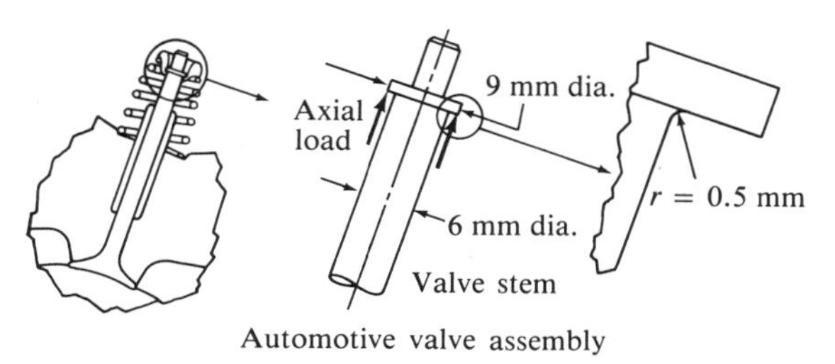

Figure 3-39 (Problem 37)

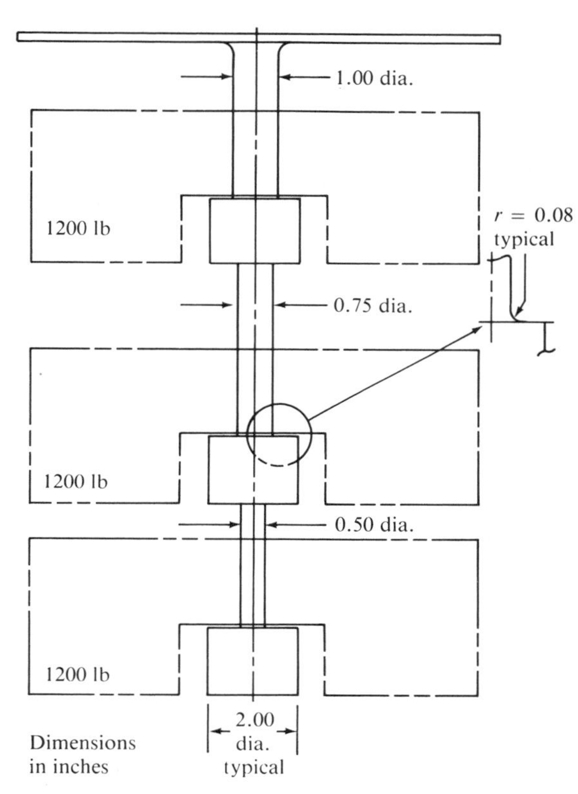

Figure 3-40 (Problem 38)

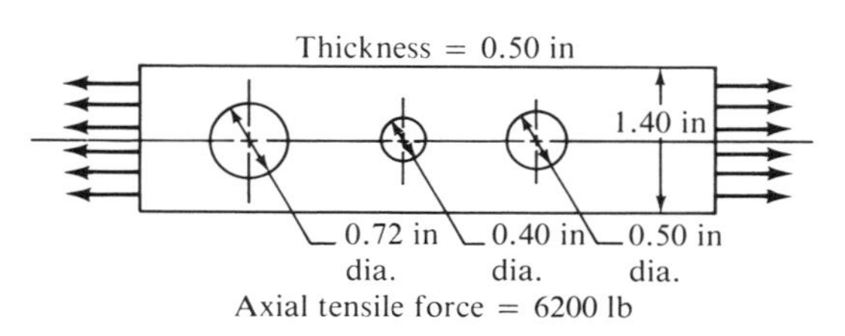

Figure 3-41 (Problem 39)

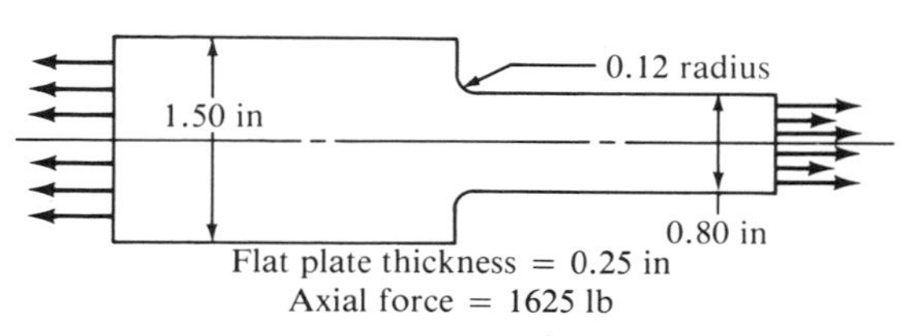

Figure 3-42 (Problem 40)

38. A conveyor fixture shown in Figure 3-40 carries three heavy assemblies (1 200 lb each). Compute the maximum stress in the fixture, considering stress concentrations at the fillets and assuming that the load acts axially.
39. For the flat plate in tension in Figure 3-41, compute the stress at each hole, assuming the holes are sufficiently far apart that their effects do not interact.

For problems 40 through 44 compute the maximum stress in the member, considering stress concentrations.

40. Use Figure 3-42.
41. Use Figure 3-43.
42. Use Figure 3-44.
43. Use Figure 3-45.
44. Use Figure 3-46.

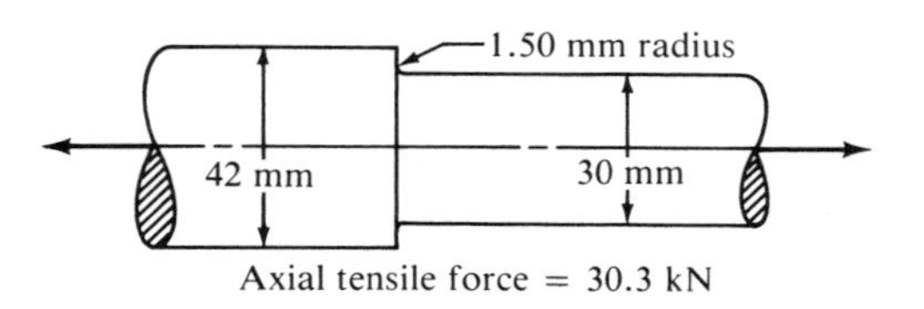

Figure 3-43 (Problem 41)

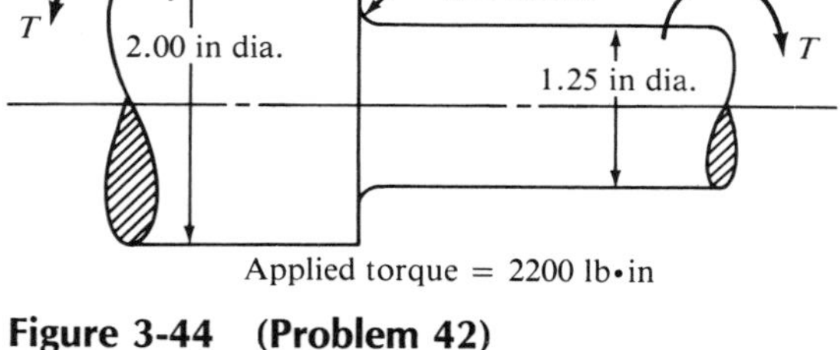

Figure 3-44 (Problem 42)

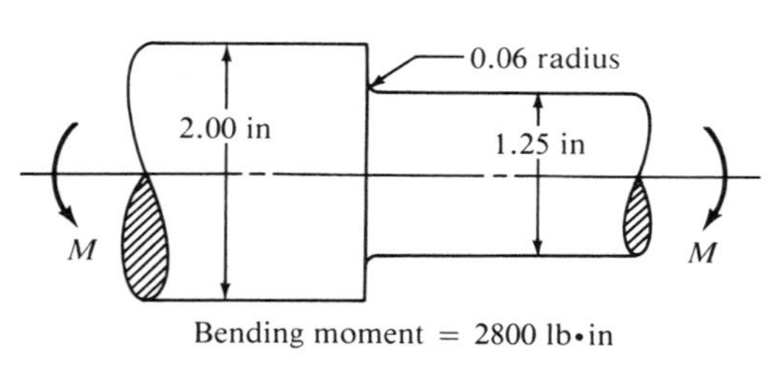

Figure 3-45 (Problem 43)

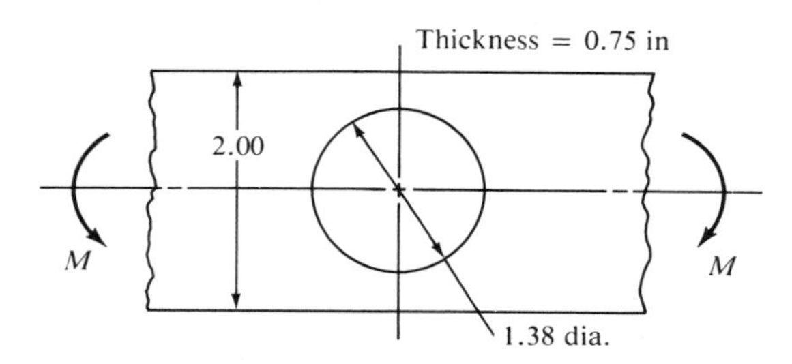

Figure 3-46 (Problem 44)

4 Combined Stresses and Mohr's Circle

4-1 OVERVIEW

Chapter 3 reviewed the basic stress analysis relationships for which only one type of stress occurs at a time. In many machine design situations, an element is subjected to two or more types of stress simultaneously. This chapter will discuss three approaches to analyzing such cases.

The first method applies when all the components of the load tend to produce normal stresses in the same direction, a very special case. The second method deals with combined torsional shear stress and bending stress, as would be encountered by a power transmission shaft. The general case, in which normal stresses in two directions can be considered along with shear stresses, is then presented in an analytical form. Lastly, the general case is solved by using a graphic aid called *Mohr's circle*.

4-2 COMBINED NORMAL STRESSES

When the same cross section of a load-carrying member is subjected to both a direct tensile or compressive stress and a stress due to bending, the resulting normal stress can be computed by the method of superposition. The formula would be

$$\sigma = \pm M/Z \pm F/A \tag{4-1}$$

where tensile stresses are positive and compressive stresses are negative.

Figure 4-1 shows a beam subjected to a load applied downward and to the right through a bracket below the beam. Resolving the load into horizontal and vertical components shows that its effect can be broken into three parts:

1. The vertical component tends to place the beam in bending with tension on the top and compression on the bottom.

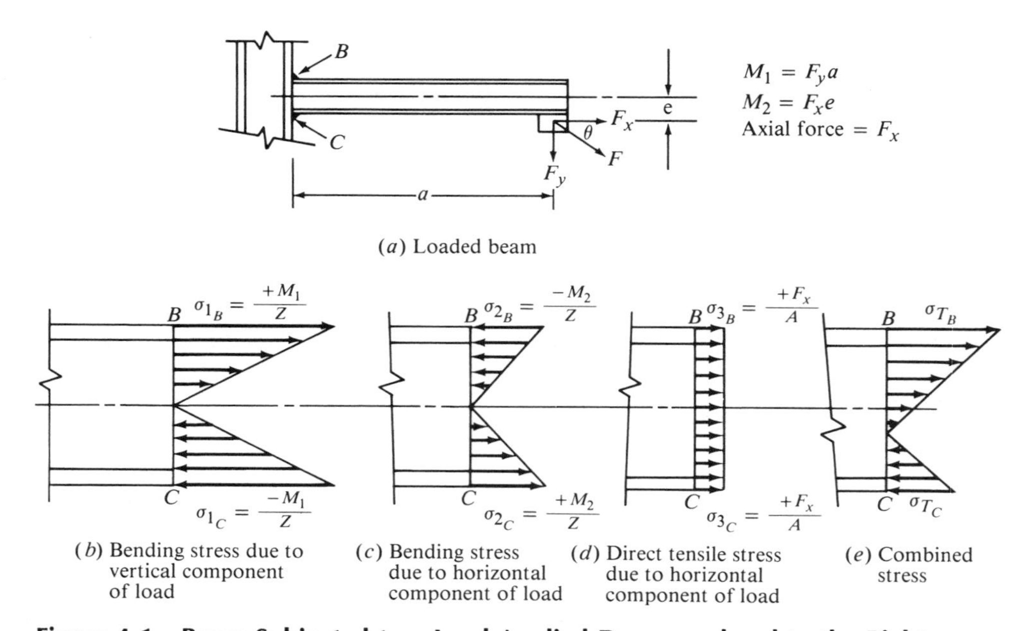

Figure 4-1 Beam Subjected to a Load Applied Downward and to the Right

2. The horizontal component, because it acts away from the neutral axis of the beam, causes bending with tension on the bottom and compression on the top.
3. The horizontal component causes direct tensile stress across the entire cross section.

The stress analysis can proceed by considering each of these cases individually and then combining the stress at any point.

Example Problem 4-1. The cantilever beam in Figure 4-1 is a steel American Standard beam, S6X12.5 (section modulus $= Z = 7.37 \text{ in}^3$; area $= A = 3.67 \text{ in}^2$). The force F is 10 000 pounds and acts at an angle of 30 degrees (°) below the horizontal, as shown. Use $a = 24$ in and $e = 6.0$ in. Compute the normal stress at B and C, where the beam joins the rigid column.

Solution. First, resolve the force into horizontal and vertical components.

$$F_x = F\cos(30°) = 10\,000\cos(30°) = 8\,660 \text{ lb}$$
$$F_y = F\sin(30°) = 10\,000\sin(30°) = 5\,000 \text{ lb}$$

Now, compute the bending moments. Due to F_y:

$$M_1 = F_y a = (5\,000)(24) = 120\,000 \text{ lb}\cdot\text{in}$$

Due to F_x:

$$M_2 = F_x e = (8\,660)(6.0) = 51\,960 \text{ lb}\cdot\text{in}$$

Compute the individual stress components.
Bending stress due to M_1:

$$\sigma_1 = \pm M_1/Z = \pm(120\,000 \text{ lb}\cdot\text{in})/(7.37 \text{ in}^3) = \pm 16\,280 \text{ psi}$$

This is tensile at B and compressive at C [Fig. 4-1(b)].
Bending stress due to M_2:

$$\sigma_2 = \pm M_2/Z = \pm(51\,960 \text{ lb}\cdot\text{in})/(7.37 \text{ in}^3) = \pm 7\,050 \text{ psi}$$

This is compressive at B and tensile at C [Fig. 4-1(c)].
Direct tensile stress due to F_x:

$$\sigma_3 = +F_x/a = (+8\,660 \text{ lb})/3.67 \text{ in}^2 = +2\,360 \text{ psi}$$

This is uniform across the entire cross section [Fig. 4-1(d)].
Now combine the stresses at B:

$$\begin{aligned}\sigma_B &= +\sigma_1 - \sigma_2 + \sigma_3 = 16\,280 - 7\,050 + 2\,360\\ &= +11\,590 \text{ psi} \quad \text{(tensile)}\end{aligned}$$

Now combine the stresses at C:

$$\begin{aligned}\sigma_C &= -\sigma_1 + \sigma_2 + \sigma_3 = -16\,280 + 7\,050 + 2\,360\\ &= -6\,870 \text{ psi} \quad \text{(compressive)}\end{aligned}$$

The resulting combined stress is shown in Figure 4-1(e).

4-3 COMBINED BENDING AND TORSION ON CIRCULAR SHAFTS

Combined bending and torsion on circular shafts is a very important concept because bending and torsion are combined any time a shaft transmits power and carries loads transverse to its axis. Power transmission elements such as gears, chain sprockets, and belt sheaves all exert such transverse forces. Thus, virtually any shaft is subjected to combined bending and torsion. Figure 4-2, which shows such a case, illustrates that the maximum bending stress and the maximum torsional shear stress do occur at the same point in the shaft, namely at the top and bottom surfaces.

Because bending produces normal stresses, whereas torsion produces shear stresses, simple superposition cannot be used. The *maximum shear stress theory of failure,* discussed in detail in Chapter 5, is frequently used for such cases. This theory states that a ductile material fails when the maximum shear stress exceeds the yield strength of the material in shear. For an element of material subjected to a normal stress, σ, and a shear stress, τ, the maximum shear stress on the element is

$$\tau_{max} = \sqrt{(\sigma/2)^2 + \tau^2} \tag{4-2}$$

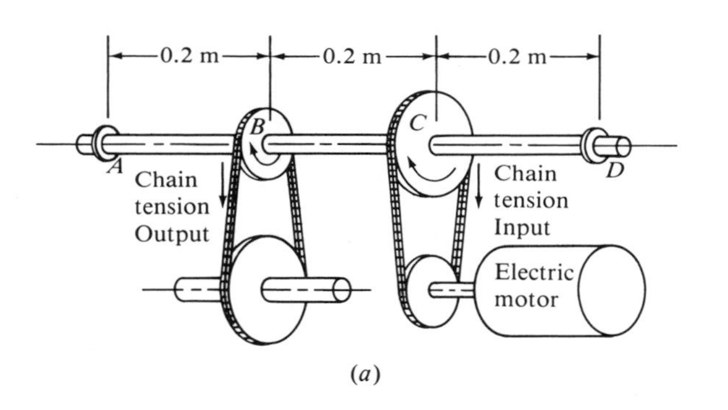

(*a*)

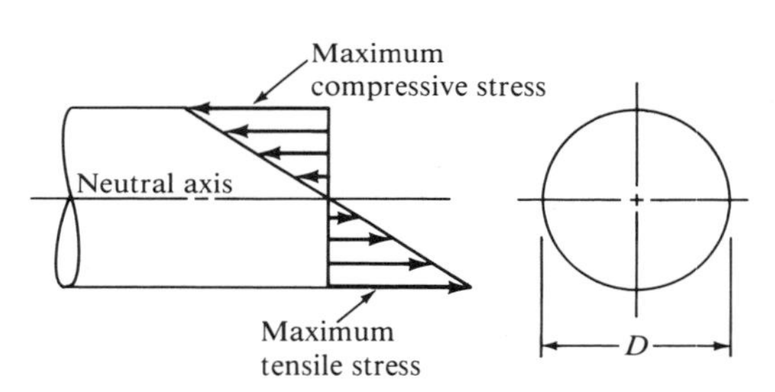

(*b*) Bending stress distribution in a round shaft

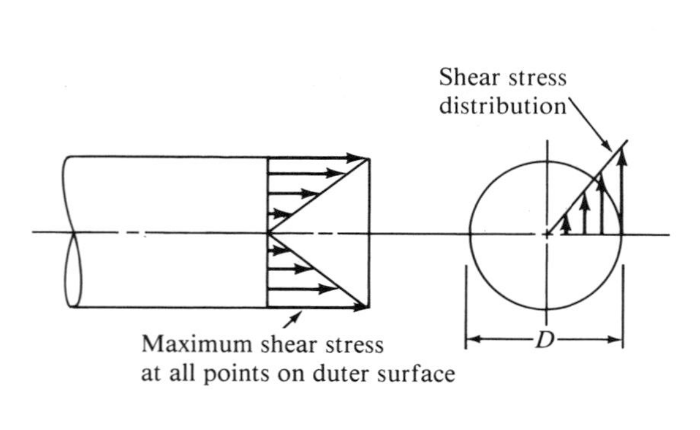

(*c*) Torsional shear stress distribution in a round shaft

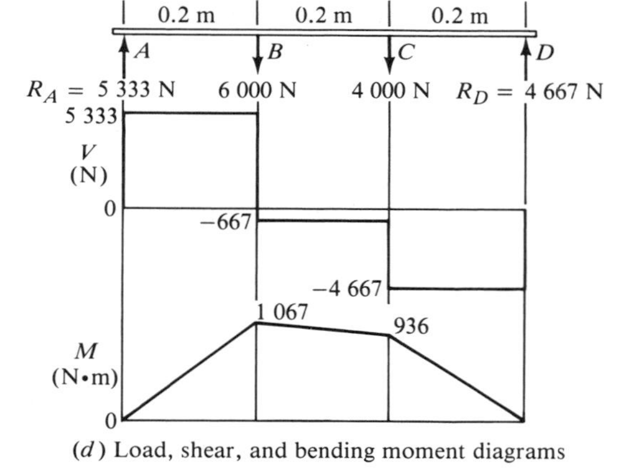

(*d*) Load, shear, and bending moment diagrams

Figure 4-2 Combined Bending and Torsion on Circular Shafts

Because in this special case, $\sigma = M/Z$, and $\tau = T/Z_p$, equation (4-2) can be written

$$\tau_{\max} = \sqrt{(M/2Z)^2 + (T/Z_p)^2}$$

And because $2Z = Z_p$ for circular shafts,

$$\tau_{\max} = \frac{\sqrt{(M^2 + T^2)}}{Z_p}$$

We can define an *equivalent torque*, T_e:

$$T_e = \sqrt{M^2 + T^2} \qquad (4\text{-}3)$$

Finally, then, we can compute the maximum shear stress from

$$\tau_{\max} = T_e/Z_p \qquad (4\text{-}4)$$

Typically, the design shear stress, τ_d, would be defined for given conditions. Then equation (4-4) would be solved for the required polar section modulus. From that the required dimensions of the cross section, such as the diameter of a shaft, could be determined.

If a stress concentration occurs at the section of interest, equation (4-4) would be modified to the form

$$\tau_{\max} = K_t T_e/Z_p \qquad (4\text{-}5)$$

If the stress concentration factor is different for bending and torsion, the following form can be used:

$$\tau_{\max} = \frac{\sqrt{(K_{tM} M)^2 + (K_{tT} T^2)}}{Z_p} \qquad (4\text{-}6)$$

Example Problem 4-2. The shaft shown in Figure 4-2 transmits 750 N · m of torque from the sprocket at B to that at C. The chain pull places downward forces on the shaft of 6 000 N at B and 4 000 N at C. Sled runner keyseats are used at the sprockets. If the shaft is to be of uniform diameter throughout its length, compute the required diameter to limit the shear stress to 54.0 MPa.

Solution. We will use the equivalent torque technique because the shaft is subjected to a combination of bending and torsion only. The stress concentration factor of 1.6 for the keyseat will be applied to the equivalent torque at both B and C. First, we can see from the bending moment diagram that the maximum bending moment is 1 067 N · m at B. Because the maximum torque and the stress concentration factor are also located at B, it is the critical point for design. The equivalent torque is

$$T_{eB} = \sqrt{(1\,067)^2 + (750)^2} = 1\,304 \text{ N} \cdot \text{m}$$

The required polar section modulus is

$$Z_p = \frac{K_t T_{eB}}{\tau_d} = \frac{1.6(1\,304 \text{ N} \cdot \text{m})}{54.0 \text{ N/mm}^2} \frac{10^3 \text{ mm}}{\text{m}} = 38\,640 \text{ mm}^3$$

From $Z_p = \pi D^3/16$, we can solve for the required diameter:

$$D = \sqrt[3]{16Z_p/\pi} = \sqrt[3]{[(16)(38\,640)]/(\pi)} = 58.2 \text{ mm}$$

We would specify the convenient size of 60 mm.

4-4 GENERAL CASE OF COMBINED STRESS

To visualize the general case of combined stress, it is helpful to consider a small element of the load-carrying member on which combined normal and shear stresses act. For this discussion we will consider a two-dimensional stress condition, as illustrated in Figure 4-3. The x and y axes are aligned with corresponding axes on the member being analyzed.

The normal stresses, σ_x and σ_y, could be due to a direct tensile force or to bending. If the normal stresses were compressive (negative), the vectors would be pointing in the opposite sense, into the stress element.

The shear stress could be due to direct shear, torsional shear, or the vertical shear stress. The double subscript notation helps to orient the direction of shear stresses. For example, τ_{xy} indicates the shear stress acting on the element face that is perpendicular to the x axis and parallel to the y axis. A *positive shear stress* is one that tends to rotate the stress element clockwise. In Figure 4-3, τ_{xy} is positive and τ_{yx} is negative. Their magnitudes must be equal to maintain the element in equilibrium.

It is necessary to determine the magnitudes and signs of each of these stresses in order to show them properly on the stress element. The example problem following the definition of principal stresses illustrates the process.

With the stress element defined, the objectives of the remaining analysis are to determine the maximum normal stress, the maximum shear stress, and the planes on which these stresses occur. The following are the governing formulas.

Maximum Normal Stresses: Principal Stresses

The combination of the applied normal and shear stresses that produces the maximum normal stress is called the *maximum principal stress*, σ_1. The magnitude of σ_1 can be computed from the following equation:

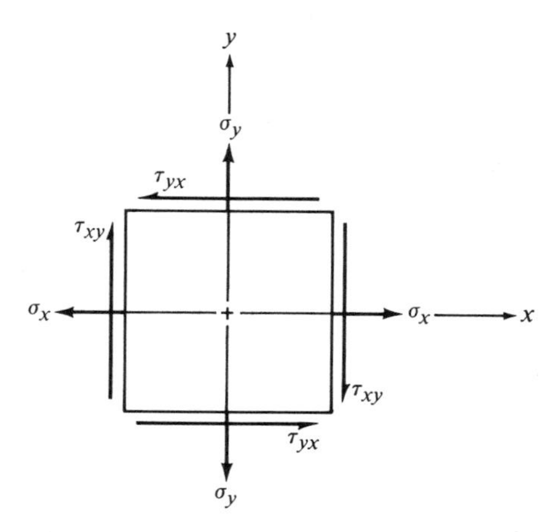

Figure 4-3 General Two-dimensional Stress Element

$$\sigma_1 = \frac{\sigma_x + \sigma_y}{2} + \sqrt{\left(\frac{\sigma_x - \sigma_y}{2}\right)^2 + \tau_{xy}^2} \tag{4-7}$$

The combination of the applied stress that produces the minimum normal stress is called the *minimum principal stress,* σ_2. Its magnitude can be computed from

$$\sigma_2 = \frac{\sigma_x + \sigma_y}{2} - \sqrt{\left(\frac{\sigma_x - \sigma_y}{2}\right)^2 + \tau_{xy}^2} \tag{4-8}$$

Particularly in experimental stress analysis, it is important to know the orientation of the principal stresses. The angle of inclination of the planes on which the principal stresses act, called the *principal planes,* can be found from equation (4-9).

$$\phi_\sigma = \tfrac{1}{2}\arctan[2\tau_{xy}/(\sigma_x - \sigma_y)] \tag{4-9}$$

The angle ϕ_σ is measured from the positive x axis of the original stress element to the maximum principal stress, σ_1. Then, the minimum principal stress, σ_2, is on the plane 90° from σ_1.

When the stress element is oriented as discussed so that the principal stresses are acting on it, the shear stress is zero. The resulting stress element is shown in Figure 4-4.

Maximum Shear Stress

On a different orientation of the stress element the maximum shear stress will occur. Its magnitude can be computed from equation (4-10).

$$\tau_{max} = \sqrt{\left(\frac{\sigma_x - \sigma_y}{2}\right)^2 + \tau_{xy}^2} \tag{4-10}$$

The angle of inclination of the element on which the maximum shear stress occurs is computed as follows:

$$\phi_\tau = \tfrac{1}{2}\arctan[-(\sigma_x - \sigma_y)/2\tau_{xy}] \tag{4-11}$$

The angle between the principal stress element and the maximum shear stress element is always 45°.

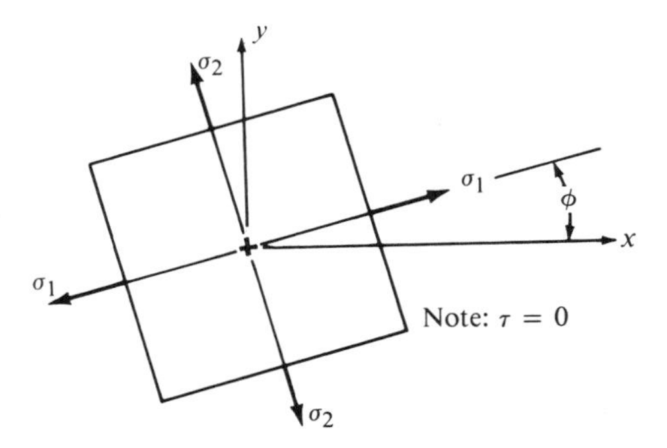

Figure 4-4 Principal Stress Element

On the maximum shear stress element there will be normal stresses of equal magnitude acting perpendicular to the planes on which the maximum shear stresses are acting. These normal stresses have the value

$$\sigma = (\sigma_x + \sigma_y)/2 \tag{4-12}$$

Note that this is the *average* of the two applied normal stresses.

The following example will illustrate the use of the equations for the principal stresses and the maximum shear stress.

Example Problem 4-3. A shaft is supported between two bearings and carries two chain sprockets, as shown in Figure 4-5. The tensions in the chains exert horizontal forces on the shaft, tending to bend it in the x-z plane. Sprocket B exerts a clockwise torque on the shaft when viewed toward the origin of the coordinate system along the x axis. Sprocket C exerts an equal but opposite torque on the shaft.

For the loading condition shown, determine the stress condition on element K on the front surface of the shaft (on the positive z side) just to the right of sprocket B.

1. Determine the stresses on element K in the x-y plane and show the stresses on the stress element.
2. Compute the principal stresses on the element and the directions in which they act.
3. Draw the stress element on which the principal stresses act and show its orientation to the original x axis.
4. Compute the maximum shear stress on the element and the orientation of the plane on which it acts.

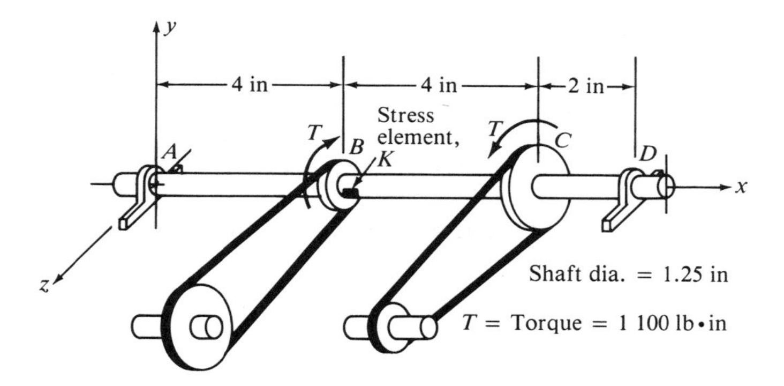

(a) Pictorial view of shaft

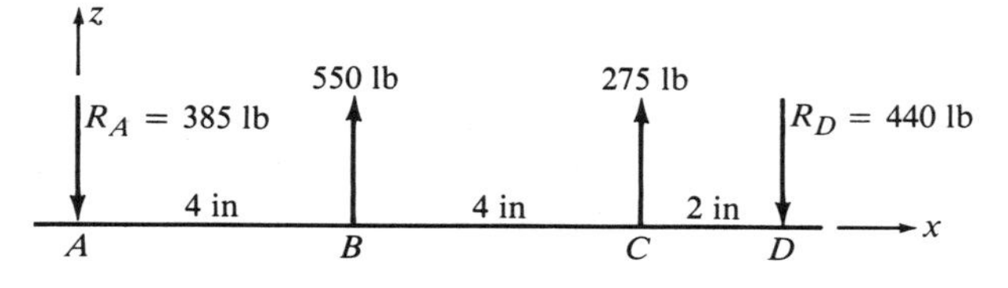

(b) Forces on shaft in x–z plane

Figure 4-5 Shaft for Example Problem 4-3

5. Draw the stress element on which the maximum shear stress acts and show its orientation to the original x axis.

Solution. Element K is subjected to bending that produces a tensile stress acting in the x direction. Also, there is a torsional shear stress acting at K. Figure 4-6 shows the shear and bending moment diagrams for the shaft and indicates that the bending moment at K is 1 540 lb · in. The bending stress is therefore

$$\sigma_x = M/Z$$
$$Z = \pi D^3/32 = [\pi(1.25\text{ in})^3]/32 = 0.192\text{ in}^3$$
$$\sigma_x = (1\,540\text{ lb}\cdot\text{in})/(0.192\text{ in}^3) = 8\,030\text{ psi}$$

The torsional shear stress acts on element K in a way that causes a downward shear stress on the right side of the element and an upward shear stress on the left side. This action results in a tendency to rotate the element in a *clockwise* direction; this is the *positive* direction for shear stresses, according to the standard convention. Also, the notation for shear stresses also uses double subscripts. For example, τ_{xy} indicates the shear stress acting on the face of an element that is perpendicular to the x axis and parallel to the y axis. Thus for element K:

$$\tau_{xy} = T/Z_p$$
$$Z_p = \pi D^3/16 = \pi(1.25\text{ in})^3/16 = 0.383\text{ in}^3$$
$$\tau_{xy} = (1\,100\text{ lb}\cdot\text{in})/0.383\text{ in}^3 = 2\,870\text{ psi}$$

The values of the normal stress, σ_x, and the shear stress, τ_{xy}, are shown on the stress element K in Figure 4-7. Note that the stress in the y direction is zero for this loading. Also, the value of the shear stress, τ_{yx}, must be equal to τ_{xy}, and it must act as shown in order for the element to be in equilibrium.

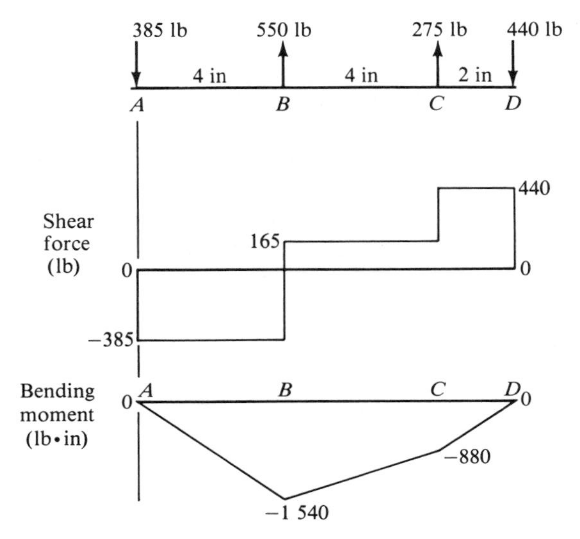

Figure 4-6 Shear and Bending Moment Diagrams for the Shaft

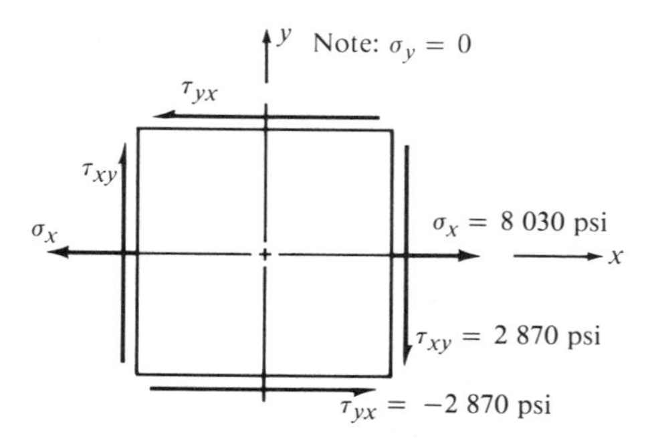

Figure 4-7 Stresses on Element *K*

We can now compute the principal stresses on the element, using equations (4-7) through (4-9). The maximum principal stress is

$$\sigma_1 = \frac{\sigma_x + \sigma_y}{2} + \sqrt{\left(\frac{\sigma_x - \sigma_y}{2}\right)^2 + \tau_{xy}^2} \tag{4-7}$$

$$\sigma_1 = (8\,030/2) + \sqrt{(8\,030/2)^2 + (2\,870)^2}$$

$$\sigma_1 = 4\,015 + 4\,935 = 8\,950 \text{ psi}$$

The minimum principal stress is

$$\sigma_2 = \frac{\sigma_x + \sigma_y}{2} - \sqrt{\left(\frac{\sigma_x - \sigma_y}{2}\right)^2 + \tau_{xy}^2} \tag{4-8}$$

$$\sigma_2 = (8\,030/2) - \sqrt{(8\,030/2)^2 + (2\,870)^2}$$

$$\sigma_2 = 4\,015 - 4\,935 = -920 \text{ psi} \quad \text{(compression)}$$

The direction in which the maximum principal stress acts is

$$\phi_\sigma = \tfrac{1}{2}\arctan[2\tau_{xy}/(\sigma_x - \sigma_y)] \tag{4-9}$$

$$\phi_\sigma = \tfrac{1}{2}\arctan[(2)(2\,870)/8\,030] = 17.8°$$

The principal stresses can be shown on a stress element as illustrated in Figure 4-8. Note that the element is shown in relation to the original element to emphasize the direction of the principal stresses in relation to the original x axis.

Now the maximum shear stress element can be defined, using equations (4-10) through (4-12):

$$\tau_{max} = \sqrt{\left(\frac{\sigma_x - \sigma_y}{2}\right)^2 + \tau_{xy}^2} \tag{4-10}$$

$$\tau_{max} = \sqrt{(8\,030/2)^2 + (2\,870)^2}$$

$$\tau_{max} = \pm 4\,935 \text{ psi}$$

The two pairs of shear stresses, $+\tau_{max}$ and $-\tau_{max}$, are equal in magnitude but opposite in direction.

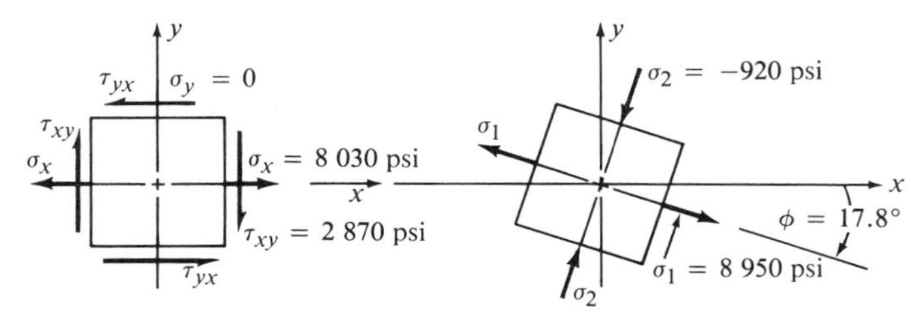

(*a*) Original stress element (*b*) Principal stress element

Figure 4-8 Principal Stresses on a Stress Element

The orientation of the element on which the maximum shear stress acts is found from Equation 4-11.

$$\phi_\tau = \tfrac{1}{2}\arctan[-(\sigma_x - \sigma_y)/2\tau_{xy}] \tag{4-11}$$

$$\phi_\tau = \tfrac{1}{2}\arctan(-8\,030/[(2)(2\,870)]) = -27.2°$$

There are equal normal stresses acting on the faces of this stress element, which have the value of

$$\sigma = (\sigma_x + \sigma_y)/2 \tag{4-12}$$

$$\sigma = 8\,030/2 = 4\,015 \text{ psi}$$

Figure 4-9 shows the stress element on which the maximum shear stress acts in relation to the original stress element. Note that the angle between this element and the principal stress element shown in Figure 4-8, is 45°.

This concludes the example problem. All of the values necessary to draw the requested stress elements were obtained from calculations. Although these values are fairly straightforward, *Mohr's circle,* which is discussed next, is somewhat easier to use and offers added insight into the actual stress condition at a point.

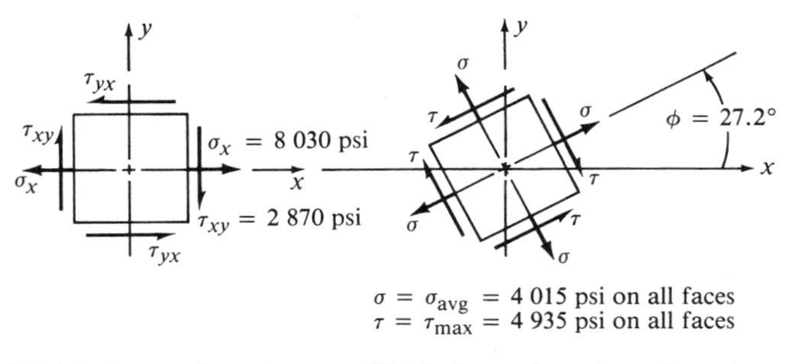

(*a*) Original stress element (*b*) Maximum shear stress element

Figure 4-9 Relation of Maximum Shear Stress Element to Original Stress Element

4-5 MOHR'S CIRCLE

Because of the many terms and signs involved, and the many calculations required in the computation of the principal stresses and the maximum shear stress, there is a rather high probability of error. Using the graphic aid Mohr's circle helps to minimize errors and gives a better "feel" for the stress condition at the point of interest.

After Mohr's circle is constructed, it can be used for the following:

1. Finding the maximum and minimum principal stresses and the directions in which they act
2. Finding the maximum shear stresses and the orientation of the planes on which they act
3. Finding the value of the normal stresses that act on the planes where the maximum shear stresses act
4. Finding the values of the normal and shear stresses that act on an element with any orientation

The data needed to construct Mohr's circle are, of course, the same as those needed to compute the above values, because the graphical approach is an exact analogy to the computations. If the normal and shear stresses that act on any two mutually perpendicular planes of an element are known, the circle can be constructed and any of items 1 through 4 can be found. This is particularly valuable in experimental stress analysis work because the results obtained from many types of standard strain gage instrumentation techniques give the necessary inputs for the creation of Mohr's circle. When the principal stresses and the maximum shear stress are known, the complete design and analysis can be done, using the various theories of failure discussed in the next chapter.

Mohr's circle is constructed as follows:

1. Perform the stress analysis to determine the magnitudes and directions of the normal and shear stresses acting at the point of interest.
2. Draw the stress element at the point of interest. Normal stresses on any two mutually perpendicular planes are drawn with tensile stresses positive: projecting outward from the element. Compressive stresses are negative: directed inward on the face. Note that the *resultants* of all normal stresses acting in the chosen directions are plotted. Shear stresses are considered to be positive if they tend to rotate the element in a *clockwise* direction, and negative otherwise.
3. Set up a rectangular coordinate system in which the positive horizontal axis represents positive (tensile) normal stresses, and the positive vertical axis represents positive (clockwise) shear stresses. Thus the plane created will be referred to as the *σ-τ plane*.
4. Plot points on the σ-τ plane corresponding to the stresses acting on the faces of the stress element. If the element is drawn in the x-y plane, the two points to be plotted are σ_x, τ_{xy} and σ_y, τ_{yx}.
5. Draw the line connecting the two points.
6. The resulting line crosses the σ axis at the center of Mohr's circle at the *average* normal stress, where

$$\sigma_{\text{avg}} = (\sigma_x + \sigma_y)/2$$

The following steps can be better visualized if the first six steps are drawn as shown in Figure 4-10. Note that on the stress element illustrated σ_x is positive, σ_y is negative,

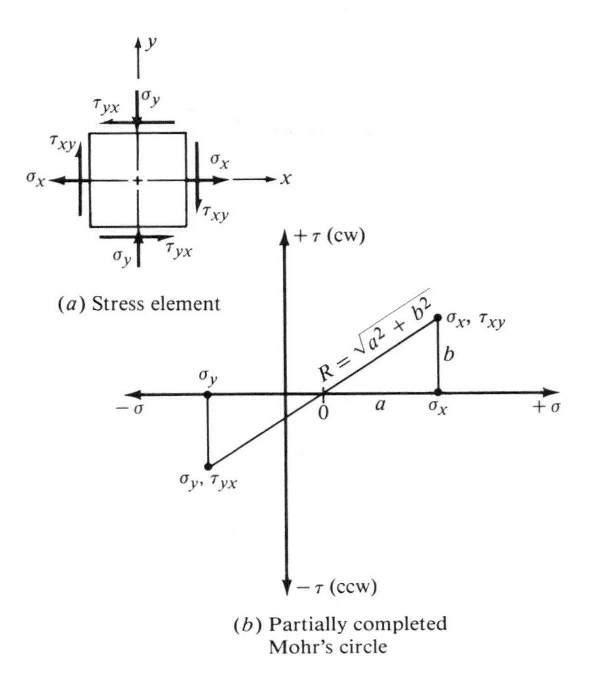

Figure 4-10 Partially Completed Mohr's Circle, Steps 1 through 6

τ_{xy} is positive, and τ_{yx} is negative. This is arbitrary for the purpose of illustration. In general, any combination of positive and negative values could exist.

Note in Figure 4-10 that a right triangle has been formed, having the sides a, b, and R, where

$$R = \sqrt{a^2 + b^2}$$

By inspection, it can be seen that

$$a = (\sigma_x - \sigma_y)/2$$
$$b = \tau_{xy}$$

The point labeled O is at a distance of $\sigma_x - a$ from the origin of the coordinate system. We can now proceed with the construction of the circle.

7. Draw the complete circle with the center at O and a radius of R, as shown in Figure 4-11.
8. The point where the circle crosses the σ axis at the right gives the value of the maximum principal stress, σ_1.
9. The point where the circle crosses the σ axis at the left gives the minimum principal stress, σ_2.
10. The coordinates of the top of the circle give the maximum shear stress and the normal stress that acts on the element having the maximum shear stress.

The following steps relate to determining the angles of inclination of the principal stress element and the maximum shear stress element in relation to the original x axis.

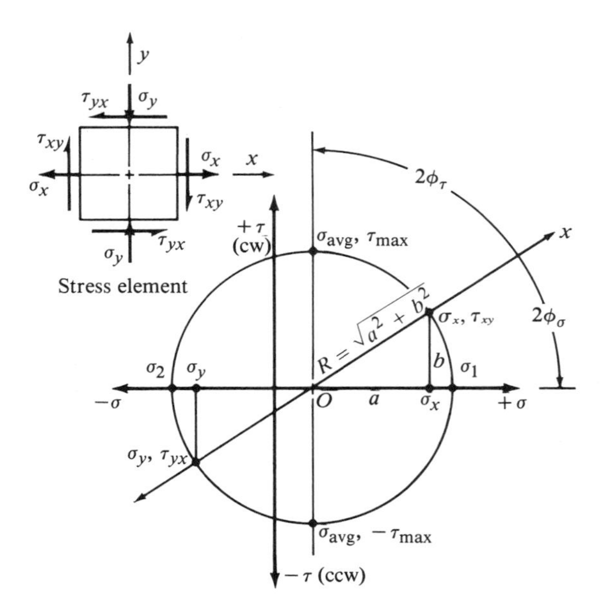

Figure 4-11 Complete Mohr's Circle

It is important to note that angles on Mohr's circle are actually *double* the true angles. Referring to Figure 4-11, the line from O through the first point plotted, σ_x, τ_{xy}, represents the original x axis, as noted in the figure. The line from O through the point σ_y, τ_{yx} represents the original y axis. Of course, on the original element, these axes are 90° apart, not 180°, illustrating the double-angle feature of Mohr's circle. Having made this observation, we can continue with the development of the process.

11. The angle $2\phi_\sigma$ is measured from the x axis on the circle as defined to the σ_1 axis. Note that

$$2\phi_\sigma = \arctan(b/a)$$

It is also important to note the direction from the x axis to the σ axis (clockwise or counterclockwise). This is necessary for representing the relation of the principal stress element to the original stress element properly.

12. The angle from the x axis on the circle to the vertical line through τ_{max} gives $2\phi_\tau$. From the geometry of the circle, it can be seen that

$$2\phi_\tau = 90° - 2\phi_\sigma$$

Again it is important to note the direction from the x axis to the τ_{max} axis for use in orienting the maximum shear stress element. You should also note that the σ_1 axis and the τ_{max} axis are always 90° apart on the circle and therefore 45° apart on the actual element.

13. The final step in the process of using Mohr's circle is to draw the resulting stress elements in their proper relation to the original element, as shown in Figure 4-12.

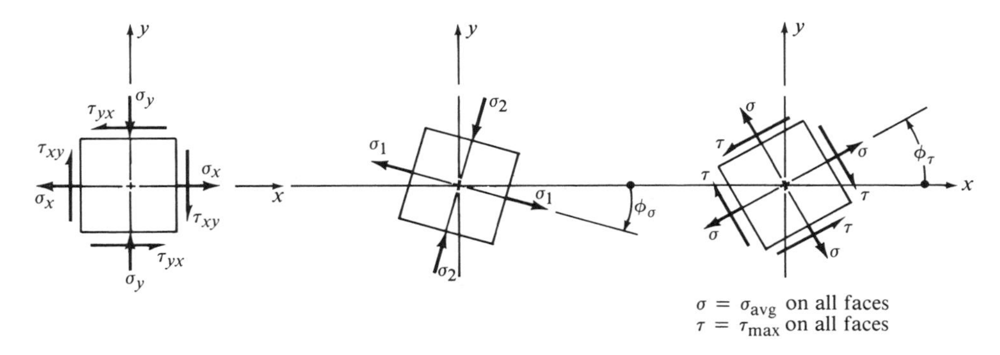

(*a*) Original stress element (*b*) Principal stress element (*c*) Maximum shear stress element

Figure 4-12 Display of Results From Mohr's Circle

Example Problem 4-4. The construction of Mohr's circle will now be illustrated by using the same data as in Example Problem 4-3, in which the principal stresses and the maximum shear stress were computed directly from the equations. You should refer to the statement of the problem dealing with the shaft carrying two chain sprockets and to Figures 4-5 through 4-9.

Figure 4-13 is identical to Figure 4-7, which was used earlier to show the stresses acting on the element K in the original x-y plane. From these data, Figure 4-14 was constructed according to steps 1–6 of the process for drawing Mohr's circle. Before proceeding with step 7, we must compute the values for a, b, and R.

$$a = (\sigma_x - \sigma_y)/2 = (8\,030 - 0)/2 = 4\,015 \text{ psi}$$

$$b = \tau_{xy} = 2\,870 \text{ psi}$$

$$R = \sqrt{a^2 + b^2} = \sqrt{(4\,015)^2 + (2\,870)^2} = 4\,935 \text{ psi}$$

Figure 4-15 shows the complete Mohr's circle. Step 7 was the drawing of the circle itself with its center at O and the radius R. Note that the circle passes through the two

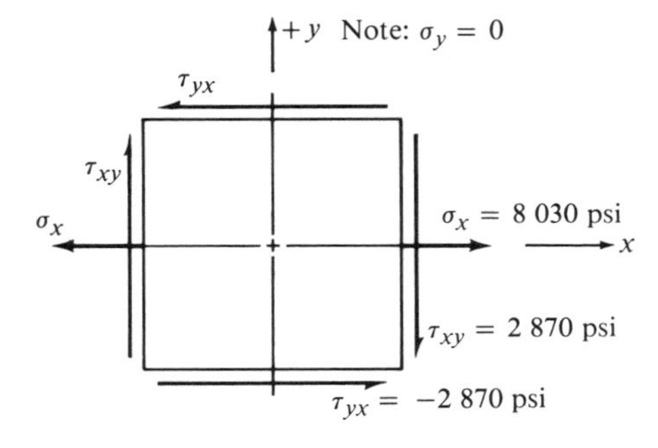

Figure 4-13 Stresses on Element *K*

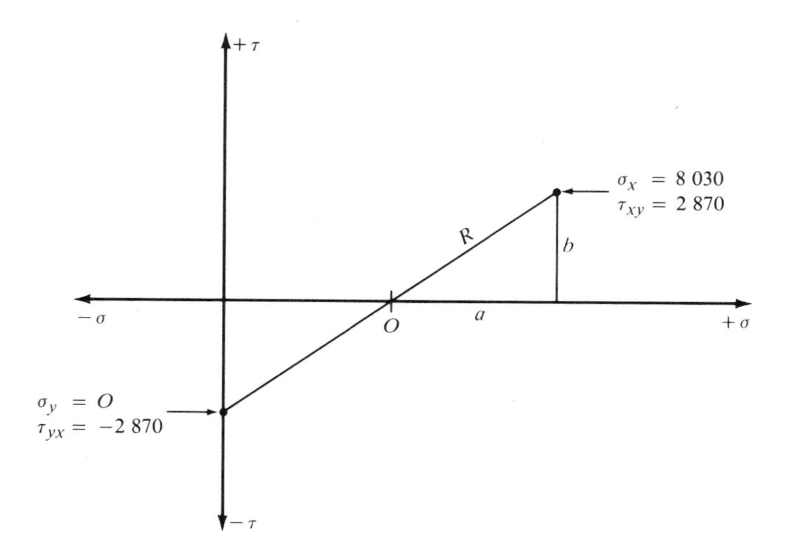

Figure 4-14 Partially Completed Mohr's Circle for Example Problem 4-4

points originally plotted. This it must do because the circle represents all possible states of stress on the element K.

8.

$$\sigma_1 = 4015 + 4935 = 8\,950 \text{ psi} \quad \text{at the right side of the circle}$$

9.

$$\sigma_2 = 4015 - 4935 = -920 \text{ psi} \quad \text{at the left side of the circle}$$

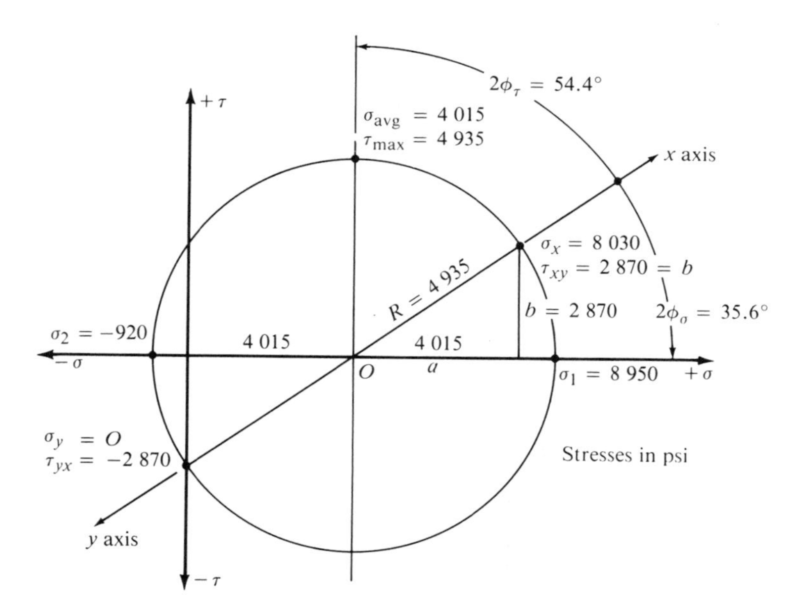

Figure 4-15 Completed Mohr's Circle

10.

$$\tau_{max} = R = 4\,935 \text{ psi}$$

The value of the normal stress on the element that carries the maximum shear stress is the same as the coordinate of O, the center of the circle. Recall that this is the average normal stress, 4 015 psi.

11.

$$2\phi_\sigma = \arctan(b/a) = \arctan(2\,870/4\,015) = 35.6°$$
$$\phi_\sigma = 35.6°/2 = 17.8°$$

Note that ϕ_σ must be measured *clockwise* from the original x axis to the direction of the line of action of σ_1.

12.

$$2\phi_\tau = 90° - 2\phi_\sigma = 90° - 35.6° = 54.4°$$

Then $$\phi_\tau = 54.4°/2 = 27.2°$$

Note that the stress element on which the maximum shear stress acts must be rotated *counterclockwise* from the orientation of the original element.

13. Figure 4-16 shows the required stress elements. They are identical to those shown in Figure 4-8 and 4-9.

Mohr's Circle Practice Problems

To a person seeing Mohr's circle for the first time, it may seem confusing. But with practice under a variety of combinations of normal and shear stresses, you should be able to execute the ten steps quickly and accurately. Below is a table giving six sets of data for normal and shear stresses in the x-y plane. You are advised to complete the Mohr's circle for each before looking at the solutions in Figures 4-17 through 4-22. From the circle, determine the two principal stresses, the maximum shear stress, and the planes on which these stresses act. Then draw the given stress element, the principal stress ele-

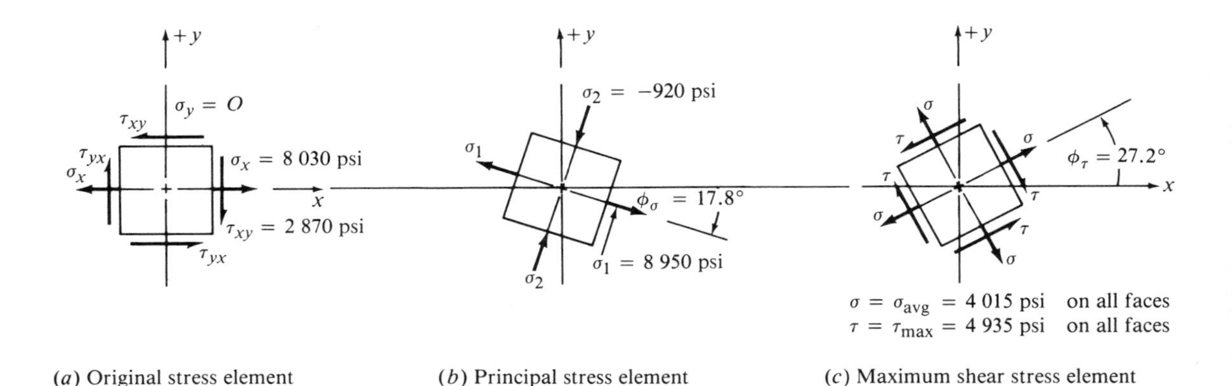

Figure 4-16 Results from Mohr's Circle Analysis of Example Problem 4-4

ment, and the maximum shear stress element, all oriented properly with respect to the x and y directions.

Practice Problems for Mohr's Circle

Example Problem	σ_x	σ_y	τ_{xy}	*Fig. No.*
4-5	+10.0 Ksi	−4.0 Ksi	+5.0 Ksi	4-17
4-6	+10.0 Ksi	−2.0 Ksi	−4.0 Ksi	4-18
4-7	+4.0 Ksi	−10.0 Ksi	+4.0 Ksi	4-19
4-8	+120 MPa	−30 MPa	+60 MPa	4-20
4-9	−80 MPa	+20 MPa	−50 MPa	4-21
4-10	−80 MPa	+20 MPa	+50 MPa	4-22

4-6 CARE WHEN BOTH PRINCIPAL STRESSES HAVE THE SAME SIGN

Remember that all of the problems presented thus far are plane stress problems, also called *biaxial stress* problems, because stresses are acting in only two directions within one plane. Obviously, real load-carrying members are three-dimensional objects. The assumption here is that if no stress is given for the third direction, it is zero. In most

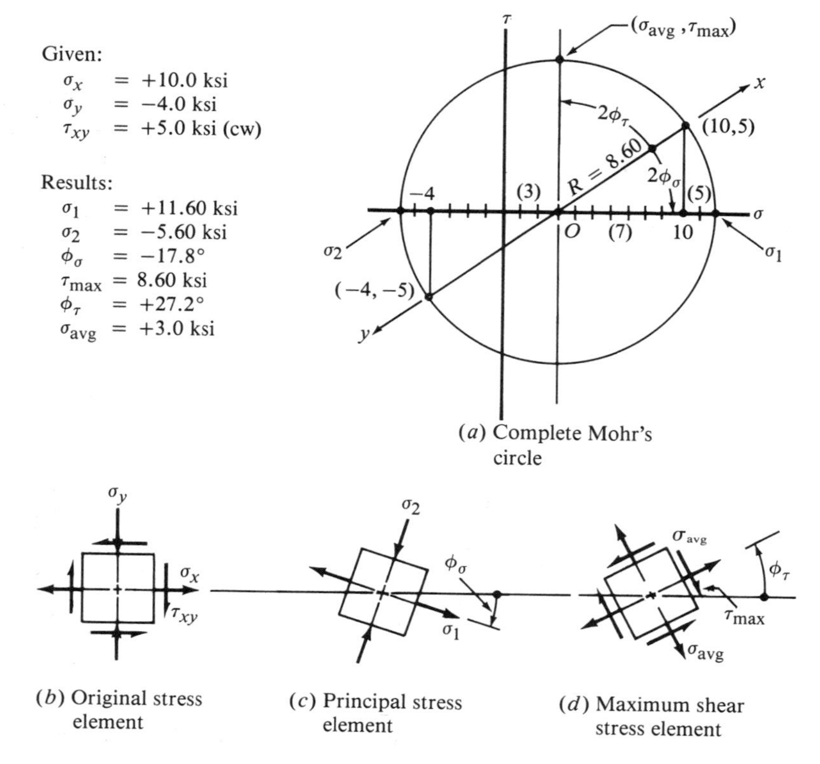

Figure 4-17 Solution for Example Problem 4-5

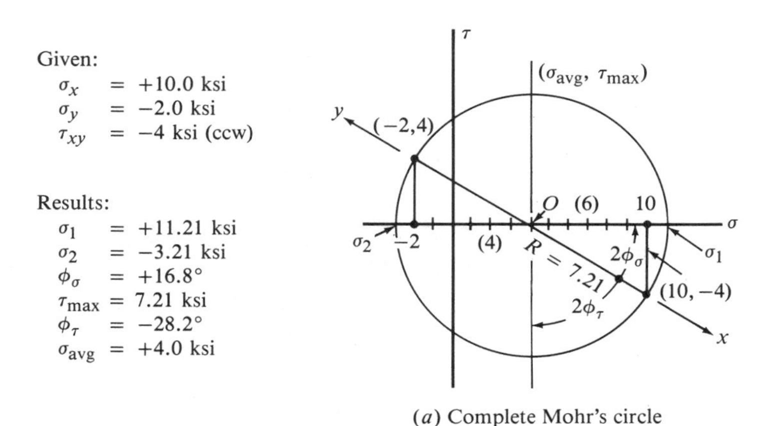

(a) Complete Mohr's circle

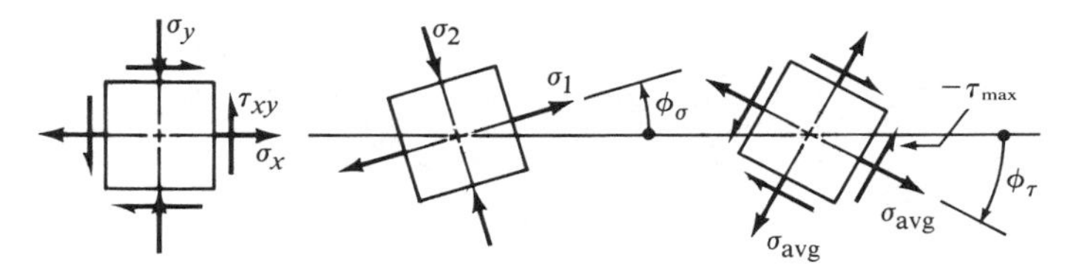

(b) Original stress element (c) Principal stress element (d) Maximum shear stress element

Figure 4-18 Solution for Example Problem 4-6

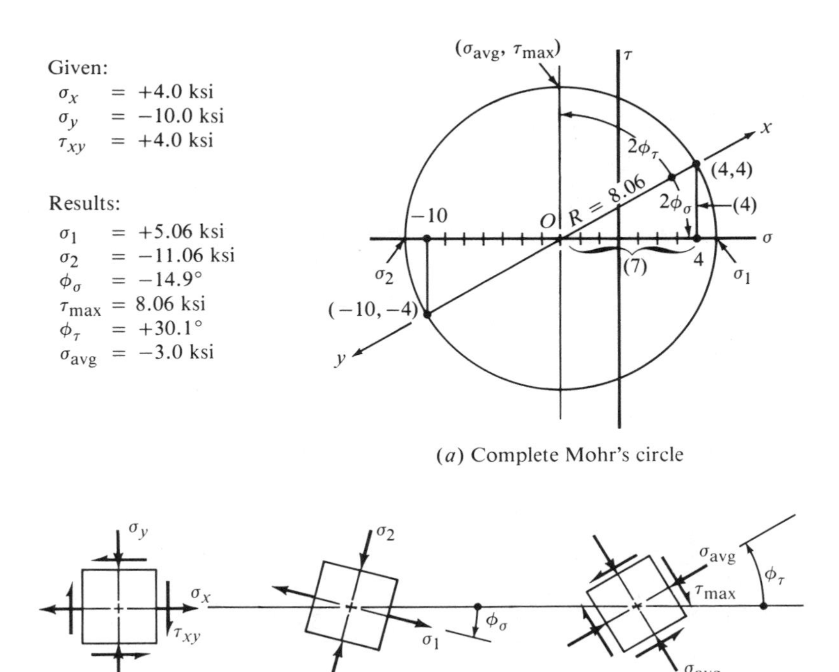

(a) Complete Mohr's circle

(b) Original stress element (c) Principal stress element (d) Maximum shear stress element

Figure 4-19 Solution for Example Problem 4-7

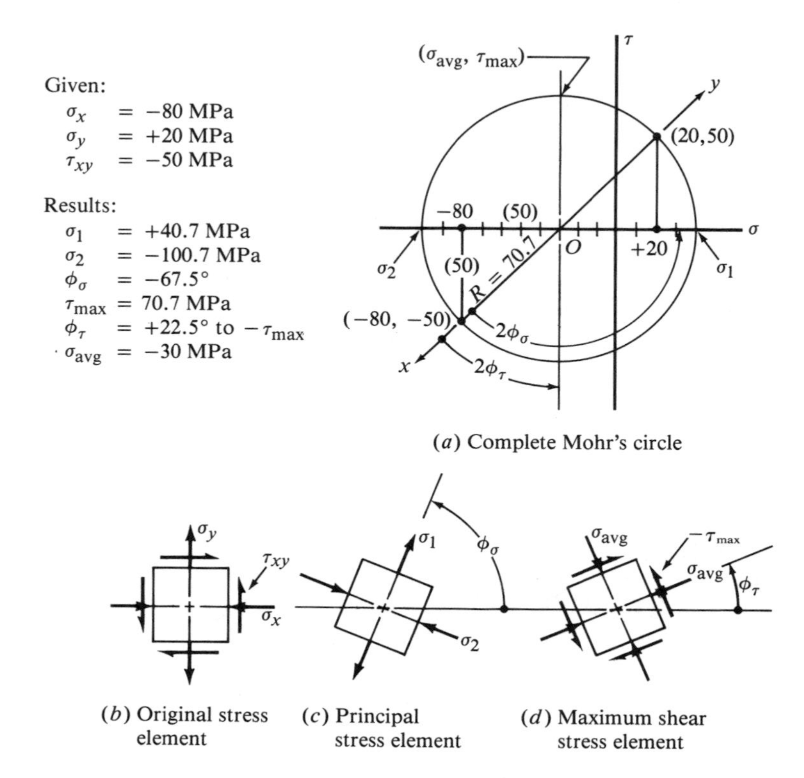

Figure 4-20 Solution for Example Problem 4-8

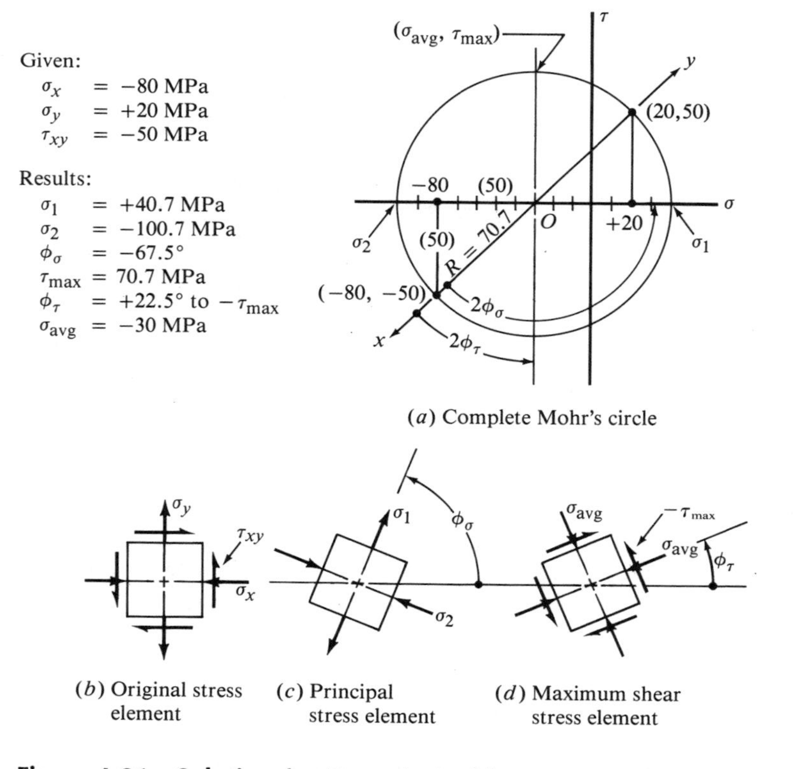

Figure 4-21 Solution for Example Problem 4-9

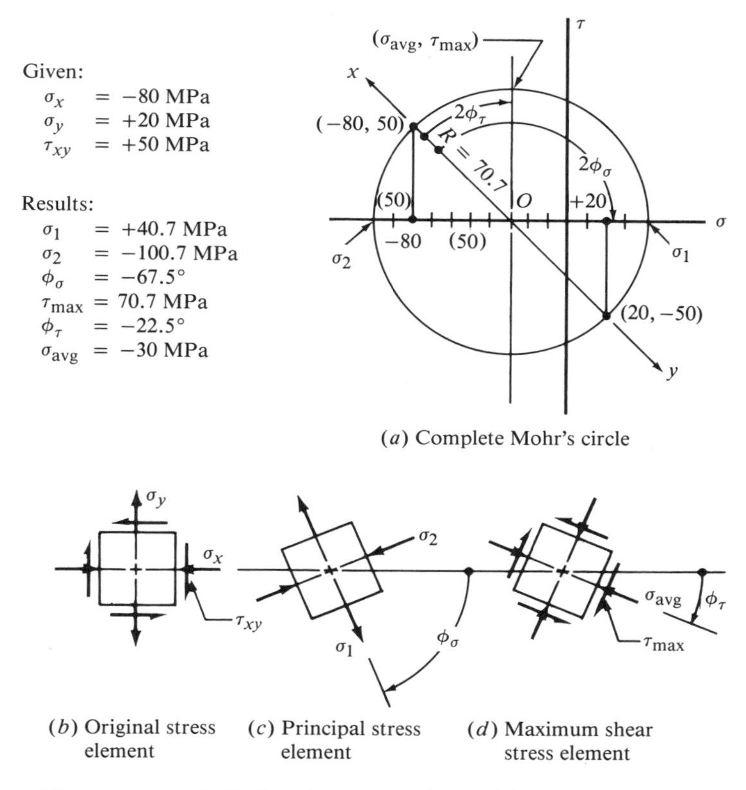

(*b*) Original stress element (*c*) Principal stress element (*d*) Maximum shear stress element

Figure 4-22 Solution for Example Problem 4-10

cases, the solutions as given will produce the true maximum shear stress, along with the two principal stresses for the given plane. This will always be true if the two principal stresses have opposite signs; that is, one is tensile and the other is compressive.

But the true maximum shear stress on the element will not be found if the two principal stresses are of the same sign. In such cases, you must consider the three-dimensional case. A familiar example will illustrate this situation.

Example Problem 4-11. Figure 4-23 shows a thin-walled cylinder subjected to an internal pressure with its ends closed. Determine the principal stresses, the maximum shear stress, and the orientation of the maximum shear stress element. The internal pressure is 500 psi, the wall thickness is 0.080 in, and the diameter of the cylinder is 4.0 in.

Solution. This is an example of a biaxial stress problem from elementary strength of materials. If we consider a small element on the surface of the cylinder, the stress aligned with the length of the cylinder (the y direction in the figure) is the *longitudinal stress* and has the value

$$\sigma_y = \frac{pD}{4t} = \frac{(500\text{ psi})(4.0\text{ in})}{(4)(0.080\text{ in})} = 6\,250\text{ psi} \quad \text{(tension)}$$

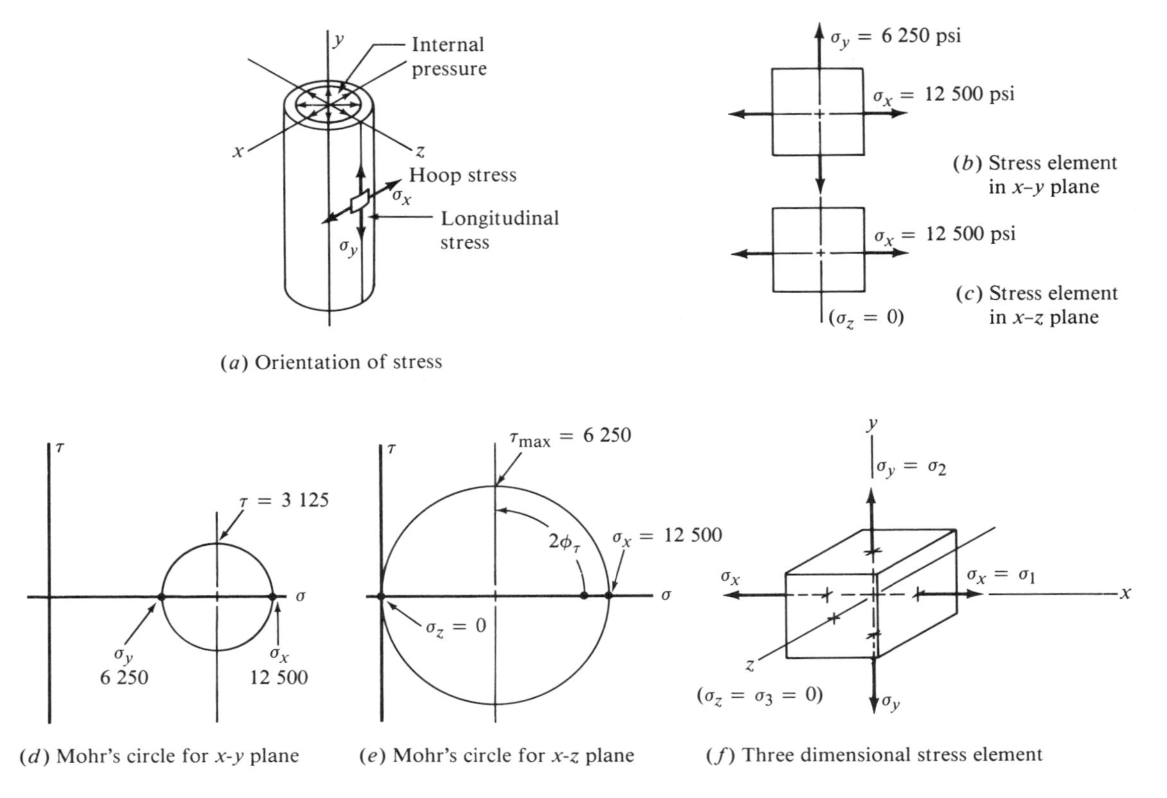

Figure 4-23 Stress Analysis for a Thin-walled Cylinder Subjected to Pressure with Its Ends Closed

The stress in the circumferential direction is the *hoop stress,* and it is aligned with the x axis on the infinitesimal element.

$$\sigma_x = \frac{pD}{2t} = \frac{(500 \text{ psi})(4.0 \text{ in})}{(2)(0.080 \text{ in})} = 12\,500 \text{ psi} \quad \text{(tension)}$$

There are no shear stresses applied in the x and y directions.

Part (b) of the figure shows the stress element for the x-y plane, and part (d) shows the corresponding Mohr's circle. Because there are no applied shear stresses, σ_x and σ_y are the principal stresses for the plane. The circle would predict the maximum shear stress to be equal to the radius of the circle, 3 125 psi.

But notice part (c) of the figure. We could have chosen the x-z plane for analysis, instead of the x-y plane. The stress in the z direction is zero because it is perpendicular to the free face of the element. Likewise, there are no shear stresses on this face. The Mohr's circle for this plane is shown in part (e) of the figure. The maximum shear stress is equal to the radius of the circle, 6 250 psi, or *twice* as much as would be predicted from the x-y plane.

This approach should be used any time the two principal stresses in a biaxial stress problem have the same sign. In summary, on a general three-dimensional stress element, there will be one orientation of the element in which there are no shear stresses acting. The normal stresses on the three perpendicular faces are then the three principal stresses.

If we call these stresses σ_1, σ_2, and σ_3, taking care to order them such that $\sigma_1 > \sigma_2 > \sigma_3$, then the maximum shear stress on the element will always be

$$\tau_{max} = \frac{\sigma_1 - \sigma_3}{2}$$

For the cylinder of Figure 4-23, we can conclude that

$$\sigma_1 = \sigma_x = 12\,500 \text{ psi}$$
$$\sigma_2 = \sigma_y = 6\,250 \text{ psi}$$
$$\sigma_3 = \sigma_z = 0$$
$$\tau_{max} = (\sigma_1 - \sigma_3)/2 = (12\,500 - 0)/2 = 6\,250 \text{ psi}$$

Figure 4-24 shows two additional examples in which the two principal stresses in the given plane have the same sign. Then the zero stress in the third direction is added to the diagram and the new Mohr's circle is superimposed on the original one. This serves to illustrate that the maximum shear stress will occur on the Mohr's circle having the largest radius.

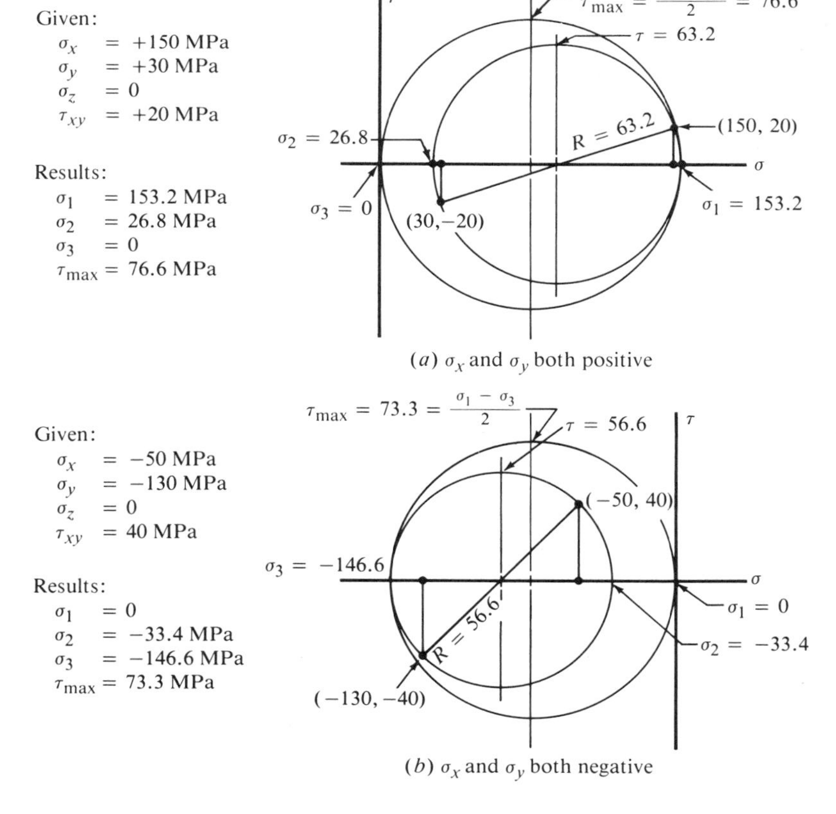

Figure 4-24 Mohr's Circle for Cases in Which Two Principal Stresses Have the Same Sign

PROBLEMS

1. Compute the maximum tensile stress in the bracket shown in Figure 4-25.
2. Compute the maximum tensile and compressive stresses in the horizontal beam shown in Figure 4-26.
3. For the lever shown in Figure 4-27(a), compute the stress at section *A* near the fixed end. Then redesign the lever to the tapered form shown in part (b) of the figure by adjusting only the height of the cross section at sections *B* and *C* so they have no greater stress than section *A*.
4. Compute the maximum tensile stress at sections *A* and *B* on the crane boom shown in Figure 4-28.
5. Refer to Figure 3-15 in Chapter 3. Compute the maximum tensile stress in the print head just to the right of the right guide. The head has a rectangular cross section, 5.0 mm high in the plane of the paper and 2.4 mm thick.
6. Refer to Figure 3-19 in Chapter 3. Compute the maximum tensile and compressive stresses in the member *B-C* if the load *F* is 1 800 pounds. The cross section of *B-C* is shown in Figure 4-29.
7. Refer to Figure 3-27 in Chapter 3. The vertical member is to be made from steel with a maximum allowable stress of 12 000 psi. Specify the required size of a standard square cross section if sizes are available in increments of $^1/_{16}$ in.
8. Refer to Figure 3-30 in Chapter 3. Compute the maximum stress in the horizontal portion of the bar and tell where it occurs on the cross section. The left support resists the axial force.
9. Refer to Figure 3-31 in Chapter 3. Compute the

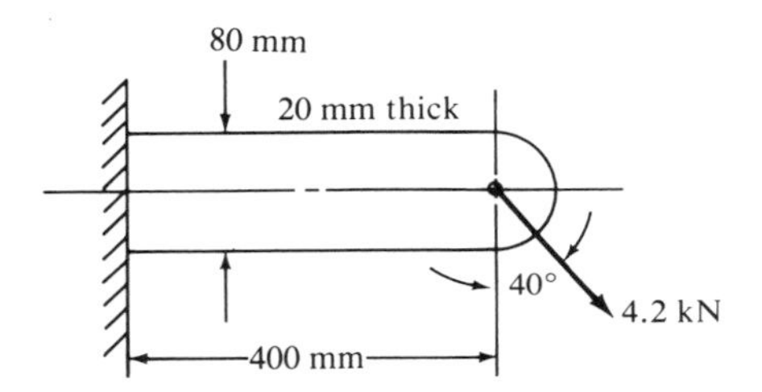

Figure 4-25 (Problem 1)

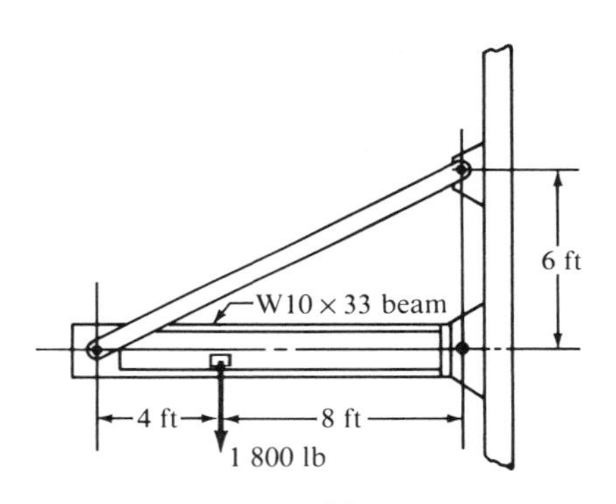

Figure 4-26 (Problem 2)

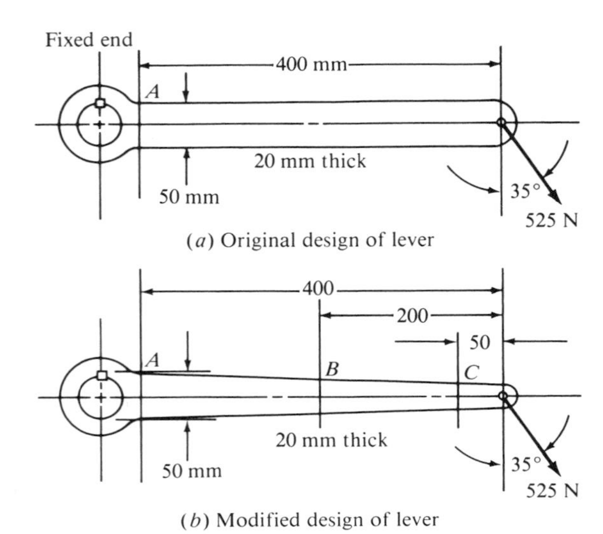

(*a*) Original design of lever

(*b*) Modified design of lever

Figure 4-27 (Problem 3)

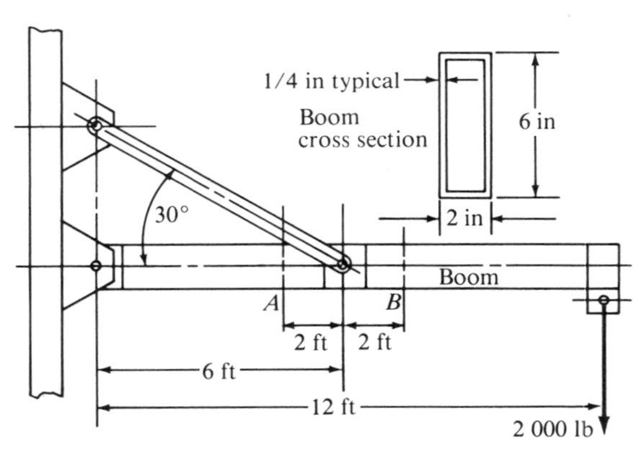

Figure 4-28 (Problem 4)

maximum stress in the horizontal portion of the bar and indicate where it occurs on the cross section. The right support resists the unbalanced axial force.

10. Refer to Figure 3-38 in Chapter 3. Specify a suitable diameter for a solid circular bar to be used for the top horizontal member, which is supported in the bearings. The left bearing resists the axial load. The allowable normal stress is 25 000 psi.
11. Refer to Figure 3-16 in Chapter 3. Using the equivalent torque method, compute the required diameter for the shaft *ABC* to keep the maximum shear stress less than 7 500 psi.
12. Refer to Figure 3-32 in Chapter 3. Using the equivalent torque method, compute the required diameter for the shaft *ABC* to keep the maximum shear stress less than 7 500 psi. The unbalanced torque is resisted at support *A* only.
13. Refer to Figure 3-33 in Chapter 3. Using the equivalent torque method, compute the required diameter for the shaft *ABC* to keep the maximum shear stress less than 7 500 psi. The unbalanced torque is resisted at support *A* only.
14. Refer to Figure 4-30. Using the equivalent torque method, compute the maximum shear stress on the top surface of the circular bar near the support.
15. Refer to Figure 4-31. Using the equivalent torque method, compute the maximum shear stress in the shaft between sections *B* and *C*. Assume that there are stress concentrations near both *B* and *C*. Use $K_t = 2.2$ for bending and $K_t = 1.7$ for torsion.

Problems 16–31.

For the following sets of given stresses on an element (see table below), draw a complete Mohr's circle, find the principal stresses and the maximum shear stress, and draw the principal stress element and the maximum shear stress element. Any stress components not shown are assumed to be zero.

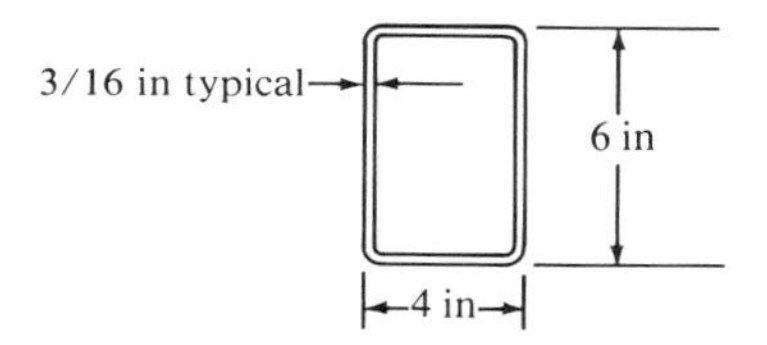

Figure 4-29 (Problem 6)

Figure 4-30 (Problem 14)

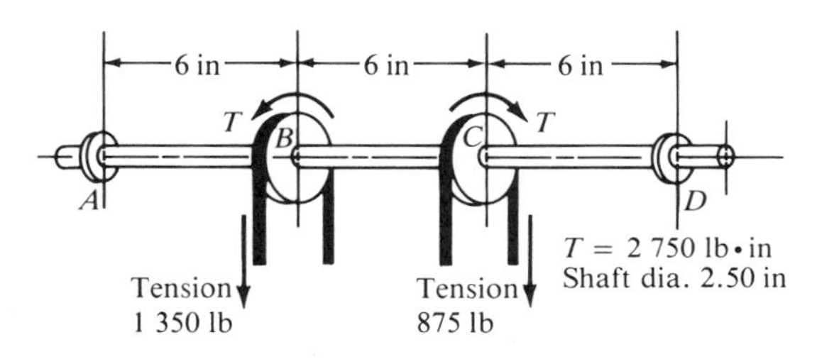

Figure 4-31 (Problem 15)

Problem	σ_x	σ_y	τ_{xy}	*Problem*	σ_x	σ_y	τ_{xy}
16	20 Ksi	0	10 Ksi	24	100 MPa	0	80 MPa
17	85 Ksi	−40 Ksi	30 Ksi	25	250 MPa	−80 MPa	110 MPa
18	40 Ksi	−40 Ksi	−30 Ksi	26	50 MPa	−80 MPa	40 MPa
19	−80 Ksi	−40 Ksi	−30 Ksi	27	−150 MPa	−80 MPa	−40 MPa
20	120 Ksi	40 Ksi	20 Ksi	28	150 MPa	80 MPa	−40 MPa
21	20 Ksi	140 Ksi	20 Ksi	29	50 MPa	180 MPa	40 MPa
22	20 Ksi	−40 Ksi	0	30	250 MPa	−80 MPa	0
23	120 Ksi	−40 Ksi	100 Ksi	31	50 MPa	−80 MPa	−30 MPa

5 Design for Different Modes of Failure

5-1 OVERVIEW

Fatigue is a condition of loading in which many cycles of stress are encountered by the part during its expected life, a situation most common in machine design.

The behavior of materials under fatigue loading conditions is the subject of much continuing study. There are many variables involved, so statistical methods are used to test parts and materials to obtain data from which reliability and life can be estimated. In general, any repeated load is called a *fatigue load*. But there are different types. The most fundamental type, called *reversed bending,* is produced by placing a shaft in bending and then rotating the shaft. Figure 5-1 shows one such loading device, in which the entire middle section experiences a uniform bending moment of $Wa/2$. A point on the surface of the shaft at the bottom and in the middle section will see the maximum tensile bending stress from the loading shown. But as the shaft is rotated 180° that point moves to the top, where it sees the maximum compressive stress. Thus in each rotation of the shaft, a given point sees the cyclic loading from maximum tension to maximum compression, as illustrated in Figure 5-2.

The loading pattern can be described by its *R-value,* where R is the ratio of the minimum stress to the maximum stress. For reversed stress, the magnitudes of the maximum and minimum stresses are equal, but they have opposite signs. Therefore,

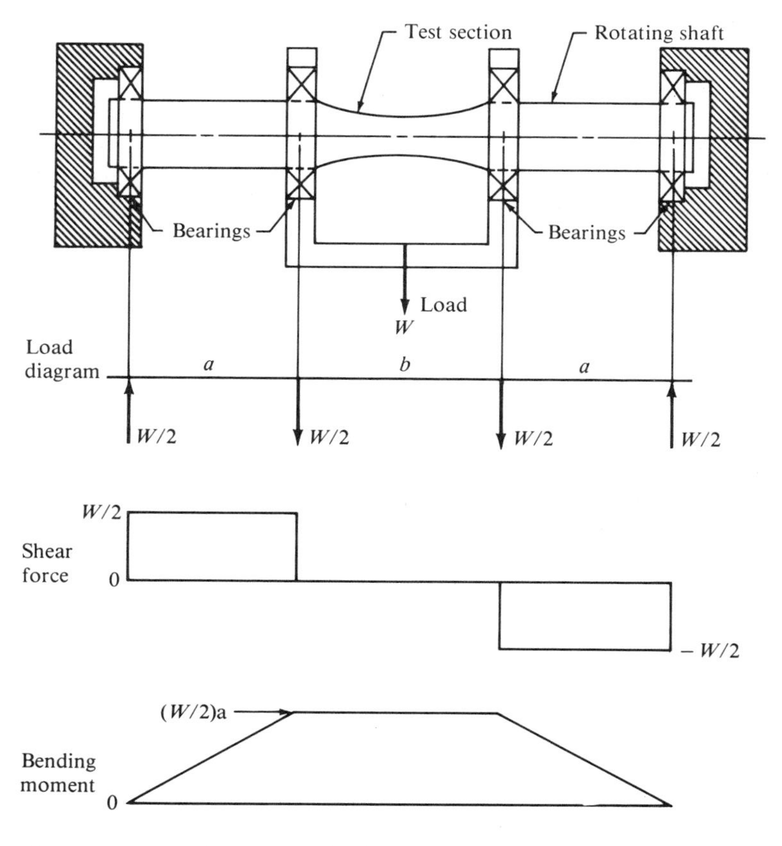

Figure 5-1 Fatigue Test Device

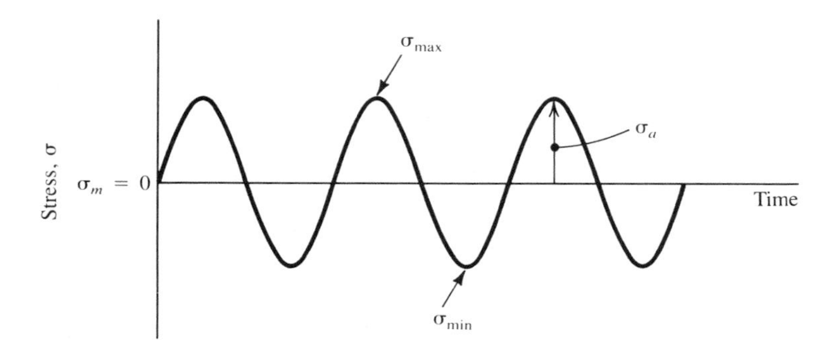

Figure 5-2 Repeated, Reversed Stress

$$R = \sigma_{\min}/\sigma_{\max} \tag{5-1}$$
$$R = (-\sigma)/(+\sigma) = -1$$

We can also define the mean stress, σ_m, and the alternating stress, σ_a (sometimes called the *stress amplitude*) from

$$\sigma_m = (\sigma_{\max} + \sigma_{\min})/2 \tag{5-2}$$

and

$$\sigma_a = (\sigma_{\max} - \sigma_{\min})/2 \tag{5-3}$$

Then for the reversed stress, $\sigma_m = 0$, and $\sigma_a = \sigma_{\max}$.

A second type of fatigue loading is the repeated, one-direction load, such as is produced in a tension link in a mechanism that starts with zero load, then increases the load and the stress as the linkage goes through its cycle. Figure 5-3 shows a graph of this type of stress. Note that the minimum stress is zero, giving the R-value of

$$R = 0/\sigma_{\max} = 0$$

The mean and alternating stresses are

$$\sigma_m = (\sigma_{\max} + 0)/2 = \sigma_{\max}/2$$
$$\sigma_a = (\sigma_{\max} - 0)/2 = \sigma_{\max}/2$$

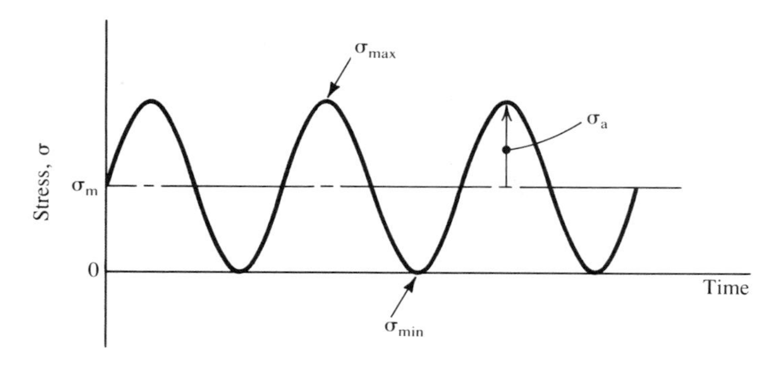

Figure 5-3 Repeated, One-direction Stress

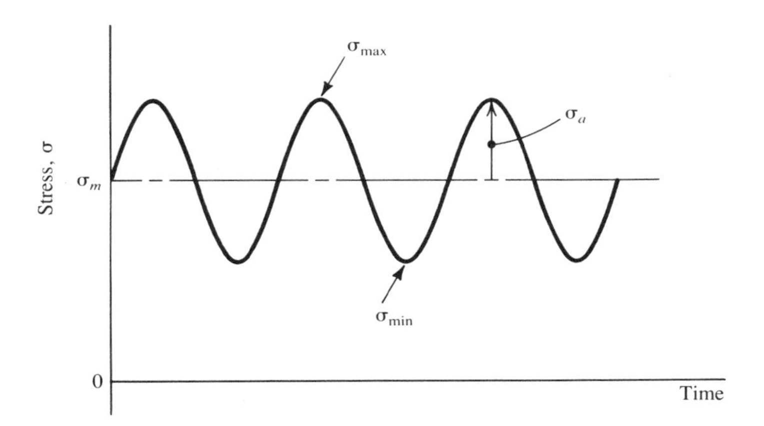

Figure 5-4 Fluctuating Stress

The third type of fatigue loading is *fluctuating stress,* in which there is a nonzero mean stress with an alternating component superimposed, as sketched in Figure 5-4. If the stresses are all tensile, as shown in Figure 5-4, the R-value will be somewhat less than 1, with the specific value depending on the magnitudes of the applied stresses.

5-2 ENDURANCE STRENGTH

The *endurance strength* of a material is its ability to withstand fatigue loads. In general it is the stress level that a material can survive for a given number of cycles of loading. If the number of cycles is infinite, the stress level is called the *endurance limit.*

Endurance strengths are usually charted on a graph like that shown in Figure 5-5, called an *S-N diagram.* Curves A, B, and D are representative of a material that does exhibit an endurance limit, such as a plain carbon steel. Curve C is typical of most nonferrous metals, such as aluminum, which do not exhibit an endurance limit. For such materials, the number of cycles to failure should be reported for the given endurance strength.

Published data for endurance strength are determined by special fatigue testing devices, which typically use a polished specimen subjected to a reversed bending load, similar to that sketched in Figure 5-1. If the actual operating conditions of a part in a machine are different, and they usually are, the fatigue strength must be reduced from the reported value. Some of the factors that decrease the endurance strength are discussed next.

Size of the Section

The test specimen is usually 0.30 in (7.6 mm) in diameter. Larger section sizes exhibit lower strengths. A size factor, $C_s = 0.9$, can be applied to the test data for sizes greater than 0.4 in (10 mm) and less than 2.0 in (50 mm). The larger sizes have a less favorable stress distribution, and the uniformity of properties is somewhat less also, particularly with heat-treated parts. Sizes larger than 2.0 in (50 mm) should use an even smaller factor, and testing is recommended.

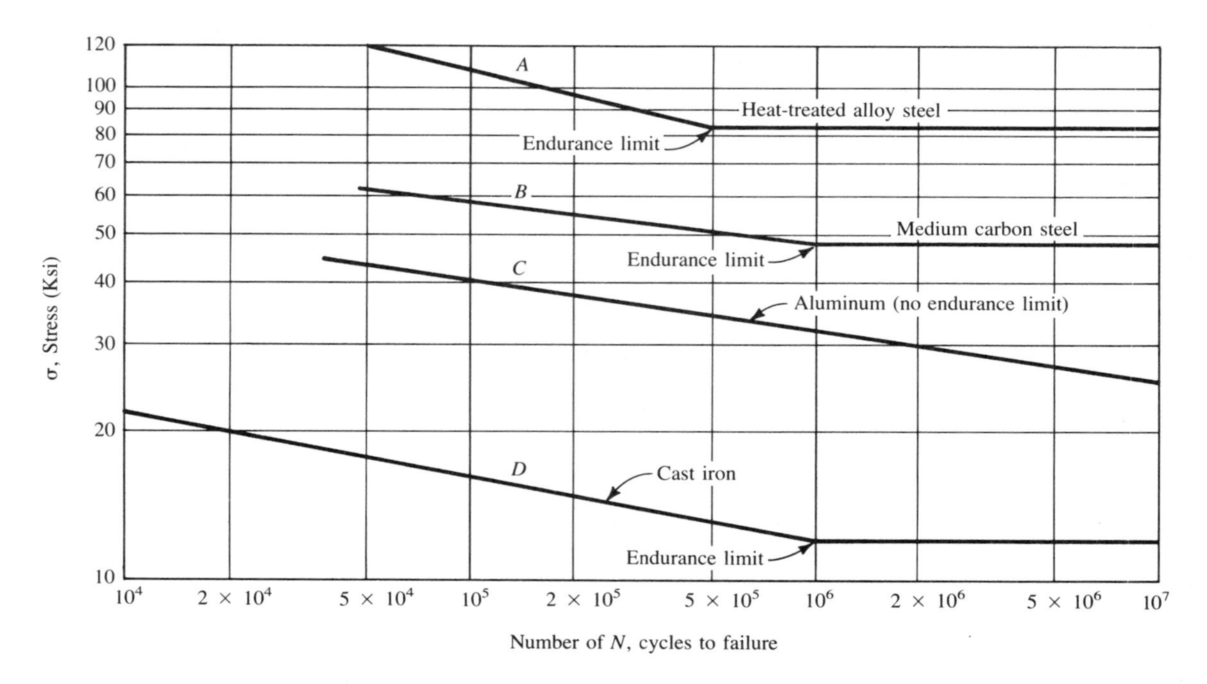

Figure 5-5 Representative Endurance Strengths

Surface Finish

Any deviation from a polished surface reduces endurance strength. Figure 5-6 shows rough estimates for the endurance strength compared with the ultimate tensile strength of steels for several practical surface conditions. It is critical that parts subjected to fatigue loading be protected from nicks, scratches, and corrosion because they drastically reduce fatigue strength.

Stress Concentrations

Sudden changes in geometry, especially sharp grooves and notches where high stress concentrations occur, are likely places for fatigue failures to occur. Care should be taken in the design and manufacture of cyclically loaded parts to keep stress concentration factors to a low value. We will apply the stress concentration factors, as found from the methods of section 3-9, to the computed stresses, rather than to the allowable strengths.

Flaws

Internal flaws of the material, especially likely in cast parts, are places in which fatigue cracks initiate. Critical parts can be inspected by x-ray techniques for internal flaws. If they are not inspected, a higher than average design factor should be specified for cast parts, and a lower endurance strength should be used.

Temperature

Most materials have a lower endurance strength at high temperatures. The reported values are for room temperatures. Operation above 160°F will reduce the endurance strength of

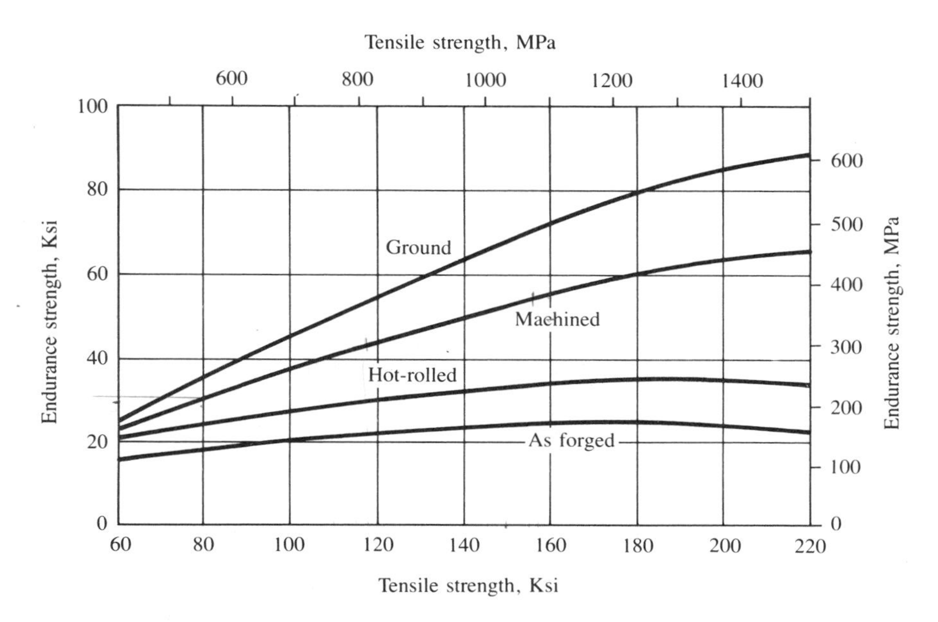

Figure 5-6 Endurance Strength versus Tensile Strength for Wrought Steel for Various Surface Conditions

most ductile materials. However, some alloys such as AISI 4340 retain a high percentage of room temperature strength up to 750°F (400°C). Specific data on the alloy used should be obtained.

Type of Stress

The reported values are for reversed bending. If axial stresses are encountered, the endurance strength should be reduced as follows:

Let s_n = Endurance strength from rotating beam test

$0.8s_n$ = Endurance strength for reversed axial load

The endurance strength in shear will be called s_{sn}. As discussed later, two theories are used for design with shear stress. The maximum shear stress theory is simpler and is used most often. For this theory,

$$s_{sn} = 0.5s_n = \text{Endurance strength for shear}$$

For the distortion energy theory,

$$s_{sn} = 0.577\ s_n$$

One difficulty facing designers is that data for endurance strengths are frequently not reported. In the absence of data, the following estimates can be used.

For wrought steel:

$$s_n = 0.50s_u$$

$S_n' = S_n C_S$ — C_S usually .9 (see Juvinall pg. 214)

(Note: This is similar to the "ground" curve in Figure 5-6.)

For cast steel:

$$s_n = 0.40s_{ut}$$

(This is 0.8 times the value for wrought steel. See item 5 below.)

For cast iron:

$$s_n = 0.35s_{ut}$$

(This is 0.7 times the value for wrought steel. See item 5 below.)

All of the factors presented are guidelines and estimates. It is highly recommended that testing be done on actual parts subjected to fatigue loading under conditions as close as possible to the actual service expected.

In problem solutions, we will use the following procedure for estimating endurance strengths for steel parts, called s_n'.

1. Determine the ultimate strength of the steel from tests or published data.
2. After specifying the manufacturing process, obtain an estimate of the endurance limit from Figure 5-6.
3. Apply a size factor if the size of the critical section is greater than 0.40 in (10 mm). Use $C_s = 0.9$ up to 2.0 in (50 mm). Above that size, reduce the size factor.
4. Apply a factor for the type of stress if it is other than rotating bending.
5. Apply a factor of 0.8 for cast steel or 0.7 for cast iron.

Example. Estimate the endurance strength for a shaft subjected to rotating bending if it is machined from AISI 1050 cold-drawn steel and has a diameter of 1.75 in.

Solution. From Appendix A-3, the ultimate strength is 100 Ksi. Then from Figure 5-6, the endurance limit is 38 Ksi for a machined part. Using a size factor of 0.9 gives

$$s_n' = (38 \text{ Ksi})(0.90) = 34.2 \text{ Ksi}$$

Example. Estimate the endurance strength for a tensile rod in a linkage subjected to reversed, repeated axial load. It is to be machined from cast steel having an ultimate strength of 120 Ksi. Its cross section is to be nominally 1.50 in square.

Solution. The machined surface would result in an endurance limit of 43 Ksi (from Figure 5-6). Using a size factor of 0.9, a load factor of 0.8 for the axial stress, and a factor of 0.8 because the part is cast,

$$s_n' = (43 \text{ Ksi})(0.9)(0.8)(0.8) = 24.8 \text{ Ksi}$$

Example. A bar has a diameter of 25 mm and is made from hot-rolled AISI 1137 steel. It is to be used as a torsion bar carrying a reversed torsion load. Estimate the endurance limit in MPa for the maximum shear stress theory.

Solution. From Appendix A-3, the ultimate strength is 607 MPa, or 88 Ksi. From Figure 5-6, $s_n = 24$ Ksi. Converting to MPa:

$$s_n = 24\ \text{Ksi}\frac{6.895\ \text{MPa}}{1.0\ \text{Ksi}} = 165\ \text{MPa}$$

Using a size factor of 0.9 and a load factor of 0.5 gives

$$s'_{sn} = (165\ \text{MPa})(0.9)(0.5) = 74.3\ \text{MPa}$$

5-3 DESIGN FOR DIFFERENT MODES OF FAILURE

It is the designer's responsibility to ensure that a machine part is safe for operation under reasonably foreseeable conditions. This is done by performing a stress analysis, either analytically or experimentally, and comparing the predicted stress with a design stress. The *design stress* is the level of stress that the part will be permitted to see under operating conditions. Some designers prefer the terms *working stress, allowable stress,* or *safe stress*. We will use *design stress* to emphasize the role of the designer in determining the appropriate value to use.

The decision on what design stress to use involves, at least, the consideration of the type of load on the part and the material from which it is made. The types of loads are the following:

Static

Repeated and reversed

Fluctuating

Shock or impact

It is important to consider the ductility of the material, in addition to its strength and stiffness. The modes of failure are dramatically different for brittle materials and ductile materials. (Review the discussion of ductility in Chapter 2.) Most machine parts are made from ductile materials, especially if they are subjected to fatigue loads or to shock or impact. Wrought metals, such as steel, aluminum, and copper, are usually ductile. For fatigue loading, it is advised that the percent elongation of the material be significantly higher than 5 percent, say 12 percent or higher, to ensure against brittle-type fracture. Gray cast iron, some highly heat-treated steels, some plastics, and some parts made from powder metal are brittle.

Predictions of Failure

Four different ways of predicting failure are considered in this section:

Maximum normal stress theory

Maximum shear stress theory

Distortion energy theory (also called the *von Mises theory* or the *Mises-Hencky theory*)

Soderberg criterion

Comparing the actual expected stress in a part with the stress that would predict failure allows the determination of a *factor of safety,* or *design factor, N*. These two terms are used interchangeably in this book.

Recommended Design Factors

The following discussion presents several ways of computing the resulting design factor, N, for given loading and material combinations. Of course, in design the use of such relationships is to compute the allowable stress to achieve an adequate design factor. At times, the design factor is specified in codes promulgated by organizations such as the American Society of Mechanical Engineers and the American Institute of Steel Construction, or in building codes or company policy, or by past experience with similar conditions.

In the absence of codes, and for problems in this book, we will use design factors as follows:

$N = 2$: Where a high level of confidence exists that the loads, material properties, and operating conditions are well known, and ductile materials are used

$N = 3$: For brittle materials where a high level of confidence exists that the loads, material properties, and operating conditions are well known

$N = 3$: For ductile materials with some doubt about the adequacy of material properties data, loads, or the stress analysis

$N = 4$: For uncertain conditions about some *combination* of material properties data, loads, and the stress analysis

Shock and impact involve not only strength considerations but also the ability of the member to absorb energy (4, 5,7).

Methods of Computing Design Factor

The methods are presented in the form that is used to compute the design factor, N, for given stress conditions and material properties. In design, many of the relationships can be solved for a design stress in terms of the design factor. If this is not convenient, an iterative procedure can be used (see the example problems).

Case A: Brittle materials under static loads
- A1: Uniaxial tensile stress
- A2: Uniaxial compressive stress
- A3: Biaxial stresses

Case B: Brittle materials under fatigue loads

Case C: Ductile materials under static loads
- C1: Maximum shear stress theory
- C2: Distortion energy theory

Case D: Reversed, repeated normal stress

Case E: Reversed torsional shear stress
- E1: Maximum shear stress theory
- E2: Distortion energy theory

Case F: Reversed combined stress
- F1: Maximum shear stress theory
- F2: Distortion energy theory

Case G: Fluctuating normal stresses

Case H: Fluctuating shear stresses

Case I: Fluctuating combined stresses
I1: Maximum shear stress theory
I2: Distortion energy theory

Case A: Brittle Materials under Static Loads

When the stress is simple tension or compression in only one direction, use the maximum normal stress theory of failure, where

$$N = s_{ut}/\sigma \quad \text{(Case A1: If } \sigma \text{ is tensile)} \tag{5-4}$$

or

$$N = s_{uc}/\sigma \quad \text{(Case A2: If } \sigma \text{ is compressive)} \tag{5-5}$$

Case A3

When a biaxial stress condition exists, use the Mohr circle to determine the principal stresses. If both principal stresses are tensile or both are compressive, use the preceding equations. If they are of different signs, the combined effect can be estimated from

$$\frac{1}{N} = \frac{\sigma_1}{s_{ut}} + \frac{\sigma_2}{s_{uc}} \tag{5-6}$$

Any stress concentration factors should be applied to the computed stresses.

Case B: Brittle Materials under Fatigue Loads

No specific recommendation will be given for brittle materials under fatigue loads because it is usually not desirable to use a brittle material in such cases. When it is necessary to do so, testing should be done to ensure safety under actual conditions of service.

Case C: Ductile Materials under Static Loads

Two failure theories will be discussed here. The maximum shear stress theory of failure is the more conservative of the two and is somewhat easier to use. The distortion energy theory is the more accurate predictor of impending failure.

Case C1: Maximum Shear Stress Theory

Failure occurs when the maximum shear stress exceeds the yield strength of the material in shear, s_{ys}. Determine the maximum shear stress from the Mohr circle. Then

$$N = s_{ys}/\tau_{max} \tag{5-7}$$

We will use $s_{ys} = 0.5s_y$. Then

$$N = 0.5s_y/\tau_{max} \quad \text{or} \quad \tau_d = 0.5s_y/N \tag{5-8}$$

It can be shown from a Mohr's circle analysis that, for the special case of uniaxial tension or compression with no applied shear stresses, equation 5-8 is the same as

$$N = s_y/\sigma_{max} \quad \text{or} \quad \sigma_d = s_y/N = \text{Design stress}$$

Case C2: Distortion Energy Theory

Failure occurs when the von Mises stress, σ', exceeds the tensile yield strength of the material. The von Mises stress is found by first using the Mohr circle to determine the two principal stresses. Then

$$\sigma' = \sqrt{\sigma_1^2 + \sigma_2^2 - \sigma_1\sigma_2} \tag{5-9}$$

The design factor is then found from

$$N = s_y/\sigma' \tag{5-10}$$

The advantage of this theory, in addition to its being a more accurate predictor of failure, is that it allows the use of the tensile yield strength for analysis.

For static loading on ductile materials it is not necessary to include a stress concentration factor when geometric discontinuities occur. If the local stress at a point of stress concentration reaches the yield strength of the material, the material will, in fact, yield, causing the stress to be redistributed and bringing the level back below the yield strength.

Case D: Reversed, Repeated Normal Stress

Determine the peak amplitude of the repeated stress (as σ_{max} in Figure 5-2) and the endurance strength for the material, s_n', adjusted for actual conditions as in section 5-2. Then

$$N = s_n'/\sigma_{max} \qquad \text{or} \qquad \sigma_d = s_n'/N \tag{5-11}$$

Stress concentration factors must be considered in computing the maximum stress.

Case E: Reversed Torsional Shear Stress

Again the maximum shear stress theory or the distortion energy theory can be used. First compute the maximum torsional shear stress, τ_{max}. Then

Case E1: For the Maximum Shear Stress Theory

$$s_{sn}' = 0.5s_n' \quad \text{(Estimate for endurance strength in shear)}$$

$$N = s_{sn}'/\tau_{max} \qquad \text{or} \qquad \tau_d = s_{sn}'/N \tag{5-12}$$

Case E2: For the Distortion Energy Theory

$$s_{sn}' = 0.577s_n'$$

$$N = s_{sn}'/\tau_{max} \tag{5-13}$$

The factor 0.577 can be derived from the definition of the von Mises stress for the special case of pure torsion. Drawing the Mohr circle, you will find that $\sigma_1 = +\tau_{max}$, and $\sigma_2 = -\tau_{max}$. Then, from equation (5-9), $\sigma' = \sqrt{3}\tau_{max} = 1.732\tau_{max}$. Putting this into the basic form of the distortion energy theory,

$$N = s_n'/\sigma' = s_n'/(1.732\tau_{max}) = 0.577s_n'/\tau_{max}$$

as used in equation (5-13).

Case F: Reversed Combined Stress

Use the Mohr circle to find the maximum shear stress and the two principal stresses by using the maximum values of the applied stresses. Then

Case F1: For the Maximum Shear Stress Theory

$$N = 0.5s_n'/\tau_{max} \tag{5-14}$$

Case F2: For the Distortion Energy Theory

$$N = 0.577s_n'/\tau_{max} \tag{5-15}$$

Or compute σ' from equation (5-9) and use

$$N = s_n'/\sigma' \tag{5-16}$$

Case G: Fluctuating Normal Stresses: The Soderberg Criterion

Recall from section 5-1 that the term *fluctuating stress* refers to the condition in which the part is subjected to a mean stress greater than zero with an alternating stress superimposed on the mean stress (see Figure 5-4). One method of failure prediction uses the Soderberg diagram, as shown in Figure 5-7, which is drawn for the case of normal stresses only. Mean stress is plotted along the horizontal axis, and alternating stress is plotted along the vertical axis. Failure when the alternating stress is zero (steady stress, case C), would be predicted as the stress reaches the yield strength of the material. Failure under alternating stress with a zero mean stress (reversed, repeated load, case D), would be predicted as the stress reaches the endurance strength of the material. A line drawn between these two points is called the *Soderberg line*. Then for fluctuating stress, if the point representing the actual combination of mean and alternating stress plots on the diagram below the Soderberg line, it should be safe. Conversely, a combination outside the line would be predicted to cause failure.

Applying a design factor to the Soderberg line produces the design stress line on the diagram. Actual operating stress levels should be kept below this line for safety. When stress concentrations are encountered, it is found that they have virtually no effect

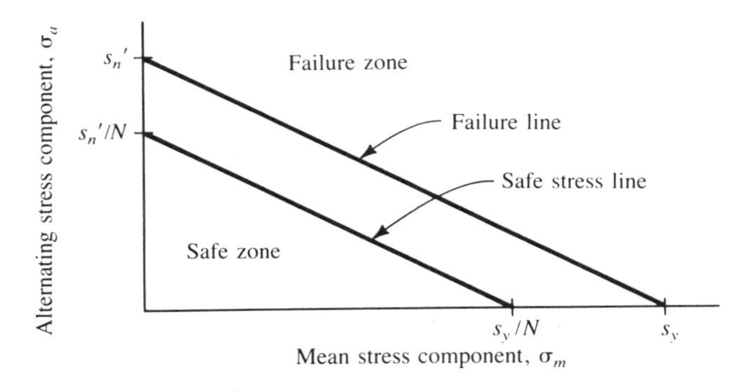

Figure 5-7 Soderberg Criterion for Fluctuating Stresses

on the mean stress component, but that the stress concentration factor should be applied to the alternating stress component. Also, the endurance strength plotted on the Soderberg diagram should be adjusted for surface finish and other modifying factors.

An analytical determination of the design factor using the Soderberg diagram can be made from the following equation:

$$\frac{1}{N} = \frac{\sigma_m}{s_y} + \frac{K_t \sigma_a}{s_n'} \tag{5-17}$$

Case H: Fluctuating Shear Stresses

The preceding development of the Soderberg criterion can also be done for fluctuating shear stresses instead of normal stresses. The design factor equation would then be

$$\frac{1}{N} = \frac{\tau_m}{s_{sy}} + \frac{K_t \tau_a}{s_{sn}'} \tag{5-18}$$

Experimental work indicates that this form of the Soderberg equation is very conservative. If the maximum shear stress theory of failure is used, $s_{sy} = 0.5s_y$, and $s_{sn}' = 0.5s_n'$. For the distortion energy theory, $s_{sy} = 0.577s_y$, and $s_{sn}' = 0.577s_n'$.

Case I: Fluctuating Combined Stresses

The approach presented here is similar to the Soderberg criterion described previously. But in this case, the effect of the combined stresses is first determined by the use of the Mohr circle.

Case I1

For the maximum shear stress theory, draw two Mohr's circles, one for the mean stresses and one for the alternating stresses. From the first circle, determine maximum mean shear stress, $(\tau_m)_{max}$. From the second circle, determine the maximum alternating shear stress, $(\tau_a)_{max}$. Then use these values in the design equation

$$\frac{1}{N} = \frac{(\tau_m)_{max}}{s_{sy}} + \frac{K_t(\tau_a)_{max}}{s_{sn}'} \tag{5-19}$$

where $s_{sy} = 0.5s_y$ and $s_{sn}' = 0.5s_n'$.

Case I2

For the distortion energy theory, draw two Mohr's circles, one for the mean stresses and one for the alternating stresses. From these circles, determine the maximum and minimum principal stresses. Then compute the von Mises stresses for both the mean and alternating components from

$$\sigma_m' = \sqrt{\sigma_{1m}^2 + \sigma_{2m}^2 - \sigma_{1m}\sigma_{2m}}$$

$$\sigma_a' = \sqrt{\sigma_{1a}^2 + \sigma_{2a}^2 - \sigma_{1a}\sigma_{2a}}$$

The Soderberg equation then becomes

$$\frac{1}{N} = \frac{\sigma_m'}{s_y} + \frac{K_t \sigma_a'}{s_n'} \tag{5-20}$$

5-4 EXAMPLE DESIGN PROBLEMS

Design Example 5-1. A rectangular bar is welded to a circular rod to form a bracket, as shown in Figure 5-8. Design the bar and the rod to carry a dead load of 250 lb. The welds would also have to be designed as explained in Chapter 19.

Solution. The bracket has two components, each subjected to a different type of stress, as shown in the free body diagrams in the figure.

Rectangular Bar

See part (b) of Figure 5-8. This is a cantilever and will be designed to be safe at point A, where the maximum tensile stress occurs. The design decisions follow.

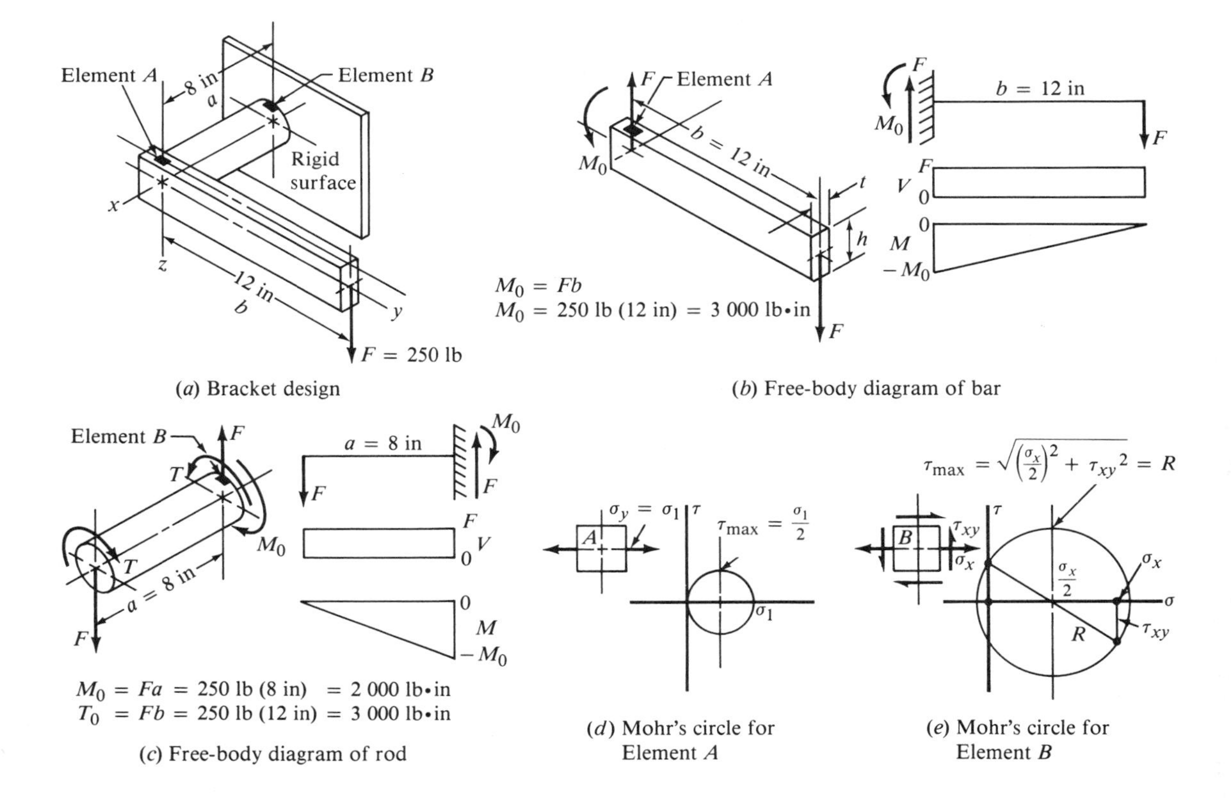

Figure 5-8 Rectangular Bar Welded to a Circular Rod to Form a Bracket

1. Material: AISI 1020 HR, $s_y = 30\,000$ psi (Appendix A-3).
2. Design factor: Use $N = 2$. Assume well-known conditions.
3. Design stress: Using case C1,

$$\sigma_d = s_y/N = (30\,000 \text{ psi})/2 = 15\,000 \text{ psi}$$

4. Stress analysis: For bending, $\sigma = M/Z$. Required section modulus is

$$Z = \frac{M}{\sigma_d} = \frac{3\,000 \text{ lb} \cdot \text{in}}{15\,000 \text{ lb/in}^2} = 0.200 \text{ in}^3$$

5. Choose the arbitrary cross-section proportions, $h = 2t$. Then

$$Z = th^2/6 = t(2t)^2/6 = 4t^3/6 = 0.667t^3$$

Then the required thickness is

$$t = \sqrt[3]{Z/0.667} = 0.669 \text{ in}$$

The nominal height would be

$$h = 2t = 1.338 \text{ in}$$

6. A vendor lists bars available in AISI 1020 HR in thicknesses of ⅝ in or ¾ in. Try ⅝ × 1½:

$$Z = \frac{th^2}{6} = \frac{(0.625)(1.50)^2}{6} = 0.234 \text{ in}^3 \quad \text{(OK)}$$

Circular Rod

See part (c) of Figure 5-8. The rod is subjected to a combination of normal stress due to bending and torsional shear stress. Both stresses are maximum at point B on the top of the rod, where it is attached to the support surface. Using the same material and design factor as for the rectangular bar, the remaining design decisions are as follows:

1. Design stress: Using the maximum shear stress theory of failure, case C1,

$$\tau_d = 0.5s_y/N = (0.5)(30\,000 \text{ psi})/2 = 7\,500 \text{ psi}$$

2. Stress analysis: Because only bending and torsion occur at B, the equivalent torque approach as outlined in equations (4-3) to (4-5) of section 4-3 can be used. The Mohr circle shown in Figure 5-8 illustrates the validity of this approach.

$$T_e = \sqrt{M^2 + T^2} = \sqrt{(2\,000)^2 + (3\,000)^2} = 3\,606 \text{ lb} \cdot \text{in}$$

Then the required polar section modulus is

$$Z_p = \frac{T_e}{\tau_d} = \frac{3\,606 \text{ lb} \cdot \text{in}}{7\,500 \text{ lb/in}^2} = 0.481 \text{ in}^3$$

But $Z_p = \pi D^3/16$. Then, solving for D gives

$$D = \sqrt[3]{\frac{16Z_p}{\pi}} = \sqrt[3]{\frac{(16)(0.481)}{\pi}} = 1.35 \text{ in}$$

3. Specify a 1.50-in-diameter rod to match the size of the rectangular bar.

Design Example 5-2. Complete the design of the bracket described in design example 5-1 if the load is repeated and reversed rather than static.

Solution. We will use the same basic geometry, but the configuration must be more completely described because of the effects of surface finish and stress concentrations on endurance.

Rectangular Bar

1. Material: Choose AISI 1020 HR: $s_u = 55\,000$ psi; $s_y = 30\,000$ psi.
2. Endurance strength: The members will have the as-rolled surface. From Figure 5-6, $s_n = 20\,000$ psi. Then, applying a size factor of 0.9,

$$s_n' = (20\,000 \text{ psi})(0.9) = 18\,000 \text{ psi}$$

3. Design stress: We will choose $N = 3$ because of the difficulty of controlling the geometry around the welds. From case D:

$$\sigma_d = s_n'/N = (18\,000 \text{ psi})/3 = 6\,000 \text{ psi}$$

4. Stress analysis: A stress concentration factor is applied to the stress due to bending because of the attachment of the rectangular bar to the circular rod. This case is not specifically covered in the available charts. We will estimate $K_t = 2.0$ and specify a well-rounded fillet at the joint. The required section modulus is then

$$Z = \frac{K_t M}{\sigma_d} = \frac{(2.0)(3\,000 \text{ lb} \cdot \text{in})}{6\,000 \text{ lb/in}^2} = 1.00 \text{ in}^3$$

Solving for t as in problem 5-1,

$$t = \sqrt[3]{\frac{Z}{0.667}} = \sqrt[3]{\frac{1.00 \text{ in}^3}{0.667}} = 1.15 \text{ in}$$

5. A standard bar, $1\frac{1}{4} \times 2\frac{1}{4}$, is available from a steel supplier.

$$Z = \frac{th^2}{6} = \frac{(1.25)(2.25)^2}{6} = 1.06 \text{ in}^3 \quad \text{(OK)}$$

Circular Rod

Using the equivalent torque approach again and the design stress from case F1:

1. Design stress:

$$\tau_d = 0.5s_n'/N = 0.5(18\,000 \text{ psi})/3 = 3\,000 \text{ psi}$$

2. Required polar section modulus:

$$Z_p = \frac{K_t T_e}{\tau_d} = \frac{(2.0)(3\,606 \text{ lb} \cdot \text{in})}{3\,000 \text{ lb/in}^2} = 2.40 \text{ in}^3$$

3. Required diameter:

$$D = \sqrt[3]{\frac{16Z_p}{\pi}} = \sqrt[3]{\frac{16(2.40)}{\pi}} = 2.30 \text{ in}$$

4. A standard 2⅜-in (2.375-in) rod is available.

Comments on Design Details

In order to obtain a "well-rounded" fillet at the welds, we can refer to the charts for stress concentration factors for a stepped round shaft and a stepped flat bar. To have $K_t < 2.0$, a ratio of $r/d = 0.15$ is desirable. Using the rod diameter for d, $r = 0.15d = 0.36$ in is recommended. The welds should be blended smoothly with the rod and the bar. The welds themselves must be analyzed by the methods discussed in Chapter 19.

Design Example 5-3. A small cam drives a reciprocating follower that ejects spherical balls at a controlled rate into an assembly machine, as shown in Figure 5-9. Design the flat spring that holds the follower in contact with the cam. The following design parameters have been set:

1. Spring deflection: $y_{max} = 8.0$ mm; $y_{min} = 3.0$ mm.

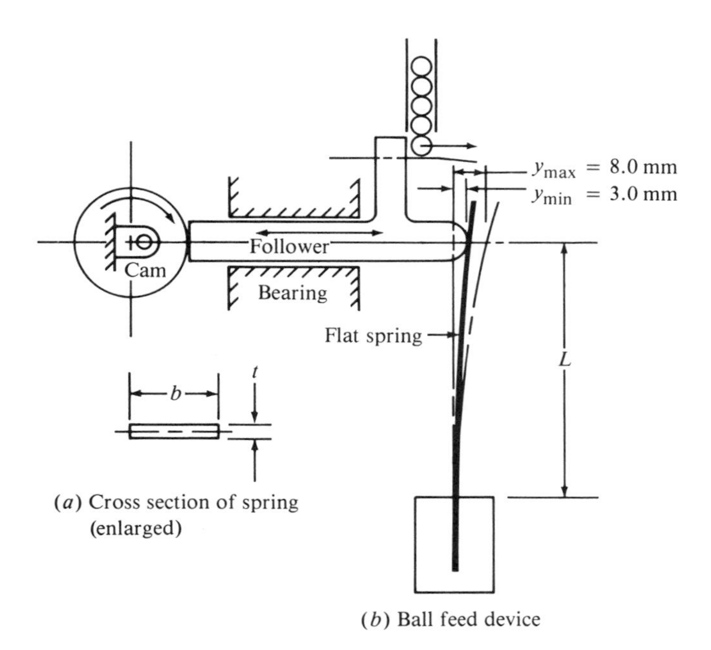

Figure 5-9 Design Example for Cyclic Loading

2. Spring length: less than 100 mm.
3. Width of spring, b: 5 mm $< b <$ 8 mm.
4. Thickness of spring, t: 0.50 mm $< t <$ 2.0 mm. Spring materials available in 0.10-mm increments.
5. Force of spring on follower at y_{max}: 4.0 $N < F_{max} <$ 8.0 N.
6. Desired design factor for static loading at F_{max}: $N = 3$. This permits some careless handling at assembly.
7. Desired design factor for cyclic loading: $N = 2$.

Solution. The logic of the solution is to consider the static loading at the maximum deflection to determine the trial size of the spring to be safe and to generate a satisfactory force on the follower. Then the cyclic loading will be evaluated as a fluctuating normal stress, using case G, the Soderberg criterion. Note that there are a great many variables in this problem and that design decisions must be made throughout the process.

1. Select a material. A high strength and high modulus of elasticity are desirable for good spring characteristics. Because of the cyclic loading, a high ductility (percent elongation) and a good surface finish are desirable. Try AISI 8650 OQT 700 steel with 12 percent elongation:

$$s_u = 240 \text{ Ksi} \quad (1\,650 \text{ MPa}); \qquad s_y = 222 \text{ Ksi} \quad (1\,530 \text{ MPa})$$

2. Compute design stresses. For static loading, case C:

$$\sigma_d = s_y/N = 1\,530 \text{ MPa}/3 = 510 \text{ MPa}$$

3. Consider the load-deflection characteristics to determine the factors affecting the force exerted by the spring on the follower. From Appendix A-14 for beam deflections, case 11 gives the formula for a cantilever:

$$y = \frac{FL^3}{3EI}$$

Solving for F and letting $I = bt^3/12$ for the flat strip gives

$$F = \frac{3EIy}{L^3} = \frac{3Ey}{L^3} \cdot \frac{bt^3}{12} = \frac{Eybt^3}{4L^3} \tag{a}$$

4. Consider the stress-deflection characteristics. The flexure formula will be used with $c = t/2$ and $M = FL$:

$$\sigma = \frac{Mc}{I} = \frac{FL}{I} \cdot \frac{t}{2}$$

Substituting equation (a),

$$\sigma = \frac{3EIy}{L^3} \cdot \frac{L}{I} \cdot \frac{t}{2} = \frac{3Eyt}{2L^2} \tag{b}$$

5. The results of steps 3 and 4 indicate that the spring thickness and length have the greatest effect on its performance. Also notice from equation (b) that the stress *increases* with an increase in the thickness, t. This is a bit unusual. Let's solve for t as a function of L from both equations (a) and (b):

$$t = L\left[\frac{4F}{Eyb}\right]^{1/3} \quad \text{[From (a)]}$$

$$t = L^2\left[\frac{2\sigma}{3Ey}\right] \quad \text{[From (b)]}$$

From design decisions already made we know for steel $E = 207$ GPa (207×10^3 N/mm²); $y = y_{max} = 8.0$ mm; and $\sigma = \sigma_d = 510$ MPa (510 N/mm²). If we specify a trial value for F and b, we can solve for t and L. Let's use $F = 6.0$ N and $b = 6.0$ mm near the middle of the given ranges. Then

$$t = L\left[\frac{4(6.0)}{(207 \times 10^3)(8.0)(6.0)}\right]^{1/3} = (0.0134)L \quad \text{[From (a)]}$$

$$t = L^2\left[\frac{2(510)}{3(207 \times 10^3)(8.0)}\right] = (2.05 \times 10^{-4})L^2 \quad \text{[From (b)]}$$

6. Equating these two expressions for t and solving for L gives $L = 65.3$ mm. From equation (a), $t = (0.0134)L = 0.875$ mm. These are nominal design values that can be rounded off to convenient values. Let's try $t = 0.80$ mm and $L = 65$ mm.
7. Now we can compute the actual expected static stress in the spring at maximum deflection from the first form of equation (b):

$$\sigma = \frac{3Eyt}{2L^2} = \frac{3(207 \times 10^3)(8.0)(0.80)}{2(65)^2} = 470 \text{ N/mm}^2 = 470 \text{ MPa}$$

This is acceptable for static loading because it is less than the design stress.

8. Now compute the force exerted by the spring from the first form of equation (b).

$$F = \frac{Eybt^3}{4L^3} = \frac{(207 \times 10^3)(8.0)(6.0)(0.80)^3}{4(65)^3} = 4.63 \text{ N} \quad \text{(OK)}$$

9. The design is satisfactory for static conditions. To evaluate cyclic conditions by using equation (5-17), we need the mean stress, the alternating stress, and the endurance strength for the spring material. From Figure 5-6, using an ultimate strength of 240 Ksi and specifying a ground surface for the spring gives $s_n' = 90$ Ksi. Converting to MPa,

$$s_n' = 90 \text{ Ksi} \cdot \frac{6.895 \text{ MPa}}{1.0 \text{ Ksi}} = 621 \text{ MPa}$$

The minimum stress during service corresponds to a deflection of 3.0 mm. Because the stress is proportional to the deflection,

$$\sigma_{min} = \sigma_{max}(3.0\text{ mm})/(8.0\text{ mm}) = (470\text{ MPa})\left(\tfrac{3}{8}\right) = 176\text{ MPa}$$

$$\sigma_m = \frac{\sigma_{max} + \sigma_{min}}{2} = \frac{470 + 176}{2} = 323\text{ MPa}$$

$$\sigma_a = \frac{\sigma_{max} - \sigma_{min}}{2} = \frac{470 - 176}{2} = 147\text{ MPa}$$

10. We can now evaluate N from equation (5-17).

$$\frac{1}{N} = \frac{\sigma_m}{s_y} + \frac{\sigma_a}{s_n'} = \frac{323}{1\,530} + \frac{147}{621} = 0.448$$

Then $N = 2.23$ (OK).
The summary for the design is as follows:

Material: AISI 8650 OQT 700 steel, ground

Size: 0.80 mm thick; 6.0 mm wide; 65 mm long outside of the support, plus some extension beyond the contact point

Force on follower at maximum deflection: $F = 4.63\ N$

REFERENCES

1. Boyer, H. E., ed. *Metals Handbook No. 10: Failure Analysis and Prevention*. 8th ed. Metals Park, Ohio: American Society for Metals, 1975.
2. Fuchs, H. O., and Stephens, R. I. *Metal Fatigue in Engineering*. New York: John Wiley & Sons, 1980.
3. Graham, J. A., ed. *Fatigue Design Handbook*. New York: Society of Automotive Engineers, 1968.
4. Harris, C. M., and Crede, C. E. *Shock and Vibration Handbook*. New York: McGraw-Hill Book Company, 1976.
5. Krulick, T. G. "A Road Map for Stress Analysis." *Machine Design Magazine*. (August 1981).
6. Juvinall, R. C. *Engineering Considerations of Stress, Strain, and Strength*. New York: McGraw-Hill Book Company, 1967.
7. Juvinall, R. C. *Fundamentals of Machine Component Design*. New York: John Wiley & Sons, 1983.
8. Madayag, A. F. *Metal Fatigue: Theory and Design*. New York: John Wiley & Sons, 1969.
9. Rinehart, J. S., and Pearson, J. *Behavior of Metals under Impulsive Loads*. Metals Park, Ohio: The American Society for Metals, 1954.
10. Shigley, J. E. *Mechanical Engineering Design*. 4th ed. New York: McGraw-Hill Book Company, 1983.
11. Sines, G., and Waisman, J. L., eds. *Metal Fatigue*. New York: McGraw-Hill Book Company, 1959.
12. Spotts, M. F. *Design of Machine Elements*. 5th ed. Englewood Cliffs, N.J.: Prentice-Hall, 1978.
13. Vigness, Irwin, and Welch, W. P. "Shock and Impact Considerations in Design." *ASME Handbook: Metals Engineering—Design*, Oscar J. Horger, ed. New York: McGraw-Hill Book Company, 1965.

PROBLEMS

1. The cast iron cylinder shown in Figure 5-10 carries only an axial compressive load of 75 000 lb. (The torque $T = 0$.) Compute the design factor if it is made from gray cast iron, Grade 40, having a tensile ultimate strength of 40 Ksi and a compressive ultimate strength of 140 Ksi.
2. Repeat problem 1, except that the load is tensile with a magnitude of 12 000 lb.
3. Repeat problem 1, except that the load is a combination of an axial compressive load of 75 000 lb and a torsion of 20 000 lb · in.
4. The shaft shown in Figure 5-11 is supported by bearings at each end, which have bores of 15.0 mm. Design the shaft to carry the given load if it is steady and the shaft is stationary. Make the dimension a as large as possible while keeping the stress safe. Determine the required diameter in the middle portion. The maximum fillet permissible is 2.0 mm.

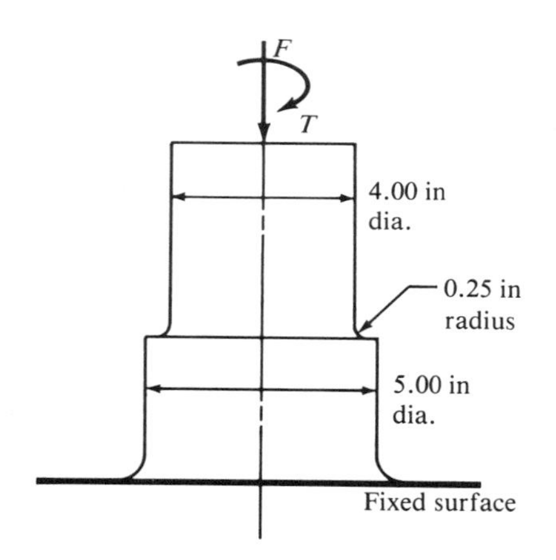

Figure 5-10 (Problems 1, 2, 3)

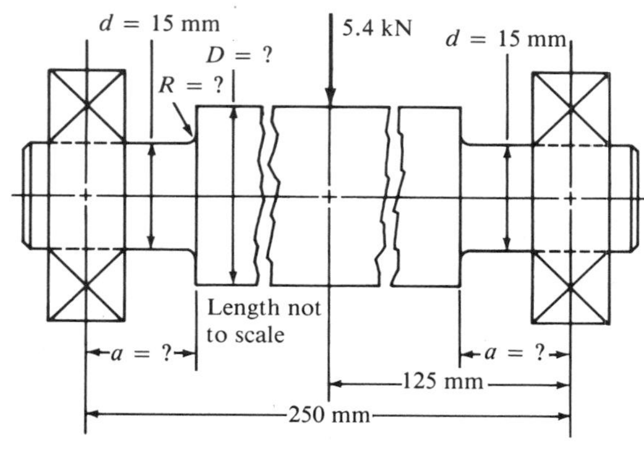

Figure 5-11 (Problems 4, 5, 6)

Use AISI 1137 cold-drawn steel. Use a design factor of 3.

5. Repeat problem 4, except that the shaft is rotating.
6. Repeat problem 4, except that the shaft is rotating and it transmits a torque of 350 N · m from the left bearing to the middle of the shaft. Also, there is a profile keyseat at the middle under the load.
7. Figure 5-12 shows a proposed design for a seat support. The vertical member is to be a standard structural tube (see Appendix A-16). Specify a suitable tube to resist static loads simultaneously in the vertical and horizontal directions, as shown. The tube has properties similar to AISI 1020 hot-rolled steel. Use a design factor of 3.
8. A torsion bar is to have a solid circular cross section. It is to carry a fluctuating torque from 30 N · m to 65 N · m. Use AISI 4140 OQT 1000 for the bar and determine the required diameter for a design factor of 2. Attachments produce a stress concentration of 2.5 near the ends of the bar.
9. Determine the required size for a square bar to be made from AISI 1213 cold-drawn steel. It carries a constant axial tensile load of 1 500 lb and a bending load that varies from zero to a maximum of 800 lb at the center of the 48-in length of the bar. Use a design factor of 3.
10. Repeat problem 9 but add a constant torsional moment of 1 200 lb · in to the other loads.

In some of the following problems, you are asked to compute the design factor resulting from the proposed design for the given loading. Unless stated otherwise, assume that the element being analyzed has a machined surface. If the design factor is significantly different from $N = 3$, redesign the component to achieve approximately $N = 3$. (See figures in Chapter 3).

11. A tensile member in a machine structure is subjected to a steady load of 4.50 kN. It has a length of 750 mm and is made from a steel tube, AISI 1040 hot-rolled, having an outside diameter of 18 mm and an inside diameter of 12 mm. Compute the resulting design factor.
12. A steady tensile load of 5.00 kN is applied to a square bar, 12 mm on a side, and having a length of 1.65 m. Compute the stress in the bar and the resulting design factor if it is made from (a) AISI 1020 hot-rolled steel, (b) AISI 8650 OQT 1000 steel, (c) ductile iron A536-77 (60-40-18), (d) aluminum alloy 6061-T6, (e) titanium alloy Ti-6Al-4V, annealed, (f) rigid PVC plastic, and (g) phenolic plastic.
13. An aluminum rod, made from alloy 6061-T6, is made in the form of a hollow square tube, 2.25 in outside with a wall thickness of 0.125 in. Its length is 16.0 in. It carries an axial compressive force of 12 600 lb. Compute the resulting design factor. Assume the tube does not buckle.
14. Compute the design factor in the middle portion only of the rod AC in Figure 3-19 if the steady vertical force on the boom is 2 500 lb. The rod is rectangular, 1.50 in by 3.50 in, and is made from AISI 1144 cold-drawn steel.
15. Compute the forces in the two angled rods in Figure 3-20 for a steady applied force, $F =$ 1 500 lb if the angle θ is 45°. Then design the middle portion of each rod to be circular and made from AISI 1040 hot-rolled steel. Specify a suitable diameter.
16. Repeat problem 15 if the angle θ is 15°.
17. Figure 3-18 shows a portion of a circular bar that is subjected to a repeated and reversed force of

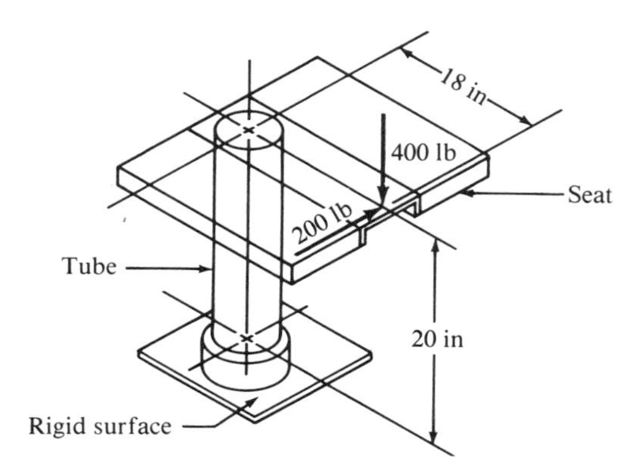

Figure 5-12 (Problem 7)

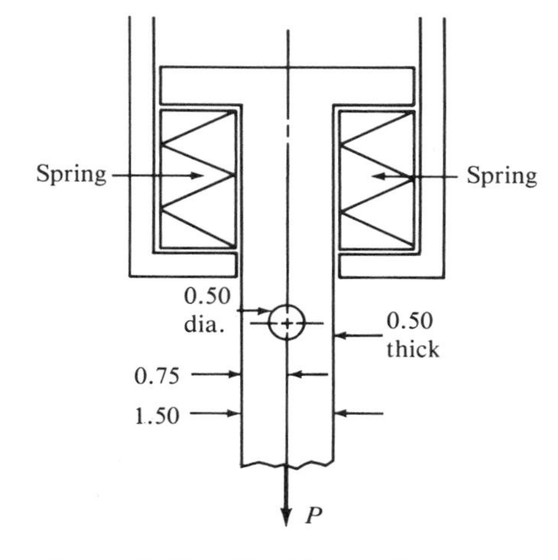

Figure 5-13 (Problem 24)

7 500 N. If the bar is made from AISI 4140 OQT 1000, compute the resulting design factor.

18. Compute the torsional shear stress in a circular shaft having a diameter of 50 mm when subjected to a torque of 800 N · m. If the torque is completely reversed and repeated, compute the resulting design factor. The material is AISI 1040 OQT 1000.
19. If the torque in problem 18 fluctuates from zero to the maximum of 800 N · m, compute the resulting design factor.
20. Compute the torsional shear stress in a circular shaft 0.40 in in diameter that is due to a steady torque of 88.0 lb · in. Specify a suitable aluminum alloy for the rod.
21. Compute the required diameter for a solid circular shaft if it is transmitting a maximum of 110 hp at a speed of 560 rpm. The torque varies from zero to the maximum. There are no other significant loads on the shaft. Use AISI 4130 OQT 700.
22. Specify a suitable material for a hollow shaft with an outside diameter of 40 mm and an inside diameter of 30 mm when transmitting 28 kilowatts (kW) of steady power at a speed of 45 radians per second (rad/s).
23. Repeat problem 22 if the power fluctuates from 15 kW to 28 kW.
24. Figure 5-13 shows part of a support bar for a heavy machine, suspended on springs to soften applied loads. The tensile load on the bar varies from 12 500 lb to a minimum of 7 500 lb. Rapid cycling for many million cycles is expected. The bar is made from AISI 6150 OQT 1300 steel. Compute the design factor for the bar in the vicinity of the hole.
25. Figure 3-39 shows a valve stem from an engine subjected to an axial tensile load applied by the valve spring. The force varies from 0.80 kN to 1.25 kN. Compute the resulting design factor at the fillet under the shoulder. The valve is made from AISI 8650 OQT 1300 steel.
26. A conveyor fixture shown in Figure 3-40 carries three heavy assemblies (1 200 lb each). The fixture is machined from AISI 1144 OQT 900 steel. Compute the resulting design factor in the fixture, considering stress concentrations at the fillets and assuming that the load acts axially. The load will vary from zero to the maximum as the conveyor is loaded and unloaded.
27. For the flat plate in tension in Figure 3-41, compute the minimum resulting design factor, assuming the holes are sufficiently far apart that their effects do not interact. The plate is machined from stainless steel, UNS S17400 in condition H1150. The load is repeated and varies from 4 000 lb to 6 200 lb.

For the following problems, select a suitable material for the member, considering stress concentrations, for the given loading to produce a design factor of $N = 3$.

28. Use Figure 3-42. The load is steady. The material is to be some grade of gray cast iron, ASTM A48-76.
29. Use Figure 3-43. The load varies from 20.0 to 30.3 kN. The material is to be titanium.
30. Use Figure 3-44. The torque varies from zero to 2 200 lb · in. The material is to be steel.
31. Use Figure 3-45. The bending moment is steady. The material is to be ductile iron, ASTM A536-77.
32. Use Figure 3-46. The bending moment is completely reversed. The material is to be stainless steel.

6 Columns

6-1 OVERVIEW

A *column* is a member carrying an axial compressive load that tends to fail by elastic instability, or buckling, rather than by crushing of the material. *Elastic instability* is the condition in which the shape of the column is insufficiently rigid to hold it straight. A radical deflection of the axis of the column occurs very suddenly, and if the load is not reduced, the column will collapse.

Figure 6-1 shows that after a column has begun to buckle, a bending moment is developed in it in addition to the axial compressive load. The combination leads to progressively worse conditions and ultimate failure.

Four approaches are presented in this chapter for analyzing columns: The Euler formula for straight, centrally loaded, long, slender columns; the J. B. Johnson formula for straight, centrally loaded short columns; the crooked column formula; and the formula for eccentrically loaded columns. This chapter does not develop the theoretical bases for these methods; rather, it describes the factors needed to use them properly.

6-2 RADIUS OF GYRATION

The geometric properties of the cross section of a column are characterized by its *area, moment of inertia,* and *radius of gyration.* Formulas for these quantities for a variety of shapes can be found in Appendix A-1. Unless there is special restraint on the column at its ends or near its midlength, it will tend to buckle first about an axis for which the moment of inertia is the least. In Figure 6-1, that would be axis Y-Y.

The radius of gyration is defined in terms of the area and moment of inertia of the cross section:

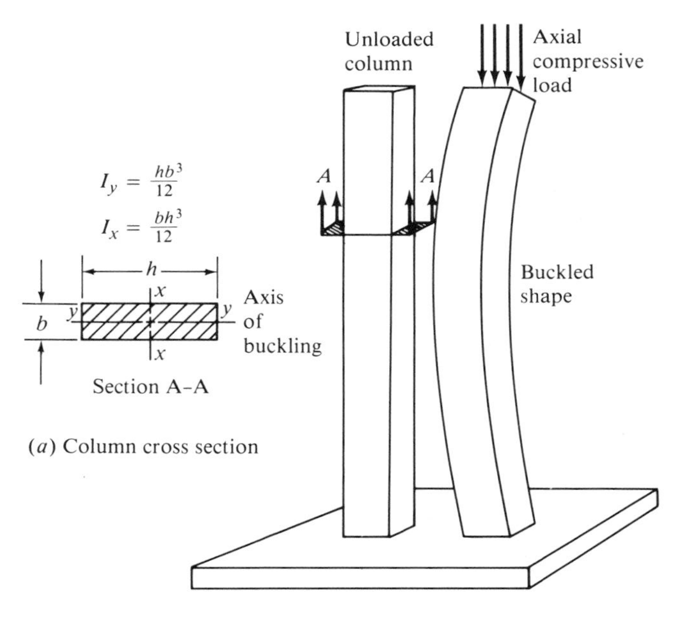

Figure 6-1 Buckling in a Column Exhibiting Elastic Instability

$$r = \sqrt{I/A} \tag{6-1}$$

Normally, a column will tend to buckle about an axis with the least radius of gyration.

6-3 END FIXITY AND EFFECTIVE LENGTH

The term *end fixity* refers to the manner in which the ends of the column are supported. The most important variable is the amount of restraint offered at the ends of the column to the tendency for rotation. Three forms of end restraint are *pinned, fixed,* and *free*.

A *pinned-end* column is guided so the end cannot sway from side to side, but it offers no resistance to rotation of the end. The best approximation of the pinned end would be a frictionless ball and socket joint. A cylindrical pin joint offers little resistance about one axis, but it may restrain the axis perpendicular to the pin axis.

A *fixed end* is one that is held against rotation at the support. An example is a cylindrical column inserted into a tight-fitting sleeve that itself is rigidly supported. The sleeve prohibits any tendency for the fixed end of the column to rotate. A column end securely welded to a rigid base plate is also a good approximation of a fixed-end column.

The *free end* can be visualized by the example of a flagpole. The top end of a flagpole is unrestrained and unguided, the worst case for column loading.

The manner of support of both ends of the column affects the *effective length* of the column, defined as

$$L_e = KL \tag{6-2}$$

where L is the actual length of the column between supports and K is a constant dependent on the end fixity, as illustrated in Figure 6-2.

6-4 SLENDERNESS RATIO

The *slenderness ratio* is the ratio of the effective length of the column to its least radius of gyration. That is,

$$\text{Slenderness ratio} = L_e/r_{\min} = KL/r_{\min} \tag{6-3}$$

We will use the slenderness ratio to aid in the selection of the method of performing the analysis of straight, centrally loaded columns.

6-5 TRANSITION SLENDERNESS RATIO

In the following sections, two methods for analyzing straight, centrally loaded columns are presented: the Euler formula for long, slender columns; and the J. B. Johnson formula for short columns.

The choice of which method to use depends on the value of the actual slenderness ratio for the column being analyzed in relation to the *transition slenderness ratio,* or *column constant,* C_c, defined as

$$C_c = \sqrt{\frac{2\pi^2 E}{S_y}} \tag{6-4}$$

where E is the modulus of elasticity of the material of the column and S_y is the yield strength of the material.

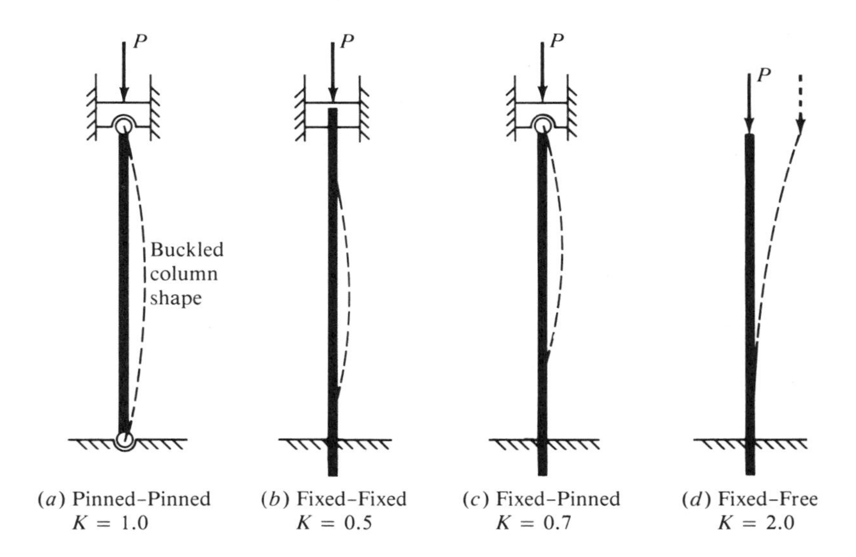

Figure 6-2 Schematic Representations of End Fixity

The use of the column constant is illustrated in the following procedure for analyzing straight, centrally loaded columns.

1. For the given column, compute its actual slenderness ratio.
2. Compute the value of C_c.
3. Compare C_c with KL/r. Because C_c represents the value of the slenderness ratio that separates a long column from a short one, the result of the comparison indicates which type of analysis should be used:
4. If the actual KL/r is greater than C_c, the column is *long*. Use Euler's equation, as described in section 6-6.
5. If KL/r is less than C_c, the column is *short*. Use the J. B. Johnson formula, described in section 6-7.

Figure 6-3 is a logical flowchart for this procedure.

The value of the column constant, or transition slenderness ratio, is dependent on the material properties of modulus of elasticity and yield strength. For any given class of material, say steel, the modulus of elasticity is nearly constant. Thus, the value of C_c varies inversely as the square root of the yield strength. Figures 6-4 and 6-5 show the resulting values for steel and aluminum, respectively, for the range of yield strengths expected for each material. The figures show that the value of C_c decreases as the yield strength increases. The importance of this observation is discussed in the following section.

6-6 LONG COLUMN ANALYSIS: THE EULER FORMULA

The analysis of a long column employs the Euler formula,

$$P_{cr} = \frac{\pi^2 EA}{(KL/r)^2} \tag{6-5}$$

The equation gives the critical load, P_{cr}, at which the column would begin to buckle.

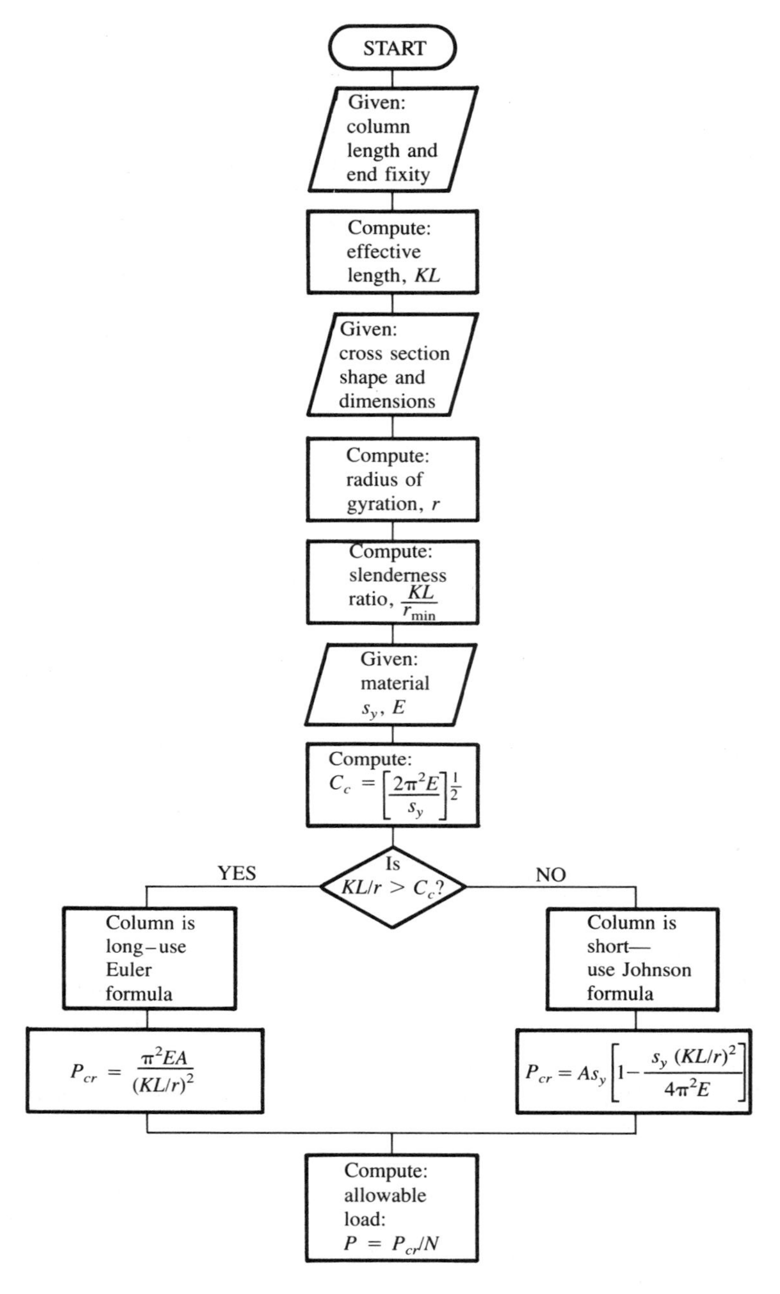

Figure 6-3 Analysis of a Straight, Centrally Loaded Column

An alternate form of the Euler formula is often desirable. Note that, from equation (6-5),

$$P_{cr} = \frac{\pi^2 EA}{(KL/r)^2} = \frac{\pi^2 EA}{(KL)^2/r^2} = \frac{\pi^2 EAr^2}{(KL)^2}$$

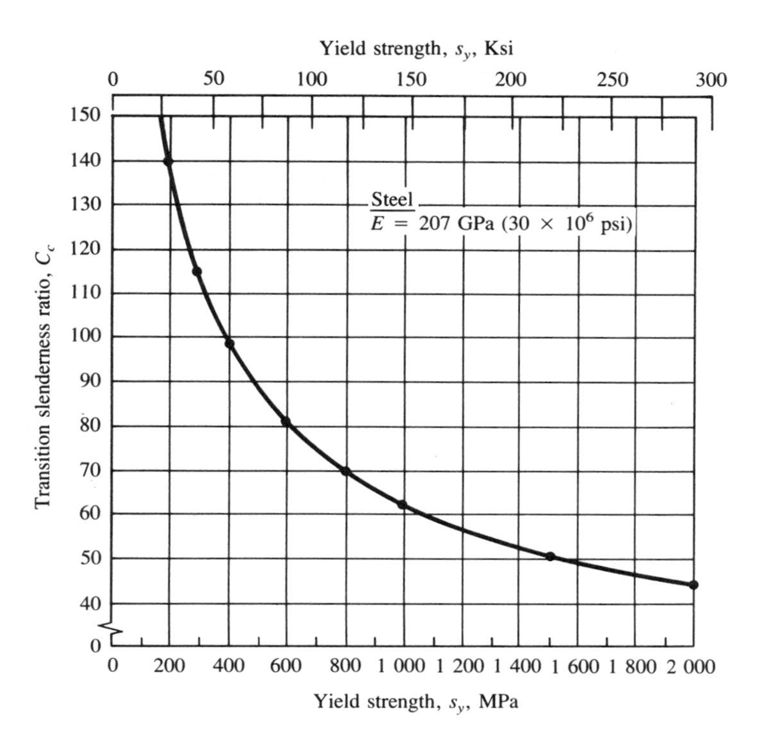

Figure 6-4 Transition Slenderness Ratio C_c Vs. Yield Strength for Steel

But, from the definition of the radius of gyration, r,

$$r = \sqrt{I/A}$$

$$r^2 = I/A$$

Then,

$$P_{cr} = \frac{\pi^2 EA}{(KL)^2}\frac{I}{A} = \frac{\pi^2 EI}{(KL)^2} \tag{6-6}$$

This form of the Euler equation aids in a design problem in which the objective is to specify a size and shape of a column cross section to carry a certain load. The moment of inertia for the required cross section can be easily determined from equation (6-6).

Notice that the buckling load is dependent only on the geometry (length and cross section) of the column and the stiffness of the material represented by the modulus of elasticity. The strength of the material is not involved at all. For these reasons, it is often of no benefit to specify a high-strength material in a long-column application. A lower strength material having the same stiffness, E, would perform as well.

Because the failure is predicted to occur at a limiting load, rather than a stress, the concept of a design factor is applied differently than it is for most other load-carrying members. Rather than applying the design factor to the yield strength or the ultimate strength of the material, it is applied to the critical load, from equation (6-5). For typical machine design applications, a design factor of 3 is used. For stationary columns with

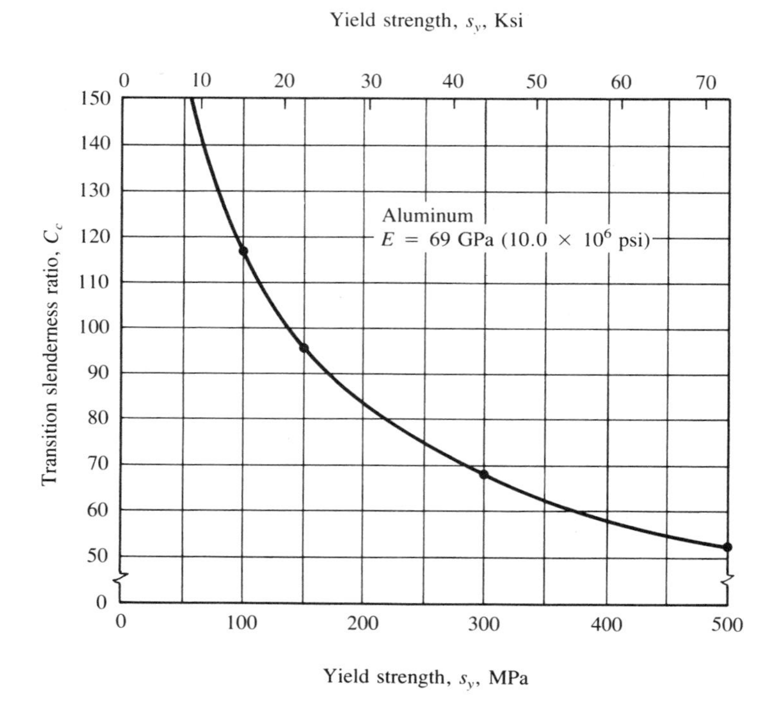

Figure 6-5 Transition Slenderness Ratio C_c Vs. Yield Strength for Aluminum

well-known loads and end fixity, a lower factor can be used, such as 2.0. A factor of 1.92 is used in some construction applications. Conversely, for very long columns, where there is some uncertainty about the loads or the end fixity, or where special dangers are presented, larger factors are advised.

Example Problem 6-1. A column has a solid circular cross section, 1.25 in in diameter; has a length of 4.50 ft; and is pinned at both ends. If it is made from AISI 1020 CD steel, what would be a safe column loading?

Solution. We will follow the procedure in Figure 6-3.

1. For the pinned-end column, the end fixity factor is $K = 1.0$. The effective length equals the actual length. $KL = 4.50 \text{ ft} = 54.0 \text{ in}$.
2. From Appendix A-1, for a solid round section,

$$r = D/4 = 1.25/4 = 0.3125 \text{ in}$$

3. Compute the slenderness ratio:

$$\frac{KL}{r} = \frac{1.0(54)}{0.3125} = 173$$

4. Compute the column constant from equation (6-4). For AISI 1020 CD steel, the yield strength is 51 000 psi and the modulus of elasticity is 30×10^6 psi. Then,

$$C_c = \sqrt{\frac{2\pi^2 E}{S_y}} = \sqrt{\frac{2\pi^2 (30 \times 10^6)}{51\,000}} = 108$$

5. Because KL/r is greater than C_c, the column is long and Euler's formula should be used. The area is

$$A = \pi D^2/4 = \pi(1.25)^2/4 = 1.23 \text{ in}^2$$

Then the critical load is

$$P_{cr} = \frac{\pi^2 EA}{(KL/r)^2} = \frac{\pi^2(30 \times 10^6)(1.23)}{(173)^2} = 12\,200 \text{ lb}$$

At this load, the column should just begin to buckle. A safe load would be a reduced value, found by applying the design factor to the critical load. Let's use $N = 3$ to compute the *allowable load*, $P = P_{cr}/N$:

$$P = (12\,200)/3 = 4\,067 \text{ lb}$$

6-7 SHORT COLUMN ANALYSIS: THE J. B. JOHNSON FORMULA

When the actual slenderness ratio for a column, KL/r, is less than the transition value, C_c, then the column is short and the J. B. Johnson formula should be used. Use of the Euler formula in this range would predict a critical load greater than it really is.

The J. B. Johnson formula is written as follows:

$$P_{cr} = As_y\left[1 - \frac{s_y(KL/r)^2}{4\pi^2 E}\right] \tag{6-7}$$

Figure 6-6 shows a plot of the results of this equation as a function of the slenderness ratio, KL/r. Notice that it becomes tangent to the result of the Euler formula at the transition slenderness ratio, the limit of its application. Also, at very low values for the slenderness ratio, the second term of the equation approaches zero and the critical load approaches the yield load. Curves for three different materials are included in the figure to illustrate the effect of E and s_y on the critical load and the transition slenderness ratio.

The critical load for a short column is affected by the strength of the material in addition to its stiffness, E. As shown in the preceding section, strength is not a factor for a long column when using the Euler formula.

Example Problem 6-2. Determine the critical load on a steel column having a rectangular cross section, 12 mm by 18 mm, and a length of 280 mm. It is proposed to use AISI 1040 HR steel. The lower end of the column is inserted into a close-fitting socket and welded securely. The upper end is pinned (see Figure 6-7).

Solution

1. Compute the slenderness ratio. The radius of gyration must be computed about the axis that gives the least value. This is the Y-Y axis, for which

$$r = \frac{B}{\sqrt{12}} = \frac{12 \text{ mm}}{\sqrt{12}} = 3.46 \text{ mm}$$

The column has a fixed-pinned end fixity for which $K = 0.7$. Then,

$$KL/r = [(0.7)(280)]/3.46 = 56.6$$

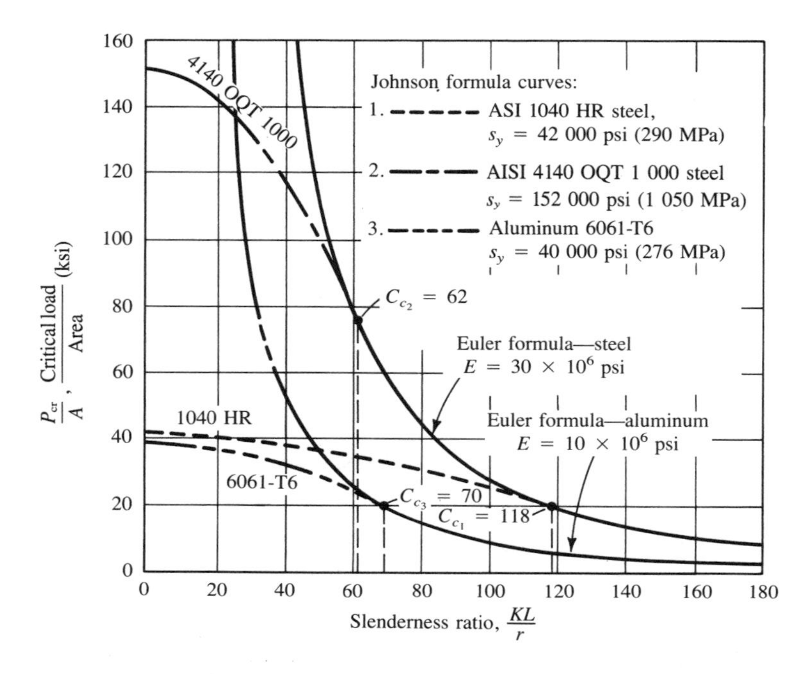

Figure 6-6 Johnson Formula Curves

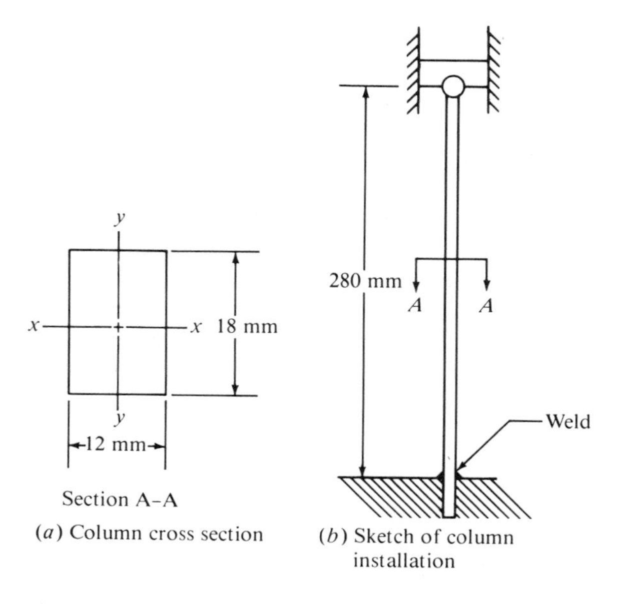

(*a*) Column cross section

(*b*) Sketch of column installation

Figure 6-7 Column For Example Problem 6-2

2. Compute the transition slenderness ratio. For the AISI 1040 HR steel, $E = 207$ GPa and $S_y = 290$ MPa. Then, from equation (6-4),

$$C_c = \sqrt{\frac{2\pi^2(207 \times 10^9 \text{ Pa})}{290 \times 10^6 \text{ Pa}}} = 119$$

3. Then $KL/r < C_c$: The column is short. Use the J. B. Johnson formula to compute the critical load:

$$P_{cr} = As_y\left[1 - \frac{s_y(KL/r)^2}{4\pi^2 E}\right] \tag{6-7}$$

$$P_{cr} = (216 \text{ mm}^2)(290 \text{ N/mm}^2)\left[1 - \frac{(290 \times 10^6 \text{ Pa})(56.6)^2}{4\pi^2(207 \times 10^9 \text{ Pa})}\right]$$

$$P_{cr} = 55.5 \times 10^3 \text{ N} = 55.5 \text{ kN}$$

This is the buckling load. A design factor would have to be applied to determine the allowable load.

6-8 EFFICIENT SHAPES FOR COLUMN CROSS SECTIONS

An *efficient shape* is one that provides good performance with a small amount of material. In the case of columns, the shape of the cross section and its dimensions determine the value of the radius of gyration, r. From the definition of the slenderness ratio, KL/r, we can see that as r gets larger, the slenderness ratio gets smaller. In the critical load equations, a smaller slenderness ratio results in a larger critical load, the most desirable situation. Therefore, it is desirable to maximize the radius of gyration to design an efficient column cross section.

Unless end fixity varies with respect to the axes of the cross section, the column would tend to buckle with respect to the axis with the *least* radius of gyration. So a column with equal values for the radius of gyration in any direction is desirable.

Review again the definition of the radius of gyration:

$$r = \sqrt{I/A}$$

This indicates that for a given area of material we should try to maximize the moment of inertia to maximize the radius of gyration. A shape with a high moment of inertia has its area distributed far away from its centroidal axis.

Shapes that have the desirable characteristics described include circular hollow pipes and tubes, square hollow tubing, and fabricated column sections made from structural shapes placed at the outer boundaries of the section. Solid circular sections and solid square sections are also good, although not as efficient as the hollow sections. Figure 6-8 illustrates some of these shapes. In the case of the section in Figure 6-8(c), the angle sections at the corners provide the greatest contribution to the moment of inertia. The lacing bars merely hold the angles in position. The built-up section in (d) gives a rigid boxlike section approximating the hollow square tube in larger sizes. The H-column in (e) has an equal depth and width and relatively heavy flanges and web. The moment of inertia with respect to the Y-Y axis is still smaller than for the X-X axis, but they are more nearly equal than for most other I-sections designed to be used as beams with bending in only one direction. Thus this shape would be more desirable for columns.

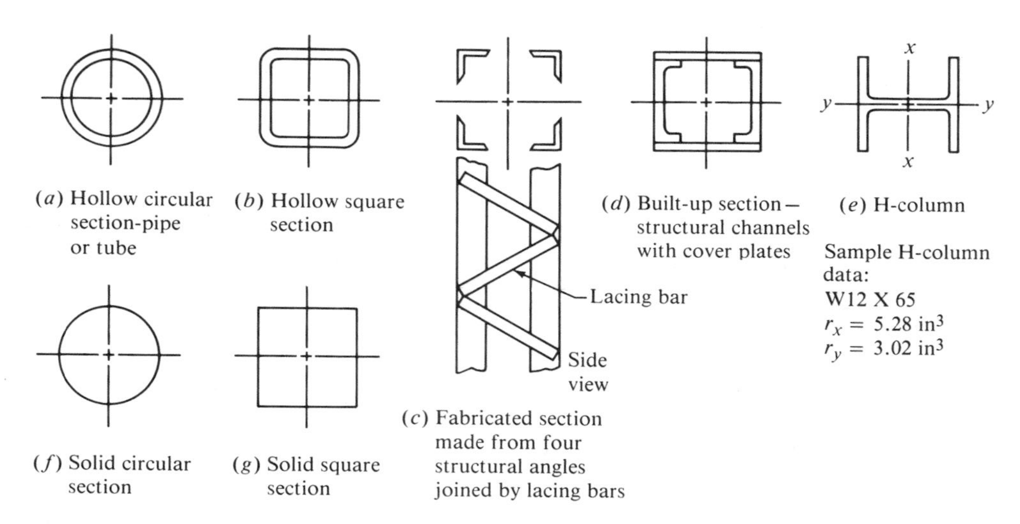

Figure 6-8 Column Cross Sections

6-9 THE DESIGN OF COLUMNS

In a design situation, the expected load on the column would be known along with the length required by the application. The designer would then specify the following:

1. The manner of attaching the ends to the structure that affects the end fixity
2. The general shape of the column cross section (for example, round, square, rectangular, and hollow tube)
3. The material for the column
4. The design factor, considering the application
5. The final dimensions for the column

It may be desirable to propose and analyze several different designs to approach an optimum for the application. A computer program facilitates the process.

It is assumed that items 1 through 4 are specified by the designer for any given trial. For some simple shapes, such as the solid round or square section, the final dimensions are computed from the appropriate formula: the Euler formula, equation (6-5) or (6-6); or the J. B. Johnson formula, equation (6-7). If an algebraic solution is not possible, iteration can be done.

In a design situation, the unknown cross-sectional dimensions make computing the radius of gyration and therefore the slenderness ratio, KL/r, impossible. Without the slenderness ratio, it cannot be determined whether the column is long (Euler) or short (Johnson). Thus, the proper formula to use is not known.

We overcome this difficulty by making an assumption that the column is either long or short and proceeding with the corresponding formula. Then, after the dimensions are determined for the cross section, the actual value of KL/r will be computed and compared with C_c. This will show whether or not the correct formula has been used. If so, the computed answer is correct. If not, the alternate formula must be used and the computation repeated to determine new dimensions. Figure 6-9 shows a flowchart for the design logic described here.

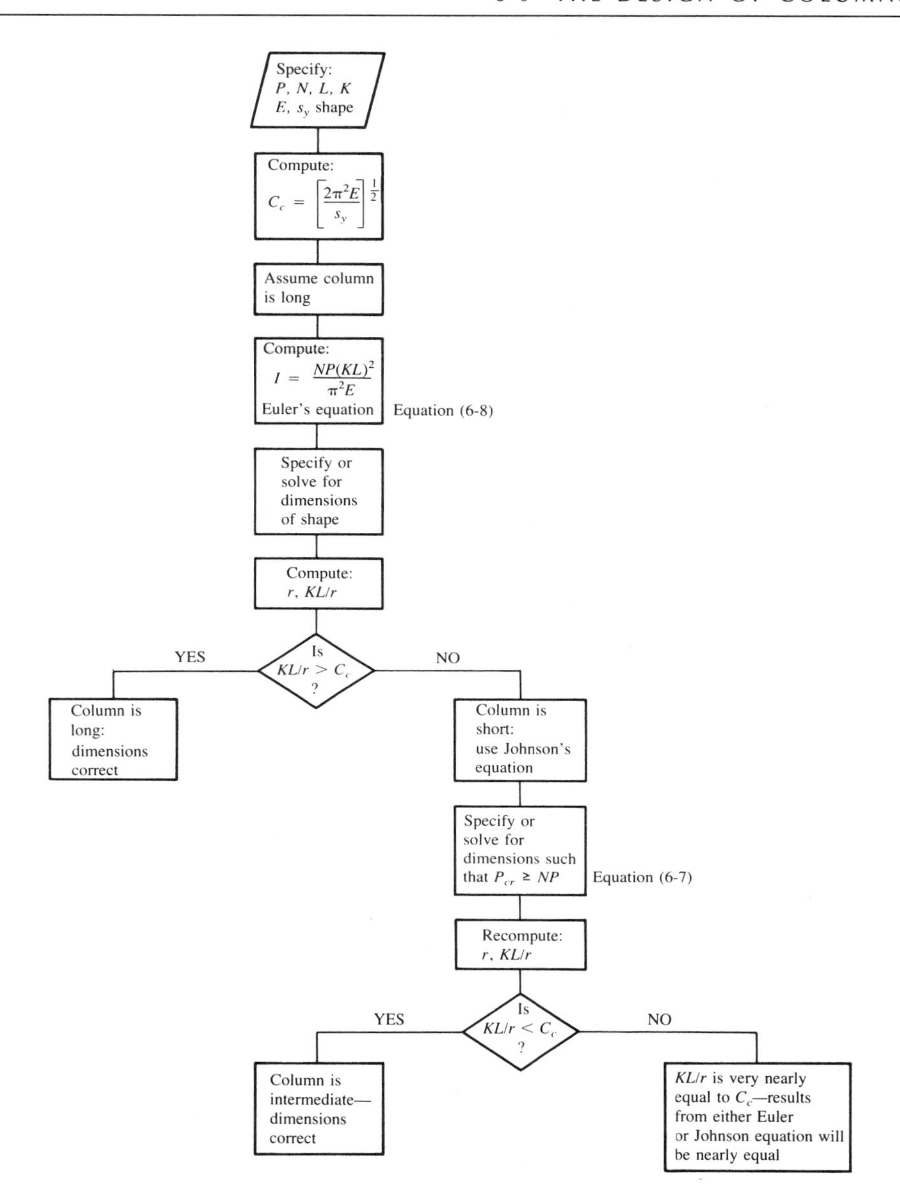

Figure 6-9 Design of a Straight, Centrally Loaded Column

Design: Assuming the Column Is Long

Euler's formula is used if the assumption is that the column is long. Equation (6-6) would be the most convenient form because it can be solved for the moment of inertia, I:

$$I = \frac{P_{cr}(KL)^2}{\pi^2 E} = \frac{NP(KL)^2}{\pi^2 E} \tag{6-8}$$

where P is the allowable load, usually set equal to the actual maximum expected load. Having the required value for I, the dimensions for the shape can be determined by additional computations or by scanning tables of data of the properties of commercially available sections.

The solid circular section is one for which it is possible to derive a final equation for the characteristic dimension, the diameter. The moment of inertia is

$$I = \pi D^4/64$$

Substituting this into equation (6-8) gives

$$I = \frac{\pi D^4}{64} = \frac{NP(KL)^2}{\pi^2 E}$$

Solving for D,

$$D = \left[\frac{64\,NP(KL)^2}{\pi^3 E}\right]^{1/4} \tag{6-9}$$

Design: Assuming a Short Column

The J. B. Johnson formula is used to analyze a short column. It is difficult to derive a convenient form for use in design. In the general case, then, trial and error is used.

For some special cases, including the solid circular section, it is possible to solve the Johnson formula for the characteristic dimension, the diameter:

$$P_{cr} = As_y\left[1 - \frac{s_y(KL/r)^2}{4\pi^2 E}\right] \tag{6-7}$$

But

$$A = \pi D^2/4$$
$$r = D/4 \quad \text{(From Appendix A-1)}$$
$$P_{cr} = NP$$

Then,

$$NP = \frac{\pi D^2}{4} s_y\left[1 - \frac{s_y(KL)^2}{4\pi^2 E(D/4)^2}\right]$$

$$\frac{4NP}{\pi s_y} = D^2\left[1 - \frac{s_y(KL)^2(16)}{4\pi^2 ED^2}\right]$$

Solving for D gives

$$D = \left[\frac{4NP}{\pi s_y} + \frac{4s_y(KL)^2}{\pi^2 E}\right]^{1/2} \tag{6-10}$$

Example Problem 6-3. Determine the required diameter of a solid round cross section for a machine link if it is to carry 9 800 lb of axial compressive load. The length will be 25 in and the ends will be pinned. Use a design factor of 3.

Solution. Using the logic shown in Figure 6-9, let's assume that the link will behave as a long column. Then, from equation (6-9),

$$D = \left[\frac{64NP(KL)^2}{\pi^3 E}\right]^{1/4} = \left[\frac{64(3)(9\,800)(25)^2}{\pi^3(30 \times 10^6)}\right]^{1/4}$$

$$D = 1.06 \text{ in}$$

The radius of gyration can now be found:

$$r = D/4 = 1.06/4 = 0.265 \text{ in}$$

The slenderness ratio is

$$KL/r = [(1.0)(25)]/0.265 = 94.3$$

For the AISI 1020 hot-rolled steel, $s_y = 30\,000$ psi. The graph in Figure 6-4 shows C_c to be approximately 138. Thus, the actual KL/r is less than the transition value, and the column must be redesigned as a short column, using equation (6-10) derived from the Johnson formula:

$$D = \left[\frac{4NP}{\pi s_y} + \frac{4s_y(KL)^2}{\pi^2 E}\right]^{1/2} \tag{6-10}$$

$$D = \left[\frac{4(3)(9\,800)}{(\pi)(30\,000)} + \frac{4(30\,000)(25)^2}{\pi^2(30 \times 10^6)}\right]^{1/2} = 1.23 \text{ in}$$

Checking the slenderness ratio again,

$$KL/r = [(1.0)(25)]/(1.23/4) = 81.3$$

This is still less than the transition value, so our analysis is acceptable.

6-10 CROOKED COLUMNS

The Euler and Johnson formulas assume that the column is straight and that the load acts in line with the centroid of the cross section of the column. If the column is somewhat crooked, bending occurs in addition to the column action (see Figure 6-10).

The crooked column formula allows an initial crookedness, a, to be considered (7):

$$P^2 - \frac{1}{N}\left[s_y A + \left(1 + \frac{ac}{r^2}\right)P_{cr}\right]P + \frac{s_y A P_{cr}}{N^2} = 0 \tag{6-11}$$

In this formula, c is the distance from the neutral axis of the cross section to its outer edge. P_{cr} is defined to be the critical load found from the *Euler formula*. Although this formula may become increasingly inaccurate for shorter columns, it is not appropriate to switch to the Johnson formula as it is for straight columns.

The crooked column formula is a quadratic with respect to the allowable load P. Evaluating all constant terms in equation (6-11) produces an equation of the form

$$P^2 + C_1 P + C_2 = 0$$

Then, from the solution for a quadratic equation,

$$P = 0.5[-C_1 - \sqrt{C_1^2 - 4C_2}]$$

The smaller of the two possible solutions is selected.

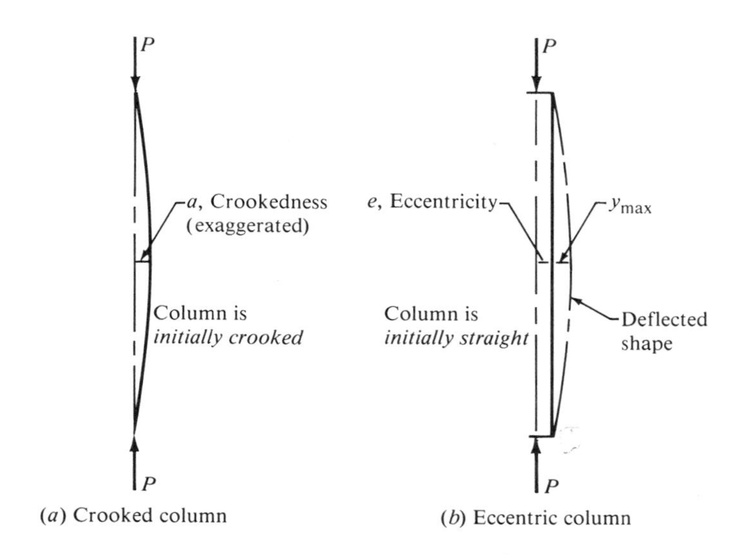

Figure 6-10 Illustration of Crooked and Eccentric Columns

Example Problem 6-4. A column has both ends pinned and has a length of 32 in. It has a circular cross section, 0.75 in in diameter, and an initial crookedness of 0.125 in. The material is AISI 1040 hot-rolled. Compute the allowable load for a design factor of 3.

Solution. Let's evaluate all factors in equation (6-11).

$$A = \pi D^2/4 = [(\pi)(0.75)^2/4 = 0.442 \text{ in}^2$$

$$r = D/4 = 0.75/4 = 0.188 \text{ in}$$

$$c = D/2 = 0.75/2 = 0.375 \text{ in}$$

$$KL/r = [(1.0)(32)]/0.188 = 171$$

$$P_{cr} = \frac{\pi^2 EA}{(KL/r)^2} = \frac{\pi^2(30\,000\,000)(0.442)}{(171)^2} = 4\,476 \text{ lb}$$

$$C_1 = -9\,649$$

$$C_2 = 9.232 \times 10^6$$

The quadratic is therefore

$$P^2 - 9\,649P + 9.232 \times 10^6 = 0$$

From this, $P = 1\,077$ lb is the allowable load.

6-11 ECCENTRICALLY LOADED COLUMNS

An *eccentric load* is one that is applied away from the centroidal axis of the cross section (see Figure 6-10). Such a load exerts bending in addition to the column action. The eccentrically loaded column formula, sometimes called the *secant formula,* is

$$\sigma_{max} = \frac{P}{A}\left[1 + \frac{ec}{r^2}\sec\left(\frac{L}{2r}\sqrt{\frac{P}{AE}}\right)\right] \tag{6-12}$$

Note that the form of this formula computes the maximum normal stress in the column for a given load. This stress can be compared with the yield strength of the material to determine the resulting design factor.

Another critical factor may be the amount of deflection of the axis of the column due to the eccentric load:

$$y_{max} = e\left[\sec\left(\frac{L}{2}\sqrt{\frac{P}{EI}}\right) - 1\right] \tag{6-13}$$

Note that the argument of the secant is in radians. Also, the secant is equal to 1/cosine.

Example Problem 6-5. For the column of example problem 6-4, compute the maximum stress and deflection if a load of 1 075 lb is applied with an eccentricity of 0.75 in.

Solution. All terms have been evaluated before, except the moment of inertia, I:

$$I = \pi D^4/64 = (\pi)(0.75)^4/4 = 0.015\,5 \text{ in}^4$$

Then the maximum stress is found from equation (6-12).

$$\sigma_{max} = \frac{1\,075}{0.442}\left[1 + \frac{(0.75)(0.375)}{(0.188)^2}\sec\left(\frac{32}{2(0.188)}\sqrt{\frac{1\,075}{(0.442)(30 \times 10^6)}}\right)\right]$$

$$\sigma_{max} = 29\,300 \text{ psi}$$

The maximum deflection is found from equation (6-13).

$$y_{max} = 0.75\left[\sec\left(\frac{32}{2}\sqrt{\frac{1\,075}{(30 \times 10^6)(0.015\,5)}}\right) - 1\right] = 0.293 \text{ in}$$

This stress and deflection are quite high. The resulting design factor, based on the yield strength of the AISI 1040 hot-rolled, is

$$N = s_y/\sigma_{max} = (42\,000)/(29\,300) = 1.43 \quad \text{(Low)}$$

6-12 COLUMN ANALYSIS PROGRAM

The following computer program can be used to analyze columns according to the methods described in this chapter for straight, centrally loaded long and short columns. The program executes the logic shown in the flowchart in Figure 6-3. Either U.S. Customary or SI units can be used, with the program asking the user to make that choice at the start of each run. Then the subroutine at statements 430–590 sets up string variable names for the units for the various quantities. It is essential that input data be given in the consistent unit sets identified in statements 440 and 450. Then, after the analysis is complete, the results are printed with the appropriate units on each part of the output.

A sample of the execution of the program is shown after the program listing. Note that the two runs duplicate the data from example problems 6-1 and 6-2, that one long and one short column have been run, and that one is in U.S. customary units and the other is in SI units.

Column Analysis: Program Listing

```
  10 PRINT "ANALYSIS OF COLUMNS"
  20 PRINT
  30 GOSUB 430                                              Subroutine for
  40 PRINT                                                  units
 [50 PRINT "INPUT ACTUAL LENGTH (L) AND END FIXITY (K)"
▶ 60 INPUT L,K
  70 REM L1 = EFFECTIVE LENGTH = KL
  80 L1 = K*L
  90 PRINT "INPUT RADIUS OF GYRATION (R) AND AREA (A)"      Input problem data
▶ 100 INPUT R,A
  110 REM S = SLENDERNESS RATIO (KL/R)
  120 S = L1/R
  130 PRINT "INPUT MATERIAL PROPERTIES, E, SY"
▶ 140 INPUT E, S1
  150 REM C = COLUMN CONSTANT (TRANSITION SLENDERNESS RATIO)
  160 C = SQR(19.74*E/S1)
  170 IF S < C THEN 220
 [180 PRINT "COLUMN IS LONG"
  190 PRINT                                                 Critical load from
  200 P = 9.8696*E*A/(S^2)                                  Euler's equation
  210 GOTO 260
 [220 PRINT "COLUMN IS SHORT"
  230 PRINT                                                 Critical load from
  240 P = A*S1*(1 - (S1*S^2)/(4 * E*9.8696))                Johnson's equation
  250 REM P1 = ALLOWABLE LOAD = P/N
  260 PRINT "INPUT DESIGN FACTOR (N)"
▶ 270 INPUT N
  280 PRINT
  290 P1 = P/N
 [300 PRINT "ACTUAL LENGTH =";L;L$
  310 PRINT "EFFECTIVE LENGTH = ";L1;L$
  320 PRINT
  330 PRINT "COLUMN CONSTANT =";INT(C)
  340 PRINT "ACTUAL SLENDERNESS RATIO =";INT(S)              Print results
  350 PRINT
  360 PRINT "CRITICAL LOAD =";INT(P);P$
  370 PRINT "ALLOWABLE LOAD ="; INT(P1);P$; "FOR DESIGN FACTOR
    =";N
  380 PRINT
 [390 PRINT "TYPE Y TO RUN NEW DATA"
▶ 400 INPUT A$                                               Decision on how to
  410 IF A$ = "Y" THEN 10                                    continue
  420 END
 [430 REM: SUBROUTINE FOR UNITS
  440 PRINT "TYPE 1 FOR U.S. CUSTOMARY: LB, IN, IN^2, PSI"   Decision on unit
  450 PRINT "TYPE 2 FOR SI UNITS: N, m, m^2, Pa"             system
▶ 460 INPUT U
  470 IF U = 2 THEN 540
  480 PRINT "U.S. CUSTOMARY UNITS"
 [490 P$ = "LB"                                              Set units in U.S.
  500 L$ = "IN"                                              Customary system
  510 A$ = "IN^2"
  520 S$ = "PSI"
  530 RETURN
  540 PRINT "SI UNITS"
  550 P$ = "N"
  560 L$ = "m"                                               Set units in SI
  570 A$ = "m^2"                                             system
  580 S$ = "Pa"
  590 RETURN
```

Lines in the program calling for input data and lines in the output that correspond to the input statements are highlighted with a ▶.

Column Analysis: Sample Output 1[1]

```
    ANALYSIS OF COLUMNS

    TYPE 1 FOR U.S. CUSTOMARY: LB, IN, IN^2, PSI
    TYPE 2 FOR SI UNITS : N, m, m^2, Pa
►   ? 1                                                   Statement 460
    U.S. CUSTOMARY UNITS

    INPUT ACTUAL LENGTH (L) AND END FIXITY (K)
►   ? 54 , 1                                              Statement 60
    INPUT RADIUS OF GYRATION (R) AND AREA (A)
►   ? .3125 , 1.23                                        Statement 100
    INPUT MATERIAL PROPERTIES, E, SY
►   ? 3E+07 , 51000                                       Statement 140
    COLUMN IS LONG

    INPUT DESIGN FACTOR (N)
►   ? 3                                                   Statement 270

   ┌ACTUAL LENGTH = 54 IN
   │EFFECTIVE LENGTH = 54 IN
   │
   │COLUMN CONSTANT = 107                                 Results in U.S.
   │ACTUAL SLENDERNESS RATIO = 172                        Customary units
   │
   │CRITICAL LOAD = 12196 LB
   └ALLOWABLE LOAD = 4065 LB FOR DESIGN FACTOR = 3

    TYPE Y TO RUN NEW DATA
►   ? Y                                                   Statement 400
```

Column Analysis: Sample Output 2[2]

```
    ANALYSIS OF COLUMNS

    TYPE 1 FOR U.S. CUSTOMARY: LB, IN, IN^2, PSI
    TYPE 2 FOR SI UNITS: N, m, m^2, Pa
►   ? 2                                                   Statement 460
    SI UNITS

    INPUT ACTUAL LENGTH (L) AND END FIXITY (K)
►   ? .28 , .7                                            Statement 60
    INPUT RADIUS OF GYRATION (R) AND AREA (A)
►   ? .00346 , .000216                                    Statement 100
    INPUT MATERIAL PROPERTIES, E, SY
►   ? 2.07E+11 , 2.9E+08                                  Statement 140
    COLUMN IS SHORT

    INPUT DESIGN FACTOR (N)
►   ? 3                                                   Statement 270

   ┌ACTUAL LENGTH = .28 m
   │EFFECTIVE LENGTH = .196 m
   │
   │COLUMN CONSTANT = 118                                 Results in SI units
   │ACTUAL SLENDERNESS RATIO = 56
   │
   │CRITICAL LOAD = 55506 N
   └ALLOWABLE LOAD = 18502 N FOR DESIGN FACTOR = 3

    TYPE Y TO RUN NEW DATA
►   ? N                                                   Statement 400
```

[1]Data are from example problem 6-1.
[2]Data are from example problem 6-2.

REFERENCES

1. Aluminum Association. *Specifications for Aluminum Structures,* 4th ed. New York, 1982.
2. American Institute of Steel Construction. *Manual of Steel Construction*. New York, 1982.
3. Muvdi, B. B., and McNabb, J. W. *Engineering Mechanics of Materials*. New York: Macmillan Publishing Company, 1980.
4. Roark, R. J., and Young, W. C. *Formulas for Stress and Strain,* 5th ed. New York: McGraw-Hill Book Company, 1975.
5. Shigley, J. E., and Mitchell, L. D. *Mechanical Engineering Design,* 4th ed. New York: McGraw-Hill Book Company, 1983.
6. Spotts, M. F. *Design of Machine Elements,* 5th ed. Englewood Cliffs, N.J.: Prentice-Hall, 1978.
7. Timoshenko, S. *Strength of Materials,* 2d ed. New York: Van Nostrand Reinhold Company, vol. 2, 1941.
8. Timoshenko, S., and Gere, J. M. *Theory of Elastic Stability,* 2d ed. New York: McGraw-Hill Book Company, 1961.

PROBLEMS

1. A column has both ends pinned and has a length of 32 in. It is made of AISI 1040 HR steel and has a circular shape with a diameter of 0.75 in. Determine the critical load.
2. Repeat problem 1 with a length of 15 in.
3. Repeat problem 1 with the bar made of aluminum 6061-T4.
4. Repeat problem 1 with both ends fixed.
5. Repeat problem 1 with a square cross section, 0.65 in on a side, instead of the circular cross section.
6. Repeat problem 1 with the bar made from high-impact acrylic plastic.
7. A rectangular steel bar has a cross section 0.50 by 1.00 in and is 8.5 in long. The bar has pinned ends and is made of AISI 4150 OQT 1000 steel. Compute the critical load.
8. A steel pipe has an outside diameter of 1.60 in, a wall thickness of 0.109 in, and a length of 6.25 ft. Compute the critical load for each of the end conditions shown in Figure 6-2. Use AISI 1020 HR steel.
9. Compute the required diameter of a circular bar to be used as a column carrying a load of 8 500 lb with pinned ends. The length is 50 in. Use AISI 4140 OQT 1000 steel and a design factor of 3.0.
10. Repeat problem 9 with AISI 1020 HR steel.
11. Repeat problem 9 with aluminum 2014-T4.
12. In section 6-9, equations were derived for the design of a solid circular column, either long or short. Perform the derivation for a solid square cross section.
13. Repeat the derivations called for in problem 12 for a hollow circular tube for any ratio of inside to outside diameter. That is, let the ratio $R = ID/OD$, and solve for the required OD for a given load, material, design factor, and end fixity.
14. Determine the required dimensions of a column with a square cross section to carry an axial compressive load of 6 500 lb if its length is 64 in and its ends are fixed. Use a design factor of 3.0. Use aluminum 6061-T6.
15. Repeat problem 14 for a hollow aluminum tube (6061-T6) with the ratio of $ID/OD = 0.80$. Compare the weight of this column with that of problem 14.
16. A toggle device is being used to compact scrap steel shavings, as illustrated in Figure 6-11. Design the two links of the toggle to be steel, AISI 5160 OQT 1000, with a circular cross section and pinned ends. The force P required to crush the shavings is 5 000 lb. Use $N = 3.50$.
17. Repeat problem 16 but propose a design that will be lighter than the solid circular cross section.
18. A sling, sketched in Figure 6-12, is to carry 18 000 lb. Design the spreader.
19. For the sling in Figure 6-12, design the spreader if the angle shown is changed from 30° to 15°.
20. A rod for a certain hydraulic cylinder behaves as a fixed-free column when used to actuate a compactor of industrial waste. Its maximum extended length will be 10.75 ft. If it is to be made of AISI 1144 OQT 1300 steel, determine the required diameter of the rod for a design factor of 2.5 for an axial load of 25 000 lb.
21. Design a column to carry 40 000 lb. One end is pinned and the other is fixed. The length is 12.75 ft.
22. Repeat problem 21 with a length of 4.25 ft.
23. Repeat problem 1 if the column has an initial crookedness of 0.08 in. Determine the allowable load for a design factor of 3.
24. Repeat problem 7 if the column has an initial crookedness of 0.04 in. Determine the allowable load for a design factor of 3.
25. Repeat Problem 8 if the column has an initial crookedness of 0.15 in. Determine the allowable load for a design factor of 3 and pinned ends only.
26. An aluminum (6063-T4) column is 42 in long and has a square cross section, 1.25 in on a side. If it carries a compressive load of 1 250 lb, applied with an eccentricity of 0.60 in, compute the maximum

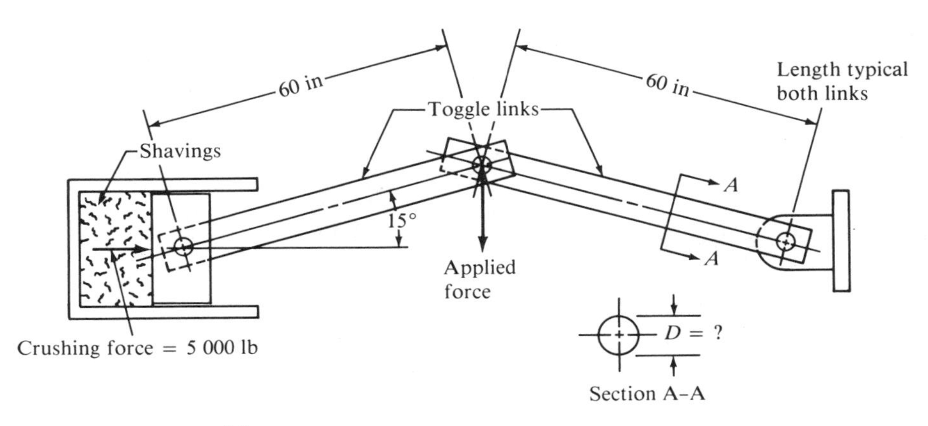

Figure 6-11 (Problem 16)

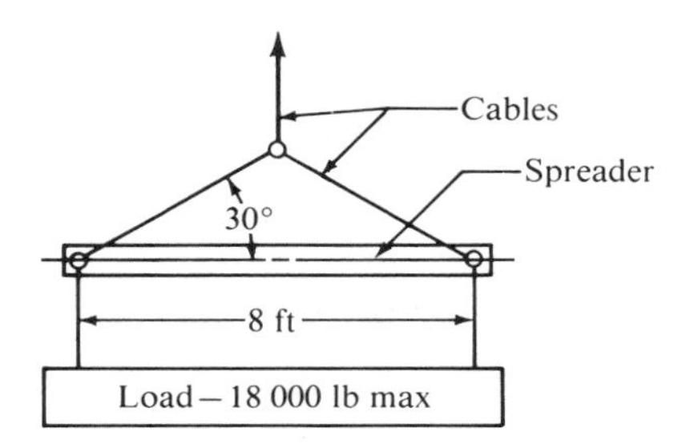

Figure 6-12 (Problem 18)

stress in the column and the maximum deflection.

27. A steel (AISI 1020 hot-rolled) column is 3.2 m long and is made from TS3X0.216 structural tubing (see Appendix A-16). If a compressive load of 30.5 kN is applied with an eccentricity of 150 mm, compute the maximum stress in the column and the maximum deflection.
28. A connecting link in a mechanism is 14.75 in long and has a square cross section, 0.250 in on a side. It is made from annealed AISI 410 stainless steel. Use $E = 28\,000\,000$ psi. If it carries a compressive load of 45 lb with an eccentricity of 0.30 in, compute the maximum stress and maximum deflection.
29. A hollow square steel tube, 40 in long, is proposed for use as a prop to hold up the ram of a punch press during installation of new dies. The ram weighs 75 000 lb. The prop has an outside dimension of 4.00 in and a wall thickness of 0.250 in. It is made from steel similar to structural steel, ASTM A242. If the load applied by the ram could have an eccentricity of 0.50 in, would the prop be safe?
30. Determine the allowable load on a column 16.0 ft long made from a wide flange beam shape, W6X20. The load will be centrally applied. The end conditions are somewhat between fixed and hinged, say $K = 0.8$. Use a design factor of 3.

7 Springs

7-1 OVERVIEW

A *spring* is a flexible element used to exert a force or a torque and, at the same time, to store energy. The force can be a linear push or pull, or it can be radial, acting similarly to a rubber band around a roll of drawings. The torque can be used to cause a rotation, for example, to close a door on a cabinet or to provide a counterbalance force for a machine element pivoting on a hinge.

Springs inherently store energy when they are deflected and return the energy when the force that causes the deflection is removed. When this is the primary design objective, the spring is frequently referred to as a *power spring* or *motor spring*.

Table 7-1 lists several types of springs and shows their uses.

7-2 HELICAL COMPRESSION SPRINGS

In the most common form of helical compression spring round wire is wrapped into a cylindrical form with a constant pitch between adjacent coils. This basic form is completed by a variety of end treatments, as shown in Figure 7-1.

For medium- to large-size springs used in machinery, the squared and ground end treatment provides a flat surface on which to seat the spring. The end coil is collapsed against the adjacent coil (squared), and the surface is ground until at least 270° of the last coil is in contact with the bearing surface. Springs made from smaller wire (less than approximately 0.020 in, or 0.50 mm) are usually squared only, without grinding. In unusual cases the ends may be ground without squaring or left with plain ends, simply cut to length after coiling.

You are probably familiar with many uses of helical compression springs. The retractable ballpoint pen depends on the helical compression spring, usually installed around the ink supply barrel. Suspension systems for cars, trucks, and motorcycles frequently incorporate these springs. Other automotive applications include the valve springs in engines, hood linkage counterbalancing, and the clutch pressure plate springs. In manufacturing, springs are used in dies to actuate stripper plates, in hydraulic control valves, as pneumatic cylinder return springs, and in the mounting of heavy equipment for shock isolation. Many small devices such as electrical switches and ball check valves

Table 7-1 Types of Springs

Uses	*Types of Springs*
Push	Helical compression spring (Figures 7-1 and 7-26) Belleville spring (Figure 7-27) Torsion spring (Figure 7-19): force acting at the end of the torque arm Flat spring, such as a cantilever or leaf spring
Pull	Helical extension spring (Figures 7-14 and 7-17) Torsion spring (Figure 7-19): force acting at the end of the torque arm Flat spring, such as a cantilever or leaf spring Drawbar spring (special case of the compression spring) Constant force spring
Radial	Garter spring, elastomeric band, spring clamp
Torque	Torsion spring (Figure 7-19), power spring

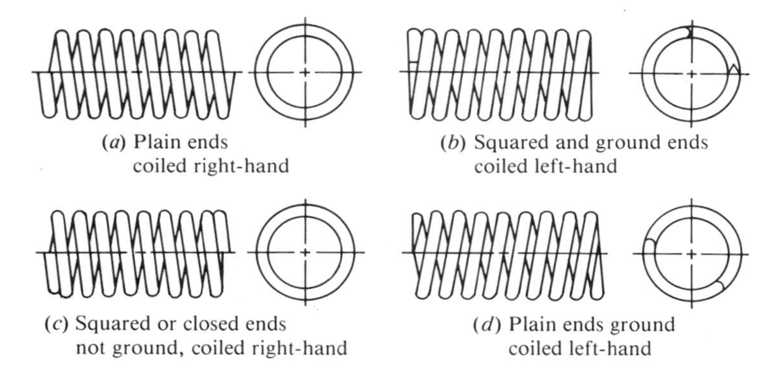

Figure 7-1 Appearance of Helical Compression Springs Showing End Treatments

incorporate helical compression springs. Desk chairs have stout springs to return the chair seat to its upright position. And don't forget the venerable pogo stick!

The following paragraphs define the many variables used to describe and to analyze the performance of helical compression springs.

Diameters

Figure 7-2 shows the notation used in referring to the characteristic diameters of helical compression springs. The outside diameter (OD), the inside diameter (ID), and the wire diameter (D_w) are obvious and can be measured with standard measuring instruments. In calculating the stress and deflection of a spring, the mean diameter, D_m, is used. Notice that

$$OD = D_m + D_w$$

$$ID = D_m - D_w$$

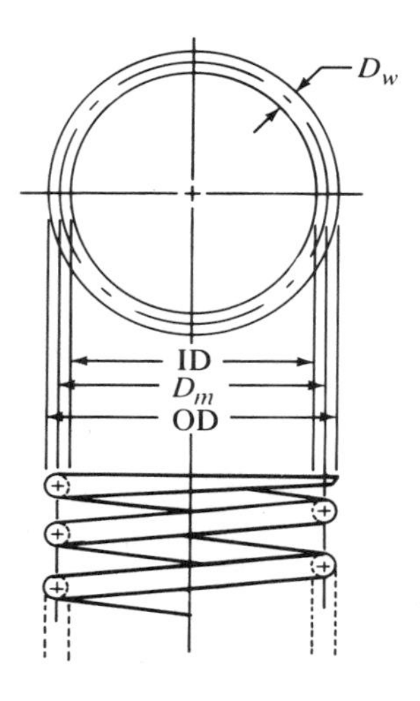

Figure 7-2 Notation for Diameters

Standard Wire Diameters

The specification of the required wire diameter is one of the most important outcomes of the design of springs. There are several types of materials typically used for spring wire, and the wire is produced in sets of standard diameters covering a broad range. Table 7-2 lists the most common standard wire gages. Notice that, except for music wire, the wire size gets smaller as the gage number gets larger. Also see the notes to the table.

Lengths

It is important to understand the relationship between the length of the spring and the force exerted by it (see Figure 7-3). The *free length,* L_f, is the length that the spring assumes when it is exerting no force as if it were simply sitting on a table. The *solid length,* L_s, is found when the spring is collapsed to the point where all coils are touching. This is obviously the shortest possible length the spring can have. The spring is usually not compressed to the solid length during operation.

The shortest length for the spring during normal operation is the *operating length,* L_o. At times, a spring will be designed to operate between two limits of deflection. Consider the valve spring for an engine, for example, as shown in Figure 7-4. When the valve is open, the spring assumes its shortest length, which is L_o. Then, when the valve is closed, the spring gets longer but still exerts a force to keep the valve securely on its seat. The length at this condition is called the *installed length,* L_i. So the valve spring length changes from L_o to L_i during normal operation as the valve itself reciprocates.

Forces

We will use the symbol F to indicate forces exerted by a spring with various subscripts to specify which level of force is being considered. The subscripts correspond to those used for the lengths. Thus

F_s = Force at solid length, L_s: The maximum force that the spring ever sees.

F_o = Force at operating length, L_o: The maximum force the spring sees in *normal operation*.

F_i = Force at installed length, L_i: The force varies between F_o and F_i for a reciprocating spring.

F_f = Force at free length, L_f: This force is zero.

Spring Rate

The relationship between the force exerted by a spring and its deflection is called its *spring rate, k.* Any change in force divided by the corresponding change in deflection can be used to compute the spring rate.

$$k = \Delta F/\Delta L \tag{7-1}$$

For example,

$$k = \frac{F_o - F_i}{L_i - L_o}$$

Table 7-2 Wire Gages and Diameters for Springs

Gage No.	U.S. Steel Wire Gage (in)[a]	Music Wire Gage (in)[b]	Brown & Sharpe Gage (in)[c]	Preferred Metric Diameters (mm)[d]
7/0	0.4900	—	—	13.0
6/0	0.4615	0.004	0.5800	12.0
5/0	0.4305	0.005	0.5165	11.0
4/0	0.3938	0.006	0.4600	10.0
3/0	0.3625	0.007	0.4096	9.0
2/0	0.3310	0.008	0.3648	8.5
0	0.3065	0.009	0.3249	8.0
1	0.2830	0.010	0.2893	7.0
2	0.2625	0.011	0.2576	6.5
3	0.2437	0.012	0.2294	6.0
4	0.2253	0.013	0.2043	5.5
5	0.2070	0.014	0.1819	5.0
6	0.1920	0.016	0.1620	4.8
7	0.1770	0.018	0.1443	4.5
8	0.1620	0.020	0.1285	4.0
9	0.1483	0.022	0.1144	3.8
10	0.1350	0.024	0.1019	3.5
11	0.1205	0.026	0.0907	3.0
12	0.1055	0.029	0.0808	2.8
13	0.0915	0.031	0.0720	2.5
14	0.0800	0.033	0.0641	2.0
15	0.0720	0.035	0.0571	1.8
16	0.0625	0.037	0.0508	1.6
17	0.0540	0.039	0.0453	1.4
18	0.0475	0.041	0.0403	1.2
19	0.0410	0.043	0.0359	1.0
20	0.0348	0.045	0.0320	0.90
21	0.0317	0.047	0.0285	0.80
22	0.0286	0.049	0.0253	0.70
23	0.0258	0.051	0.0226	0.65
24	0.0230	0.055	0.0201	0.60 or 0.55
25	0.0204	0.059	0.0179	0.50 or 0.55
26	0.0181	0.063	0.0159	0.45
27	0.0173	0.067	0.0142	0.45
28	0.0162	0.071	0.0126	0.40
29	0.0150	0.075	0.0113	0.40
30	0.0140	0.080	0.0100	0.35
31	0.0132	0.085	0.00893	0.35
32	0.0128	0.090	0.00795	0.30 or 0.35
33	0.0118	0.095	0.00708	0.30
34	0.0104	0.100	0.00630	0.28
35	0.0095	0.106	0.00501	0.25
36	0.0090	0.112	0.00500	0.22
37	0.0085	0.118	0.00445	0.22
38	0.0080	0.124	0.00396	0.20
39	0.0075	0.130	0.00353	0.20
40	0.0070	0.138	0.00314	0.18

[a]Use the U.S. Steel Wire Gage for steel wire, except music wire. This gage has also been called the *Washburn and Moen Gage (W&M),* the *American Steel Wire Co. Gage,* and the *Roebling Wire Gage.*

[b]Use the Music Wire Gage only for music wire (ASTM A228).

[c]Use the Brown & Sharpe Gage for nonferrous wires such as brass, phosphor bronze, etc.

[d]The preferred metric sizes are from Associated Spring, Barnes Group, Inc., and are listed as the nearest preferred metric size to the U.S. Steel Wire Gage. The gage numbers do not apply.

Source: Associated Spring, Barnes Group, Inc. *Engineering Guide to Spring Design.* Bristol, Conn., 1981; Carlson, Harold. *Spring Designer's Handbook.* New York: Marcel Dekker, 1978; Oberg, E., et al. *Machinery's Handbook,* 22d ed. New York: Industrial Press, 1984.

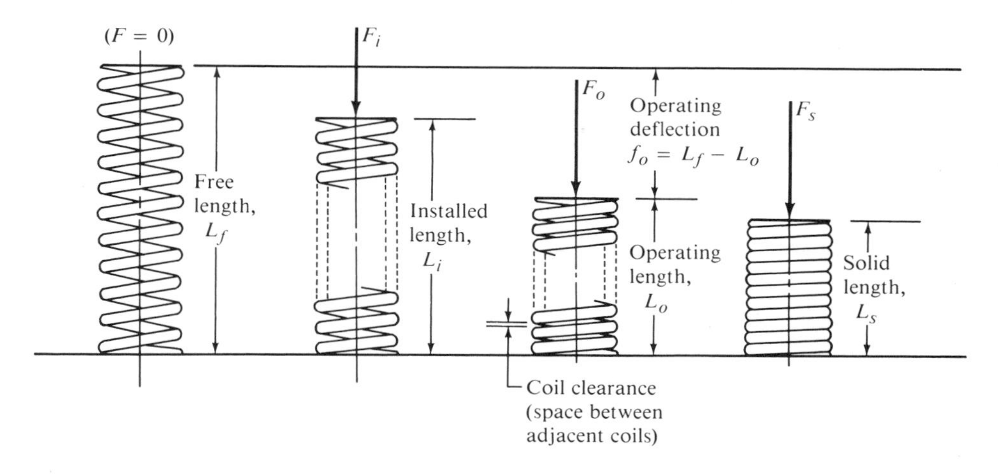

Figure 7-3 Notation for Lengths and Forces

or

$$k = \frac{F_o}{L_f - L_o}$$

or

$$k = \frac{F_i}{L_f - L_i}$$

In addition, if the spring rate is known, the force at any deflection can be computed. For example, if a spring has a rate of 42.0 pounds per inch (lb/in), the force exerted at

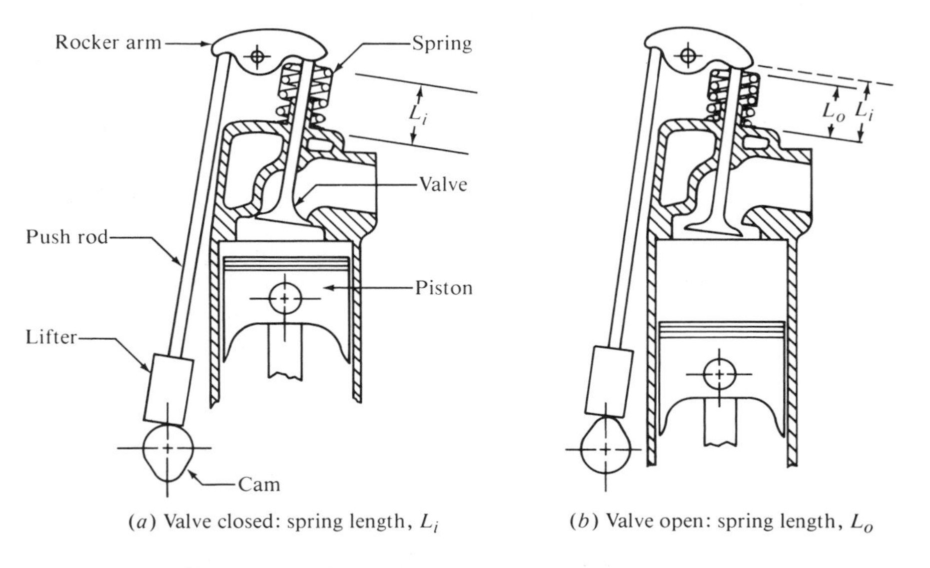

(*a*) Valve closed: spring length, L_i

(*b*) Valve open: spring length, L_o

Figure 7-4 Illustration of Installed Length and Operating Length

a deflection from free length of 2.25 in would be

$$F = k(L_f - L) = (42.0 \text{ lb/in})(2.25 \text{ in}) = 94.5 \text{ lb}$$

Spring Index

The ratio of the mean diameter of the spring to the wire diameter is called the *spring index, C*.

$$C = D_m/D_w$$

It is recommended that C be greater than 5.0, with typical machinery springs having C values ranging from 5 to 12. For C less than 5, the forming of the spring will be very difficult, and the severe deformation required may create cracks in the wire. The stresses and deflections in springs are dependent on C, and a larger C will help to eliminate the tendency for a spring to buckle.

Number of Coils

The total number of coils in a spring will be called N. But in the calculation of stress and deflections for a spring, some of the coils are inactive and are neglected. For example, in a spring with squared and ground ends or simply squared ends, each end coil is inactive and the number of *active coils*, N_a, is $N - 2$. For plain ends all coils are active: $N_a = N$. For plain coils with ground ends, $N_a = N - 1$.

Pitch

Pitch, p, refers to the axial distance from a point on one coil to the corresponding point on the next adjacent coil. The relationships among the pitch, free length, wire diameter, and number of active coils are given next.

Squared and ground ends:	$L_f = pN_a + 2D_w$
Squared ends only:	$L_f = pN_a + 3D_w$
Plain and ground ends:	$L_f = p(N_a + 1)$
Plain ends:	$L_f = pN_a + D_w$

Pitch Angle

Figure 7-5 shows the pitch angle, λ; it can be seen that the larger the pitch angle, the steeper the coils appear to be. Most practical spring designs produce a pitch angle less than about 12°. If the angle is greater than 12°, undesirable compressive stresses develop in the wire and the formulas presented later are inaccurate. The pitch angle can be computed by the formula

$$\lambda = \tan^{-1}\left[\frac{p}{\pi D_m}\right] \qquad (7\text{-}2)$$

The logic of this formula can be seen by taking one coil of a spring and unwrapping it onto a flat surface, as illustrated in Figure 7-5. The horizontal line is the mean circumference of the spring, and the vertical line is the pitch, p.

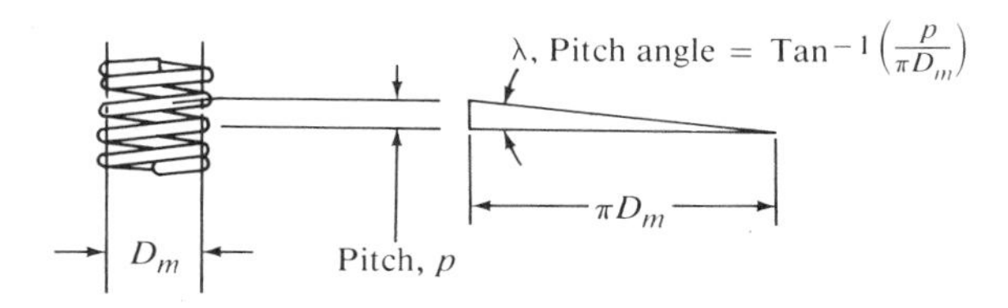

Figure 7-5 Pitch Angle

Installation Considerations

Frequently, a spring is installed in a cylindrical hole or around a rod. When it is, adequate clearances must be provided. When a compression spring is compressed, its diameter gets larger. Thus the inside diameter of a hole enclosing the spring must be greater than the outside diameter of the spring to eliminate rubbing. An initial diametral clearance of one-tenth of the wire diameter is recommended for springs having a diameter of 0.50 in (12 mm) or greater. If a more precise estimate of the actual outside diameter of the spring is required, the following formula can be used for the OD at the solid length condition:

$$OD_s = \sqrt{D_m^2 + \frac{p - D_w^2}{\pi^2}} + D_w \tag{7-3}$$

Even though the spring ID gets larger, it is also recommended that the clearance at the ID be approximately $0.1D_w$.

Springs with squared ends or squared-and-ground ends are frequently mounted on a button-type seat or in a socket with a depth equal to the height of just a few coils for the purpose of locating the spring.

Coil Clearance

The term *coil clearance* refers to the space between adjacent coils when the spring is compressed to its operating length, L_o. The actual coil clearance can be estimated from

$$cc = (L_o - L_s)/N_a$$

One guideline is that the coil clearance should be greater than $D_w/10$, especially in springs loaded cyclically. Another recommendation relates to the overall deflection of the spring:

$$(L_o - L_s) > 0.15(L_f - L_s)$$

Materials Used for Springs

Virtually any elastic material can be used for a spring. However, most mechanical applications use metallic wire, either high-carbon steel (most common), alloy steel, stainless steel, brass, bronze, beryllium copper, or nickel base alloys. Most spring materials are made according to specifications of the ASTM. Table 7-3 lists some common types.

Types of Loading and Allowable Stresses

The allowable stress to be used for a spring depends on the type of loading, the material, and size of the wire. Loading is usually classified into three types:

Light service: Static loads or up to 10 000 cycles of loading with a low rate of loading (nonimpact).

Average service: Typical machine design situations; moderate rate of loading and up to 1 million cycles.

Table 7-3 Spring Materials

Material Type	*ASTM No.*	*Relative Cost*	*Temperature Limits, °F*
High-carbon steels:			
Hard drawn	A227	1.0	0–250
General-purpose steel with 0.60–0.70% carbon; low cost			
Music wire	A228	2.6	0–250
High-quality steel with 0.80–0.95% carbon; very high strength; excellent surface finish; hard drawn; good fatigue performance; used mostly in smaller sizes up to 0.125 in			
Oil-tempered	A229	1.3	0–350
General-purpose steel with 0.60–0.70% carbon; used mostly in larger sizes above 0.125 in; not good for shock or impact			
Alloy steels:			
Chromium-vanadium	A231	3.1	0–425
Good strength, fatigue resistance, impact strength, high-temperature performance; valve spring quality			
Chromium-silicon	A401	4.0	0–475
Very high strength and good fatigue and shock resistance			
Stainless steels:			
Type 302	A313(302)	7.6	<0–550
Very good corrosion resistance and high-temperature performance; nearly nonmagnetic; cold drawn; types 304 and 316 also fall under this ASTM class and have improved workability but lower strength			
Type 17-7 PH	A313(631)	11.0	0–600
Good high-temperature performance			
Copper alloys: All have good corrosion resistance and electrical conductivity			
Spring brass	B134	High	0–150
Phosphor bronze	B159	8.0	<0–212
Beryllium copper	B197	27.0	0–300
Nickel-base alloys: All are corrosion-resistant, have good high- and low-temperature properties, are nonmagnetic or nearly nonmagnetic (trade names of the International Nickel Company)			
Monel	—	—	−100–425
K-Monel	—	—	−100–450
Inconel	—	—	Up to 700
Inconel-X	—	44.0	Up to 850

Source: Associated Spring, Barnes Group, Inc. *Engineering Guide to Spring Design.* Bristol, Conn., 1981; Carlson, Harold. *Spring Designer's Handbook.* New York: Marcel Dekker, 1978; Oberg, E., et al. *Machinery's Handbook,* 22d ed. New York: Industrial Press, 1984.

Severe service: Rapid cycling for above 1 million cycles; possibility of shock or impact loading; engine valve springs are a good example.

The strength of a given material is greater for the smaller sizes. Figures 7-6 through 7-11 show the allowable working stresses for six different materials. Note that some curves can be used for more than one material by the application of a factor. As a conservative approach to design, we will use the average service curve for most design examples, unless true high cycling conditions exist. We will use the light service curve as the upper limit on stress when the spring is compressed to its solid height. If the stress exceeds the light service value by a small amount, the spring will undergo permanent set because of yielding.

7-3 STRESSES AND DEFLECTION FOR HELICAL COMPRESSION SPRINGS

As a compression spring is compressed under an axial load, the wire is twisted. Therefore, the stress developed in the wire is *torsional shear stress,* and it can be derived from the classical equation $\tau = Tc/J$.

When applied specifically to a helical compression spring, some modifying factors are needed to account for the curvature of the spring wire and for the direct shear stress created as the coils resist the vertical load. Also, it is convenient to express the shear stress in terms of the design variables encountered in springs. The resulting equation for stress is attributed to Wahl (7). The maximum shear stress, which occurs at the inner surface of the wire, is

$$\tau = \frac{8KFD_m}{\pi D_w^3} = \frac{8KFC}{\pi D_w^2} \tag{7-4}$$

These are two forms of the same equation as the definition of $C = D_m/D_w$ demonstrates. The shear stress for any applied force, F, can be computed. Normally we will be concerned about the stress when the spring is compressed to solid length under the influence of F_s and when the spring is operating at its normal maximum load, F_o. Notice that the stress is inversely proportional to the *cube* of the wire diameter. This illustrates the great effect that variation in the wire size has on the performance of the spring.

The Wahl factor, K, in equation (7-4) is the term that accounts for the curvature of the wire and the direct shear stress. Analytically, K is related to C:

$$K = \frac{4C - 1}{4C - 4} + \frac{0.615}{C} \tag{7-5}$$

Figure 7-12 shows a plot of K versus C for round wire. Recall that $C = 5$ is the recommended minimum value of C. The value of K rises rapidly for $C < 5$.

Deflection

Because the primary manner of loading on the wire of a helical compression spring is torsion, the deflection is computed from the angle of twist formula,

$$\theta = TL/GJ$$

where θ is the angle of twist in radians, T is the applied torque, L is the length of the wire, G is the modulus of elasticity of the material in shear, and J is the polar moment of inertia

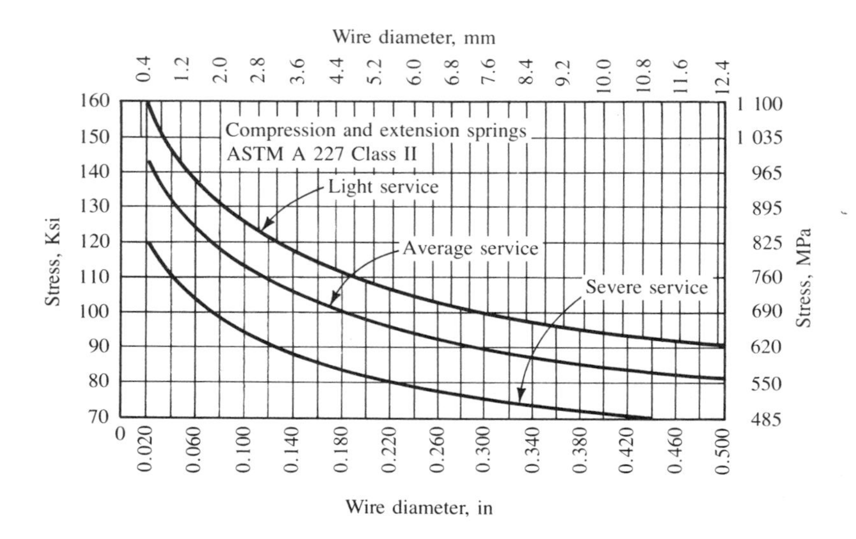

Figure 7-6 Design Stresses, ASTM A227 Steel Wire, Hard Drawn (Reprinted from Carlson, Harold. *Spring Designer's Handbook,* p. 144, by courtesy of Marcel Dekker, Inc.)

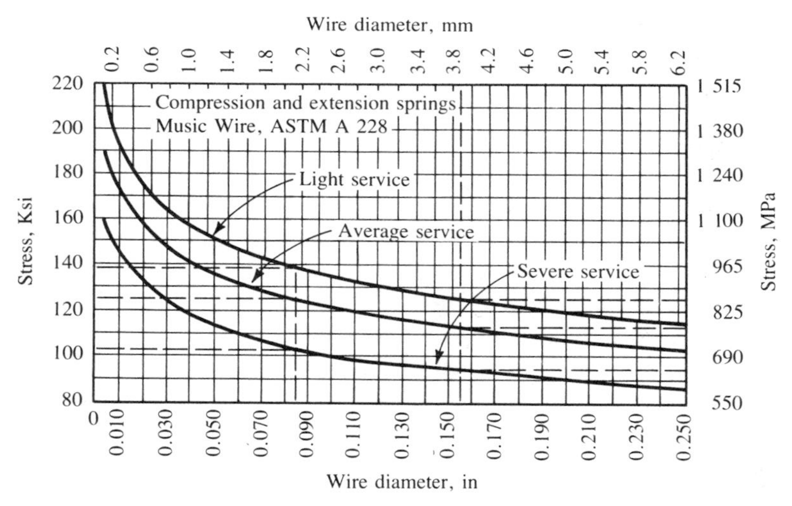

Similar design stresses for ASTM A 313, type 631 stainless steel wire, 17-7 PH

Figure 7-7 Design Stresses, ASTM A228 Steel Wire (Music Wire) (Reprinted from Carlson, Harold. *Spring Designer's Handbook,* p. 143, by courtesy of Marcel Dekker, Inc.)

of the wire. Again, for convenience, we will use a different form of the equation in order to calculate the linear deflection, f, of the spring from the typical design variables of the spring. The resulting equation is

$$f = \frac{8FD_m^3N_a}{GD_w^4} = \frac{8FC^3N_a}{GD_w} \qquad (7\text{-}6)$$

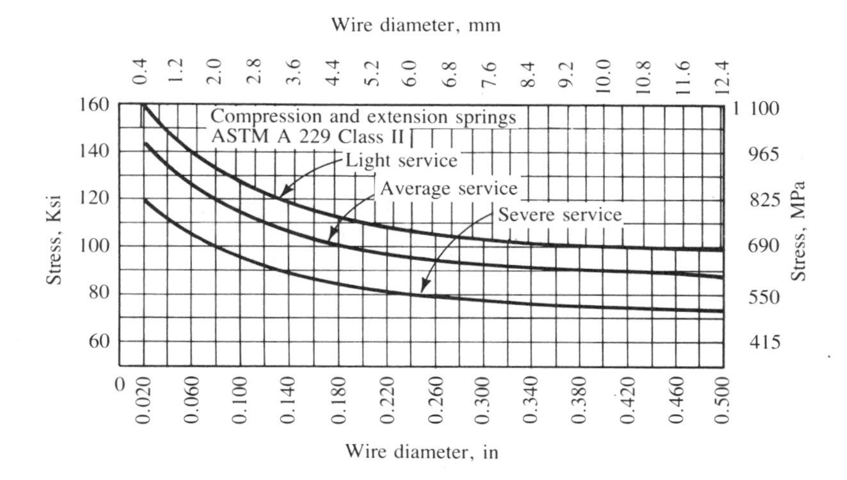

Figure 7-8 Design Stresses, ASTM A229 Steel Wire, Oil-tempered (Reprinted from Carlson, Harold. *Spring Designer's Handbook,* p. 146, by courtesy of Marcel Dekker, Inc.)

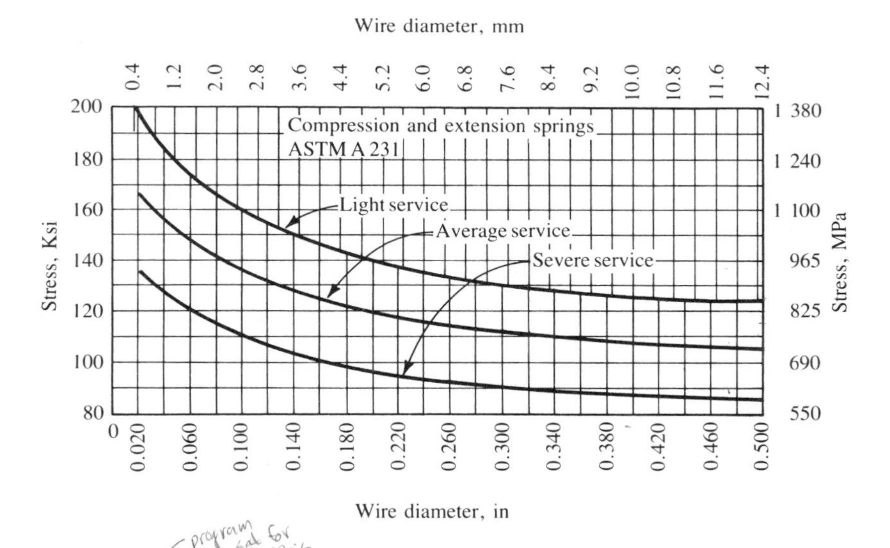

program set for this

Figure 7-9 Design Stresses, ASTM A231 Steel Wire, Chromium-vanadium Alloy, Valve Spring Quality (Reprinted from Carlson, Harold. *Spring Designer's Handbook,* p. 147, by courtesy of Marcel Dekker, Inc.)

Recall that N_a is the number of *active* coils, as discussed in section 7-2. Table 7-4 lists the values for G for typical spring materials. Note again, in equation (7-6), that the wire diameter has a strong effect on the performance of the spring.

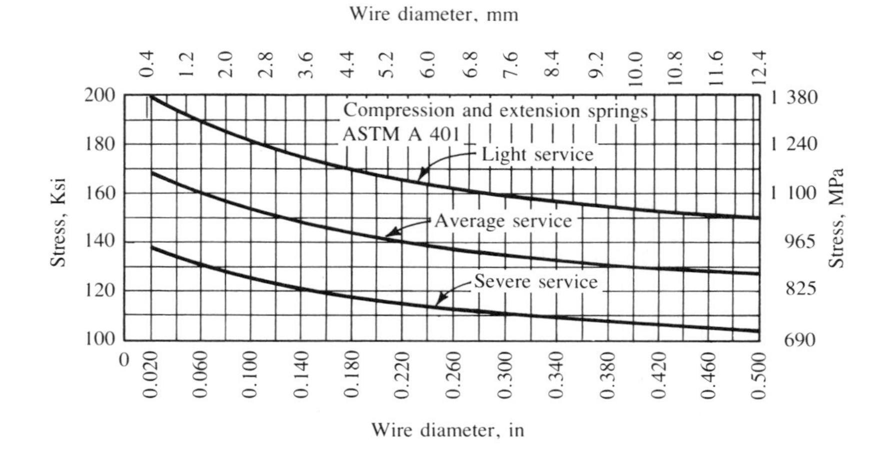

Figure 7-10 Design Stresses, ASTM A401 Steel Wire, Chromium-silicon Alloy, Oil-tempered (Reprinted from Carlson, Harold. *Spring Designer's Handbook,* p. 148, by courtesy of Marcel Dekker, Inc.)

Buckling

The tendency for a spring to buckle increases as the spring becomes tall and slender, much as for a column. Figure 7-13 shows plots of the critical ratio of deflection to the free length, versus the ratio of free length to the mean diameter for the spring. Three different end conditions are described in the figure. As an example of the use of this figure, consider a spring having squared and ground ends, a free length of 6.0 in, and a mean diameter of 0.75 in. We want to know what deflection would cause the spring to buckle. First compute

$$\frac{L_f}{D_m} = \frac{6.0}{0.75} = 8.0$$

Then, from Figure 7-13, the critical deflection ratio is 0.20. From this we can compute the critical deflection.

$$\frac{f_o}{L_f} = 0.20 \qquad \text{or} \qquad f_o = 0.20(L_f) = 0.20(6.0 \text{ in}) = 1.20 \text{ in}$$

That is, if the spring is deflected more than 1.20 in, the spring should buckle.

7-4 DESIGN OF HELICAL COMPRESSION SPRINGS

The objective of the design of helical compression springs is to specify the geometry for the spring to operate under specified limits of load and deflection, possibly with space limitations, also. The material and the type of service will be specified by considering the environment and the application.

A typical problem statement follows.

Example Problem 7-1. A helical compression spring is to exert a force of 8.0 lb when compressed to a length of 1.75 in. At a length of 1.25 in, the force must be 12.0 lb. The

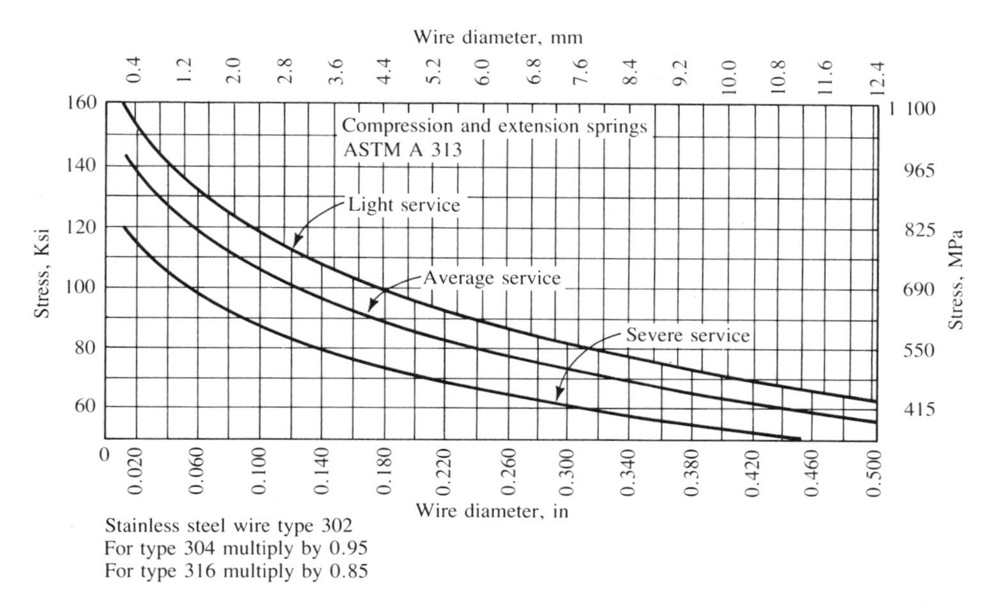

Figure 7-11 Design Stresses, ASTM A313 Corrosion-resistant Stainless Steel Wire (Reprinted from Carlson, Harold. *Spring Designer's Handbook,* p. 150, by courtesy of Marcel Dekker, Inc.)

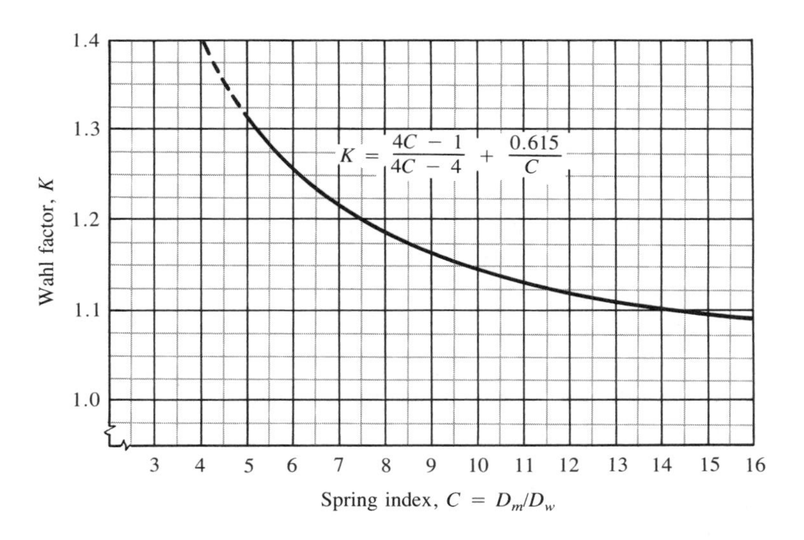

Figure 7-12 Wahl Factor vs. Spring Index for Round Wire

spring will be installed in a machine that cycles slowly, and approximately 200 000 cycles total are expected. The temperature will not exceed 200°F. It is planned to install the spring in a hole having a diameter of 0.75 in.

For this application, specify a suitable material, wire diameter, mean diameter, OD, ID, free length, solid length, number of coils, and type of end condition. Check the stress at the maximum operating load and at the solid length condition.

Table 7-4 Modulus of Elasticity: G (Shear); E (Tension)

MATERIAL AND ASTM NO.	SHEAR MODULUS, G (psi)	SHEAR MODULUS, G (GPa)	TENSION MODULUS, E (psi)	TENSION MODULUS, E (GPa)
Hard drawn steel: A227	11.5×10^6	79.3	28.6×10^6	197
Music wire: A228	11.85×10^6	81.7	29.0×10^6	200
Oil-tempered: A229	11.2×10^6	77.2	28.5×10^6	196
Chromium-vanadium: A231	11.2×10^6	77.2	28.5×10^6	196
Chromium-silicon: A401	11.2×10^6	77.2	29.5×10^6	203
Stainless steels: A313				
Types 302, 304, 316	10.0×10^6	69.0	28.0×10^6	193
Type 17-7 PH	10.5×10^6	72.4	29.5×10^6	203
Spring brass: B134	5.0×10^6	34.5	15.0×10^6	103
Phosphor bronze: B159	6.0×10^6	41.4	15.0×10^6	103
Beryllium copper: B197	7.0×10^6	48.3	17.0×10^6	117
Monel and K-Monel	9.5×10^6	65.5	26.0×10^6	179
Inconel and Inconel-X	10.5×10^6	72.4	31.0×10^6	214

Note: Data are average values. Slight variation with wire size and treatment may occur.

Two solution procedures will be shown, and each will be implemented in a computer program and worked out in the solution. The numbered steps can be used as a guide for future problems and as a kind of algorithm for the computer programs.

Solution Method 1. The procedure works directly toward the overall geometry of the spring by specifying the mean diameter to meet the space limitations. The process requires that the designer have tables of data available for wire diameters (such as Table 7-2) and design stresses for the material from which the spring will be made (such as Figures 7-6 to 7-11). An initial estimate for the design stress for the material must be made by consulting the charts of design stress versus wire diameter to make a reasonable choice. In general, more than one trial must be made, but the results of early trials will help you to decide the values to use for later trials.

Step 1. Specify a material and its shear modulus of elasticity, G.

For this problem, several standard spring materials can be used. Let's select ASTM A231 chromium-vanadium steel wire, having a value of $G = 11\,200\,000$ psi.

Step 2. From the problem statement, identify the operating force, F_o; the operating length at which that force must be exerted, L_o; the force at some other length, called the *installed force*, F_i; and the installed length, L_i.

Remember, F_o is the maximum force the spring experiences under normal operating conditions. Many times, the second force level is not specified. In that case, let F_i = zero and specify a design value for the free length, L_f, in place of L_i.

For this problem, $F_o = 12.0$ lb; $L_o = 1.25$ in; $F_i = 8.0$ lb; and $L_i = 1.75$ in.

Step 3. Compute the spring rate, k.

$$k = \frac{F_o - F_i}{L_i - L_o} = \frac{12.0 - 8.0}{1.75 - 1.25} = 8.00 \text{ lb/in}$$

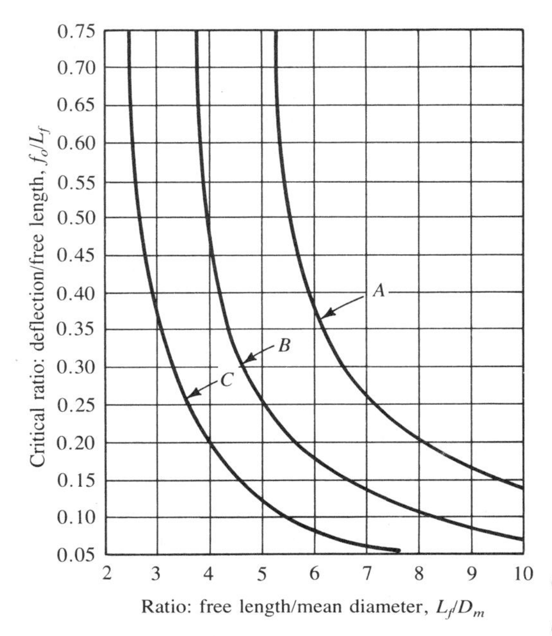

Curve *A*: Fixed ends (e.g., squared and ground ends on guided, flat, parallel surfaces)
Curve *B*: One fixed end; one pinned end (e.g., one end on flat surface, one in contact with a spherical ball)
Curve *C*: Both ends pinned (e.g., ends in contact with surfaces which are pinned to the structure and permitted to rotate)

Figure 7-13 Spring Buckling Criteria. If actual ratio of L_O/L_f is greater than critical ratio, spring will buckle at operating deflection.

Step 4. Compute the free length, L_f.

$$L_f = L_i + F_i/k = 1.75 \text{ in} + [(8.00 \text{ lb})/(8.00 \text{ lb/in})] = 2.75 \text{ in}$$

The second term in the preceding equation is the amount of deflection from free length to the installed length in order to develop the installed force, F_i. Of course, this step becomes unnecessary if the free length is specified in the original data.

Step 5. Specify an initial estimate for the mean diameter, D_m.

Keeping in mind that the mean diameter will be smaller than the outside diameter and larger than the inside diameter, judgment is necessary to get started. For this problem, let's specify $D_m = 0.60$ in. This should permit the installation into the 0.75-in diameter hole.

Step 6. Specify an initial design stress.

The charts for the design stresses for the selected material can be consulted, considering also the service. In this problem, we should use average service. Then, for the ASTM A231 steel, as shown in Figure 7-9, a nominal design stress would be 130 000 psi. This is strictly an estimate based on the strength of the material. The process includes a check on stress later.

Step 7. Compute the trial wire diameter by solving equation (7-4) for D_w. Notice that everything else in the equation is known except the Wahl factor, K, because it depends on the wire diameter itself. But K varies only little over the normal range of spring indexes, C. From Figure 7-12, it can be seen that $K = 1.2$ is a nominal value. This, too, will be checked later. With the assumed value of K, some simplification can be done.

$$D_w = \left[\frac{8KF_oD_m}{\pi\tau_d}\right]^{1/3} = \left[\frac{(8)(1.2)(F_o)(D_m)}{(\pi)(\tau_d)}\right]^{1/3}$$

Combining constants gives

$$D_w = \left[\frac{8KF_oD_m}{\pi\tau_d}\right]^{1/3} = \left[\frac{(3.06)(F_o)(D_m)}{\tau_d}\right]^{1/3} \qquad (7\text{-}7)$$

For this problem,

$$D_w = \left[\frac{(3.06)(F_o)(D_m)}{\tau_d}\right]^{1/3} = \left[\frac{(3.06)(12)(0.6)}{130\,000}\right]^{0.333}$$

$$D_w = 0.055\,3 \text{ in}$$

Step 8. Select a standard wire diameter from the tables and then determine the design stress and maximum allowable stress for the material at that diameter. The design stress will normally be for average service, unless high cycling rates or shock indicate that severe service is warranted. The light service curve should be used with care because it is very near to the yield strength. In fact, we will use the light service curve as an estimate of the maximum allowable stress.

For this problem, the next larger standard wire size is 0.062 5 in, no. 16 on the U.S. Steel Wire Gage chart. For this size, the curves for ASTM A231 steel wire show the design stress to be approximately 145 000 psi for average service, and the maximum allowable stress to be 170 000 psi from the light service curve.

Step 9. Compute the actual values of C and K, the spring index, and the Wahl factor.

$$C = D_m/D_w = 0.60/0.062\,5 = 9.60$$

$$K = \frac{4C - 1}{4C - 4} + \frac{0.615}{C} = \frac{4(9.60) - 1}{4(9.60) - 4} + \frac{0.615}{9.60} = 1.15$$

Step 10. Compute the actual expected stress due to the operating force, F_o, from equation (7-4).

$$\tau_o = \frac{8KF_oD_m}{\pi D_w^3} = \frac{(8)(1.15)(12.0)(0.60)}{(\pi)(0.062\,5)^3} = 86\,450 \text{ psi}$$

Comparing this with the design stress of 145 000 psi, it is safe.

Step 11. Compute the number of active coils required to give the proper deflection characteristics for the spring. Using equation (7-6) and solving for N_a,

$$f = \frac{8FC^3N_a}{GD_w}$$

$$N_a = \frac{fGD_w}{8FC^3} = \frac{GD_w}{8kC^3} \quad \text{(Note: } F/f = k\text{, the spring rate)} \qquad (7\text{-}8)$$

Then, for this problem,

$$N_a = \frac{GD_w}{8kC^3} = \frac{(11\,200\,000)\,(0.062\,5)}{(8)\,(8.0)\,(9.60)^3} = 12.36 \text{ coils}$$

Notice that $k = 8.0$ lb/in is the spring rate. Do not confuse this with K, the Wahl factor.

Step 12. Compute the solid length, L_s; the force on the spring at solid length, F_s; and the stress in the spring at solid length, τ_s. This computation will give the maximum stress the spring will receive.

The solid length occurs when all the coils are touching, but recall that there are two inactive coils.

$$L_s = D_w(N_a + 2) = 0.062\,5(14.36) = 0.898 \text{ in}$$

The force at solid length is the product of the spring rate times the deflection to solid length $(L_f - L_s)$.

$$F_s = k(L_f - L_s) = (8.0 \text{ lb/in})\,(2.75 - 0.898) \text{ in} = 14.8 \text{ lb}$$

Because the stress in the spring is directly proportional to the force, a simple method of computing the solid length stress is

$$\tau_s = (\tau_o)\,(F_s/F_o) = (86\,450 \text{ psi})\,(14.8/12.0) = 106\,750 \text{ psi}$$

When this is compared with the maximum allowable stress of 170 000 psi, it is safe and the spring will not yield when compressed to solid length.

Step 13. Complete the computations of geometric features and compare them with space and operational limitations.

$$\text{OD} = D_m + D_w = 0.60 + 0.062\,5 = 0.663 \text{ in}$$
$$\text{ID} = D_m - D_w = 0.60 - 0.062\,5 = 0.538 \text{ in}$$

These dimensions are satisfactory for installation in a hole having a diameter of 0.75 in.

This procedure completes the design of one satisfactory spring for this application. It may be desirable to make other trials to attempt to find a more nearly optimum spring. Also, the tendency to buckle should be checked along with the coil clearance. These items will be discussed within the development of method 2.

Computer Program for Spring Design Method 1

In the following discussion, the listing of the code for a computer program in the BASIC language performs the 13-step design procedure developed previously. Figure 7-14 is a flowchart for the program. The *REM*, or remark, statements in the program listing should help you follow the program. Notice that several checks are made and messages are printed if (1) the stresses are too high, (2) the spring index is less than 5.0, or (3) the computed solid length happens to exceed the specified operating length, a clearly impossible situation. Statements 890–940 allow the operator to elect several options on how to proceed after completing any given design. Many designs can be completed in a short time with such a program.

Following the program is a set of output matching the example just completed. In addition, a second trial, made with a slightly larger mean diameter, results in a spring that fits better into the specified hole.

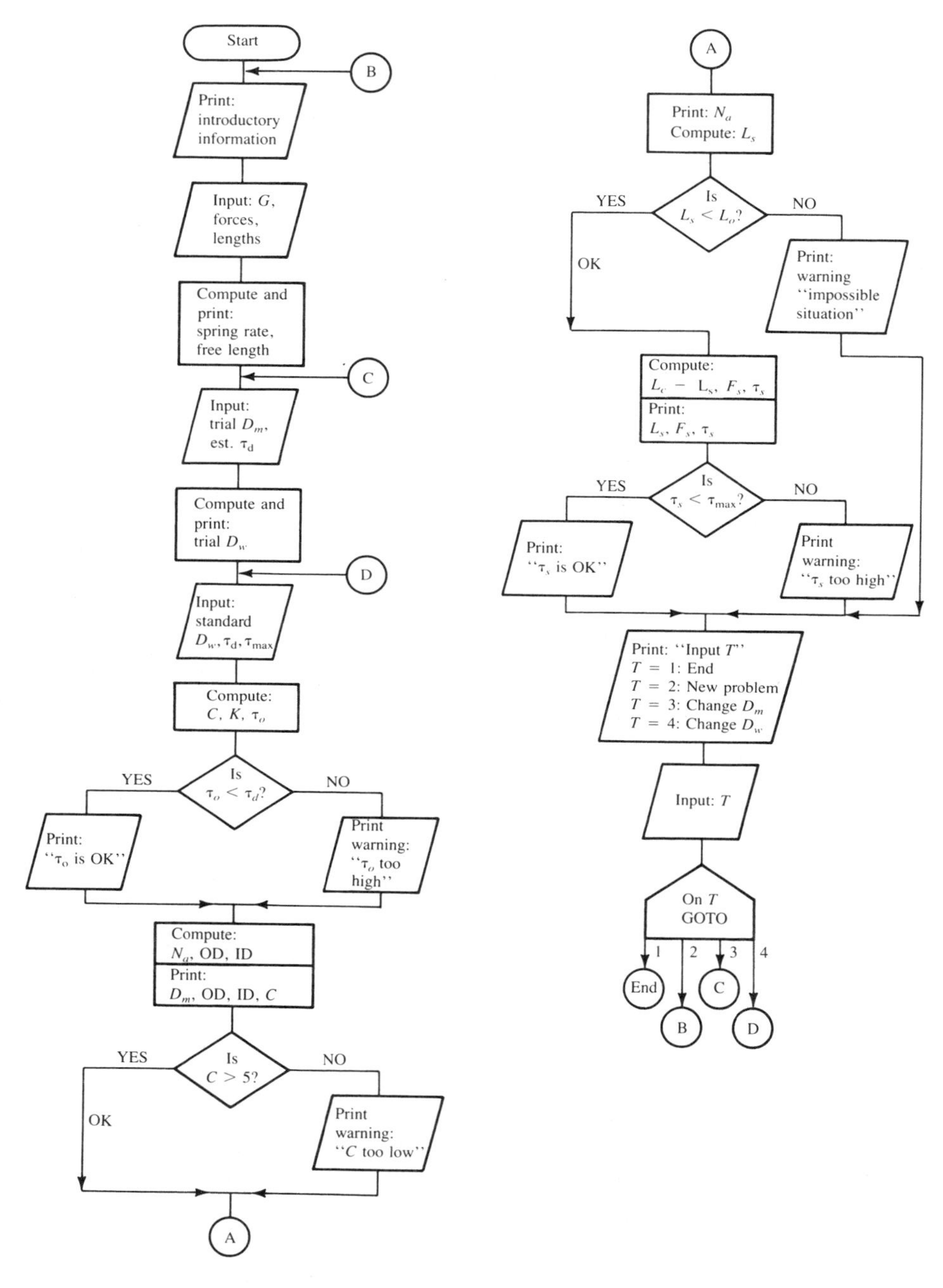

Figure 7-14 Flow Chart for Spring Design Method 1

Helical Spring Design (Method 1): Program Listing

```
10 PRINT "PROGRAM TO ASSIST IN THE DESIGN OF"
20 PRINT "  HELICAL COMPRESSION SPRINGS"
30 PRINT "  WITH A SPECIFIED MEAN DIAMETER"
40 PRINT
50 PRINT "INPUT DATA REQUIRED ARE:"
60 PRINT "  SHEAR MODULUS OF ELASTICITY OF SPRING WIRE,
  G, PSI"
70 PRINT "  MAXIMUM OPERATING FORCE, FO, LB"
80 PRINT "  OPERATING LENGTH, LO, INCHES"
90 PRINT "  INSTALLED FORCE, FI, LB"
100 PRINT "  INSTALLED LENGTH, LI, INCHES"
110 PRINT
120 PRINT "NOTE: INSTALLED LENGTH CAN BE FREE LENGTH -"
130 PRINT "       THEN INSTALLED FORCE = ZERO"
140 PRINT
150 PRINT
160 PRINT "NOW INPUT G, FO, LO, FI, LI"
▶ 170 INPUT G,FO,LO,F1,L1
180 PRINT
190 REM : K = SPRING RATE
200 K = (FO - F1)/(L1 - LO)
210 PRINT USING "SPRING RATE = ###.## LB/IN";K
220 REM : L2 = FREE LENGTH
230 L2 = L1 + F1/K
240 PRINT USING "FREE LENGTH = ##.### INCHES";L2
250 REM : DO = DEFLECTION AT OPERATING LENGTH
260 PRINT
270 PRINT "INPUT TRIAL VALUE FOR MEAN DIAMETER, D2"
▶ 280 INPUT D2
290 PRINT
300 PRINT "INPUT INITIAL ESTIMATE OF DESIGN STRESS, S1"
▶ 310 INPUT S1
320 REM : D1 = INITIAL TRIAL VALUE OF WIRE DIAMETER
330 D1 = (3.06*FO*D2/S1)^(1/3)
340 PRINT
350 PRINT USING "TRIAL WIRE DIAMETER = #.#### INCHES";D1
360 PRINT
370 PRINT "INPUT STANDARD WIRE SIZE, DESIGN STRESS, AND MAX
  STRESS"
▶ 380 INPUT D1, S1, S4
390 REM : C = SPRING INDEX
400 C = D2/D1
410 REM : K1 = WAHL FACTOR
420 K1 = (4*C - 1)/(4*C - 4) + .615/C
430 S2 = 2.546*K1*FO*D2/(D1^3)
440 PRINT
450 PRINT USING "STRESS DUE TO OPERATING FORCE = ######.
  PSI";S2
460 IF S2 > S1 THEN 490
470 PRINT " -OPERATING STRESS OK- "
480 GOTO 500
490 PRINT " * * * OPERATING STRESS EXCEEDS DESIGN STRESS * * *"
500 PRINT
510 REM : N1 = NUMBER OF ACTIVE COILS
520 N1 = G*D1/(8*K*C^3)
530 REM : D3 = OUTSIDE DIAMETER
540 REM : D4 = INSIDE DIAMETER
550 D3 = D2 + D1
```

Introduction (lines 10–130)

Input design data; compute spring rate and trial D_w (lines 160–330)

Select standard wire size and allowable stress (lines 340–380)

Stress due to operating force

Compute C, K, and operating stress (lines 390–490)

Compute and print geometry (lines 510–550)

Lines in the program calling for input of data and lines in the output that correspond to the input statements are highlighted with a ▶.

```
560 D4 = D2 - D1
570 PRINT USING "FOR SPECIFIED MEAN DIAMETER = ##.###
  INCHES:";D2
580 PRINT USING "   OUTSIDE DIAMETER = ##.### INCHES";D3
590 PRINT USING "   INSIDE DIAMETER = ##.### INCHES";D4
600 PRINT
610 PRINT USING "   SPRING INDEX = ##.##";C
620 IF C > 5 THEN 640
630 PRINT " * * * SPRING INDEX LESS THAN 5.0 * * *"
640 PRINT
650 PRINT USING "   NUMBER OF ACTIVE COILS = ##.##";N1
660 PRINT
670 REM : L3 = SOLID LENGTH
680 L3 = D1*(N1 + 2)
690 IF L3 < L0 THEN 740
700 PRINT USING " * * * SOLID LENGTH = ##.### INCHES WHICH
  IS";L3
710 PRINT USING " * * * GREATER THAN OPERATING LENGTH, ##.###
  INCHES";L0
720 PRINT " * * * * IMPOSSIBLE SITUATION * * * *"
730 GOTO 870
740 REM : D5 = DEFLECTION AT SOLID LENGTH
750 D5 = L2 - L3
760 REM : F2 = FORCE AT SOLID LENGTH
770 F2 = K*D5
780 REM : S3 = STRESS AT SOLID LENGTH
790 S3 = S2*F2/F0
800 PRINT USING "SOLID LENGTH = ##.### INCHES";L3
810 PRINT USING "FORCE AT SOLID LENGTH = ###.## LB";F2
820 PRINT USING "STRESS AT SOLID LENGTH = ######. PSI";S3
830 IF S3 > S4 THEN 860
840 PRINT " -STRESS AT SOLID LENGTH OK- "
850 GOTO 870
860 PRINT " * * * SOLID LENGTH STRESS EXCEEDS MAX ALLOWABLE * *
  * "
870 PRINT
880 PRINT
890 PRINT "TYPE  1  TO END"
900 PRINT "TYPE  2  TO RUN A NEW PROBLEM"
910 PRINT "TYPE  3  TO CHANGE MEAN DIAMETER"
920 PRINT "TYPE  4  TO CHANGE WIRE DIAMETER"
930 INPUT T
940 ON T GOTO 950, 10, 260, 360
950 END
```

Force and stress at solid length

Decision on how to continue

Helical Spring Design (Method 1): Sample Output

```
PROGRAM TO ASSIST IN THE DESIGN OF
  HELICAL COMPRESSION SPRINGS
  WITH A SPECIFIED MEAN DIAMETER

INPUT DATA REQUIRED ARE:
  SHEAR MODULUS OF ELASTICITY OF SPRING WIRE, G, PSI
  MAXIMUM OPERATING FORCE, F0, LB
  OPERATING LENGTH, L0, INCHES
  INSTALLED FORCE, FI, LB
  INSTALLED LENGTH, LI, INCHES

NOTE: INSTALLED LENGTH CAN BE FREE LENGTH -
      THEN INSTALLED FORCE = ZERO

NOW INPUT G, F0, L0, FI, LI
? 1.12E + 07 , 12 , 1.25 , 8 , 1.75

SPRING RATE = 8.00 LB/IN
FREE LENGTH = 2.750 INCHES
```

Statement 170

```
   INPUT TRIAL VALUE FOR MEAN DIAMETER, D2
▶  ? .6                                                        Statement 280

   INPUT INITIAL ESTIMATE OF DESIGN STRESS, S1
▶  ? 130000                                                    Statement 310

   TRIAL WIRE DIAMETER = 0.0553 INCHES

   INPUT STANDARD WIRE SIZE, DESIGN STRESS, AND MAX STRESS
▶  ? .0625 , 145000 , 170000                                   Statement 380

   STRESS DUE TO OPERATING FORCE = 86443. PSI
     - OPERATING STRESS OK -

   FOR SPECIFIED MEAN DIAMETER = 0.600 INCHES:
     OUTSIDE DIAMETER = 0.663 INCHES
     INSIDE DIAMETER = 0.538 INCHES

     SPRING INDEX = 9.60

     NUMBER OF ACTIVE COILS = 12.36

   SOLID LENGTH = 0.898 INCHES
   FORCE AT SOLID LENGTH = 14.82 LB
   STRESS AT SOLID LENGTH = 106748. PSI
     - STRESS AT SOLID LENGTH OK -

   TYPE 1 TO END
   TYPE 2 TO RUN A NEW PROBLEM
   TYPE 3 TO CHANGE MEAN DIAMETER
   TYPE 4 TO CHANGE WIRE DIAMETER
▶  ? 3                                                         Statement 930

   INPUT TRIAL VALUE FOR MEAN DIAMETER, D2                     Begin new design
▶  ? .65                                                       with new mean
                                                               diameter.
   INPUT INITIAL ESTIMATE OF DESIGN STRESS, S1                 Statement 280
▶  ? 145000                                                    Statement 310

   TRIAL WIRE DIAMETER = 0.0548 INCHES

   INPUT STANDARD WIRE SIZE, DESIGN STRESS, AND MAX STRESS
▶  ? .0625 , 145000 , 170000                                   Statement 380

   STRESS DUE TO OPERATING FORCE = 92642. PSI
     - OPERATING STRESS OK -

   FOR SPECIFIED MEAN DIAMETER = 0.650 INCHES:
     OUTSIDE DIAMETER = 0.713 INCHES
     INSIDE DIAMETER = 0.588 INCHES

     SPRING INDEX = 10.40

     NUMBER OF ACTIVE COILS = 9.72

   SOLID LENGTH = 0.733 INCHES
   FORCE AT SOLID LENGTH = 16.14 LB
   STRESS AT SOLID LENGTH = 124590. PSI
     - STRESS AT SOLID LENGTH OK -

   TYPE 1 TO END
   TYPE 2 TO RUN A NEW PROBLEM
   TYPE 3 TO CHANGE MEAN DIAMETER
   TYPE 4 TO CHANGE WIRE DIAMETER
▶  ? 1                                                         Statement 930
```

Computer Program for Spring Design Method 2

The next program presents the code for a second spring design method. Figure 7-15 is the flowchart for this program. This design procedure is much more open-ended in that it allows a wide latitude for the designer to create designs that satisfy the basic

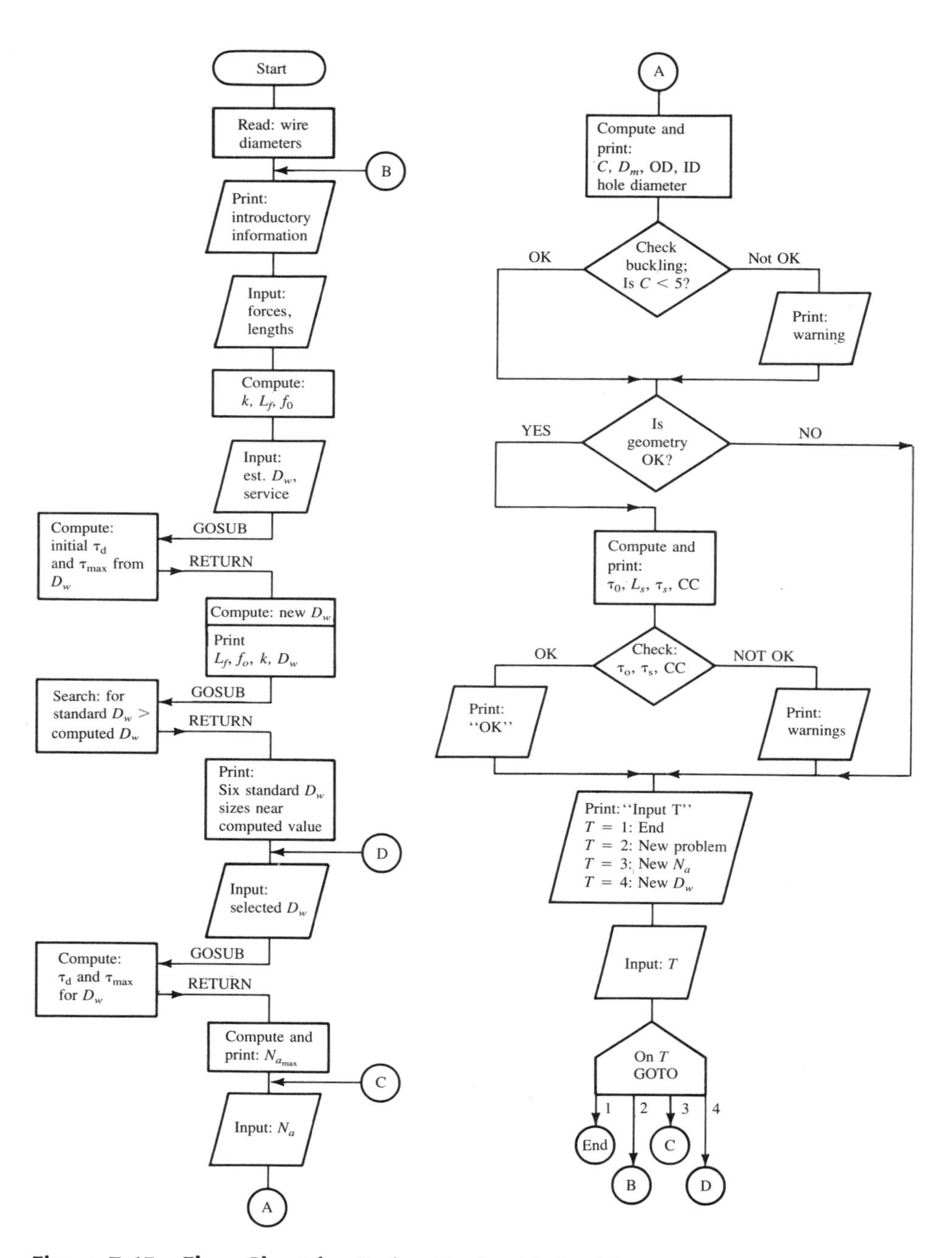

Figure 7-15 Flow Chart for Spring Design Method 2

Helical Spring Design (Method 2): Program Listing

```
10 DIM D(35)
20 FOR I = 1 TO 32
30 READ D(I)
40 NEXT I
50 DATA .0204, .0230, .0258, .0286, .0317, .0348, .0410
55 DATA .0475, .0540, .0625, .0720, .0800, .0915, .1055
60 DATA .1205, .1350, .1483, .1620, .1770, .1920, .2070
70 DATA .2253, .2437, .2625, .2830, .3065, .3310, .3625
80 DATA .3938, .4305, .4615, .4900
```

Read standard wire sizes into memory

```
90 PRINT "PROGRAM TO ASSIST IN THE DESIGN OF"
100 PRINT "   HELICAL COMPRESSION SPRINGS"
110 PRINT "   MADE FROM ASTM A231 WIRE"
120 PRINT
130 PRINT "SQUARED AND GROUND ENDS ASSUMED"
140 REM : FOR A231 WIRE, G = 11,200,000 PSI
150 G = 1.12E + 07
160 PRINT
170 PRINT "INPUT DATA REQUIRED ARE:"
180 PRINT "   MAXIMUM OPERATING FORCE, FO, LB."
190 PRINT "   OPERATING LENGTH, LO, INCHES"
200 PRINT "   INSTALLED FORCE, FI, LB."
210 PRINT "   INSTALLED LENGTH, LI, INCHES"
220 PRINT
230 PRINT "NOTE: INSTALLED LENGTH CAN BE FREE LENGTH -"
240 PRINT "         THEN INSTALLED FORCE = ZERO"
250 PRINT
```

Introduction

```
260 PRINT
270 PRINT "NOW INPUT FO,LO,FI,LI"
▶ 280 INPUT FO,LO,F1,L1
290 REM : K = SPRING RATE
300 K = (FO - F1)/(L1 - LO)
310 REM : L2 = FREE LENGTH
320 L2 = L1 + F1/K
330 REM : DO = DEFLECTION AT OPERATING LENGTH
340 DO = L2 - LO
```

Input design data; compute spring rate and operating deflection

```
350 PRINT
360 PRINT "INPUT INITIAL ESTIMATE OF WIRE DIAMETER, INCHES"
▶ 370 INPUT D1
380 PRINT
390 PRINT "WHAT TYPE OF SERVICE?"
400 PRINT "TYPE 1 FOR LIGHT; 2 FOR AVERAGE; 3 FOR SEVERE"
▶ 410 INPUT S
```

Rough estimate for D_w; specify service class

```
420 PRINT
430 GOSUB 1620
```

Subroutine for allowable stresses

```
440 PRINT USING "INITIAL DESIGN STRESS = ######. PSI";S1
450 PRINT USING "MAX ALLOWABLE STRESS = ######. PSI";S4
```

Results from subroutine

```
460 REM : NEW D1 = COMPUTED TRIAL WIRE DIAMETER
470 D1 = SQR(21.4*FO/S1)
```

Compute trial D_w

```
480 PRINT
490 PRINT
500 PRINT "TRIAL RESULTS:"
510 PRINT
520 PRINT USING "FREE LENGTH = ##.#### INCHES";L2
530 PRINT USING "DEFLECTION AT OPERATING LENGTH = ##.####
   INCHES";DO
540 PRINT USING "SPRING RATE = ###.## LB/IN";K
550 PRINT
560 PRINT USING "TRIAL WIRE DIAMETER = #.#### INCH";D1
```

Trial results

```
570 GOSUB 1700
```

Subroutine: Search for standard wire size

```
580 PRINT
590 PRINT "NEAREST STANDARD WIRE DIAMETERS ARE:"
600 PRINT D(I - 1), D(I), D(I + 1)
610 PRINT D(I + 2), D(I + 3), D(I + 4)
```

Results from subroutine; operator selects wire size

```
620 PRINT
630 PRINT "INPUT SELECTED WIRE DIAMETER
640 INPUT D1
650 GOSUB 1620
660 N2 = (L0 - 2*D1)/D1
670 PRINT
680 PRINT USING "INPUT SELECTED NUMBER OF COILS LESS THAN
  ##.##";N2
690 INPUT N1
700 IF N1 > N2 THEN 680
710 REM : C = SPRING INDEX
720 C = (G*D1/(8*K*N1))^(1/3)
730 IF C > 5 THEN 760
740 PRINT USING "C = #.## - TOO SMALL";C
750 REM : D2 = MEAN DIAMETER
760 D2 = C*D1
770 REM : D3 = OUTSIDE DIAMETER
780 D3 = D2 + D1
790 REM : D4 = INSIDE DIAMETER
800 D4 = D2 - D1
810 PRINT
820 PRINT "TRIAL GEOMETRY:"
830 PRINT
840 PRINT USING "WIRE DIAMETER = #.#### INCH";D1
850 PRINT
860 PRINT USING "MEAN DIAMETER = ##.### INCHES";D2
870 PRINT USING "OUTSIDE DIAMETER = ##.### INCHES";D3
880 PRINT USING "INSIDE DIAMETER = ##.### INCHES";D4
890 PRINT
900 H = D3 + D1/2
910 PRINT USING "NOTE: HOLE DIAMETER SHOULD BE > ##.###
  INCHES";H
915 PRINT
920 PRINT USING "SPRING INDEX = ##.##";C
925 PRINT
930 PRINT USING "FREE LENGTH/MEAN DIAMETER = ##.##";L2/D2
940 IF L2/D2 > 5.4 THEN 970
950 PRINT "BUCKLING SHOULD NOT BE A PROBLEM"
960 GOTO 1030
970 PRINT " * * * BUCKLING MAY BE A PROBLEM * * *"
980 PRINT USING " RATIO: OPERATING DEFLECTION/FREE LENGTH =
  #.###";D0/L2
990 R = 14.2748/(L2/D2)^2.03558
1000 PRINT USING "    MAX RATIO = #.###";R
1010 PRINT "    TRY TO INCREASE THE MEAN DIAMETER"
1020 PRINT "       OR DECREASE THE FREE LENGTH"
1030 PRINT
1040 PRINT USING "DESIGN STRESS = ######. PSI";S1
1050 PRINT USING "MAX ALLOWABLE STRESS = ######. PSI";S4
1060 PRINT
1070 PRINT "IS GEOMETRY OK? YES/NO"
1080 INPUT A$
1090 IF A$ <> "YES" THEN 1540
1100 PRINT
1110 PRINT "GEOMETRY OK! CHECK STRESSES"
1120 PRINT
1130 REM : K1 = WAHL FACTOR
1140 K1 = (4*C - 1)/(4*C - 4) + .615/C
1150 REM : S2 = ACTUAL EXPECTED STRESS AT OPERATING FORCE
1160 S2 = 2.546*K1*C*F0/(D1^2)
1170 REM : L3 = SOLID LENGTH
1180 L3 = D1*(N1 + 2)
1190 REM : D5 = DEFLECTION AT SOLID LENGTH
1200 D5 = L2 - L3
1210 REM : F2 = FORCE AT SOLID LENGTH
1220 F2 = K*D5
```

Subroutine for allowable stresses

Number of coils

Compute geometry factors

Check buckling

Check geometry

Evaluate operating stress and stress at solid length

```
1230 REM : S3 = STRESS AT SOLID LENGTH
1240 S3 = S2*F2/F0
1250 PRINT USING "STRESS DUE TO OPERATING FORCE = ######.
  PSI";S2
1260 IF S2 > S1 THEN 1290
1270 PRINT "OPERATING STRESS OK"
1280 GOTO 1310
1290 PRINT " * * * OPERATING STRESS TOO HIGH * * *"
1300 PRINT USING "        DESIGN STRESS = ######. PSI";S1
1310 PRINT
1320 PRINT USING "SOLID LENGTH = ##.### INCHES";L3
1330 PRINT USING "STRESS AT SOLID LENGTH = ######. PSI";S3
1340 IF S3 > S4 THEN 1380
1350 PRINT "SOLID LENGTH STRESS OK"
1360 PRINT
1370 GOTO 1430
1380 PRINT " * * * SOLID LENGTH STRESS TOO HIGH * * *"
1390 PRINT USING "        MAX SOLID STRESS = ######. PSI";S4
1400 PRINT "PERMANENT SET WILL OCCUR AT SOLID LENGTH"
1410 PRINT
1420 REM : C1 = COIL CLEARANCE AT OPERATING LENGTH
1430 C1 = (L0 - L3)/N1
1440 REM : C2 = RECOMMENDED COIL CLEARANCE
1450 C2 = .1*D1
1460 IF C1 < C2 THEN 1500
1470 PRINT USING "COIL CLEARANCE = #.#### INCH - OK -";C1
1480 PRINT
1490 GOTO 1540
1500 PRINT USING "COIL CLEARANCE = #.#### INCH: 0.1*D1 =
  #.#### INCH";C1,C2
1510 PRINT "DECREASE NUMBER OF COILS OR INCREASE WIRE
  DIAMETER"
1520 PRINT "   TO INCREASE COIL CLEARANCE"
1530 PRINT
1540 PRINT "TYPE  1  TO END"
1550 PRINT "TYPE  2  TO RUN A NEW PROBLEM"
1560 PRINT "TYPE  3  TO CHANGE THE NUMBER OF COILS"
1570 PRINT "TYPE  4  TO CHANGE THE WIRE DIAMETER"
1580 INPUT T
1590 ON T GOTO 1610,90,680,580
1600 GOTO 1540
1610 END
1620 S4 = 110843K/D1^(.1486)
1630 ON S GOTO 1640,1660,1680
1640 S1 = S4
1650 RETURN
1660 S1 = 94487K/D1^(.1484)
1670 RETURN
1680 S1 = 75942K/D1^(.1532)
1690 RETURN
1700 FOR I = 1 TO 32
1710 IF D(I) > D1 THEN 1730
1720 NEXT I
1730 RETURN
```

Evaluate coil clearance

Decide how to continue

Subroutine: Compute allowable stresses for given D_w

Subroutine: Search for standard D_w greater than computed trial D_w

force/length/stress requirements of a given problem. But the geometry of the spring is not tightly controlled; it is an output from the procedure, rather than a factor to be specified at the start, as in method 1.

As written, the program is limited to the design of springs from ASTM A231 steel wire and to springs with squared and ground ends. However, other materials and other end conditions can be added by the reader, using the techniques shown in the program. The following is a discussion of the various parts of the program.

Lines 10–80. The standard wire diameters from the U.S. Steel Wire Gage system are read into memory, each being identified by a subscripted variable. Later, the list will be searched to prompt the designer on the appropriate wire sizes.

Lines 90–260. General introductory information.

Lines 270–350. The designer types in the basic load/length data desired for the spring. From this, the spring rate, free length, and operating deflection from free length to operating length are computed.

Lines 360–420. The process requires an initial estimate of the design stress in the wire. However, the designer need not look up any data. From the magnitude of the forces required from the spring, a very rough estimate is input for the wire size that the designer judges will handle the loads.

Lines 430–490 plus Subroutine at Lines 1620–1690. The subroutine is a feature of the program that precludes the need for the designer to consult a chart or table of data. The formulas at lines 1620, 1660, and 1680 compute the design stress as a function of wire diameter for light, average, or severe service. They are curve-fit equations for the data shown in Figure 7-9 for the ASTM A231 steel wire. The equations compute the stress levels within about 2 percent of what can be read from the graphs directly. The subroutine also selects the appropriate design stress, depending on the value of the variable S, which the designer chose earlier in line 410. In any case, the stress called *S4* is from the light service curve and is used as the maximum allowable stress.

Upon return from the subroutine, the design stress and the maximum allowable stress are printed for the designer to see, and the trial wire diameter is computed at line 470. This formula is derived from equation (7-4) in section 7-3.

$$\tau = \frac{8KFC}{\pi D_w^2} \qquad (7\text{-}4)$$

Letting $F = F_o$ and $\tau = \tau_d$ (the design stress), and solving for the wire diameter gives

$$D_w = \sqrt{\frac{8KF_oC}{\pi\tau_d}} \qquad (7\text{-}8)$$

The values of K and C are not yet known. But a good estimate for the wire diameter can be computed if the spring index is assumed to be approximately 7.0, a reasonable value. The corresponding value for the Wahl factor is $K = 1.2$, from equation (7-5). Combining these assumed values with the other constants in the above equation gives

$$D_w = \sqrt{21.4(F_o)/(\tau_d)} \qquad (7\text{-}9)$$

This is the formula in line 470 of the program. It is a reasonable estimate of the required wire size to limit the stress to the design value.

Lines 480–560. The trial data are printed.

Lines 570–620 plus Subroutine at Lines 1700–1730. The list of standard wire sizes is searched to find the size just larger than the computed trial wire diameter. In order to give the designer some flexibility in the selection of the wire size, lines 590–610 list several sizes near the value found from the search.

Lines 630–650. The designer selects the wire size to be used in subsequent analyses. Then the program uses the subroutine at line 1620 again to compute the design stress and maximum allowable stress for this size of wire.

Lines 660–700. Several geometry features are computed, starting with line 660, where the maximum permissible number of active coils for the spring is computed. The logic here is that *the solid length must be less than the operating length*. The solid length, for squared and ground ends, is

$$L_s = D_w(N_a + 2)$$

Solving for the number of coils and letting $L_s = L_o$ as a limit,

$$(N_a)_{\max} = (L_o - 2D_w)/D_w \tag{7-10}$$

This is line 660. The result is printed at line 680 and the designer is asked to select a value less than the maximum. The smaller the selected value of N_a, the more clearance there will be between adjacent coils, but the higher the stresses will be. These factors will be checked later, but the designer should be aware of the effects of the decision for the number of coils.

Lines 710–920. Everything in equation (7-6) relating to spring deflection is now known except the spring index, C. Thus we can solve for C:

$$C = \left[\frac{GD_w}{8kN_a}\right]^{(1/3)} \tag{7-11}$$

This is line 720. The mean diameter, outside diameter, inside diameter, and recommended hole diameter to contain the spring are then computed and printed.

Lines 930–1030. The tendency for the spring to buckle is checked automatically, using the curve in Figure 7-13 (review the discussion of buckling in section 7-3). For squared and ground ends, when $L_f/D_m < 5.4$ (approximately), there is no tendency to buckle. At larger values of this ratio, the critical ratio of operating deflection to free length, at which buckling would be expected, is found from the curve. In the program, line 990 computes this critical ratio from an equation, fitted to the data in Figure 7-13.

Lines 1070–1100 plus Lines 1540–1610. The designer is asked to evaluate the design thus far on the basis of geometry only. There is no need to complete the stress analysis if the spring has an unsatisfactory geometry. Four options are presented in lines 1540–1570, with the appropriate branching logic following in lines 1580–1610. If the geometry is satisfactory ("yes" in line 1080), the stress analysis is begun.

Lines 1110–1410. The actual expected stress in the spring due to the operating force and the force at solid length are computed and compared with the allowable values. Messages tell the designer whether or not the stresses are acceptable.

Lines 1420–1530. The coil clearance at the operating height is checked.

Lines 1540–1610. Again, the designer is permitted to take one of four actions. Choosing option 3 or 4 starts a new spring design for the same conditions of force and length. In this way, several candidate designs can be generated quickly. Then an optimum design can be selected.

The following is a sample output from this program, using the data from example problem 7-1. The first trial results in a long slender spring, which would tend to buckle. The second trial uses a larger wire diameter, which produces an acceptable design on all counts: geometry, operating stress, solid length stress, buckling, and coil clearance.

Helical Spring Design (Method 2): Sample Output

```
PROGRAM TO ASSIST IN THE DESIGN OF
  HELICAL COMPRESSION SPRINGS
  MADE FROM ASTM A231 WIRE

SQUARED AND GROUND ENDS ASSUMED

INPUT DATA REQUIRED ARE:
  MAXIMUM OPERATING FORCE, FO, LB.
  OPERATING LENGTH, LO, INCHES
  INSTALLED FORCE, FI, LB.
  INSTALLED LENGTH, LI, INCHES

NOTE: INSTALLED LENGTH CAN BE FREE LENGTH -
      THEN INSTALLED FORCE = ZERO

  NOW INPUT FO,LO,FI,LI
► ? 12 , 1.25 , 8 , 1.75                              Statement 280

  INPUT INITIAL ESTIMATE OF WIRE DIAMETER, INCHES
► ? .06                                               Statement 370

  WHAT TYPE OF SERVICE?
  TYPE 1 FOR LIGHT; 2 FOR AVERAGE; 3 FOR SEVERE
► ? 2                                                 Statement 410

  INITIAL DESIGN STRESS = 143448. PSI
  MAX ALLOWABLE STRESS = 168374. PSI

  TRIAL RESULTS:                                      Preliminary analysis

  FREE LENGTH = 2.7500 INCHES
  DEFLECTION AT OPERATING LENGTH = 1.5000 INCHES
  SPRING RATE = 8.00 LB/IN

  TRIAL WIRE DIAMETER = 0.0423 INCH

  NEAREST STANDARD WIRE DIAMETERS ARE:
    .041        .0475       .054
    .0625       .072        .08

  INPUT SELECTED WIRE DIAMETER
► ? .0475                                             Statement 640

  INPUT SELECTED NUMBER OF COILS LESS THAN 24.32
► ? 22                                                Statement 690

  TRIAL GEOMETRY:

  WIRE DIAMETER = 0.0475 INCH

  MEAN DIAMETER = 0.343 INCHES
  OUTSIDE DIAMETER = 0.391 INCHES
  INSIDE DIAMETER = 0.296 INCHES

  NOTE: HOLE DIAMETER SHOULD BE > 0.415 INCHES
```

Lines in the program calling for input of data and lines in the output that correspond to the input statements are highlighted with a ►.

```
SPRING INDEX = 7.23                                          Initial trial
                                                             unsuccessful!
FREE LENGTH/MEAN DIAMETER = 8.01
  * * * BUCKLING MAY BE A PROBLEM * * *
    RATIO: OPERATING DEFLECTION/FREE LENGTH = 0.545
    MAX RATIO = 0.207
    TRY TO INCREASE THE MEAN DIAMETER
      OR DECREASE THE FREE LENGTH

DESIGN STRESS = 148508. PSI
MAX ALLOWABLE STRESS = 174322. PSI

IS GEOMETRY OK? YES/NO                                       Statement 1080
? NO
TYPE  1  TO END
TYPE  2  TO RUN A NEW PROBLEM
TYPE  3  TO CHANGE THE NUMBER OF COILS
TYPE  4  TO CHANGE THE WIRE DIAMETER
? 4                                                          Statement 1580

NEAREST STANDARD WIRE DIAMETERS ARE:                         Begin second trial
  .041        .0475       .054                               with a new wire
  .0625       .072        .08                                size

INPUT SELECTED WIRE DIAMETER
? .0625                                                      Statement 640

INPUT SELECTED NUMBER OF COILS LESS THAN 18.00
? 16                                                         Statement 690

TRIAL GEOMETRY:

WIRE DIAMETER = 0.0625 INCH

MEAN DIAMETER = 0.551 INCHES
OUTSIDE DIAMETER = 0.613 INCHES
INSIDE DIAMETER = 0.488 INCHES

NOTE: HOLE DIAMETER SHOULD BE > 0.644 INCHES

SPRING INDEX = 8.81

FREE LENGTH/MEAN DIAMETER = 4.99
BUCKLING SHOULD NOT BE A PROBLEM

DESIGN STRESS = 142582. PSI                                  Second trial —
MAX ALLOWABLE STRESS = 167356. PSI                           successful!

IS GEOMETRY OK? YES/NO
? YES                                                        Statement 1080

GEOMETRY OK! CHECK STRESSES

STRESS DUE TO OPERATING FORCE = 80326. PSI
OPERATING STRESS OK

SOLID LENGTH = 1.125 INCHES
STRESS AT SOLID LENGTH = 87020. PSI
SOLID LENGTH STRESS OK

COIL CLEARANCE = 0.0078 INCH - OK -

TYPE  1  TO END
TYPE  2  TO RUN A NEW PROBLEM
TYPE  3  TO CHANGE THE NUMBER OF COILS
TYPE  4  TO CHANGE THE WIRE DIAMETER
? 1                                                          Statement 1580
```

7-5 EXTENSION SPRINGS

Extension springs are designed to exert a pulling force and to store energy. They are made from closely coiled helical coils similar in appearance to helical compression springs. Most extension springs are made with adjacent coils touching in such a manner that an initial force must be applied to separate the coils. Once separated, the force is linearly proportional to the deflection, as it is for helical compression springs. Figure 7-16 shows a typical extension spring, and Figure 7-17 shows the characteristic type of load-deflection curve. By convention, the initial force is found by projecting the straight-line portion of the curve back to zero deflection.

The stresses and deflections for an extension spring can be computed by using the formulas used for compression springs. Equation (7-4) is used for the torsional shear stress, equation (7-5) for the Wahl factor to account for the curvature of the wire and the direct shear stress, and equation (7-6) for the deflection characteristics. All coils in an extension spring are active. In addition, since the end loops or hooks deflect, their deflection may affect the actual spring rate.

The initial tension in an extension spring is typically 10–25 percent of the maximum design force. Figure 7-18 shows one manufacturer's recommendation of the preferred torsional stress due to initial tension as a function of the spring index.

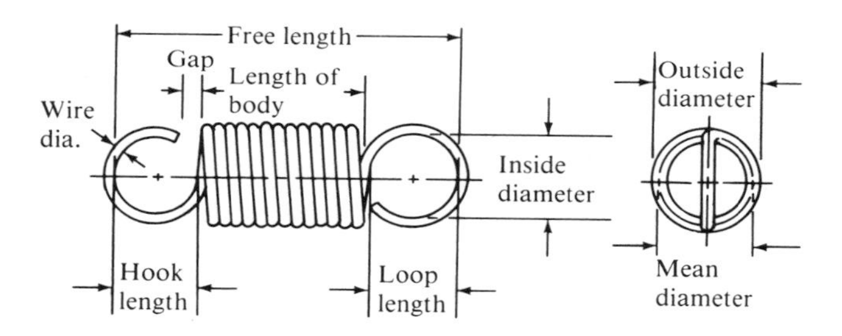

Figure 7-16 Extension Spring

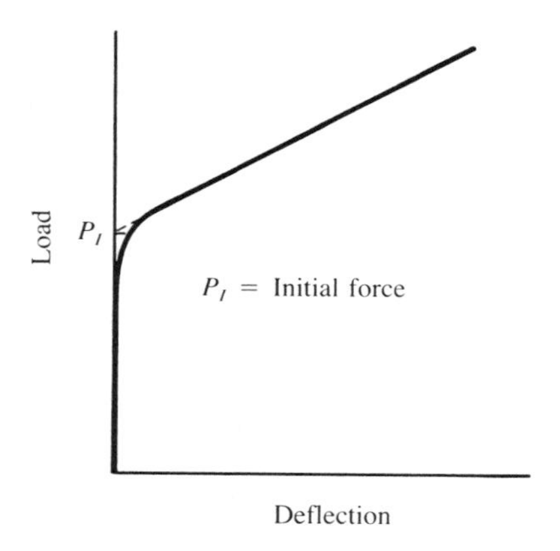

Figure 7-17 Load-deflection Curve for an Extension Spring

End Configurations for Extension Springs

A wide variety of end configurations may be obtained for attaching the spring to mating machine elements, some of which are shown in Figure 7-19. The cost of the spring can be greatly affected by its end type, and it is recommended that the manufacturer be consulted before specifying ends.

Frequently the weakest part of an extension spring is its end, especially in fatigue loading cases. The loop end shown in Figure 7-20, for example, has a high bending stress at point A and a torsional shear stress at point B. Approximations for the stresses at these points can be computed as follows:

Bending stress at A:

$$\sigma_A = \frac{16D_m F_o K_1}{\pi D_w^3} + \frac{4F_o}{\pi D_w^2} \tag{7-12}$$

$$K_1 = \frac{4C_1^2 - C_1 - 1}{4C_1(C_1 - 1)} \tag{7-13}$$

$$C_1 = 2R_1/D_w$$

Torsional stress at B:

$$\tau_B = \frac{8D_m F_o K_2}{\pi D_w^3} \tag{7-14}$$

$$K_2 = \frac{4C_2 - 1}{4C_2 - 4} \tag{7-15}$$

$$C_2 = 2R_2/D_w$$

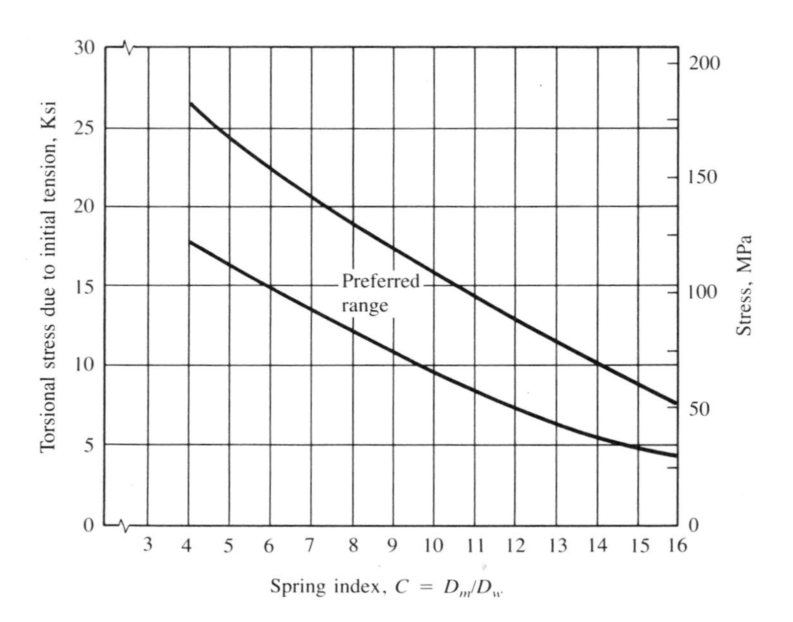

Figure 7-18 Recommended Torsional Shear Stress Due to Initial Tension (Data from Associated Spring, Barnes Group, Inc.)

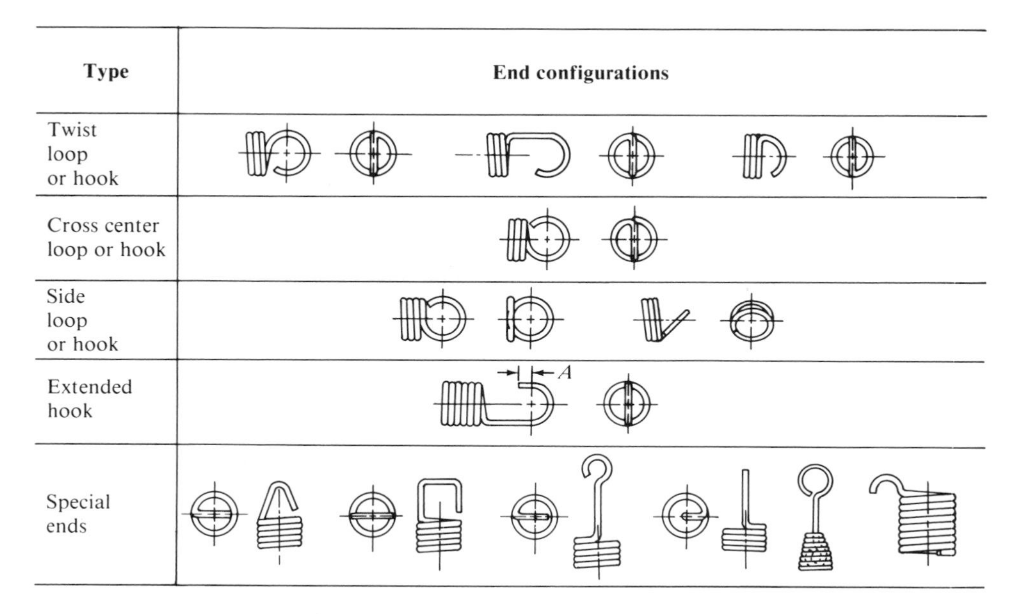

Figure 7-19 End Configurations for Extension Springs

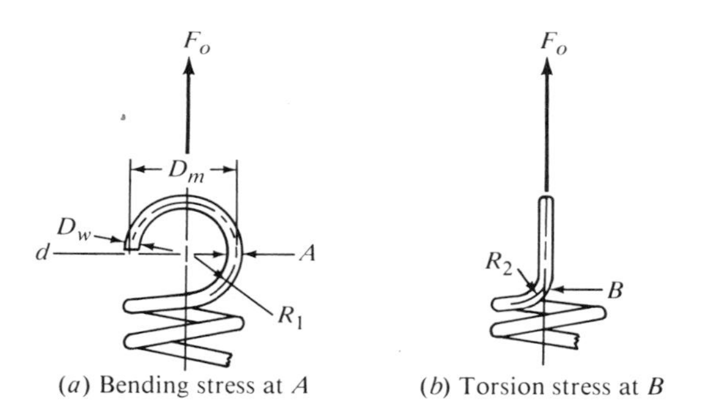

Figure 7-20 Stresses in Ends of Extension Spring

The ratios C_1 and C_2 relate to the curvature of the wire and should be large, typically greater than 4, to avoid high stresses.

Allowable Stresses for Extension Springs

The torsional shear stress in the coils of the spring and in the end loops can be compared to the curves of Figures 7-6 to 7-11. Some designers reduce these allowable stresses by about 10 percent. The bending stress in the end loops, such as that from equation (7-12), should be compared to the allowable bending stresses for torsion springs, as presented in the next section.

Example Problem 7-2. A helical extension spring is to be installed in a latch for a large commercial laundry machine. When the latch is closed, the spring must exert a force of 16.25 lb at a length between attachment points of 3.50 in. As the latch is opened, the spring is pulled to a length of 4.25 in with a maximum force of 26.75 lb. An outside diameter of ⅝ in (0.625 in) is desired. The latch will be cycled only about 10 times per day, and thus the design stress will be based on average service. Use ASTM A227 steel wire.

Solution. The suggested design procedure will be given as numbered steps, followed by the calculations for this set of data.

Step 1. Assume a trial mean diameter and a trial design stress for the spring.
Let the mean diameter be 0.500 in. For ASTM A227 wire under average service, a design stress of 110 000 psi is reasonable (from Figure 7-6).

Step 2. Compute a trial wire diameter from equation (7-4) for the maximum operating force, the assumed mean diameter and design stress, and an assumed value for K of about 1.20.

$$D_w = \left[\frac{8KF_oD_m}{\pi\tau_d}\right]^{1/3} = \left[\frac{8(1.20)(26.75)(0.50)}{(\pi)(110\,000)}\right]^{1/3} = 0.072 \text{ in}$$

A standard wire size in the U.S. Steel Wire Gage system is available in this size. Use 15-gage.

Step 3. Determine the actual design stress for the selected wire size.
From Figure 7-6, at a wire size of 0.072 in, the design stress is 120 000 psi.

Step 4. Compute the actual values for outside diameter, mean diameter, inside diameter, spring index, and the Wahl factor, K. These factors are the same as those defined for helical compression springs.
Let the outside diameter be as specified, 0.625 in.

$$D_m = OD - D_w = 0.625 - 0.072 = 0.553 \text{ in}$$

$$ID = OD - 2D_w = 0.625 - 2(0.072) = 0.481 \text{ in}$$

$$C = D_m/D_w = 0.553/0.072 = 7.68$$

$$K = \frac{4C-1}{4C-4} + \frac{0.615}{C} = \frac{4(7.68)-1}{4(7.68)-4} + \frac{0.615}{7.68} = 1.19$$

Step 5. Compute the actual expected stress in the spring wire under the operating load from equation (7-4).

$$\tau_o = \frac{8KF_oD_m}{\pi D_w^3} = \frac{8(1.19)(26.75)(0.553)}{\pi(0.072)^3} = 120\,000 \text{ psi} \quad \text{(OK)}$$

Step 6. Compute the required number of coils to produce the desired deflection characteristics. Solve equation (7-6) for the number of coils and substitute k = force/deflection = F/f.

$$k = \frac{26.75 - 16.25}{4.25 - 3.50} = 14.0 \text{ lb/in}$$

$$N_a = \frac{GD_w}{8C^3k} = \frac{(11.5 \times 10^6)(0.072)}{8(7.68)^3(14.0)} = 16.3 \text{ coils}$$

Step 7. Compute the body length for the spring and propose a trial design for the ends.

$$\text{Body length} = D_w(N_a + 1) = (0.072)(16.3 + 1) = 1.25 \text{ in}$$

Let's propose to use a full loop at each end of the spring, adding a length equal to the ID of the spring at each end. Then the total free length is

$$L_f = \text{body length} + 2(ID) = 1.25 + 2(0.481) = 2.21 \text{ in}$$

Step 8. Compute the deflection from free length to operating length.

$$f_o = L_o - L_f = 4.25 - 2.21 = 2.04 \text{ in}$$

Step 9. Compute the initial force in the spring at which the coils just begin to separate. This is done by subtracting the amount of force due to the deflection, f_o.

$$F_I = F_o - kf_o = 26.75 - (14.0)(2.04) = -1.81 \text{ lb}$$

The negative force resulting from a free length that is too small for the specified conditions, is clearly impossible.

Let's try $L_f = 2.50$ in, which will require a redesign of the end loops. Then,

$$f_o = 4.25 - 2.50 = 1.75 \text{ in}$$

$$F_I = 26.75 - (14.0)(1.75) = 2.25 \text{ lb} \quad \text{(Reasonable)}$$

Step 10. Compute the stress in the spring under the initial tension and compare with the recommended levels in Figure 7-18.

Because the stress is proportional to the load,

$$\tau_I = \tau_o(F_I/F_o) = (120\,000)(2.25/26.75) = 10\,100 \text{ psi}$$

For $C = 7.68$, this stress is in the preferred range from Figure 7-18.

At this point, the coiled portion of the spring is satisfactory. The final configuration of the end loops must be completed and analyzed for stress.

7-6 HELICAL TORSION SPRINGS

Many machine elements require a spring that exerts a rotational moment, or torque, instead of a push or pull force. The helical torsion spring is designed to satisfy this requirement. The spring has the same general appearance as either the helical compression or extension spring, with round wire wrapped into a cylindrical shape. Usually the coils are close together but with a small clearance allowing no initial tension in the spring, as there is for extension springs. Figure 7-21 shows a few examples of torsion springs with a variety of end treatments.

The familiar clip type of clothes pin uses a torsion spring to provide the gripping force. Many cabinet doors are designed to close automatically under the influence of a torsion spring. Some timers and switches use torsion springs to actuate mechanisms or close contacts. Torsion springs often provide counterbalancing of machine elements that are mounted to a hinged plate.

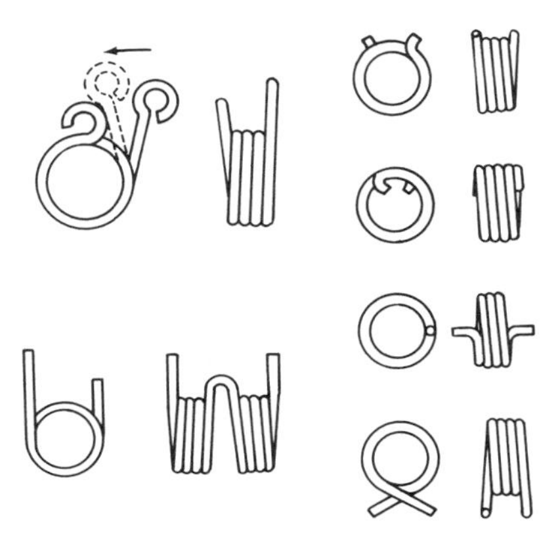

Figure 7-21 Torsion Springs Showing a Variety of End Types

The following are some of the special features and design guides for torsion springs:

1. The moment applied to a torsion spring should always act in a direction that causes the coils to wind tighter, rather than to open the spring up. This takes advantage of favorable residual stresses in the wire after forming.
2. In the free condition (no load) the definition of mean diameter, outside diameter, inside diameter, wire diameter, and spring index are the same as those used for compression springs.
3. As the load on a torsion spring increases, its mean diameter, D_m, decreases, and its length, L, increases, according to the following relations:

$$D_m = D_{ml} N_a/(N_a + \theta) \tag{7-16}$$

where D_{ml} is the initial mean diameter at the free condition, and N_a is the number of active coils in the spring (to be defined later). The term θ is the angular deflection of the spring from the free condition, expressed in revolutions or a fraction of a revolution.

$$L = D_w(N_a + 1 + \theta) \tag{7-17}$$

This equation assumes that all coils are touching. If any clearance is provided, as is often desirable to reduce friction, a length of N_a times the clearance must be added.

4. Torsion springs must be supported at three or more points. They are usually installed around a rod to provide location and to transfer reaction forces to the structure. The rod diameter should be approximately 90 percent of the ID of the spring at the maximum load.

Stress Calculations

The stress in the coils of a helical torsion spring is *bending stress* created as the applied moment tends to bend each coil into a smaller diameter. Thus the stress is computed from a form of the flexure formula, $\sigma = Mc/I$, modified to account for the curved wire.

Also, because most torsion springs are made from round wire, the section modulus I/c is $Z = \pi D_w^3/32$. Then,

$$\sigma = \frac{McK_b}{I} = \frac{MK_b}{Z} = \frac{MK_b}{\pi D_w^3/32} = \frac{32MK_b}{\pi D_w^3} \tag{7-18}$$

K_b is the curvature correction factor and is reported by Wahl (7) to be

$$K_b = \frac{4C^2 - C - 1}{4C(C - 1)} \tag{7-19}$$

where C is the spring index.

Deflection, Spring Rate, and Number of Coils

The basic equation governing deflection is

$$\theta' = ML_w/EI$$

where θ' is the angular deformation of the spring in radians (rad); M is the applied moment, or torque; L_w is the length of wire in the spring; E is the tensile modulus of elasticity; and I is the moment of inertia of the spring wire. We can substitute equations for L_w and I and convert θ' (radians) to θ (revolutions) to produce a more convenient form for application to torsion springs.

$$\theta = \frac{ML_w}{EI} = \frac{M(\pi D_m N_a)}{E(\pi D_w^4/64)} \frac{1 \text{ rev}}{2\pi \text{ rad}} = \frac{10.2MD_m N_a}{ED_w^4} \tag{7-20}$$

To compute the spring rate, k_θ (moment per revolution), solve for M/θ.

$$k_\theta = \frac{M}{\theta} = \frac{ED_w^4}{10.2D_m N_a} \tag{7-21}$$

Friction between coils and between the ID of the spring and the guide rod may decrease the rate slightly from this value.

The number of coils, N_a, is composed of a combination of the number of coils in the body of the spring, called N_b, and the contribution of the ends as they are subjected to bending, also. Calling N_e the contribution of the ends having lengths L_1 and L_2,

$$N_e = (L_1 + L_2)/(3\pi D_m) \tag{7-22}$$

Then compute $N_a = N_b + N_e$.

Design Stresses

Because the stress in a torsion spring is bending and not torsional shear, the design stresses are different from those used for compression and extension springs. Figures 7-22 through 7-27 show graphs of the design stress versus wire diameter for the same alloys used earlier for compression springs.

Design Procedure and Example Problem 7-3. A general procedure for designing torsion springs is presented and illustrated with an example next.

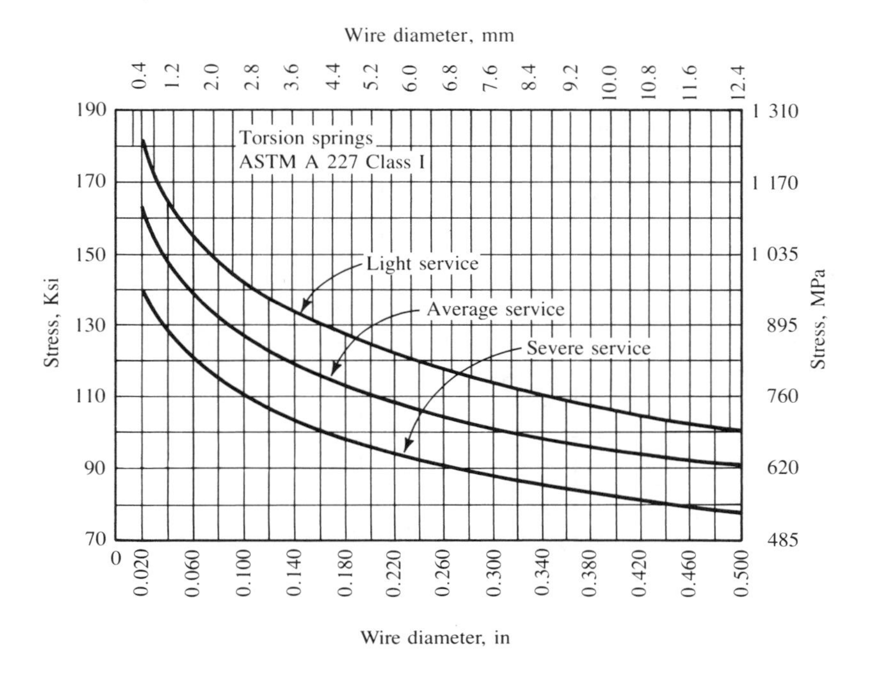

Figure 7-22 Design Stresses for Torsion Springs ASTM A227 Steel Wire, Hard Drawn (Reprinted from Carlson, Harold. *Spring Designer's Handbook,* p. 144, by courtesy of Marcel Dekker, Inc.)

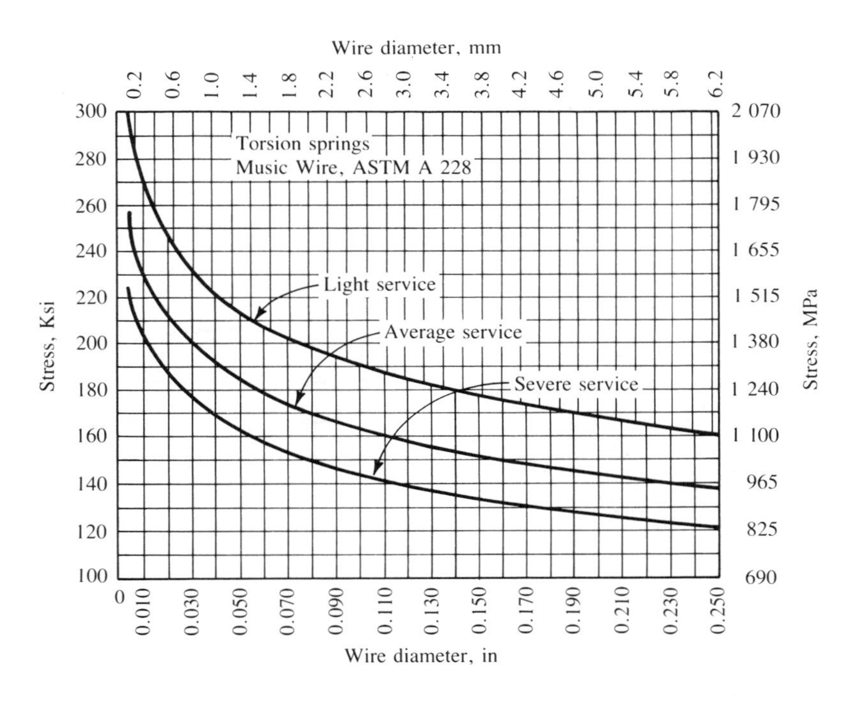

Figure 7-23 Design Stresses for Torsion Springs ASTM A228 Music Wire (Reprinted from Carlson, Harold. *Spring Designer's Handbook,* p. 143, by courtesy of Marcel Dekker, Inc.)

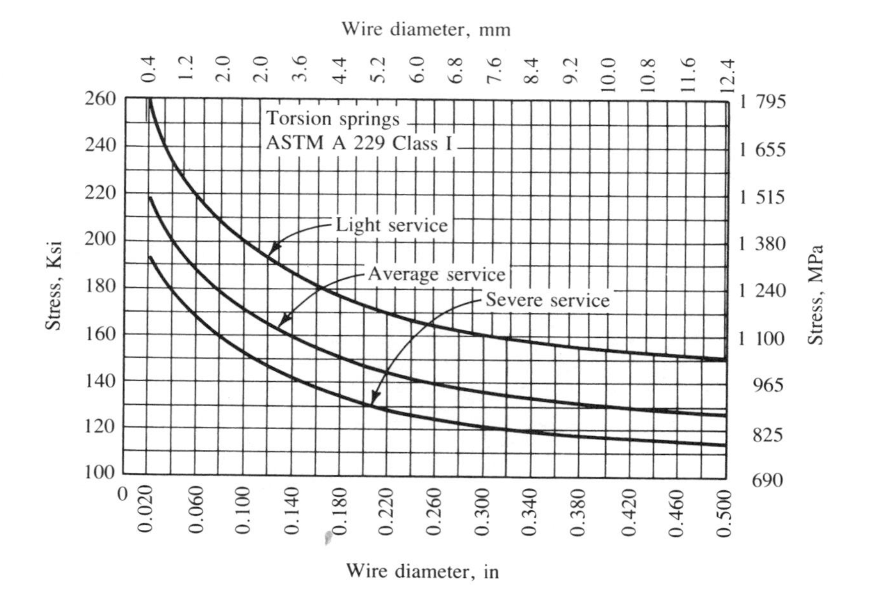

Figure 7-24 Design Stresses for Torsion Springs ASTM A229 Steel Wire, Oil-tempered, MB grade (Reprinted from Carlson, Harold. *Spring Designer's Handbook,* p. 146, by courtesy of Marcel Dekker, Inc.)

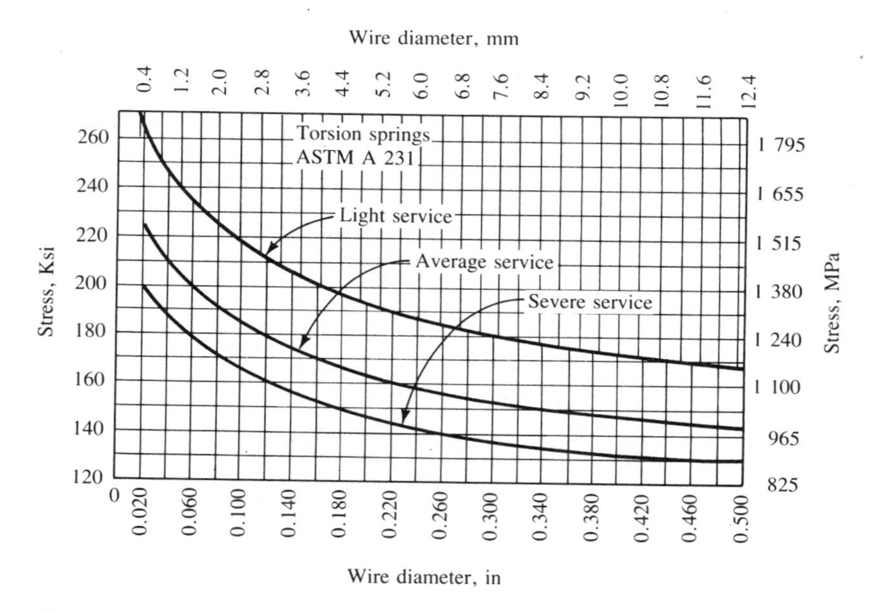

Figure 7-25 Design Stresses for Torsion Springs ASTM A231 Steel Wire, Chromium-vanadium Alloy, Valve Spring Quality (Reprinted from Carlson, Harold. *Spring Designer's Handbook,* p. 147, by courtesy of Marcel Dekker, Inc.)

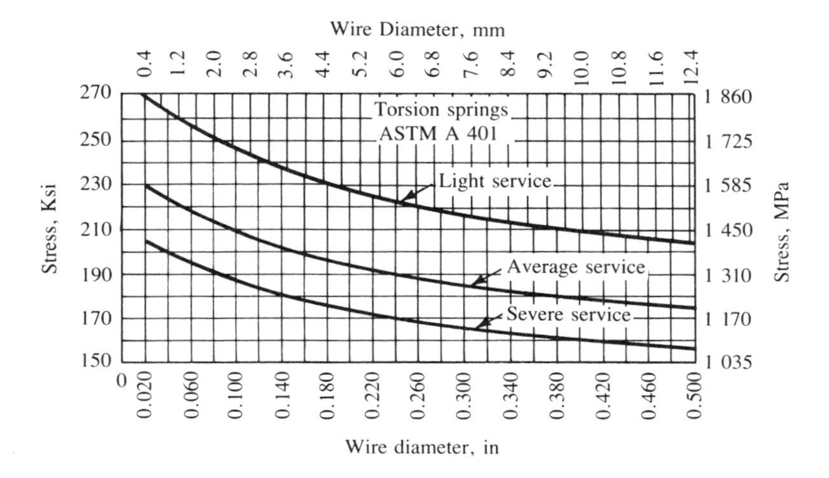

Figure 7-26 Design Stresses for Torsion Springs ASTM A401 Steel Wire, Chromium-silicon Alloy, Oil-tempered (Reprinted from Carlson, Harold. *Spring Designer's Handbook,* p. 148, by courtesy of Marcel Dekker, Inc.)

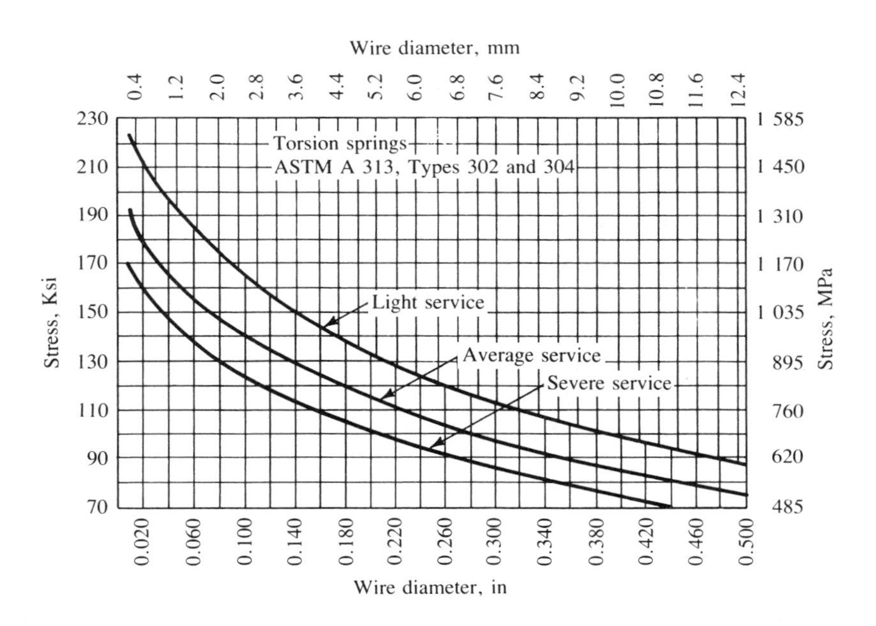

Figure 7-27 Design Stresses for Torsion Springs ASTM A313 Stainless Steel Wire, Types 302 and 304, Corrosion-resistant (Reprinted from Carlson, Harold. *Spring Designer's Handbook,* p. 150, by courtesy of Marcel Dekker, Inc.)

A timer incorporates a mechanism to close a switch after the timer rotates one complete revolution. The switch contacts are actuated by a torsion spring that is to be designed. A cam on the timer shaft slowly moves a lever attached to one end of the spring to a point where the maximum torque on the spring is 3.00 lb · in. At the end of the revolution, the cam permits the lever to rotate 60° suddenly with the motion produced by the energy stored in the spring. At this new position, the torque on the spring is 1.60 lb · in. Because of space limitations, the OD of the spring should not be greater than 0.50 in and the length should not be greater than 0.75 in. Use music wire, ASTM A228 steel wire. The number of cycles of the spring will be moderate, so use design stresses for average service.

Step 1. Assume a trial value for the mean diameter and an estimate for the design stress.

Let's use a mean diameter of 0.400 in and estimate the design stress for A228 music wire, average service, to be 180 000 psi (Figure 7-23).

Step 2. Solve equation (7-18) for the wire diameter, compute a trial size, and select a standard wire size. Let $K_b = 1.15$ as an estimate. Also use the largest applied torque.

$$D_w = \left[\frac{32MK_b}{\pi\sigma_d}\right]^{1/3} = \left[\frac{32(3.0)(1.15)}{\pi(180\,000)}\right]^{1/3} = 0.058 \text{ in}$$

From Table 7-2, we can choose 25-gage music wire with a diameter of 0.059 in. For this size wire, the actual design stress for average service is 178 000 psi.

Step 3. Compute the OD, ID, spring index, and the new K_b.

$$OD = D_m + D_w = 0.400 + 0.059 = 0.459 \text{ in} \quad \text{(OK)}$$

$$ID = D_m - D_w = 0.400 - 0.059 = 0.341 \text{ in}$$

$$C = D_m/D_w = 0.400/0.059 = 6.78$$

$$K_b = \frac{4C^2 - C - 1}{4C(C - 1)} = \frac{4(6.78)^2 - 6.78 - 1}{4(6.78)(6.78 - 1)} = 1.123$$

Step 4. Compute the actual expected stress from equation (7-18).

$$\sigma = \frac{32MK_b}{\pi D_w^3} = \frac{32(3.0)(1.123)}{(\pi)(0.059)^3} = 167\,000 \text{ psi} \quad \text{(OK)}$$

Step 5. Compute the spring rate from the given data.

The torque exerted by the spring decreases from 3.00 to 1.60 lb · in as the spring rotates 60°. Convert 60° to a fraction of a revolution (rev).

$$\theta = 60/360 = 0.167 \text{ rev}$$

$$k_\theta = \frac{M}{\theta} = \frac{3.00 - 1.60}{0.167} = 8.40 \text{ lb} \cdot \text{in/rev}$$

Step 6. Compute the required number of coils by solving for N_a from equation (7-21).

$$N_a = \frac{ED_w^4}{10.2D_m k_\theta} = \frac{(29 \times 10^6)(0.059)^4}{(10.2)(0.400)(8.40)} = 10.3 \text{ coils}$$

Step 7. Compute the equivalent number of coils due to the ends of the spring from equation (7-22).

This requires some design decisions. For the example problem, let's use straight ends, 2.0 in long on one side and 1.0 in long on the other. These ends will be attached to the structure of the timer during operation. Then,

$$N_e = (L_1 + L_2)/(3\pi D_m) = (2.0 + 1.0)/[3\pi(0.400)] = 0.80 \text{ coil}$$

Step 8. Compute the required number of coils in the body of the spring.

$$N_b = N_a - N_e = 10.3 - 0.8 = 9.5 \text{ coils}$$

Step 9. Complete the geometric design of the spring, including the size of the rod on which it will be mounted.

We first need the total angular deflection for the spring from the free condition to the maximum load. In this case, we know that the spring rotates 60° during operation. To this we must add the rotation from the free condition to the initial torque, 1.60 lb · in.

$$\theta_I = M_I/k_\theta = 1.60 \text{ lb} \cdot \text{in}/(8.4 \text{ lb} \cdot \text{in/rev}) = 0.19 \text{ rev}$$

Then the total rotation is

$$\theta_t = \theta_I + \theta_o = 0.19 + 0.167 = 0.357 \text{ rev}$$

From equation (7-16), the mean diameter at the maximum operating torque is

$$D_m = D_{mI} N_a/(N_a + \theta_t) = [(0.400)(10.3)]/[(10.3 + 0.357)] = 0.387 \text{ in}$$

The minimum inside diameter is

$$ID_{\min} = 0.387 - D_w = 0.387 - 0.059 = 0.328 \text{ in}$$

The rod diameter on which the spring is mounted should be approximately 0.90 times this value.

$$D_r = 0.9(0.328) = 0.295 \text{ in} \quad (\text{say } 0.30 \text{ in})$$

The length of the spring, assuming all coils are originally touching, is

$$L_{\max} = D_w(N_a + 1 + \theta_t) = (0.059)(10.3 + 1 + 0.356) = 0.688 \text{ in} \quad (\text{OK})$$

If the spring is made 0.75 in long, a small clearance will exist between adjacent coils.

7-7 OTHER TYPES OF SPRINGS

This chapter has discussed the most popular types of springs: helical compression, extension, and torsion springs made from round wire. However, many other types are available.

Square or rectangular wire can be used to make springs similar to those already discussed and usually at some increase in performance. However, the cost of such springs is typically greater, and the analysis is more difficult.

Compression and extension springs can be made in a wide variety of styles in addition to the familiar cylindrical shape with a uniform pitch of the coils. They can have a conical, barrel, or hourglass shape; have nonuniform pitch; or be connected together in series or parallel to create special performance characteristics (see Figure 7-28).

Flat spring materials, brass, bronze, steel, and others can be used to make *leaf springs* that can be analyzed as beams. These are frequently loaded as cantilevers or simple beams and sometimes are produced with an initial curvature.

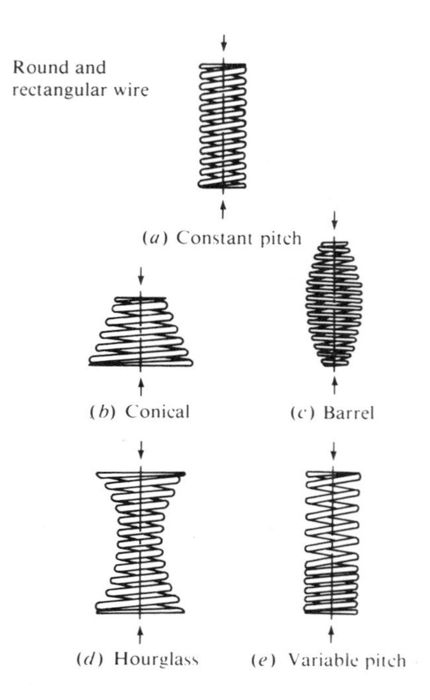

Figure 7-28 Variations of Helical Compression Springs. Each style can be made with round or rectangular wire.

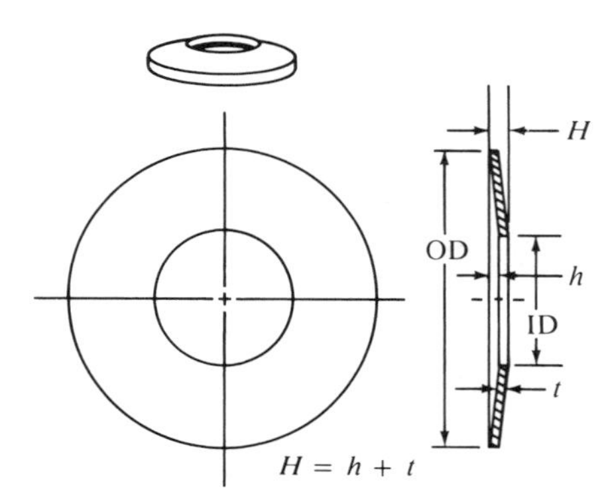

Figure 7-29 Belleville Spring

A *Belleville spring* has the shape of a shallow conical disk with a central hole, as shown in Figure 7-29. A very high spring rate or spring force can be developed in a very small axial space with such springs. A variety of load-deflection characteristics can be obtained by varying the height of the cone to the thickness of the disk.

A *torsion bar,* as its name implies, is a round bar loaded in torsion. The analysis is similar to that presented for circular shafts in which the torsional shear stress and the angular deformation are computed.

Power springs, sometimes referred to as *motor* or *clock springs,* are made from flat spring steel stock, wound in a spiral shape. The strip is placed in bending and a torque is exerted by the spring as it tends to unwrap the spiral.

Constant force springs are flat strips wound into a spiral and loaded in tension similar to extension springs. However, the force exerted by the spring at any amount of deflection is nearly the same (thus the name).

Garter springs are actually long helical springs, either extension or compression, formed into a circular shape with the ends joined. They exert a radial force nearly uniformly around the circumference of an object. One typical use is in lip seals, where the garter spring forces the seal into uniform intimate contact with a shaft.

REFERENCES

1. Associated Spring, Barnes Group, Inc. *Engineering Guide to Spring Design*. Bristol, Conn., 1981.
2. Carlson, Harold. *Spring Designer's Handbook*. New York: Marcel Dekker, 1978.
3. Carlson, Harold. *Springs—Troubleshooting and Failure Analysis*. New York: Marcel Dekker, 1980.
4. Faires, V. M. *Design of Machine Elements,* 4th ed. New York: Macmillan Publishing Company, 1965.
5. Oberg, E., et al. *Machinery's Handbook,* 22d ed. New York: Industrial Press, 1984.
6. Shigley, J. E., and L. D. Mitchell. *Mechanical Engineering Design,* 4th ed. New York: McGraw-Hill Book Company, 1983.
7. Wahl, A. M. *Mechanical Springs*. New York: McGraw-Hill Book Company, 1963.

PROBLEMS

1. A spring has an overall length of 2.75 in when it is not loaded and a length of 1.85 in when carrying a load of 12.0 lb. Compute its spring rate.
2. A spring is loaded initially with a load of 4.65 lb and has a length of 1.25 in. The spring rate is given to be 18.8 lb/in. What is the free length of the spring?
3. A spring has a spring rate of 76.7 lb/in. At a load of 32.2 lb, it has a length of 0.830 in. Its solid length is 0.626 in. Compute the force required to compress the spring to solid height. Also compute the free length of the spring.
4. A spring has an overall length of 63.5 mm when it is not loaded and a length of 37.1 mm when carrying a load of 99.2 N. Compute its spring rate.
5. A spring is loaded initially with a load of 54.05 N and it has a length of 39.47 mm. The spring rate is given to be 1.47 N/mm. What is the free length of the spring?
6. A spring has a spring rate of 8.95 N/mm. At a load of 134 N it has a length of 29.4 mm. Its solid length is 21.4 mm. Compute the force required to compress the spring to solid height. Also compute the free length of the spring.
7. A helical compression spring with squared and ground ends has an outside diameter of 1.100 in, a wire diameter of 0.085 in, and a solid height of 0.563 in. Compute the ID, the mean diameter, the spring index, and the approximate number of coils.
8. The following data are known for a spring:
   Total number of coils = 19
   Squared and ground ends
   Outside diameter = 0.560 in
   Wire diameter = 0.059 in (25-gage music wire)
   Free length = 4.22 in
   For this spring, compute the spring index, the pitch, the pitch angle, and the solid length.
9. For the spring of problem 8, compute the force required to reduce its length to 3.00 in. At that force, compute the stress in the spring. Would that stress be satisfactory for average service?
10. Would the spring of problems 8 and 9 tend to buckle when compressed to the length of 3.00 in?
11. For the spring of problem 8, compute the estimate for the outside diameter when it is compressed to solid length.
12. The spring of problem 8 must be compressed to solid length to install it. What force is required to do this? Compute the stress at solid height. Is that stress satisfactory?
13. A support bar for a machine component is suspended to soften applied loads. During operation, the load on each spring varies from 180 lb to 220 lb. The position of the rod is to move no more than 0.500 in as the load varies. Design a compression spring for this application. Several million cycles of load application are expected. Use ASTM A229 steel wire.
14. Design a helical compression spring to exert a force of 22.0 lb when compressed to a length of 1.75 in. When its length is 3.00 in, it must exert a force of 5.0 lb. The spring will be cycled rapidly, with severe service required. Use ASTM A401 steel wire.
15. Design a helical compression spring for a pressure relief valve. When the valve is closed, the spring length is 2.0 in, and the spring force is to be 1.50 lb.

As the pressure on the valve increases, a force of 14.0 lb causes the valve to open and compress the spring to a length of 1.25 in. Use corrosion-resistant ASTM A313, type 302 steel wire, and design for average service.

16. Design a helical compression spring to be used to return a pneumatic cylinder to its original position after being actuated. At a length of 10.50 in, the spring must exert a force of 60 lb. At a length of 4.00 in, it must exert a force of 250 lb. Severe service is expected. Use ASTM A231 steel wire.
17. Design a helical compression spring using music wire that will exert a force of 14.0 lb when its length is 0.68 in. The free length is to be 1.75 in. Use average service.
18. Design a helical compression spring using stainless steel wire, ASTM A313, type 316, for average service, which will exert a force of 8.00 lb after deflecting 1.75 in from a free length of 2.75 in.
19. Repeat problem 18 with the additional requirement that the spring must operate around a rod having a diameter of 0.625 in.
20. Repeat problem 17 with the additional requirement that the spring is to be installed in a hole having a diameter of 0.750 in.
21. Design a helical compression spring using ASTM A231 steel wire for severe service, which will exert a force of 45.0 lb at a length of 3.05 in and a force of 22.0 lb at a length of 3.50 in.
22. Design a helical extension spring using music wire to exert a force of 7.75 lb when the length between attachment points is 2.75 in, and 5.25 lb at a length of 2.25 in. The outside diameter must be less than 0.300 in. Use severe service. Be sure the stress in the spring under initial tension is within the range suggested in Figure 7-18.
23. Design a helical extension spring for average service using music wire to exert a force of 15.0 lb when the length between attachment points is 5.00 in, and 5.20 lb at a length of 3.75 in. The outside diameter must be less than 0.75 in.
24. Design a helical extension spring for severe service using music wire to exert a maximum force of 10.0 lb at a length of 3.00 in. The spring rate should be 6.80 lb/in. The outside diameter must be less than 0.75 in.
25. Design a helical extension spring for severe service using music wire to exert a maximum force of 10.0 lb at a length of 6.00 in. The spring rate should be 2.60 lb/in. The outside diameter must be less than 0.75 in.
26. Design a helical extension spring for average service using music wire to exert a maximum force of 10.0 lb at a length of 9.61 in. The spring rate should be 1.50 lb/in. The outside diameter must be less than 0.75 in.
27. Design a helical extension spring for average service using stainless steel wire, ASTM A313, type 302, to exert a maximum force of 162 lb at a length of 10.80 in. The spring rate should be 38.0 lb/in. The outside diameter should be approximately 1.75 in.
28. An extension spring has an end similar to that shown in Figure 7-20. The pertinent data are as follows: U.S. Wire gage no. 19; mean diameter = 0.28 in; $R_1 = 0.25$ in; $R_2 = 0.094$ in. Compute the expected stresses at points A and B in the figure for a force of 5.0 lb. Would those stresses be satisfactory for ASTM A227 steel wire for average service?
29. Design a helical torsion spring for average service using stainless steel wire, ASTM A313, type 302, to exert a torque of 1.00 lb · in after a deflection of 180° from the free condition. The outside diameter of the coil should be no more than 0.500 in. Specify the diameter of a rod on which to mount the spring.
30. Design a helical torsion spring for severe service using stainless steel wire, ASTM A313, type 302, to exert a torque of 12.0 lb · in after a deflection of 270° from the free condition. The outside diameter of the coil should be no more than 1.250 in. Specify the diameter of a rod on which to mount the spring.
31. Design a helical torsion spring for severe service using music wire to exert a maximum torque of 2.50 lb · in after a deflection of 360° from the free condition. The outside diameter of the coil should be no more than 0.750 in. Specify the diameter of a rod on which to mount the spring.
32. A helical torsion spring has a wire diameter of 0.038 in; an outside diameter of 0.368 in; 9.5 coils in the body; one end 0.50 in long; the other end 1.125 in long; and a material of ASTM A401 steel. What torque would cause the spring to rotate 180°? What would be the stress then? Would it be safe?

8 Tolerances and Fits

8-1 OVERVIEW

Most of the analysis procedures discussed in this book have had the objective of determining the minimum acceptable geometric size at which a component is safe or performs properly under specified conditions. As a designer, you must also then specify the final dimensions for the components, including the *tolerances* on those dimensions. Often overlooked by students studying machine design and avoided by novice designers, this is a most important step. The proper performance of a machine can depend on the tolerances specified for its parts, particularly those that must fit together for location or for suitable relative motion.

The term *tolerance* refers to the permissible deviation of a dimension from the specified basic size. A *unilateral tolerance* deviates in only one direction from the basic size. A *bilateral tolerance* deviates both above and below the basic size. The *total tolerance* is the difference between the maximum and minimum permissible dimensions.

Geometric tolerancing is often used instead of the traditional method in controlling size, form, location, and finish of manufactured parts. It is especially useful in quantity production of interchangeable parts. The complete definition of the standards for geometric dimensioning and tolerancing is found in the Standard ANSI Y14.5M—1982.

The term *allowance* refers to an intentional difference between the maximum material limits of mating parts. For example, a positive allowance for a hole/shaft pair would define the minimum *clearance* between the mating parts from the largest shaft mating with the smallest hole. A negative allowance would result in the shaft's being larger than the hole (*interference*).

The term *fit* refers to the relative looseness (clearance fit) or tightness (interference fit) of mating parts, especially as it affects the motion of the parts or the force between them after assembly. Specifying the degree of clearance or interference is one of the tasks of the designer.

Consider the plain surface bearings designed in Chapter 14. A critical part of the design is the specification of the diametral clearance between the journal and the bearing. The typical value is just a few thousandths of an inch. But there must be some variation allowed on both the journal outside diameter and the bearing inside diameter for reasons of economy of manufacture. Thus there will be a variation of the actual clearance in production devices depending on where the individual mating components fall within their own tolerance bands. Such variations must be accounted for in the analysis of the bearing performance. Too small a clearance could cause seizing. Conversely, too large a clearance would reduce the precision of the machine and adversely affect the lubrication.

The mounting of power transmission elements onto shafts is another situation in which tolerances and fits must be considered. A chain sprocket for general-purpose mechanical drives is usually produced with a bore that will easily slide into position on the shaft during assembly. But once in place it will transmit power smoothly and quietly only if it is not excessively loose. A high-speed turbine rotor must be installed on its shaft with an interference fit, eliminating any looseness that would cause vibrations at the high rotational speeds.

When relative motion between two parts is required, a clearance fit is necessary. But here, too, there are differences. Some measuring instruments have parts that must move with no perceptible looseness (sometimes called *play*) between mating parts that would adversely affect the accuracy of measurement. An idler sheave in a belt drive system must rotate on its shaft reliably, without the tendency to seize, but with only a small amount of play. The requirement for the mounting of a wheel of a child's wagon on its axle is much different. A loose clearance fit is satisfactory for the expected use of the wheel, and it permits wide tolerances on the wheel bore and on the axle diameter for economy.

In applying tolerances and the allowance to the dimensions of mating parts, the basic hole system and the basic shaft system are used. In the *basic hole system,* the design size of the hole is the basic size and the allowance is applied to the shaft; the basic size is the minimum size of the hole. In the *basic shaft system,* the design size of the shaft is the basic size and the allowance is applied to the hole; the basic size is the maximum size of the shaft. These concepts are illustrated in section 8-4 in the discussion of clearance fits.

8-2 TOLERANCES, PRODUCTION PROCESSES, AND COST

It is costly to produce components with very small tolerances on dimensions. It is the designer's responsibility to set the tolerances at the highest possible level that results in satisfactory operation of the machine. Certainly judgment and experience must be exercised in such a process. In quantity production situations, it may be cost-effective to test prototypes with a range of tolerances to observe the limits of acceptable performance.

In general, the production of parts with small tolerances on their dimensions requires multiple processing steps. A shaft may first have to be turned on a lathe and then ground to produce the final dimensions and surface finish. In extreme cases, lapping may be required. Each subsequent step in the manufacture adds cost. Even if different operations are not required, the holding of small tolerances on a single machine, such as a lathe, may require several passes, ending with a finish cut. Cutting tool changes must be more frequent, also, as wear of the tool takes the part out of tolerance.

The production of part features with small tolerances usually involves finer surface finishes, as well. Figure 8-1 shows the general relationship between the surface finish and the cost of producing a part. The typical total tolerance produced by the processes described is included in the figure. The increase in cost is dramatic for the small tolerances.

Figure 8-2 presents the relationship between the surface finish and the machining operations available to produce it.

The basic reference for tolerances and fits is the USA Standard B4.1 — 1967 (1974), *Preferred Limits and Fits for Cylindrical Parts,* sponsored by the ASME (1).

8-3 PREFERRED BASIC SIZES

The first step in specifying a dimension for a part is to decide on the basic size, that dimension to which the tolerances are applied. The analysis for strength, deflection, or performance of the part determines the nominal or minimum size required. Unless special conditions exist, the basic size should be chosen from the following lists of preferred basic sizes: Table 8-1 for fractional inch sizes, Table 8-2 for decimal inch sizes, and Table 8-3 for metric sizes from the SI system. If possible, select from the "first"-choice column. If a size between two first-choice sizes is required, the "second"-choice column should be used. This will limit the number of sizes typically encountered in the manufacture of products and lead to cost-effective standardization. The choice of system depends on the policies of the company and the market for the product.

8-4 CLEARANCE FITS

When there must always be a clearance between mating parts, a clearance fit is specified. The designation for standard clearance fits from USA Standard B4.1 for members that must move together is the class called *Running or sliding clearance fit* (*RC*). Within this standard, there are nine classes, RC1 through RC9, with RC1 providing the smallest

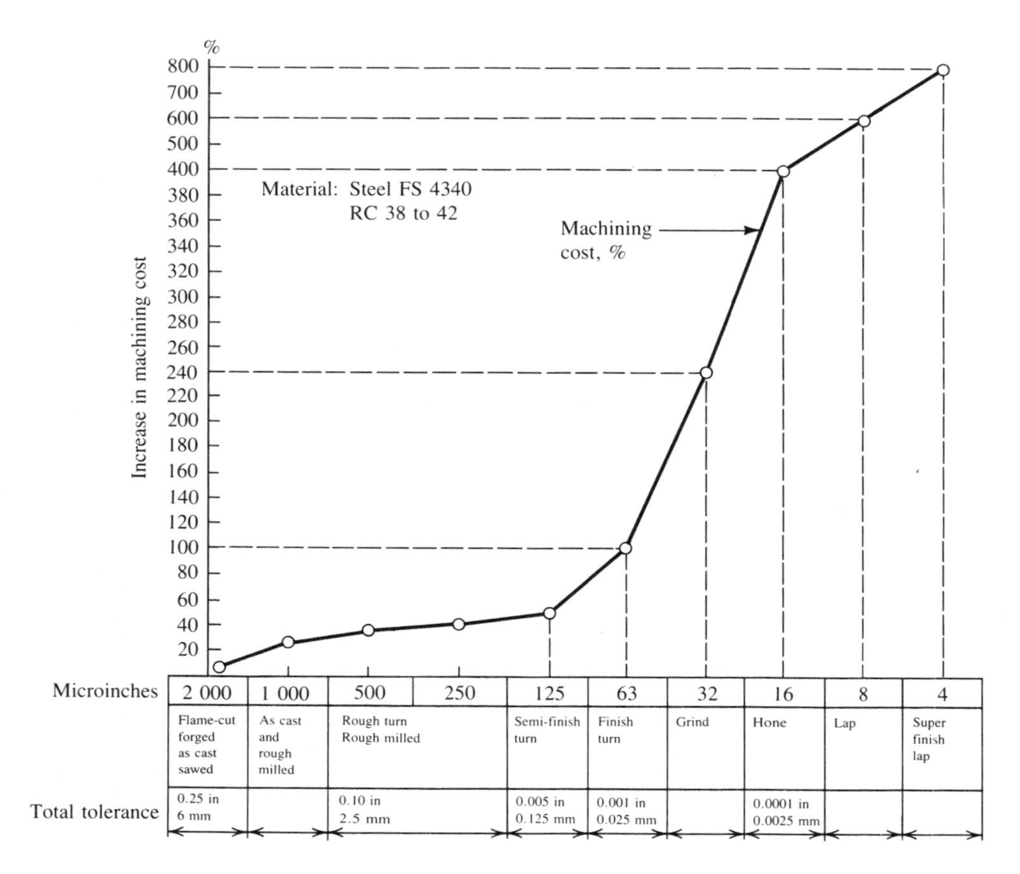

Figure 8-1 Machining Costs versus Surface Finish Specified (Reprinted with permission from the Association for Integrated Manufacturing Technology [formerly the Numerical Control Society])

clearance, and RC9 the largest. The descriptions for the individual members of this class should help you to decide which is most appropriate for a given application.

RC1 (close sliding fit): Accurate location of parts that must assemble without perceptible play.

RC2 (sliding fit): Parts will move and turn easily but are not intended to run freely. Parts may seize with small temperature changes, especially in the larger sizes.

RC3 (precision running fit): Precision parts operating at slow speeds and light loads that must run freely. Changes in temperature may cause difficulties.

RC4 (close running fit): Provides accurate location with minimum play for use under moderate loads and speeds. A good choice for accurate machinery.

RC5 (medium running fit): Accurate machine parts for higher speeds and/or loads than for RC4.

RC6 (medium running fit): Similar to RC5 for applications in which larger clearance is desired.

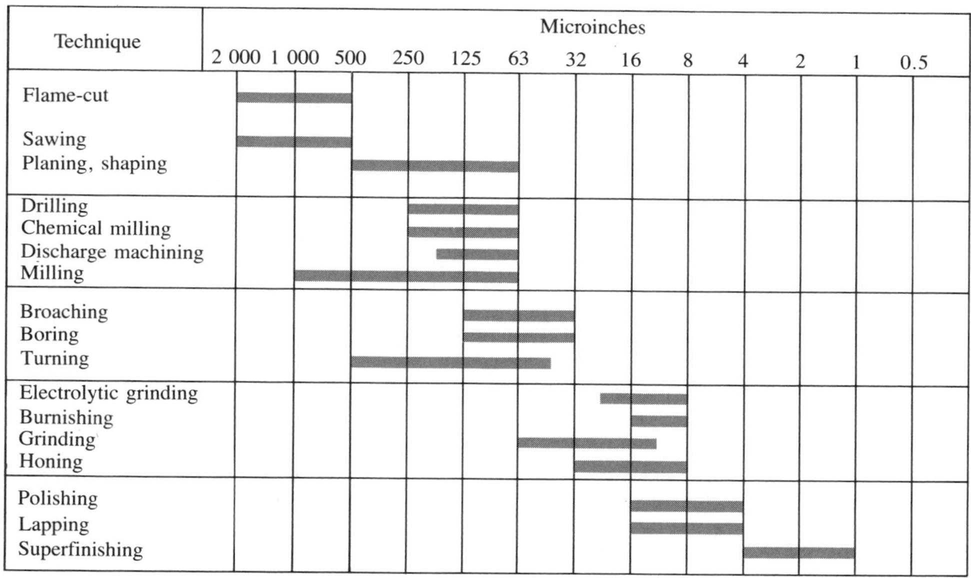

Figure 8-2 Finishes Produced by Various Techniques (Reprinted with permission from the Association for Integrated Manufacturing Technology [formerly the Numerical Control Society])

RC7 (free running fit): Reliable relative motion under wide temperature variations in applications where accuracy is not critical.

RC8 (loose running fit): Large clearances are permitted, allowing the use of parts with commercial, "as received" tolerances.

RC9 (loose running fit): Similar to RC8, with approximately 50 percent larger clearances.

The complete standard USA B4.1 lists the tolerances on the mating parts and the resulting limits of clearances for all nine classes and for sizes from zero to 200 in. Table 8-4 is abstracted from the standard. Let RC2 represent the precision fits (RC1, RC2, RC3); let RC5 represent the accurate, reliable motion fits (RC4, RC5, RC6); and let RC8 represent the loose fits (RC7, RC8, RC9).

The numbers in the table are in thousandths of an inch. Thus a clearance of 2.8 from the table means a difference in size between the inside and outside parts of 0.002 8 in. The tolerances on the hole and the shaft are to be applied to the basic size to determine the limits of size for that dimension.

Example Problem 8-1. A shaft carrying an idler sheave for a belt drive system is to have a nominal size of 2.00 in. The sheave must rotate reliably on the shaft but with the smoothness characteristic of accurate machinery. Specify the limits of size for the shaft and the sheave bore and list the limits of clearance that will result. Use the basic hole system.

Table 8-1 Preferred Basic Sizes

Fractional (Inch)			
1/64	0.015 625	5	5.000 0
1/32	0.031 25	5 1/4	5.250 0
1/16	0.062 5	5 1/2	5.500 0
3/32	0.093 75	5 3/4	5.750 0
1/8	0.125 0	6	6.000 0
5/32	0.156 25	6 1/2	6.500 0
3/16	0.187 5	7	7.000 0
1/4	0.250 0	7 1/2	7.500 0
5/16	0.312 5	8	8.000 0
3/8	0.375 0	8 1/2	8.500 0
7/16	0.437 5	9	9.000 0
1/2	0.500 0	9 1/2	9.500 0
9/16	0.562 5	10	10.000 0
5/8	0.625 0	10 1/2	10.500 0
11/16	0.687 5	11	11.000 0
3/4	0.750 0	11 1/2	11.500 0
7/8	0.875 0	12	12.000 0
1	1.000 0	12 1/2	12.500 0
1 1/4	1.250 0	13	13.000 0
1 1/2	1.500 0	13 1/2	13.500 0
1 3/4	1.750 0	14	14.000 0
2	2.000 0	14 1/2	14.500 0
2 1/4	2.250 0	15	15.000 0
2 1/2	2.500 0	15 1/2	15.500 0
2 3/4	2.750 0	16	16.000 0
3	3.000 0	16 1/2	16.500 0
3 1/4	3.250 0	17	17.000 0
3 1/2	3.500 0	17 1/2	17.500 0
3 3/4	3.750 0	18	18.000 0
4	4.000 0	18 1/2	18.500 0
4 1/4	4.250 0	19	19.000 0
4 1/2	4.500 0	19 1/2	19.500 0
4 3/4	4.750 0	20	20.000 0

Table 8-2 Preferred Basic Sizes

Decimal (Inch)		
0.010	2.00	8.50
0.012	2.20	9.00
0.016	2.40	9.50
0.020	2.60	10.00
0.025	2.80	10.50
0.032	3.00	11.00
0.040	3.20	11.50
0.05	3.40	12.00
0.06	3.60	12.50
0.08	3.80	13.00
0.10	4.00	13.50
0.12	4.20	14.00
0.16	4.40	14.50
0.20	4.60	15.00
0.24	4.80	15.50
0.30	5.00	16.00
0.40	5.20	16.50
0.50	5.40	17.00
0.60	5.60	17.50
0.80	5.80	18.00
1.00	6.00	18.50
1.20	6.50	19.00
1.40	7.00	19.50
1.60	7.50	20.00
1.80	8.00	

Source: USA Standard B4.1—1967 (rev. 1974), *Preferred Limits and Fits for Cylindrical Parts* (American Society of Mechanical Engineers, New York).

Solution. An RC5 fit should be satisfactory in this application. From Table 8-4, the hole tolerance limits are +1.8 and −0. The sheave hole then should be within the following limits:

Sheave hole:

$$2.000\,0 + 0.001\,8 = 2.001\,8 \text{ in} \quad \text{(largest)}$$

$$2.000\,0 - 0.000\,0 = 2.000\,0 \text{ in} \quad \text{(smallest)}$$

Notice that the smallest hole is the basic size.

Table 8-3 Preferred Metric Sizes

NOMINAL SIZE, mm		NOMINAL SIZE, mm		NOMINAL SIZE, mm	
First	*Second*	*First*	*Second*	*First*	*Second*
1		10		100	
	1.1		11		110
1.2		12		120	
	1.4		14		140
1.6		16		160	
	1.8		18		180
2		20		200	
	2.2		22		220
2.5		25		250	
	2.8		28		280
3		30		300	
	3.5		35		350
4		40		400	
	4.5		45		450
5		50		500	
	5.5		55		550
6		60		600	
	7		70		700
8		80		800	
	9		90		900
				1 000	

The shaft tolerance limits are −2.5 and−3.7. The resulting size limits are as follows. Shaft diameter:

$$2.0000 - 0.0025 = 1.9975 \text{ in} \quad \text{(largest)}$$

$$2.0000 - 0.0037 = 1.9963 \text{ in} \quad \text{(smallest)}$$

Figure 8-3 illustrates these results.

Combining the smallest shaft with the largest hole gives the largest clearance. Conversely, combining the largest shaft with the smallest hole gives the smallest clearance. Therefore, the limits of clearance are

$$2.0018 - 1.9963 = 0.0055 \text{ in} \quad \text{(largest)}$$

$$2.0000 - 1.9975 = 0.0025 \text{ in} \quad \text{(smallest)}$$

These values check with the limits of clearance in Table 8-4. Notice that the total tolerance for the shaft is 0.001 2 in and for the hole 0.001 8 in, both relatively small values.

Locational Clearance Fits

Another clearance fit system is available for parts for which control of the location is desired although the parts will not normally move in relation to each other in operation.

Table 8-4 Clearance Fits (RC)

NOMINAL SIZE RANGE (INCHES)	CLASS RC2			CLASS RC5			CLASS RC8			NOMINAL SIZE RANGE (INCHES)
	Limits of Clearance	STANDARD LIMITS		*Limits of Clearance*	STANDARD LIMITS		*Limits of Clearance*	STANDARD LIMITS		
Over To		*Hole*	*Shaft*		*Hole*	*Shaft*		*Hole*	*Shaft*	*Over To*
0–0.12	0.1 0.55	+0.25 0	−0.1 −0.3	0.6 1.6	+0.6 −0	−0.6 −1.0	2.5 5.1	+1.6 0	−2.5 −3.5	0–0.12
0.12–0.24	0.15 0.65	+0.3 0	−0.15 −0.35	0.8 2.0	+0.7 −0	−0.8 −1.3	2.8 5.8	+1.8 0	−2.8 −4.0	0.12–0.24
0.24–0.40	0.2 0.85	+0.4 0	−0.2 −0.45	1.0 2.5	+0.9 −0	−1.0 −1.6	3.0 6.6	+2.2 0	−3.0 −4.4	0.24–0.40
0.40–0.71	0.25 0.95	+0.4 0	−0.25 −0.55	1.2 2.9	+1.0 −0	−1.2 −1.9	3.5 7.9	+2.8 0	−3.5 −5.1	0.40–0.71
0.71–1.19	0.3 1.2	+0.5 0	−0.3 −0.7	1.6 3.6	+1.2 −0	−1.6 −2.4	4.5 10.0	+3.5 0	−4.5 −6.5	0.71–1.19
1.19–1.97	0.4 1.4	+0.6 0	−0.4 −0.8	2.0 4.6	+1.6 −0	−2.0 −3.0	5.0 11.5	+4.0 0	−5.0 −7.5	1.19–1.97
1.97–3.15	0.4 1.6	+0.7 0	−0.4 −0.9	2.5 5.5	+1.8 −0	−2.5 −3.7	6.0 13.5	+4.5 0	−6.0 −9.0	1.97–3.15
3.15–4.73	0.5 2.0	+0.9 0	−0.5 −1.1	3.0 6.6	+2.2 −0	−3.0 −4.4	7.0 15.5	+5.0 0	−7.0 −10.5	3.15–4.73
4.73–7.09	0.6 2.3	+1.0 0	−0.6 −1.3	3.5 7.6	+2.5 −0	−3.5 −5.1	8.0 18.0	+6.0 0	−8.0 −12.0	4.73–7.09
7.09–9.85	0.6 2.6	+1.2 0	−0.6 −1.4	4.0 8.6	+2.8 −0	−4.0 −5.8	10.0 21.5	+7.0 0	−10.0 −14.5	7.09–9.85
9.85–12.41	0.7 2.8	+1.2 0	−0.7 −1.6	5.0 10.0	+3.0 0	−5.0 −7.0	12.0 25.0	+8.0 0	−12.0 −17.0	9.85–12.41

Note: Limits are in thousandths of an inch.

Source: USA Standard B4.1—1967 (rev. 1974), *Preferred Limits and Fits for Cylindrical Parts* (American Society of Mechanical Engineers, New York, N.Y.).

Called *locational clearance* (*LC*) fits, they include eleven classes. The first four, LC1 to LC4, have a zero clearance (size to size) as the lower limit of the fit, regardless of size or class. The upper limit of fit increases with both the size of the parts and the class number. Classes LC5 through LC11 provide some positive clearance for all sizes, increasing with the size of the parts and with the class. Numerical values for the tolerances and fits for these classes have been published (1).

8-5 INTERFERENCE FITS

Interference fits are those in which the inside member is larger than the outside member, requiring the application of force during assembly. There is some deformation of the parts after assembly and a pressure exists at the mating surfaces.

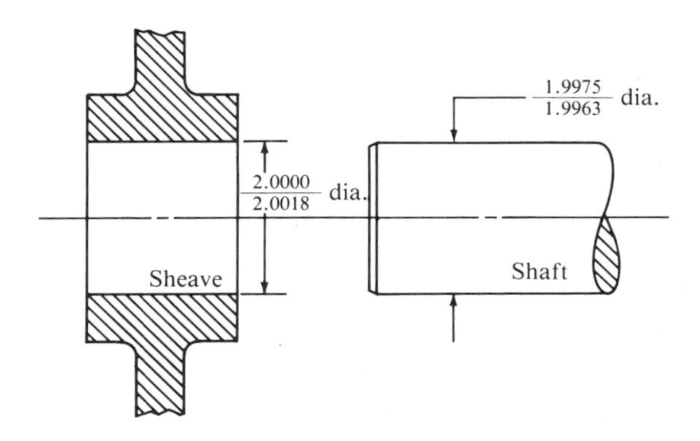

Figure 8-3 An RC5 Fit Using the Basic Hole System

Force fits are designed to provide a controlled pressure between mating parts throughout the range of sizes for a given class. They are used where forces or torques are transmitted through the joint. Instead of assembling through the application of force, similar fits are obtained by *shrink fitting,* in which one member is heated to expand it while the other remains cool. The parts are then assembled with little or no force. After cooling, the same dimensional interference exists as for the force fit. *Locational interference fits* are used for location only. There is no movement between parts after assembly, but there is no special requirement for the resulting pressure between mating parts.

Force Fits (FN)

Five classes of force fits are defined in USA Standard B4.1 (1). (See Table 8-5.)

FN1 (light drive fit): Only light pressure is required to assemble mating parts. Used for fragile parts and where no large forces must be transmitted across the joint.

FN2 (medium drive fit): General-purpose class used frequently for steel parts of moderate cross section.

FN3 (heavy drive fit): Used for heavy steel parts.

FN4 (force fit): Used for high-strength assemblies where high resulting pressures are required.

FN5 (force fit): Similar to FN4 for higher pressures.

The use of shrink fit methods is desirable in most cases of interference fits and is virtually required in the heavier classes and larger-size parts. The temperature increase required to produce a given expansion for assembly can be computed from the basic definition of the coefficient of thermal expansion:

$$\delta = \alpha L(\Delta t) \tag{8-1}$$

where

δ = Total deformation desired (in or mm)

α = Coefficient of thermal expansion (in/in · °F or mm/mm · °C)

L = Nominal length of the member being heated (in or mm)

Δt = Temperature difference (°F or °C)

Table 8-5 Force and Shrink Fits (FN)

NOMINAL SIZE RANGE (INCHES)	CLASS FN1			CLASS FN2			CLASS FN3			CLASS FN4			CLASS FN5		
	Limits of Interference	STANDARD LIMITS		*Limits of Interference*	STANDARD LIMITS		*Limits of Interference*	STANDARD LIMITS		*Limits of Interference*	STANDARD LIMITS		*Limits of Interference*	STANDARD LIMITS	
Over To		*Hole*	*Shaft*		*Hole*	*Shaft*		*Hole*	*Shaft*		*Hole*	*Shaft*		*Hole*	*Shaft*
0–0.12	0.05 0.5	+0.25 −0	+0.5 +0.3	0.2 0.85	+0.4 −0	+0.85 +0.6				0.3 0.95	+0.4 −0	+0.95 +0.7	0.3 1.3	+0.6 −0	+1.3 +0.9
0.12–0.24	0.1 0.6	+0.3 −0	+0.6 +0.4	0.2 1.0	+0.5 −0	+1.0 +0.7				0.4 1.2	+0.5 −0	+1.2 +0.9	0.5 1.7	+0.7 −0	+1.7 +1.2
0.24–0.40	0.1 0.75	+0.4 −0	+0.75 +0.5	0.4 1.4	+0.6 −0	+1.4 +1.0				0.6 1.6	+0.6 −0	+1.6 +1.2	0.5 2.0	+0.9 −0	+2.0 +1.4
0.40–0.56	0.1 0.8	+0.4 −0	+0.8 +0.5	0.5 1.6	+0.7 −0	+1.6 +1.2				0.7 1.8	+0.7 −0	+1.8 +1.4	0.6 2.3	+1.0 −0	+2.3 +1.6
0.56–0.71	0.2 0.9	+0.4 −0	+0.9 +0.6	0.5 1.6	+0.7 −0	+1.6 +1.2				0.7 1.8	+0.7 −0	+1.8 +1.4	0.8 2.5	+1.0 −0	+2.5 +1.8
0.71–0.95	0.2 1.1	+0.5 −0	+1.1 +0.7	0.6 1.9	+0.8 −0	+1.9 +1.4				0.8 2.1	+0.8 −0	+2.1 +1.6	1.0 3.0	+1.2 −0	+3.0 +2.2
0.95–1.19	0.3 1.2	+0.5 −0	+1.2 +0.8	0.6 1.9	+0.8 −0	+1.9 +1.4	0.8 2.1	+0.8 −0	+2.1 +1.6	1.0 2.3	+0.8 −0	+2.3 +1.8	1.3 3.3	+1.2 −0	+3.3 +2.5
1.19–1.58	0.3 1.3	+0.6 −0	+1.3 +0.9	0.8 2.4	+1.0 −0	+2.4 +1.8	1.0 2.6	+1.0 −0	+2.6 +2.0	1.5 3.1	+1.0 −0	+3.1 +2.5	1.4 4.0	+1.6 −0	+4.0 +3.0
1.58–1.97	0.4 1.4	+0.6 −0	+1.4 +1.0	0.8 2.4	+1.0 −0	+2.4 +1.8	1.2 2.8	+1.0 −0	+2.8 +2.2	1.8 3.4	+1.0 −0	+3.4 +2.8	2.4 5.0	+1.6 −0	+5.0 +4.0
1.97–2.56	0.6 1.8	+0.7 −0	+1.8 +1.3	0.8 2.7	+1.2 −0	+2.7 +2.0	1.3 3.2	+1.2 −0	+3.2 +2.5	2.3 4.2	+1.2 −0	+4.2 +3.5	3.2 6.2	+1.8 −0	+6.2 +5.0
2.56–3.15	0.7 1.9	+0.7 −0	+1.9 +1.4	1.0 2.9	+1.2 −0	+2.9 +2.2	1.8 3.7	+1.2 −0	+3.7 +3.0	2.8 4.7	+1.2 −0	+4.7 +4.0	4.2 7.2	+1.8 −0	+7.2 +6.0
3.15–3.94	0.9 2.4	+0.9 −0	+2.4 +1.8	1.4 3.7	+1.4 −0	+3.7 +2.8	2.1 4.4	+1.4 −0	+4.4 +3.5	3.6 5.9	+1.4 −0	+5.9 +5.0	4.8 8.4	+2.2 −0	+8.4 +7.0
3.94–4.73	1.1 2.6	+0.9 −0	+2.6 +2.0	1.6 3.9	+1.4 −0	+3.9 +3.0	2.6 4.9	+1.4 −0	+4.9 +4.0	4.6 6.9	+1.4 −0	+6.9 +6.0	5.8 9.4	+2.2 −0	+9.4 +8.0
4.73–5.52	1.2 2.9	+1.0 −0	+2.9 +2.2	1.9 4.5	+1.6 −0	+4.5 +3.5	3.4 6.0	+1.6 −0	+6.0 +5.0	5.4 8.0	+1.6 −0	+8.0 +7.0	7.5 11.6	+2.5 −0	+11.6 +10.0
5.52–6.30	1.5 3.2	+1.0 −0	+3.2 +2.5	2.4 5.0	+1.6 −0	+5.0 +4.0	3.4 6.0	+1.6 −0	+6.0 +5.0	5.4 8.0	+1.6 −0	+8.0 +7.0	9.5 13.6	+2.5 −0	+13.6 +12.0

Note: Limits are in thousandths of an inch.

Source: USA Standard B4.1—1967 (rev. 1974), *Preferred Limits and Fits for Cylindrical Parts* (American Society of Mechanical Engineers, New York).

For cylindrical parts, L is the diameter and δ is the change in diameter required. Table 8-6 gives the values for α for several materials.

8-6 TRANSITION FITS

The *location transition fit (LT)* is used where accuracy of location is important but where a small amount of clearance or a small amount of interference is acceptable. There are six classes, LT1 to LT6. In any class there is an overlap in the tolerance limits for both the hole and the shaft so that possible combinations produce a small clearance, a small interference, or even a size-to-size fit. Complete tables of data for these fits are published (1).

8-7 STRESSES FOR FORCE FITS

When force fits are used to secure mechanical parts, the interference creates a pressure acting at the mating surfaces. The pressure causes stresses in each part. Under heavy force fits or even with lighter fits in fragile parts, the stresses developed can be great enough to yield ductile materials. A permanent set results, which normally destroys the usefulness of the assembly. With brittle materials such as cast iron, actual fracture may result.

The stress analysis applicable to force fits is related to the analysis of thick-walled cylinders. The outer member expands under the influence of the pressure at the mating surface, with the tangential tensile stress developed being a maximum at the mating surface. There is a radial stress equal to the pressure itself. Also, the inner member contracts because of the pressure and is subjected to a tangential compressive stress along with the radial compressive stress equal to the pressure.

The usual objective of the analysis is to determine the magnitude of the pressure due to a given interference fit that would be developed at the mating surfaces. Then the

Table 8-6 Coefficient of Thermal Expansion

	COEFFICIENT OF THERMAL EXPANSION, α	
MATERIAL	$in/in \cdot °F$	$mm/mm \cdot °C$
Steel:		
AISI 1020	6.5×10^{-6}	11.7×10^{-6}
AISI 1050	6.1×10^{-6}	11.0×10^{-6}
AISI 4140	6.2×10^{-6}	11.2×10^{-6}
Stainless steel:		
AISI 301	9.4×10^{-6}	16.9×10^{-6}
AISI 430	5.8×10^{-6}	10.4×10^{-6}
Aluminum:		
2014	12.8×10^{-6}	23.0×10^{-6}
6061	13.0×10^{-6}	23.4×10^{-6}
Bronze:	10.0×10^{-6}	18.0×10^{-6}

stresses due to this pressure in the mating members are computed. The following procedure can be used:

1. Determine the amount of interference from the design of the parts. For standard force fits, Table 8-5 can be used. The maximum limit of interference would, of course, give the maximum stresses for the parts. Note that the interference values are based on the total interference on the diameter, which is the sum of the expansion of the outer ring plus the contraction of the inner member (see Figure 8-4).
2. Compute the pressure at the mating surface from equation (8-2) if both members are of the same material.

$$p = \frac{E\delta}{2b}\left[\frac{(c^2 - b^2)(b^2 - a^2)}{2b^2(c^2 - a^2)}\right] \tag{8-2}$$

   Use equation (8-3) if they are of different materials.

$$p = \frac{\delta}{2b\left[\dfrac{1}{E_o}\left(\dfrac{c^2 + b^2}{c^2 - b^2} + \nu_o\right) + \dfrac{1}{E_i}\left(\dfrac{b^2 + a^2}{b^2 - a^2} - \nu_i\right)\right]} \tag{8-3}$$

   where

   p = Pressure at the mating surface
   δ = Total diametral interference
   E = Modulus of elasticity of each member if they are the same
   E_o = Modulus of elasticity of outer member
   E_i = Modulus of elasticity of inner member
   ν_o = Poisson's ratio for outer member
   ν_i = Poisson's ratio for inner member

3. Compute the tensile stress in the outer member from

$$\sigma_o = p\left(\frac{c^2 + b^2}{c^2 - b^2}\right) \quad \text{(tensile at the inner surface in the tangential direction)} \tag{8-4}$$

4. Compute the compressive stress in the inner member from

$$\sigma_i = -p\left(\frac{b^2 + a^2}{b^2 - a^2}\right) \quad \text{(compressive at the outer surface in the tangential direction)} \tag{8-5}$$

5. If desired, the increase in the diameter of the outer member due to the tensile stress can be computed from

$$\delta_o = \frac{2bp}{E_o}\left[\frac{c^2 + b^2}{c^2 - b^2} + \nu_o\right] \tag{8-6}$$

6. If desired, the decrease in the diameter of the inner member due to the compressive stress can be computed from

$$\delta_i = -\frac{2bp}{E_i}\left[\frac{b^2 + a^2}{b^2 - a^2} - \nu_i\right] \tag{8-7}$$

The stresses computed from equations (8-4) and (8-5) were derived assuming that the two cylinders are of equal length. If the outer member is shorter than the inner member, the stresses are higher at its ends by as much as a factor of 2.0. Such a factor should be applied like a stress concentration factor.

In the absence of applied shear stresses, the tensile stress in the tangential direction in the outer member is the maximum principal stress and can be compared with the yield strength of the material to determine the resulting design factor.

An example problem that shows the application of these relationships follows. Then a computer program is listed to solve the same equations, with sample output showing the solution to example problem 8-2, followed by the same data but using a steel sleeve on a steel bushing.

Example Problem 8-2. A bronze bushing is to be installed into a steel sleeve as indicated in Figure 8-4. The bushing has an inside diameter of 2.000 in and a nominal outside diameter of 2.500 in. The steel sleeve has a nominal inside diameter of 2.500 in and an outside diameter of 3.500 in.

1. Specify the limits of size for the outside diameter of the bushing and the inside diameter of the sleeve in order to obtain a heavy drive fit, FN3. Determine the limits of interference that would result.
2. For the maximum interference from 1, compute the pressure that would be developed between the bushing and the sleeve, the stress in the bushing and the sleeve, and the deformation of the bushing and the sleeve. Use $E = 30 \times 10^6$ psi for the steel and $E = 17 \times 10^6$ for the bronze. Use $\nu = 0.27$ for both materials.

Solution. For 1, from Table 8-5, for a part size of 2.50 in at the mating surface, the tolerance limits on the hole in the outer member are +1.2 and −0. Applying these limits to the basic size gives the dimension limits for the hole in the steel sleeve:

2.501 2 in

2.500 0 in

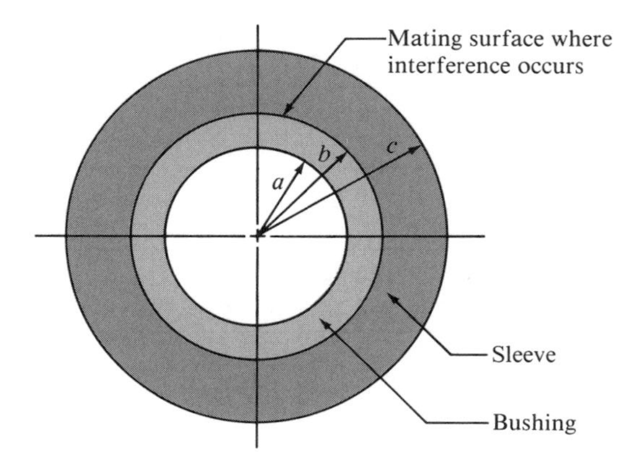

Figure 8-4 Terminology for Interference Fit

For the bronze insert, the tolerance limits are +3.2 and +2.5. Then the size limits for the outside diameter of the bushing are:

2.503 2 in

2.502 5 in

The limits of interference would be 0.001 3 to 0.003 2 in.

For 2, the maximum pressure would be produced by the maximum interference, 0.003 2 in. Then, using $a = 1.00$ in, $b = 1.25$ in, $c = 1.75$ in, $E_o = 30 \times 10^6$ psi, $E_i = 17 \times 10^6$ psi, and $\nu_o = \nu_i = 0.27$ from equation (8-3),

$$p = \frac{\delta}{2b\left[\dfrac{1}{E_o}\left(\dfrac{c^2 + b^2}{c^2 - b^2} + \nu_o\right) + \dfrac{1}{E_i}\left(\dfrac{b^2 + a^2}{b^2 - a^2} - \nu_i\right)\right]}$$

$$p = \frac{0.003\,2}{2(1.25)\left[\dfrac{1}{30 \times 10^6}\left(\dfrac{1.75^2 + 1.25^2}{1.75^2 - 1.25^2} + 0.27\right) + \dfrac{1}{17 \times 10^6}\left(\dfrac{1.25^2 + 1.00^2}{1.25^2 - 1.00^2} - 0.27\right)\right]}$$

$$p = 3\,517 \text{ psi}$$

The tensile stress in the steel sleeve is

$$\sigma_o = p\left(\frac{c^2 + b^2}{c^2 - b^2}\right) = 3\,517\left(\frac{1.75^2 + 1.25^2}{1.75^2 - 1.25^2}\right) = 10\,846 \text{ psi}$$

The compressive stress in the bronze bushing is

$$\sigma_i = -p\left(\frac{b^2 + a^2}{b^2 - a^2}\right) = -3\,517\left(\frac{1.25^2 + 1.00^2}{1.25^2 - 1.00^2}\right) = -16\,025 \text{ psi}$$

The increase in the diameter of the sleeve is

$$\delta_o = \frac{2bp}{E_o}\left[\frac{c^2 + b^2}{c^2 - b^2} + \nu_o\right]$$

$$\delta_o = \frac{2(1.25)(3\,517)}{30 \times 10^6}\left[\frac{1.75^2 + 1.25^2}{1.75^2 - 1.25^2} + 0.27\right] = 0.000\,98 \text{ in}$$

The decrease in the diameter of the bushing is

$$\delta_i = -\frac{2bp}{E_i}\left[\frac{b^2 + a^2}{b^2 - a^2} - \nu_i\right]$$

$$\delta_i = \frac{2(1.25)(3\,517)}{17 \times 10^6}\left[\frac{1.25^2 + 1.00^2}{1.25^2 - 1.00^2} + 0.27\right] = 0.002\,22 \text{ in}$$

Note that the sum of δ_o and δ_i equals 0.003 2, the total interference, δ.

The computer program follows. Figure 8-5 is a flowchart for the program.

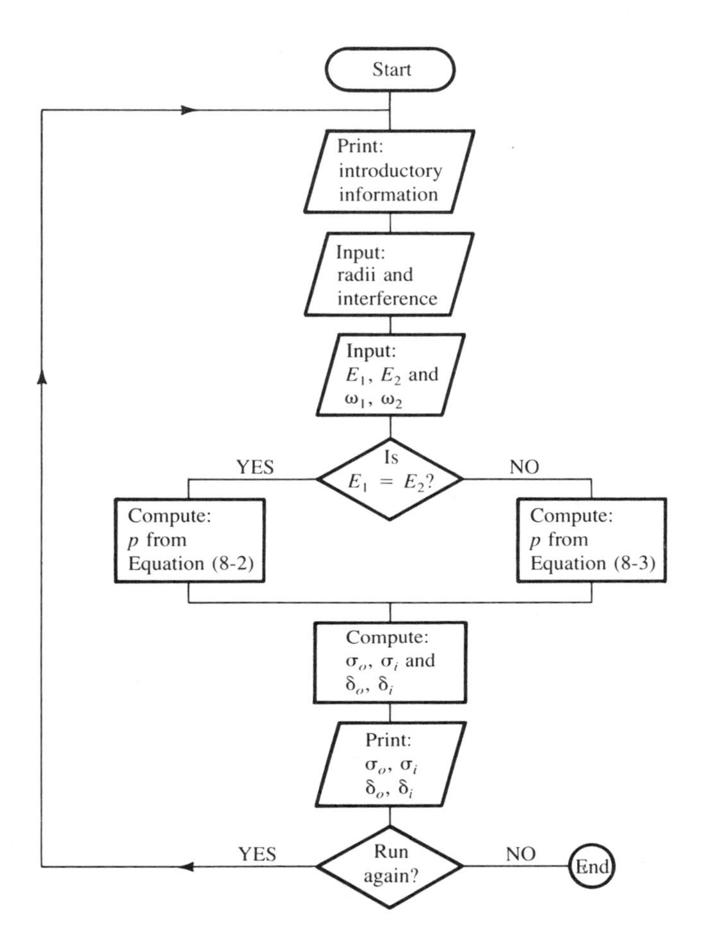

Figure 8-5 Flowchart for Program for Force Fits

Stresses for Force Fits: Program Listing

Introduction

```
10 PRINT "PROGRAM TO COMPUTE PRESSURE, STRESSES"
20 PRINT "   AND DEFLECTIONS DUE TO A FORCE FIT"
30 PRINT
40 PRINT "DATA REQUIRED ARE:"
50 PRINT "   D = DIAMETRAL INTERFERENCE AT MATING SURFACE"
60 PRINT "   A1 = INSIDE DIAMETER OF INNER MEMBER"
70 PRINT "   B1 = NOMINAL DIAMETER AT MATING SURFACE"
80 PRINT "   C1 = OUTSIDE DIAMETER OF OUTER MEMBER"
90 PRINT "   E1 = MODULUS OF ELASTICITY OF OUTER MEMBER"
100 PRINT "   E2 = MODULUS OF ELASTICITY OF INNER MEMBER"
110 PRINT "   M1 = POISSON'S RATIO FOR OUTER MEMBER"
120 PRINT "   M2 = POISSON'S RATIO FOR INNER MEMBER"
130 PRINT
```

Input data

```
140 PRINT "INPUT D, A1, B1, C1"
► 150 INPUT D,A1,B1,C1
160 PRINT
170 PRINT "INPUT E1, E2, M1, M2"
► 180 INPUT E1,E2,M1,M2
```

Lines in the program calling for input of data and lines in the output that correspond to the input statements are highlighted with a ►.

```
190 A = (A1/2)^2
200 B = (B1/2)^2
210 C = (C1/2)^2
220 F1 = (C + B)/(C - B)
230 F2 = (B + A)/(B - A)
240 IF E1 <> E2 THEN 270
250 P = E1*D*((C - B)*(B - A)/(2*B*(C - A)))/B1
260 GOTO 280
270 P = D/(B1*((F1 + M1)/E1 + (F2 - M2)/E2))
280 S1 = P*F1
290 S2 = -P*F2
300 D1 = B1*P*(F1 + M1)/E1
310 D2 = -B1*P*(F2 - M2)/E2
320 PRINT "PRESSURE AT MATING SURFACE =";P;"PSI"
330 PRINT "STRESSES:"
340 PRINT "   INNER SURFACE OF OUTER MEMBER: S1 =";S1;"PSI"
350 PRINT "   OUTER SURFACE OF INNER MEMBER: S2 =";S2;"PSI"
360 PRINT "DIAMETER CHANGES DUE TO PRESSURE;"
370 PRINT "   INCREASE IN OUTER DIAMETER: D1 =";D1;"IN."
380 PRINT "   DECREASE IN INNER DIAMETER: D2 =";D2;"IN."
390 PRINT
400 PRINT "DO YOU WANT TO RUN AGAIN? Y/N"
410 INPUT X$
420 IF X$ = "Y" THEN 40
430 END
```

Compute pressure:

Equation (8-2)

Equation (8-3)

Stresses and deformations

Results

Decision on how to continue

Stresses For Force Fits: Sample Output 1

```
PROGRAM TO COMPUTE PRESSURE, STRESSES
  AND DEFLECTIONS DUE TO A FORCE FIT

DATA REQUIRED ARE:
  D = DIAMETRAL INTERFERENCE AT MATING SURFACE
  A1 = INSIDE DIAMETER OF INNER MEMBER
  B1 = NOMINAL DIAMETER AT MATING SURFACE
  C1 = OUTSIDE DIAMETER OF OUTER MEMBER
  E1 = MODULUS OF ELASTICITY OF OUTER MEMBER
  E2 = MODULUS OF ELASTICITY OF INNER MEMBER
  M1 = POISSON'S RATIO FOR OUTER MEMBER
  M2 = POISSON'S RATIO FOR INNER MEMBER

INPUT D, A1, B1, C1
? .0032 , 2 , 2.5 , 3.5

INPUT E1, E2, M1, M2
? 3E + 07 , 1.7E + 07 , .27 , .27

PRESSURE AT MATING SURFACE = 3517.75 PSI

STRESSES:
  INNER SURFACE OF OUTER MEMBER: S1 = 10846.4 PSI
  OUTER SURFACE OF INNER MEMBER: S2 = -16025.3 PSI

DIAMETER CHANGES DUE TO PRESSURE:
  INCREASE IN OUTER DIAMETER: D1 = 9.83014E-04 IN.
  DECREASE IN INNER DIAMETER: D2 = -2.21698E-03 IN.

DO YOU WANT TO RUN AGAIN? Y/N
? Y
```

Data from example problem 8-2

Outer member: steel

Inner member: bronze

Statement 150

Statement 180

Statement 410

Stresses For Force Fits: Sample Output 2

Data as above but inner member is steel

```
DATA REQUIRED ARE:
  D = DIAMETRAL INTERFERENCE AT MATING SURFACE
  A1 = INSIDE DIAMETER OF INNER MEMBER
  B1 = NOMINAL DIAMETER AT MATING SURFACE
  C1 = OUTSIDE DIAMETER OF OUTER MEMBER
  E1 = MODULUS OF ELASTICITY OF OUTER MEMBER
  E2 = MODULUS OF ELASTICITY OF INNER MEMBER
  M1 = POISSON'S RATIO FOR OUTER MEMBER
  M2 = POISSON'S RATIO FOR INNER MEMBER

INPUT D, A1, B1, C1
► ? .0032 , 2 , 2.5 , 3.5                          Statement 150

INPUT E1, E2, M1, M2
► ? 3E + 07 , 3E + 07 , .27 , .27                  Statement 180

PRESSURE AT MATING SURFACE = 5026.91 PSI

STRESSES:
  INNER SURFACE OF OUTER MEMBER: S1 = 15499.6 PSI
  OUTER SURFACE OF INNER MEMBER: S2 = -22900.4 PSI

DIAMETER CHANGES DUE TO PRESSURE:
  INCREASE IN OUTER DIAMETER: D1 = 1.40474E-03 IN.
  DECREASE IN INNER DIAMETER: D2 = -1.79526E-03 IN.

DO YOU WANT TO RUN AGAIN? Y/N
► ? N                                              Statement 410
```

REFERENCES

1. The American Society of Mechanical Engineers. *Preferred Limits and Fits for Cylindrical Parts.* USA Standard B4.1—1967 (rev. 1974). New York.
2. The American Society of Mechanical Engineers. *Dimensioning and Tolerancing.* ANSI Standard Y14.5—1982. New York.
3. Shigley, J. E., and Mitchell, L. D. *Mechanical Engineering Design,* 4th ed. New York: McGraw-Hill Book Company, 1983.

PROBLEMS

1. Specify the class of fit, the limits of size, and the limits of clearance for the bore of a slow-moving but heavily loaded conveyor roller that must rotate freely on a stationary shaft under heavy load. The nominal diameter of the shaft is 3.500 in. Use the basic hole system.
2. A sine plate is a measuring device that pivots at its base, allowing the plate to assume different angles that are set with gage blocks for precision. The pin in the pivot has a nominal diameter of 0.5000 in. Specify the class of fit, the limits of size, and the limits of clearance for the pivot. Use the basic hole system.
3. A child's toy wagon has an axle with a nominal diameter of ⅝ in. For the wheel bore and axle, specify the class of fit, the limits of size, and the limits of clearance. Use the basic shaft system.
4. The planet gear of an epicyclic gear train must rotate reliably on its shaft while maintaining accurate position with respect to the mating gears. For the planet gear bore and its shaft, specify the class of fit, the limits of size, and the limits of clearance. Use the basic hole system. The nominal shaft diameter is 0.800 in.
5. The base of a hydraulic cylinder is mounted to the frame of a machine by means of a clevis joint, which allows the cylinder to oscillate during operation. The clevis must provide reliable motion, but some play is acceptable. For the clevis holes and the pin, which have a nominal diameter of 1.25 in, specify the class of fit, the limits of size, and the limits of clearance. Use the basic hole system.
6. A heavy door on a furnace swings upward to permit access to the interior of the furnace. During various

modes of operation, the door and its hinge assembly see temperatures of 50–500°F. The nominal diameter of each hinge pin is 4.00 in. For the hinge and its pin, specify the class of fit, the limits of size, and the limits of clearance. Use the basic shaft system.

7. The stage of an industrial microscope pivots to permit the mounting of a variety of shapes. The stage must move with precision and reliably under widely varying temperatures. For the pin mount for the stage, having a nominal diameter of 0.750 in, specify the class of fit, the limits of size, and the limits of clearance. Use the basic hole system.
8. An advertising sign is suspended from a horizontal rod and is allowed to swing under wind forces. The rod is to be commercial bar stock, nominally 1.50 in in diameter. For the mating hinges on the sign, specify the class of fit, the limits of size, and the limits of clearance. Use the basic shaft system.
9. A spacer made of AISI 1020 hot-rolled steel is in the form of a hollow cylinder with a nominal inside diameter of 3.25 in and an outside diameter of 4.000 in. It is to be mounted on a solid steel shaft with a heavy force fit. Specify the dimensions for the shaft and sleeve and compute the stress in the sleeve after installation. Use the basic hole system.
10. A bronze bushing, with an inside diameter of 3.50 in and a nominal outside diameter of 4.00 in is pressed into a steel sleeve having an outside diameter of 4.50 in. For an FN3 class of fit, specify the dimensions of the parts, the limits of interference, and the stresses created in the bushing and the sleeve. Use the basic hole system.
11. It is proposed to install a steel rod with a nominal diameter of 3.00 in in a hole of an aluminum cylinder with an outside diameter of 5.00 in, with an FN5 force fit. Would this be satisfactory?
12. The allowable compressive stress in the wall of an aluminum tube is 8 500 psi. Its outside diameter is 2.000 in, and the wall thickness is 0.065 in. What is the maximum amount of interference between this tube and a steel sleeve that can be tolerated? The sleeve has an outside diameter of 3.00 in.
13. To what temperature would the spacer of problem 9 have to be heated so it would slide over the shaft with a clearance of 0.002 in? The ambient temperature is 75°F.
14. For the bronze bushing and the steel sleeve of problem 10, the ambient temperature is 75°F. How much would the bronze bushing shrink if placed in a freezer at −20°F? Then to what temperature would the steel sleeve have to be heated to provide a clearance of 0.004 in for assembly on the cold bushing?
15. For the bronze bushing of problem 10, what would be the final inside diameter after assembly into the sleeve if it started with an ID of 3.5000 in?

9 Shaft Design

9-1 OVERVIEW

A shaft is the component of mechanical devices that transmits rotational motion and power. It is integral to such devices as gear type speed reducers, belt or chain drives, conveyors, pumps, fans, agitators, and many types of automation equipment. In the process of transmitting power at a given rotational speed, the shaft is inherently subjected to a torsional moment, or torque. Thus, torsional shear stress is developed in the shaft. Also, a shaft usually carries power-transmitting components such as gears, belt sheaves, or chain sprockets, which exert forces on the shaft in the *transverse direction* (perpendicular to its axis). These transverse forces cause bending moments to be developed in the shaft, requiring analysis of the stress due to bending.

Because of the simultaneous occurrence of torsional shear stresses and normal stresses due to bending, the stress analysis virtually always involves the use of a combined stress approach. The approach most often used for shaft design and analysis is the *maximum shear stress theory of failure*. This theory was introduced in Chapter 5 and will be discussed more fully in section 9-4. Vertical shear stresses and direct normal stresses due to axial loads also occur at times, but they typically have such a small effect that they can be neglected. On very short shafts or on portions of shafts where no bending or torsion occur, such stresses may be dominant. For a shaft that is not rotating, the discussions in Chapters 3, 4, and 5 explain the appropriate analysis.

The specific tasks to be performed in the design and analysis of a shaft depend on the shaft's proposed design in addition to the manner of loading and support. With this in mind, the following is a recommended procedure for the design of a shaft.

1. Determine the rotational speed of the shaft.
2. Determine the power or the torque to be transmitted by the shaft.
3. Determine the design of the power-transmitting components or other devices that will be mounted on the shaft and specify the required location of each device.
4. Specify the location of bearings to support the shaft. The reactions on bearings supporting radial loads are assumed to act at the midpoint of the bearings. For example, if a single-row ball bearing is used, the load path is assumed to pass directly through the balls. If thrust (axial) loads exist in the shaft, you must specify which bearing is to be designed to react against the thrust load. Then the bearing that does not resist the thrust should be permitted to move slightly in the axial direction to ensure that no unexpected and undesirable thrust load is exerted on that bearing.

   Another important concept is that normally two and only two bearings are used to support a shaft. They should be placed on either side of the power-transmitting elements if possible to provide stable support for the shaft and to produce reasonably well-balanced loading of the bearings. The bearings should be placed close to the power-transmitting elements to minimize bending moments. Also, the overall length of the shaft should be kept small to keep deflections at reasonable levels.
5. Propose the general form of the geometry for the shaft, considering how each element on the shaft will be held in position axially and how power transmission from each element to the shaft is to take place. For example, consider the shaft in Figure 9-1, which is to carry two gears as the intermediate shaft in a double-reduction spur gear type speed reducer. Gear A accepts power from gear P by way of the input shaft. The power is transmitted from gear A to the shaft through the key at the interface between the gear hub and the shaft. The power is then transmitted down the shaft to point C,

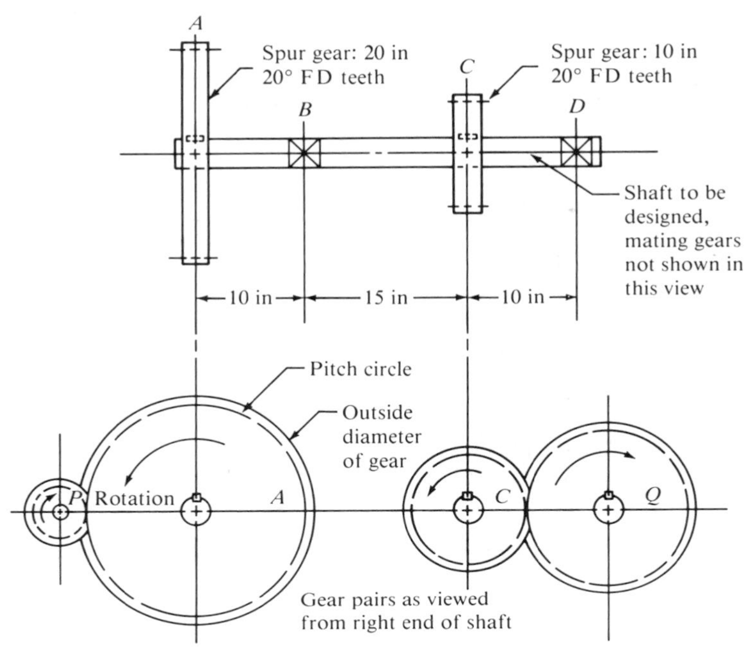

Figure 9-1 Intermediate Shaft for a Double-reduction Spur Gear Type Speed Reducer

where it passes through another key into gear C. Gear C then transmits the power to gear Q and thus to the output shaft.

It is now decided that the bearings will be placed at points B and D on the shaft to be designed. But how will the bearings and the gears be located so as to ensure that they stay in position during operation, handling, shipping, and so forth? Of course, there are many ways to do this. One way is proposed in Figure 9-2. Shoulders are to be machined in the shaft to provide surfaces against which to seat the bearings and the gears on one side of each element. The gears are restrained on the other side by retaining rings snapped into grooves in the shaft. The bearings will be held in position by the housing acting on the outer races of the bearings. Keyseats will be machined in the shaft at the location of each gear. This proposed geometry provides for positive location of each element.

6. Determine the magnitude of torque that the shaft sees at all points. It is recommended that a torque diagram be prepared, as will be shown later.
7. Determine the forces that are exerted on the shaft, both radially and axially.
8. Resolve the forces into components in perpendicular directions, usually vertically and horizontally.
9. Solve for the reactions on all support bearings in each plane.
10. Produce the complete shearing force and bending moment diagrams to determine the distribution of bending moments in the shaft.

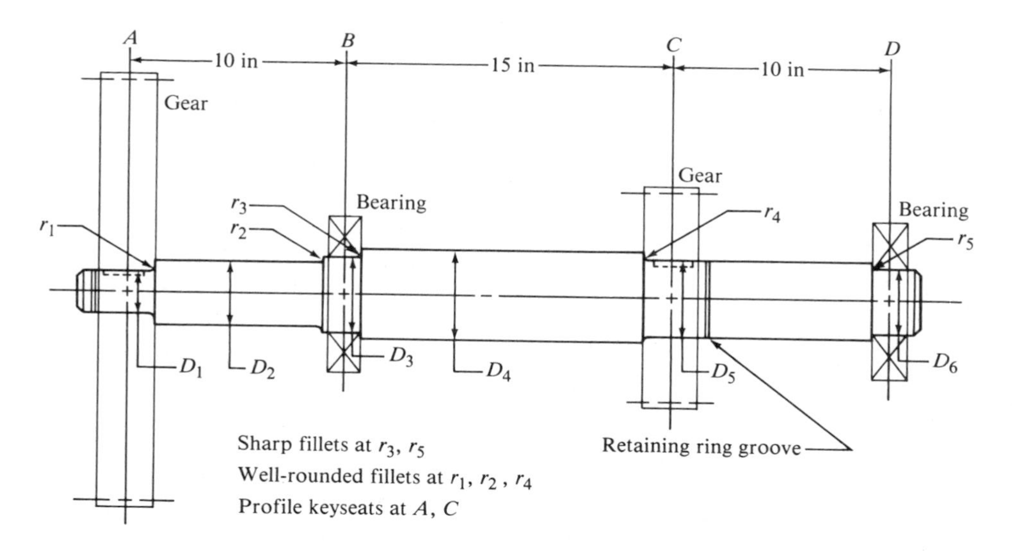

Figure 9-2 Proposed Geometry for Shaft in Figure 9-1. Sharp fillets at r_3, r_5; well-rounded fillets at r_1, r_2, r_4; profile keyseats at A, C.

11. Select the material from which the shaft will be made and specify its condition: cold-drawn, heat-treated, and so on.
12. Determine an appropriate design stress, considering the manner of loading (smooth, shock, repeated and reversed, or other).
13. Analyze each critical point of the shaft to determine the minimum acceptable diameter of the shaft at that point in order to ensure safety under the loading at that point. In general, the critical points are several and include those where a change of diameter takes place, where the higher values of torque and bending moment occur, and where stress concentrations occur.
14. Specify the final dimensions for each point on the shaft. Normally, the results from step 13 are used as a guide, and convenient values are then chosen. Design details such as tolerances, fillet radii, shoulder heights, and keyseat dimensions must also be specified. Sometimes the size and tolerance for a shaft diameter are dictated by the element to be mounted there. For example, ball bearing manufacturers' catalogs give recommended limits for bearing seat diameters on shafts.

This process will be demonstrated after the concepts of force and stress analysis are presented.

9-2 FORCES EXERTED ON SHAFTS BY MACHINE ELEMENTS

Gears, belt sheaves, chain sprockets, and other elements typically carried by shafts exert forces on the shaft that cause bending moments. The following is a discussion of the methods for computing these forces for some cases. In general, you will have to use the principles of statics and dynamics to determine the forces for any particular element.

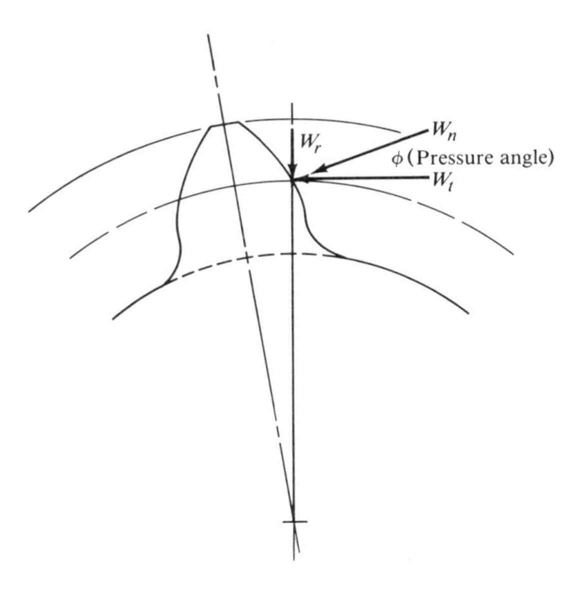

Figure 9-3 Forces on Gear Teeth

Spur Gears

The force exerted on a gear tooth during power transmission acts normal (perpendicular) to the involute tooth profile, as discussed in Chapter 11 and shown in Figure 9-3. It is convenient for the analysis of shafts to consider the rectangular components of this force acting in the radial and tangential directions. It is most convenient to compute the tangential force, W_t, directly from the known torque being transmitted by the gear.

$$T = 63\,000(P)/n \tag{9-1}$$

$$W_t = T/(D/2) \tag{9-2}$$

where P is the power being transmitted in hp, n is the rotational speed in rpm, T is the torque on the gear in pound-inches, and D is the pitch diameter of the gear in inches.

The angle between the total force and the tangential component is equal to the pressure angle, ϕ, of the tooth form. Thus, if the tangential force is known, the radial force can be computed directly from

$$W_r = W_t \tan \phi \tag{9-3}$$

and there is no need to compute the total force at all. For gears, the pressure angle is typically 14½, 20, or 25 degrees.

Helical Gears

In addition to the tangential and radial forces encountered with spur gears, helical gears produce an axial force (as discussed in Chapter 12). First compute the tangential force from equations (9-1) and (9-2). Then, if the helix angle of the gear is ψ, and the normal pressure angle is ϕ_n, the radial load can be computed from

$$W_r = W_t \tan \phi_n / \cos \psi \tag{9-4}$$

The axial load is

$$W_a = W_t \tan \psi \tag{9-5}$$

Bevel Gears

Refer to Chapter 12 to review the formulas for the three components of the total force on bevel gear teeth in the tangential, radial, and axial directions.

Worms and Wormgears

Chapter 12 also gives the formulas for computing the forces on worms and wormgears in the tangential, radial, and axial directions.

Chain Sprockets

Figure 9-4 shows a pair of chain sprockets transmitting power. The upper part of the chain is in tension and produces the torque on either sprocket. The lower part of the chain is referred to as the *slack side* and exerts no force on either sprocket. Therefore the total bending force on the shaft carrying the sprocket is equal to the tension in the tight side of the chain. If the torque on a certain sprocket is known,

$$F_c = T/(D/2) \tag{9-6}$$

where D is the pitch diameter of that sprocket.

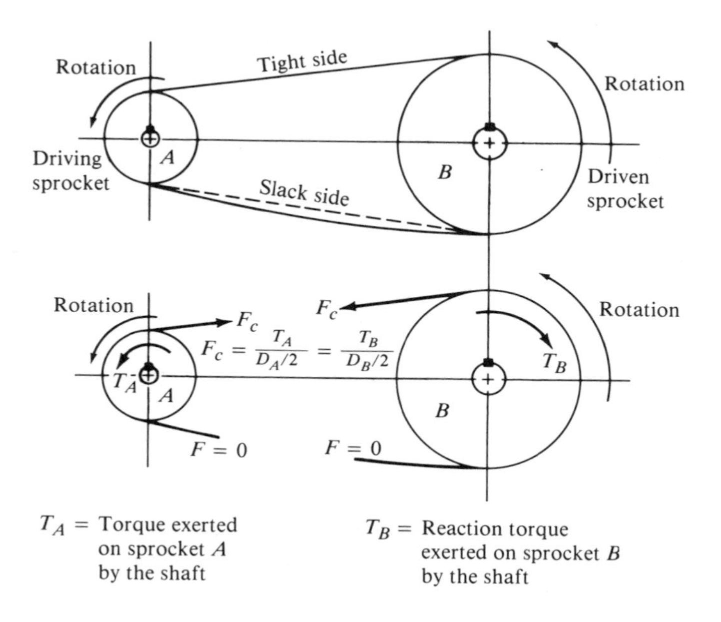

Figure 9-4 Forces on Chain Sprockets

V-belt Sheaves

The general appearance of the V-belt drive system looks similar to the chain drive system. But there is one important difference. Both sides of the V-belt are in tension, as indicated in Figure 9-5. The tight side tension, F_1, is greater than the "slack side" tension, F_2, and thus there is a net driving force on the sheaves equal to

$$F_N = F_1 - F_2 \tag{9-7}$$

The magnitude of the net driving force can be computed from the torque transmitted,

$$F_N = T/(D/2) \tag{9-8}$$

But notice that the bending force on the shaft carrying the sheave is dependent on the *sum*, $F_1 + F_2 = F_B$. To be more precise, the components of F_1 and F_2 parallel to the line of centers of the two sprockets should be used. But unless the two sprockets are radically different in diameter, little error will result from $F_B = F_1 + F_2$.

To determine the bending force, F_B, a second equation involving the two forces F_1 and F_2 is needed. This is provided by assuming a ratio of the tight side tension to the slack side tension. For V-belt drives, the ratio is normally taken to be

$$F_1/F_2 = 5 \tag{9-9}$$

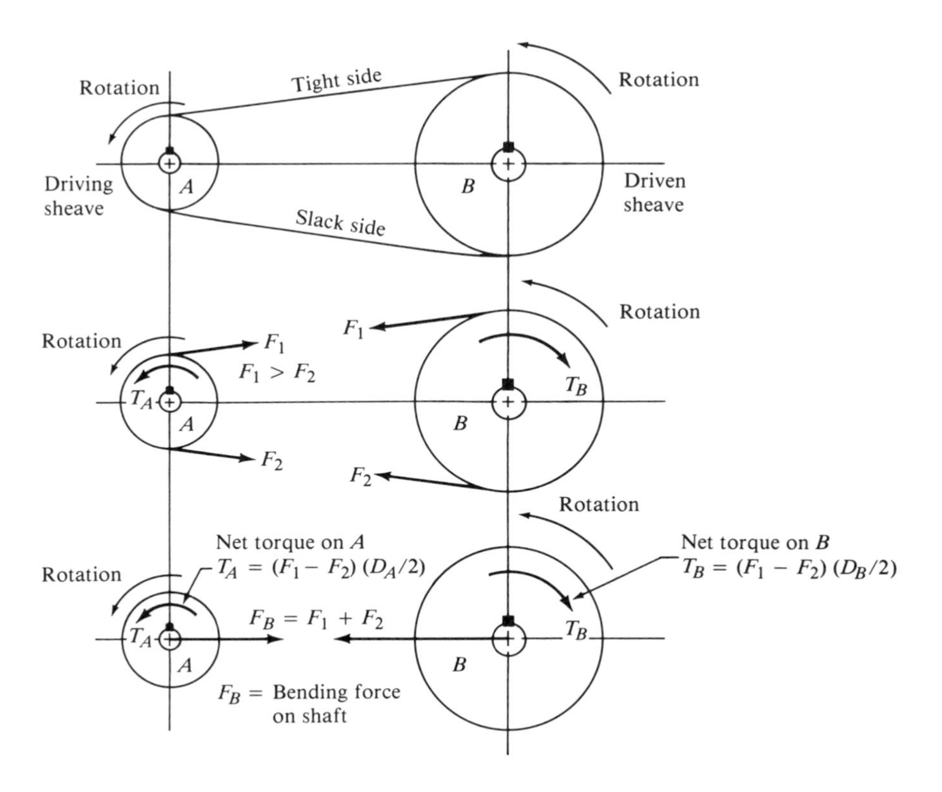

Figure 9-5 Forces on Belt Sheaves or Pulleys

It is convenient to derive a relationship between F_N and F_B of the form

$$F_B = CF_N \tag{9-10}$$

where C is a constant to be determined.

$$C = \frac{F_B}{F_N} = \frac{F_1 + F_2}{F_1 - F_2} \tag{9-11}$$

But from equation (9-9), $F_1 = 5F_2$. Then,

$$C = \frac{F_1 + F_2}{F_1 - F_2} = \frac{5F_2 + F_2}{5F_2 - F_2} = \frac{6F_2}{4F_2} = 1.5$$

Equation (9-10) then becomes, for V-belt drives,

$$F_B = 1.5F_N = 1.5T/(D/2) \tag{9-12}$$

It is customary to consider the bending force, F_B, to act as a single force in the direction along the line of centers of the two sheaves as shown in Figure 9-5.

Flat Belt Pulleys

The analysis of the bending force exerted on shafts by flat belt pulleys is identical to that for V-belt sheaves except that the ratio of the tight side to the slack side tension is typically taken to be 3 instead of 5. Using the same logic as with V-belt sheaves, we can compute the constant C to be 2.0. Then, for flat belt drives,

$$F_B = 2.0F_N = 2.0T/(D/2) \tag{9-13}$$

Flexible Couplings

More detailed discussion of flexible couplings is presented in Chapter 10, but it is important to observe here how the use of a flexible coupling affects the design of a shaft.

A flexible coupling is used to transmit power between shafts while accommodating minor misalignments in the radial, angular, or axial directions. Thus, the shafts adjacent to the couplings are subjected to torsion but the misalignments cause no axial or bending loads.

9-3 STRESS CONCENTRATIONS IN SHAFTS

In order to mount and locate the several types of machine elements on shafts properly, typically a final design contains several diameters, keyseats, ring grooves, and other geometric discontinuities that create stress concentrations. The shaft design proposed in Figure 9-2 is an example of this observation.

These stress concentrations must be taken into account during the design analysis. But a problem exists because the true design values of the stress concentration factors are unknown at the start of the design process. Most of the values are dependent on the diameters of the shaft and on the fillet and groove geometries, and these are the objectives of the design.

This dilemma can be overcome by establishing a set of preliminary design values for commonly encountered stress concentration factors, which can be used to produce initial estimates for the minimum acceptable shaft diameters. Then, after the refined dimensions are selected, the final geometry can be analyzed to determine the real values for stress concentration factors. Comparing the final values with the preliminary values will enable you to judge the acceptability of the design. The charts from which the final values can be determined are found in Appendixes A-15-1 and A-15-4.

Preliminary Design Values

Considered here are the types of geometric discontinuities most often found in power-transmitting shafts: keyseats, shoulder fillets, and retaining ring grooves. In each case, a suggested design value is relatively high in order to produce a conservative result for the first approximation to the design. Again it is emphasized that the final design should be checked for safety. That is, if the final value is less than the original design value, the design is still safe. Conversely, if the final value is higher, the stress analysis for the design has to be rechecked.

Keyseats

A *keyseat* is a longitudinal groove cut into a shaft for the mounting of a key, permitting the transfer of torque from the shaft to a power-transmitting element, or vice versa. The detail design of keys is covered in Chapter 10.

Two types of keyseats are most frequently used, profile and sled runner (see Figure 9-6). The profile keyseat is milled into the shaft, using an end mill having a

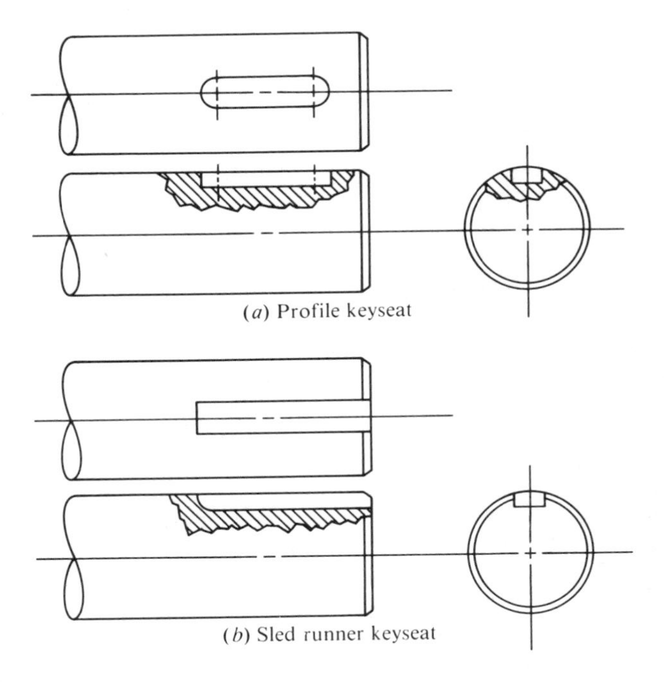

Figure 9-6 Keyseats

diameter equal to the width of the key. The resulting groove is flat-bottomed and has a sharp, square corner at its end. The sled runner keyseat is produced by a circular milling cutter having a width equal to the width of the key. As the cutter begins or ends the keyseat, it produces a smooth radius. For this reason, the stress concentration factor for the sled runner keyseat is lower than that for the profile keyseat. Normally used design values are

$$K_t = 2.0 \quad \text{(profile)}$$
$$K_t = 1.6 \quad \text{(sled runner)}$$

Each of these is to be applied to the bending stress calculation for the shaft, using the full diameter of the shaft. The factors take into account both the reduction in cross section and the effect of the discontinuity. Consult the references listed for more detail about stress concentration factors for keyseats (4). If the torsional shear stress is fluctuating rather than steady, the stress concentration factor is also applied to that.

Shoulder Fillets

When a change in diameter occurs in a shaft to create a shoulder against which to locate a machine element, a stress concentration dependent on the ratio of the two diameters and on the radius in the fillet is produced (see Figure 9-7). It is recommended that the fillet radius be as large as possible to minimize the stress concentration, but at times the design of the gear, bearing, or other element affects the radius that can be used. For the purpose of design, we will classify fillets into two categories, sharp and well-rounded.

The term *sharp* here does not mean truly sharp, without any fillet radius at all. Such a shoulder configuration would have a very high stress concentration factor and should be avoided. Instead, *sharp* describes a shoulder with a relatively small fillet radius. One situation in which this is likely to occur is where a ball or roller bearing is to be located. The inner race of the bearing has a factory-produced radius, but it is small. The fillet radius on the shaft must be smaller yet in order for the bearing to be seated properly against the shoulder. When an element with a large chamfer on its bore is located against the shoulder or when nothing at all seats against the shoulder, the fillet radius can be much larger (*well-rounded*) and the corresponding stress concentration factor is smaller. We will use the following values for design for bending:

$$K_t = 2.5 \quad \text{(sharp fillet)}$$
$$K_t = 1.5 \quad \text{(well-rounded fillet)}$$

Referring to the chart for stress concentration factors in Appendix A-15-1, you can see that these values correspond to ratios of r/d of approximately 0.03 for the sharp fillet case and 0.17 for the well-rounded fillet for a D/d ratio of 1.50.

Retaining Ring Grooves

Retaining rings are used for many types of locating tasks in shaft applications. The rings are installed in grooves in the shaft after the element to be retained is in place. The geometry of the groove is dictated by the ring manufacturer. Its usual configuration is a shallow groove with straight side walls and bottom and a small fillet at the base of the groove. The behavior of the shaft in the vicinity of the groove can be approximated by considering two sharp-filleted shoulders positioned close together. Thus the stress concentration factor for a groove is fairly high.

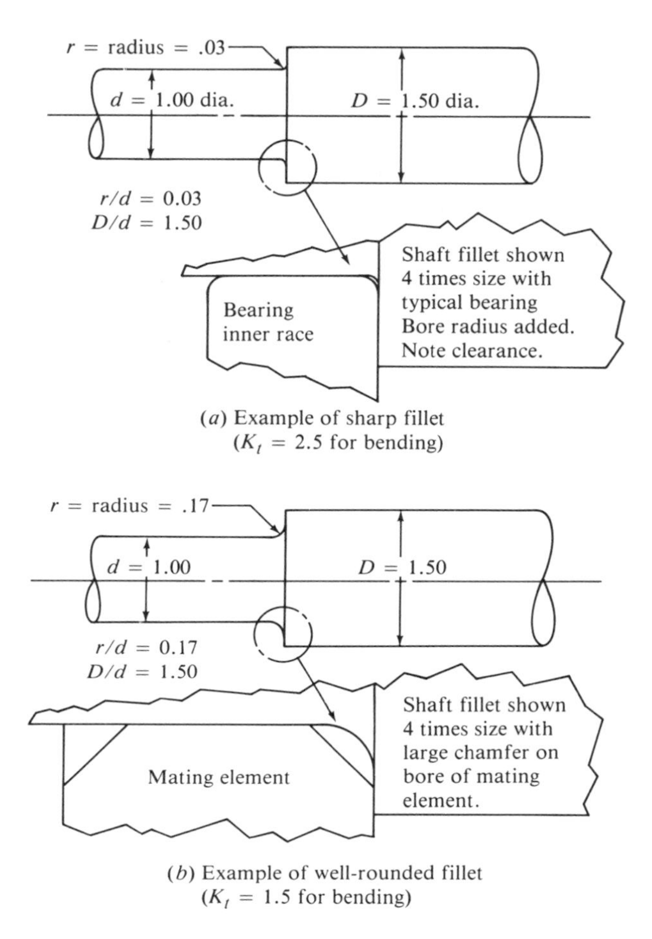

(a) Example of sharp fillet ($K_t = 2.5$ for bending)

(b) Example of well-rounded fillet ($K_t = 1.5$ for bending)

Figure 9-7 Fillets on Shafts

When bending exists, we will use $K_t = 3.0$ for preliminary design as an estimate to account for the fillets and the reduction in diameter at the groove to determine the nominal shaft diameter before the groove is cut. When torsion exists along with bending, or when only torsion exists at a section of interest, the stress concentration factor is not applied to the torsional shear stress component because it is steady. To account for the decrease in diameter at the groove, however, increase the resulting computed diameter by approximately 6 percent, a typical value for commercial retaining ring grooves. But after the final shaft diameter and groove geometry are specified, the stress in the groove should be computed with the appropriate stress concentration factor for the groove geometry.

9-4 DESIGN STRESSES FOR SHAFTS

In a given shaft, there can be several different stress conditions existing at the same time. For any part of the shaft that transmits power, there will be a torsional shear stress; a bending stress is usually present on the same parts. There may be other parts where only bending stresses occur. Some points may not be subjected to either bending or torsion but will experience vertical shearing stress. Axial tensile or compressive stresses may be superimposed on the other stresses. Then there may be some points where no significant stresses at all are created.

Thus, the decision of what design stress to use depends on the particular situation at the point of interest. In many shaft design and analysis projects, computations must be done at several points to account completely for the variety of loading and geometry conditions that exist.

In Chapter 5, several cases discussed for computing design factors, N, are useful for determining design stresses for shaft design. The bending stresses will be assumed to be completely reversed and repeated because of the rotation of the shaft. Because ductile materials perform better under such loads, it will be assumed that the material for the shaft is ductile. It will also be assumed that the torsional loading is relatively constant and acting in one direction. If other situations exist, consult the appropriate case from Chapter 5.

The symbol τ_d will be used for the design stress when a shear stress is the basis for the design. The symbol σ_d will be used when a normal stress is the basis.

For torsional load, vertical shear, or combined stress:

$$\tau_d = 0.5s_y/N \tag{9-14}$$

For reversed bending:

$$\sigma_d = s_n'/N \tag{9-15}$$

For typical steel shafts we will use the values in Figure 5-6 as a rough estimate for the endurance strength and apply a size factor for larger shafts. Another approach is to estimate the endurance strength to be $0.5s_u$, as suggested in Chapter 5. Then,

$$\sigma_d = 0.5s_u/N \tag{9-16}$$

But in this case, a higher-than-average value for N should be used to account for any less than ideal conditions.

Steady Normal Stress

Axial loads exerted on shafts, such as those carrying helical, bevel, or worm gears, produce a steady axial tensile or compressive stress. Frequently these stresses are minor, compared with bending and torsion. When this stress is significantly high and superimposed on the bending and torsion, the methods of combined stress must be used. When the steady axial stress exists alone, the design stress is

$$\sigma_d = s_y/N \tag{9-17}$$

Design Factor, N

Under typical industrial conditions the design factor of $N = 3$ is recommended. If the application is very smooth, a value as low as $N = 2$ may be justified. Under conditions of shock or impact, $N = 4$ or higher should be used, and careful testing is advised.

9-5 SHAFTS IN BENDING AND TORSION ONLY

Examples of shafts subjected to bending and torsion only are those carrying spur gears, V-belt sheaves, or chain sprockets. The power being transmitted causes the torsion, and the transverse forces on the elements cause bending. In the general case, the transverse forces do not all act in the same plane. In such cases, the bending moment diagrams for two perpendicular planes are prepared first. Then the resultant bending moment at each point of interest is determined. The process is illustrated in example problem 9-1.

At a point where both bending and torsion occur, the maximum shear stress theory of failure is used in the design process, while also applying the Soderberg criterion to account for the fluctuating bending stress.

For rotating, solid circular shafts, the bending stress due to a bending moment, M, and a stress concentration factor, K_t, is

$$\sigma = K_t M/Z \tag{9-18}$$

where $Z = \pi D^3/32$ is the rectangular section modulus. The torsional shear stress is

$$\tau = T/Z_p \tag{9-19}$$

where $Z_p = \pi D^3/16$ is the polar section modulus. Note that $Z_p = 2Z$ and that, therefore,

$$\tau = T/(2Z)$$

No stress concentration factor is included here because the torque does not vary.

By using these equations and the Soderberg approach for combining the steady torsional shear stress and the fluctuating bending stress (similar to cases G and H in section 5-3), the following design equation can be developed:

$$\frac{1}{N} = \frac{\sqrt{(K_t M/s_n')^2 + (T/s_y)^2}}{Z} \tag{9-20}$$

For design, the required section modulus is

$$Z = N\sqrt{(K_t M/s_n')^2 + (T/s_y)^2} \tag{9-21}$$

Using $Z = \pi D^3/32$, the required diameter would be

$$D = [(32N/\pi)\sqrt{(K_t M/s_n')^2 + (T/s_y)^2}]^{1/3} \tag{9-22}$$

The complete development of equation (9-20) can be found in reference 5 or 6.

9-6 SHAFT DESIGN EXAMPLE

Example Problem 9-1. Design the shaft shown in Figures 9-1 and 9-2 if it is to be machined from AISI 1144 OQT 1000 steel. The shaft is part of the drive for a large blower system supplying air to a furnace. Gear A receives 200 hp from gear P. Gear C delivers the power to gear Q. The shaft rotates at 600 rpm.

Solution. First determine the properties of the steel for the shaft. From Appendix A-4, $s_y = 83\,000$ psi, $s_u = 118\,000$ psi, and the percent elongation is 19 percent. Thus the material has good ductility. Using Figure 5-6 we can estimate $s_n = 42\,000$ psi. A size factor of 0.80 is applied because the shaft will probably be much larger than 0.30 in. Then,

$$s_n' = 0.80 s_n = 0.80(42\,000) = 33\,600 \text{ psi}$$

The design factor is taken to be $N = 3$. The blower is not expected to present any unusual shock or impact.

Now we can compute the torque in the shaft from equation (9-1):

$$T = 63\,000(P)/n = 63\,000(200)/600 = 21\,000 \text{ lb} \cdot \text{in}$$

Note that only that part of the shaft from A to C is subjected to this torque. There is zero torque from the right of gear C over to bearing D.

Forces on the gears: Figure 9-8 shows the two pairs of gears with the forces acting *on gears* A *and* C *shown*. You should observe that gear A is driven by gear P, and gear C drives gear Q. It is very important for the directions of these forces to be correct. The values of the forces are found from equations (9-2) and (9-3).

$$W_{tA} = T_A/(D_A/2) = 21\,000/(20/2) = 2\,100 \text{ lb}$$

$$W_{rA} = W_{tA} \tan(\phi) = 2\,100 \tan(20°) = 764 \text{ lb}$$

$$W_{tC} = T_C/(D_C/2) = 21\,000/(10/2) = 4\,200 \text{ lb}$$

$$W_{rC} = W_{tC} \tan(\phi) = 4\,200 \tan(20°) = 1\,529 \text{ lb}$$

The next step is to show these forces on the shaft in their proper planes of action and in the proper direction. The reactions at the bearings are computed, and the shearing force and bending moment diagrams are prepared. The results are shown in Figure 9-9.

The design will continue by computing the minimum acceptable diameter of the shaft at several points along the shaft. At each point, we will observe the magnitude of torque and bending moment that exist at the point and estimate the value of any stress

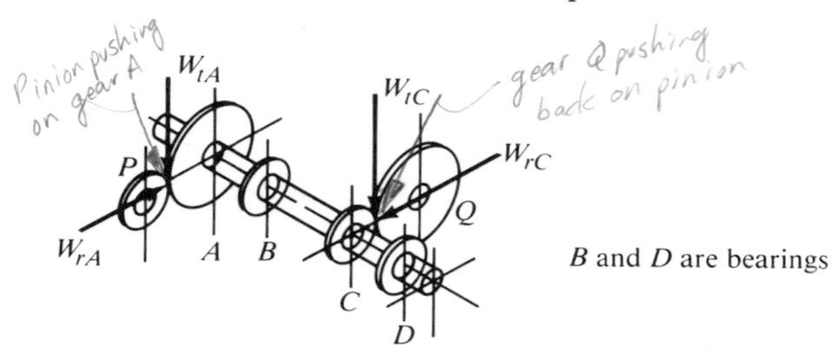

(*a*) Pictorial view of forces on gears A and C

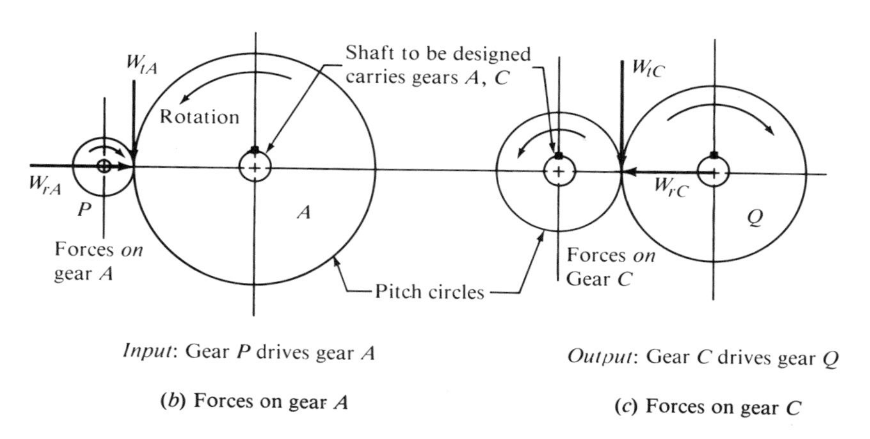

Input: Gear P drives gear A

Output: Gear C drives gear Q

(*b*) Forces on gear A

(*c*) Forces on gear C

Figure 9-8 Forces on Gears *A* and *C*

concentration factors. If more than one stress concentration exists in the vicinity of the point of interest, the larger value is used for design. This assumes that the geometric discontinuities themselves do not interact, which is good practice. For example, at point A, the keyseat should end well before the shoulder fillet begins.

Point A

Because gear A is located at point A, the shaft from A and to the right is subjected to torque. To the left of A, where there is a retaining ring, there are no forces, moments, or torques. Because it is the free end of the shaft, there is no bending moment at A. Stress concentrations can be neglected because the torque is steady. Then, from equations (9-14) and (9-19),

$$\tau_d = 0.5s_y/N = 0.5(83\,000)/3 = 13\,833 \text{ psi}$$

Letting $\tau = \tau_d$ in equation (9-19) and solving for Z_p gives

$$Z_p = T/\tau_d = 21\,000/13\,833 = 1.518 \text{ in}^3$$

But because $Z_p = \pi D_1^3/16$,

$$D_1 = \sqrt[3]{16(Z_p)/\pi} = \sqrt[3]{16(1.518)/\pi} = 1.98 \text{ in}$$

Point B

Point B is the location of a bearing with a sharp fillet to the right of B and a well-rounded fillet to the left. It is desirable to make D_2 at least slightly smaller than D_3 at the bearing

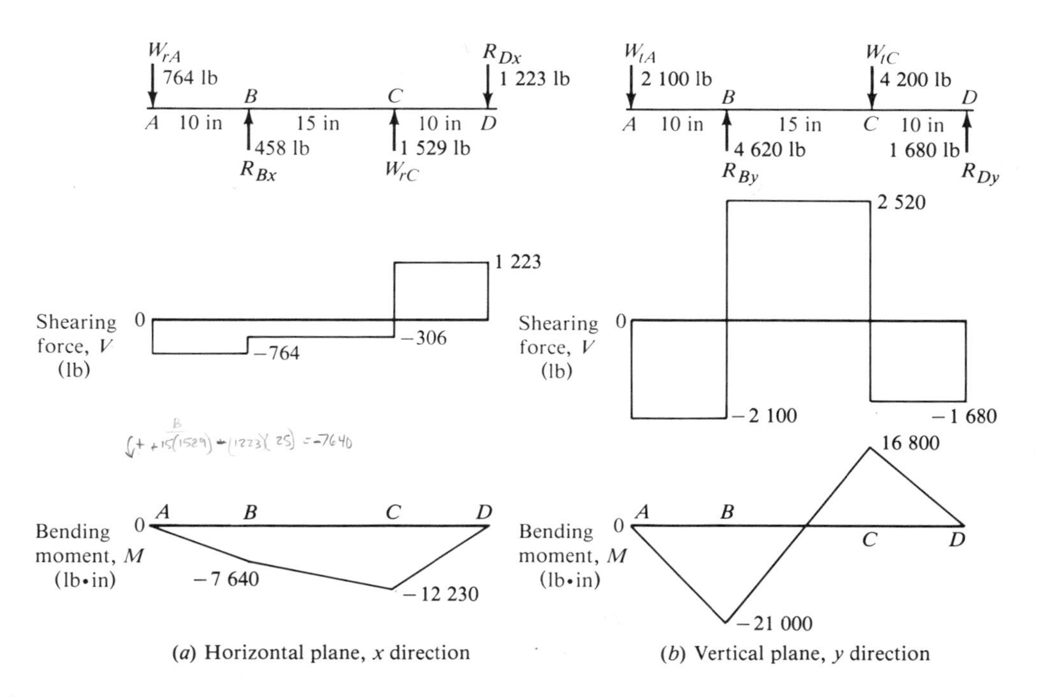

Figure 9-9 Load, Shear, and Moment Diagrams for Shaft in Figure 9-8

seat to permit the bearing to be slid easily onto the shaft up to the place where it is pressed to its final position. There is usually a light press fit between the bearing bore and the shaft seat.

To the left of B (diameter D_2):

$$T = 21\,000 \text{ lb} \cdot \text{in}$$

The bending moment at B is the resultant of the moment in the x and y planes from Figure 9-9.

$$M_B = \sqrt{M_{Bx}^2 + M_{By}^2} = \sqrt{(21\,000)^2 + (7\,640)^2} = 22\,350 \text{ lb} \cdot \text{in}$$

$$K_t = 1.5 \quad \text{(well-rounded fillet)}$$

Using equation (9-22) because of the combined stress condition,

$$D_2 = [(32N/\pi)\sqrt{(K_t M/s_n')^2 + (T/s_y)^2}]^{1/3} \tag{9-22}$$

$$D_2 = \left[\frac{32(3)}{\pi}\sqrt{\left[\frac{1.5(22\,350)}{33\,600}\right]^2 + \left[\frac{21\,000}{83\,000}\right]^2}\right]^{1/3} = 3.16 \text{ in}$$

At B and to the right of B (diameter D_3) everything is the same, except the value of $K_t = 2.5$ for the sharp fillet. Then,

$$D_3 = \left[\frac{32(3)}{\pi}\sqrt{\left[\frac{2.5(22\,350)}{33\,600}\right]^2 + \left[\frac{21\,000}{83\,000}\right]^2}\right]^{1/3} = 3.72 \text{ in}$$

Notice that D_4 will be larger than D_3 in order to provide a shoulder for the bearing. Therefore, it will be safe. Its actual diameter will be specified after completing the stress analysis and after selecting the bearing at B. The bearing manufacturer's catalog will specify the minimum acceptable diameter to the right of the bearing to provide a suitable shoulder against which to seat the bearing.

Point *C*

Point C is the location of gear C with a well-rounded fillet to the left, a profile keyseat at the gear, and a retaining ring groove to the right. The use of a well-rounded fillet at this point is actually a design decision that requires that the design of the bore of the gear accommodate a large fillet. Usually, this means that a chamfer is produced at the ends of the bore. The bending moment at C is

$$M_C = \sqrt{M_{Cx}^2 + M_{Cy}^2} = \sqrt{(12\,230)^2 + (16\,800)^2} = 20\,780 \text{ lb} \cdot \text{in}$$

To the left of C the torque of 21 000 lb · in exists with the profile keyseat giving $K_t = 2.0$. Then,

$$D_5 = \left[\frac{32(3)}{\pi}\sqrt{\left[\frac{2.0(20\,780)}{33\,600}\right]^2 + \left[\frac{21\,000}{83\,000}\right]^2}\right]^{1/3} = 3.38 \text{ in}$$

To the right of C there is no torque, but the ring groove suggests $K_t = 3.0$ for design, and there is reversed bending. Using equations (9-15) and (9-18),

$$\sigma_d = s_n'/N = 33\,600/3 = 11\,200 \text{ psi}$$

Then let $\sigma = \sigma_d$ in equation (9-18) and compute the required Z.

$$Z = K_t M_C / \sigma_d = 3.0(20\,780)/11\,200 = 5.57 \text{ in}^3$$

But $Z = \pi D_5^3/32$. Then,

$$D_5 = \sqrt[3]{32Z/\pi} = \sqrt[3]{32(5.57)/\pi} = 3.84 \text{ in}$$

This value is higher than that computed for the left of C so it governs the design at point C.

Point *D*

Point D is the seat for bearing D and there is no torque or bending moment here. However, there is a vertical shearing force equal to the reaction at the bearing. Using the resultant of the x and y plane reactions,

$$V_D = \sqrt{(1\,223)^2 + (1\,680)^2} = 2\,078 \text{ lb}$$

The vertical shear stress for a solid circular shaft is

$$\tau = 4V/3A$$

Letting $\tau = \tau_d$, the required area of the shaft is

$$A = 4V/3\tau_d = 4(2\,078)/(3)(13\,833) = 0.200 \text{ in}^2$$

From $A = \pi D_6^2/4$, and solving for D_6,

$$D_6 = \sqrt{4A/\pi} = \sqrt{4(0.200)/\pi} = 0.505 \text{ in}$$

This is very small as compared to the other computed diameters, and it will usually be so. In reality the diameter at D will probably be made much larger than this computed value because of the size of a reasonable bearing to carry the radial load of 2 078 lb.

Summary

The computed minimum required diameters for the various parts of the shaft in Figure 9-2 are

$$D_1 = 1.98 \text{ in} \qquad D_5 = 3.84 \text{ in}$$
$$D_2 = 3.16 \text{ in} \qquad D_6 = 0.51 \text{ in}$$
$$D_3 = 3.72 \text{ in}$$

Also, D_4 must be somewhat greater than 3.84 in order to provide adequate shoulders for gear C and bearing B.

9-7 RECOMMENDED BASIC SIZES FOR SHAFTS

When mounting a commercially available element, of course, follow the manufacturer's recommendation for the basic size of the shaft and the tolerance.

In the U.S. Customary Unit system, diameters are usually specified to be common fractions or their decimal equivalents. Tables 8-1 and 8-2 list the preferred basic decimal inch sizes that you can use for dimensions over which you have control.

When commercially available unmounted bearings are to be used on a shaft, it is likely that their bores will be in metric dimensions. Typical sizes available and their decimal equivalents are listed in Table 15-2. Other preferred metric dimensions are listed in Table 8-3.

Example Problem 9-2. Specify convenient decimal inch dimensions for the six diameters from example problem 9-1. Choose the bearing seat dimensions from Table 15-2. Choose all other dimensions from Table 8-2.

Solution. Of course, the diameters D_3 and D_6 have to be checked against the bearing capacity requirements. D_4 has to be checked to see that it provides an acceptable shoulder for the bearing at B and the gear at C. The following dimensions are recommended.

MATING PART	DIAMETER NO.	MINIMUM DIAMETER	SPECIFIED DIAMETER
	(From Problem 9-1)		*(Basic Size)*
Gear	D_1	1.98	2.000
Nothing	D_2	3.16	3.200
Bearing	D_3	3.72	3.740 2 (95 mm)
Nothing	D_4	$>D_3$ or D_5	4.400
Gear	D_5	3.84	4.000
Bearing	D_6	0.51	3.149 6 (80 mm)

9-8 ADDITIONAL DESIGN EXAMPLES

Two additional design examples are given in this section. The first is for a shaft that contains three different types of power transmission devices: a V-belt sheave, a chain sprocket, and a spur gear. Some of the forces are acting at angles other than vertical and horizontal, requiring the resolution of the bending forces on the shaft into components before creating the shearing force and bending moment diagrams. Example problem 9-4 involves a shaft carrying a wormgear and a chain sprocket. The axial force on the wormgear presents a slight modification of the design procedure. Except for these differences, the design procedure is the same as that for example problem 9-1. Thus much of the detailed manipulation of formulas is omitted.

Example Problem 9-3. The shaft shown in Figure 9-10 receives 110 hp from a water turbine through a chain sprocket at point C. The gear pair at E delivers 80 hp to an electrical generator. The V-belt sheave at A delivers 30 hp to a bucket elevator that carries grain to an elevated hopper. The shaft rotates at 1 700 rpm. The sprocket, sheave, and gear are located axially by retaining rings. The sheave and gear are keyed with sled runner keyseats, and there is a profile keyseat at the sprocket. Use AISI 1040 cold-drawn steel for the shaft. Compute the minimum acceptable diameters D_1 through D_7 as defined in Figure 9-10.

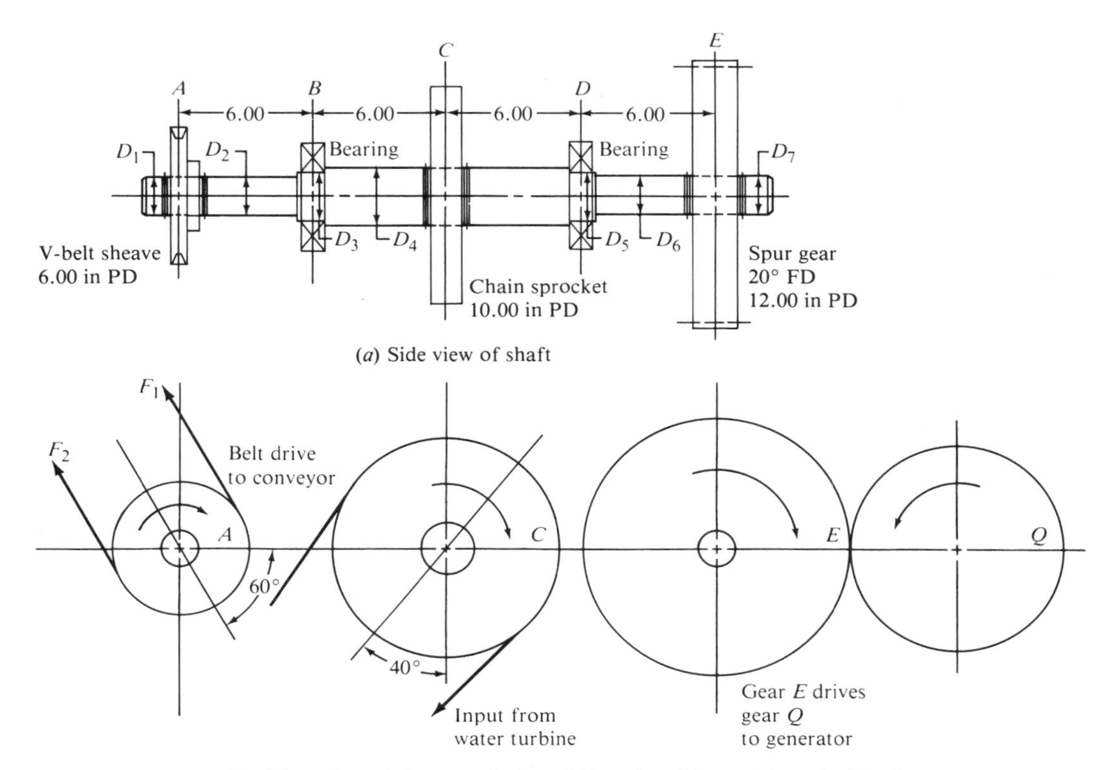

(b) Orientation of elements A, C, and E as viewed from right end of shaft

Figure 9-10 Shaft for Example Problem 9-3

Solution. First, the material properties for the AISI 1040 cold-drawn steel are found from Appendix A-3.

$$s_y = 71\,000 \text{ psi} \qquad s_u = 80\,000 \text{ psi}$$

Then from Figure 5-6, $s_n = 30\,000$ psi. Applying a size factor of 0.85 because of the moderately large size that should result gives $s_n' = 0.85(30\,000) = 25\,500$ psi.

This application is fairly smooth: a turbine drive and a generator and a conveyor at the output points. A design factor of $N = 3$ should be satisfactory.

Torque Distribution in the Shaft

Recalling that all the power comes into the shaft at C, we can then observe that 30 hp is delivered down the shaft from C to the sheave at A. Also, 80 hp is delivered down the shaft from C to the gear at E. From these observations, the torque in the shaft can be computed.

$$T = 63\,000(30)/1\,700 = 1\,112 \text{ lb} \cdot \text{in} \quad \text{from } A \text{ to } C$$

$$T = 63\,000(80)/1\,700 = 2\,965 \text{ lb} \cdot \text{in} \quad \text{from } C \text{ to } E$$

When designing the shaft at C, we will use 2 965 lb · in at C and to the right, but we can use 1 112 lb · in to the left of C. Notice that no part of the shaft is subjected to the full

110 hp that comes into the sprocket at C. The power splits into two parts as it enters the shaft. When analyzing the sprocket itself, we must use the full 110 hp and the corresponding torque.

$$T = 63\,000(110)/1\,700 = 4\,076 \text{ lb} \cdot \text{in} \quad \text{torque in the sprocket}$$

Forces

We will compute the forces at each element separately and show the component forces that act in the vertical and horizontal planes, as in example problem 9-1. Figure 9-11 shows the directions of the applied forces and their components for each element.

Forces on Sheave A

Use equations (9-7), (9-8), and (9-12).

$$F_N = F_1 - F_2 = T_A/(D_A/2) = 1\,112/3.0 = 371 \text{ lb} \quad \text{net driving force}$$

$$F_A = 1.5F_N = 1.5(371) = 556 \text{ lb} \quad \text{bending force}$$

The bending force acts upward and to the left at an angle of 60° from the horizontal. As shown in Figure 9-11, the components of the bending force are

$$F_{Ax} = F_A \cos(60°) = (556)\cos(60°) = 278 \text{ lb} \quad \text{(toward the left)}$$

$$F_{Ay} = F_A \sin(60°) = (556)\sin(60°) = 482 \text{ lb} \quad \text{(upward)}$$

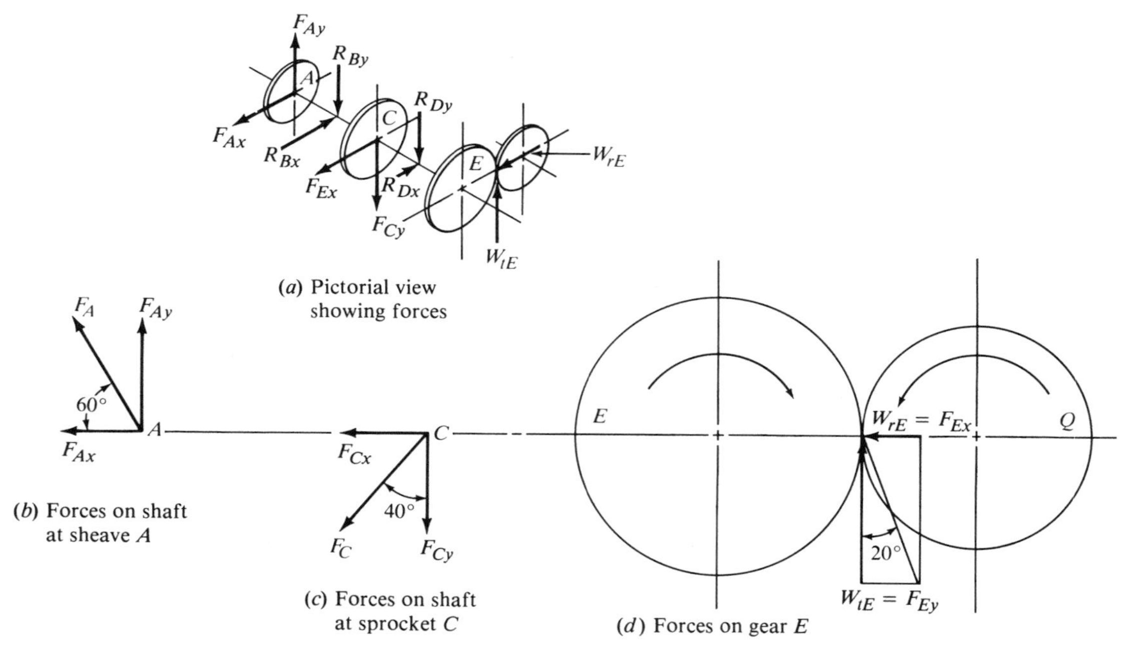

Figure 9-11 Forces Resolved into x and y Components

Forces on Sprocket C

Use equation (9-6).

$$F_C = T_C/(D_C/2) = 4076/5.0 = 815 \text{ lb}$$

This is the bending load on the shaft. The components are

$$F_{Cx} = F_C \sin(40°) = (815)\sin(40°) = 524 \text{ lb} \quad \text{(to the left)}$$
$$F_{Cy} = F_C \cos(40°) = (815)\cos(40°) = 624 \text{ lb} \quad \text{(downward)}$$

Forces on Gear E

The transmitted load is found from equation (9-2) and the radial load from equation (9-3). The directions are shown in Figure 9-11.

$$F_{Ey} = W_{tE} = T_E/(D_E/2) = 2965/6.0 = 494 \text{ lb} \quad \text{(upward)}$$
$$F_{Ex} = W_{rE} = W_{tE} \tan(\phi) = (494)\tan(20°) = 180 \text{ lb} \quad \text{(to the left)}$$

Load, Shear, and Moment Diagrams

Figure 9-12 shows the forces acting on the shaft at each element, the reactions at the bearings, and the shearing force and bending moment diagrams for both the horizontal (x) and vertical (y) planes. In the figure, the computations of the resultant bending moment at points B, C, and D are shown.

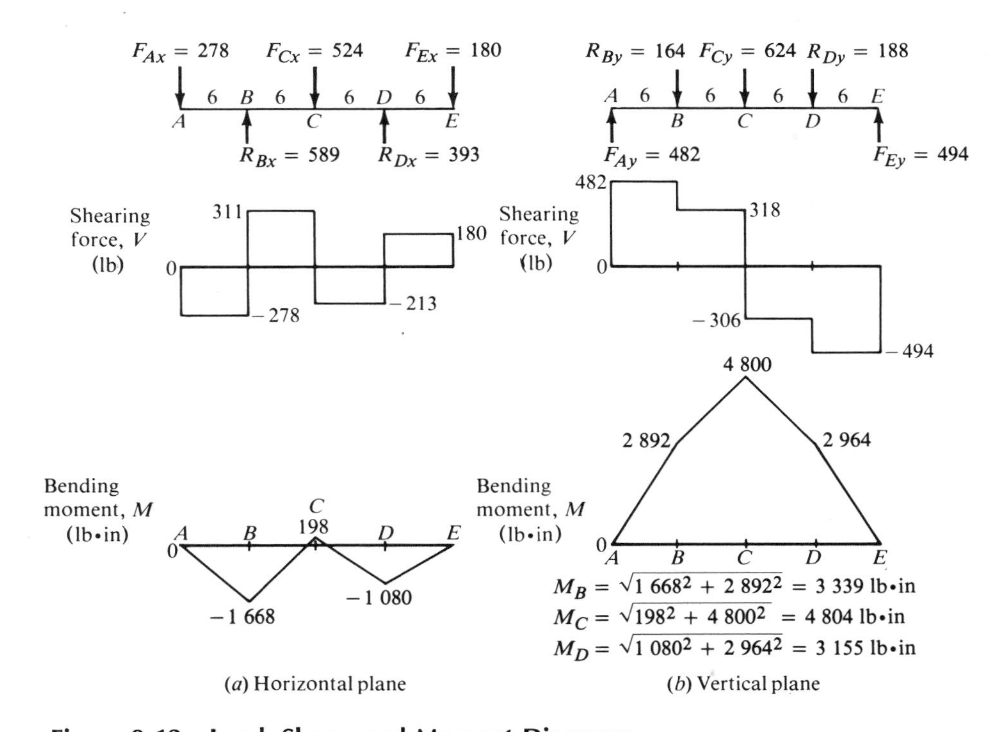

Figure 9-12 Load, Shear, and Moment Diagrams

Design of the Shaft

We will use equation (9-22) to determine the minimum acceptable diameter of the shaft at each point of interest. Because the equation requires a fairly large number of individual operations, and because we will be using it at least seven times, it may be desirable to write a computer program just for that operation. Or the use of a programmable calculator would be nearly ideal because the number of steps involved is within the range of most calculators. At the end of this chapter is the listing of such a program, which was used to compute the results given below. Note that equation (9-22) can be used even though there is only torsion or only bending by entering zero for the missing value.

Equation (9-22) is repeated here for reference. In the solution below, the data used for each design point are listed. You may want to verify the calculations for the required minimum diameters. The design factor of $N = 3$ has been used.

$$D = \left[\frac{32N}{\pi}\sqrt{\left(\frac{K_t M}{s_n'}\right)^2 + \left(\frac{T}{s_y}\right)^2}\right]^{1/3}$$

Point *A*

Torque = 1 112 lb · in; moment = 0. The sheave is located with retaining rings. Because the torque is steady, we will not use a stress concentration factor in this calculation, as discussed in section 9-3. But then the nominal diameter at the groove will be found by increasing the computed result by about 6 percent. The result should be conservative for typical groove geometries.

Using Equation 9-22, $D_1 = 0.78$ in. Increasing this by 6 percent gives $D_1 = 0.83$ in.

To the Left of Point *B*

This is the relief diameter leading up to the bearing seat. A well-rounded fillet radius will be specified for the place where D_2 joins D_3. Thus,

Torque = 1 112 lb · in Moment = 3 339 lb · in $K_t = 1.5$

Then $D_2 = 1.82$ in.

At Point *B* and to the Right

This is the bearing seat with a shoulder fillet at the right, requiring a fairly sharp fillet:

Torque = 1 112 lb · in Moment = 3 339 lb · in $K_t = 2.5$

Then $D_3 = 2.16$ in.

At Point *C*

It is planned that the diameter be the same all the way from the right of bearing *B* to the left of bearing *D*. The worst condition is at the right of *C*, where there is a ring groove and the larger torque value:

Torque = 2 965 lb · in Moment = 4 804 lb · in $K_t = 3.0$

Then $D_4 = 2.59$ in.

At Point D and to the Left

This is a bearing seat similar to that at B:

$$\text{Torque} = 2\,965 \text{ lb} \cdot \text{in} \qquad \text{Moment} = 3\,155 \text{ lb} \cdot \text{in} \qquad K_t = 2.5$$

Then $D_5 = 2.12$ in.

To the Right of Point D

This is a relief diameter similar to D_2:

$$\text{Torque} = 2\,965 \text{ lb} \cdot \text{in} \qquad \text{Moment} = 3\,155 \text{ lb} \cdot \text{in} \qquad K_t = 1.5$$

$D_6 = 1.80$ in.

At Point E

The gear is mounted with retaining rings on each side:

$$\text{Torque} = 2\,965 \text{ lb} \cdot \text{in} \qquad \text{Moment} = 0 \qquad D_7 = 1.15$$

Summary with Convenient Values Specified

Using Tables 8-1 and 8-2, we specify convenient fractions at all places, including bearing seats. It is assumed that inch-bearings of the pillow block type will be used.

MATING PART	DIAMETER NO.	MINIMUM DIAMETER	SPECIFIED DIAMETER	
			Fraction	*Decimal*
Sheave	D_1	0.83	2	2.000
Nothing	D_2	1.82	2	2.000
Bearing	D_3	2.16	2¼	2.2500
Sprocket	D_4	2.59	2¾	2.750
Bearing	D_5	2.12	2¼	2.2500
Nothing	D_6	1.80	2	2.000
Gear	D_7	1.15	2	2.000

It was decided to make the diameters D_1, D_2, D_6, and D_7 the same to minimize machining and to provide a little extra safety factor at the ring grooves. Again the bearing bore sizes would have to be checked against the load rating of the bearings. The size of D_4 would have to be checked to see that it provides a sufficient shoulder for the bearings at B and D.

Example Problem 9-4. A wormgear is mounted at the end of the shaft as shown in Figure 9-13. The gear has the same design as that discussed in Chapter 12 and delivers 6.68 hp to the shaft at a speed of 101 rpm. The magnitudes and directions of the forces on the gear are given in the figure. Notice that there is a system of three orthogonal forces

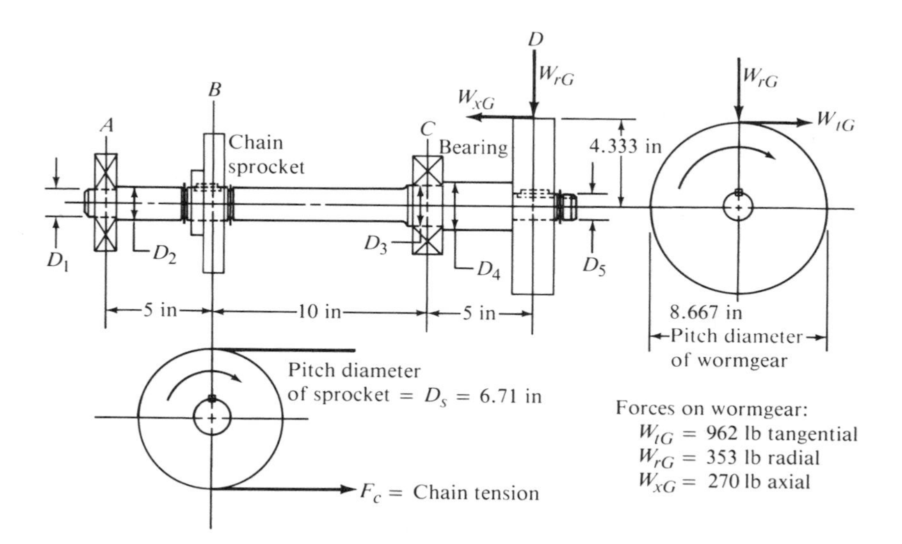

Figure 9-13 Shaft for Example Problem 9-4

acting on the gear. The power is transmitted by a chain sprocket at B to drive a conveyor removing cast iron chips from a machining system. Design the shaft.

Solution. The torque on the shaft from the wormgear at point D to the chain sprocket at B is

$$T = W_{tG}(D_G/2) = 962(4.333) = 4\,168 \text{ lb} \cdot \text{in}$$

The force on the chain sprocket is

$$F_c = T/(D_s/2) = 4\,168/(6.71/2) = 1\,242 \text{ lb}$$

This force acts horizontally toward the right as viewed from the end of the shaft.

Bending Moment Diagrams

Figure 9-14 shows the forces acting on the shaft in both the vertical and horizontal planes and the corresponding shearing force and bending moment diagrams. You should review these diagrams, especially that for the vertical plane, to grasp the effect of the axial force of 270 lb. Notice that, because it acts above the shaft, it creates a bending moment at the end of the shaft of 1 165 lb · in. It also affects the reactions at the bearings. The resultant bending moments at B, C, and D are also shown in the figure.

In the design of the entire system, it must be decided which bearing will resist the axial force. For this problem, let's specify that the bearing at C will transfer the axial thrust force to the housing. This decision places a compressive stress in the shaft from C to D and requires that means be provided to transmit the axial force from the wormgear to the bearing. The geometry proposed in Figure 9-13 accomplishes this, and it will be adopted for the following stress analysis. The procedures are the same as those used in example problems 9-1 to 9-3 and only summary results will be shown. The consideration of the axial compressive stress is discussed along with the computations at point C on the shaft.

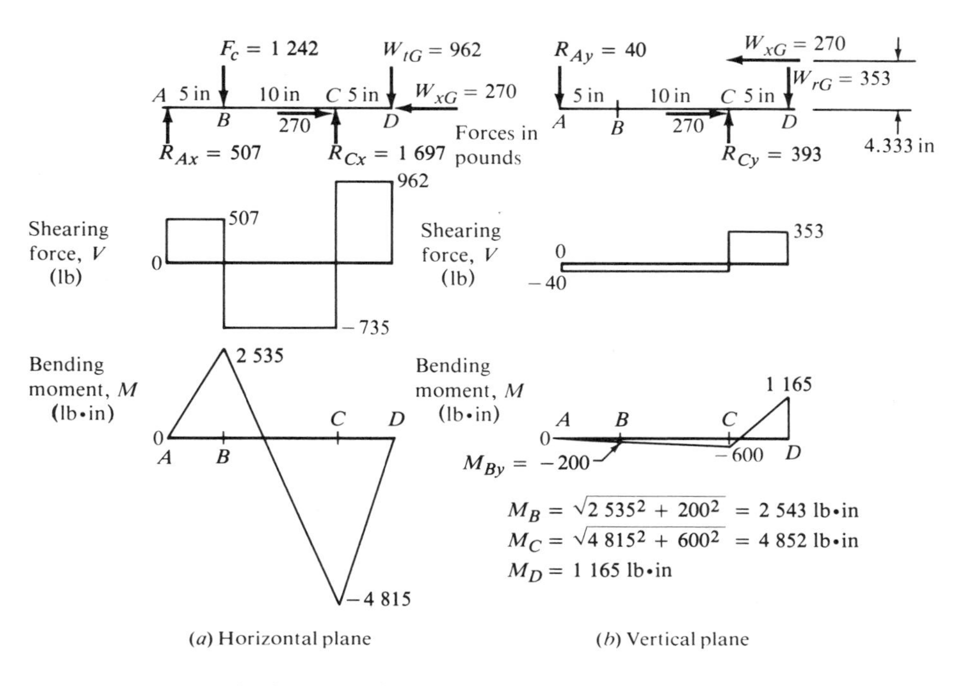

Figure 9-14 Load, Shear, and Moment Diagrams for Shaft in Figure 9-13

Material Selection and Design Strengths

A medium carbon steel with good ductility and a fairly high strength is desired for this demanding application. We will use AISI 1340 OQT 1000 (Appendix A-3), having an ultimate strength of 144 000 psi, a yield strength of 132 000 psi, and a 17 percent elongation. From Figure 5-6, and applying a size factor of 0.85, we can estimate the endurance strength to be 43 350 psi. Because the use of the conveyor is expected to be rough, we will use a design factor of $N = 4$, higher than average.

Except at point A, where only a vertical shear stress exists, the computation of the minimum required diameter is done using equation (9-22).

Point A

The left bearing mounts at point A, carrying the radial reaction force only, which acts as a vertical shearing force in the shaft. There is no torque or bending moment here.

The design shear stress is

$$\tau_d = 0.5 s_y / N = 0.5(132\,000)/4 = 16\,500 \text{ psi}$$

The required area and diameter are

$$V = \sqrt{R_{Ax}^2 + R_{Ay}^2} = \sqrt{(507)^2 + (40)^2} = 509 \text{ lb}$$

$$A = \frac{4V}{3\tau_d} = \frac{4(509)}{3(16\,500)} = 0.041\,1 \text{ in}^2$$

$$D_1 = \sqrt{4A/\pi} = \sqrt{4(0.041\,1)/\pi} = 0.229 \text{ in}$$

As seen before, this is quite small and the final specified diameter will probably be larger, depending on the bearing selected.

Point *B*

The chain sprocket mounts at point *B* and it is located axially by retaining rings on both sides. The critical point is at the right of the sprocket at the ring groove, where $T = 4\,168$ lb · in, $M = 2\,543$ lb · in, and $K_t = 3.0$ for bending.

The computed minimum diameter required is $D_2 = 1.94$ in. Because of the torsion on the groove, we should increase this by approximately 6 percent, as discussed in section 9-3. Then,

$$D_2 = 1.06(1.94 \text{ in}) = 2.05 \text{ in}$$

To the Left of Point *C*

This is the relief diameter for the bearing seat. The diameter here will be specified to be the same as that at *B*, but different conditions occur. Torque $= 4\,168$ lb · in, $M = 4\,852$ lb · in, and $K_t = 1.5$ for the well-rounded fillet for bending only. The required diameter is 1.91 in. Because this is smaller than that at *B*, the previous calculation will govern.

At Point *C* and to the Right

The bearing will seat here, and it is assumed that the fillet will be rather sharp. Thus, $T = 4\,168$ lb · in, $M = 4\,852$ lb · in, and $K_t = 2.5$ for bending only. The required diameter is $D_3 = 2.26$ in.

The axial thrust load acts between points *C* and *D*. The inclusion of this load in the computations would greatly complicate the solution for the required diameters. In most cases, the axial normal stress is relatively small compared with the bending stress. Also, the fact that the stress is compressive improves the fatigue performance of the shaft because fatigue failures normally initiate at points of tensile stress. For these reasons, the axial stress is ignored in these calculations. The computed diameters are also interpreted as nominal minimum diameters, and the final selected diameter is larger than the minimum. This, too, tends to make the shaft safe even when there is an added axial load. When in doubt, or when a relatively high axial tensile stress is encountered, the methods of Chapter 5 should be applied. Long shafts in compression should also be checked for buckling.

Point *D*

The wormgear mounts at point *D*. We will specify that a well-rounded fillet will be placed to the left of *D* and that there will be a sled runner keyseat. Thus, $T = 4\,168$ lb · in, $M = 1\,165$ lb · in, and $K_t = 1.6$ for bending only. The computed required diameter is $D_5 = 1.30$ in. Notice that D_4 must be greater than either D_3 or D_5 because it provides the means to transfer the thrust load from the wormgear to the inner race of the bearing at *C*.

Summary and Selection of Convenient Diameters

For this application we have chosen to use fractional inch dimensions from Table 8-1 except at the bearing seats, where the use of metric bearing bores from Table 15-2 are selected.

MATING PART	DIAMETER NO.	MINIMUM DIAMETER	SPECIFIED DIAMETER *Fraction (metric)*	*Decimal*
Bearing *A*	D_1	0.23	(35 mm)	1.378 0
Sprocket *B*	D_2	2.05	2¼	2.250
Bearing *C*	D_3	2.26	(60 mm)	2.362 2
(Shoulder)	D_4	$>D_3$	2¾	2.750
Wormgear *D*	D_5	1.30	1½	1.500

9-9 CALCULATOR PROGRAM FOR SHAFT DIAMETER

The calculation of the minimum required diameter of a shaft when subjected to only bending and torsion uses equation (9-22). The following is a program for a programmable calculator to perform the calculations required for that equation. It was prepared for the Texas Instruments TI-55-II calculator, which uses an algebraic mode of entering equations. For other calculators, especially those with reverse Polish notation, some modification of the program is required. The following symbols are used.

R/S (Run/stop): Execution stops to permit data entry. Striking the key after data entry resumes execution.

RCL (Recall): Recalls the value from the memory number that follows.

y^x (Exponentiation): y to the x power

Equation (9-22) has the form

$$D = [(32N/\pi)\sqrt{(K_t M/s_n')^2 + (T/s_y)^2}]^{1/3}$$

The first term can be written as a single constant and loaded into memory. Calling the constant C_0 and recalling that the value of N is usually 2, 3, or 4,

$$C_0 = 20.38 \quad \text{for } N = 2$$

$$C_0 = 30.56 \quad \text{for } N = 3$$

$$C_0 = 40.76 \quad \text{for } N = 4$$

To begin, load C_0 into memory (0). Then load the value of ⅓ = 0.333 333 into memory (1). The program follows.

Action	*Operation*	*Step No.*
Assume program has been reset.		
Input M and hit R/S		
	× (Times)	00
	R/S	01
Input K_t and hit R/S		
	÷ (Division)	02
	R/S	03

Input s_n and hit R/S		
	=	04
	x^2	05
	+	06
	(	07
	R/S	08
Input T and hit R/S		
	÷	09
	R/S	10
Input s_y and hit R/S		
	)	11
	x^2	12
	=	13
	$\sqrt{x}$	14
	×	15
	RCL	16
	00	17
	=	18
	y^x	19
	RCL	20
	01	21
	=	22
	RST	23

Now the computed value of the diameter is displayed and the program is reset to enable you to run another problem.

For checking your program you can use any of the worked-out examples, such as example problem 9-1 or 9-3. For example, for diameter D_2 in example problem 9-1, $N = 3$ and

$$M = 22\,350 \qquad T = 21\,000 \qquad K_t = 1.5 \qquad s_n' = 33\,600 \qquad s_y = 83\,000$$

The result is $D_2 = 3.16$ in.

REFERENCES

1. American Society of Mechanical Engineers. *Preferred Limits and Fits for Cylindrical Parts*. USA Standard B4.1—1967 (rev. 1974).
2. Baumeister, T., et al. *Marks' Standard Handbook for Mechanical Engineers,* 8th ed. New York: McGraw-Hill Book Company, 1978.
3. Oberg, Erik, et al. *Machinery's Handbook,* 22d ed. New York: Industrial Press, 1984.
4. Peterson, R. E. *Stress Concentration Factors,* 2d ed. New York: John Wiley & Sons, 1974.
5. Shigley, J. E., and Mitchell, L. D. *Mechanical Engineering Design,* 4th ed. New York: McGraw-Hill Book Company, 1983.
6. Soderberg, C. R. "Working Stresses." *Journal of Applied Mechanics*. 57 (1935): A-106.

PROBLEMS

For each of the following problems, it will be necessary to determine the magnitude of the torque in the shaft at all points, to compute the forces acting on the shaft at all power-transmitting elements, to compute the reactions at the bearings, and to draw the complete load, shear, and bending moment diagrams. Neglect the weight of the elements on the shafts, unless otherwise noted.

The objective of any problem, at the discretion of the instructor, could be any of the following:

Design the complete shaft, including the specification of the overall geometry and the consideration of stress concentration factors. The analysis would show the minimum acceptable diameter at each point on the shaft to be safe from the standpoint of strength.

Given a suggested geometry of one part of the shaft, specify the minimum acceptable diameter for the shaft at that point.

Specify the required geometry at any selected element on the shaft: a gear, sheave, bearing, or other.

Make a working drawing of the design for the shaft, following the appropriate stress analysis, and specify the final dimensions.

Suggest how the given shaft can be redesigned by moving or reorienting the elements on the shaft to improve the design to produce lower stresses, smaller shaft size, more convenient assembly, and so on.

Incorporate the given shaft into a more comprehensive machine and complete the design of the entire machine. In most problems, the type of machine for which the shaft is being designed is suggested.

1. The shaft in Figure 9-15 is part of a drive for an automated transfer system in a metal stamping plant. Gear *Q* delivers 30 hp to gear *B*. Sheave *D* delivers the power to its mating sheave as shown. The shaft carrying *B* and *D* rotates at 550 rpm. Use AISI 1040 cold-drawn steel.
2. The shaft in Figure 9-16 rotates at 200 rpm. Pulley *A* receives 10 hp from below. Gear *C* delivers 6 hp to the mating gear below it. Chain sprocket *D* delivers 4 hp to a shaft above. Use AISI 1117 cold-drawn steel.
3. The shaft in Figure 9-17 is part of a special machine designed to reclaim scrap aluminum cans. The gear at *B* delivers 5 hp to a chopper that cuts the cans into small pieces. The V-belt sheave at *D* delivers 3 hp to a blower that draws air through the chopper. V-belt sheave *E* delivers 3 hp to a conveyor that raises the shredded aluminum to an elevated hopper. The shaft rotates at 480 rpm. All power comes into the shaft through the chain sprocket at *C*. Use AISI 1137 OQT 1300 steel for the shaft. The elements at *B*, *C*, *D*, and *E* are held in position with retaining rings and keys in profile keyseats. The shaft is to be of uniform diameter, except at its ends, where the bearings are to be mounted. Determine the required diameter.
4. The shaft in Figure 9-18 is part of a grain-drying system. At *A* is a propeller-type fan that requires 12 hp when rotating at 475 rpm. The fan weighs 34 lb, and its weight should be included in the analysis. The flat belt pulley at *D* delivers 3.5 hp to a

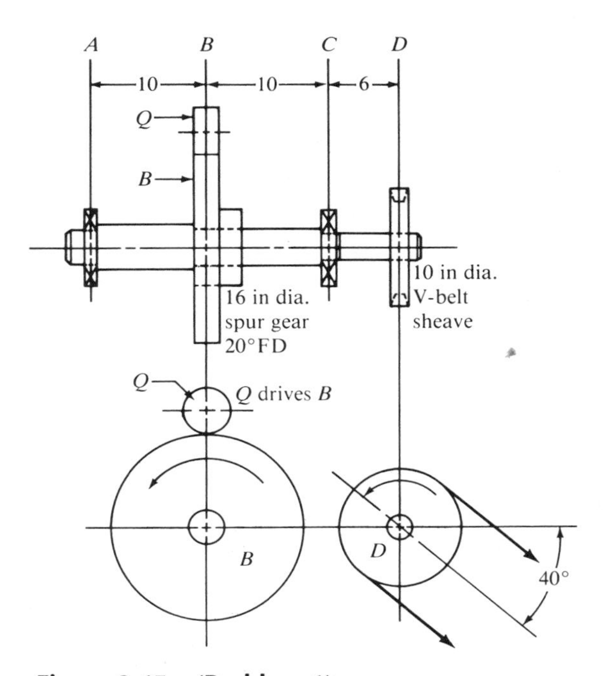

Figure 9-15 (Problem 1)

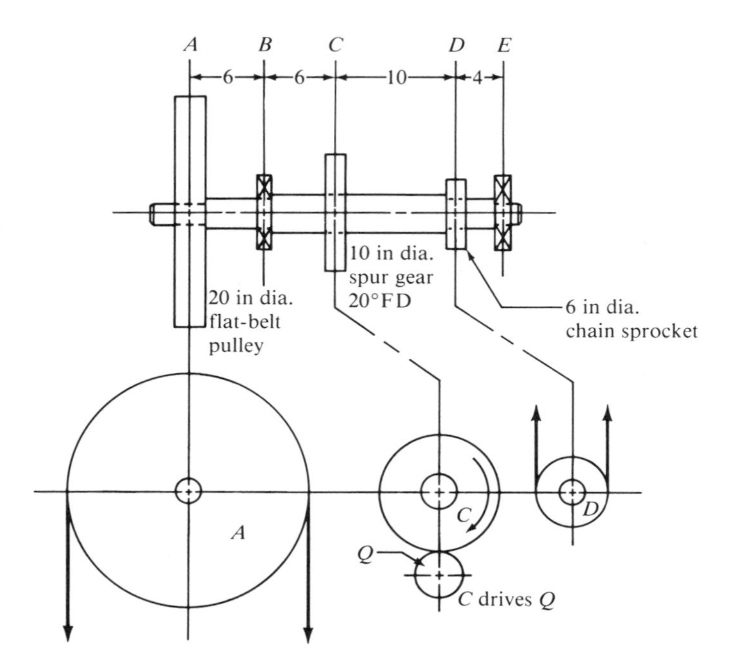

Figure 9-16 (Problem 2)

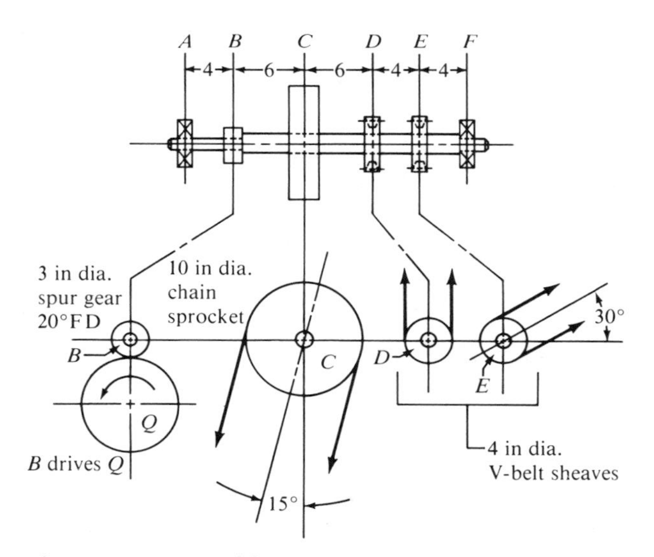

Figure 9-17 (Problem 3)

screw conveyor handling the grain. All power comes into the shaft through the V-belt sheave at C. Use AISI 1144 cold-drawn steel.

5. The shaft in Figure 9-19 is the drive shaft for a large bulk material conveyor. The gear receives 40 hp and rotates at 120 rpm. Each chain sprocket delivers 20 hp to one side of the conveyor. Use AISI 1020 cold-drawn steel.
6. The shaft in Figure 9-20 is part of a conveyor drive system that feeds crushed rock into a railroad car. The shaft rotates at 240 rpm and is subjected to moderate shock during operation. All power is input

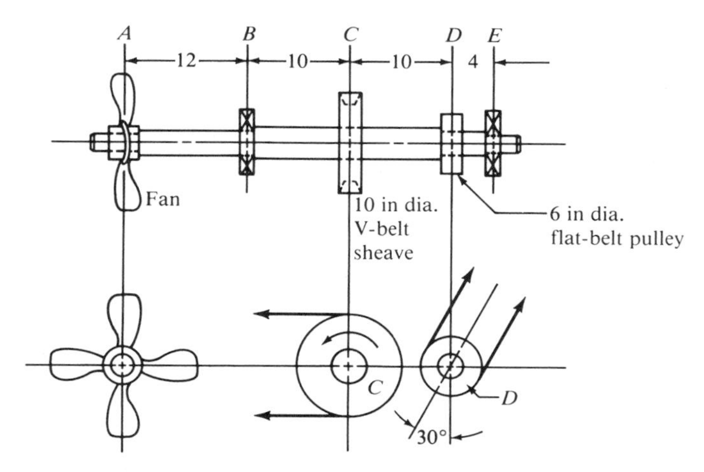

Figure 9-18 (Problem 4)

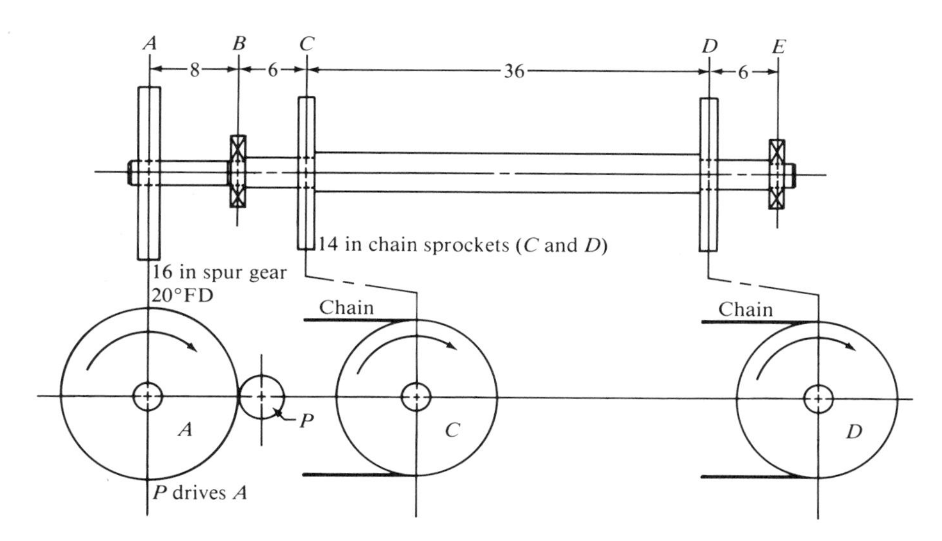

Figure 9-19 (Problem 5)

to the gear at D. The V-belt sheave at A delivers 10.0 hp vertically downward. Chain sprocket E delivers 5.0 hp. Note the position of the gear Q, which drives gear D.

7. Figure 9-21 illustrates an intermediate shaft of a punch press that rotates at 310 rpm while transmitting 20 hp from the V-belt sheave to the gear. The flywheel is not absorbing or giving up any energy at this time. Consider the weight of all elements in the analysis.
8. The shaft in Figure 9-22 is part of a material-handling system aboard a ship. All power comes into the shaft through gear C, which rotates at 480 rpm. Gear A delivers 30 hp to a hoist. V-belt sheaves D and E each deliver 10 hp to hydraulic pumps. Use AISI 3140 OQT 1000 steel.
9. The shaft in Figure 9-23 is part of an automatic machining system. All power is input through gear D. Gears C and F drive two separate tool feed devices requiring 5.0 hp each. The V-belt sheave at B requires 2.5 hp to drive a coolant pump. The shaft rotates at 220 rpm. All gears are spur gears with 20° full-depth teeth. Use AISI 1020 cold-drawn steel for the shaft.

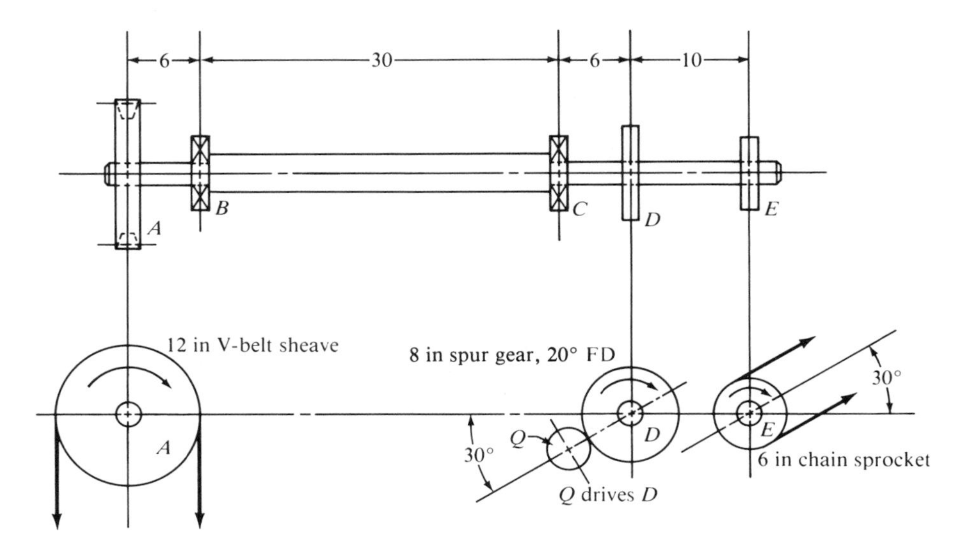

Figure 9-20 (Problem 6)

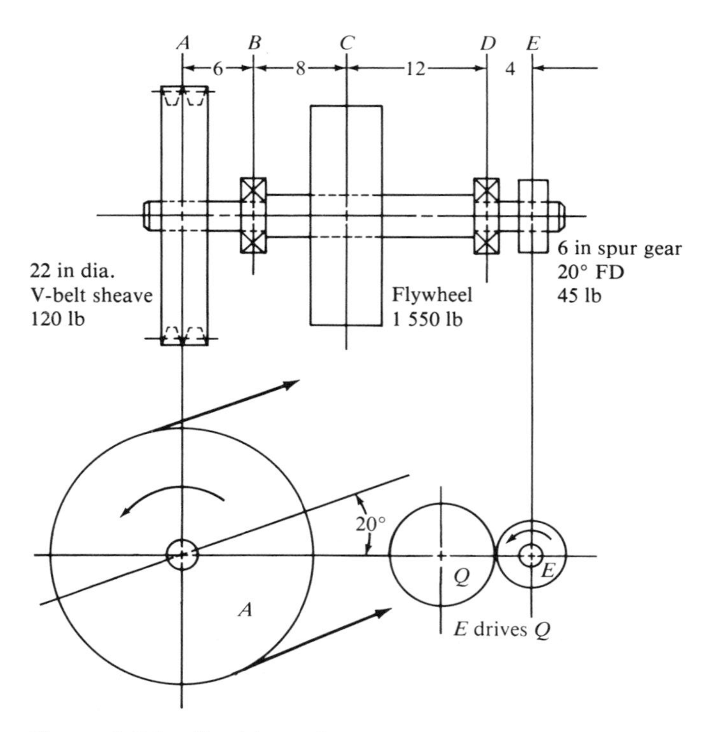

Figure 9-21 (Problem 7)

10. Figure 9-24 shows a helical gear mounted on a shaft that rotates at 650 rpm while transmitting 7.5 hp. The gear is also analyzed in example problem 12-2, and the tangential, radial, and axial forces on it are shown in the figure. The pitch diameter of the gear is 4.141 in. The power is delivered from the shaft through a flexible coupling at its right end. A spacer is used to position the gear relative to bearing C. The thrust load is taken at bearing A.

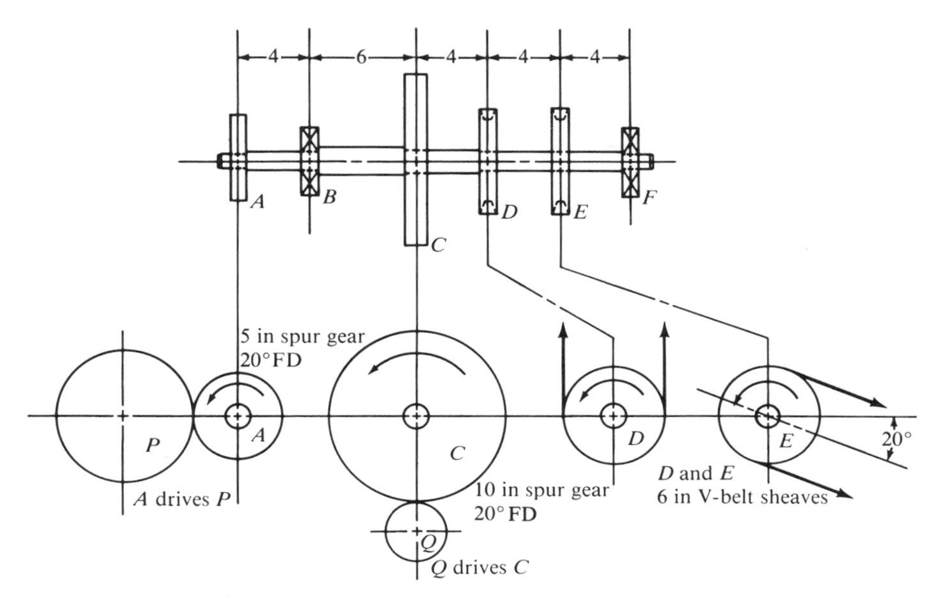

Figure 9-22 (Problem 8)

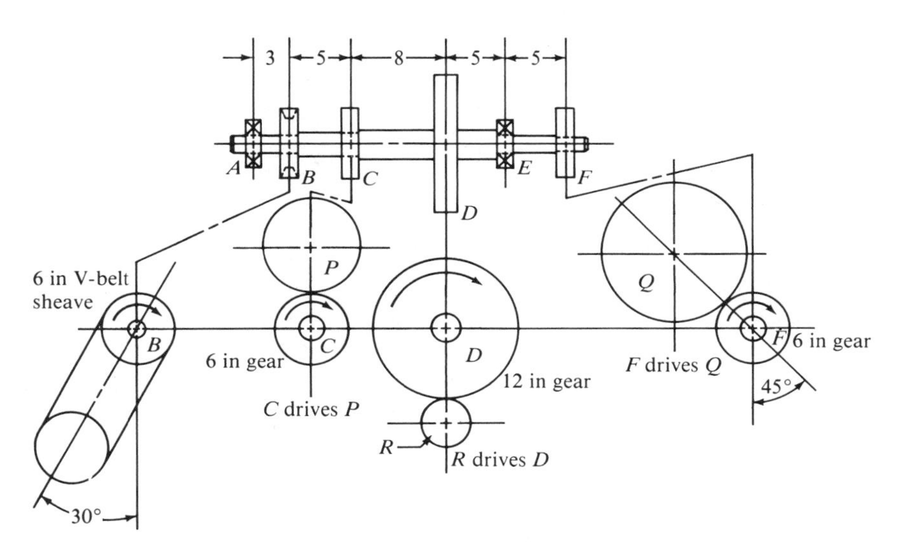

Figure 9-23 (Problem 9)

11. The double-reduction helical gear reducer shown in Figure 9-25 transmits 5.0 hp. Shaft 1 is the input, rotating at 1 800 rpm and receiving power directly from an electric motor through a flexible coupling. Shaft 2 rotates at 900 rpm. Shaft 3 is the output, rotating at 300 rpm. A chain sprocket is mounted on the output shaft as shown and delivers the power upward. The data for the gears are as follows:

Gear	*Diametral Pitch*	*Pitch Diameter*	*Number of Teeth*	*Face Width*
P	8	1.500 in	12	0.75 in
B	8	3.000 in	24	0.75 in
C	6	2.000 in	12	1.00 in
Q	6	6.000 in	36	1.00 in

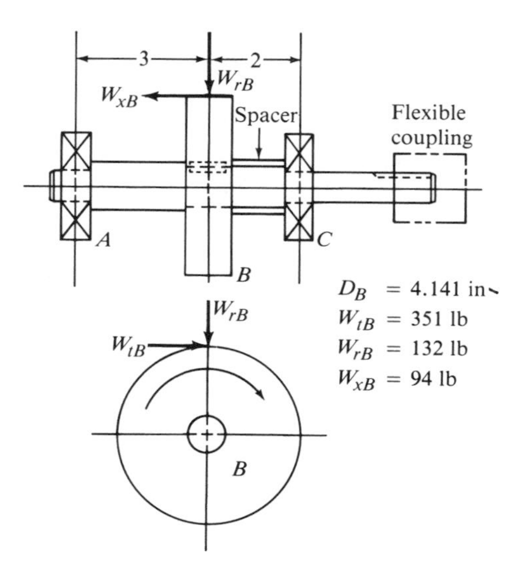

Figure 9-24 (Problem 10)

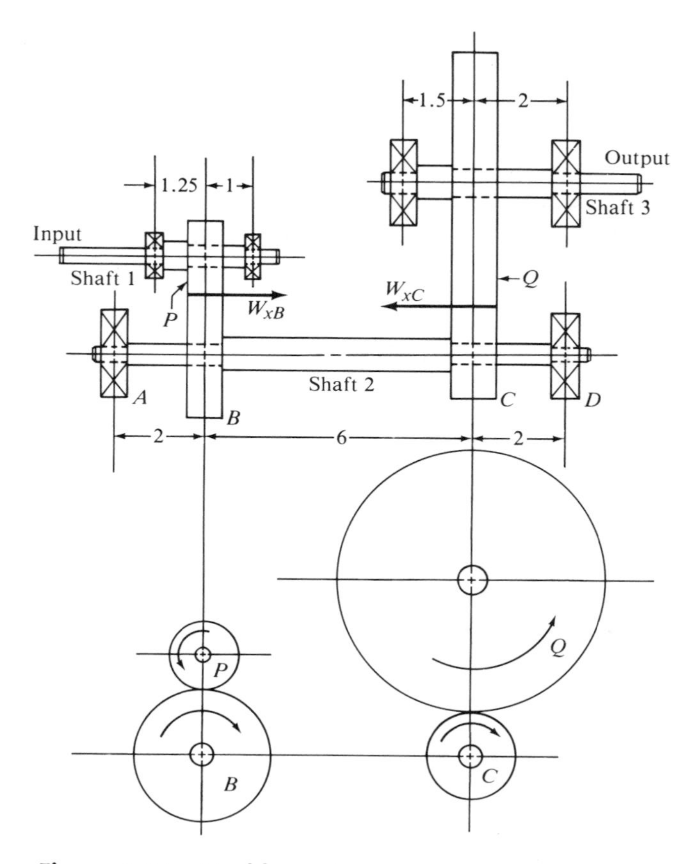

Figure 9-25 (Problem 11)

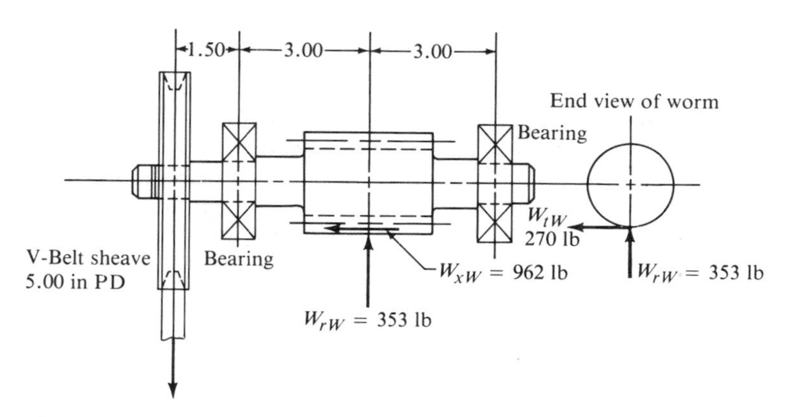

Figure 9-26 (Problem 12)

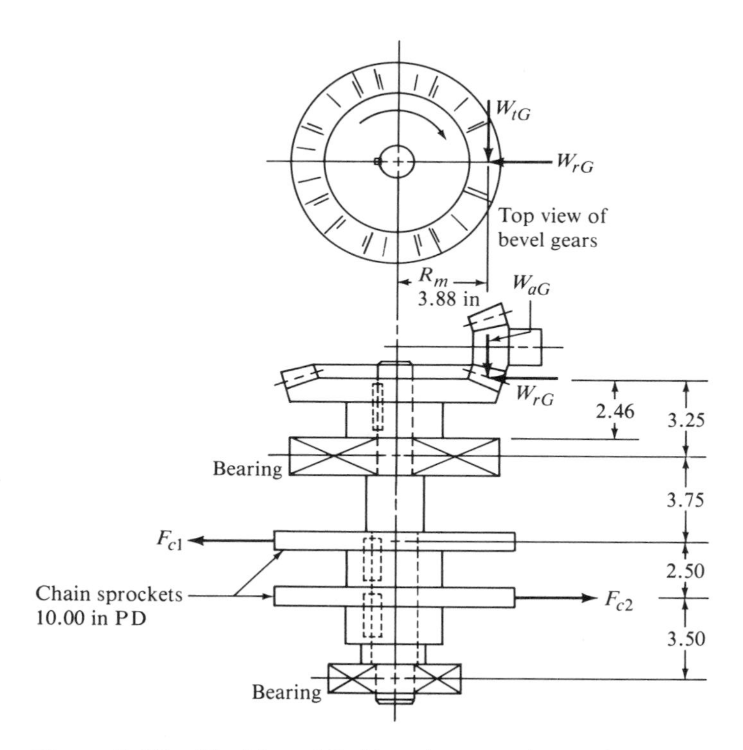

Figure 9-27 (Problem 14) (Bevel gears shown in section. See Chapter 12.)

Each gear has a 14½° normal pressure angle and a 45° helix angle. The combinations of left- and right-hand helixes are arranged so that the axial forces oppose each other on shaft 2 as shown. Use AISI 4140 OQT 1200 for the shafts.

12. The shaft shown in Figure 9-26 is the input shaft of a wormgear drive. The V-belt sheave receives 7.5 hp from directly downward. The worm rotates at 1 750 rpm and has a pitch diameter of 2.000 in. This is the driving worm for the wormgear described in example problem 9-4. The tangential, radial, and axial forces are shown in the figure. The worm is to be machined integrally with the shaft, and it has a root diameter of 1.614 in. Assume that the geometry of the root area presents a stress concentration factor of 1.5 for bending. Analyze the stress in the area of the worm thread root and specify a suitable material for the shaft.

13. Complete the design of the shafts carrying the bevel gear and pinion shown in Figure 12-9 in Chapter 12. The forces on the gears, the bearing reactions, and the bending moment diagrams are developed in example problems 12-4 to 12-6 and are shown in Figures 12-10 to 12-14. Assume that the 2.50 hp enters the pinion shaft from the right through a flexible coupling. The power is delivered by way of the lower extension of the gear shaft through another flexible coupling. Use AISI 1040 OQT 1200 for the shafts.

14. The vertical shaft shown in Figure 9-27 is driven at a speed of 600 rpm with 4.0 hp entering through the bevel gear. Each of the two chain sprockets delivers 2.0 hp to the side to drive mixer blades in a chemical reactor vessel. The bevel gear has a diametral pitch of 5, a pitch diameter of 9.000 in, a face width of 1.31 in, and a pressure angle of 20°. Use AISI 4140 OQT 1000 steel for the shaft. See Chapter 12 for the methods for computing the forces on the bevel gear.

10 Keys and Couplings

10-1 OVERVIEW

This chapter deals with the many ways in which power-transmitting elements can be attached to shafts and two shafts may be connected. The most frequent way of attaching a gear, pulley, belt sheave, chain sprocket, or similar device to a shaft for the purpose of transmitting power is by installing a key at the interface between the shaft and the hub of the power-transmitting element. The design and installation of keys and keyseats are discussed more fully in the following section. Of the several other methods used for the same function, the following list shows some of the more common.

Pinning	Taper and screw
Clamping	Press fit
Split taper bushing	Molding
Set screws	Mechanical lock
Splines	

This chapter will discuss these methods further because designers must be aware of alternative methods for fastening elements to shafts.

The term *coupling* is used for the means of joining two shafts at their ends. The coupling may be either rigid or flexible, as will be discussed in a later section.

10-2 KEYS

A *key* is a machinery component placed at the interface between a shaft and the hub of a power-transmitting element for the purpose of transmitting torque (see Figure 10-1*a*). The key is demountable to facilitate assembly and disassembly of the shaft system. It is

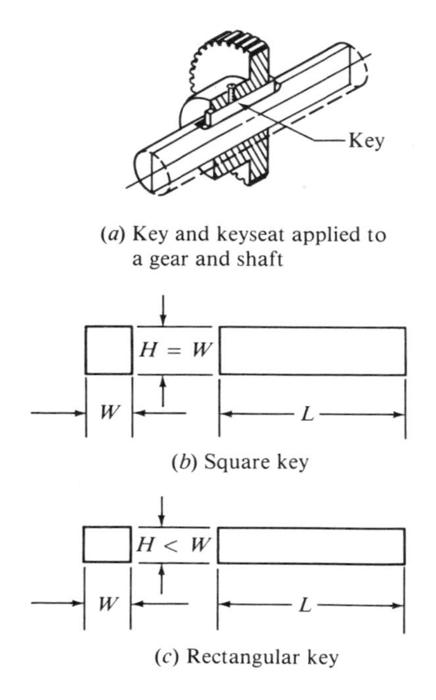

(*a*) Key and keyseat applied to a gear and shaft

(*b*) Square key

(*c*) Rectangular key

Figure 10-1 Parallel Keys

installed in an axial groove machined into the shaft, called a *keyseat*. A similar groove in the hub of the power-transmitting element is usually called a *keyway* but is more properly also a keyseat. The key is typically installed into the shaft keyseat first; then the hub keyseat is aligned with the key and the hub is slid into position.

The most common type of key for shafts up to 6½ in in diameter is the square key, as illustrated in Figure 10-1(b). The rectangular key, Figure 10-1(c), is recommended for larger shafts and is used for smaller shafts where the shorter height can be tolerated. Both the square and the rectangular keys are referred to as *parallel keys* because the top and bottom and sides of the key are parallel. Other types of keys are the taper key, the gib head key, the Woodruff key, and the pin key, as shown in Figure 10-2.

Table 10-1 gives the preferred dimensions for parallel keys as a function of shaft diameter, as specified in USA Standard B17.1-1967. The width is nominally ¼ of the diameter of the shaft.

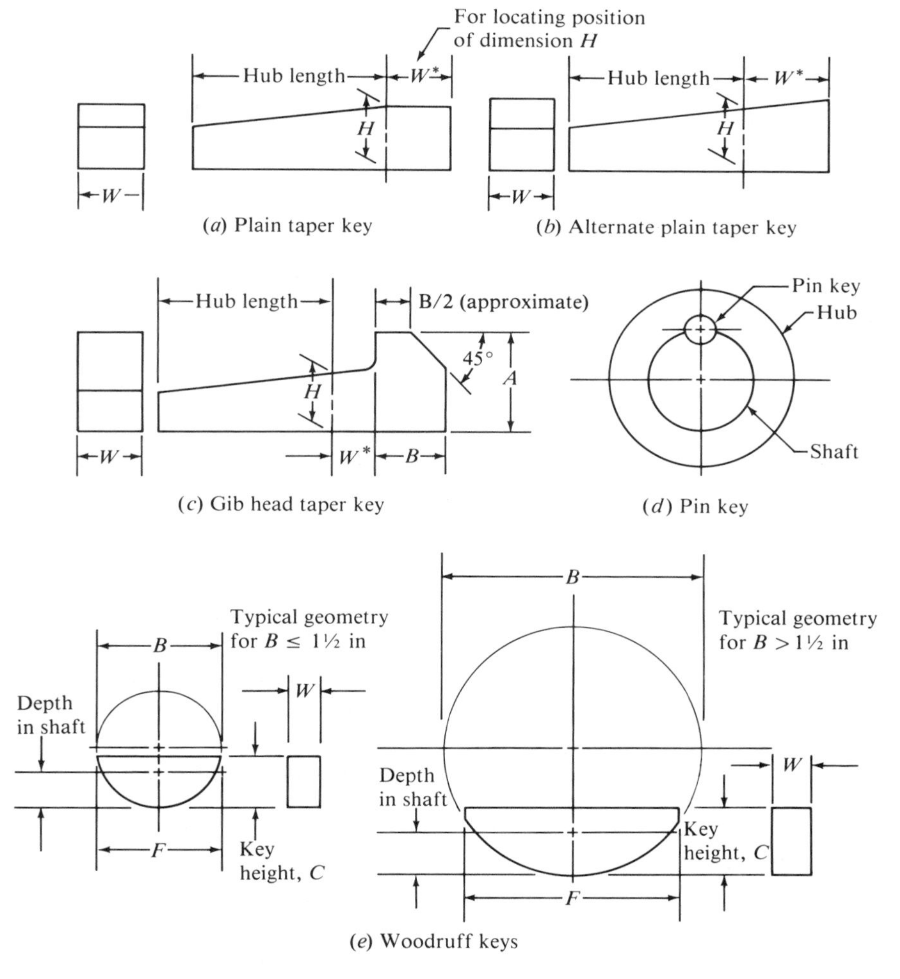

*Note: Plain and gib head taper keys have a 1/8″ taper in 12″.

Figure 10-2 Key Types

Table 10-1 Key Size Versus Shaft Diameter

NOMINAL SHAFT DIAMETER		NOMINAL KEY SIZE		
			Height, H	
Over	*To (Incl.)*	*Width,* W	*Square*	*Rectangular*
5/16	7/16	3/32	3/32	
7/16	9/16	1/8	1/8	3/32
9/16	7/8	3/16	3/16	1/8
7/8	1 1/4	1/4	1/4	3/16
1 1/4	1 3/8	5/16	5/16	1/4
1 3/8	1 3/4	3/8	3/8	1/4
1 3/4	2 1/4	1/2	1/2	3/8
2 1/4	2 3/4	5/8	5/8	7/16
2 3/4	3 1/4	3/4	3/4	1/2
3 1/4	3 3/4	7/8	7/8	5/8
3 3/4	4 1/2	1	1	3/4
4 1/2	5 1/2	1 1/4	1 1/4	7/8
5 1/2	6 1/2	1 1/2	1 1/2	1
6 1/2	7 1/2	1 3/4	1 3/4	1 1/2*
7 1/2	9	2	2	1 1/2
9	11	2 1/2	2 1/2	1 3/4
11	13	3	3	2
13	15	3 1/2	3 1/2	2 1/2
15	18	4		3
18	22	5		3 1/2
22	26	6		4
26	30	7		5

Note: Values in nonshaded areas preferred. Dimensions are in inches.
Source: USA Standard B17.1—1967, *Keys and Keyseats* (American Society of Mechanical Engineers, New York).

The keyseats in the shaft and the hub are designed so that exactly one-half of the height of the key is bearing on the side of the shaft keyseat and the other half on the side of the hub keyseat. Figure 10-3 shows the resulting geometry. The distance Y is the radial distance from the theoretical top of the shaft, before the keyseat is machined, to the top edge of the finished keyseat to produce a keyseat depth of exactly $H/2$. To assist in machining and inspecting the shaft or the hub, the dimensions S and T can be computed and shown on the part drawings. The equations are given in Figure 10-3. Tabulated values of Y, S, and T are available (1, 2).

As discussed in Chapter 9, keyseats in shafts are usually machined with either an end mill or a circular milling cutter, producing the profile or sled runner keyseat, respectively (refer to Figure 9-6). In general practice, the keyseats and keys are left with essentially square corners. But radiused keyseats and chamfered keys can be used to reduce the stress concentrations. Table 10-2 shows suggested values from USAS B17.1. When using chamfers on a key, be sure to take that factor into account when computing the bearing stress on the side of the key (see equation [10-3]).

Standard key stock conforming to the dimensions of USAS B17.1 is available in the sizes shown in Table 10-1. Typical materials used include low- to medium-carbon plain

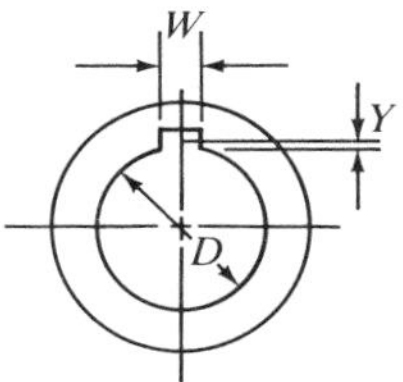

$$Y = \frac{D - \sqrt{D^2 - W^2}}{2}$$

(*a*) Chordal height

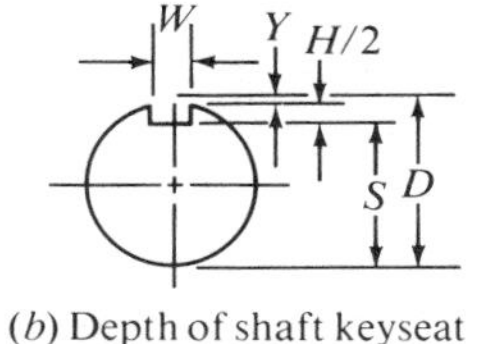

$$S = D - Y - \frac{H}{2} = \frac{D - H + \sqrt{D^2 - W^2}}{2}$$

(*b*) Depth of shaft keyseat

W Y H/2 T D

$$T = D - Y + \frac{H}{2} + C = \frac{D + H + \sqrt{D^2 - W^2}}{2} + C$$

(*c*) Depth of hub keyseat

Symbols
- C = Allowance
  - +0.005 in clearance for parallel keys
  - −0.020 in interference for taper keys
- D = Nominal shaft or bore diameter, in
- H = Nominal key height, in
- W = Nominal key width, in
- Y = Chordal height, in

Figure 10-3 Dimensions for Keyseats

Table 10-2 Suggested Fillet Radii and Key Chamfers

H/2 KEYSEAT DEPTH			
Over	*To (Incl.)*	FILLET RADIUS	45° CHAMFER
1/8	1/4	1/32	3/64
1/4	1/2	1/16	5/64
1/2	7/8	1/8	5/32
7/8	1 1/4	3/16	7/32
1 1/4	1 3/4	1/4	9/32
1 3/4	2 1/2	3/8	13/32

Note: All dimensions given in inches.
Source: USA Standard B17.1—1967, *Keys and Keyseats* (American Society of Mechanical Engineers, New York).

carbon steels, such as AISI 1020 or 1040 in the cold-drawn condition. Of course, virtually any material can be used to meet special requirements for strength, corrosion resistance, and so on.

The taper key permits the key to be inserted from the end of the shaft after the hub is in position. If the opposite end of the key is not accessible to be driven out, the gib head key provides the means of extracting the key from the same end at which it was installed. On both the plain taper and the gib head key, the taper is ⅛ in per foot. The cross-sectional dimensions of the key, W and H, are the same as those used for parallel keys, with the height, H, measured at the position specified in Figure 10-2.

Design of Keys and Keyseats

The key and keyseat for a particular application are usually designed after the shaft diameter is specified by the methods of Chapter 9. Then, with the shaft diameter as a guide, the size of the key is selected from Table 10-1. The only remaining variables are the length of the key and its material. One of these can be specified, and the requirements for the other can then be computed.

Typically the length of a key is specified to be a substantial portion of the hub length of the element in which it is installed to provide for good alignment and stable operation. But if the keyseat in the shaft is to be in the vicinity of other geometric changes such as shoulder fillets and ring grooves, it is important to provide some axial clearance between them so the effects of the stress concentrations are not compounded.

The key can be cut off square at its ends or provided with a radius at each end when installed in a profile keyseat to improve location. Square cut keys are usually used with the sled runner type keyseat.

The key is sometimes held in position with a set screw in the hub over the key. However, the reliability of this approach is questionable because of the possibility of the set screw's backing out with vibration of the assembly. Axial location of the assembly should be provided by more positive means, such as shoulders, retaining rings, or spacers.

Stress Analysis to Determine Key Length

There are two basic modes of potential failure for keys transmitting power: shear across the shaft/hub interface and compression failure due to the bearing action between the sides of the key and the shaft or hub material. The analysis for either failure mode requires an understanding of the forces that act on the key. Figure 10-4 shows the idealized case in which the torque on the shaft creates a force on the left side of the key. The key in turn exerts a force on the right side of the hub keyseat. The reaction force of the hub back on the key then produces a set of opposing forces that place the key in direct shear over its cross section, $W \times L$. The magnitude of the shearing force can be found from

$$F = T/(D/2)$$

The shearing stress is then

$$\tau = \frac{F}{A_s} = \frac{T}{(D/2)(WL)} = \frac{2T}{DWL} \tag{10-1}$$

In design, we can set the shearing stress equal to a design stress in shear for the maximum shear stress theory of failure,

$$\tau_d = 0.5s_y/N$$

Then the required length of the key is

$$L = \frac{2T}{\tau_d DW} \tag{10-2}$$

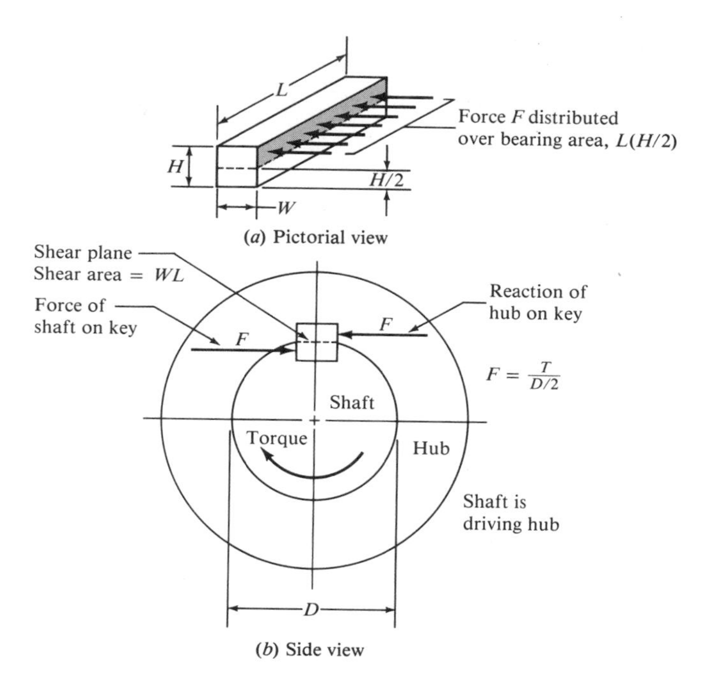

(a) Pictorial view

(b) Side view

Figure 10-4 Forces on a Key

The failure in bearing is related to the compressive stress on the side of the key, the side of the shaft keyseat, or the side of the hub keyseat. The area in compression is the same for either of these zones, $L \times (H/2)$. Thus the failure occurs on the surface with the lowest compressive yield strength. Let's define a *design stress for compression* as

$$\sigma_d = s_y/N$$

Then the compressive stress is

$$\sigma = \frac{F}{A_c} = \frac{T}{(D/2)(L)(H/2)} = \frac{4T}{DLH} \tag{10-3}$$

Letting this stress equal the design compressive stress allows the computation of the required length of the key for this mode of failure,

$$L = \frac{4T}{\sigma_d DH} \tag{10-4}$$

In typical industrial applications, $N = 3$ is adequate.

Example Problem 10-1. Example problem 9-3 specified that the shaft size at the right end of the shaft where the gear is to be mounted was to be 2.00 in. The gear transmits 2 965 lb · in of torque. The shaft is to be made of AISI 1040 cold-drawn steel. The gear is made from AISI 8650 OQT 1000 steel. Design the key.

Solution. From Table 10-1, the standard key dimension for a 2.00-in diameter shaft would be ½ in square. Choosing a standard key material, AISI 1020 cold-drawn steel, the design stress for shear is

$$\tau_d = 0.5s_y/N = [0.5(51\,000)]/3 = 8\,500 \text{ psi}$$

The required length of the key is found from equation (10-2):

$$L = \frac{2T}{\tau_d DW} = \frac{2(2\,965)}{(8\,500)\,(2.00)\,(0.500)} = 0.698 \text{ in}$$

For compression we can examine the material for the key, the shaft, and the hub and find that the key is the weakest. The design compressive stress is

$$\sigma_d = s_y/N = (51\,000)/3 = 17\,000 \text{ psi}$$

The required length of the key to resist bearing is then

$$L = \frac{4T}{\sigma_d DH} = \frac{4(2\,965)}{(17\,000)\,(2.00)\,(0.500)} = 0.698 \text{ in}$$

The fact that the two computations produced the same result is to be expected because, comparing the two equations, there is a factor of 2 involved in both the numerator and denominator if the same material strength is involved in each calculation. A convenient dimension for the specified length, consistent with the design of the hub of the gear and the geometry of the shaft in the vicinity of the gear, is chosen.

If the computed required length of the key is too long for the mating hub, a higher-strength material can be specified. Two keys placed 90° apart may also be used.

Woodruff Keys

Where light loading and relatively easy assembly and disassembly are desired, the *Woodruff key* should be considered. Figure 10-2 shows the standard configuration. The circular groove in the shaft holds the key in position while the mating part is slid over the key. The stress analysis for this type of key proceeds in the manner discussed for the parallel key, taking into consideration the particular geometry of the Woodruff key. The USA Standard B17.2-1967 lists the dimensions for a large number of standard Woodruff keys and their mating keyseats. Table 10-3 provides a sampling. Note that the *key number* indicates the nominal key dimensions. The last two digits give the nominal diameter B in eighths of an inch, and the digits preceding the last two give the nominal width W in thirty-seconds of an inch. For example, key number 1 210 has a diameter of 10/8 in (1¼ in), and a width of 12/32 in (3/8 in). The actual size of the key is slightly smaller than half of the full circle, as can be seen from dimensions C and F in the table.

10-3 SPLINES

A *spline* can be described as a series of axial ribs machined into a shaft, with corresponding ribs machined into the bore of the mating part (gear, sheave, sprocket, and so on) (see Figure 10-5). The splines perform the same function as a key in transmitting torque from the shaft to the mating element. The advantages of splines over keys are many. Because usually four or more splines are used, as compared with one or two keys, a more uniform transfer of the torque and a lower loading on a given part of the shaft/hub interface

Table 10-3 Woodruff Key Dimensions

Key Number	*Nominal Key Size,* W × B	*Actual Length,* F	*Height of Key,* C	*Shaft Keyseat Depth*	*Hub Keyseat Depth*
202	1/16 × 1/4	0.248	0.104	0.0728	0.0372
204	1/16 × 1/2	0.491	0.200	0.1668	0.0372
406	1/8 × 3/4	0.740	0.310	0.2455	0.0685
608	3/16 × 1	0.992	0.435	0.3393	0.0997
810	1/4 × 1 1/4	1.240	0.544	0.4170	0.1310
1210	3/8 × 1 1/4	1.240	0.544	0.3545	0.1935
1628	1/2 × 3 1/2	2.880	0.935	0.6830	0.2560
2428	3/4 × 3 1/2	2.880	0.935	0.5580	0.3810

Note: All dimensions given in inches.
Source: USA Standard B17.1—1967, *Keys and Keyseats* (American Society of Mechanical Engineers, New York).

result. The splines are integral with the shaft so no relative motion can occur as between a key and the shaft. Splines are accurately machined to provide a controlled fit between the mating internal and external spline. The surface of the spline is often hardened to resist wear and facilitate its use in applications in which axial motion of the mating element is

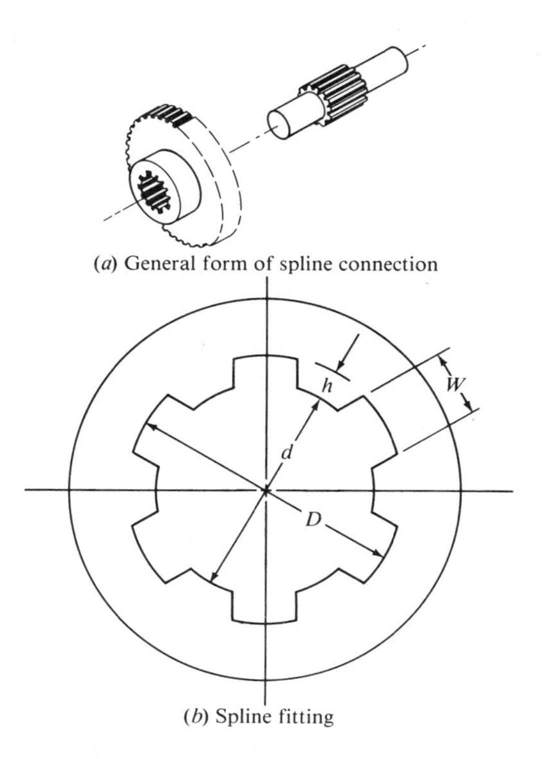

(*a*) General form of spline connection

(*b*) Spline fitting

Figure 10-5 Spline

desired. Relative motion between a key and the mating element should not be permitted. Because of the multiple splines on the shaft, the mating element can be indexed to various positions.

Splines can be either straight-sided or involute. The involute form is preferred because it provides for self-centering of the mating element and because it can be machined with standard hobs used to cut gear teeth.

Straight splines are made according to the specifications of the Society of Automotive Engineers (SAE) and usually contain 4, 6, 10, or 16 splines. Figure 10-5 shows the six-spline version, from which you can see the basic design parameters of D (major diameter), d (minor diameter), W (spline width), and h (spline depth). The dimensions for d, W, and h are related to the nominal major diameter D by the formulas given in Table 10-4. Note that the values of h and d differ according to the use of the spline. The permanent fit, A, is used when the mating part is not to be moved after installation. The B fit is used if the mating part will be moved along the shaft without a torque load. When the mating part must be moved under load, the C fit is used.

The torque capacity for SAE splines is based on the limit of 1 000 psi bearing stress on the sides of the splines, from which the following formula is derived:

$$T = 1\,000\,NRh \tag{10-5}$$

where N is the number of splines, R is the mean radius of the splines, and h is the depth of the splines (from Table 10-4). The torque capacity is per inch of length of the spline. But note that

$$R = \frac{1}{2}\left[\frac{D}{2} + \frac{d}{2}\right] = \frac{D + d}{4}$$

$$h = \frac{1}{2}(D - d)$$

Then

$$T = 1\,000N\frac{(D + d)}{4}\frac{(D - d)}{2} = 1\,000N\frac{(D^2 - d^2)}{8} \tag{10-6}$$

Table 10-4 Formulas for SAE Straight Splines

		A		B		C	
	W,	PERMANENT FIT		TO SLIDE WITHOUT LOAD		TO SLIDE UNDER LOAD	
NO. OF SPLINES	FOR ALL FITS	h	d	h	d	h	d
Four	0.241D	0.075D	0.850D	0.125D	0.750D	....	....
Six	0.250D	0.050D	0.900D	0.075D	0.850D	0.100D	0.800D
Ten	0.156D	0.045D	0.910D	0.070D	0.860D	0.095D	0.810D
Sixteen	0.098D	0.045D	0.910D	0.070D	0.860D	0.095D	0.810D

These formulas give the maximum dimensions for W, h, and d.

This equation can be further refined for each of the types of splines in Table 10-4 by substituting the appropriate relationships for N and d. For example, for the six-spline version and the B fit, $N = 6$, $d = 0.850D$, and $d^2 = 0.722\,5D^2$. Then

$$T = 1\,000(6)\frac{[D^2 - 0.722\,5D^2]}{8} = 208D^2$$

Then the required diameter to transmit a given torque would be

$$D = \sqrt{T/208}$$

In these formulas, dimensions are in inches and the torque is in pound-inches. Using this same approach, the torque capacities and required diameters for the other versions of straight splines are found (Table 10-5).

The graphs in Figure 10-6 enable you to choose an acceptable diameter for a spline to carry a given torque, depending on the desired fit, A, B, or C. The data were taken from Table 10-5. A detailed analysis of the stresses in splines has been published (2).

Involute splines are typically made with pressure angles of 30°, 37.5°, or 45°. The 30° form is illustrated in Figure 10-7, showing the two types of fit that can be specified. The *major diameter fit* produces accurate concentricity between the shaft and the mating element. In the *side fit,* contact occurs only on the sides of the teeth, but the involute form tends to center the shaft in the mating splined hub.

10-4 OTHER METHODS OF FASTENING ELEMENTS TO SHAFTS

The following discussion will acquaint you with some of the ways in which power-transmitting elements can be attached to shafts without keys or splines. In most cases the designs have not been standardized, and analysis of individual cases, considering the forces exerted on the elements and the manner of loading of the fastening means, is necessary. In several of the designs, the analysis of shear and bearing will follow a

Table 10-5 Torque Capacity for Straight Splines Per Inch of Spline Length

Number of Splines	*Fit*	*Torque Capacity*	*Required Diameter*
4	A	$139D^2$	$\sqrt{T/139}$
4	B	$219D^2$	$\sqrt{T/219}$
6	A	$143D^2$	$\sqrt{T/143}$
6	B	$208D^2$	$\sqrt{T/208}$
6	C	$270D^2$	$\sqrt{T/270}$
10	A	$215D^2$	$\sqrt{T/215}$
10	B	$326D^2$	$\sqrt{T/326}$
10	C	$430D^2$	$\sqrt{T/430}$
16	A	$344D^2$	$\sqrt{T/344}$
16	B	$521D^2$	$\sqrt{T/521}$
16	C	$688D^2$	$\sqrt{T/688}$

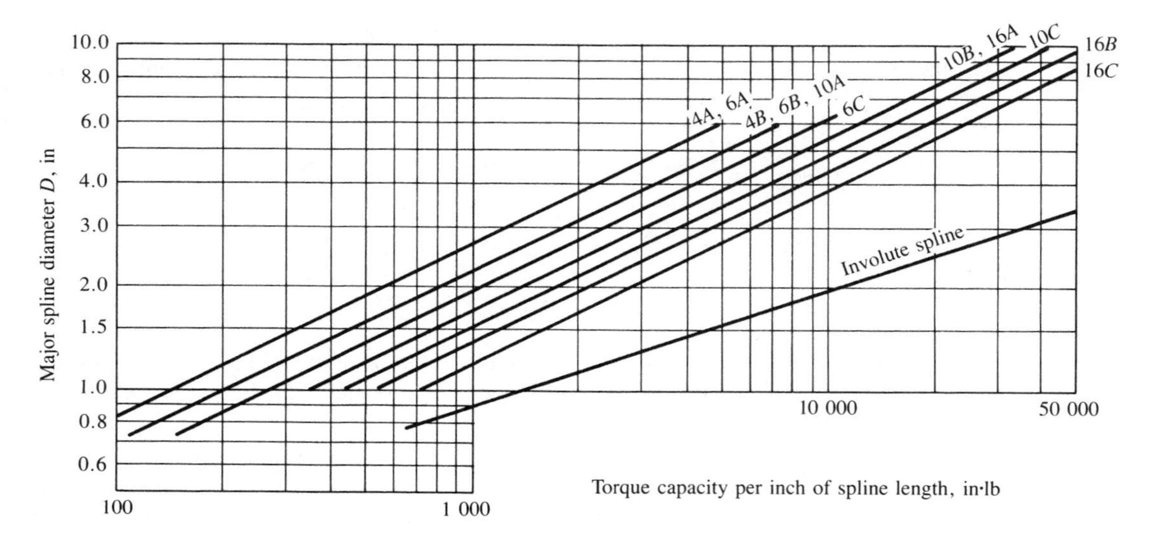

Figure 10-6 Torque Capacity per Inch of Spline Length, lb · in

procedure similar to that shown for keys. If a satisfactory analysis is not possible, testing of the assembly is recommended.

Pinning

With the element in position on the shaft, a hole can be drilled through both the hub and the shaft and a pin can be inserted in the hole. Figure 10-8 shows three examples of this approach. The straight solid cylindrical pin is subjected to shear over two cross sections. If there is a force F on each end of the pin at the shaft/hub interface, then

$$T = 2F(D/2) = FD$$

or $F = T/D$. With the symbol d representing the pin diameter, the shear stress in the pin is

$$\tau = \frac{F}{A_s} = \frac{T}{D(\pi d^2/4)} = \frac{4T}{D(\pi d^2)} \tag{10-7}$$

Letting the shear stress equal the design stress in shear as before, solving for d gives the required pin diameter:

$$d = \sqrt{\frac{4T}{D(\pi)(\tau_d)}} \tag{10-8}$$

Sometimes the diameter of the pin is purposely made small to ensure that the pin will break if a moderate overload is encountered, in order to protect critical parts of a mechanism. Such a pin is called a *shear pin*.

One problem with a cylindrical pin is that fitting it adequately to provide precise location of the hub and to prevent the pin from falling out is difficult. The taper pin

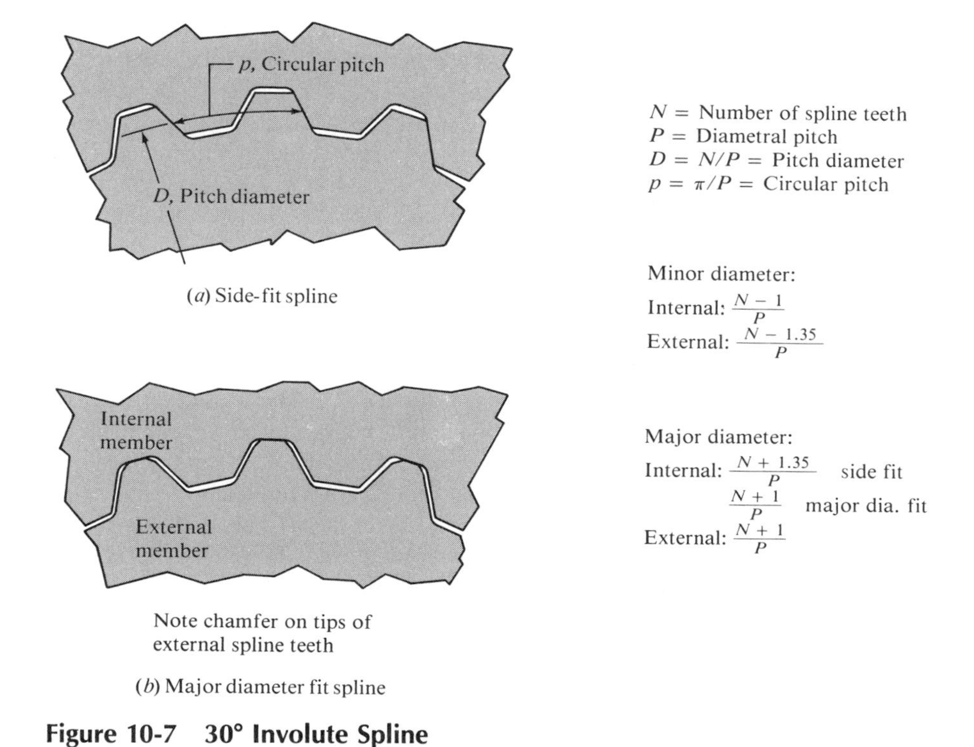

Figure 10-7 30° Involute Spline

overcomes some of these problems, as does the split spring pin shown in Figure 10-8. For the split spring pin, the hole is made slightly smaller than the pin diameter so that a light force is required to assemble the pin in the hole. The spring force retains the pin in the hole and holds the assembly in position. But, of course, the presence of any of the pin type connections produces stress concentrations in the shaft.

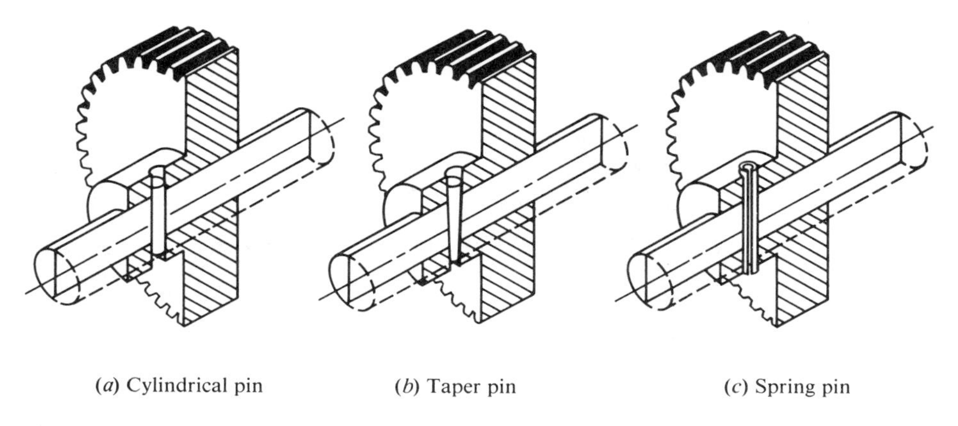

Figure 10-8 Pinning

Clamping

A hub on a gear or other element can be slotted axially, and a clamp can be placed over the hub. When the clamp is drawn down tight, it forces the split hub into contact with the shaft. The pressure of the hub on the surface of the shaft permits transmission of torque. This type of connection is difficult to control, and predicting the torque-transmitting capacity accurately is difficult.

Split Taper Bushing

The difficulties of simple clamping can be overcome by using a *split taper bushing,* available commercially on several belt sheaves and chain sprockets (see Figure 10-9). Whereas simple clamping relies on friction to transmit torque produced by pressure between the hub and the shaft, the split taper bushing uses a key that is a positive means of torque transmission; that is, no slipping can occur between the hub and the shaft. The split bushing has a small taper on the outer surface. When the bushing is pulled into a mating hub with a set of capscrews, the bushing is brought into tight contact with the shaft to hold the assembly in the proper axial position. The small taper locks the assembly together. Removal of the bushing is accomplished by removing the capscrews and using them in push-off holes to force the hub off the taper. The assembly can then be easily disassembled.

Set Screws

A *set screw* is a threaded fastener driven radially through a hub to bear on the outer surface of a shaft (see Figure 10-10). The point of the set screw is either flat, oval, cone-shaped, cupped, or any of several proprietary forms. The point bears on the shaft or digs slightly into its surface. Thus the set screw transmits torque by the friction between the point and the shaft or by the resistance of the material in shear. The capacity for torque transmission is somewhat variable, depending on the hardness of the shaft material and the clamping force created when the screw is installed. Furthermore the screw may loosen during operation because of vibration. For these reasons, set screws should be used with care. Some manufacturers provide set screws with plastic inserts in the side among the threads. When screwed into a tapped hole, the plastic is deformed by the threads and holds the screw securely, resisting vibration. Using a liquid adhesive also helps resist loosening.

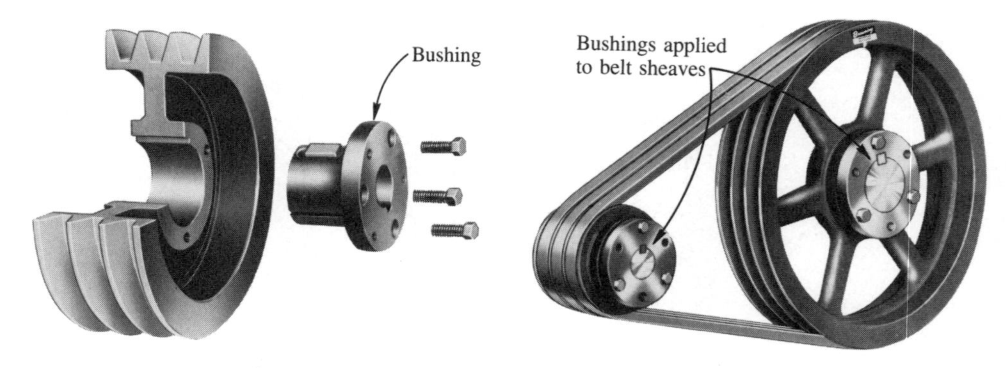

Figure 10-9 Split Taper Bushing (Browning Mfg. Division, Emerson Electric Co., Maysville, Ky.)

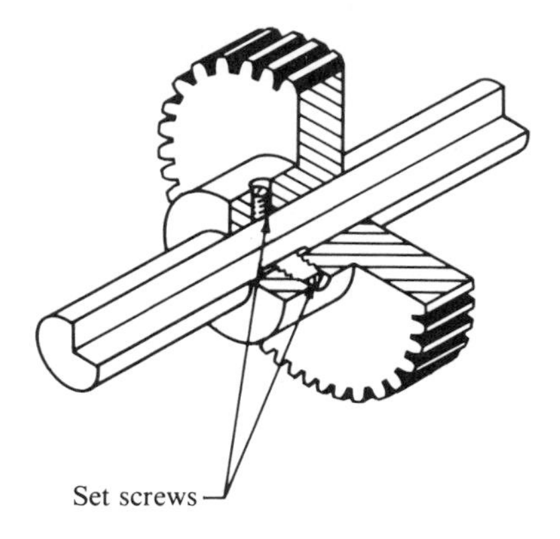

Figure 10-10 Set Screws

Another problem with using set screws is that the shaft surface is damaged by the point; this damage may make disassembly difficult. Machining a flat on the surface of the shaft may help reduce the problem and also produce a more consistent assembly.

When properly assembled on typical industrial shafting, the force capability of set screws is approximately as follows:

Screw Diameter	*Holding Force*
¼ in	100 lb
⅜ in	250 lb
½ in	500 lb
¾ in	1 300 lb
1 in	2 500 lb

Source: Oberg, Erik, et al. *Machinery's Handbook*, 22d ed. New York: Industrial Press, 1984.

Taper and Screw

The power-transmitting element (gear, sheave, sprocket, or other) that is to be mounted at the end of a shaft can be secured with a screw and washer in the manner shown in Figure 10-11(a). The taper provides good concentricity and moderate torque-transmission capacity. Because of the machining required, the connection is fairly costly. A modified form uses the tapered shaft with a threaded end for the application of a nut, as shown in Figure 10-11(b). The inclusion of a key lying in a keyseat machined parallel to the taper increases the torque-transmitting capacity greatly and ensures positive alignment.

Press Fit

Making the diameter of the shaft greater than the bore diameter of the mating element results in an interference fit. The resulting pressure between the shaft and the hub permits the transmission of torque at fairly high levels, depending on the degree of interference. This is discussed more fully in Chapter 8. Sometimes the press fit is combined with a key: the key providing the positive drive and the press fit ensuring concentricity.

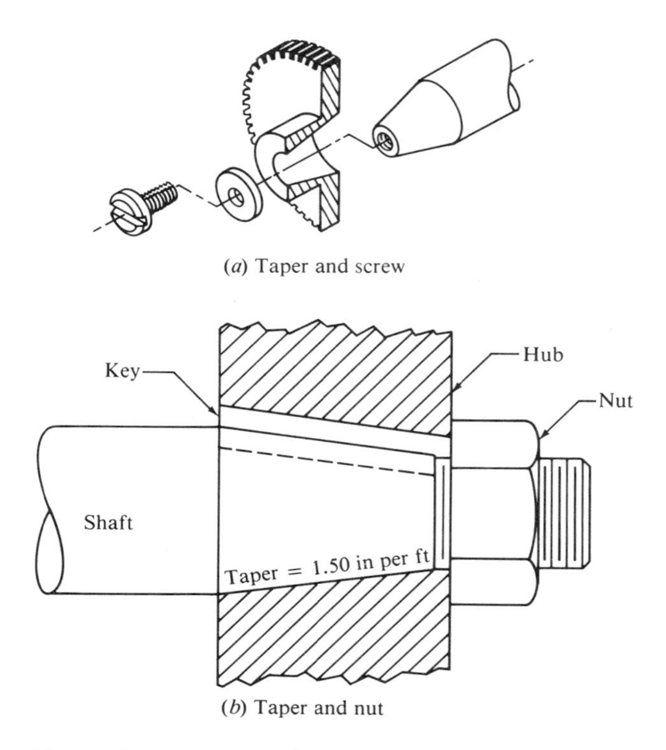

(*a*) Taper and screw

(*b*) Taper and nut

Figure 10-11 Tapered Shaft for Fastening Machine Elements to Shafts

Molding

Plastic and die cast gears can be molded directly to their shafts. Often the gear is applied to a location that is knurled to improve the ability to transmit torque. A modification of this procedure is to take a separate gear blank with a prepared hub, locate it over the proper position on a shaft, and then cast zinc into the space between the shaft and the hub to lock them together.

Mechanical Locking

Some of the variations of methods of locking gears to shafts by mechanical means are illustrated in Figure 10-12. Part (a), *staking,* is produced by deforming part of the hub into the gear. Part (b), *pegging* or *pinning,* is similar to the pin key shown in Figure 10-2. In part (c), the gear is riveted to a flange on the shaft. Here other types of fasteners, such as standard nuts and bolts, can be used to permit removal of the gear. Spinning is used in part (d) to deform part of the hub radially outward into intimate contact with the gear.

10-5 COUPLINGS

The term *coupling* refers to a device used to connect two shafts together at their ends for the purpose of transmitting power. There are two general types of couplings, rigid and flexible.

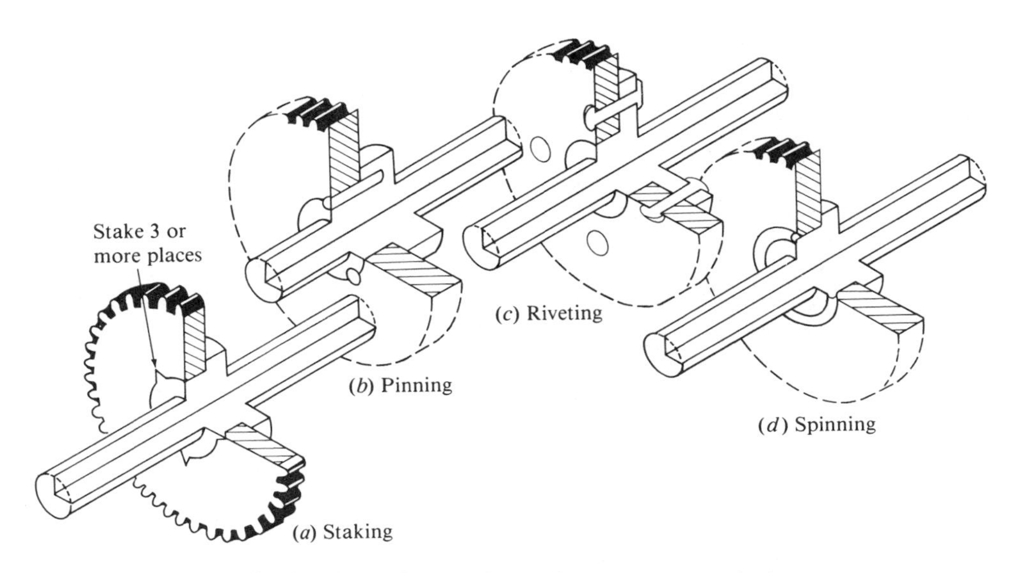

Figure 10-12 Methods of Mechanically Locking Gears to Shafts

Rigid couplings are designed to draw the two shafts together tightly so that no relative motion can occur between them. This design is desirable for certain kinds of equipment in which precise alignment of two shafts is required and can be provided. In such cases, the coupling must be designed to be capable of transmitting the torque in the shafts.

A typical rigid coupling is shown in Figure 10-13, in which flanges are mounted on the ends of each shaft and drawn together by a series of bolts. The load path is then from the driving shaft to its flange, through the bolts, into the mating flange, and out to the driven shaft. The torque places the bolts in shear. The total shear force on the bolts depends on the radius of the bolt circle, $D_{bc}/2$, and the torque, T. That is,

$$F = T/(D_{bc}/2) = 2T/D_{bc}$$

Letting N be the number of bolts, the shear stress in each bolt is

$$\tau = \frac{F}{A_s} = \frac{F}{N(\pi d^2/4)} = \frac{2T}{D_{bc}N(\pi d^2/4)} \tag{10-9}$$

Letting the stress equal the design stress in shear and solving for the bolt diameter,

$$d = \sqrt{\frac{8T}{D_{bc}N\pi\tau_d}} \tag{10-10}$$

Notice that this is a similar analysis to that for pinned connections in section 10-4. The analysis assumes that the bolts are the weakest part of the coupling.

Rigid couplings should be used only when the alignment of the two shafts can be maintained very accurately, not only at the time of installation but also during operation of the machines. If significant angular, radial, or axial misalignment occurs, stresses that are difficult to predict and may lead to early failure due to fatigue will be induced in the shafts. These difficulties can be overcome by the use of flexible couplings.

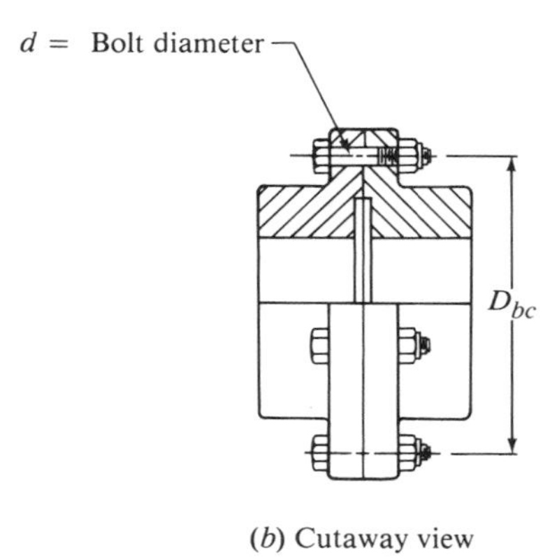

(*b*) Cutaway view

Figure 10-13 Rigid Coupling (Dodge Division, Reliance Electric Co.)

Flexible couplings are designed to transmit torque smoothly while permitting some axial, radial, and angular misalignment. The flexibility is such that when misalignment does occur, parts of the coupling move with little or no resistance. Thus no significant axial or bending stresses are developed in the shaft.

Many types of flexible couplings are available commercially, as discussed later in this section. Each is designed to transmit a given limiting torque. The manufacturer's catalog lists the design data from which you can choose a suitable coupling. Remember that torque equals power divided by rotational speed. So for a given size of coupling, as the speed of rotation increases, the amount of power the coupling can transmit also increases, although not always in direct proportion. Of course, centrifugal effects determine the upper limit of speed.

The degree of misalignment that can be accommodated by a given coupling should be obtained from the manufacturer's catalog data, with values varying with the size and design of the coupling. Small couplings may be limited to parallel misalignment of 0.005 in, although larger couplings may allow 0.030 in or more. Typical allowable angular misalignment is $\pm 3°$. Axial movement allowed, sometimes called *end float,* is up to 0.030 in for many types of couplings.

Figures 10-14 to 10-22 show several commercially available flexible couplings.

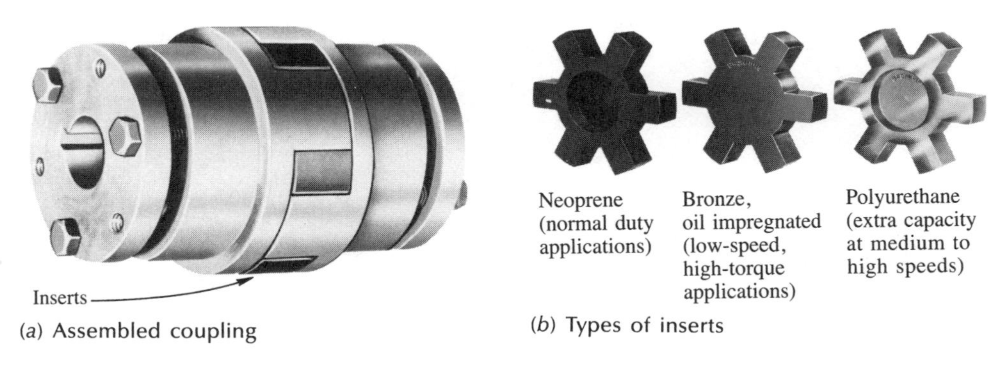

(*a*) Assembled coupling

(*b*) Types of inserts

Figure 10-14 Jaw Type Coupling (Browning Mfg. Division, Emerson Electric Co., Maysville, Ky.)

10-6 UNIVERSAL JOINTS

As noted in the preceding section, flexible couplings can accommodate up to approximately 3° of angular misalignment and approximately 0.030 in of parallel misalignment, depending on their design. Sometimes much more misalignment must be used; in these applications, the *universal joint* provides an effective coupling. Universal joints operate at angles of up to 45° at very slow speeds. At higher speeds, above approximately 10 rpm, the maximum recommended angle is 30°. Above 600 rpm, a 20° maximum angle is typical. Manufacturer's data should be consulted. Actual service conditions affect the permissible angle.

Figure 10-23 shows one of the smaller, simpler types of universal joints, consisting of two yokes, a center bearing block, and two pins that pass through the block at right angles to each other. Figure 10-24 shows a larger type, having two yokes and a center cross member, that is sometimes called a *Cardan joint*. The caps on the cross are equipped with needle bearings to permit low-friction relative motion between the members. Proper lubrication of the moving members of the joint is essential. Figure 10-25 shows another type of universal joint, which is used in heavy industrial applications.

When a single universal joint is driven by a shaft rotating at a uniform angular velocity, the output shaft has a nonuniform angular velocity. The varying angular velocity produces variable angular accelerations, which in turn cause vibrations and higher stresses. Thus the use of a single universal joint should be limited to slow-speed, low-power applications.

Using two universal joints connected by an intermediate shaft eliminates the varying angular velocity caused by one joint. With proper alignment, the angular velocity variation of each joint is cancelled by the other. Figure 10-26 shows a properly aligned double universal joint connecting a pair of parallel shafts having a large offset. If the shafts must be at a specified angle, the two joints are installed so that the angle of each joint is one-half of the total angle between the input and output shafts.

The intermediate shaft for the double universal joint is frequently made in two parts with a spline connection between them. This arrangement allows for a variation in the axial location of the parts of the drive, even while the system is in operation. This feature is particularly desirable in vehicular applications because of the movement of the chassis components. In industrial machinery, such a design would permit easier, less critical installation of drive components.

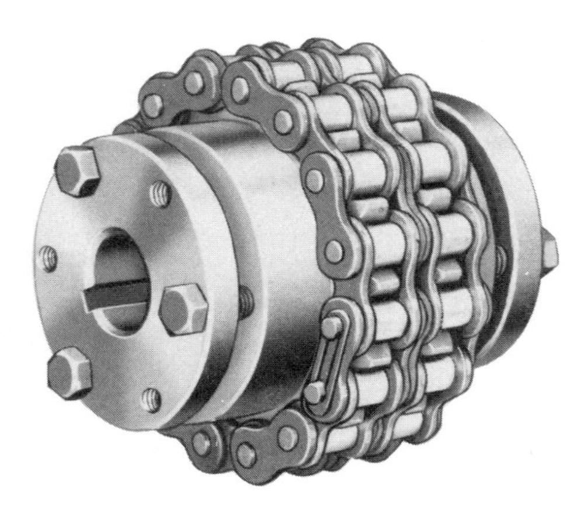

Figure 10-15 Chain Coupling. Torque is transmitted through a double roller chain. Clearances between the chain and the sprocket teeth on the two coupling halves accommodate misalignment. (Browning Mfg. Division, Emerson Electric Co., Maysville, Ky.)

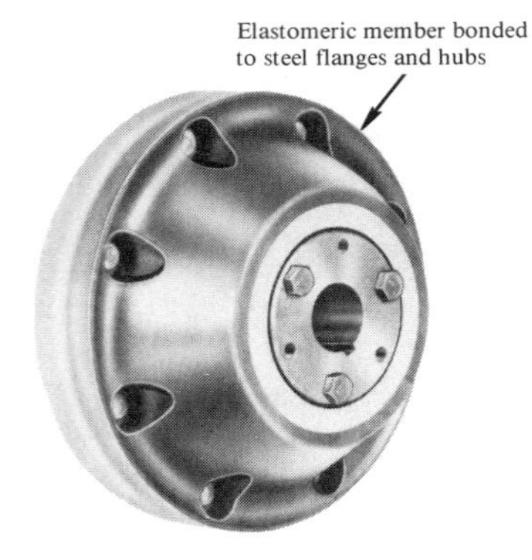

Figure 10-16 Ever-flex Coupling. Features of coupling: (1) generally minimizes torsional vibration; (2) cushions shock loads; (3) compensates for parallel misalignment up to 1/32 in; (4) accommodates angular misalignment of ±3°; (5) provides adequate end float, ±1/32 in. (Browning Mfg. Division, Emerson Electric Co., Maysville, Ky.)

Figure 10-17 Grid-flex Coupling. Torque is transmitted through flexible spring steel grid. Flexing of grid permits misalignment and is torsionally resilient to resist shock loads. (Browning Mfg. Division, Emerson Electric Co., Maysville, Ky.)

Figure 10-18 Gear Coupling. Torque is transmitted between crown-hobbed teeth from coupling half to sleeve. Crown shape on gear teeth permits misalignment. (Browning Mfg. Division, Emerson Electric Co., Maysville, Ky.)

Figure 10-19 Bellows Coupling (Stock Drive Products, New Hyde Park, N.Y.)

Figure 10-20 PARA-FLEX® Coupling Using an Elastomeric Element to Permit Misalignment and to Cushion Shocks (Dodge Division, Reliance Electric Co.)

Figure 10-21 Dynaflex® Coupling. Torque is transmitted through elastomeric material that flexes to permit misalignment and to attenuate shock loads. (Lord Corporation, Erie, Penn.)

Figure 10-22 FORM-FLEX® Coupling. Torque is transmitted from hubs through laminated flexible elements to spacer. (Formsprag Division, Dana Corp., Mt. Pleasant, Mich.)

Figure 10-23 Universal Joint Components (Curtis Universal Joint Co., Inc., Springfield, Mass.)

Several proprietary designs for universal joints that have a uniform angular velocity relationship in a single joint are available. Examples are the Rzeppa and the Bendix-Weiss joints.

10-7 RETAINING RINGS AND OTHER MEANS OF AXIAL LOCATION

The preceding sections of this chapter have focused on means of connecting machine elements to shafts for the purpose of transmitting power. Therefore, they emphasized the ability to withstand a given torque at a given speed of rotation. It must be recognized that the axial location of the machine elements must also be ensured by the designer.

The choice of the means for axial location depends heavily on whether or not there is axial thrust transmitted by the element. Spur gears, V-belt sheaves, and chain sprockets produce no significant thrust loads. Therefore, the need for axial location affects only incidental forces due to vibration, handling, and shipping. Although not severe, these forces should not be taken lightly. Movement of an element in relation to its mating

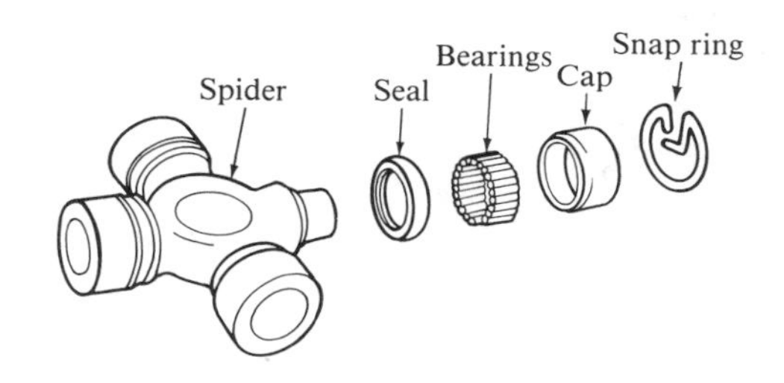

(*a*) Automotive universal joint components

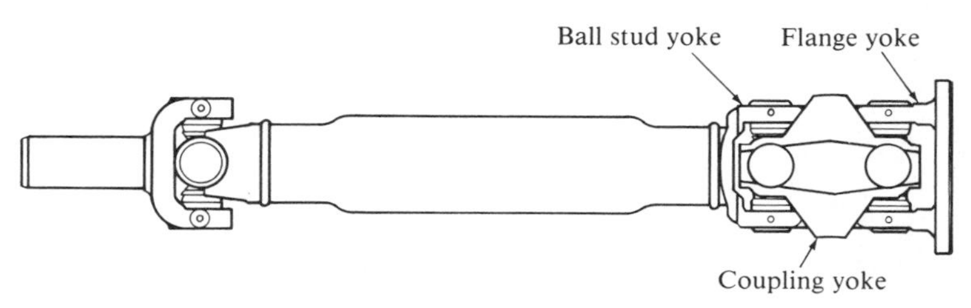

(*b*) Automotive propeller shaft with double Cardan universal joint

Figure 10-24 Automotive Universal Joints (Pontiac Motor Division, General Motors Corp., Pontiac, Mich.)

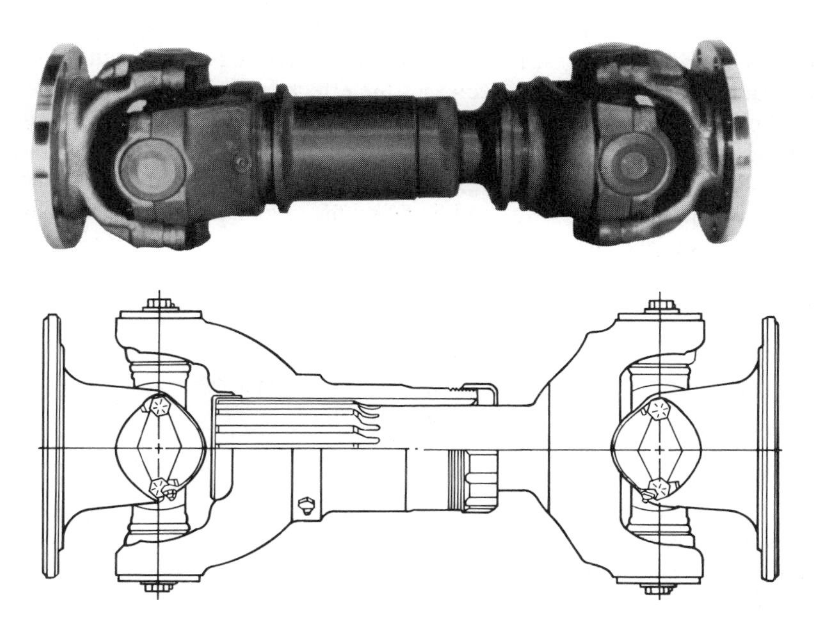

Figure 10-25 Industrial Universal Joint (Spicer Universal Joint Division, Dana Corp., Toledo, Ohio)

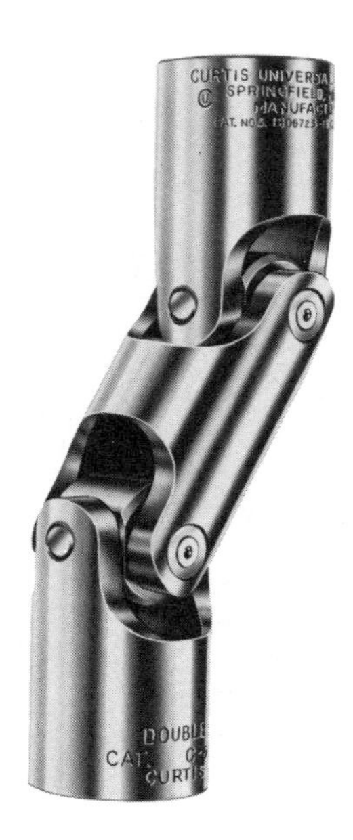

Figure 10-26 Double Universal Joint (Curtis Universal Joints Co., Inc., Springfield, Mass.)

element in an axial direction can cause noise, excessive wear, vibration, or complete disconnection of the drive. Any bicycle rider who has experienced the loss of a chain can appreciate the consequences of misalignment. Recall that for spur gears, the strength of the gear teeth and the wear resistance are both directly proportional to the face width of the gear. Axial misalignment decreases the effective face width.

Some of the methods discussed in section 10-4 for fastening elements to shafts for the purpose of transmitting power also provide some degree of axial location. Refer to the discussions of pinning, clamping, split taper bushings, set screws, the taper and screw, the taper and nut, the press fit, molding, and the several methods of mechanically locking the elements to the shaft.

Among the wide variety of other means available for axial location, we will discuss the following:

Retaining rings	Collars
Shoulders	Spacers
Locknuts	

Retaining Rings

Retaining rings are placed on a shaft, in grooves in the shaft, or in a housing to prevent the axial movement of a machine element. Figure 10-27 shows the wide variety of styles of rings available from one manufacturer. The different designs allow for either internal or external mounting of the ring. The amount of axial thrust load capacity and the shoulder height provided by the different ring styles also vary. The manufacturer's catalog provides data on the capacity and installation data for the rings. Figure 10-28 shows a few examples of retaining rings in use.

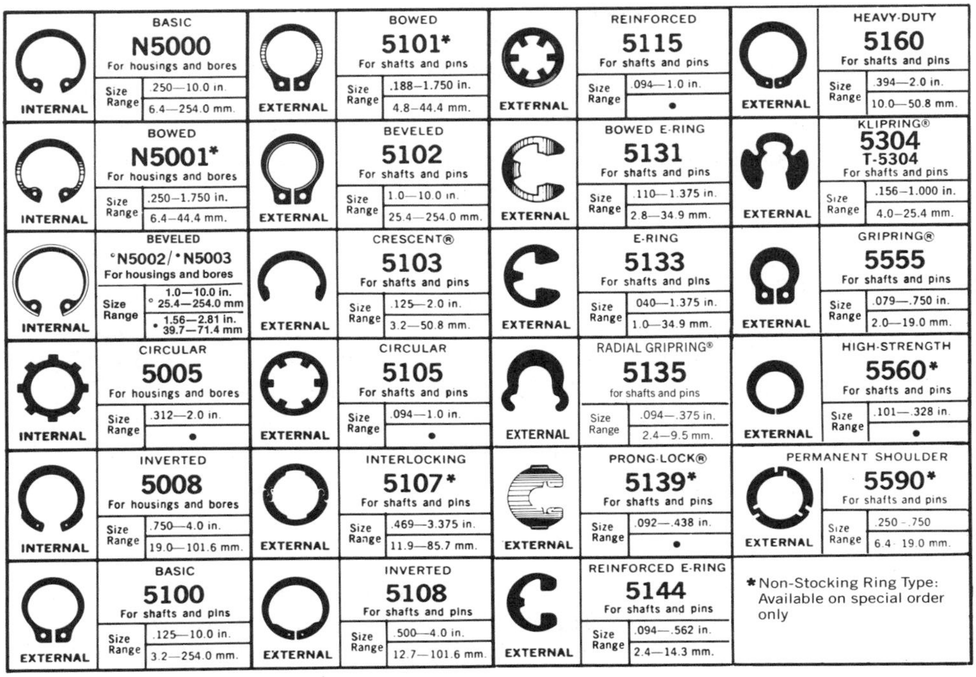

Type	Series	Ring No.	Application	Size Range (in.)	Size Range (mm)
INTERNAL	BASIC	N5000	For housings and bores	.250—10.0 in.	6.4—254.0 mm.
EXTERNAL	BOWED	5101*	For shafts and pins	.188—1.750 in.	4.8—44.4 mm.
EXTERNAL	REINFORCED	5115	For shafts and pins	.094—1.0 in.	•
EXTERNAL	HEAVY-DUTY	5160	For shafts and pins	.394—2.0 in.	10.0—50.8 mm.
INTERNAL	BOWED	N5001*	For housings and bores	.250—1.750 in.	6.4—44.4 mm.
EXTERNAL	BEVELED	5102	For shafts and pins	1.0—10.0 in.	25.4—254.0 mm.
EXTERNAL	BOWED E-RING	5131	For shafts and pins	.110—1.375 in.	2.8—34.9 mm.
EXTERNAL	KLIPRING®	5304 T-5304	For shafts and pins	.156—1.000 in.	4.0—25.4 mm.
INTERNAL	BEVELED	°N5002/•N5003	For housings and bores	° 1.0—10.0 in. • 1.56—2.81 in.	° 25.4—254.0 mm • 39.7—71.4 mm
EXTERNAL	CRESCENT®	5103	For shafts and pins	.125—2.0 in.	3.2—50.8 mm.
EXTERNAL	E-RING	5133	For shafts and pins	040—1.375 in.	1.0—34.9 mm.
EXTERNAL	GRIPRING®	5555	For shafts and pins	.079—.750 in.	2.0—19.0 mm.
INTERNAL	CIRCULAR	5005	For housings and bores	.312—2.0 in.	•
EXTERNAL	CIRCULAR	5105	For shafts and pins	.094—1.0 in.	•
EXTERNAL	RADIAL GRIPRING®	5135	for shafts and pins	.094—.375 in.	2.4—9.5 mm.
EXTERNAL	HIGH-STRENGTH	5560*	For shafts and pins	.101—.328 in.	•
INTERNAL	INVERTED	5008	For housings and bores	.750—4.0 in.	19.0—101.6 mm.
EXTERNAL	INTERLOCKING	5107*	For shafts and pins	.469—3.375 in.	11.9—85.7 mm.
EXTERNAL	PRONG-LOCK®	5139*	For shafts and pins	.092—.438 in.	•
EXTERNAL	PERMANENT SHOULDER	5590*	For shafts and pins	.250-.750	6.4-19.0 mm.
EXTERNAL	BASIC	5100	For shafts and pins	.125—10.0 in.	3.2—254.0 mm.
EXTERNAL	INVERTED	5108	For shafts and pins	.500—4.0 in.	12.7—101.6 mm.
EXTERNAL	REINFORCED E-RING	5144	For shafts and pins	.094—.562 in.	2.4—14.3 mm.

*Non-Stocking Ring Type: Available on special order only

Figure 10-27 Standard Truarc® Retaining Ring Series (Courtesy Waldes Kohinoor, Inc., Long Island City, N.Y.)

Figure 10-28 Retaining Rings Applied to a Water Conditioner Mechanism (Courtesy Waldes Kohinoor, Inc., Long Island City, N.Y.)

Collars

A *collar* is a ring slid over the shaft and positioned adjacent to a machine element for the purpose of axial location. It is held in position, typically, by set screws. Its advantage is that axial location can be set virtually anywhere along the shaft to allow adjustment of the position at the time of assembly. The disadvantages are chiefly those related to the use of set screws themselves (section 10-4).

Shoulders

A *shoulder* is the vertical surface produced when a diameter change occurs on a shaft. Such a design is an excellent method for providing for the axial location of a machine element, at least on one side. Several of the shafts illustrated in Chapter 9 incorporate shoulders. The main design considerations are providing a sufficiently large shoulder to locate the element effectively and a fillet at the base of the shoulder that produces an acceptable stress concentration factor, and is compatible with the geometry of the bore of the mating element (see Chapter 9, Figures 9-2 and 9-7).

Spacers

A *spacer* is similar to a collar in that it is slid over the shaft against the machine element that is to be located. The primary difference is that set screws and the like are not necessary because the spacer is positioned *between* two elements and thus controls only the relative position between them. Typically, one of the elements is positively located by some other means, such as a shoulder or retaining ring. Consider the shaft in Figure 10-29, on which two spacers locate each gear with respect to its adjacent bearing. The bearings are in turn located on one side by the housing supporting their outer races. The gears also seat against shoulders on one side.

Locknuts

When an element is located at the end of a shaft, a *locknut* can be used to retain it on one side. Figure 10-30 shows a bearing retainer type of locknut. These are available as stock items from bearing suppliers.

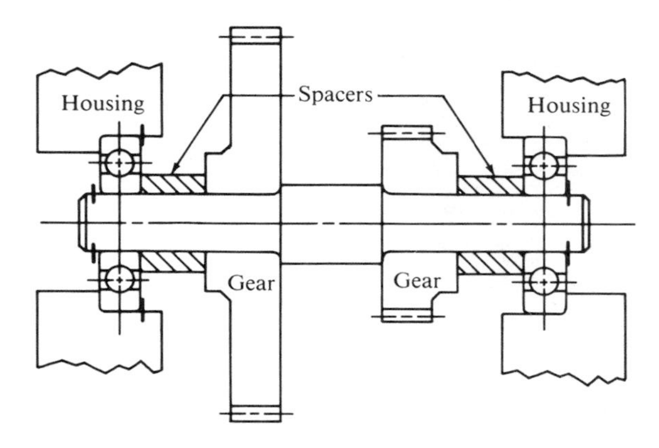

Figure 10-29 Application of Spacers

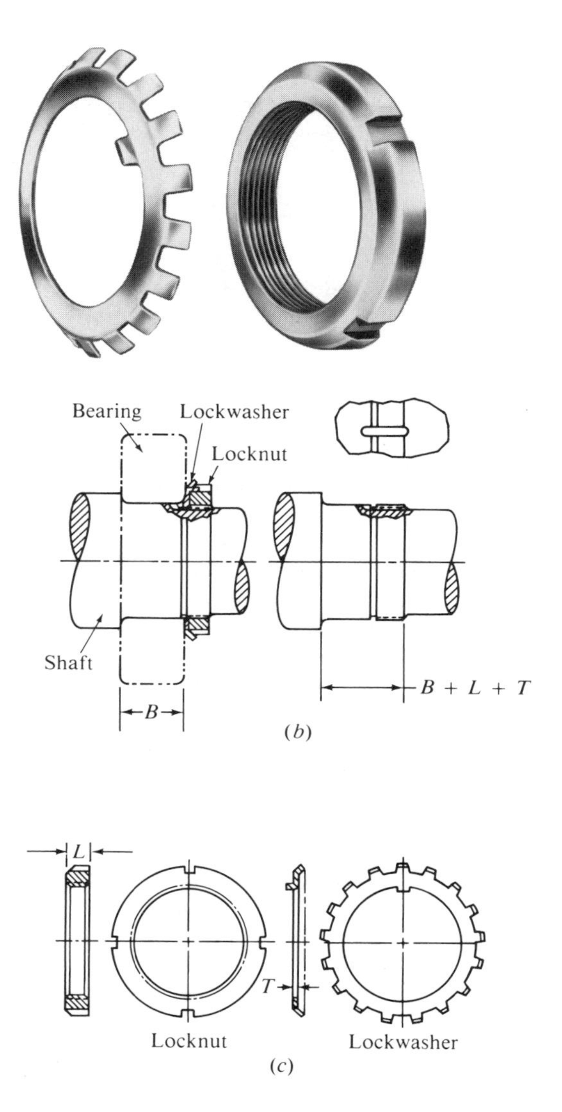

Figure 10-30 Locknut and Lockwasher for Retaining Bearings. Dimensions given in manufacturer's catalog. Part sizes are compatible with standard bearing sizes. (SKF Industries, Inc., King of Prussia, Penn.)

One practical consideration in the matter of axial location of machine elements is exercising care that elements are not *overrestrained*. Under certain conditions of differential thermal expansion or with an unfavorable tolerance stack, the elements may be forced together so tightly as to cause dangerous axial stresses. At times it may be desirable to locate only one bearing positively on a shaft and to permit the other to float slightly in the axial direction. The floating element may be held lightly with an axial spring force accommodating the thermal expansion without creating dangerous forces.

REFERENCES

1. American Society of Mechanical Engineers. *Keys and Keyseats,* USA Standard B17.1-1967 (reaffirmed 1973).
2. Oberg, Erik, et al. *Machinery's Handbook,* 22nd ed. New York: Industrial Press, 1984.

PROBLEMS

For each set of data below, determine the required key geometry: length, width, and height. Use AISI 1020 cold-drawn steel for the keys if a satisfactory design can be achieved. If not, use a higher-strength material. Assume that the key material is weakest when compared with the shaft material or the mating elements, unless otherwise stated.

1. Specify a key for a gear to be mounted on a shaft with a 2.00-in diameter. The gear transmits 21 000 lb · in of torque and has a hub length of 4.00 in.
2. Specify a key for a gear carrying 21 000 lb · in of torque if it is mounted on a 3.60-in-diameter shaft. The hub length of the gear is 4.00 in.
3. A V-belt sheave transmits 1 112 lb · in of torque to a 1.75-in-diameter shaft. The sheave is made from ASTM class 20 cast iron and has a hub length of 1.75 in.
4. A chain sprocket delivers 110 hp to a shaft at a rotational speed of 1 700 rpm. The bore of the sprocket is 2.50 in in diameter. The hub length is 3.25 in.
5. Specify a suitable spline having a B fit for each of the applications in problems 1 to 4.
6. Design a cylindrical pin to transmit the power, as in problem 4. But design it so it will fail in shear if the power exceeds 220 hp.
7. Specify a key for both the sprocket and the worm-gear from example problem 9-4 in Chapter 9. Note the specifications for the final shaft diameters at the end of the problem.
8. Describe a Woodruff key no. 204.
9. Describe a Woodruff key no. 1 628.
10. Make a detailed drawing of a Woodruff key connection between a shaft and the hub of a gear. The shaft has a diameter of 1.500 in. Use a no. 1 210 Woodruff key. Dimension the keyseat in the shaft and the hub.
11. Repeat problem 10, using a no. 406 Woodruff key in a shaft having a 0.500-in diameter.
12. Repeat problem 10, using a no. 2 428 Woodruff key in a shaft having a 3.250-in diameter.
13. Compute the torque that could be transmitted by the key of problem 10 on the basis of shear alone if the key is made from AISI 1020 cold-drawn steel with a design factor of $N = 3$.
14. Repeat problem 13 for the key of problem 11.
15. Repeat problem 13 for the key of problem 12.
16. Make a drawing of a four-spline connection having a major diameter of 1.500 in and an A fit. Show critical dimensions.
17. Make a drawing of a 10-spline connection having a major diameter of 3.500 in and a B fit. Show critical dimensions.
18. Make a drawing of a 16-spline connection having a major diameter of 2.500 in and a C fit. Show critical dimensions.
19. Determine the torque capacity of the splines in problems 16, 17, and 18.
20. Describe the manner in which a set screw transmits torque if it is used in place of a key. Discuss the disadvantages of such an arrangement.
21. Describe a press fit as it would be used to secure a power-transmission element to a shaft.
22. Describe the main differences between rigid and flexible couplings as they affect the stresses in the shafts that they connect.
23. Discuss a major disadvantage of using a single universal joint to connect two shafts with angular misalignment.
24. Describe five ways to locate power transmission elements positively on a shaft axially.

11 Spur Gears

11-1 OVERVIEW

In a *spur gear* the teeth are parallel to the axis of the shaft carrying the gear. Figure 11-1 shows a drawing of the cross section of three spur gear teeth, along with photographs of typical commercially available spur gears and sketches of gear blank styles. Their relative simplicity and ease of manufacture and installation make spur gears very popular as power-transmitting elements in mechanical design.

This chapter will describe the basic geometry of spur gears and the methods of utilizing them in trains to produce desired speed-reduction ratios. The methods of computing the loads and stresses on spur gear teeth are also developed. Wear resistance is discussed because the successful application of gears requires not only that the teeth will not break but also that they will keep their precise geometry for many hours, even years, of operation. Gears must be carried on shafts to transmit the power to the driven machinery. The forces exerted on the shaft are analyzed, along with the practical design considerations of gears and their interfaces with other machine elements such as bearings and housings.

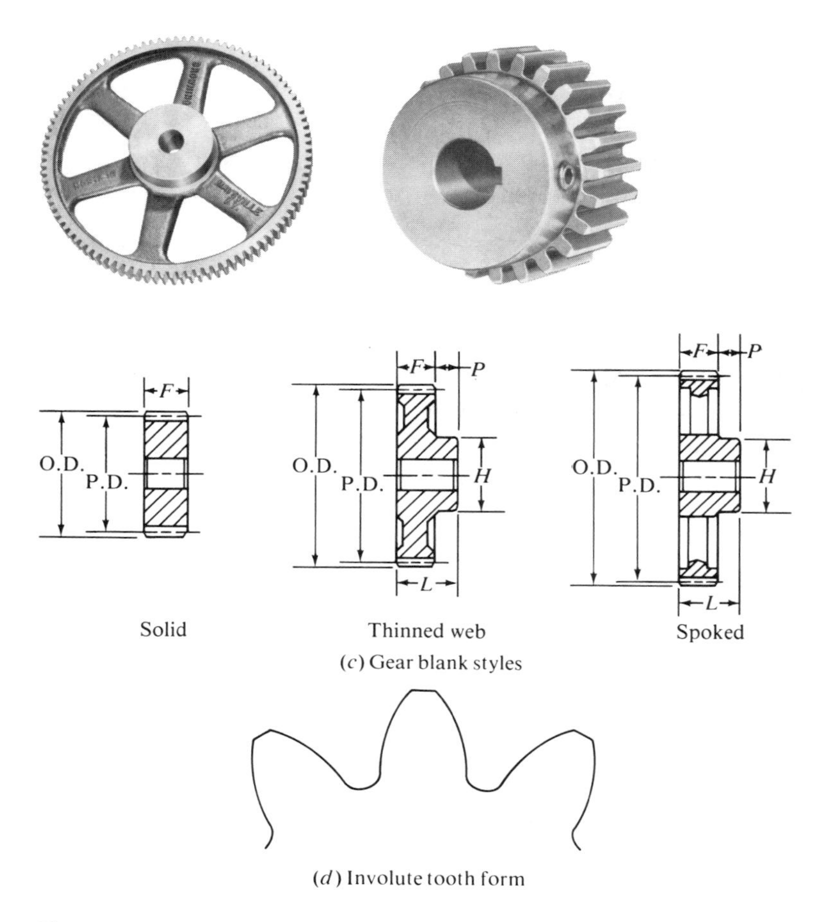

Figure 11-1 Spur Gears (Browning Mfg. Division, Emerson Electric Co., Maysville, Ky.)

11-2 GEAR GEOMETRY

Involute Tooth Form

The most widely used spur gear tooth form is the full-depth involute form. Its characteristic shape can be seen in Figure 11-1.

The involute is one of a class of geometric curves called *conjugate curves*. When two such gear teeth are in mesh and rotating, there is a *constant angular velocity* ratio between them: From the moment of initial contact to the moment of disengagement, the speed of the driving gear is in a constant proportion to the speed of the driven gear. The resulting action of the two gears is very smooth. If this were not the case, there would be some speeding up and slowing down during the engagement with the resulting accelerations causing vibration, noise, and dangerous torsional oscillations in the system.

The involute curve can be easily visualized by taking a cylinder and wrapping a string around its circumference. Tie a pencil to the end of the string. Then start with the pencil tightly against the cylinder and hold the string taut. Move the pencil away from the cylinder while keeping the string taut. The curve you will draw is an involute. Figure 11-2 is a sketch of the process.

The circle represented by the cylinder is called the *base circle*. Notice that at any position on the curve, the string represents a line tangent to the base circle and, at the same time, perpendicular to the involute. Drawing another base circle along the same centerline in such a position that the resulting involute is tangent to the first one, as shown in Figure 11-3, demonstrates that at the point of contact, the two lines tangent to the base circles are coincident and will stay in the same position as the base circles rotate. This is what happens when two gear teeth are in mesh.

It is a fundamental principle of *kinematics,* the study of motion, that if the line drawn perpendicular to the surfaces of two rotating bodies at their point of contact always crosses the centerline between the two bodies at the same place, the angular velocity ratio of the two bodies will be constant. This is a statement of the law of gearing. As demonstrated here, the gear teeth made in the involute tooth form obey the law.

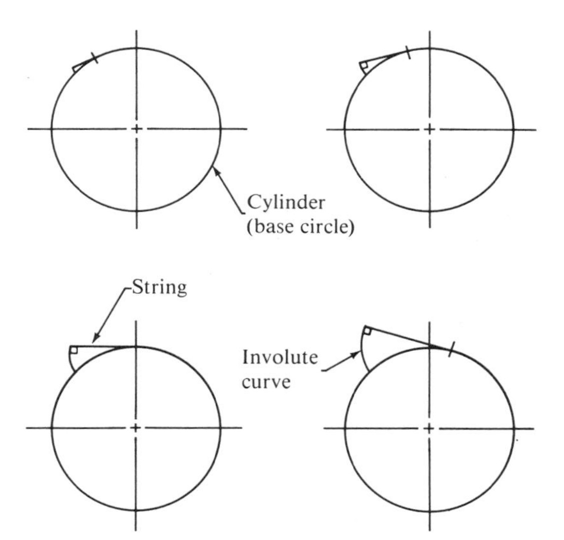

Figure 11-2 Graphical Generation of an Involute Curve

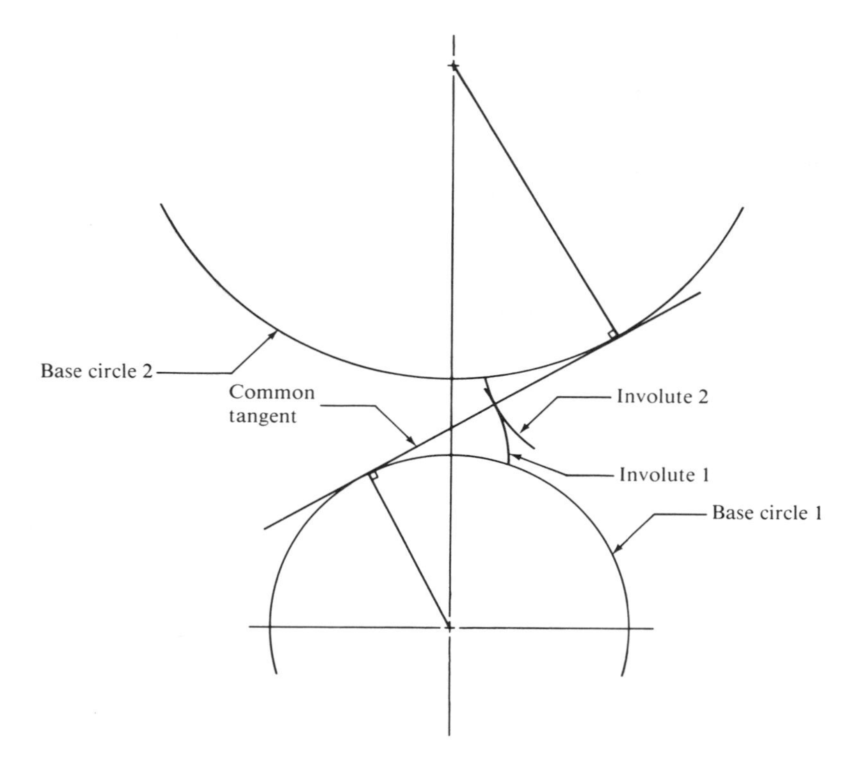

Figure 11-3 Mating Involutes

Of course, only the part of the gear tooth that actually comes into contact with the mating tooth needs to be in the involute form. Figure 11-4 shows drawings of spur gear teeth, with the symbols for the various features indicated. These features are described later in this section.

Pitch Diameter

Figure 11-5 shows teeth from two gears in mesh to demonstrate the relative positions of the teeth at several stages of engagement. One of the most important observations that can be made from Figure 11-5 is that, throughout the engagement cycle, there are two circles, one from each gear, that remain tangent. These are called the *pitch circles*. The diameter of the pitch circle of a gear is its *pitch diameter;* the point of tangency is the *pitch point*.

When two gears mesh, the smaller gear is called the *pinion* and the larger is the *gear*. We will use the symbol d to indicate the pitch diameter of the pinion and the symbol D for the pitch diameter of the gear. When referring to the number of teeth, we will use N_P for the pinion and N_G for the gear.

Notice that the pitch diameter lies somewhere within the height of the gear tooth, and thus it is not possible to measure its diameter directly. It must be calculated from other known features of the gear; this calculation depends on understanding the concept of *pitch*.

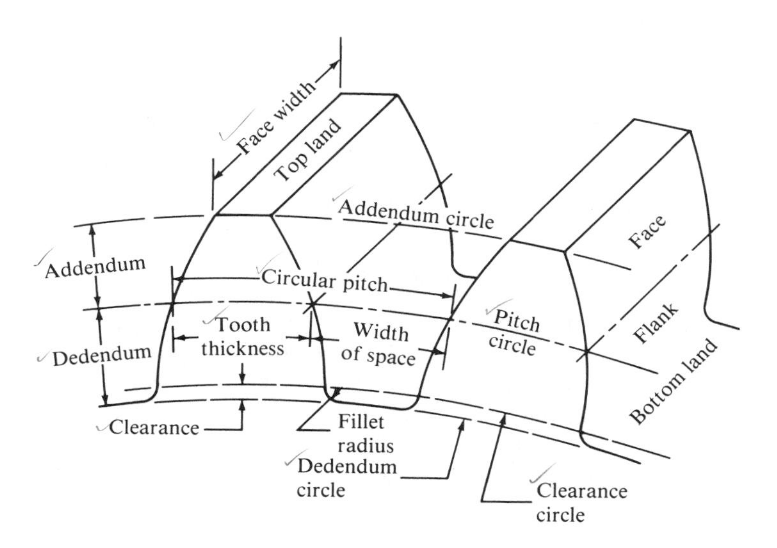

Figure 11-4 Spur Gear Teeth Features

Pitch

The spacing between adjacent teeth and the size of the teeth are controlled by the pitch of the teeth. Three types of pitch designation systems are in common use for gears: circular pitch, diametral pitch, and the metric module.

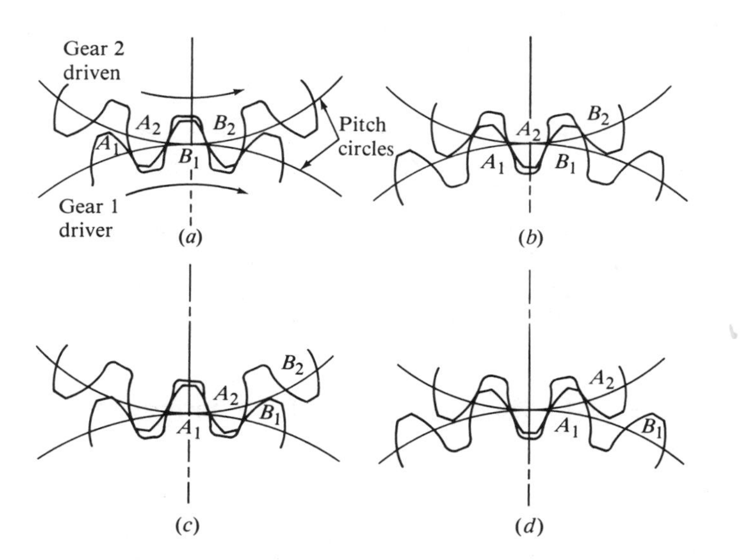

Figure 11-5 Cycle of Engagement of Gear Teeth

Circular Pitch, P_c

The distance from a point on a tooth of a gear at the pitch circle to a corresponding point on the next adjacent tooth, measured along the pitch circle, is the *circular pitch*. Note that it is an arc length, usually in inches. The value of the circular pitch can be computed by taking the circumference of the pitch circle and dividing it into a number of equal parts corresponding to the number of teeth in the gear. Using N for the number of teeth,

$$P_c = \pi D/N \tag{11-1}$$

Notice that the tooth size increases as the value of the circular pitch increases because there is a larger pitch circle for the same number of teeth. Also note that the basic sizes of the mating gear teeth must be the same for them to mesh properly. This observation leads to a very important rule: *The pitch of two gears in mesh must be identical.* This must be true whether the pitch is indicated as the circular pitch, the diametral pitch, or the metric module. Then equation (11-1) can be written in terms of either the pinion or the gear diameter:

$$P_c = \pi D/N_G = \pi d/N_P \tag{11-2}$$

Circular pitch is infrequently used now. It is sometimes an advantage to use this system when large gears are to be made by casting. The layout of the pattern for the casting is facilitated by laying off the chord of the arc length of the circular pitch. Also, some machines and product lines of machines have traditionally used circular pitch gears and continue to do so. Table 11-1 lists the recommended standard circular pitches for large gear teeth.

Diametral Pitch, P_d

The most common pitch system used today is the *diametral pitch* system, the number of teeth per inch of pitch diameter. Its basic definition is

$$P_d = N_G/D = N_P/d \tag{11-3}$$

As such, it has units of in^{-1}. However, the units are rarely reported and gears are referred to as 8-pitch or 20-pitch, for example. One of the advantages of the diametral pitch system is that there is a set list of standard pitches, and most of them have integer values. Table 11-2 lists the recommended standard pitches, with those of 20 and above called *fine pitch* and those below 20 called *coarse pitch*.

There are other intermediate values available but most manufacturers produce gears from this list of pitches. In any case, it is advisable to check availability before finally

Table 11-1 Standard Circular Pitches (Inches)

10.	7.5	5.
9.5	7.	4.5
9.	6.5	4.
8.5	6.	3.5
8.	5.5	

Table 11-2 Standard Diametral Pitches (Teeth/Inch)

Coarse Pitch ($P_d < 20$)				Fine Pitch ($P_d \geq 20$)	
1	2	5	12	20	72
1.25	2.5	6	14	24	80
1.5	3	8	16	32	96
1.75	4	10	18	48	120
				64	

specifying a pitch. In problem solutions in this book, it is expected that one of the pitches listed in Table 11-2 will be used if possible.

As stated before, the pitch of the gear teeth determines their size, and two mating gears must have the same pitch. Figure 11-6 shows the profiles of some of the standard diametral pitch gear teeth, drawn actual size. That is, you can lay a given gear down on the page and compare its size with the drawing and get a good estimate of the pitch of the teeth. Notice that as the numerical value of the diametral pitch increases, the physical size of the tooth decreases, and vice versa.

Sometimes it is necessary to convert from diametral pitch to circular pitch, or vice versa. Their definitions provide a simple means of doing this. Solving for the pitch diameter in both equations (11-1) and (11-3) gives

$$D = NP_c/\pi$$

$$D = N/P_d$$

Equating these two gives

$$N/P_d = NP_c/\pi \qquad \text{or} \qquad P_dP_c = \pi \tag{11-4}$$

From this equation, the equivalent circular pitch for a gear having a diametral pitch of 1 is $P_c = \pi/1 = 3.1416$. Referring to Tables 11-1 and 11-2, notice that the circular pitches listed are for the larger gear teeth, being preferred when the diametral pitch is less than 1. Diametral pitch is preferred for sizes equivalent to 1-pitch or smaller.

Metric Module System

In the SI system of units, the standard unit of length is the *millimeter*. The pitch of gears in the metric system is based on this unit and is designated the *module, m*. The module of a gear is found by dividing the pitch diameter of the gear in millimeters by the number of teeth. That is,

$$m = D/N_G = d/N_P \tag{11-5}$$

There is rarely a need to convert from the module system to the diametral pitch system. However, it is important to have a feel for the physical size of gear teeth. Because at this time people are more familiar with the standard diametral pitches, as shown in

Figure 11-6 Gear Tooth Size as a Function of Diametral Pitch
(Barber-Colman Company, Loves Park, Ill.)

Figure 11-6, we will develop the relationship between m and P_d. From their definitions, equations (11-3) and (11-5), we can say

$$m = 1/P_d$$

But recall that diametral pitch uses the inch unit, and module uses the millimeter. Therefore, the conversion factor of 25.4 mm per inch must be applied.

$$m = \frac{1}{P_d \text{ in}^{-1}} \cdot \frac{25.4 \text{ mm}}{\text{in}}$$

This reduces to

$$m = 25.4/P_d \tag{11-6}$$

For example, if a gear has a diametral pitch of 10, the equivalent module is

$$m = 25.4/10 = 2.54 \text{ mm}$$

This is not a standard value for module, but it is close to the standard value of 2.5. So it can be concluded that a 10-pitch gear is of similar size to a gear with module 2.5. Table 11-3 gives selected standard modules with their equivalent diametral pitches.

Table 11-3 Standard Modules

Module (mm)	*Equivalent P_d*	*Closest Standard P_d (Teeth/Inch)*
0.3	84.667	80
0.4	63.500	64
0.5	50.800	48
0.8	31.750	32
1	25.400	24
1.25	20.320	20
1.5	16.933	16
2	12.700	12
2.5	10.160	10
3	8.466	8
4	6.350	6
5	5.080	5
6	4.233	4
8	3.175	3
10	2.540	2.5
12	2.117	2
16	1.587	1.5
20	1.270	1.25
25	1.016	1

Gear Tooth Features

In design and inspection of gear teeth, several special features must be known. Figure 11-4, presented earlier, and Figure 11-7, which shows segments of two gears in mesh, identify these features. They are defined next; Table 11-4 gives the relationships needed to compute their values. Note that many of the computations involve the diametral pitch, again illustrating that the physical size of a gear tooth is determined by its diametral pitch.

Addendum (a). The radial distance from the pitch circle to the outside of a tooth.

Dedendum (b). The radial distance from the pitch circle to the bottom of the tooth space.

Clearance (c). The radial distance from the top of a tooth to the bottom of the tooth space of the mating gear when the tooth is fully engaged. Note:

$$c = b - a \tag{11-7}$$

Outside Diameter (D_o). The diameter of the circle that encloses the outside of the gear teeth. Note:

$$D_o = D + 2a \tag{11-8}$$

Also note that both the pitch diameter, D, and the addendum, a, are defined in terms of the diametral pitch, P_d. Making these substitutions produces a very useful form of the equation for outside diameter:

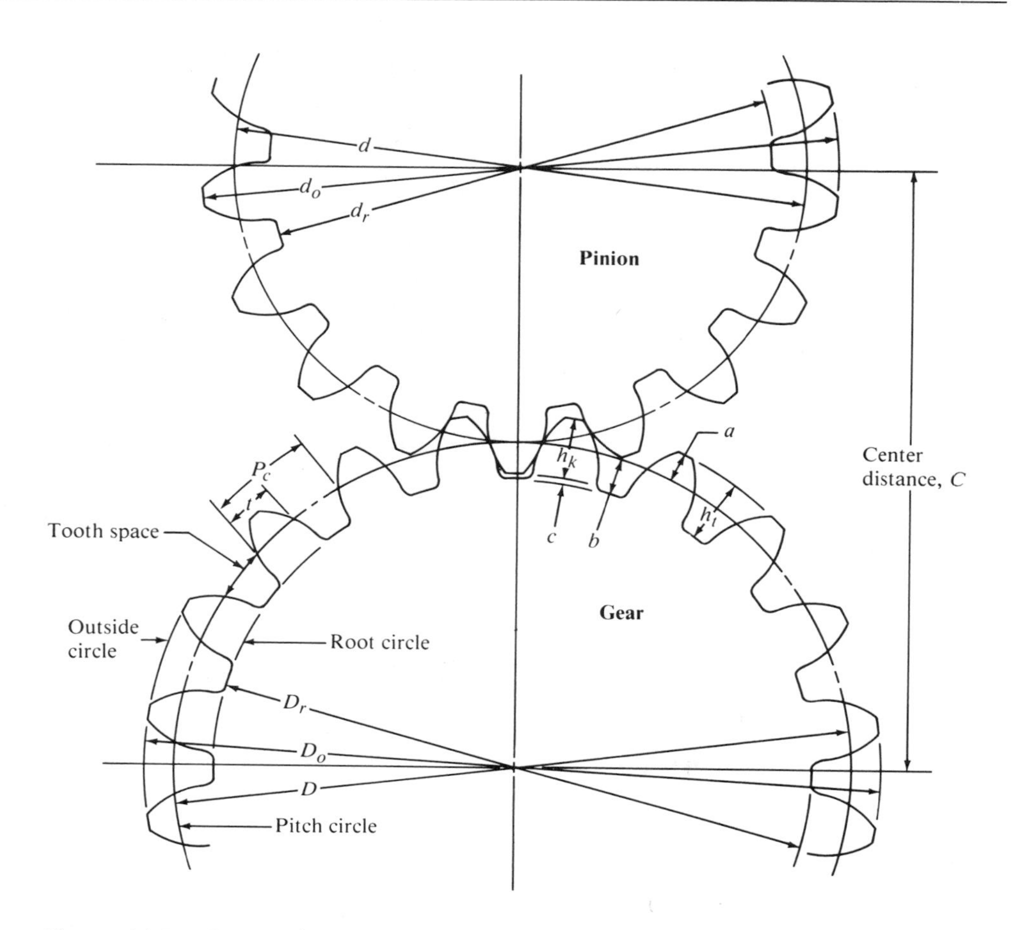

Figure 11-7 Gear Pair Features

$$D_o = \frac{N}{P_d} + 2\frac{1}{P_d} = \frac{N + 2}{P_d} \tag{11-9}$$

In the metric module system, a similar equation can be derived:

$$D_o = mN + 2m = m(N + 2) \tag{11-10}$$

Root Diameter (D_r). The diameter of the circle that contains the bottom of the tooth space; this circle is called the *root circle*. Note:

$$D_r = D - 2b \tag{11-11}$$

Whole Depth (h_t). The radial distance from the top of a tooth to the bottom of the tooth space. Note:

$$h_t = a + b \tag{11-12}$$

Table 11-4 Formulas for Gear Tooth Features for 20° Pressure Angle

		FULL-DEPTH INVOLUTE SYSTEM		METRIC MODULE SYSTEM
FEATURE	SYMBOL	*Coarse Pitch* ($P_d < 20$)	*Fine Pitch* ($P_d \geq 20$)	
Addendum	a	$1/P_d$	$1/P_d$	$1.00\ m$
Dedendum	b	$1.25/P_d$	$1.200/P_d + 0.002$	$1.25\ m$
Clearance	c	$0.25/P_d$	$0.200/P_d + 0.002$	$0.25\ m$

Working Depth (h_k). The radial distance that a gear tooth projects into the tooth space of the mating gear. Note:

$$h_k = a + a = 2a \tag{11-13}$$

and

$$h_t = h_k + c \tag{11-14}$$

Tooth Thickness (t). The arc length, measured on the pitch circle, from one side of a tooth to the other side. This is sometimes called the *circular thickness* and has the theoretical value of one-half of the circular pitch. That is,

$$t = P_c/2 = \pi/2P_d \tag{11-15}$$

Tooth Space. The arc length, measured on the pitch circle, from the right side of one tooth to the left side of the next tooth. Theoretically, the tooth space equals the tooth thickness. But for practical reasons, the tooth space is made larger (see Backlash).

Backlash. If the tooth thickness is made identical in value to the tooth space, as it theoretically is, the tooth geometry would have to be absolutely precise for the gears to operate, and there would be no space available for lubrication of the tooth surfaces. To alleviate these problems, practical gears are made with the tooth space slightly larger than the tooth thickness, the difference being called the *backlash*. To provide backlash, the cutter generating the gear teeth can be fed deeper into the gear blank than the theoretical value on either or both of the mating gears. Alternatively, backlash can be created by adjusting the center distance to a larger value than the theoretical value.

The magnitude of backlash depends on the desired precision of the gear pair and on the size and pitch of the gears. It is actually a design decision, balancing cost of production with desired performance. The American Gear Manufacturers Association (AGMA) provides recommendations for backlash in their standards. Table 11-5 lists recommended ranges for several values of pitch.

Face Width (F). The width of the tooth measured parallel to the axis of the gear.

Fillet. The arc joining the involute tooth profile to the root of the tooth space.

Face. The surface of a gear tooth from the pitch circle to the outside circle of the gear.

Table 11-5 Ranges of Recommended Backlash

Diametral Pitch P_d	*Center Distance (Coarse* P_d*)*	*Precision Class (Fine* P_d*)*	*Backlash (Inches)*
1	30 in		0.040
1	50 in		0.060
1	100 in		0.080
4	10 in		0.020
4	30 in		0.030
4	50 in		0.040
10	5 in		0.010
10	20 in		0.020
20		*A*	0.004 –0.008
20		*C*	0.001 –0.002
20		*E*	0 –0.000 5
64		*A*	0.003 –0.007
64		*C*	0.000 7–0.001 5
64		*E*	0 –0.000 4

Flank. The surface of a gear tooth from the pitch circle to the root of the tooth space, including the fillet.

Center Distance (*C*). The distance from the center of the pinion to the center of the gear; the sum of the pitch radii of two gears in mesh. That is, because radius = diameter/2:

$$C = D/2 + d/2 = (D + d)/2 \tag{11-16}$$

Also note that both pitch diameters can be expressed in terms of the diametral pitch:

$$C = \frac{1}{2}\left[\frac{N_G}{P_d} + \frac{N_P}{P_d}\right] = (N_G + N_P)/2P_d \tag{11-17}$$

It is recommended that equation (11-17) be used for center distance because all the terms are usually integers, giving greater accuracy in the computation. In the metric module system, a similar equation can be derived.

$$C = (D + d)/2 = (mN_G + mN_P)/2 = [(N_G + N_P)m]/2 \tag{11-18}$$

Pressure Angle

Figure 11-8 illustrates the *pressure angle:* the angle between the tangent to the pitch circles and the line drawn normal (perpendicular) to the surface of the gear tooth. The normal line is sometimes referred to as the *line of action*. When two gear teeth are in mesh and transmitting power, the force transferred from the driver to the driven gear tooth acts in a direction along the line of action. Also, the actual shape of the gear tooth depends on the pressure angle, as illustrated in Figure 11-9. The teeth in the figure were drawn

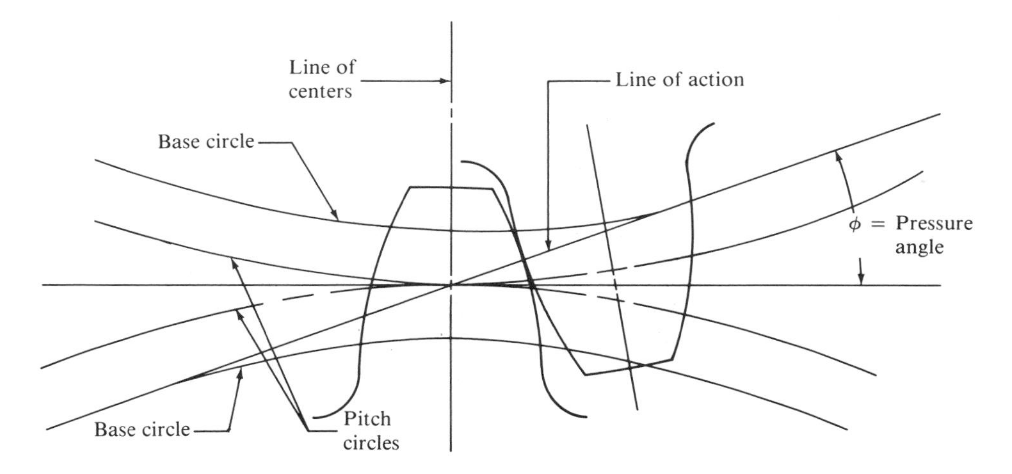

Figure 11-8 Pressure Angle

according to the proportions for a 20-tooth, 5-pitch gear having a pitch diameter of 4.000 in.

All three teeth have the same tooth thickness because, as stated in equation (11-15), the thickness at the pitch line depends only on the pitch. The difference between the three teeth shown is due to the different pressure angles because the pressure angle determines the size of the base circle. Remember that the base circle is the circle from which the involute is generated. The line of action is always tangent to the base circle. Therefore, the size of the base circle can be found from

$$D_b = D \cos \phi \tag{11-19}$$

Standard values of the pressure angle are established by gear manufacturers, and the pressure angles of two gears in mesh must be the same. Current standard pressure angles are 14½°, 20°, and 25°, as illustrated in Figure 11-9. Actually, the 14½° tooth form is

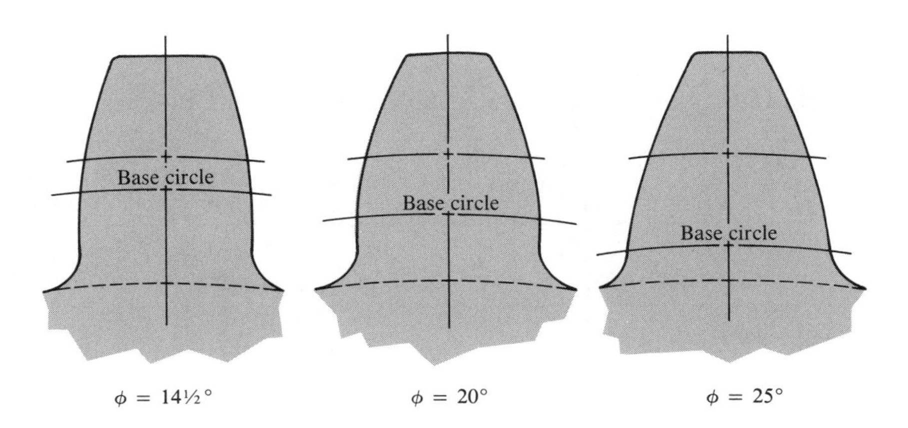

Figure 11-9 Full-Depth Involute Tooth Form for Varying Pressure Angles

considered to be obsolete. Although it is still available, it should be avoided for new designs. The 20° tooth form is the most readily available at this time. The advantages and disadvantages of the different values of pressure angle relate to the strength of the teeth, the occurrence of interference, and the magnitude of forces exerted on the shaft. Interference is discussed next. The other points are discussed later in this chapter.

Interference

For certain combinations of numbers of teeth in a gear pair there is interference between the tip of the teeth on the pinion and the fillet or root of the teeth on the gear. Obviously this cannot be tolerated because the gears simply will not mesh. The probability of interference occurring is greatest when a small pinion drives a large gear, with the worst case being a small pinion driving a rack. A *rack* is a gear with a straight pitch line and can be thought of as a gear with an infinite pitch diameter.

It is the designer's responsibility to ensure that interference does not occur in a given application. The surest way to do this is to control the minimum number of teeth in the pinion to the limiting values shown in Table 11-6. With this number of teeth or a greater number, there will be no interference with a rack or with any other gear. A designer who desires to use fewer than the listed number of teeth can use a graphical layout to test the combination of pinion and gear for interference. Texts on kinematics provide the necessary procedure. The right part of Table 11-6 indicates the maximum number of gear teeth you can use for a given number of pinion teeth to avoid interference (7).

As noted earlier, the 14½° system is considered to be obsolete. The data in Table 11-6 indicate one of the main disadvantages with that system: its potential for causing interference.

If a proposed design encounters interference, there are ways to make it work. But caution should be exercised because the tooth form or the alignment of the mating gears is changed, causing the stress and wear analysis to be inaccurate. With this in mind, the designer can provide for undercutting, modification of the addendum on the pinion or the gear, or modification of the center distance.

Undercutting is the process of cutting away the material at the fillet or root of the gear teeth, thus relieving the interference. Figure 11-10 shows the result of undercutting. It should be obvious that this process weakens the tooth; this point is discussed further in the section on stresses in gear teeth.

Table 11-6 Number of Pinion Teeth to Ensure No Interference

FOR A PINION MESHING WITH A RACK		FOR A 20° FULL-DEPTH PINION MESHING WITH A GEAR	
Tooth Form	*Minimum Number of Teeth*	*Number of Pinion Teeth*	*Maximum Number of Gear Teeth*
14½° Involute, Full-Depth	32	17	1309
20° Involute, Full-Depth	18	16	101
25° Involute, Full-Depth	12	15	45
		14	26
		13	16

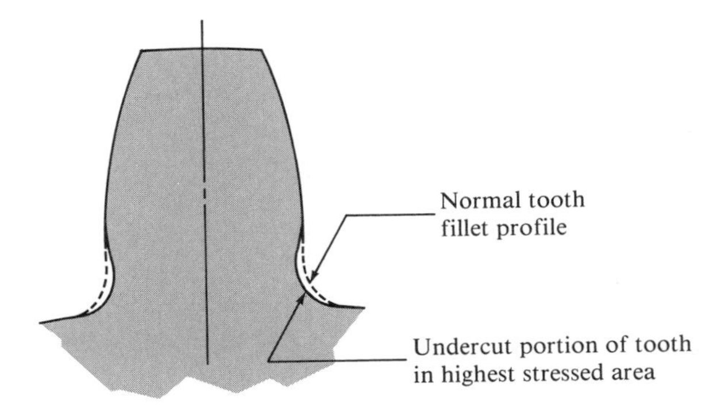

Figure 11-10 Undercutting of a Gear Tooth

The problem of interference can be alleviated by increasing the addendum of the pinion while decreasing the addendum of the gear. The center distance can remain the same as its theoretical value for the number of teeth in the pair. But the resulting gears are, of course, nonstandard (3). It is possible to make the pinion of a gear pair larger than standard while keeping the gear standard if the center distance for the pair is enlarged (4).

11-3 GEAR TRAINS

A *gear train* is one or more pairs of gears operating together to transmit power. Normally there is a speed change from one gear to the next due to the different size of the gears in mesh. The fundamental building block of the total speed change ratio in a gear train is the velocity ratio between two gears in a single pair.

Velocity Ratio

The *velocity ratio (VR)* is defined as the ratio of the rotational speed of the input gear to that of the output gear for a single pair of gears. To develop the equation for computing the velocity ratio, it is helpful to view the action of two gears in mesh, as shown in Figure 11-11. The action is equivalent to the action of two smooth wheels rolling on each other without slipping, with the diameters of the two wheels equal to the pitch diameters of the two gears. Remember that when two gears are in mesh, their pitch circles are tangent and that, obviously, the gear teeth prohibit any slipping.

As shown in Figure 11-11, without slipping there is no relative motion between the two pitch circles at the pitch point and, therefore, the linear velocity of a point on either pitch circle is the same. We will use the symbol v_t for this velocity. The linear velocity of a point that is in rotation at a distance R from its center of rotation and rotating with an angular velocity, ω, is found from

$$v_t = R\omega \tag{11-20}$$

Using the subscript P for the pinion and G for the gear for two gears in mesh,

$$v_t = R_P\,\omega_P \qquad \text{and} \qquad v_t = R_G\,\omega_G$$

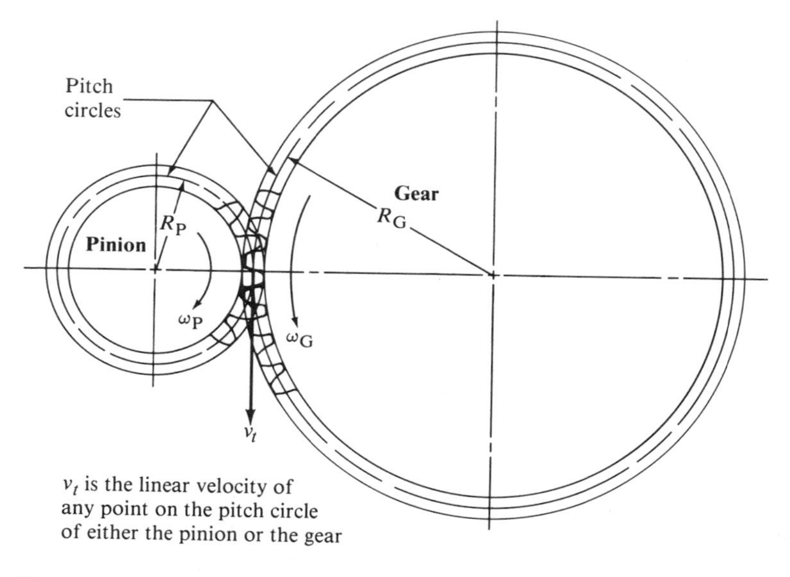

Figure 11-11 Two Gears in Mesh

This set of equations says that the pitch line speed of the pinion and the gear are the same. Equating these two and solving for ω_P/ω_G gives our definition for the velocity ratio, *VR*:

$$VR = \omega_P/\omega_G = R_G/R_P$$

In general, it is convenient to express the velocity ratio in terms of the pitch diameters, the rotational speeds, or the numbers of teeth of the two gears in mesh. Remember that

$$R_G = D/2$$
$$R_P = d/2$$
$$D = N_G/P_d$$
$$d = N_P/P_d$$
$$n_P = \text{Rotational speed of the pinion (in rpm)}$$
$$n_G = \text{Rotational speed of the gear (in rpm)}$$

The definition of velocity ratio can then be defined in any of the following ways.

$$VR = \frac{\omega_P}{\omega_G} = \frac{n_P}{n_G} = \frac{R_G}{R_P} = \frac{D}{d} = \frac{N_G}{N_P} = \frac{\text{speed}_P}{\text{speed}_G} = \frac{\text{size}_G}{\text{size}_P} \tag{11-21}$$

Most gear drives are *speed reducers,* that is, the output speed is lower than the input speed. This results in a velocity ratio greater than 1. If a *speed increaser* is desired, then *VR* is less than 1. It should be noted that not all books and articles use the same definition for velocity ratio. Some define it as the ratio of the output speed to the input speed, the inverse of our definition. It is thought that the use of *VR* greater than 1 for the reducer, that is, the majority of the time, is more convenient.

Train Value

When more than two gears are in mesh, the term *train value (TV)* refers to the ratio of the input speed (for the first gear in the train) to the output speed (the last gear in the train). Again, TV will be greater than 1 for a reducer and less than 1 for an increaser.

By definition the *train value* is the product of the values of VR for each gear pair in the train. In this definition, a gear pair is any set of two gears with a driver and a follower (driven) gear.

For example, consider the gear train sketched in Figure 11-12. The input is through the shaft carrying gear A. Gear A drives gear B. Gear C is on the same shaft with gear B and rotates at the same speed. Gear C drives gear D, which is connected to the output shaft. Then gears A and B constitute the first gear pair, and gears C and D constitute the second pair.

$$VR_1 = n_A/n_B$$

$$VR_2 = n_C/n_D$$

The train value is

$$TV = (VR_1)(VR_2) = \frac{n_A}{n_B}\frac{n_C}{n_D}$$

But, because they are on the same shaft, $n_B = n_C$, and the above equation reduces to

$$TV = n_A/n_D$$

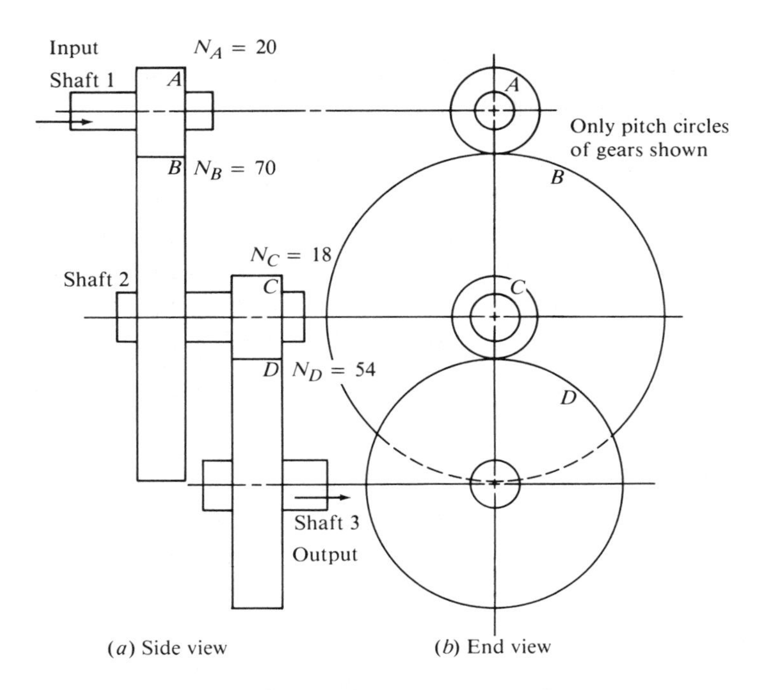

Figure 11-12 Double-Reduction Gear Train

This is the input speed divided by the output speed, the basic definition of the train value. This process can be expanded to any number of stages of reduction in a gear train. Remember that any of the forms for velocity ratio shown in equation (11-21) can be used for computing the train value. In design, it is often most convenient to express the velocity ratio in terms of the numbers of teeth in each gear because they must be integers. Then, once the diametral pitch or module is defined, the values of the diameters or radii can be determined.

Example Problem 11-1. For the gear train shown in Figure 11-12, if the input shaft rotates at 1 750 rpm clockwise, compute the speed of the output shaft and its direction of rotation.

Solution. We can find the output speed if we can determine the train value.

$$TV = n_A/n_D = \text{Input speed/output speed}$$

Then,

$$n_D = n_A/TV$$

But

$$TV = (VR_1)(VR_2) = \frac{N_B}{N_A}\frac{N_D}{N_C} = \frac{70}{20}\frac{54}{18} = \frac{3.5}{1}\frac{3.0}{1} = \frac{10.5}{1} = 10.5$$

Now

$$n_D = n_A/TV = (1\,750 \text{ rpm})/10.5 = 166.7 \text{ rpm}$$

The direction of rotation can be determined by observation, noting that there is a direction reversal for each pair of external gears.

Gear A rotates clockwise; gear B rotates counterclockwise.

Gear C rotates counterclockwise; gear D rotates clockwise.

We will use the term *positive train value* to refer to one in which the input and output gears rotate in the same direction. Conversely, if they rotate in the opposite direction, the train value will be negative. Thus the train in Figure 11-12 is a positive train.

Example Problem 11-2. Determine the train value for the train shown in Figure 11-13. If the shaft carrying gear A rotates at 1 750 rpm clockwise, compute the speed and direction of the shaft carrying gear E.

Solution. Look first at the direction of rotation. Remembering that a gear pair is defined as any two gears in mesh (a driver and a follower), there are actually three gear pairs.

Gear A drives gear B: A clockwise; B counterclockwise.

Gear C drives gear D: C counterclockwise; D clockwise.

Gear D drives gear E: D clockwise; E counterclockwise.

Because gears A and E rotate in opposite directions, the train value is negative. Now

$$TV = -(VR_1)(VR_2)(VR_3)$$

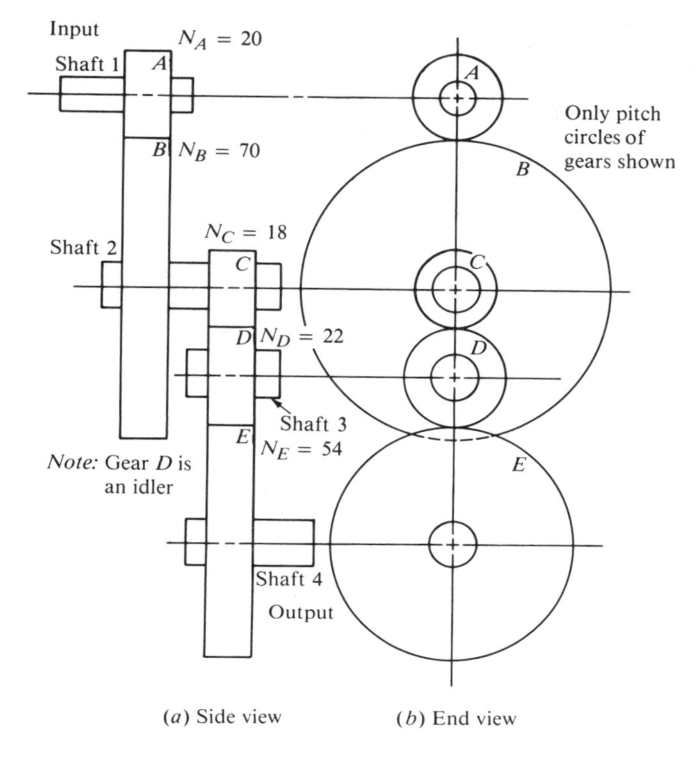

Figure 11-13 Double-Reduction Gear Train with an Idler. Gear D is an idler.

In terms of the number of teeth,

$$TV = -\frac{N_B}{N_A}\frac{N_D}{N_C}\frac{N_E}{N_D}$$

Note that the number of teeth in gear D appears in both the numerator and the denominator and thus can be cancelled. The train value then becomes

$$TV = -\frac{N_B}{N_A} \cdot \frac{N_E}{N_C} = -\frac{70}{20} \cdot \frac{54}{18} = -\frac{3.5}{1}\frac{3.0}{1} = -10.5$$

Gear D is called an *idler;* as demonstrated here, it has no effect on the magnitude of the train value, but it does cause a direction reversal. The output speed is then found from

$$TV = n_A/n_E$$

$$n_E = n_A/TV = (1\,750 \text{ rpm})/(-10.5) = -166.7 \text{ rpm} \quad \text{(counterclockwise)}$$

11-4 GEAR MATERIALS

Gears can be made from a wide variety of materials to achieve properties appropriate to the application. From a mechanical design standpoint, strength and durability (wear resistance) are the most important properties. But in general, the designer should consider the producibility of the gear, taking into account all of the manufacturing processes involved from the preparation of the gear blank, through the forming of the gear teeth, to the final assembly of the gear into a machine. Other considerations are weight, appearance, corrosion resistance, noise, and of course, cost.

Some of the more widely used materials for gears are listed in Tables 11-7, 11-8, and 11-9, along with the normal range of hardness to which they can be produced. Hardness is the major property related to the durability of the gear, and it is also related to the expected strength of the material. The values listed for tensile strength and yield strength are approximate but give an idea of the available range. The final specification of a material must include its AISI number and heat-treatment designation.

The steels are usually in the wrought form in the smaller sizes and cast in the larger sizes. Some steels used for gears include AISI 1020, 1045, 4027, 4640, 5150, 6150, and

Table 11-7 Typical Ferrous Gear Materials

MATERIAL AND CONDITION	HARDNESS RANGE*	TENSILE STRENGTH		YIELD STRENGTH	
		Ksi	MPa	Ksi	MPa
Carbon or alloy steel					
Annealed	HB 160–200	75	520	40	275
Carbon steel					
Q & T	HB 212–248	95	655	55	380
Q & T	HB 302–351	135	930	115	790
Alloy steel					
Q & T	HB 223–262	105	725	80	550
Q & T	HB 402–461	190	1 310	170	1 170
Carbon or alloy steel Surface-hardened by flame or induction	Surface HB 402–555 HRC 43–55	Depends on prior treatment			
Carbon or alloy steel Surface-hardened by carburizing	Surface HB 461–653 HRC 48–60	Depends on prior treatment			
Gray cast iron					
Class 20	HB 155	20	138	—	—
Class 60	HB 223	60	414	—	—
Nodular iron					
Annealed	HB 179	60	414	40	276
Q & T	HB 212	80	550	55	380
Q & T	HB 311	120	830	90	620

*Note: HB indicates Brinell hardness number with a 3 000-kg load. HRC indicates hardness on the Rockwell C scale.

Table 11-8 Typical Nonferrous Gear Materials

MATERIAL AND CONDITION	HARDNESS RANGE*	TENSILE STRENGTH		YIELD STRENGTH	
		Ksi	MPa	Ksi	MPa
Aluminum Bronze					
As cast	HB 116	65	448	25	172
Heat-treated	HB 121	80	550	40	276
Heat-treated	HB 202	110	758	60	414
Manganese Bronze					
As cast	HB 85	60	414	20	138
As cast	HB 200	110	758	60	414
Wrought: soft	HB 150				
Wrought: hard	HB 210				
Tin Bronze					
As cast	HB 70	40	276	20	138
Chill cast	HB 80	45	310	24	165
Centrifugal cast	HB 90	50	345	26	179
Aluminum					
2017 T4	HB 105	62	427	40	276
2024 T4	HB 120	68	469	47	324
6061 T6	HB 95	45	310	40	276

*Note: HB indicates Brinell hardness number, with 3 000-kg load for aluminum bronze and 500-kg load for all others.

Table 11-9 Typical Plastic Gear Materials

MATERIAL	HARDNESS*	TENSILE STRENGTH		FLEXURAL STRENGTH	
		Ksi	MPa	Ksi	MPa
Acetal copolymer	HRM 80	8	55	13	90
Phenolic	HRE 80	7	48	10	69
Nylon	HRR 115	12	83	—	—
Polycarbonates					
General-purpose	HRM 65	10	69	12	83
20 percent glass reinforced	HRM 91	16	110	19	131
Polyurethane elastomer	Shore 80D	7	48	1	7

*Note: HRM, HRE, and HRR indicate hardness on the Rockwell M, E, and R scale, respectively.

8650. These would normally be through-heat-treated by quenching and tempering to the desired levels of strength and hardness. The tables of properties of carbon and alloy steels in Appendixes A-3 and A-4 give the ranges of properties that can be expected.

When using through-hardened steels for gears, it is normal to make the pinion harder than the gear by 40 to 50 points on the Brinell scale. This gives more optimum wear performance of the pair when the hardness is in the range of about HB 180 to HB 400. Above this hardness, surface hardening such as flame hardening, induction hardening, or carburizing is normally used, with equal hardness given to both the pinion and the gear.

Specifications for through hardness of gear teeth are usually given over a range of 40 to 50 HB points to permit some latitude in the heat-treatment processes. Because of the many variables involved, such as alloy composition, carbon content, mass of the gear, and severity of the quench, precise values of hardness cannot be expected. Figure 11-14 gives the recommended ranges of hardness specifications for several classifications of carbon and alloy steels as defined by the AGMA.

For satisfactory surface hardening using the flame- or induction-hardening processes, the material must have a good hardenability and at least 0.30 percent carbon but less than 0.50 percent. Figure 11-15 shows a graph of the approximate surface hardness that can be expected, with flame or induction hardening of steel, as a function of carbon content.

Carburizing steels are plain carbon or alloy steels having a basic carbon content usually in the range of 0.10 percent to 0.20 percent. The surface is brought to a very high hardness by the infusion of carbon into the surface. Light-duty carburized gears may be made from AISI 1015, 1020, 1117, or 1118 carbon steel. Moderate- to heavy-duty gears may be made from AISI 4615, 4620, 8620, 2320, 4320, 4820, or 9310. The properties of many of these carburized steels are listed in Appendix A-5.

In addition to surface hardness, the depth of the case should be controlled and specified for flame-hardened, induction-hardened, or carburized gears. The wear resistance is provided by the surface itself, but because of the very high local contact stress, the case must have sufficient thickness to provide support for the load. When the case is too thin, failure sometimes occurs at the interface between the case and the core because of shear stresses developed near the point of contact.

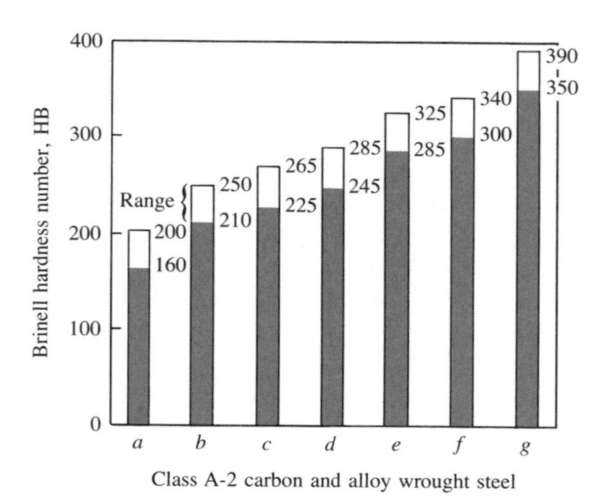

Figure 11-14 Recommended Hardness Ranges (AGMA Standard 240.01)

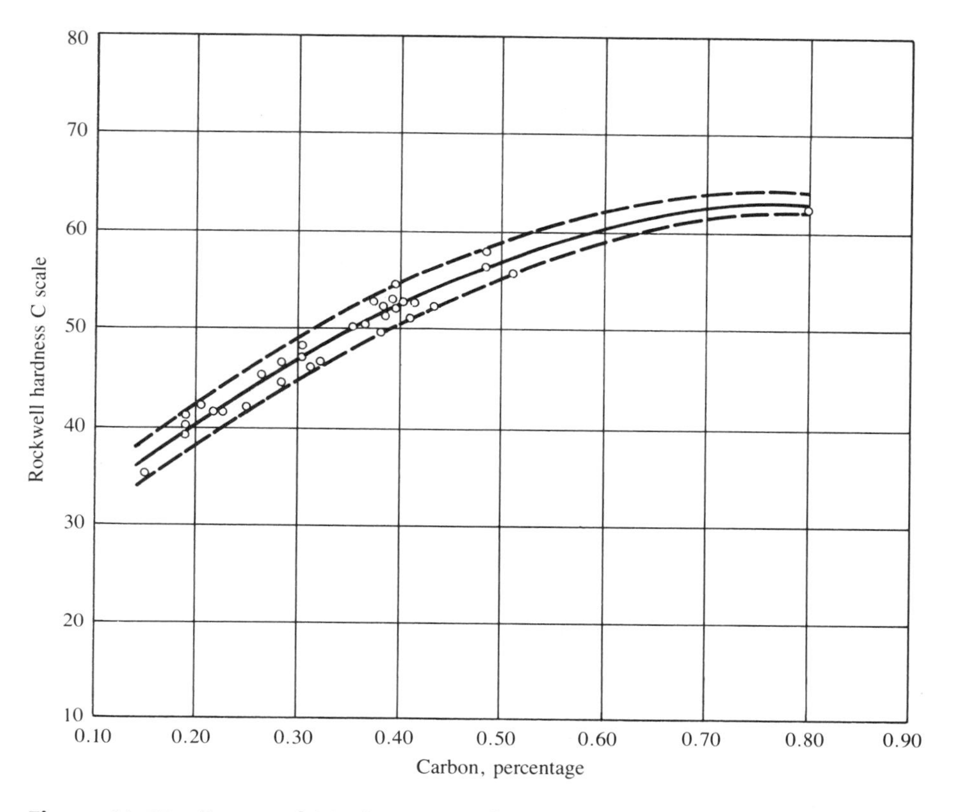

Figure 11-15 Expected Maximum Hardness of Steel as a Function of Carbon Content (Extracted from *AGMA Recommended Procedure for Induction Hardened Gears and Pinions, AGMA 248.01,* with the permission of the publisher, American Gear Manufacturers Association, 1901 North Fort Myer Drive, Suite 1000, Arlington, Virginia 22209)

Various means are used to define case depth. Sometimes the depth at which the HRC hardness reaches 5 points lower than the surface is considered to be the case depth. For gears hardened to surface hardnesses of HRC 55 and higher, the case depth is sometimes that at which the hardness reaches HRC 50.

Several guidelines exist to estimate the required case depth. Some recommend a depth equal to approximately one-sixth of the chordal thickness of the tooth. Kern (10) gives the following formula for the required depth to HRC 50:

$$\text{Depth} = \frac{12 \times 10^{-6}\, W_t}{F \cos \phi} \quad \text{in} \tag{11-22}$$

where

W_t = Tangential load on the tooth (lb)

F = Face width (in)

ϕ = Pressure angle

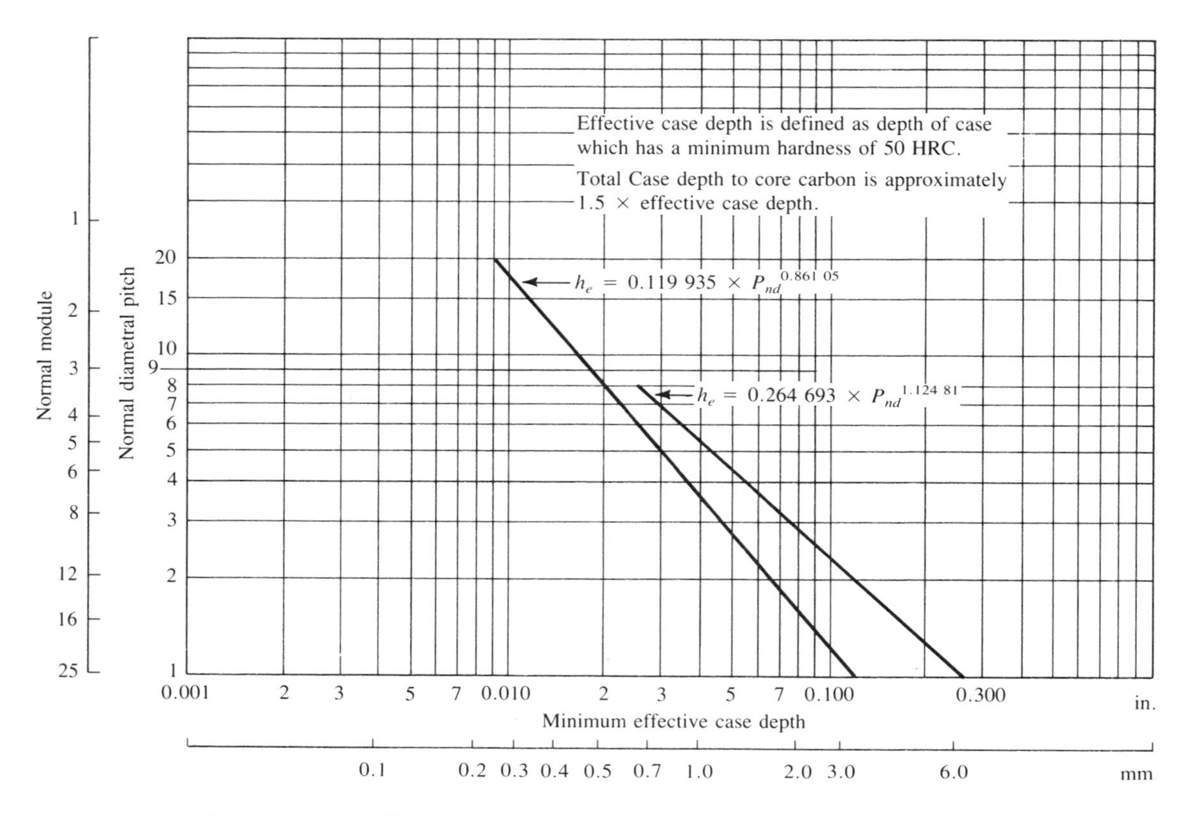

Figure 11-16 Effective Case Depth for Carburized Gears, h_e (Extracted from *AGMA Standard for Rating the Pitting Resistance and Bending Strength of Spur and Helical Involute Gear Teeth, AGMA 218.01,* with the permission of the publisher, American Gear Manufacturers Association, 1901 North Fort Myer Drive, Suite 1000, Arlington, Virginia 22209)

The AGMA has produced a graph giving recommended case depths as a function of the diametral pitch of the gear teeth, as shown in Figure 11-16. If possible, the case depth should be tested to verify its suitability in a given application.

The bronzes are frequently used for wormgears because of their favorable wear characteristics when operating with sliding contact against hardened steel worms.

Aluminum gears are used in some light-duty instrument applications. Plastics are used in timers, toys, business machines, and similar applications.

Powder metallurgy is often used for gears in which the gear teeth are formed to their finished dimensions during the pressing and sintering processes. Such gears are used in some appliance and business machine applications.

11-5 GEAR MANUFACTURE AND QUALITY

The discussion of gear manufacture will begin with the method of producing the gear blank. Small gears are frequently made from wrought plate or bar, with the hub, web, spokes, and rim machined to final or near final dimensions before the gear teeth are produced. The face width and the outside diameter of the gear teeth are also produced at this stage. Other gear blanks may be forged, sand cast, or die cast to achieve the basic form prior to machining. A few gears in which only moderate precision is required may be die cast with the teeth in virtually final form.

Large gears are frequently fabricated from components. The rim and the portion into which the teeth are machined may be rolled into a ring shape from a flat bar and then welded. The web or spokes and the hub are then welded inside the ring. Very large gears may be made in segments with the final assembly of the segments by welding or by mechanical fasteners.

The popular methods of machining the gear teeth are form milling, shaping, and hobbing, as shown in Figures 11-17, 11-18, 11-19, and 11-20.

In *form milling,* a milling cutter that has the shape of the tooth space is used, and each space is cut completely before the gear blank is turned to the position of the next adjacent space. This method is used mostly for large gears, and great care is required to achieve accurate results.

Shaping is a process in which the cutter reciprocates, usually on a vertical spindle. The shaping cutter rotates as it reciprocates and is fed into the gear blank. Thus the involute tooth form is generated gradually. This process is frequently used for internal gears.

Hobbing is a process similar to milling except that the workpiece (the gear blank) and the cutter (the hob) rotate in a coordinated fashion. Here also, the tooth form is generated gradually as the hob is fed into the blank.

The gear teeth are finished to greater precision after form milling, shaping, or hobbing by the processes of grinding, shaving, and honing. Being secondary processes, they are expensive and should only be used where the operation requires high accuracy in the tooth form and spacing.

Gear Quality

Quality in gearing is the precision of the individual gear teeth and the precision with which two gears rotate in relation to one another. The factors normally measured to determine quality are the following:

Figure 11-17 Form Milling of Gears (Illinois Tool Works, Chicago, Ill.)

(a) Barber-Colman gear shaper

(b) Close-up of gear shaping

(c) Herringbone gear shaper cutters

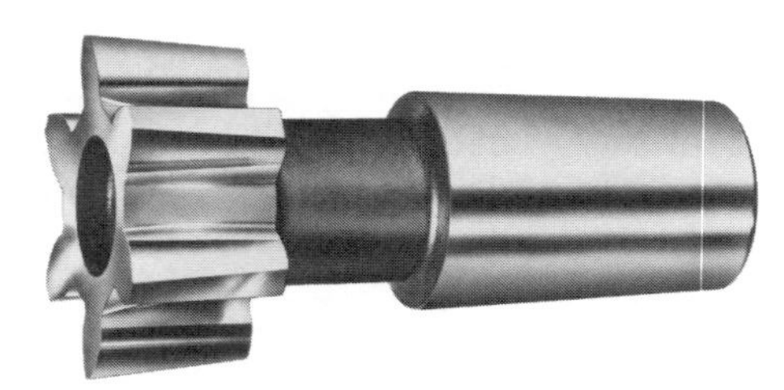

(*d*) Spur gear shaper cutter

Figure 11-18 Gear Shaping (Barber-Colman Company, Loves Park, Ill.), (a,b) (Illinois Tool Works, Chicago, Ill.), (c,d)

Runout: A measure of eccentricity and out-of-roundness

Tooth-to-tooth spacing: The difference in spacing between corresponding points on adjacent teeth

Profile: The variation of the actual tooth profile from the theoretically precise profile

Composite error is measured on a special device that places the test gear in tight mesh with a master gear of known precision. The two gears are rotated while in tight mesh and the center of one gear is free to move. The variation in center distance for a complete revolution of the test gear is recorded. Figure 11-21 shows a typical recording with both *tooth-to-tooth composite error* and *total composite error* indicated. Thus, composite error is a measure of the combined effects of several types of gear tooth errors.

The allowable amount of variation of the actual tooth form from the theoretical form, or the composite error, is specified by defining an AGMA quality number. Detailed charts giving the tolerances for many features are included in AGMA Standard 390.03. The quality numbers range from 5 to 16 with increasing precision. The actual tolerances are a function of the diametral pitch of the gear teeth and of the pitch diameter of the gear. Table 11-10 and Figure 11-22 show representative data for the total composite tolerance for several quality numbers.

(*a*) Gear hobbing machine

(*b*) Closeup of gear hobbing process

Figure 11-19 Gear Hobbing (Barber-Colman Company, Loves Park, Ill.)

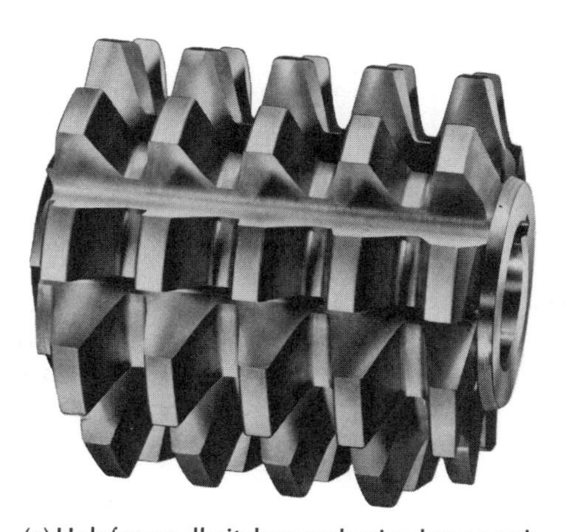

(*a*) Hob for small pitch gears having large teeth

(*b*) Hob for high pitch gears having small teeth

Figure 11-20 Gear Hobs (Illinois Tool Works, Chicago, Ill.), (a) (Barber-Colman Company, Loves Park, Ill.), (b)

These data should impress you with the precision that is normally exercised in the manufacture and installation of gears. The design of the entire gear system, including the shafts, bearings, and housing, must be consistent with this precision. Of course, the system should not be made more precise than necessary because of cost. For this reason, manufacturers have recommended quality numbers that will give satisfactory performance at a reasonable cost for a variety of applications. Table 11-11 lists several of these recommendations; AGMA Standard 390.03 gives many more.

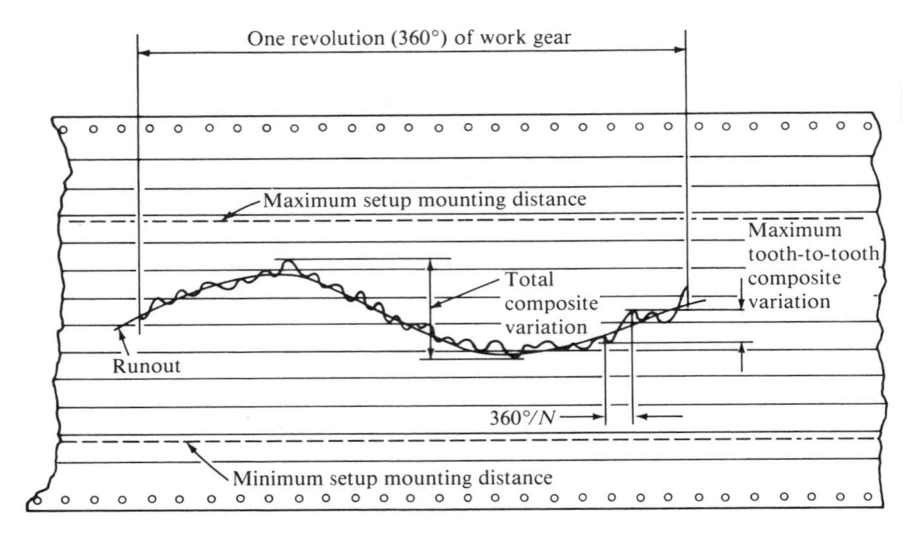

(*a*) Chart of gear-tooth errors of a typical gear when run with a specified gear in a rolling fixture

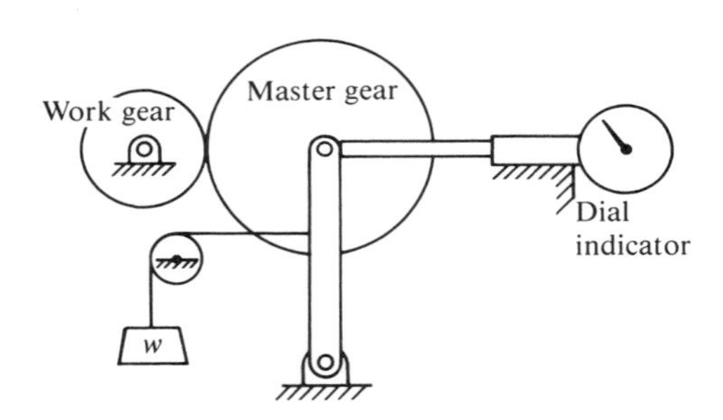

(*b*) Schematic diagram of a typical gear rolling fixture

Figure 11-21 Recording of Errors in Gear Geometry (Extracted from *AGMA Handbook for Unassembled Gears,* Volume 1, *Gear Classification Materials, and Inspection, AGMA 390.03,* with the permission of the publisher, American Gear Manufacturers Association, 1901 North Fort Myer Drive, Arlington, Virginia 22209)

In addition to the applications described in Table 11-11, it should be helpful to note the recommendation given for machine tool power drives. Because this is such a wide range of specific applications, the recommended quality numbers are related to the speed of rotation of the gears. The important factor, as discussed later in this chapter, is the *pitch line speed,* defined as the linear velocity of a point on the pitch circle of the gear.

$$v_t = \pi D n_G / 12 = \pi d n_P / 12$$

where

v_t = Pitch line speed in feet per minute (fpm)

n_G = Rotational speed of the gear (rpm)

Table 11-10 Total Composite Tolerance

AGMA Quality Number	*Diametral Pitch* P_d	*Pitch Diameter (Inches)* 2.5	4.0	10.0	25.0	50.0
		TOTAL COMPOSITE TOLERANCE (INCHES)				
6	1	—	—	0.027 38	0.029 30	0.034 61
	4	0.011 94	0.012 12	0.013 52	0.016 17	0.018 50
	12	0.007 03	0.007 60	0.008 84	0.010 29	0.011 54
	20	0.005 89	0.006 33	0.007 28	0.008 38	0.009 32
8	1	—	—	0.013 27	0.014 95	0.017 66
	4	0.005 95	0.006 13	0.006 90	0.008 25	0.009 44
	12	0.003 59	0.003 88	0.004 51	0.005 25	0.005 89
	20	0.003 01	0.003 23	0.003 72	0.004 28	0.004 76
10	1	—	—	0.006 07	0.007 63	0.009 01
	4	0.002 89	0.003 07	0.003 52	0.004 21	0.004 82
	12	0.001 83	0.001 98	0.002 30	0.002 68	0.003 00
	20	0.001 53	0.001 65	0.001 90	0.002 18	0.002 43
12	1	—	—	0.002 40	0.003 89	0.004 60
	4	0.001 33	0.001 51	0.001 80	0.002 15	0.002 46
	12	0.000 93	0.001 01	0.001 17	0.001 37	0.001 53
	20	0.000 78	0.000 84	0.000 97	0.001 11	0.001 24
14	1	—	—	0.000 53	0.001 99	0.002 35
	4	0.000 53	0.000 71	0.000 92	0.001 10	0.001 25
	12	0.000 48	0.000 51	0.000 60	0.000 70	0.000 78
	20	0.000 40	0.000 43	0.000 49	0.000 57	0.000 63

n_P = Rotational speed of the pinion (rpm)
D = Pitch diameter of the gear (in)
d = Pitch diameter of the pinion (in)

The recommended quality numbers for machine tool drives are as follows:

Pitch Line Speed	*Quality Number*
0–800 fpm	6–8
800–2 000 fpm	8–10
2 000–4 000 fpm	10–12
Over 4 000 fpm	12–14

This set of data can serve as a model for specifying quality numbers for many mechanical drive applications.

11-6 FORCES ON GEAR TEETH

To understand the method of computing stresses in gear teeth, it is helpful to consider the way power is transmitted by a gear system. Consider the simple single-reduction gear pair shown in Figure 11-23. Power is received from the motor by the input shaft rotating at

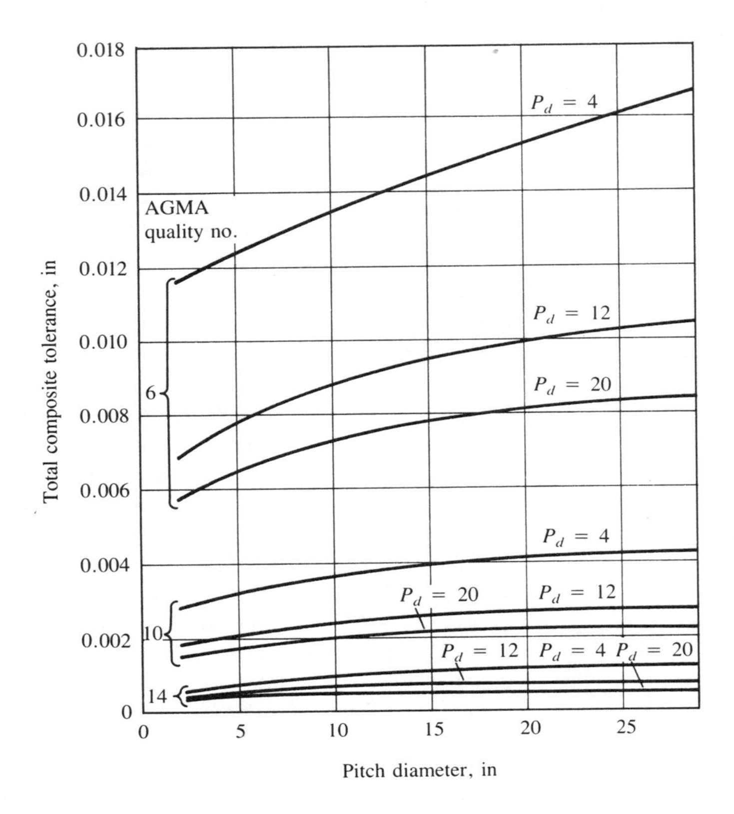

Figure 11-22 Total Composite Tolerance vs. Pitch Diameter

motor speed. Thus there is a torque in the shaft that can be computed from the following equation:

$$\text{Torque} = \text{Power/Rotational speed}$$

The input shaft transmits the power from the coupling to the point where the pinion is mounted. The power is transmitted from the shaft to the pinion through the key. The teeth

Table 11-11 Recommended AGMA Quality Numbers

Application	*Quality Number*	*Application*	*Quality Number*
Cement mixer drum drive	3–5	Small power drill	7–9
Cement kiln	5–6	Clothes washing machine	8–10
Steel mill drives	5–6	Printing press	9–11
Corn picker	5–7	Computing mechanism	10–11
Cranes	5–7	Automotive transmission	10–11
Punch press	5–7	Radar antenna drive	10–12
Mining conveyor	5–7	Marine propulsion drive	10–12
Paper-box-making machine	6–8	Aircraft engine drive	10–13
Gas meter mechanism	7–9	Gyroscope	12–14

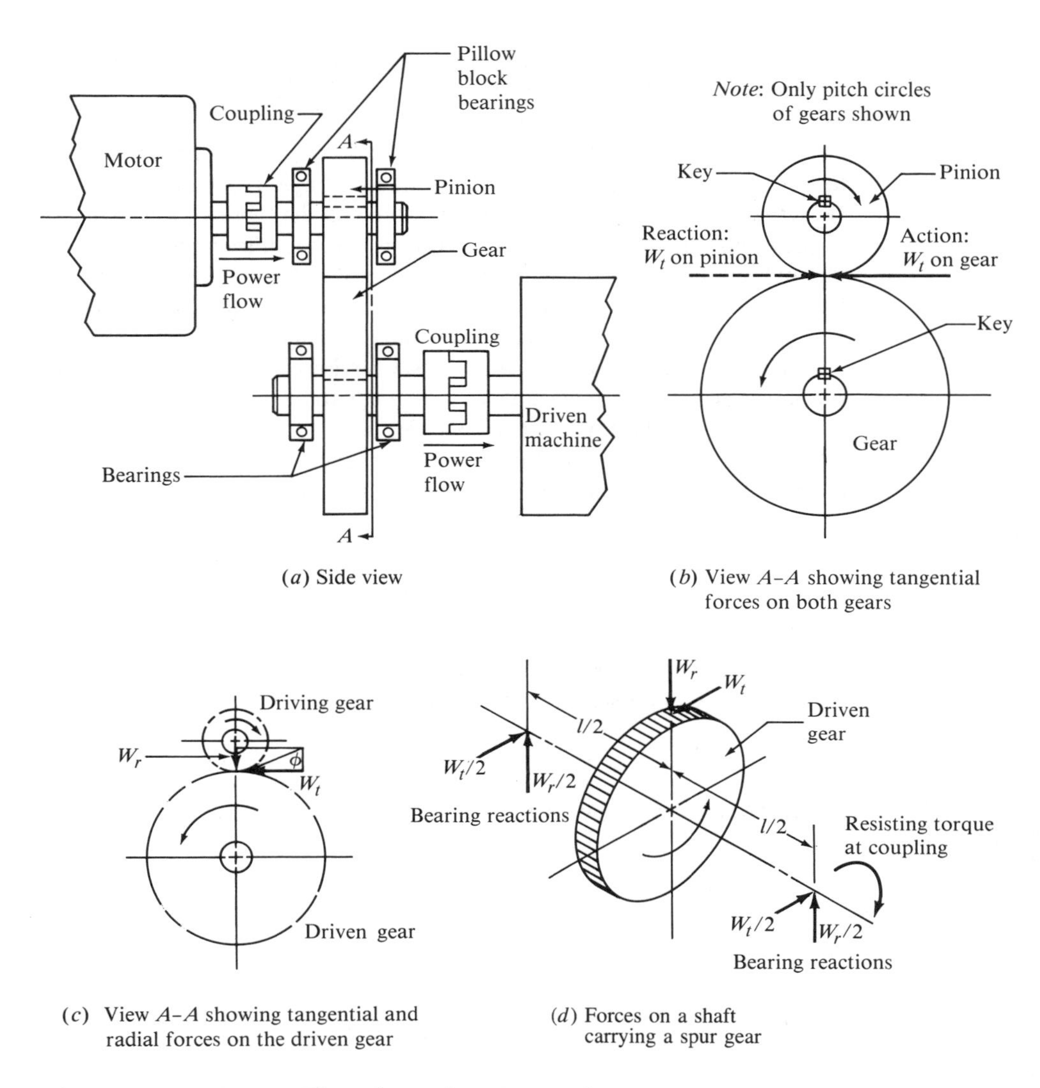

(*a*) Side view

(*b*) View *A–A* showing tangential forces on both gears

(*c*) View *A–A* showing tangential and radial forces on the driven gear

(*d*) Forces on a shaft carrying a spur gear

Figure 11-23 Power Flow through a Gear Pair

of the pinion drive the teeth of the gear and thus transmit the power to the gear. But again, power transmission actually involves the application of a torque during rotation at a given speed. The torque is the product of the force acting tangent to the pitch circle of the pinion times the pitch radius of the pinion. We will use the symbol W_t to indicate the *tangential force*. As described, W_t is the force exerted *by the pinion teeth on the gear teeth*. But if the gears are rotating at a constant speed and transmitting a uniform level of power, the system is in equilibrium. Therefore, there must be an equal and opposite tangential force exerted by the gear teeth back on the pinion teeth. This is an application of the principle of action and reaction. To complete the description of the power flow, the tangential force on the gear teeth produces a torque on the gear equal to the product of W_t times the pitch radius of the gear. Because W_t is the same on the pinion and the gear, but the pitch radius of the gear is larger than that of the pinion, the torque on the gear (the output torque) is greater than the input torque. But note that the power transmitted is the same or slightly less because of mechanical inefficiencies. The power then flows from the gear through the key to the output shaft and finally to the driven machine.

From this description of power flow, we can see that gears transmit power by exerting a force by the driving teeth on the driven teeth while the reaction force acts back on the teeth of the driving gear. Figure 11-24 shows a single gear tooth with the tangential force W_t acting on it. But this is not the total force on the tooth. Because of the involute form of the tooth, the total force transferred from one tooth to the mating tooth acts normal to the involute profile. This action is shown as W_n. The tangential force W_t is actually the horizontal component of the total force. To complete the picture, note that there is a vertical component of the total force acting radially on the gear tooth, indicated by W_r.

The computation of forces usually starts with W_t because it is based on the given data for power and speed. In general, power equals force times velocity. In gearing, as shown in Figure 11-23, the transmitted force, W_t, acts at the pitch line with a velocity v_t. The pitch line velocity is derived from the basic relationship, $v = R\omega$, for a point moving in a circle. For the typical data for gearing in U.S. Customary units,

$$v_t = \frac{\pi d n_P}{12} \text{ ft/min}$$

where d is the pinion diameter in inches and n_P is the rotational speed of the pinion in rpm.

The transmitted force can be found by dividing the power transmitted by the pitch line speed. If P is in horsepower and v_t is in ft/min,

$$W_t = 33\,000(P)/v_t \text{ lb}$$

Alternatively, the torque on the pinion can be computed from

$$T_P = 63\,000(P)/n_P$$

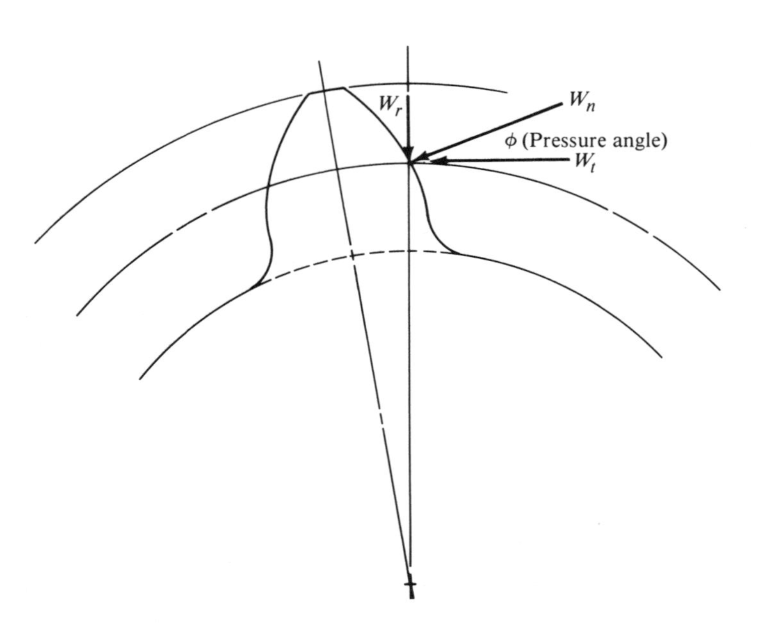

Figure 11-24 Forces on Gear Teeth

Then, because $T_P = W_t(d/2)$,

$$W_t = T_P/(d/2)$$

These values can be computed for the gear by appropriate substitutions. Remember that the pitch line speed is the same for the pinion and the gear. And the transmitted load on the pinion and the gear are the same, except they act in opposite directions.

The normal force, W_n, and the radial force, W_r, can be computed from the known W_t by using the right triangle relations evident in Figure 11-24.

$$W_r = W_t \tan \phi$$

$$W_n = W_t/\cos \phi$$

where ϕ is the pressure angle of the tooth form.

In addition to causing the stresses in the gear teeth, these forces act on the shaft. And in order to maintain equilibrium, the bearings that support the shaft must provide the reactions. Figure 11-23(d) shows the free body diagram of the output shaft of the reducer.

11-7 STRESSES IN GEAR TEETH

The stress analysis of gear teeth is facilitated by considering the orthogonal force components, W_t and W_r, as shown in Figure 11-24.

The tangential force, W_t, produces a bending moment on the gear tooth similar to that on a cantilever beam. The resulting bending stress is maximum at the base of the tooth in the fillet that joins the involute profile to the bottom of the tooth space. Taking the detailed geometry of the tooth into account, Wilfred Lewis developed the equation for the stress at the base of the involute profile, which is now called the *Lewis equation*.

$$\sigma_t = \frac{W_t P_d}{FY} \tag{11-23}$$

In the Lewis equation, W_t is the tangential force, P_d is the diametral pitch of the tooth, F is the face width of the tooth; and Y is the *Lewis form factor,* which depends on the tooth form, the pressure angle, the diametral pitch, the number of teeth in the gear, and the place where W_t acts.

While presenting the theoretical basis for the stress analysis of gear teeth, the Lewis equation must be modified for practical design and analysis. One important limitation is that it doesn't take into account the stress concentration that exists in the fillet of the tooth. Figure 11-25 is a photograph of a photoelastic stress analysis of a model of a gear tooth. It indicates a stress concentration in the fillet at the root of the tooth as well as high contact stresses at the mating surface (the contact stress is discussed in the following section). Comparing the actual stress at the root with that predicted by the Lewis equation enables one to determine the stress concentration factor, K_t, for the fillet area. Placing this into equation (11-23) gives

$$\sigma_t = \frac{W_t P_d K_t}{FY} \tag{11-24}$$

The value of the stress concentration factor is dependent on the form of the tooth, the shape and size of the fillet at the root of the tooth, and the point of application of the

Figure 11-25 Photoelastic Study of Gear Teeth Under Load (Measurements Group, Inc., Raleigh, North Carolina)

force on the tooth. Note that the value of the Lewis form factor, Y, also depends on the tooth geometry. Therefore, the two factors are combined into one term, the *geometry factor, J*, where $J = Y/K_t$. The value of J also, of course, varies with the location of the point of application of the force on the tooth because Y and K_t do.

Figure 11-26 shows graphs giving the values for the geometry factor for 20° and 25° full-depth involute teeth. The safest value to use is the one for the load applied at the tip of the tooth. However, this value is overly conservative because there is some load sharing by another tooth at the time the load is initially applied at the tip of a tooth. The critical load on a given tooth occurs when the load is at the highest point of single-tooth contact, when the tooth carries the entire load. The upper curves in Figure 11-26 give the values for J for this condition.

Using the geometry factor, J, in the stress equation gives

$$\sigma_t = \frac{W_t P_d}{FJ} \tag{11-25}$$

Another modification required for an accurate stress analysis is the inclusion of a *dynamic factor, K_v*.

$$\sigma_t = \frac{W_t P_d}{K_v FJ}$$

The dynamic factor accounts for the fact that the load is assumed by a tooth with some degree of impact and that the actual load subjected to the tooth is higher than the transmitted load alone. Sometimes the term *dynamic load,* $W_d = W_t/K_v$, is used. The value of K_v depends on the accuracy of the tooth profile, the elastic properties of the tooth, and the speed with which the teeth come into contact. Figure 11-27 shows a graph of the AGMA recommended values for K_v, where the Q_v numbers are the AGMA quality numbers referred to earlier. Gears in typical machine design would fall into the classes represented by curves 5, 6, or 7, which are for gears made by hobbing or shaping with average to good tooling. If the teeth are finish-ground or shaved to improve the accuracy of the tooth profile and spacing, curves 8, 9, 10, or 11 should be used. Under very special conditions where teeth of high precision are used in applications where there is little chance of developing external dynamic loads, the shaded area can be used. If the teeth are cut by form milling, factors lower than those found from curve 5 should be used. Note that the quality 5 gears should not be used at pitch line speeds above 2 500 ft/min. It should be understood that the dynamic factors are approximate. For severe applications, especially those operating above 4 000 ft/min, approaches taking into account the material properties, the mass and inertia of the gears, and the actual error in the tooth form should be used to predict the dynamic load (6, 9, 15, 16, 17).

So far in our discussion of the stress in gear teeth, it has been assumed that the load is uniformly distributed across the face of the tooth. Because of inaccuracies in the alignment of shafts, the placement of bearings, and the deflection of the several members in a gear set, this is rarely true. For this reason, it is recommended that a *load distribution factor,* K_m, be used (17). Table 11-12 lists conservative values.

A service factor or application factor, K_a, should be applied to the stress analysis equation to account for load variations and shock presented to the gear system by the power source or the driven machine (see Table 11-13). Specific industry data should be used where possible. Examples of the various classes of driven machines follow.

Uniform: Fans and blowers, uniformly loaded conveyors, liquid agitators, generators, centrifugal pumps

Moderate shock: Reciprocating compressors or pumps, heavy-duty conveyors, meat grinders, machine tool drives, concrete mixers, textile machinery, chain saws

Heavy shock: Rock crushers, railroad car dumpers, hammer mills, punch press drives, ball mills, rod mills, tumbling barrels, wood chippers, vibrating screens

The stress analysis equation then becomes

$$\sigma_t = \frac{K_a W_t}{K_v} \frac{P_d}{F} \frac{K_m}{J} \tag{11-26}$$

This is the form we will use in problem solutions. It accounts for all of the factors usually encountered in general machine design work. Additional factors should be taken into account if the application involves any of the following conditions:

Temperatures above 250°F

Very large teeth ($P_d < 1.0$)

Rough tooth surfaces

Nonstandard tooth forms or fillets

Limited design life

Load not carried by the full face of the tooth

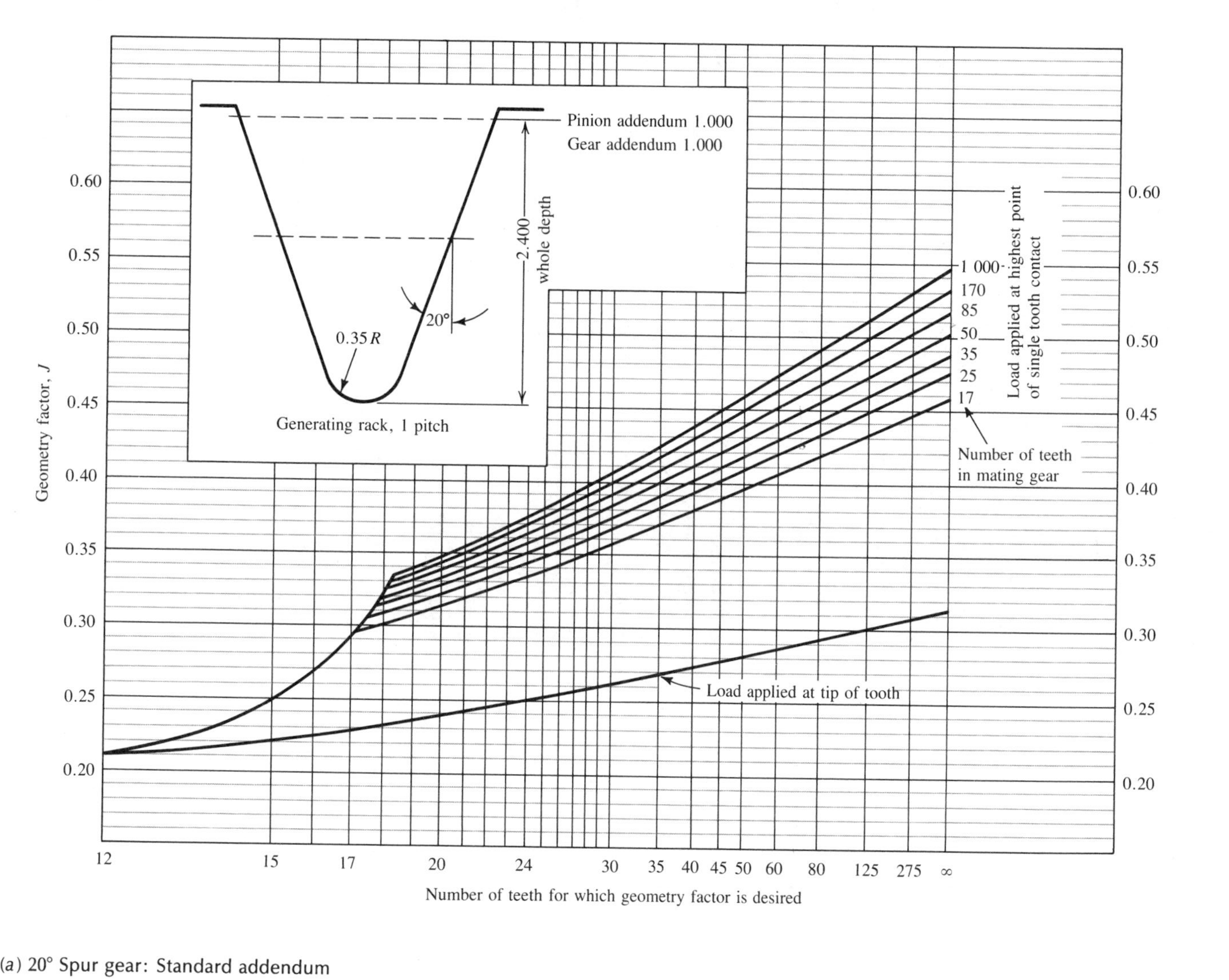

(a) 20° Spur gear: Standard addendum

Figure 11-26 Geometry Factor, J (Extracted from *AGMA Standard for Rating the Pitting Resistance and Bending Strength of Spur and Helical Involute Gear Teeth, AGMA 218.01,* with the permission of the publisher, American Gear Manufacturers Association, 1901 North Fort Myer Drive, Suite 1000, Arlington, Virginia 22209)

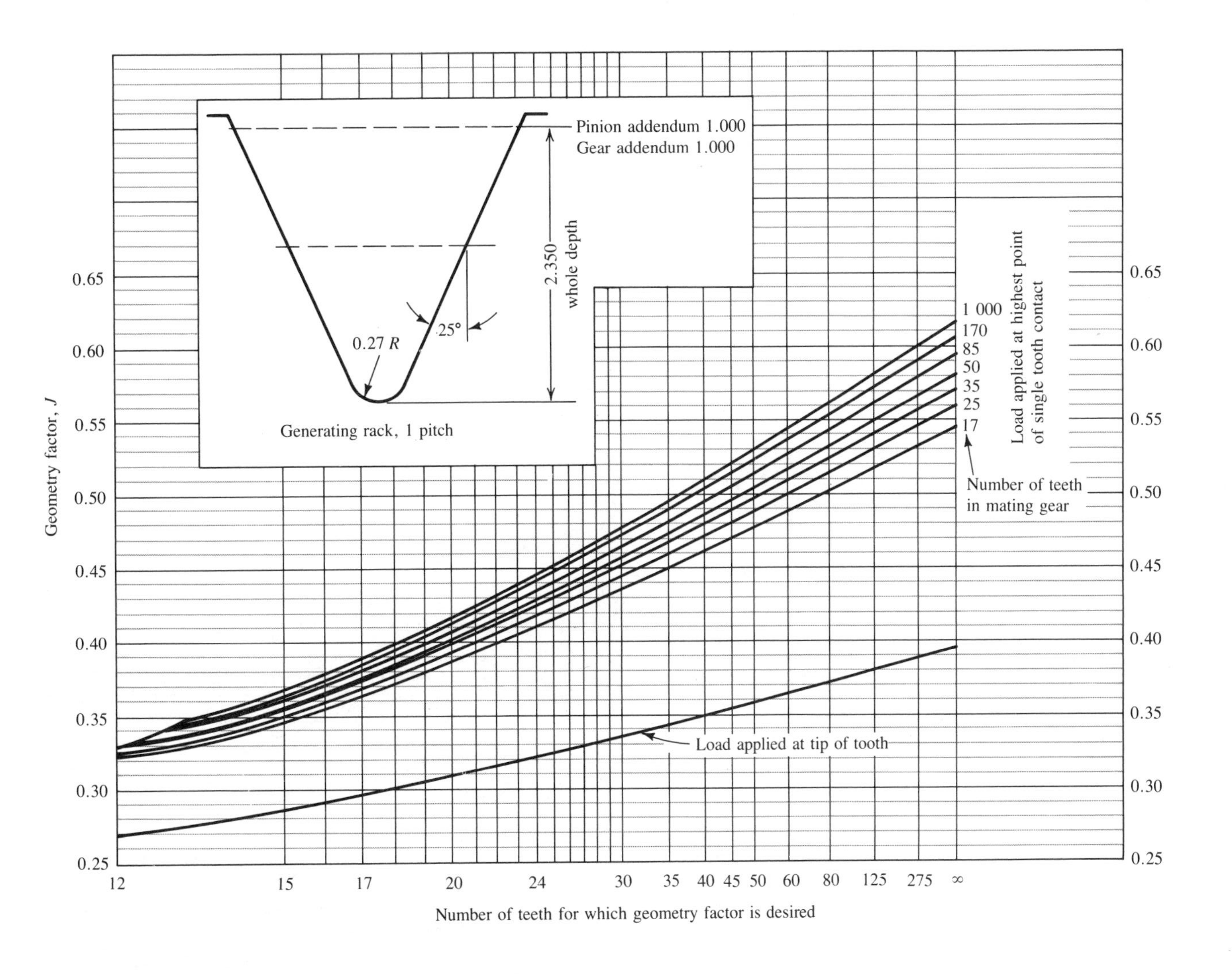

(*b*) 25° spur gear: standard addendum

Figure 11-26 (Continued)

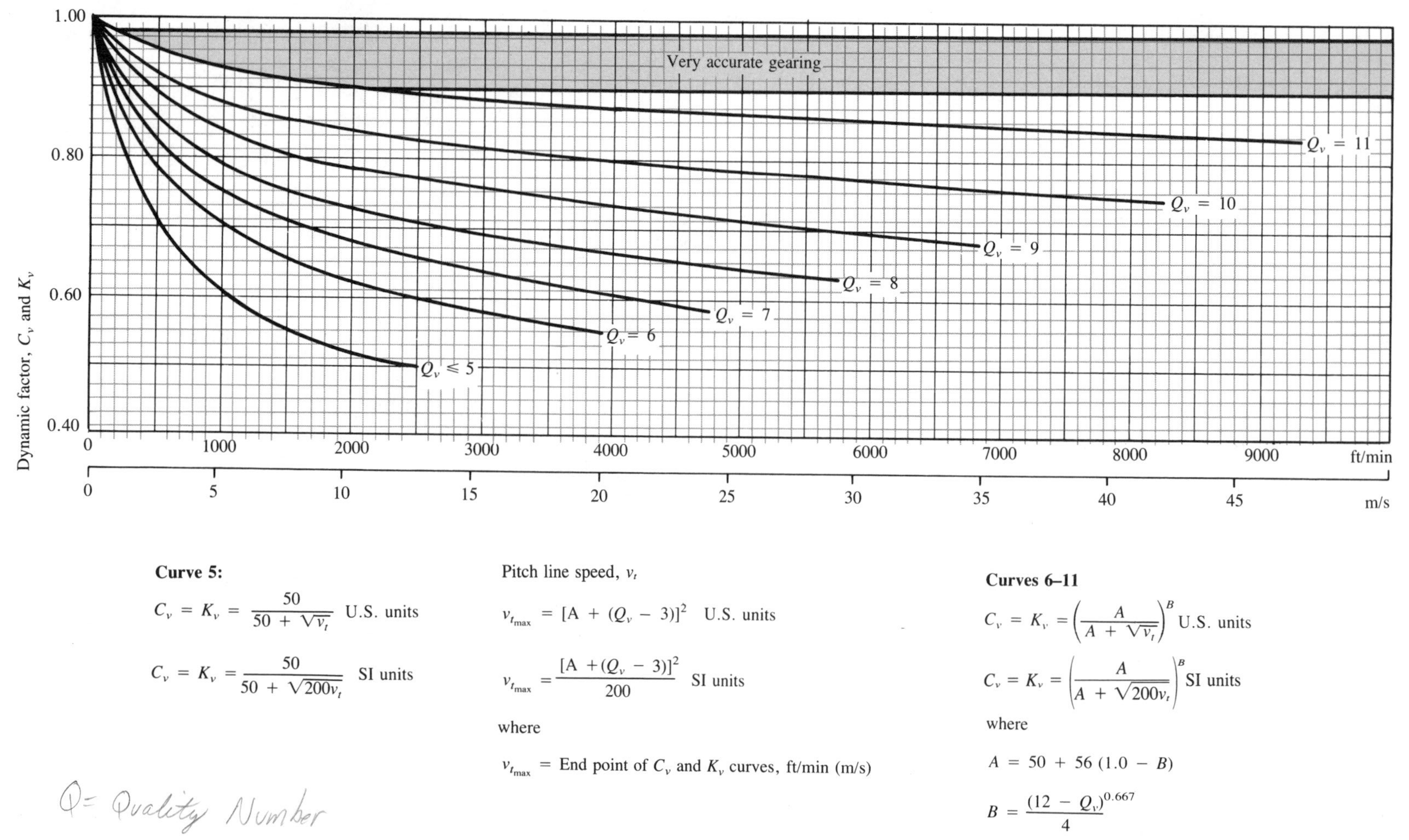

Curve 5:

$$C_v = K_v = \frac{50}{50 + \sqrt{v_t}} \quad \text{U.S. units}$$

$$C_v = K_v = \frac{50}{50 + \sqrt{200 v_t}} \quad \text{SI units}$$

Pitch line speed, v_t

$$v_{t_{\max}} = [A + (Q_v - 3)]^2 \quad \text{U.S. units}$$

$$v_{t_{\max}} = \frac{[A + (Q_v - 3)]^2}{200} \quad \text{SI units}$$

where

$v_{t_{\max}}$ = End point of C_v and K_v curves, ft/min (m/s)

Curves 6–11

$$C_v = K_v = \left(\frac{A}{A + \sqrt{v_t}}\right)^B \quad \text{U.S. units}$$

$$C_v = K_v = \left(\frac{A}{A + \sqrt{200 v_t}}\right)^B \quad \text{SI units}$$

where

$$A = 50 + 56\,(1.0 - B)$$

$$B = \frac{(12 - Q_v)^{0.667}}{4}$$

Q_v = Transmission accuracy level number

Figure 11-27 Dynamic Factor, C_V and K_V (Extracted from *AGMA Standard for Rating the Pitting Resistance and Bending Strength of Spur and Helical Involute Gear Teeth, AGMA 218.01,* with the permission of the publisher, American Gear Manufacturers Association, 1901 North Fort Myer Drive, Suite 1000, Arlington, Virginia 22209)

Table 11-12 Load Distribution Factors, K_m

Gear Face Width, F (Inches)			
≤2	6	9	≤20
1.6	1.7	1.8	2.0

Table 11-13 Application Factors, K_a

	DRIVEN MACHINE		
POWER SOURCE	*Uniform*	*Moderate Shock*	*Heavy Shock*
Uniform (Electric motor, turbine)	1.00	1.25	1.75 or higher
Light shock (Multicylinder engine)	1.25	1.50	2.00 or higher
Medium shock (Single-cylinder engine)	1.50	1.75	2.25 or higher

Pitch line speed greater than approximately 4 000 ft/min

Reliability factor greater than 0.99 required

The references at the end of the chapter provide guidance for these cases (1, 6, 15).

After computing the stress from equation (11-26), the designer must select a material that has an allowable stress, s_{at}, greater than that value. Figure 11-28 shows the AGMA recommended allowable stresses for steel and cast iron gears. The value for carburized steels assumes that the case is continuous throughout the root and fillet areas, where the bending stress is maximum.

If the computed stress is less than the allowable stress, the gears should not fail by fatigue within 10 million cycles with a reliability factor of approximately 0.99.

11-8 WEAR RESISTANCE OF GEAR TEETH

Gear teeth must be safe from failure of the teeth by breakage, and they must also be capable of running for the desired life without significant wear of the tooth form. *Wear* is the phenomenon in which small particles are removed from the surface of the tooth because of the high contact forces that are present between mating teeth. Wear is actually the fatigue failure of the tooth surface. It starts as pitting, leaving a roughened surface. Prolonged operation after pitting begins causes the tooth form to change dramatically, causing vibration and noise. Thus the design process should seek to prevent the start of pitting.

The primary property of the gear tooth that provides resistance to wear by pitting is the hardness of the tooth surface. The harder the surface, the higher is the compressive strength of the material. The force exerted by the driving tooth on the mating tooth (and the reaction back on the driving tooth) is theoretically applied to the line of contact between the tooth profiles. Actually, because of the elasticity of the material of the teeth, the force is spread over a small rectangular area. The resulting stress is highly concentrated around the point of application of the force and is called a *contact stress* or a *Hertz stress*.

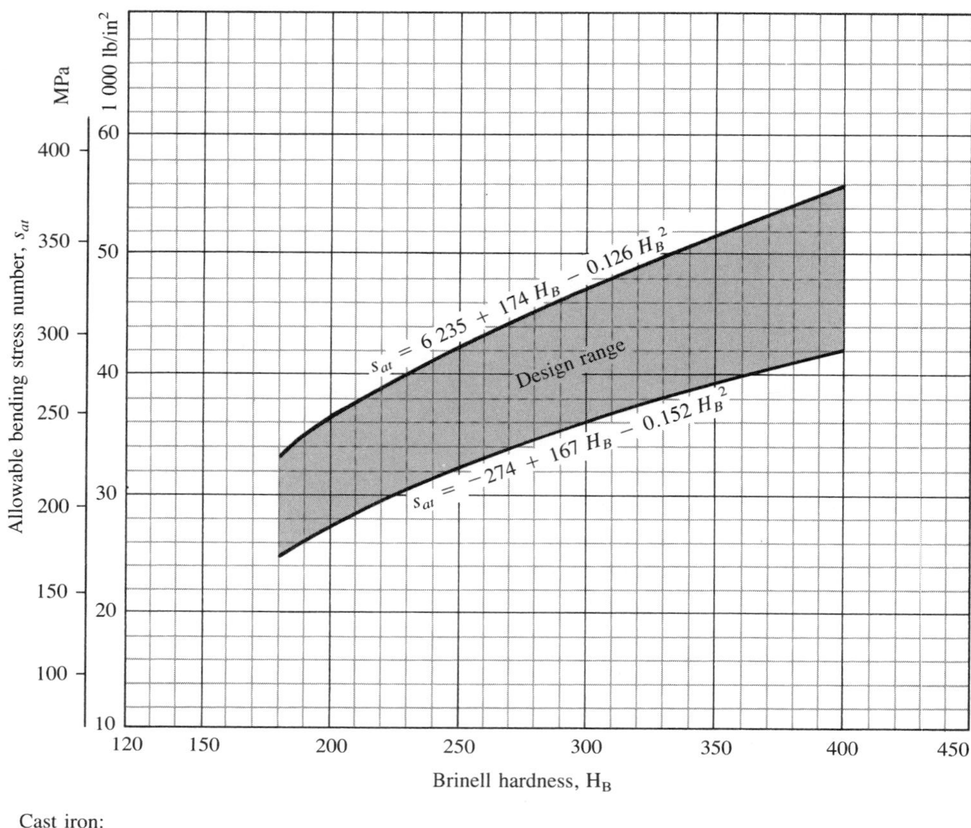

Cast iron:
Class 20 — 5 000 psi (35 MPa) = s_{at}
Class 30 — 8 500 psi (59 MPa) = s_{at}
Class 40 — 13 000 psi (90 MPa) = s_{at}
Case-hardened steel:
Flame- or induction-hardened
50–54 HRC — 45–55 000 psi (310–380 MPa) = s_{at}
Carburized and case-hardened
55 HRC — 65 000 psi (380–450 MPa) = s_{at}
60 HRC — 55–70 000 psi (380–480 MPa) = s_{at}
Bronze:
ASTM B-148-52, Heat-treated to $s_{u_{\min}}$ = 90 000 psi (620 MPa)
s_{at} = 23 000 psi (160 MPa)

Figure 11-28 Allowable Bending Stress Number for Steel Gears, s_{at} (Extracted from *AGMA Standard for Rating the Pitting Resistance and Bending Strength of Spur and Helical Involute Gear Teeth, AGMA 218.01,* with the permission of the publisher, American Gear Manufacturers Association, 1901 North Fort Myer Drive, Suite 1000, Arlington, Virginia 22209)

The development of the equation for the contact stress on gear teeth is based on the analysis for two cylinders under a radial load. The radii of the cylinders are taken to be the radii of curvature of the involute tooth forms of the mating teeth at the point of contact. The load on the teeth is the total normal load, found from

$$W_{dn} = W_t/(C_v \cos \phi) \tag{11-27}$$

where

W_{dn} = Dynamic load acting normal to the tooth surface

W_t = Transmitted load acting tangential to the pitch line

C_v = Dynamic factor

ϕ = Pressure angle

The resulting Hertz stress due to this load is

$$\sigma_h = \sqrt{\frac{W_t}{C_v F d} \frac{1}{\pi\left(\frac{1-\nu_P^2}{E_P} + \frac{1-\nu_G^2}{E_G}\right)} \frac{1}{\frac{\cos\phi \sin\phi}{2} \frac{m_G}{m_G+1}}} \tag{11-28}$$

where, in addition to the terms already defined,

d = Pitch diameter of the pinion

ν_P = Poisson's ratio for the pinion material

ν_G = Poisson's ratio for the gear material

E_P = Modulus of elasticity for the pinion material

E_G = Modulus of elasticity for the gear material

m_G = Gear ratio = N_G/N_P

The second term under the radical is dependent on the material properties and is given the name *elastic coefficient,* C_p. That is,

$$C_p = \sqrt{\frac{1}{\pi\left(\frac{1-\nu_P^2}{E_P} + \frac{1-\nu_G^2}{E_G}\right)}} \tag{11-29}$$

For commonly used gear materials, the value of C_p can be found in Table 11-14.

The last term under the radical in equation (11-28) is called the *geometry factor, I,* and is dependent on the tooth geometry and on the gear ratio. Values can be found from Figure 11-29.

Making these substitutions into equation (11-28) gives

$$\sigma_h = C_p \sqrt{\frac{W_t}{C_v F d I}} \tag{11-30}$$

As with the equation for stress in gear teeth, an application factor, C_a, and a load distribution factor, C_m, should be included in the contact stress equation. Equation (11-30) becomes

$$\sigma_h = C_p \sqrt{\frac{C_a C_m W_t}{C_v F d I}} \tag{11-31}$$

This is the form of the contact stress equation that we will use in problem solutions. As stated for the bending stress equation, equation (11-26), additional factors should be taken into account if unusual conditions exist. The values for the application factor,

Table 11-14 Elastic Coefficient, C_p

PINION MATERIAL AND MODULUS OF ELASTICITY E_P	lb/in² (MPa)	GEAR MATERIAL AND MODULUS OF ELASTICITY E_G, lb/in² (MPa)					
		Steel 30×10^6 (2×10^5)	*Malleable Iron* 25×10^6 (1.7×10^5)	*Nodular Iron* 24×10^6 (1.7×10^5)	*Cast Iron* 22×10^6 (1.5×10^5)	*Aluminum Bronze* 17.5×10^6 (1.2×10^5)	*Tin Bronze* 16×10^6 (1.1×10^5)
Steel	30×10^6 (2×10^5)	2 300 (191)	2 180 (181)	2 160 (179)	2 100 (174)	1 950 (162)	1 900 (158)
Mall. iron	25×10^6 (1.7×10^5)	2 180 (181)	2 090 (174)	2 070 (172)	2 020 (168)	1 900 (158)	1 850 (154)
Nod. iron	24×10^6 (1.7×10^5)	2 160 (179)	2 070 (172)	2 050 (170)	2 000 (166)	1 880 (156)	1 830 (152)
Cast iron	22×10^6 (1.5×10^5)	2 100 (174)	2 020 (168)	2 000 (166)	1 960 (163)	1 850 (154)	1 800 (149)
Al. bronze	17.5×10^6 (1.2×10^5)	1 950 (162)	1 900 (158)	1 880 (156)	1 850 (154)	1 750 (145)	1 700 (141)
Tin bronze	16×10^6 (1.1×10^5)	1 900 (158)	1 850 (154)	1 830 (152)	1 800 (149)	1 700 (141)	1 650 (137)

Note: Poisson's ratio = 0.30.

Source: Extracted from *AGMA Standard for Rating the Pitting Resistance and Bending Strength of Spur and Helical Involute Gear Teeth, AGMA 218.01*, with the permission of the publisher, American Gear Manufacturers Association, 1901 North Fort Myer Drive, Suite 1000, Arlington, Virginia 22209.

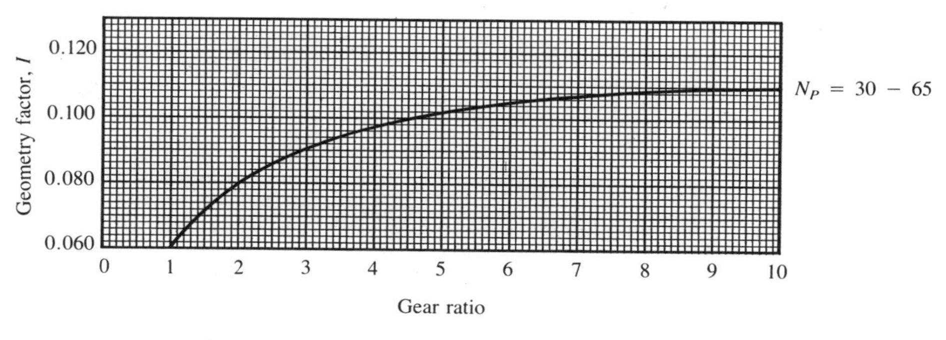

(a) $14\frac{1}{2}°$ pressure angle, full-depth teeth, standard addendum $= \frac{1}{P_d}$

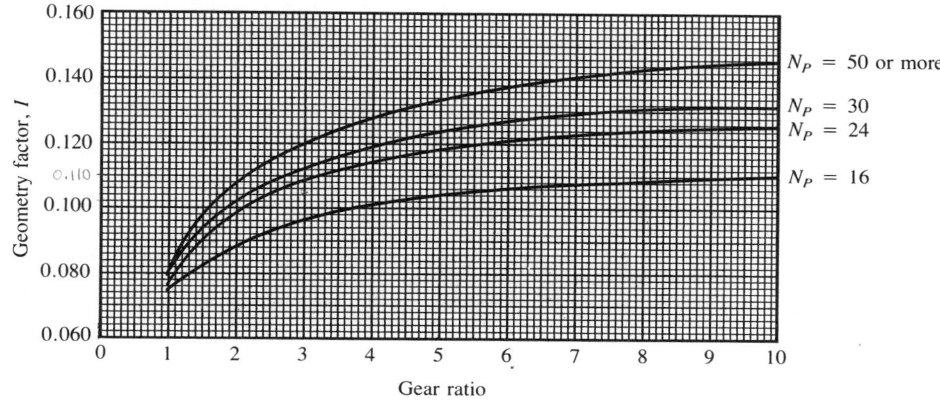

(b) 20° pressure angle, full-depth teeth, standard addendum $= \frac{1}{P_d}$

Figure 11-29 External Spur Pinion Geometry Factor, *I*, for Standard Center Distances. All curves are for the lowest point of single tooth contact on the pinion. (Extracted from *AGMA Standard for Rating the Pitting Resistance and Bending Strength of Spur and Helical Involute Gear Teeth, AGMA 218.01,* with the permission of the publisher, American Gear Manufacturers Association, 1901 North Fort Myer Drive, Suite 1000, Arlington, Virginia 22209)

C_a, and the load distribution factor, C_m, can be taken to be the same as K_a and K_m used for the bending stress analysis and presented in Tables 11-12 and 11-13.

The contact stress computed from equation (11-31) should be compared with the allowable contact stress number, s_{ac}, as shown in Figure 11-30. As can be seen, the allowable stress is dependent on the hardness of the tooth surface.

11-9 DESIGN OF SPUR GEARS

In designs involving gear drives, normally the required speeds of rotation of the pinion and the gear and the amount of power that the drive must transmit are known. These factors are determined from the application. Also, the environment and operating conditions to

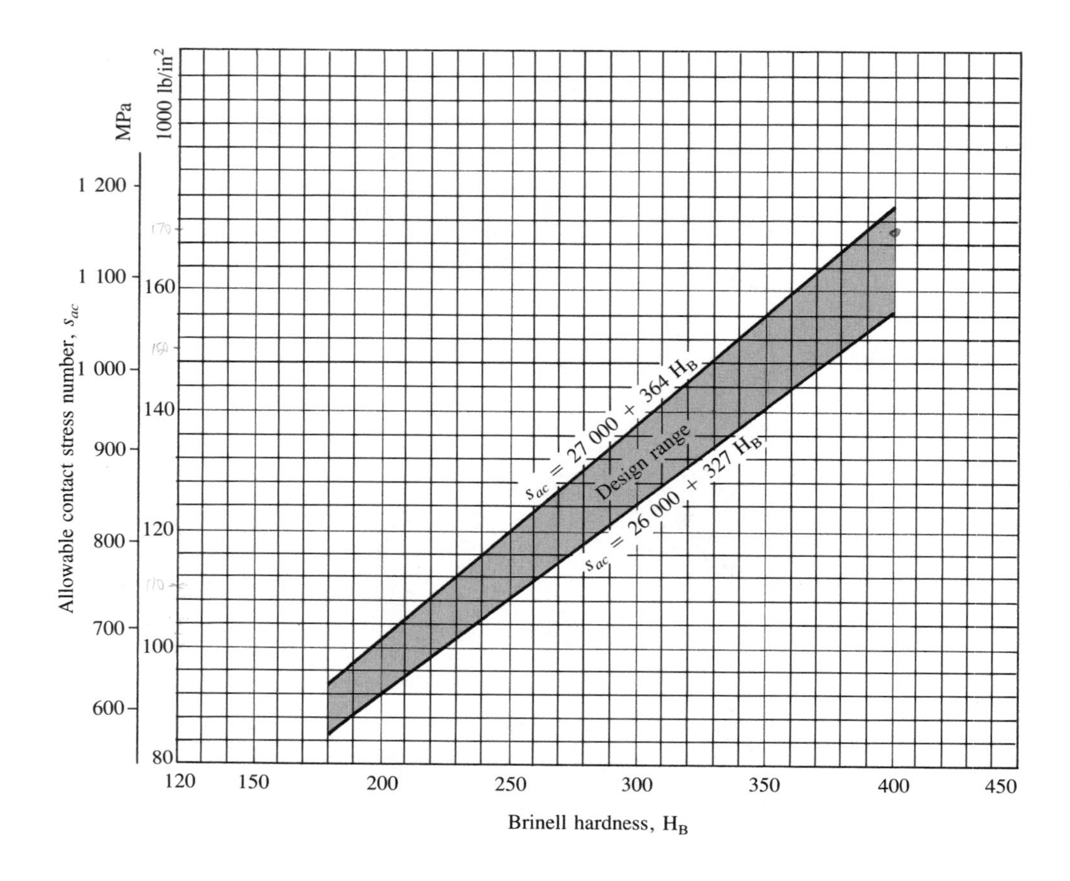

Cast iron:

- Class 20—50 000–60 000 psi (340–410 MPa) = s_{ac}
- Class 30—65 000–75 000 psi (450–520 MPa) = s_{ac}
- Class 40—75 000–85 000 psi (520–590 MPa) = s_{ac}

Case-hardened steel:

- Flame or induction-hardened
  - 50 HRC—170 000–190 000 psi (1 200–1 300 MPa) = s_{ac}
  - 54 HRC—175 000–195 000 psi (1 200–1 300 MPa) = s_{ac}

Carburized and case-hardened

- 55 HRC–180 000–200 000 psi (1 250–1 400 MPa) = s_{ac}
- 60 HRC–200 000–225 000 psi (1 400–1 550 MPa) = s_{ac}

Bronze:

ASTM B-148-52, heat-treated to $s_{u_{\min}}$ = 90 000 psi (620 MPa)
s_{ac} = 65 000 psi (450 MPa)

Figure 11-30 Allowable Contact Stress Number for Steel Gears, s_{ac} (Extracted from *AGMA Standard for Rating the Pitting Resistance and Bending Strength of Spur and Helical Involute Gear Teeth, AGMA 218.01,* with the permission of the publisher, American Gear Manufacturers Association, 1901 North Fort Myer Drive, Suite 1000, Arlington, Virginia 22209)

which the drive will be subjected must be understood. It is particularly important to know the type of driving device and the driven machine, in order to judge the proper value for the service factor or the application factor.

The designer must decide the type of gears to use; their arrangement on their shafts; materials, including their heat treatment; and the geometry of the gears: numbers of teeth, diametral pitch, pitch diameters, tooth form, face width, and quality numbers.

This section presents a design procedure that accounts for the bending fatigue strength of the gear teeth and the wear resistance, called *surface durability*. This procedure makes extensive use of the design equations presented in the preceding sections of the chapter and of the tables of material properties in Appendixes A-3 through A-5, A-8, and A-12.

It should be understood that there is no one best solution to a gear design problem; several good designs are possible. Your judgment and creativity and the specific requirements of the application will greatly affect the final design selected. The purpose here is to provide a means of approaching the problem to create a reasonable design.

Some overall objectives of a design are listed below. The resulting drive should—

Be compact and small

Operate smoothly and quietly

Have long life

Be low in cost

Be easy to manufacture

Be compatible with the other elements in the machine such as bearings, shafts, the housing, the driver, and the driven machine.

The design procedure has as its major objective the determination of a safe and long-lasting gear drive. The procedure has these primary parts.

1. A geometry that satisfies the required velocity ratio and application limitations such as center distance and physical size is proposed.
2. A tentative choice of the type of material to be used (steel, cast iron, and so on) is made.
3. A trial diametral pitch is chosen. Because of its strong effect on strength, wear resistance, and geometry, the choice of diametral pitch is critical. The remaining parts of the procedure are directed at confirming that a reasonable design can be completed. If not, the results will help to choose the next trial pitch.
4. The loads, face width, and design factors are determined.
5. The bending stress on the pinion teeth is computed. If a reasonable value results, the procedure continues. Otherwise a new pitch or revised geometry is selected.
6. The contact stress on the surface of the teeth is computed and the required material properties determined to ensure against pitting wear.
7. The final specifications of the materials for the pinion and the gear are made to satisfy the requirements of both strength and wear.

The detailed steps involved in the design of spur gears will be illustrated in an example problem. It is not unusual to make several trials before settling on a design. If the process is set up in a computer program or on a programmable calculator, successive trials can be made very quickly (see section 11-11).

Figure 11-31 shows a graph of power capacity of a pair of steel gears versus the speed of rotation of the pinion, with several values of diametral pitch shown. The hardnesses for the pinion and the gear used for these curves are in the middle range of possible values for steel (approximately HB 300). This should give you some feel of where to start. But note that the curves assume a uniform load ($K_a = 1.0$) and good alignment ($K_m = 1.0$).

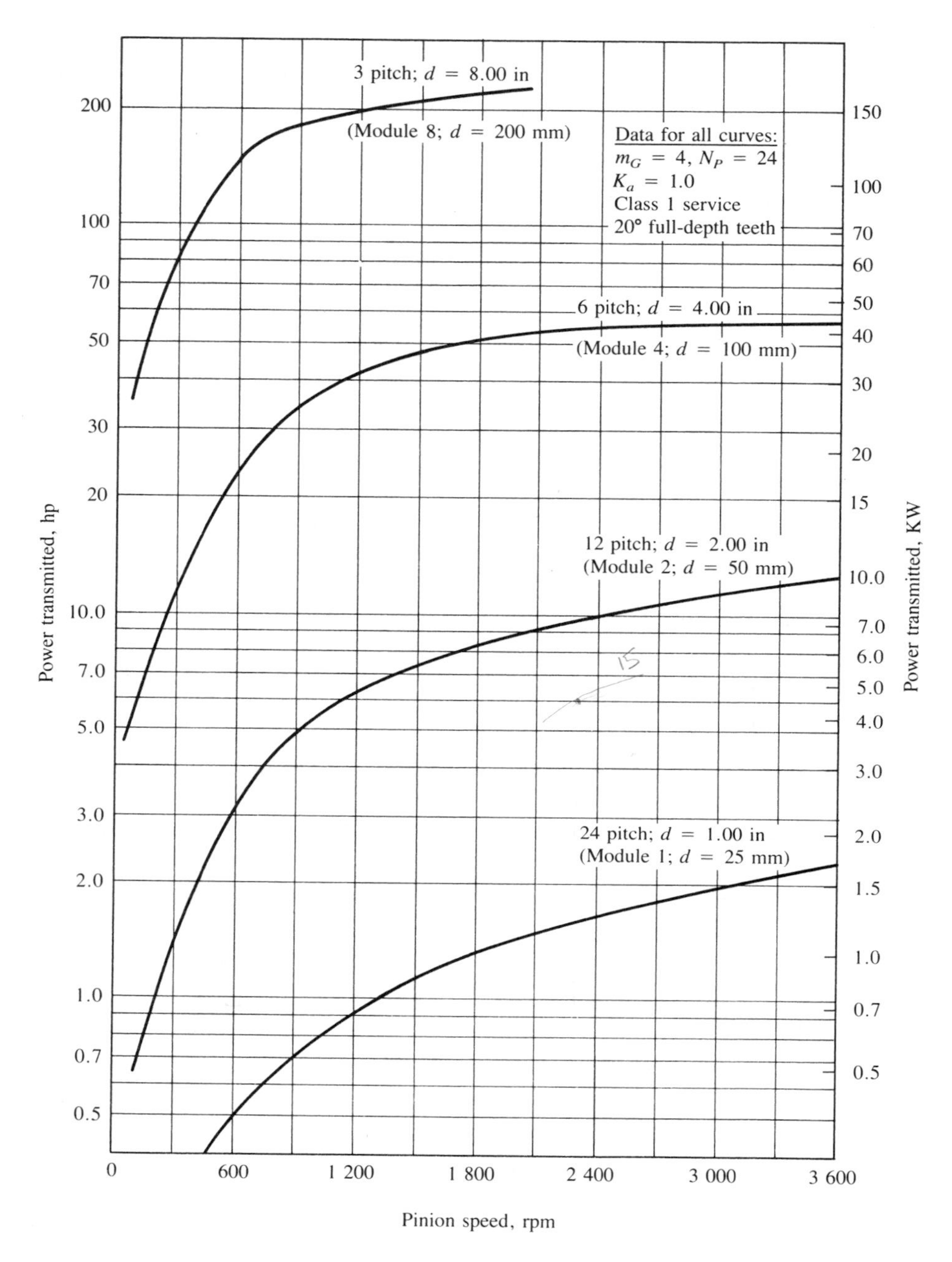

Figure 11-31 Power vs. Pinion Speed

If your application has higher values for either factor, the entire set of curves is shifted downward. That is, you should choose a lower value for diametral pitch than indicated on the graph.

The face width, F, can be specified once the diametral pitch is chosen. Although a wide range of face widths is possible, the following limits are used for general machine drive gears:

$$8/P_d < F < 16/P_d$$

$$\text{Nominal value of } F = 12/P_d \tag{11-32}$$

Also, the face width should not be greater than the pitch diameter of the pinion. An upper limit is placed on the face width to minimize problems with alignment. A very wide face width increases the chance for less than full face loading of the teeth. When the face width is less than the lower limit of equation (11-32), it is probable that a more compact design can be achieved with a different pitch.

The following relationships should help to determine what changes in your design assumptions you should make after the first set of calculations.

Decreasing the numerical value of the diametral pitch results in larger teeth and generally lower stresses. Also, the lower value of the pitch usually means a larger face width, which decreases stress and increases surface durability.

Increasing the diameter of the pinion decreases the transmitted load and generally lowers the stresses and improves the surface durability.

Increasing the face width lowers the stress and improves the surface durability, but to a generally lesser extent than either the pitch or pitch diameter changes discussed previously.

Gears with more and smaller teeth tend to run more smoothly and quietly than gears with fewer and larger teeth.

Standard values of diametral pitch should be used for ease of manufacture and lower cost (see Table 11-2).

Using high-alloy steels with high surface hardness results in the most compact system, but the cost is higher.

Using very accurate gears (ground or shaved teeth) results in lower dynamic loads and consequently lower stresses and improved surface durability, but the cost is higher.

The number of teeth in the pinion should generally be as small as possible to make the system compact. But the possibility of interference is greater with fewer teeth. Check Table 11-6 to ensure that no interference will occur.

Example Problem 11-3. A pair of spur gears with 20° full-depth teeth is to be designed as a part of the drive for a wood chipper to cut pulpwood for use in a paper mill. The power source is an electric motor that drives the pinion at 1 750 rpm. The gear must rotate between 460 and 465 rpm. The gears must transmit 3.0 hp. A compact design is desired, but in no case can the center distance exceed 5.00 in. The gears will be hobbed and will conform to an AGMA quality number of 5.

Complete the design that will be safe with a fatigue life of greater than 10 million cycles and with an expected infinite life with regard to surface durability. Specify the diametral pitch, the number of teeth in the pinion and the gear, the pitch diameters of the pinion and the gear, the face width, and the material with its heat treatment.

Solution.

1. Compute the nominal velocity ratio.
   At the midpoint of the given range of output speeds:

$$VR = n_P/n_G = (1\,750 \text{ rpm})/(462.5 \text{ rpm}) = 3.78$$

2. Specify the geometry for the first trial design.
   Let's start by specifying the diametral pitch and the number of teeth in the pinion. Referring to Figure 11-31, it would appear that a diametral pitch somewhat greater than 12 would work. But remember that those graphs are for uniform loads and good alignment. This design, for a wood chipper, has a high application factor and will probably have fairly high misalignment. Therefore, let's choose $P_d = 12$. Because we want a compact design, let's choose the smallest number of teeth for the pinion that will ensure no interference. From Table 11-6, we can use $N_P = 18$.
   Now, using the basic equations for gear pair geometry from section 11-2:

$$N_G = N_P(VR) = (18)(3.78) = 68.04$$

   But, because N_G must be an integer, use $N_G = 68$. The final velocity ratio is $VR = N_G/N_P = 68/18 = 3.778$. The actual output speed is $n_G = n_P/VR = (1\,750 \text{ rpm})/(3.778) = 463$ rpm. This is within the specified range.
   Compute the pitch diameters.

$$d = N_P/P_d = 18/12 = 1.500 \text{ in}$$
$$D = N_G/P_d = 68/12 = 5.667 \text{ in}$$

   Compute the center distance.

$$C = (N_P + N_G)/(2P_d) = (18 + 68)/(24) = 3.583 \text{ in}$$

   The geometry is acceptable.
3. Compute the pitch line speed.

$$v_t = \pi d n_P/12 = [\pi(1.500)(1\,750)]/12 = 687 \text{ ft/min}$$

4. Compute the transmitted load for 3.0 hp.

$$W_t = 33\,000(P)/v_t = 33\,000(3.0)/687 = 144 \text{ lb}$$

5. Determine the dynamic factor, K_v. For the commercially hobbed gear teeth, use curve 5 of Figure 11-27. The value can be read from the graph or computed from

$$K_v = \frac{50}{50 + \sqrt{v_t}} = \frac{50}{50 + \sqrt{687}} = 0.656$$

6. Select the application factor, K_a. From Table 11-13 with a uniform power source (electric motor) and a heavy shock load (wood chipper), use $K_a = 1.75$.
7. Select the load distribution factor, K_m. From Table 11-12, and assuming only average precision, use $K_m = 1.6$.
8. Determine the geometry factor, J. From Figure 11-26, $J = 0.330$.

9. Specify the face width, F. Using the nominal value:

$$F = 12/P_d = 12/12 = 1.00 \text{ in}$$

10. Compute the bending stress on the pinion teeth. Using equation (11-26):

$$\sigma_t = \frac{K_a W_t}{K_v} \frac{P_d}{F} \frac{K_m}{J} = \frac{(1.75)(144)(12)(1.6)}{(0.656)(1.00)(0.330)} = 22\,350 \text{ psi}$$

For the normal reliability factor of 0.99, this computed stress can be compared directly with the allowable stresses shown in Figure 11-28. An approximate Brinell hardness of 160 is required. This would allow the choice of virtually any steel. Therefore, the design is satisfactory from the standpoint of strength. In fact, it appears that even higher stresses could be tolerated for most gear materials. However, the design should first be checked for surface durability, as this often is the limiting case.

11. To evaluate the stress on the gear teeth, notice that all terms in equation (11-26) would be the same except the geometry factor, J. Having the value of the stress for the pinion, the stress for the gear can be found from a simple ratio.

$$\sigma_{tG} = \sigma_{tP}(J_P)/(J_G) \tag{11-33}$$

where

σ_{tG} = Bending stress in the gear teeth

σ_{tP} = Bending stress in the pinion teeth

J_G = Geometry factor for the gear teeth

J_P = Geometry factor for the pinion teeth

Because the value of J_P is always less than J_G, the pinion stress will always be the larger. However, frequently the gear is made of a different, and weaker, material. If so, it must be checked for safety.

In this example, $J_P = 0.330$ and $J_G = 0.420$. Then,

$$\sigma_{tG} = 22\,350 \text{ psi}(0.330)/0.420 = 17\,600 \text{ psi}$$

This requires a hardness of only about HB 140.

12. Compute the contact stress on the teeth from equation (11-31). This requires that we decide on the types of materials to use for the pinion and the gear. In this case, let's use steel for both. Then, from Table 11-14, $C_p = 2\,300$. We must also determine the geometry factor for surface durability, I, from Figure 11-29. We read $I = 0.104$.

The other factors are the same as those used for the stress analysis.

$$C_v = K_v = 0.656$$

$$C_a = K_a = 1.75$$

$$C_m = K_m = 1.6$$

Then in equation (11-31):

$$\sigma_h = C_p \sqrt{\frac{C_a C_m W_t}{C_v F d I}} = 2\,300 \sqrt{\frac{(1.75)(1.6)(144)}{(0.656)(1.00)(1.500)(0.104)}}$$

$$\sigma_h = 144\,400 \text{ psi}$$

Comparing the computed contact stress with the allowable contact stress number, s_{ac}, from Figure 11-30 indicates that a surface hardness of HB 350 is required for the teeth.

13. Specify the material for the gears. In general, the requirements of both strength and wear must be satisfied. In this case, the wear resistance requires the harder tooth and therefore it governs the choice of materials. It is typical for the pinion to be made about 16 percent harder than the gear for optimum performance. Table 11-15 shows some typical gear-and-pinion hardness combinations.

    We can now select the materials. In summary, we need a pinion material with a surface hardness of at least HB 350. The gear should have a hardness about 50 points lower than the pinion, approximately HB 300.

    Referring to the appendix tables and graphs of properties of steels, we can see that none of the listed carbon steels has the proper hardness level. Several of the alloy steels can be used, however. For example, from Appendix A-4-4, we can use

    For the pinion (AISI 4140, OQT 950):

    $$s_u = 175 \text{ Ksi} \quad \text{Hardness} = \text{HB } 365 \quad 16 \text{ percent elongation}$$

    For the gear (AISI 4140, OQT 1100):

    $$s_u = 148 \text{ Ksi} \quad \text{Hardness} = \text{HB } 311 \quad 19 \text{ percent elongation}$$

Alternate Solution

Although the solution may be quite satisfactory, another possible solution is presented here to show the effects of making changes in assumptions and objectives of the design. Note that one of the original objectives was that the gear drive have a small size. Let's see whether a smaller system can be successfully designed.

The diameters of the gears must be reduced to achieve a smaller size. It may also be desirable, or necessary, to change the diametral pitch. Our next trial will use

$$P_d = 14 \quad \text{and} \quad N_P = 18$$

Then, pinion pitch diameter is

$$d = N_P/P_d = 18/14 = 1.286 \text{ in}$$

Number of teeth in the gear is

$$N_G = N_P VR = (18)(3.78) = 68.04 \quad \text{or } 68$$

Table 11-15 Hardness Combinations for Gear and Pinion

Gear HB	*Pinion HB*	*Gear HB*	*Pinion HB*
180	210	255	300
210	245	270	315
225	265	285	335
245	285	300	350

Thus the final velocity ratio and the output speed are the same as for the first trial: $VR = 3.778$ and $n_G = 463$ rpm.

Gear pitch diameter:

$$D = N_G/P_d = 68/14 = 4.857 \text{ in}$$

Center distance:

$$C = (N_P + N_G)/(2P_d) = (18 + 68)/(28) = 3.071 \text{ in}$$

This represents about a 14 percent reduction in size. Much of the remaining analysis is similar to that for the first trial. Only those places where significant changes occur will be discussed.

Pitch line speed: $v_t = \pi d n_P/12 = \pi(1.286)(1\,750)/12 = 589$ ft/min

Transmitted load: $W_t = [33\,000(P)]/v_t = (33\,000)(3.0)/589 = 168$ lb

Dynamic factor for $v_t = 589$ ft/min: $K_v = 0.673$.

Values from trial 1: $K_a = 1.75$; $K_m = 1.6$; $J_P = 0.330$; $J_G = 0.420$

Face width: nominal $F = 12/P_d = 12/14 = 0.857$ in. Because the face width is not a critical dimension, it is recommended that common, convenient values be used for F. Let's use $F = 0.75$ in.

Bending stress in pinion teeth:

$$\sigma_{tP} = \frac{K_a W_t}{K_v}\frac{P_d}{F}\frac{K_m}{J_P} = \frac{(1.75)(168)(14)(1.6)}{(0.673)(0.75)(0.330)} = 39\,500 \text{ psi}$$

Bending stress in gear teeth:

$$\sigma_{tG} = \sigma_{tP}(J_P)/J_G = 39\,500 \text{ psi}(0.330)/0.420 = 31\,000 \text{ psi}$$

Required hardness of pinion material (from Figure 11-28):

$$\text{Hardness} = \text{HB } 350$$

Required hardness of gear material (from Figure 11-28):

$$\text{Hardness} = \text{HB } 230$$

Contact stress:

$$\sigma_h = C_p\sqrt{\frac{C_a C_m W_t}{C_v F d I}} = 2\,300\sqrt{\frac{(1.75)(1.6)(168)}{(0.673)(0.75)(1.286)(0.104)}}$$

$$\sigma_h = 192\,000 \text{ psi}$$

Required surface hardness from Figure 11-30: Case hardening by carburizing is required, with a resulting hardness of HRC 55. Note that for the case-hardened options, the pinion and the gear are given equivalent treatments.

Material selection: To satisfy both the strength and hardness requirements, refer to the table of properties for carburized steels in Appendix A-5. The safest approach would be to select a material whose core properties satisfy the strength requirement and whose case hardness is adequate for surface durability. One possible selection, from Appendix A-5, is AISI E9310 DOQT 300, case-hardened by carburizing:

Core: $s_u = 174$ Ksi Hardness = HB 363

Case: Hardness = HRC 60

This is conservative because the maximum stress occurs in the fillet of the tooth at its root, where the carburizing is effective, thus increasing the strength above that of the core. Note in Figure 11-28 that an allowable bending stress of 55 Ksi can be used for carburized teeth. However, the case is relatively thin, and it is possible that the material just under the case will be subjected to a high stress. Thus if the core properties are sufficient to withstand the maximum bending stress, the design will be safe.

Two different solutions have been shown for the same problem. Others are certainly possible. The procedure outlined here should enable you to try several designs from which an optimum design can be selected.

11-10 GEAR DESIGN FOR THE METRIC MODULE SYSTEM

Section 11-2 on Gear Geometry described the metric module system of gearing and its relation to the diametral pitch system. As the design process was being developed in section 11-9, the data for stress analysis and surface durability analysis were taken from charts using U.S. Customary Units (inches, pounds, horsepower, feet per minute, and Kips per square inch). Data for the metric module system were also available in the charts in units of millimeters, newtons, kilowatts, meters per second, and megapascals. But to use the SI data, some of the formulas must be modified.

The following is an example problem using SI units. The procedure will be virtually the same as that used to design with U.S. Customary Units. Those formulas that are converted to SI units are identified.

Example Problem 11-4. A gear pair is to be designed to transmit 15.0 kilowatts (kW) of power to a large meat grinder in a commercial meat processing plant. The pinion is attached to the shaft of an electric motor rotating at 575 rpm. The gear must operate at 270 to 280 rpm. Commercially hobbed (quality number 5), 20° full-depth involute gears are to be used in the metric module system. The maximum center distance is to be 200 mm. Specify the design of the gears.

Solution. The nominal velocity ratio is

$$VR = 575/275 = 2.09$$

From Figure 11-31, $m = 4$ is a reasonable trial module. Then

$$N_P = 18 \quad \text{(design decision)}$$

$$d = N_P m = (18)(4) = 72 \text{ mm}$$

$$N_G = N_P VR = (18)(2.09) = 37.6 \quad \text{(use 38)}$$

$$D = N_G m = (38)(4) = 152 \text{ mm}$$

$$\text{Final output speed} = n_G = n_P(N_P/N_G)$$

$$n_G = 575 \text{ rpm} \times (18/38) = 272 \text{ rpm} \quad \text{(OK)}$$

$$\text{Center distance} = C = (N_P + N_G)m/2 \quad \text{[Eq. (11-18)]}$$
$$C = (18 + 38)(4)/2 = 112 \text{ mm} \quad \text{(OK)}$$

In SI units, the pitch line speed in meters per second (m/s) is

$$v_t = \pi d n_P/(60\,000)$$

where d is in mm and n_P is in revolutions per minute (rpm). Then

$$v_t = [(\pi)(72)(575)]/(60\,000) = 2.17 \text{ m/s}$$

In SI units, the transmitted load, W_t, is in the unit of newtons (N). If the power, P, is in kW, and v_t is in m/s,

$$W_t = 1\,000(P)/v_t = (1\,000)(15)/(2.17) = 6\,920 \text{ N}$$

In the U.S. Customary Unit system, it was recommended that the face width be approximately $F = 12/P_d$ in. The equivalent SI values are $F = 12(m)$ mm. For this problem, $F = 12(4) = 48$ mm. Let's use $F = 50$ mm.

Other factors are found as before.

$$K_v = 0.74 \quad \text{(Figure 11-27)}$$
$$K_a = 1.25 \quad \text{(Table 11-13)}$$
$$K_m = 1.6 \quad \text{(Table 11-12)}$$
$$J_P = 0.315 \qquad J_G = 0.380 \quad \text{[Figure 11-26(a)]}$$

Then the stress in the pinion is found from equation (11-26), modified by letting $P_d = 1/m$:

$$\sigma_{tP} = \frac{K_a W_t}{K_v} \frac{1}{Fm} \frac{K_m}{J_P} = \frac{(1.25)(6\,920)(1)(1.6)}{(0.74)(50)(4)(0.315)} = 297 \text{ MPa}$$

This is a reasonable stress level. The required hardness of the tooth material is HB 400, as found in Figure 11-28. Proceed with wear design.

For two steel gears:

$$C_p = 191 \quad \text{(Table 11-14)}$$
$$I = 0.092 \quad \text{(Figure 11-29)}$$
$$C_v = K_v = 0.74$$
$$C_a = K_a = 1.25$$
$$C_m = K_m = 1.6$$

Contact stress [equation (11-31)]:

$$\sigma_h = C_p \sqrt{\frac{C_a C_m W_t}{C_v F d I}} = 191 \sqrt{\frac{(1.25)(1.6)(6\,920)}{(0.74)(50)(72)(0.092)}} = 1\,435 \text{ MPa}$$

From Figure 11-30, the required surface hardness is HRC 60, case carburized.

Material selection from Appendix A-5, for carburized steels:

AISI 4320 SOQT 300; s_u = 1 500 MPa; 13 percent elongation

Case harden by carburizing to HRC 60 minimum

Case depth: 0.6 mm minimum (Figure 11-16)

11-11 COMPUTER-ASSISTED GEAR DESIGN

The gear design process described in this chapter has been written into a program shown at the end of this section. Figure 11-32 is the flowchart. A sample output, using the data from example problem 11-3, immediately follows the listing. The correlation of the variable names with the terms used in the gear design process is noted in the following list.

P:	Power transmitted by the gear pair (hp)
N1:	Rotational speed of the pinion (rpm)
N2:	Rotational speed of the gear (rpm)
R:	Velocity ratio of the gear pair
P1:	Diametral pitch
T1:	Number of teeth in the pinion
T2:	Number of teeth in the gear
D1:	Pitch diameter of the pinion (in)
D2:	Pitch diameter of the gear (in)
C:	Center distance (in)
V:	Pitch line speed of the pinion or the gear (ft/min)
W:	Transmitted load on the teeth (lb)
F:	Face width of the teeth (in)
F1, F2, F3:	Nominal, minimum, and maximum recommended face widths, respectively (in)
K1:	Dynamic factor, K_v (computed from equation in Figure 11-27)
K2:	Application factor, K_a (from Table 11-13)
K3:	Alignment factor, K_m (from Table 11-12)
J1:	Geometry factor, J_P, for the pinion (from Figure 11-26)
J2:	Geometry factor, J_G, for the gear (from Figure 11-26)
S1:	Computed stress in the pinion teeth (psi)
S2:	Computed stress in the gear teeth (psi)
S3:	Computed contact stress on the teeth (psi)
S4:	Required tensile strength of the pinion or gear material (Ksi)
H1:	Required Brinell hardness number
C1:	Elastic coefficient, C_p (from Table 11-14)
E1, E2:	Modulus of elasticity for the pinion and gear (psi)
I1:	Geometry factor for surface durability (Figure 11-29)

Other variable names are used as "dummy variables" to allow the operator of the program to input control values. These are variables G (line 400), S (line 800), M (line 930), and Q (line 1510).

Lines 100 to 260 are introductory and call for the basic power and speed data to be input by the designer. The initial design decisions made by the designer are the first trial values of the diametral pitch and the number of teeth in the pinion. Recall that Figure 11-31 is an aid to these choices.

Lines 270 to 410 compute the various geometric features from the given data and allow the designer to evaluate the suitability of the design at this early point. If it is unsatisfactory, new choices can be made for P_d and N_P.

Values for pitch line speed, v_t; transmitted load, W_t; and the dynamic factor, K_v, are then computed. The dynamic factor assumes commercial gears with quality number 5 to be conservative. Lines for the other quality numbers can be added to enhance the program. The final selection of the face width is up to the designer at line 550. However, the program computes the guidelines, giving the nominal, minimum, and maximum recommended values, as discussed in section 11-9 and computed by equation (11-32).

The application factor, alignment factor, and geometry factors are determined by the designer and input to the program in steps 560 to 610. Lines 620 to 820 perform the computation of the stresses in the teeth and the required strength and hardness of the materials for the pinion and the gear.

Lines 680 and 690 call for special note. Recall that after the computation of the actual stress in the pinion teeth, the required tensile strength and hardness of the material are determined from Figure 11-28. The need for the designer to look up this information has been eliminated by creating an equation for the required hardness of the material as a function of the allowable stress. A polynomial curve fit routine was used to determine the equation. Many computer systems have such curve fit routines, and the technique could have been used for other similar operations in this program. But in the interest of keeping the program to a moderate level of complexity and length, only this one operation uses a curve fit.

Lines 830 to 1470 constitute the analysis for surface durability. First (lines 840 to 930), the operator selects the material combination for the pinion and the gear. From this, the value of C_p is determined from the values in Table 11-14. Here many of the tabulated values are looked up automatically. But for special cases, lines 1070 to 1100 allow computation of C_p if the operator inputs the values of the modulus of elasticity.

After the value of the geometry factor for surface durability, I, is input, the contact stress is computed. Lines 1180 to 1440 determine the required surface hardness to withstand the computed contact stress, by automatically interpreting Table 11-15 and Figure 11-30.

Finally, the program allows several options on how to proceed. If satisfied with the design just completed, the operator can end the program. If other designs for the same requirements are desired, the program is designed to branch back to the point where the diametral pitch and the number of teeth in the pinion are specified. For a new set of requirements, the program goes back to line 180, allowing input of new values for power and speeds.

With such a program, the designer can produce a complete new trial design in approximately 30 sec, with little chance for error. The design decisions that require thought and judgment are performed by the designer, while the computer performs computations and repetitive tasks.

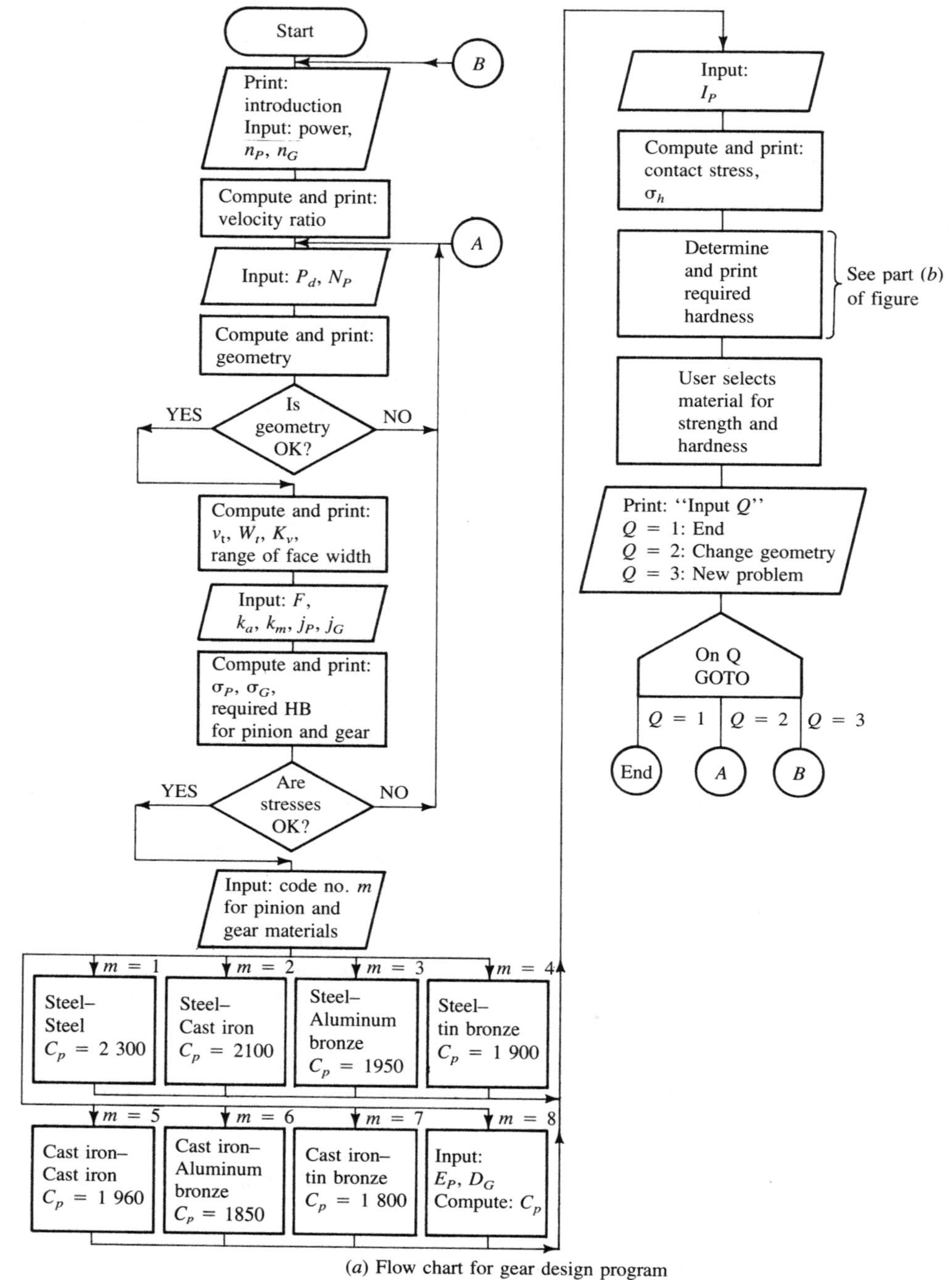

(a) Flow chart for gear design program

Figure 11-32 Gear Design Process

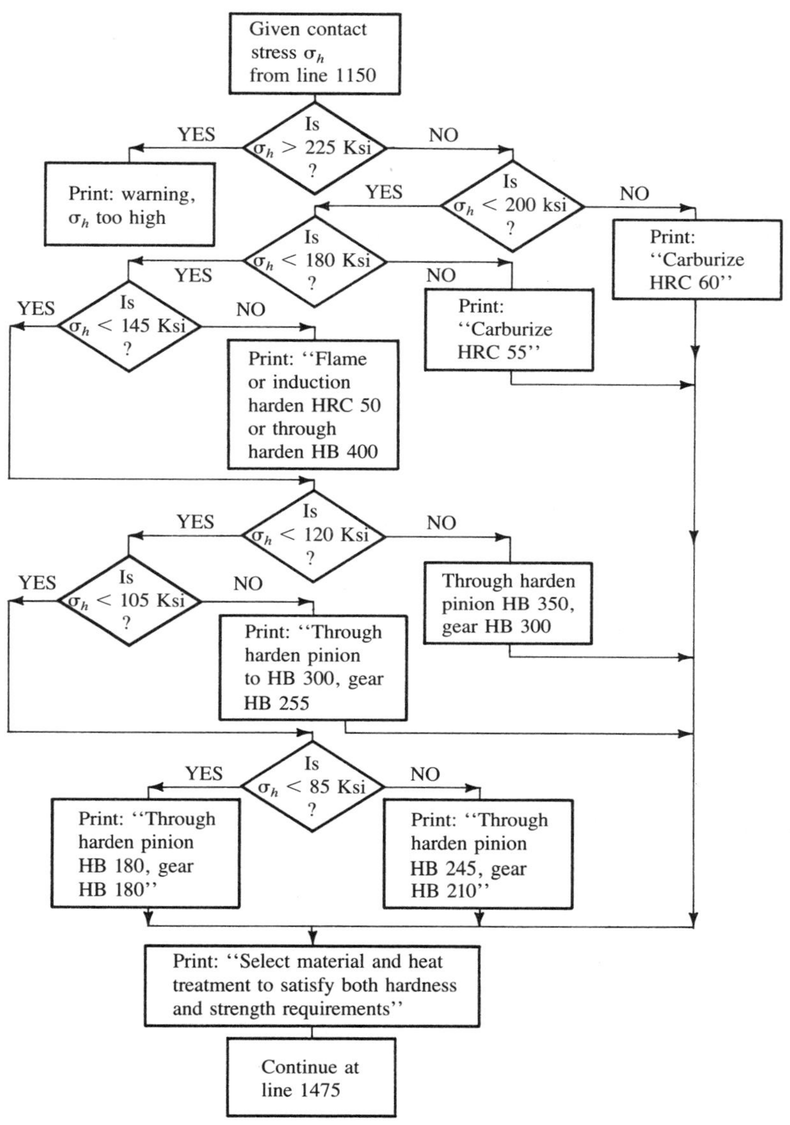

(b) Logic to determine required hardness

(*b*) Logic to determine required hardness

Figure 11-32 (Continued)

Gear Design: Program Listing

Introduction

```
  100 PRINT "THIS PROGRAM ASSISTS IN THE DESIGN OF SPUR GEARS"
  110 PRINT
  120 PRINT "THE INPUT DATA REQUIRED ARE:"
  130 PRINT "   POWER TRANSMITTED IN HORSEPOWER"
  140 PRINT "   PINION SPEED IN RPM"
  150 PRINT "   DESIRED GEAR SPEED IN RPM"
  160 PRINT
  170 PRINT "OTHER DATA REQUESTED AS NEEDED"
  180 PRINT
```

Input data

```
  190 PRINT "INPUT POWER, PINION SPEED, GEAR SPEED"
► 200 INPUT P,N1,N2
  210 R = N1/N2
  215 PRINT
  216 PRINT
  220 PRINT "DESIRED VELOCITY RATIO =";R
  225 PRINT
  226 PRINT
  230 PRINT "INPUT TRIAL DIAMETRAL PITCH"
► 240 INPUT P1
  250 PRINT "INPUT NUMBER OF TEETH IN THE PINION"
► 260 INPUT T1
```

Compute geometry

```
  270 D1 = T1/P1
  280 T2 = INT(T1*R + .5)
  290 D2 = T2/P1
  300 C = (T1 + T2)/(2*P1)
  310 R = T2/T1
  320 N2 = N1*T1/T2
  325 PRINT
  326 PRINT
```

Geometry results

```
  330 PRINT "DIAMETRAL PITCH =";P1
  340 PRINT "PINION: ";T1;"TEETH       PITCH DIAMETER
    =";D1;"INCHES"
  350 PRINT "GEAR: ";T2;"TEETH       PITCH DIAMETER
    =";D2;"INCHES"
  360 PRINT "CENTER DISTANCE =";C;"INCHES"
  370 PRINT "ACTUAL VELOCITY RATIO =";R
  380 PRINT "ACTUAL GEAR SPEED =";N2;"RPM"
  385 PRINT
  386 PRINT
```

Decision on how to continue

```
  390 PRINT "IF GEOMETRY IS OK, TYPE 1; TYPE 2 TO TRY AGAIN"
► 400 INPUT G
  410 IF G <> 1 THEN 230
  415 PRINT
  416 PRINT
  420 PRINT "GEOMETRY OK"
```

Transmitted load and face width

```
  430 V = 3.1415926 *D1*N1/12
  440 W = 33000 *P/V
  450 K1 = 50/(50 + SQR(V))
  460 F1 = 12/P1
  470 F2 = 8/P1
  480 F3 = 16/P1
  490 PRINT "PITCH LINE SPEED =";V;"FT/MIN"
  500 PRINT "TRANSMITTED LOAD =";W;"POUNDS"
  510 PRINT "DYNAMIC FACTOR, KV =";K1
  515 PRINT
  516 PRINT
  520 PRINT "NOMINAL FACE WIDTH =";F1;"IN."
  530 PRINT "ACCEPTABLE RANGE OF FACE WIDTH:
    ";F2;"TO";F3;"IN."
  540 PRINT "INPUT SELECTED FACE WIDTH"
► 550 INPUT F
  555 PRINT
  556 PRINT
```

Lines in the program calling for input of data and lines in the output that correspond to the input statements are highlighted with a ►.

Operator input

```
560 PRINT "INPUT APPLICATION FACTOR (KA) AND ALIGNMENT
  FACTOR (KM)"
570 INPUT K2,K3
580 PRINT "INPUT GEOMETRY FACTOR, JP, FOR ";T1;"TEETH IN
  PINION"
590 INPUT J1
600 PRINT "INPUT GEOMETRY FACTOR, JG, FOR ";T2;"TEETH IN
  GEAR"
610 INPUT J2
615 PRINT
616 PRINT
```

Stress analysis

```
620 S1 = K2*W*P1*K3/(K1*F*J1)
630 S2 = S1*J1/J2
640 PRINT "BENDING STRESS IN PINION TEETH =";S1;"PSI"
650 PRINT "BENDING STRESS IN GEAR TEETH =";S2;"PSI"
660 PRINT
670 PRINT
680 H1 = 549.5 - 0.5*SQR(1.2E6 - 26.32*S1)
690 H1 = INT(H1)
700 IF S1 = S2 THEN 740
710 PRINT "REQUIRED CORE HARDNESS FOR PINION = HB";H1
720 S1 = S2
730 GOTO 670
740 PRINT "REQUIRED CORE HARDNESS FOR GEAR = HB";H1
750 PRINT
760 PRINT
770 PRINT
780 PRINT
```

Handwritten annotations (line 680 struck out):

```
671 Z= 1.2E6 -26.32*1
672 IF Z≤Ø GOTO 674
673 IF Z >Ø GOTO 680
674 PRINT " TOOTH STRESSES ARE TOO HIGH, TRY ANOTHER DIAM. PITCH OR NO. PINION TEETH"
675 GOTO 230
680 H1 = 549.5 -0.5 *SQR(Z)
```

Decision on how to continue

```
790 PRINT "IF STRESSES ARE OK, TYPE 1; TYPE 2 TO TRY AGAIN"
800 INPUT S
810 IF S <> 1 THEN 220
820 PRINT "STRESSES ARE OK"
825 PRINT
826 PRINT
830 PRINT "DESIGN FOR WEAR NOW BEGINS"
```

Material selection

```
840 PRINT "WHAT MATERIALS ARE TO BE USED?"
850 PRINT "TYPE  1  FOR STEEL PINION, STEEL GEAR"
860 PRINT "TYPE  2  FOR STEEL PINION, CAST IRON GEAR"
870 PRINT "TYPE  3  FOR STEEL PINION, ALUM. BRONZE GEAR"
880 PRINT "TYPE  4  FOR STEEL PINION, TIN BRONZE GEAR"
890 PRINT "TYPE  5  FOR CAST IRON PINION, CAST IRON GEAR"
900 PRINT "TYPE  6  FOR CAST IRON PINION, ALUM. BRONZE GEAR"
910 PRINT "TYPE  7  FOR CAST IRON PINION, TIN BRONZE GEAR"
920 PRINT "TYPE  8  FOR ANY OTHER MATERIAL COMBINATION"
930 INPUT M
```

Determine value of C_p

```
940 ON M GOTO 950,970,990,1010,1030,1050,1065,1070
950 C1 = 2300
960 GOTO 1110
970 C1 = 2100
980 GOTO 1110
990 C1 = 1950
1000 GOTO 1110
1010 C1 = 1900
1020 GOTO 1110
1030 C1 = 1960
1040 GOTO 1110
1050 C1 = 1850
1060 GOTO 1110
1065 C1 = 1800
1066 GOTO 1110
1070 PRINT "INPUT E FOR PINION AND E FOR GEAR IN PSI"
1080 REM POISSON'S RATIO ASSUMED TO BE 0.3
1090 INPUT E1,E2
1100 C1 = SQR(1/(3.1415926*(.91/E1 + .91/E2)))
1110 PRINT "ELASTIC COEFFICIENT, CP = ";C1
```

```
1115 PRINT
1116 PRINT
1120 PRINT "INPUT I, THE GEOMETRY FACTOR FOR SURFACE
  DURABILITY"
1130 PRINT "USE I FOR ";T1;"TEETH"
1140 INPUT I1
1145 PRINT
1146 PRINT
1150 S3 = C1*SQR(K2*K3*W/(K1*F*D1*I1))
1160 PRINT "CONTACT STRESS =";S3;"PSI"
1170 IF M <> 1 THEN 1460
1180 PRINT "FOR STEEL PINION AND GEAR:"
1190 IF S3 > 225000 THEN 1440
1200 IF S3 < 200000 THEN 1240
1210 PRINT "CASE CARBURIZED TEETH REQUIRED, HARDNESS =
  HRC 60"
1220 PRINT "BOTH PINION AND GEAR"
1230 GOTO 1460
1240 IF S3 < 180000 THEN 1280
1250 PRINT "CASE CARBURIZED TEETH REQUIRED, HARDNESS =
  HRC 55"
1260 PRINT "BOTH PINION AND GEAR"
1270 GOTO 1460
1280 IF S3 < 145000 THEN 1330
1290 PRINT "FLAME OR INDUCTION HARDEN, HARDNESS HRC 50"
1300 PRINT " - OR - THROUGH HARDEN TO HB 400"
1310 PRINT "BOTH PINION AND GEAR"
1320 GOTO 1460
1330 IF S3 < 120000 THEN 1360
1340 PRINT "THROUGH HARDEN PINION TO HB 350; GEAR TO HB 300"
1350 GOTO 1460
1360 IF S3 < 105000 THEN 1390
1370 PRINT "THROUGH HARDEN PINION TO HB 300; GEAR TO HB 255"
1380 GOTO 1460
1390 IF S3 < 85000 THEN 1420
1400 PRINT "THROUGH HARDEN PINION TO HB 245; GEAR TO HB 210"
1410 GOTO 1460
1420 PRINT "THROUGH HARDEN PINION AND GEAR TO HB 180"
1430 GOTO 1460
1440 PRINT "CONTACT STRESS TOO HIGH; REDESIGN"
1450 GOTO 1480
1460 PRINT "SELECT MATERIAL AND HEAT TREATMENT TO SATISFY"
1470 PRINT "   BOTH SURFACE HARDNESS AND CORE STRENGTH
  REQUIREMENTS"
1475 PRINT
1476 PRINT
1480 PRINT "TYPE  1  TO END"
1490 PRINT "TYPE  2  TO TRY NEW GEOMETRY FOR SAME POWER AND
  SPEED"
1500 PRINT "TYPE  3  TO TRY WHOLE NEW PROBLEM"
1510 INPUT Q
1520 ON Q GOTO 1530, 220, 180
1530 END
```

Geometry factor

Contact stress

Material selection based on contact stress

Decision on how to continue

Gear Design: Sample Output[1]

```
THIS PROGRAM ASSISTS IN THE DESIGN OF SPUR GEARS

THE INPUT DATA REQUIRED ARE:
  POWER TRANSMITTED IN HORSEPOWER
  PINION SPEED IN RPM
  DESIRED GEAR SPEED IN RPM
```

[1]Data from Example Problem 11-3

```
    OTHER DATA REQUESTED AS NEEDED

    INPUT POWER, PINION SPEED, GEAR SPEED
 ►  ? 3 , 1750 , 462.5                                      Statement 200

    DESIRED VELOCITY RATIO = 3.78378

    INPUT TRIAL DIAMETRAL PITCH
 ►  ? 12                                                    Statement 240
    INPUT NUMBER OF TEETH IN THE PINION
 ►  ? 18                                                    Statement 260

    DIAMETRAL PITCH = 12
    PINION:    18 TEETH    PITCH DIAMETER = 1.5 INCHES
    GEAR:    68 TEETH    PITCH DIAMETER = 5.66667 INCHES
    CENTER DISTANCE = 3.58333 INCHES
    ACTUAL VELOCITY RATIO = 3.77778
    ACTUAL GEAR SPEED = 463.235 RPM

    IF GEOMETRY IS OK, TYPE 1; TYPE 2 TO TRY AGAIN
 ►  ? 1                                                     Statement 400

    GEOMETRY OK
    PITCH LINE SPEED = 687.223 FT/MIN
    TRANSMITTED LOAD = 144.058 POUNDS
    DYNAMIC FACTOR, KV = .656039

    NOMINAL FACE WIDTH = 1 IN.
    ACCEPTABLE RANGE OF FACE WIDTH: .666667 TO 1.33333 IN.
    INPUT SELECTED FACE WIDTH
 ►  ? 1                                                     Statement 550

    INPUT APPLICATION FACTOR (KA) AND ALIGNMENT FACTOR (KM)
 ►  ? 1.75 , 1.6                                            Statement 570
    INPUT GEOMETRY FACTOR, JP, FOR 18 TEETH IN PINION
 ►  ? .33                                                   Statement 590
    INPUT GEOMETRY FACTOR, JG, FOR 68 TEETH IN GEAR
 ►  ? .42                                                   Statement 610
    BENDING STRESS IN PINION TEETH = 22358 PSI
    BENDING STRESS IN GEAR TEETH = 17567 PSI

    REQUIRED CORE HARDNESS FOR PINION = HB 158

    REQUIRED CORE HARDNESS FOR GEAR = HB 120

    IF STRESSES ARE OK, TYPE 1; TYPE 2 TO TRY AGAIN
 ►  ? 1                                                     Statement 800
    STRESSES ARE OK

    DESIGN FOR WEAR NOW BEGINS
    WHAT MATERIALS ARE TO BE USED?
    TYPE  1  FOR STEEL PINION, STEEL GEAR
    TYPE  2  FOR STEEL PINION, CAST IRON GEAR
    TYPE  3  FOR STEEL PINION, ALUM. BRONZE GEAR
    TYPE  4  FOR STEEL PINION, TIN BRONZE GEAR
    TYPE  5  FOR CAST IRON PINION, CAST IRON GEAR
    TYPE  6  FOR CAST IRON PINION, ALUM. BRONZE GEAR
    TYPE  7  FOR CAST IRON PINION, TIN BRONZE GEAR
    TYPE  8  FOR ANY OTHER MATERIAL COMBINATION
 ►  ? 1                                                     Statement 930
    ELASTIC COEFFICIENT, CP = 2300
```

```
INPUT I , THE GEOMETRY FACTOR FOR SURFACE DURABILITY
USE I FOR 18 TEETH
▶ ? .104                                                  Statement 1140

CONTACT STRESS = 144394 PSI
FOR STEEL PINION AND GEAR:
THROUGH HARDEN PINION TO HB 350; GEAR TO HB 300
SELECT MATERIAL AND HEAT TREATMENT TO SATISFY
  BOTH SURFACE HARDNESS AND CORE STRENGTH REQUIREMENTS

TYPE  1  TO END
TYPE  2  TO TRY NEW GEOMETRY FOR SAME POWER AND SPEED
TYPE  3  TO TRY WHOLE NEW PROBLEM
▶ ? 1                                                     Statement 1510
```

11-12 PRACTICAL CONSIDERATIONS FOR GEARS AND INTERFACES WITH OTHER ELEMENTS

It is important to consider the design of the entire gear system when designing the gears because they must work in harmony with the other elements in the system. This section will discuss briefly some of these practical considerations and show commercially available speed reducers.

Our discussion so far has been primarily on the gear teeth, including the tooth form, pitch, face width, material selection, and heat treatment. Also to be considered is the type of gear blank. Figure 11-1 shows three styles of blanks. Smaller gears and lightly loaded gears are typically made in the plain style. Gears with pitch diameters of approximately 5.0 in through 8.0 in are frequently made with thinned webs between the rim and the hub for lightening, with some having holes bored in the webs for additional lightening. Larger gears, typically with pitch diameters greater than 8.0 in, are made from cast blanks with spokes between the rim and the hub.

In many precision special machines and gear systems produced in large quantities, the gears are machined integral with the shaft carrying the gears. This, of course, eliminates some of the problems associated with mounting and location of the gears, but it may complicate the machining operations.

In general machine design, gears are usually mounted on separate shafts, with the torque transmitted from the shaft to the gear through a key. This setup provides a positive means of transmitting the torque while permitting easy assembly and disassembly. The axial location of the gear must be provided by another means, such as a shoulder on the shaft, a retaining ring, or a spacer (see Figures 9-2 and 10-1).

Sections 10-3 and 10-4 and Figures 10-5 and 10-8 through 10-12 illustrate other means of fastening gears to shafts.

Other considerations include the forces exerted on the shaft and the bearings that are due to the action of the gears. These subjects are discussed in section 11-6. The housing design must provide adequate support for the bearings and protection of the interior components. Normally, it must also provide a means of lubricating the gears.

Lubrication

The action of spur gear teeth is a combination of rolling and sliding. Because of the relative motion, and because of the high local forces exerted at the gear faces, adequate lubrication is critical to smoothness of operation and gear life. A continuous supply of oil at the pitch line is desirable for most gears unless they are lightly loaded or operate only intermittently.

In splash-type lubrication, one of the gears in a pair dips into an oil supply sump and carries the oil to the pitch line. At higher speeds, the oil may be thrown onto the inside surfaces of the case; then it flows down, in a controlled fashion, onto the pitch line. Simultaneously, the oil can be directed to the bearings that support the shafts. One difficulty with the splash type of lubrication is that the oil is churned; at high gear speeds, excessive heat can be generated and foaming can occur.

A positive oil circulation system is used for high-speed and high-capacity systems. A separate pump draws the oil from the sump and delivers it at a controlled rate to the meshing teeth.

The primary functions of gear lubricants are to reduce friction at the mesh and to keep operating temperatures at acceptable levels. It is essential that a continuous film of lubricant be maintained between the mating tooth surfaces of highly loaded gears and that there be a sufficient flow rate and total quantity of oil to maintain cool temperatures. Heat is generated by the meshing gear teeth, by the bearings, and by the churning of the oil. This heat must be dissipated from the oil to the case or to some other external heat exchange device in order to keep the oil itself below 160–180°F (approximately 80°C). Above this temperature, the lubricating ability of the oil, as indicated by its viscosity, is severely decreased. Also, there can be chemical changes produced in the oil that decrease its lubricity. Because of the wide variety of lubricants available and the many different conditions under which they must operate, it is recommended that suppliers of lubricants be consulted for proper selection.

Commercially Available Gear-Type Speed Reducers

By studying the design of commercially available gear-type speed reducers, you should get a better feel for these design details and the relationships among the component parts: the gears, the shafts, the bearings, the housing, the means of providing lubrication, and the coupling to the driving and driven machines.

Figure 11-33 Double-reduction Spur Gear Reducer (Bison Gear & Engineering Corp., Downers Grove, Ill.)

Figure 11-34 Double-reduction Gear Reducer. First stage, helical gears; second stage, spur gears. (Bison Gear & Engineering Corp., Downers Grove, Ill.)

(*a*) Cutaway of a concentric helical gear reducer

(*b*) Complete reducer

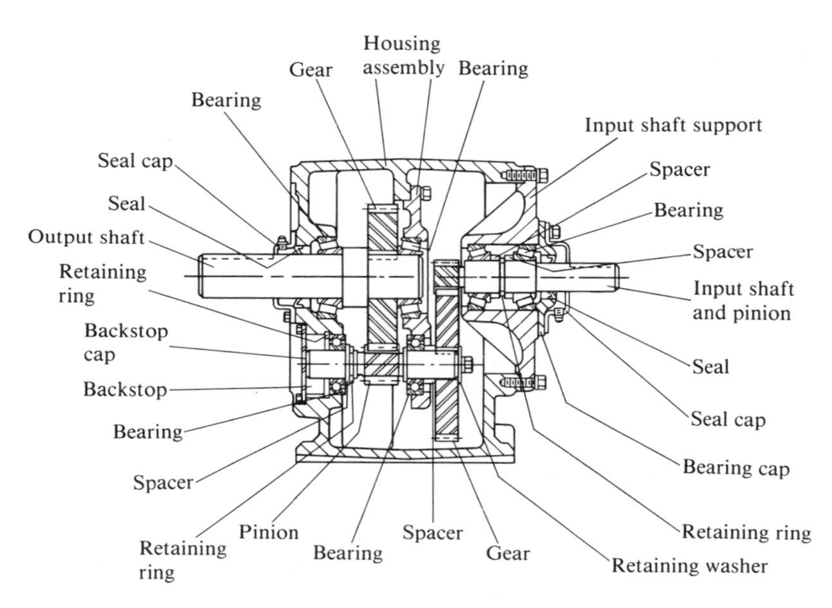

(*c*) Parts index

Figure 11-35 Concentric Helical Gear Reducer (Winsmith Division, UMC Industries, Springville, N.Y.)

Figure 11-33 shows a double-reduction spur gear speed reducer with an electric motor rigidly attached. Such a unit is often called a *gearmotor*. Figure 11-34 is similar, except that one of the stages of reduction uses helical gears (discussed in the next chapter). The cross-section drawing shown with Figure 11-35 gives a clear picture of the several components of a reducer.

The planetary reducer in Figure 11-36 has quite a different design to accommodate the placement of the sun, planet, and ring gears. Figure 11-37 shows the eight-speed transmission from a large farm tractor and illustrates the high degree of complexity that may be involved in the design of transmissions.

(*a*)

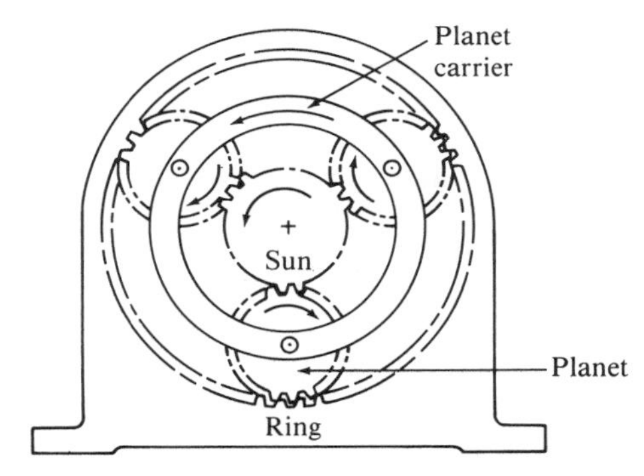

(*b*) Schematic arrangement of planetary gearing

Figure 11-36 Planetary Gear Reducer (Rexnord, Milwaukee, Wis.)

Figure 11-37 Eight-Speed Tractor Transmission (Ford Tractor Operations, Troy, Mich.)

REFERENCES

1. American Gear Manufacturers Association. *AGMA Standard for Rating the Pitting Resistance and Bending Strength of Spur and Helical Involute Gear Teeth*. Standard AGMA 218.01. Washington, D.C.: December 1982.
2. American Gear Manufacturers Association. *Nomenclature of Gear Tooth Failure Modes*. Standard 110.04-1980. Arlington, Va.: 1980.
3. American National Standards Institute. *Tooth Proportions for Coarse-Pitch Involute Spur Gears*. Standard ANSI B6.1. New York: 1974.
4. American National Standards Institute. *Tooth Proportions for Fine Pitch Involute Spur and Helical Gears*. Standard ANSI B6.7-1977. New York: 1977.
5. Colbourne, J. R. *The Design of Undercut Involute Gears*. ASME Technical Paper No. 80-C2/DET-67, 1980.
6. Dudley, Darle W. *Practical Gear Design*. New York: McGraw-Hill Book Company, 1954.
7. Hebenstreit, Herbert. "Guidelines for Quality Carburized Gearing." *Machine Design*, 54, no. 20, Sept. 9, 1982, 67.
8. Hirt, Manfred. *German and American Quality System of Spur, Helical and Bevel Gears; Influence of Gear Quality on Costs and Load Capacity*. ASME Technical Paper No. 80-C2/DET-18, 1980.
9. Imwalle, D. E., Labath, O.A., and Hutchinson, R. N., *A Review of Recent Gear Rating Developments: ISO/AGMA Comparison Study*. ASME Technical Paper No. 80-C2/DET-25, 1980.
10. Kern, R. F., and Suess, M. E. *Steel Selection*. New York: John Wiley & Sons, 1979.
11. Lipp, Robert. "Avoiding Tooth Interference in Gears." *Machine Design*. 54, no. 1, Jan. 7, 1982.
12. Martin, K. F. *The Efficiency of Involute Spur Gears*. ASME Technical Paper No. 80-C2/DET-16, 1980.
13. Michalec, G. W. *Precision Gearing: Theory and Practice*. New York: John Wiley & Sons, 1966.
14. Oberg, Erik, et al. *Machinery's Handbook*, 22d ed. New York: Industrial Press, 1984.
15. Shigley, Joseph E. *Mechanical Engineering Design*, 4th ed. New York: McGraw-Hill Book Company, 1983.
16. Tucker, A. I. "Gear Design: Dynamic Loads." *Mechanical Engineering*. 93, no. 10, October 1971.
17. Tucker, A. I. *The Gear Design Process*. ASME Technical Paper No. 80-C2/DET-13, 1980.

PROBLEMS

1. A gear has 44 teeth of the 20° full-depth involute form and a diametral pitch of 12. Compute the following:
   a. Pitch diameter
   b. Circular pitch
   c. Equivalent module
   d. Nearest standard module
   e. Addendum
   f. Dedendum
   g. Clearance
   h. Whole depth
   i. Working depth
   j. Tooth thickness
   k. Outside diameter

Problems 2–9: Repeat problem 1 for the following gears:

2. $N = 34;\ P_d = 24$
3. $N = 45;\ P_d = 2$
4. $N = 18;\ P_d = 8$
5. $N = 22;\ P_d = 1.75$
6. $N = 20;\ P_d = 64$
7. $N = 180;\ P_d = 80$
8. $N = 28;\ P_d = 18$
9. $N = 28;\ P_d = 20$

Problems 10–17: Repeat problem 1 for the following gears in the metric module system. Replace part c with equivalent P_d and part d with nearest standard P_d.

10. $N = 34;\ m = 3$
11. $N = 45;\ m = 1.25$
12. $N = 18;\ m = 12$
13. $N = 22;\ m = 20$
14. $N = 20;\ m = 1$
15. $N = 180;\ m = 0.4$
16. $N = 28;\ m = 1.5$
17. $N = 28;\ m = 0.8$

18. Define *backlash* and discuss the methods used to produce it.
19. For the gears of problems 6 and 9, recommend the amount of backlash to provide for class C precision.
20. An 8-pitch pinion with 18 teeth mates with a gear having 64 teeth. The pinion rotates at 2 450 rpm. Compute the following:
    a. Center distance
    b. Velocity ratio
    c. Speed of gear
    d. Pitch line speed

Problems 21–28: Repeat problem 20 for the following data:

21. $P_d = 4;\ N_P = 20;\ N_G = 92;\ n_P = 225$ rpm
22. $P_d = 20;\ N_P = 30;\ N_G = 68;\ n_P = 850$ rpm
23. $P_d = 64;\ N_P = 40;\ N_G = 250;\ n_P = 3\,450$ rpm
24. $P_d = 12;\ N_P = 24;\ N_G = 88;\ n_P = 1\,750$ rpm
25. $m = 2;\ N_P = 22;\ N_G = 68;\ n_P = 1\,750$ rpm
26. $m = 0.8;\ N_P = 18;\ N_G = 48;\ n_P = 1\,150$ rpm
27. $m = 4;\ N_P = 36;\ N_G = 45;\ n_P = 150$ rpm
28. $m = 12;\ N_P = 15;\ N_G = 36;\ n_P = 480$ rpm

Problems 29–32. Tell what is wrong with the following statements:

29. An 8-pitch pinion having 24 teeth mates with a 10-pitch gear having 88 teeth. The pinion rotates at 1 750 rpm and the gear at approximately 477 rpm. The center distance is 5.900 in.

30. A 6-pitch pinion having 18 teeth mates with a 6-pitch gear having 82 teeth. The pinion rotates at 1 750 rpm and the gear at approximately 384 rpm. The center distance is 8.3 in.
31. A 20-pitch pinion having 12 teeth mates with a 20-pitch gear having 62 teeth. The pinion rotates at 825 rpm and the gear at approximately 160 rpm. The center distance is 1.850 in.
32. A 16-pitch pinion having 24 teeth mates with a 16-pitch gear having 45 teeth. The outside diameter of the pinion is 1.625 in. The outside diameter of the gear is 2.938 in. The center distance is 2.281 in.
33. The gear pair described in problem 20 is to be installed in a rectangular housing. Specify the dimensions X and Y as sketched in Figure 11-38 that would provide a minimum clearance of 0.10 in.
34. Repeat problem 33 for the data of problem 23.
35. Repeat problem 33 for the data of problem 26 but make the clearance 2.0 mm.
36. Repeat problem 33 for the data of problem 27 but make the clearance 2.0 mm.

Problems 37–40: For the gear trains sketched in the given figures, compute the output speed and direction of rotation of the output shaft if the input shaft rotates at 1 750 rpm clockwise.

37. Use Figure 11-39.
38. Use Figure 11-40.
39. Use Figure 11-41.
40. Use Figure 11-42.

Problems 41 through 47 describe design situations. For each, design a pair of spur gears, specifying (at least) the diametral pitch, the number of teeth in each gear, the pitch diameters of each gear, the center distance, the face width, and the material from which the gears are to be made. Design for indefinite life with regard to both strength and wear. Work toward designs that are compact. Use standard values of diametral pitch and avoid designs for which interference could occur. Assume the input to the gear pair is from an electric motor unless otherwise stated.

If the data are given in SI units, complete the design in the metric module system with dimensions in millimeters, forces in newtons, and stresses in megapascals.

41. A pair of spur gears is to be designed to transmit 5.0 hp while the pinion rotates at 1 200 rpm. The gear must rotate between 385 and 390 rpm. The gear drives a reciprocating compressor.

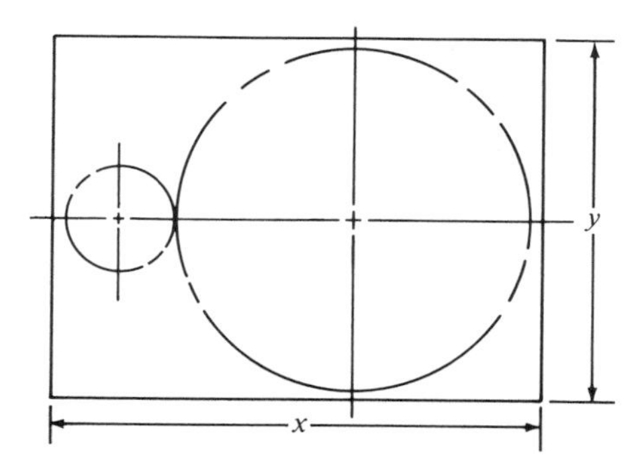

Figure 11-38 (Problem 33)

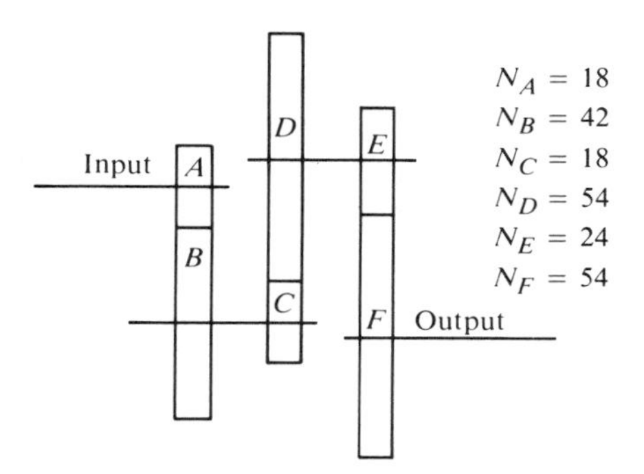

Figure 11-39 (Problem 37)

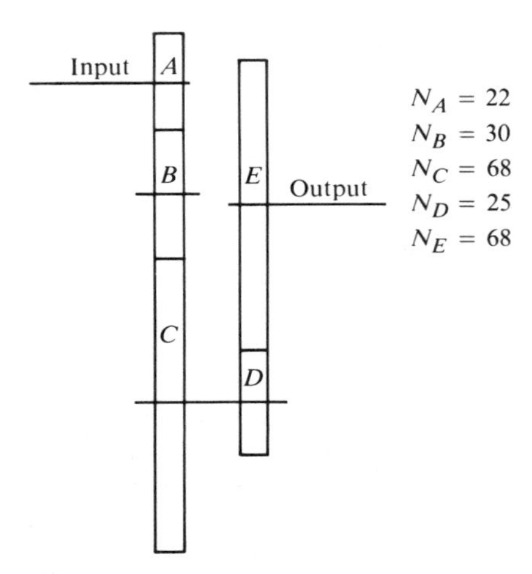

Figure 11-40 (Problem 38)

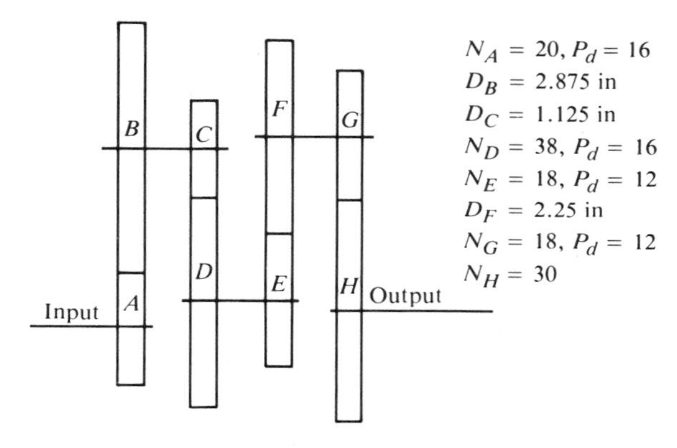

Figure 11-41 (Problem 39)

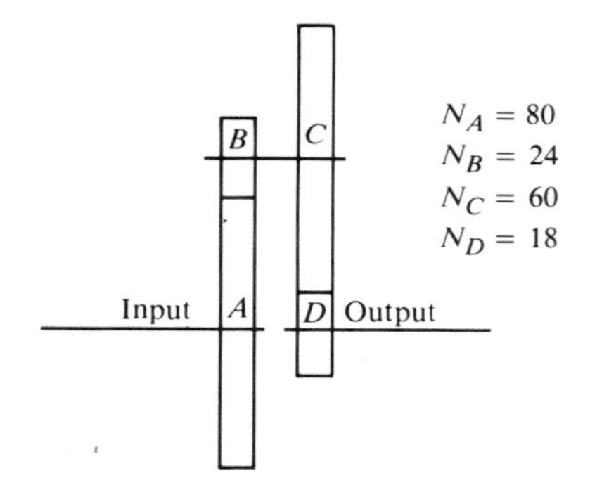

Figure 11-42 (Problem 40)

42. A gear pair is to be a part of the drive for a milling machine requiring 20.0 hp with the pinion speed at 550 rpm and the gear speed to be between 180 and 190 rpm.
43. A drive for a punch press requires 50.0 hp with the pinion speed of 900 rpm and the gear speed of 225 to 230 rpm.
44. A single-cylinder gasoline engine has the pinion of a gear pair on its output shaft. The gear is attached to the shaft of a small cement mixer. The mixer requires 2.5 hp while rotating at approximately 75 rpm. The engine is governed to run at approximately 900 rpm.
45. A four-cylinder industrial engine runs at 2 200 rpm and delivers 75 hp to the input gear of a drive for a large wood chipper used to prepare pulpwood chips for paper making. The output gear must run between 4 500 and 4 600 rpm.
46. A small commercial tractor is being designed for chores such as lawn mowing and snow removal. The wheel drive system is to be through a gear pair in which the pinion runs at 600 rpm while the gear, mounted on the hub of the wheel, runs at 170 to 180 rpm. The wheel is 300 mm in diameter. The gasoline engine delivers 3.0 kW of power to the gear pair.
47. A water turbine transmits 75 kW of power to a pair of gears at 4 500 rpm. The output of the gear pair must drive an electric power generator at 3 600 rpm. The center distance for the gear pair must not exceed 150 mm.

48. Determine the power-transmitting capacity for a pair of spur gears having 20° full-depth teeth, a diametral pitch of 10, a face width of 1.25 in, 25 teeth in the pinion, 60 teeth in the gear, and an AGMA quality class of 8. The pinion is made from AISI 4140 OQT 1000, and the gear is made from AISI 4140 OQT 1100. The pinion will rotate at 1 725 rpm on the shaft of an electric motor. The gear will drive a centrifugal pump.
49. Determine the power-transmitting capacity for a pair of spur gears having 20° full-depth teeth, a diametral pitch of 6, 35 teeth in the pinion, 100 teeth in the gear, a face width of 2.00 in, and an AGMA quality class of 6. A gasoline engine drives the pinion at 1 500 rpm. The gear drives a conveyor for crushed rock in a quarry. The pinion is made from AISI 1040 OQT 800. The gear is made from gray cast iron, ASTM A48-76, class 30.
50. It was found that the gear pair described in problem 49 wore out when driven by a 25-hp engine. Propose a redesign that would be expected to give indefinite life under the conditions described.
51. Design a double-reduction gear train that will transmit 10.0 hp from an electric motor running at 1 750 rpm to an assembly conveyor whose drive shaft must rotate between 146 and 150 rpm. Note that this will require the design of two pairs of gears. Sketch the arrangement of the train and compute the actual output speed.
52. A commercial food waste grinder in which the final shaft rotates at between 40 and 44 rpm is to be designed. The input is from an electric motor running at 850 rpm and delivering 0.50 hp. Design a double-reduction spur gear train for the grinder.
53. A small powered hand drill is driven by an electric motor running at 3 000 rpm. The drill speed is to be approximately 550 rpm. Design the speed reduction for the drill. The power transmitted is 0.25 hp.
54. The output from the drill described in problem 53 provides the drive for a small bench-scale band saw similar to the one in Figure 1-1 in Chapter 1. The saw blade is to move with a linear velocity of 375 ft/min. Design a spur gear reduction to drive the band saw. Consider using plastic gears.

12 Helical Gears, Bevel Gears, and Wormgearing

12-1 OVERVIEW

Following the presentation of the design of spur gears in Chapter 11, this chapter extends the technology of gears to include helical gears, bevel gears, and wormgearing. Each type of gear has its special geometry and operation. But many of the concepts used in Chapter 11 carry over to this chapter. Each type of gearing must be designed to have adequate gear tooth strength and good surface durability.

The design for strength and wear is presented for each type of gear, with an analysis of the geometry and the forces exerted on the gears.

12-2 HELICAL GEAR GEOMETRY

Helical and spur gears are distinguished by the orientation of the teeth. On spur gears, the teeth are straight and aligned with the axis of the gear. On helical gears, the teeth are inclined at an angle with the axis, that angle being called the *helix angle*. If the gear were very wide, it would appear that the teeth wind around the gear blank in a continuous helical path. However, practical considerations limit the width of the gears so that the teeth normally appear to be merely inclined with respect to the axis. Figure 12-1 shows some examples of commercially available helical gears. Unless otherwise stated, we will assume that the gears are mounted on parallel shafts. With special geometry, it is possible to have nonparallel shafts. When helical gears operate on shafts at an angle of 90° to each other, they are called *crossed helical gears*.

The forms of helical gear teeth, and therefore the methods of analyzing them for strength and wear, are very similar to those discussed for spur gears in Chapter 11. The basic task is to account for the effect of the helix angle.

Figure 12-2 shows the pertinent geometry of the helical gear teeth. The *helix angle*, called ψ (the Greek letter *psi*), is the angle between the plane through the axis of the gear

(*a*) Helical gears with parallel shafts

(*b*) Crossed helical gears, shafts at right angle

Figure 12-1 Helical Gears. These gears have a 45° helix angle. (Browning Mfg. Division, Emerson Electric Co., Maysville, Ky.)

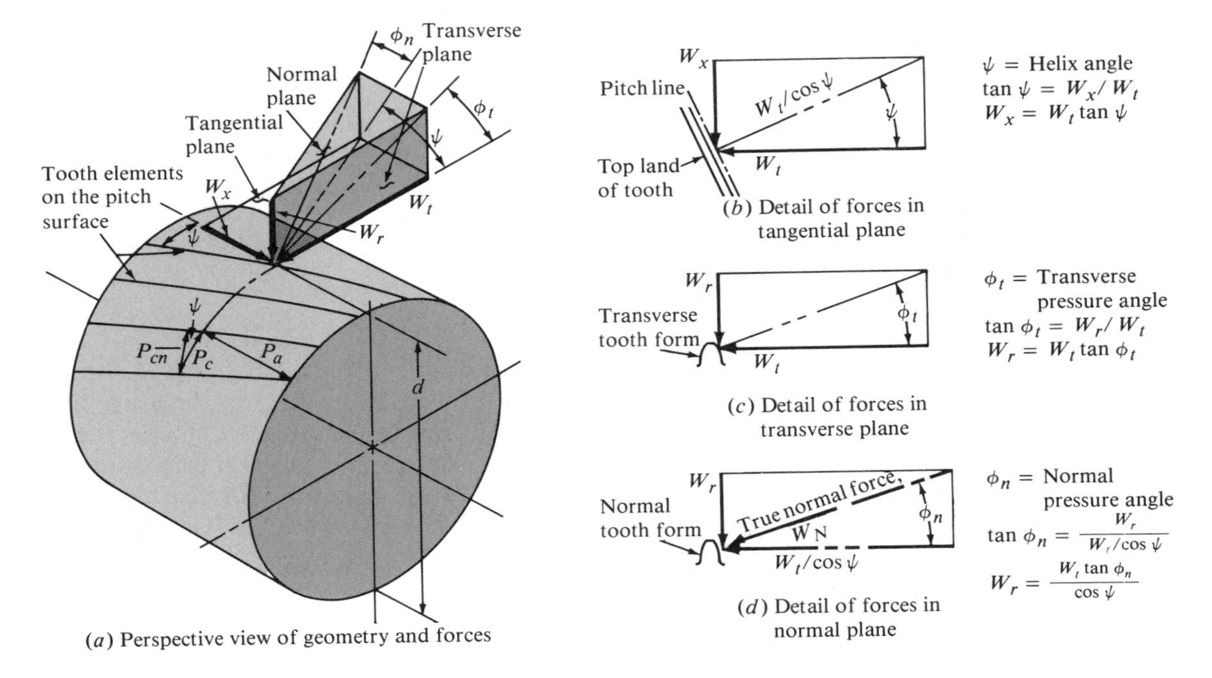

Figure 12-2 Helical Gear Geometry and Forces

and the tangent to the helix that follows the pitch surface of the teeth. When two helical gears operate together, one must have a right-hand helix and the other a left-hand helix. The appearance of the right-hand helix is similar to that of the standard right-hand screw thread.

The main advantage of helical gears over spur gears is that a given tooth assumes the load gradually instead of suddenly. Contact starts at one end of a tooth near the tip and progresses across the face in a path downward across the pitch line to the lower flank of the tooth, where it leaves engagement. Because more than one tooth is in contact at a given time a lower average load per tooth and smoother operation result.

The main disadvantage of helical gears is that there is a *thrust,* or axial load, produced because of the inclined orientation of the teeth. The force system on a helical gear tooth is best visualized in terms of its three rectangular components, as defined next and illustrated in Figure 12-2. Because the transmitted load, W_t, is derived directly from the known parameters of the gear system, it is the most convenient to start with in computing the forces on the teeth.

Transmitted Load, W_t

The force that acts tangential to the pitch surface of the gear is called the *transmitted load* and is the force that actually transmits torque and power from the driver to the driven gear. It acts in a direction perpendicular to the axis of the shaft carrying the gear. This is the same definition used for W_t in the chapter on spur gears, and the methods of computing it are the same. If the torque transmitted is known,

$$W_t = T/(d/2) \tag{12-1}$$

where d is the pitch diameter of the gear. If the power transmitted (in hp) and the rotational speed (in rpm) are known, the torque in inch·pounds can be computed from

$$T = 63\,000(P)/n \tag{12-2}$$

If the power transmitted (in hp) and the pitch line speed (in ft/min) are known,

$$W_t = 33\,000(P)/v_t \tag{12-3}$$

Axial Load, W_x

The *axial load* is the force directed parallel to the axis of the shaft carrying the gear. Also called a *thrust load,* this is the generally undesirable force that must be resisted by shaft bearings having a thrust capacity. This topic will be discussed in Chapter 15.

Referring to Figure 12-2, consider a plane tangent to the pitch circle of the gear, which thus contains the transmitted load, W_t. This tangential plane would also contain the axial load acting perpendicular to the transmitted load. The helix angle ψ establishes the relationship between W_t and W_x:

$$W_x = W_t \tan \psi \tag{12-4}$$

Figure 12-2(b) shows a detail of the tangential plane. Note that the axial load increases as the value of the helix angle increases. Helix angles typically range from 15° to 45°.

Radial Load, W_r

The *radial load* is the force that acts toward the center of the gear, that is, radially. The direction of the force is always such that it tends to push the two gears apart. To determine the value of the radial load, it is first necessary to discuss the nature of the pressure angle for helical gear teeth. Refer again to Figure 12-2(a) and (b).

Two different pressure angles must be considered to understand the geometry of the teeth on helical gears. The plane passed through the gear teeth in a direction perpendicular (transverse) to the axis of the gear is called the *transverse plane*. The pressure angle in this plane is referred to as ϕ_t.

A plane passed through the gears in a direction normal to the teeth themselves is called the *normal plane*. The pressure angle of the tooth form thus obtained is called ϕ_n.

Some helical gears are made with the transverse pressure angle set at a given angle, say 20° or 14½°. Others are made with the normal pressure angle set. Equation (12-5) relates these two pressure angles for a given helix angle.

$$\tan \phi_n = \tan \phi_t \cos \psi \tag{12-5}$$

This equation is developed after further discussion.

Now we can see in Figure 12-2 that the angle between the transmitted load, W_t, and the normal to the gear tooth in the transverse plane is the transverse pressure angle, ϕ_t. It then follows that the radial load can be computed from

$$W_r = W_t \tan \phi_t \tag{12-6}$$

Also shown in Figure 12-2 is the normal plane. The force acting perpendicular to the tooth form is the true normal force, W_N. Note that it is the resultant of the three components, W_t, W_x, and W_r. From the detail of the normal plane in Figure 12-2(b), you can see that

$$W_N = \frac{W_t}{\cos \phi_n \cos \psi} \tag{12-7}$$

The details of both the transverse plane and the normal plane contain formulas for the radial load, W_r. By equating them, you can derive equation (12-5).

Pitches for Helical Gears

To get a clear picture of the geometry of helical gears, it is necessary to understand five different pitches.

Circular Pitch, P_c

Circular pitch is the distance from a point on one tooth to the corresponding point on the next adjacent tooth, measured at the pitch line in the transverse plane. This is the same definition used for spur gears. Then

$$P_c = \pi d / N \tag{12-8}$$

Normal Circular Pitch, P_{cn}

Normal circular pitch is the distance between corresponding points on adjacent teeth measured on the pitch surface in the normal direction. P_c and P_{cn} are related by the following equation:

$$P_{cn} = P_c \cos \psi \tag{12-9}$$

Diametral Pitch, P_d

Diametral pitch is the ratio of the number of teeth in the gear to the pitch diameter. This is the same definition as the one for spur gears and applies in considerations of the form of the teeth in the diametral or transverse plane. Thus it is sometimes called the *transverse diametral pitch*.

$$P_d = N/d \tag{12-10}$$

Normal Diametral Pitch, P_{dn}

Normal diametral pitch is the equivalent diametral pitch in the plane normal to the teeth.

$$P_{dn} = P_d/\cos \psi \tag{12-11}$$

It is helpful to remember these relationships:

$$P_d P_c = \pi \tag{12-12}$$

$$P_{dn} P_{cn} = \pi \tag{12-13}$$

Axial Pitch, P_x

Axial pitch is the distance between corresponding points on adjacent teeth, measured on the pitch surface in the axial direction.

$$P_x = P_c/\tan \psi = \pi/(P_d \tan \psi) \quad (12\text{-}14)$$

It is necessary to have at least two axial pitches in the face width to have the benefit of full helical action and its smooth transfer of the load from tooth to tooth.

The use of equations (12-1) through (12-14) is now illustrated in two example problems.

Example Problem 12-1. A helical gear has a transverse diametral pitch of 12, a transverse pressure angle of 14½°, 28 teeth, a face width of 1.25 in, and a helix angle of 30°. Compute the circular pitch, normal circular pitch, normal diametral pitch, axial pitch, pitch diameter, and the normal pressure angle. Compute the number of axial pitches in the face width.

Solution.

1. Circular pitch [equation (12-8)]:

$$P_c = \pi/P_d = \pi/12 = 0.262 \text{ in}$$

2. Normal circular pitch [equation (12-9)]:

$$P_{cn} = P_c \cos \psi = 0.262 \cos(30) = 0.227 \text{ in}$$

3. Normal diametral pitch [equation (12-11)]:

$$P_{dn} = P_d/\cos \psi = 12/\cos(30) = 13.856$$

4. Axial pitch [equation (12-14)]:

$$P_x = P_c/\tan \psi = 0.262/\tan(30) = 0.453 \text{ in}$$

5. Pitch diameter [equation (12-9)]:

$$d = N/P_d = 28/12 = 2.333 \text{ in}$$

6. Normal pressure angle [equation (12-5)]:

$$\phi_n = \tan^{-1}(\tan \phi_t \cos \psi)$$

$$\phi_n = \tan^{-1}[\tan(14\tfrac{1}{2}) \cos(30)] = 12.62°$$

7. Number of axial pitches in the face width:

$$F/P_x = 1.25/0.453 = 2.76 \text{ pitches}$$

Since this is greater than 2.0, there will be full helical action.

Example Problem 12-2. A helical gear has a normal diametral pitch of 8, a normal pressure angle of 20°, 32 teeth, a face width of 3.00 in, and a helix angle of 15°. Compute the diametral pitch, circular pitch, normal circular pitch, axial pitch, number of axial pitches in the face width, transverse pressure angle, and pitch diameter. If the gear is rotating at 650 rpm while transmitting 7.50 hp, compute the pitch line speed, the transmitted load, the axial load, the radial load, and the true normal force.

Solution.

1. Diametral pitch [equation (12-11)]:

$$P_d = P_{dn} \cos \psi = 8 \cos(15) = 7.727$$

2. Circular pitch [equation (12-12)]:

$$P_c = \pi/P_d = \pi/7.727 = 0.407 \text{ in}$$

3. Normal circular pitch [equation (12-9)]:

$$P_{cn} = P_c \cos \psi = 0.407 \cos(15) = 0.393 \text{ in}$$

4. Axial pitch [equation (12-14)]:

$$P_x = \pi/(P_d \tan \psi) = \pi/[7.727 \tan(15)] = 1.517 \text{ in}$$

5. Number of axial pitches in the face width:

$$F/P_x = 3.00/1.517 = 1.98 \text{ pitches}$$

6. Transverse pressure angle [equation (12-5)]:

$$\phi_t = \tan^{-1}(\tan \phi_n/\cos \psi)$$
$$\phi_t = \tan^{-1}[\tan(20)/\cos(15)] = 20.65°$$

7. Pitch diameter [equation (12-10)]:

$$d = N/P_d = 32/7.727 = 4.141 \text{ in}$$

8. Pitch line speed, v_t:

$$v_t = \pi dn/12 = \pi(4.141)(650)/12 = 704.7 \text{ ft/min}$$

9. Transmitted load, W_t [equation (12-3)]:

$$W_t = 33\,000(P)/v_t = 33\,000(7.5)/704.7 = 351 \text{ lb}$$

10. Axial load, W_x [equation (12-4)]:

$$W_x = W_t \tan \psi = 351 \tan(15) = 94 \text{ lb}$$

11. Radial load, W_r [equation (12-6)]:

$$W_r = W_t \tan \phi_t = 351 \tan(20.65) = 132 \text{ lb}$$

12. True normal load, W_N [equation (12-7)]:

$$W_N = \frac{W_t}{\cos \phi_n \cos \psi} = \frac{351}{\cos(20) \cos(15)} = 387 \text{ lb}$$

12-3 STRESSES IN HELICAL GEAR TEETH

We will use the same basic equation for computing stresses in helical gear teeth as we did for spur gear teeth in Chapter 11, given in equation (11-26) and repeated here.

$$\sigma_t = \frac{K_a W_t}{K_v} \frac{P_d}{F} \frac{K_m}{J} \qquad (11\text{-}26)$$

Refer to the following tables and figures for the factors in this equation.

K_a: Application factor (Table 11-13)

K_v: Dynamic factor (Figure 11-27); use the curve for the appropriate quality number.

K_m: Load distribution factor (Table 11-12)

Figures 12-3, 12-4, and 12-5 show the values for the geometry factor, J, for helical gear teeth with 15°, 20°, and 22° normal pressure angles, respectively.

The stress computed from equation (11-26) should be compared with the allowable fatigue stress for steel, as shown in Figure 11-28. The conditions discussed in Chapter 11, section 11-7, should be followed.

12-4 WEAR RESISTANCE OF HELICAL GEAR TEETH

As with spur gears, the wear resistance of helical gear teeth depends on the contact stress. The AGMA standard for helical gears includes formulas and data for a detailed analysis of the contact stress. The data, especially the determination of the geometry factor I, are too complex to repeat here.

Therefore, a different method is shown here to allow the estimation of the wear load capacity of helical gears. A wear load equation can be written

$$F_w = dFQK/\cos^2 \psi \qquad (12\text{-}15)$$

where F_w is called the *limiting load for wear.* We can define the dynamic load, W_d, to be

$$W_d = W_t/C_v$$

Then, if F_w is greater than W_d, the teeth should not wear out. The factor Q is related to the size ratio between the pinion and the gear, but it is most convenient to compute it from

$$Q = 2N_G/(N_P + N_G) \qquad (12\text{-}16)$$

K, the *wear load factor,* depends on the form of the gear teeth and the hardness of the surface of the teeth in contact. Remember, wear is a fatigue failure of the surface of the teeth. Table 12-1 gives some values of K for steel gears.

Figure 12-6 is a graph of K versus surface hardness. The hardness number refers to the nominal value for the pair. But, as discussed in Chapter 11, the pinion should be 40 to 50 points higher on the Brinell scale than the gear. Thus, for the pinion, you may add about 20 points to the nominal value and subtract about 20 points for the gear.

In design we set $F_w = W_d = W_t/C_v$ and solve equation (12-15) for the required value of K. Then, by referring to Table 12-1, a satisfactory surface hardness for the teeth can be found. From this a material can be specified as it was for spur gears.

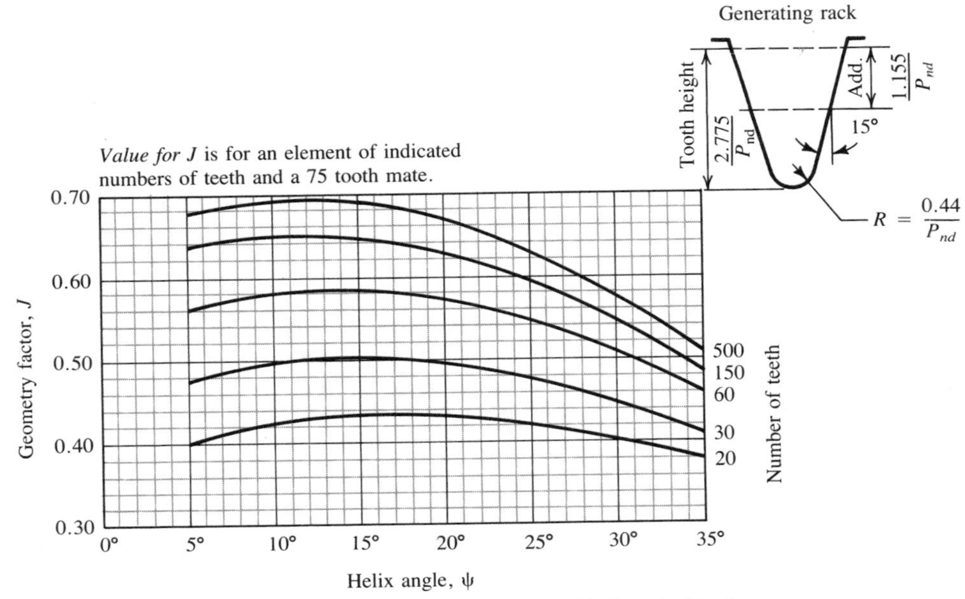

(a) Geometry factor (J), for 15° normal pressure angle and indicated addendum

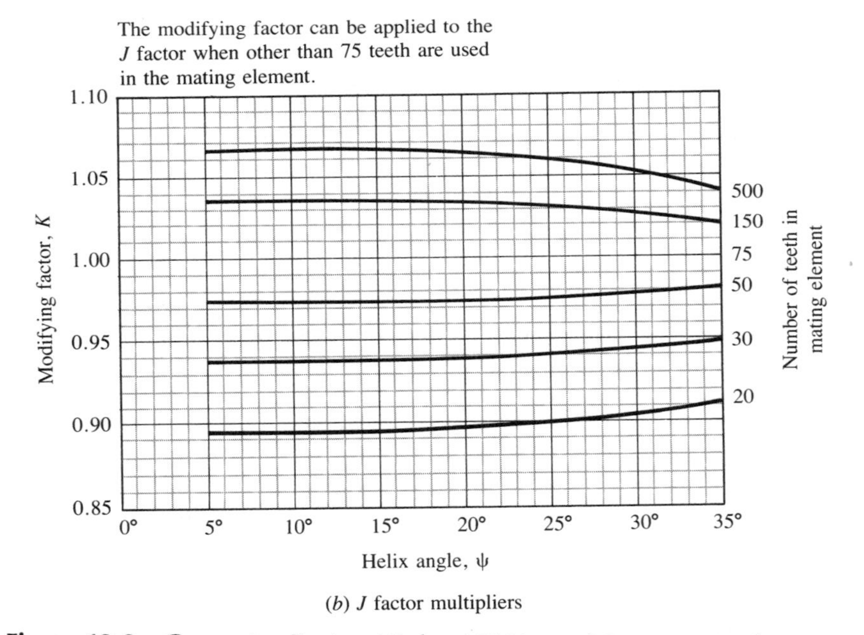

(b) J factor multipliers

Figure 12-3 Geometry Factor (J) for 15° Normal Pressure Angle (Extracted from AGMA with the permission of the publisher, the American Gear Manufacturers Association, 1901 North Fort Myer Drive, Suite 1000, Arlington, Virginia 22209)

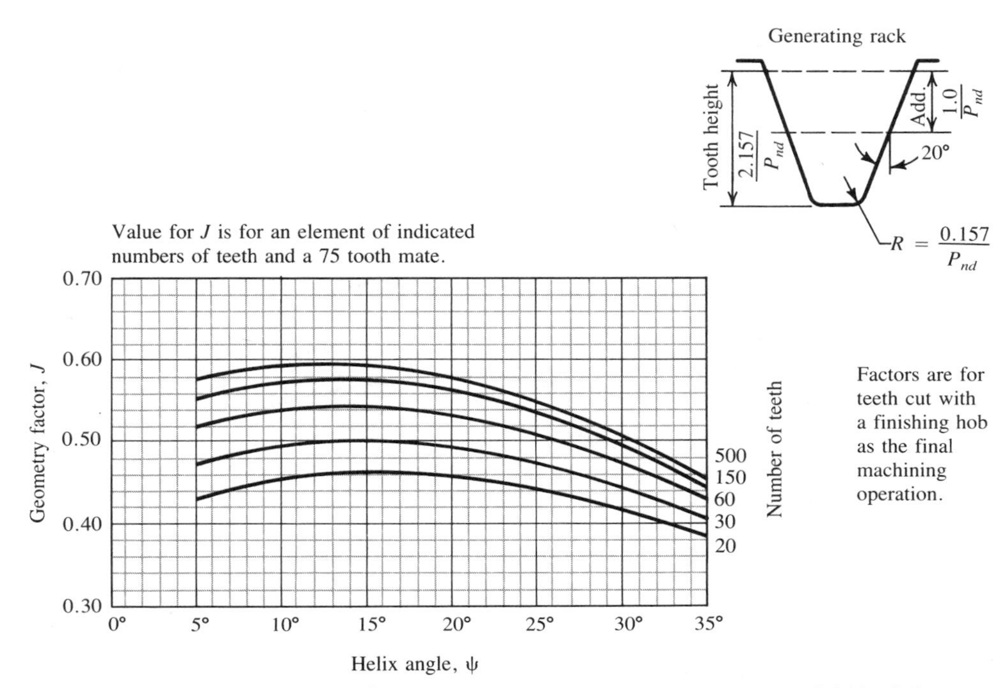

(*a*) Geometry factor (*J*), for 20° normal pressure angle, standard addendum, and finishing hob

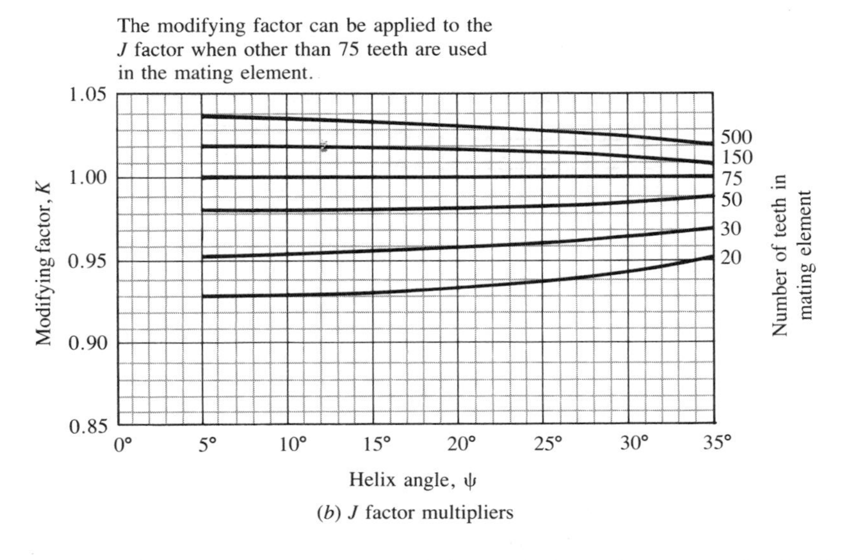

(*b*) *J* factor multipliers

Figure 12-4 Geometry Factor (*J*) for 20° Normal Pressure Angle (Extracted from AGMA with the permission of the publisher, the American Gear Manufacturers Association, 1901 North Fort Myer Drive, Suite 1000, Arlington, Virginia 22209)

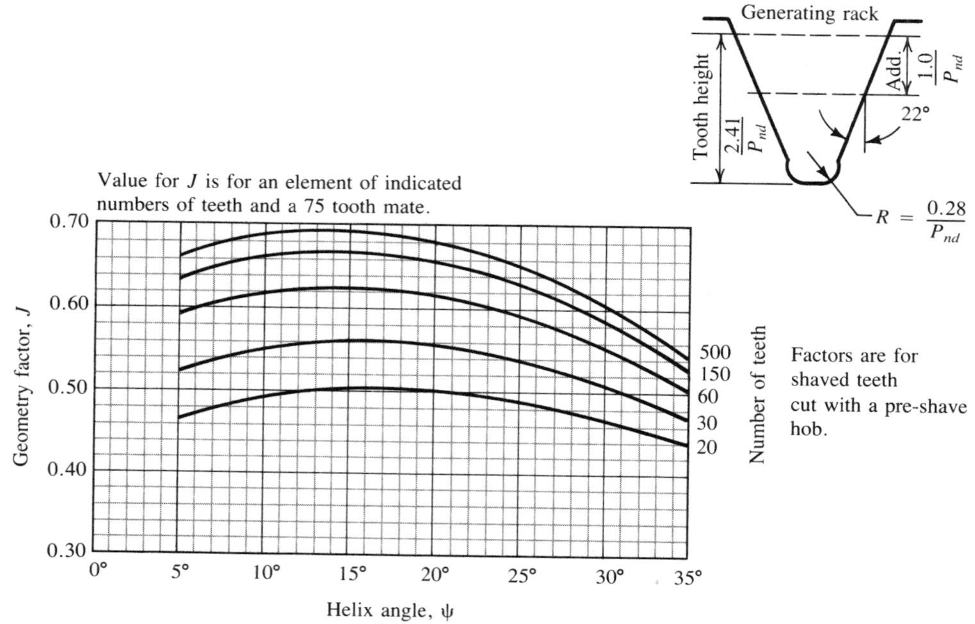

(a) Geometry factor (J), for 22° normal pressure angle, standard addendum, and pre-shave hob

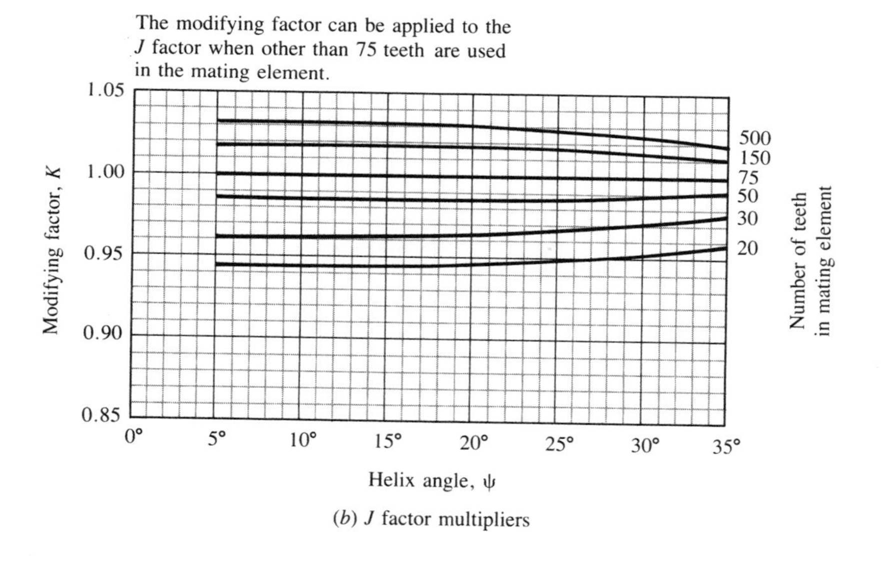

(b) J factor multipliers

Figure 12-5 Geometry Factor (J) for 22° Normal Pressure Angle (Extracted from AGMA with the permission of the publisher, the American Gear Manufacturers Association, 1901 North Fort Myer Drive, Suite 1000, Arlington, Virginia 22209)

Table 12-1 Wear Load Factors K for Steel Gear Teeth

NOMINAL SURFACE HARDNESS, HB	PRESSURE ANGLE, ϕ_t 14½°	20°	25°
	Wear Load Factor, K		
150	30	41	51
175	43	58	72
200	58	79	98
225	76	103	127
250	96	131	162
275	119	162	200
300	144	196	242
325	171	233	288
350	196	270	333
375	233	318	384
400	268	366	453
450	344	470	560
500 (Induction-hardened)	405	555	675
600 (Carburized)	550	750	910

The required K is

$$K = \frac{W_t \cos^2 \psi}{C_v dFQ} \tag{12-17}$$

12-5 DESIGN OF HELICAL GEARS

Figures 12-7 and 12-8 show commercially available reducers employing helical gears. The example problem that follows illustrates the procedure to design such gears.

Example Problem 12-3. A pair of helical gears for a milling machine drive is to transmit 65 hp with a pinion speed of 3 450 rpm and a gear speed of 1 100 rpm. The power is from an electric motor. Design the gears.

Solution. Of course, there are several possible solutions. Here is one. Let's try a normal diametral pitch of 12, 24 teeth in the pinion, a helix angle of 15°, a normal pressure angle of 20°, and a quality number of 8.

Now compute the transverse diametral pitch, the axial pitch, the transverse pressure angle, and the pitch diameter. Then we will choose a face width that will give at least two axial pitches to ensure true helical action.

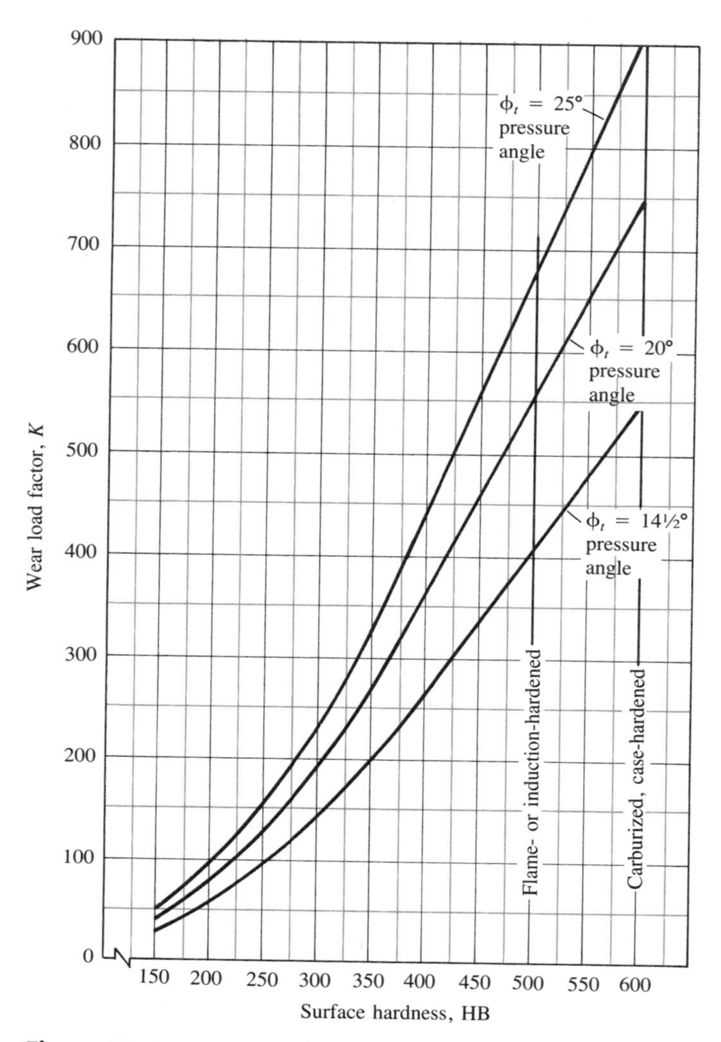

Figure 12-6 Wear Load Factor K vs. Hardness

Figure 12-7 Parallel Shaft Reducer (Browning Mfg. Division, Emerson Electric Co., Maysville, Ky.)

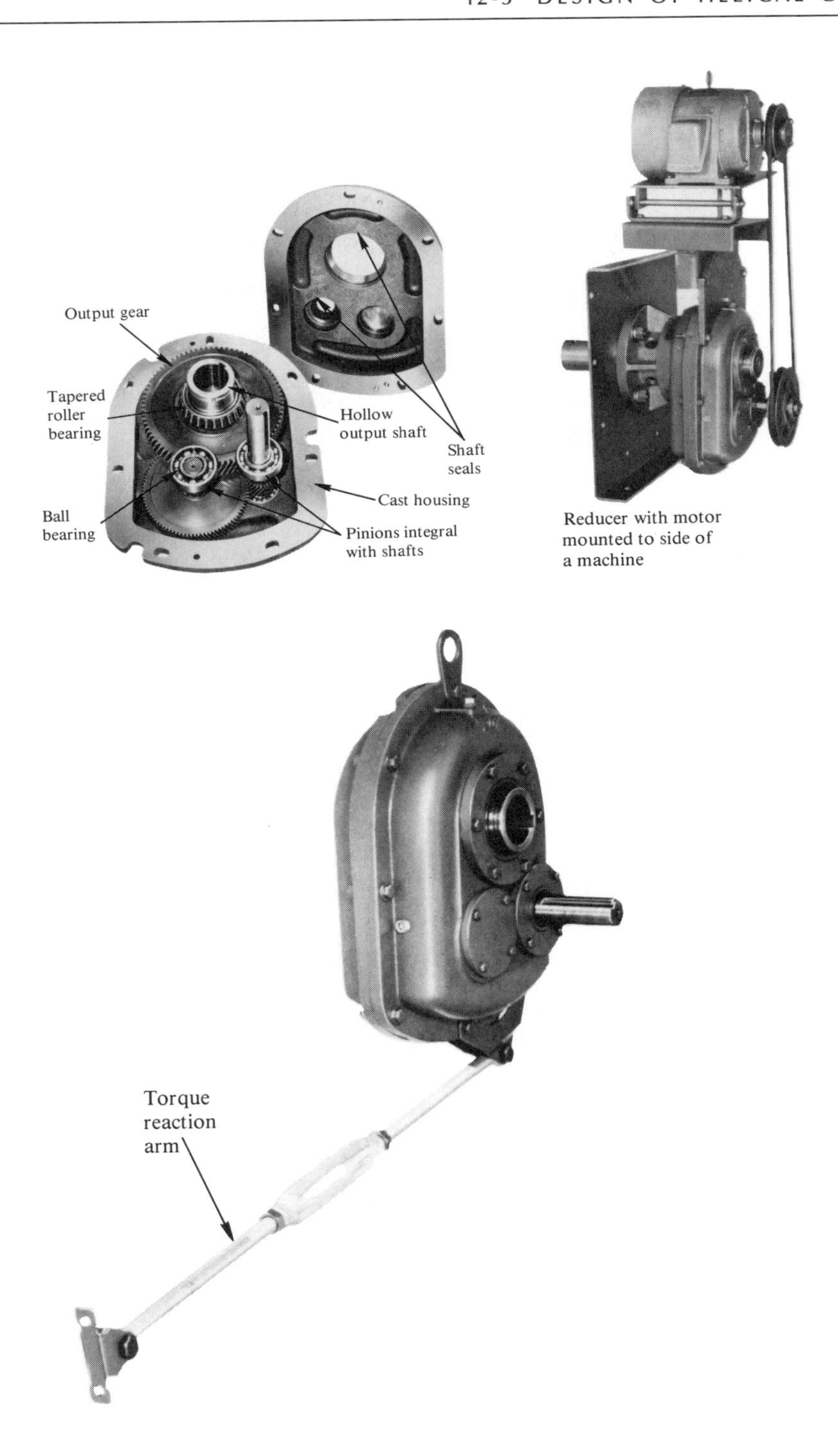

Figure 12-8 Helical Shaft Mount Reducer (Browning Mfg. Division, Emerson Electric Co., Maysville, Ky.)

$$P_d = P_{dn} \cos \psi = 12 \cos(15°) = 11.59$$

$$P_x = \frac{\pi}{P_d \tan \psi} = \frac{\pi}{11.59 \tan(15°)} = 1.012 \text{ in}$$

$$\phi_t = \tan^{-1}(\tan \phi_n/\cos \psi) = \tan^{-1}[\tan(20°)/\cos(15°)] = 20.65°$$

$$d = N_P/P_d = 24/11.59 = 2.071 \text{ in}$$

$$F = 2P_x = 2(1.012) = 2.024 \text{ in} \quad \text{Nominal face width}$$

Let's use 2.25 in, a more convenient value. Pitch line speed and transmitted load:

$$v_t = \pi dn/12 = \pi(2.071)(3\,450)/12 = 1\,871 \text{ ft/min}$$

$$W_t = 33\,000 \text{ (hp)}/v_t = 33\,000(65)/1\,871 = 1\,146 \text{ lb}$$

Factors affecting bending stress:

$J = 0.48$ (from Figure 12-4 for 24 teeth)

$K_v = 0.73$ (from Figure 11-27, curve 8)

$K_a = 1.25$ (from Table 11-13, moderate shock, uniform power)

$K_m = 1.6$ (from Table 11-12, conservative)

We can now compute the bending stress in the teeth.

$$\sigma_t = \frac{K_a W_t}{K_v} \frac{P_d}{F} \frac{K_m}{J} \qquad (11\text{-}26)$$

$$\sigma_t = \frac{(1.25)(1\,146)}{0.73} \frac{11.59}{2.25} \frac{1.6}{0.48} = 33\,700 \text{ psi}$$

From Figure 11-28, the required hardness of the tooth material is 270. This is a reasonable value for steel gears.

Equation (12-17) is now used to evaluate the wear load factor. We need the number of teeth in the gear to evaluate Q. From the required rotational speeds of the pinion and the gear,

$$VR = n_P/n_G = 3\,450/1\,100 = 3.14$$

Then

$$N_G = N_P(VR) = (24)(3.14) = 75 \quad \text{(integer value)}$$

$$Q = 2N_G/(N_P + N_G) = 2(75)/(24 + 75) = 1.52$$

Then

$$K = \frac{W_t \cos^2 \psi}{C_v dFQ} = \frac{(1\,146)\cos^2(15)}{(0.73)(2.071)(2.25)(1.52)} = 207$$

Figure 12-6 indicates that a surface hardness of approximately HB 310 is required. Because this is higher than that required for strength, it governs the selection of the material and its heat treatment. From the Appendix tables for properties of steels, several are available to satisfy these requirements. Appendix A-4-6 shows that AISI 6150 OQT 1100 gives a hardness of HB 341 with a percent elongation of 15 percent that would be satisfactory for the pinion. The same steel, but tempered at 1200 °F, would give HB 293, which is good for the gear.

12-6 BEVEL GEAR GEOMETRY

Bevel gears are used to transfer motion between nonparallel shafts, usually at 90° to one another. Some of the several types that are available commercially are the straight bevel, the Zerol bevel, the spiral bevel, and the hypoid. The details of the straight bevel will be described.

As the name implies, the teeth of the straight bevel gear are straight and are located along the elements of a cone. The surface that would pass through the pitch line of all the teeth would be part of a right circular cone. When two bevel gears mesh, the axes of their pitch cones intersect, with the vertices of the two pitch cones at the same point. This is illustrated in Figure 12-9.

The angle of the pitch cone surface is dependent on the ratio of the number of teeth in the two mating gears. For the smaller of the two gears, the *pinion,* the pitch cone angle is

$$\gamma = \tan^{-1}(N_P/N_G) \quad \text{(Lowercase Greek } gamma\text{)} \tag{12-18}$$

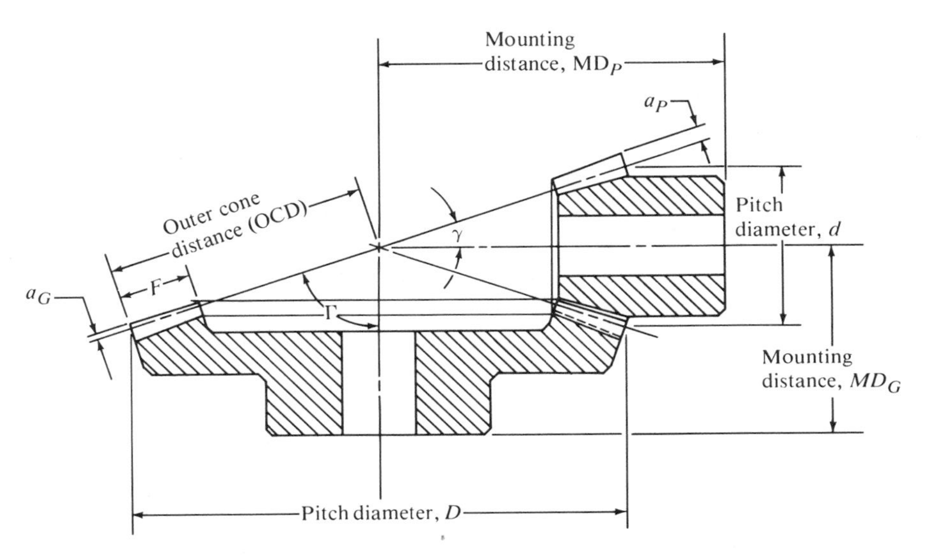

Figure 12-9 Bevel Gear Geometry (Browning Mfg. Division, Emerson Electric Co., Maysville, Ky.)

For the larger gear:

$$\Gamma = \tan^{-1}(N_G/N_P) \quad \text{(Uppercase Greek } gamma\text{)} \tag{12-19}$$

These equations apply if the two shafts carrying the gears are 90° apart. In this case, $\gamma + \Gamma = 90°$. If the ratio is unity, the two gears have the same number of teeth, and the angles of the two pitch cones are each 45°. This type of bevel gear pair is called a set of *miter gears*.

Note that the teeth of straight bevel gears are tapered from the outside toward the middle. The tooth form for bevel gears is characterized by the form of the large end of the teeth. Most are made in the diametral pitch system with 20° involute teeth, although other pressure angles are used. The pitch diameter is defined for the large end of the teeth and is computed in the same way as it is for spur gears:

$$P_d = N_P/d = N_G/D \tag{12-20}$$

where d is the pitch diameter of the pinion, D is the pitch diameter of the gear, and P_d is the diametral pitch.

Other geometrical features of straight bevel gears are typically made according to the relationships shown in Table 12-2. Again, other variations are used.

The mounting of bevel gears is critical if satisfactory performance is to be achieved. Most commercial gears have a defined mounting distance similar to that shown in Figure 12-9. It is the distance from some reference surface, usually the back of the hub of the gear, to the apex of the pitch cone. Because the pitch cones of the mating gears have coincident apexes, the mounting distance also locates the axis of the mating gear. If the gear is mounted at a distance smaller than the recommended mounting distance, the teeth will likely bind. If mounted at a greater distance, there will be excessive backlash, causing noisy and rough operation.

Example Problem 12-4. Compute the values for the geometrical features of a pair of straight bevel gears having a diametral pitch of 8, a 20° pressure angle, 16 teeth in the pinion, and 48 teeth in the gear. The shafts are at 90°.

Solution.

1. Pitch diameters at the large end of the gears
   a. Pinion: $d = N_P/P_d = 16/8 = 2.000$ in
   b. Gear: $D = N_G/P_d = 48/8 = 6.000$ in
2. Pitch cone angles
   a. Pinion: $\gamma = \tan^{-1}(N_P/N_G) = \tan^{-1}(16/48) = 18.43°$
   b. Gear: $\Gamma = \tan^{-1}(N_G/N_P) = \tan^{-1}(48/16) = 71.57°$
   c. Check: $\gamma + \Gamma = 18.43° + 71.57° = 90°$ (OK)
3. Whole depth: $h_t = 2.188/P_d + 0.002 = 2.188/8 + 0.002 = 0.2755$ in
4. Working depth: $h_k = 2.000/P_d = 2.000/8 = 0.2500$ in
5. Clearance: $c = 0.188/P_d + 0.002 = 0.188/8 + 0.002 = 0.0255$ in
6. Addendum: gear

$$a_G = \frac{0.54}{P_d} + \frac{0.460}{P_d(N_G/N_P)^2} = \frac{0.54}{8} + \frac{0.460}{8(48/16)^2} = 0.0739 \text{ in}$$

Table 12-2 Geometrical Features of Straight Bevel Gear Teeth

Whole depth	$h_t = 2.188/P_d + 0.002$
Working depth	$h_k = 2.000/P_d$
Clearance	$c = 0.188/P_d + 0.002$
Addendum: Gear	$a_G = \dfrac{0.54}{P_d} + \dfrac{0.460}{P_d(N_G/N_P)^2}$
Addendum: Pinion	$a_P = h_k - a_G$
Outside diameter: Gear	$D_o = D + 2a_G \cos \Gamma$
Outside diameter: Pinion	$d_o = d + 2a_P \cos \gamma$
Outer cone distance	$OCD = D/(2 \sin \Gamma) = d/(2 \sin \gamma)$
Preferred face width	$F = OCD/3$ or less (maximum $F = 10/P_d$)

7. Addendum: pinion: $a_P = h_k - a_G = 0.2500 - 0.0739 = 0.1761$ in
8. Outside diameter: gear: $D_o = D + 2a_G \cos(\Gamma)$

$$D_o = 6.000 + 2(0.0739)\cos(71.57°) = 6.047 \text{ in}$$

9. Outside diameter: pinion: $d_o = d + 2a_P \cos(\gamma)$

$$d_o = 2.000 + 2(0.1761)\cos(18.43°) = 2.334 \text{ in}$$

10. Outer cone distance

$$OCD = D/(2 \sin \Gamma) = 6.000/[2 \sin(71.57°)] = 3.162 \text{ in}$$

11. Preferred face width: $F = OCD/3 = 3.162/3 = 1.054$ in (or less)
12. Maximum face width: $F = 10/P_d = 10/8 = 1.250$ in

12-7 FORCES ON STRAIGHT BEVEL GEARS

Because of the conical shape of bevel gears and because of the involute tooth form, a three-component set of forces acts on bevel gear teeth. Using notation similar to that for helical gears, we will compute the transmitted load, W_t, the radial load, W_r, and the axial load, W_x. It is assumed that the three forces act concurrently at the mid-face of the teeth and on the pitch cone (see Figure 12-10). Although the actual point of application of the resultant force is a little displaced from the middle, no serious error results.

The transmitted load acts tangential to the pitch cone and is the force that produces the torque on the pinion and the gear. The torque can be computed from the known power transmitted and the rotational speed:

$$T = 63\,000\ P/n$$

Then, using the pinion, for example, the transmitted load is

$$W_{tP} = T/r_m \qquad (12\text{-}21)$$

where r_m is the mean radius of the pinion. The value of r_m can be computed from

$$r_m = d/2 - (F/2)\sin \gamma \qquad (12\text{-}22)$$

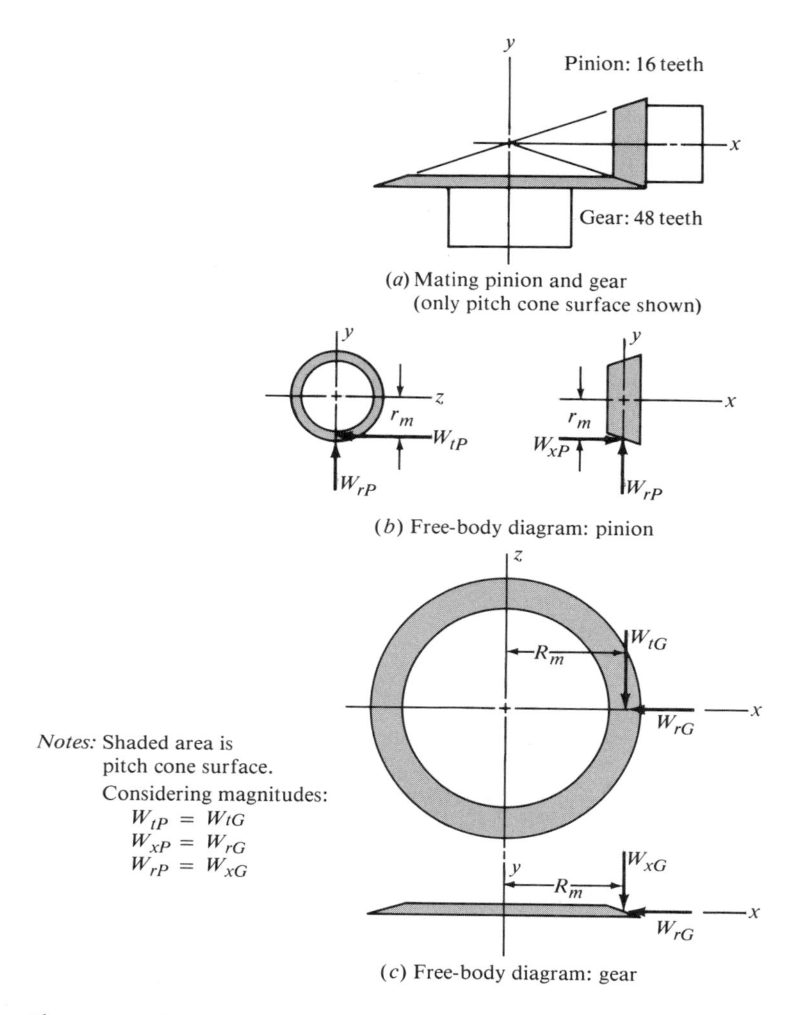

Figure 12-10 Forces on Bevel Gears

Remember that the pitch diameter, d, is measured to the pitch line of the tooth at its large end.

The radial load acts toward the center of the pinion, perpendicular to its axis, causing bending of the pinion shaft.

$$W_{rP} = W_t \tan \phi \cos \gamma \tag{12-23}$$

The axial load acts parallel to the axis of the pinion, tending to push it away from the mating gear. It causes a thrust load on the shaft bearings. It also produces a bending moment on the shaft because it acts at the distance from the axis equal to the mean radius of the gear.

$$W_{xP} = W_t \tan \phi \sin \gamma \tag{12-24}$$

The values for the forces on the gear can be calculated by the same equations shown here for the pinion, if the geometry for the gear is substituted for that of the pinion. Refer to Figure 12-10 for the relationships between the forces on the pinion and the gear in both magnitude and direction.

Example Problem 12-5. For the gear pair described in example problem 12-4, calculate the forces on the pinion and the gear if they are transmitting 2.50 hp with a pinion speed of 600 rpm. The face width for both gears is 0.84 in. The other geometry factors computed in example problem 12-4 apply.

Solution. Forces on the pinion are described by the following equation:

$$W_t = T/r_m$$

But

$$T = 63\,000\ (P)/n_p = [63\,000(2.50)]/600 = 263\ \text{lb}\cdot\text{in}$$
$$r_m = d/2 - (F/2)\sin\gamma$$
$$r_m = (2.000/2) - (0.84/2)\sin(18.43°) = 0.87\ \text{in}$$

Then

$$W_t = 263\ \text{lb}\cdot\text{in}/0.87\ \text{in} = 302\ \text{lb}$$
$$W_r = W_t \tan\phi \cos\gamma = 302\ \text{lb}\tan(20°)\cos(18.43°) = 104\ \text{lb}$$
$$W_x = W_t \tan\phi \sin\gamma = 302\ \text{lb}\tan(20°)\sin(18.43°) = 35\ \text{lb}$$

To determine the forces on the gear, first let's calculate the rotational speed of the gear:

$$n_G = n_P(N_P/N_G) = 600\ \text{rpm}(16/48) = 200\ \text{rpm}$$

Then

$$T = 63\,000(2.50)/200 = 788\ \text{lb}\cdot\text{in}$$
$$R_m = D/2 - (F/2)\sin\Gamma$$
$$R_m = 6.000/2 - (0.84/2)\sin(71.57°) = 2.60\ \text{in}$$
$$W_t = T/R_m = (788\ \text{lb}\cdot\text{in})/(2.60\ \text{in}) = 302\ \text{lb}$$
$$W_r = W_t \tan\phi \cos\Gamma = 302\ \text{lb}\tan(20°)\cos(71.57°) = 35\ \text{lb}$$
$$W_x = W_t \tan\phi \sin\Gamma = 302\ \text{lb}\tan(20°)\sin(71.57°) = 104\ \text{lb}$$

Note from Figure 12-10 that the forces on the pinion and the gear form an *action-reaction pair*. That is, the forces on the gear are equal to those on the pinion, but they act in the opposite direction. Also, because of the 90° orientation of the shafts, the radial force on the pinion becomes the axial thrust load on the gear and the axial thrust load on the pinion becomes the radial load on the gear.

12-8 BEARING FORCES ON SHAFTS CARRYING BEVEL GEARS

Because of the three-dimensional force system that acts on bevel gears, the calculation of the forces on shaft bearings can be cumbersome. An example is worked out here to show the procedure. In order to obtain numerical data, the arrangement shown in Figure 12-11 is proposed for the bevel gear pair that was the subject of example problems 12-4 and

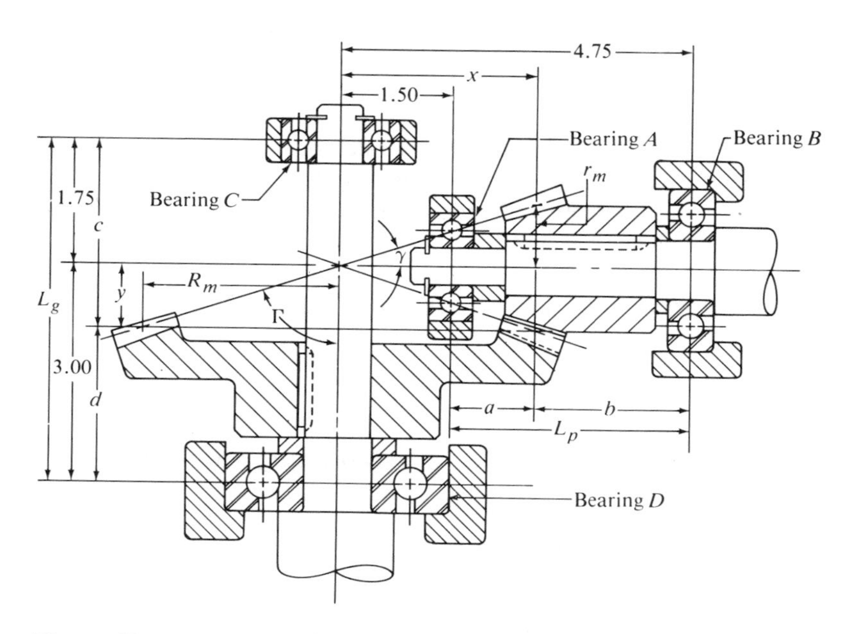

Figure 12-11 Layout of Bevel Gear Pair For Example Problem 12-6

12-5. The locations for the bearings are given with respect to the apex of the two pitch cones where the shaft axes intersect.

Note that both the pinion and the gear are *straddle-mounted;* that is, each gear is positioned between the supporting bearings. This is the most preferred arrangement because it usually provides the greatest rigidity and maintains the alignment of the teeth during power transmission. Care should be exercised to provide rigid mountings and stiff shafts when using bevel gears.

The arrangement of Figure 12-11 is designed so that the bearing on the right resists the axial thrust load on the pinion, and the lower bearing resists the axial thrust load on the gear.

Example Problem 12-6. Compute the reaction forces on the bearings that support the shafts carrying the bevel gear pair shown in Figure 12-11. The values of example problems 12-4 and 12-5 apply.

Solution. Referring to the results of example problem 12-5 and Figure 12-10, the forces acting on the gears are listed in the following table:

	Pinion	*Gear*
Tangential	$W_{tP} = 302$ lb	$W_{tG} = 302$ lb
Radial	$W_{rP} = 104$ lb	$W_{rG} = 35$ lb
Axial	$W_{xP} = 35$ lb	$W_{xG} = 104$ lb

It is critical to be able to visualize the directions in which these forces are acting because of the three-dimensional force system. Notice in Figure 12-10 that a rectangular coordinate system has been set up. Figure 12-12 is an isometric sketch of the free body diagrams of the pinion and the gear, simplified to represent the concurrent forces acting

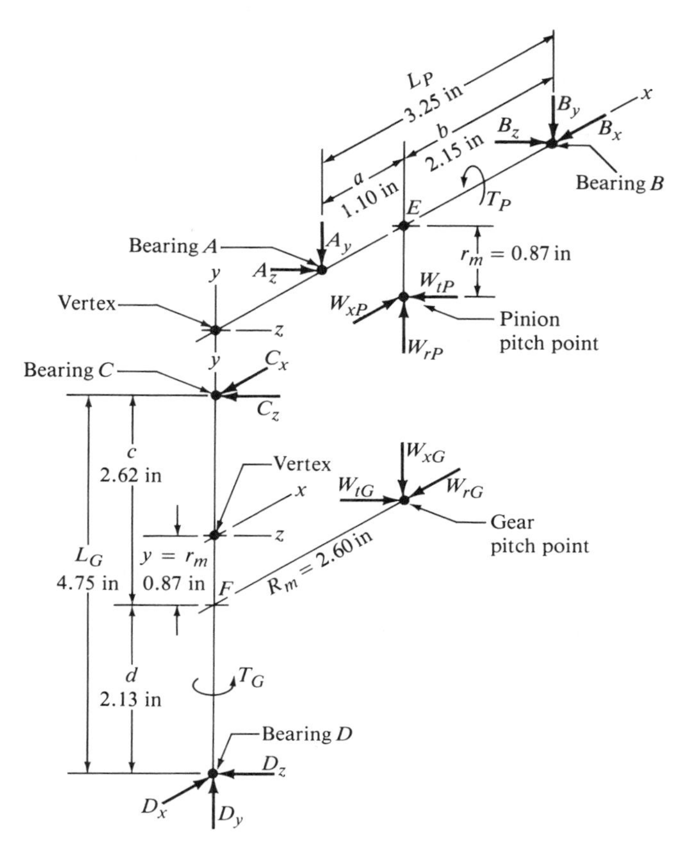

Figure 12-12 Free Body Diagrams for Pinion and Gear Shafts

at the pinion/gear interface and at the bearing locations. Although the two free body diagrams are separated for clarity, notice that they can be brought together by moving the point called *vertex* on each sketch together. This is the point in the actual gear system where the vertices of the two pitch cones lie at the same point. The two pitch points also coincide.

For setting up the equations of static equilibrium needed to solve for the bearing reactions, the distances a, b, c, d, L_P, and L_G are needed, as shown in Figure 12-11. These require the two dimensions labeled x and y. Note from example problem 12-5,

$$x = R_m = 2.60 \text{ in}$$

$$y = r_m = 0.87 \text{ in}$$

Then

$$a = x - 1.50 = 2.60 - 1.50 = 1.10 \text{ in}$$

$$b = 4.75 - x = 4.75 - 2.60 = 2.15 \text{ in}$$

$$c = 1.75 + y = 1.75 + 0.87 = 2.62 \text{ in}$$

$$d = 3.00 - y = 3.00 - 0.87 = 2.13 \text{ in}$$

$$L_P = 4.75 - 1.50 = 3.25 \text{ in}$$

$$L_G = 1.75 + 3.00 = 4.75 \text{ in}$$

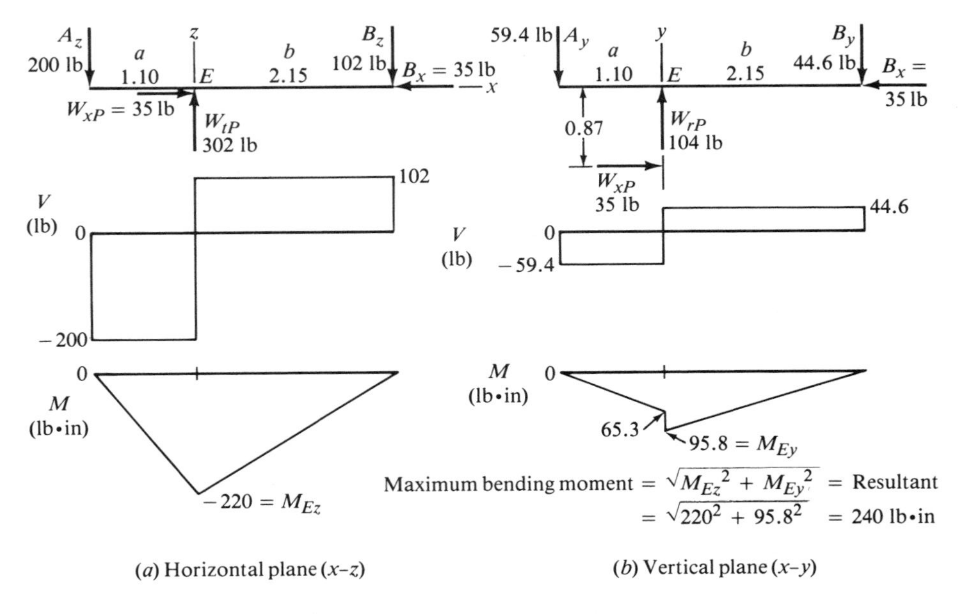

(*a*) Horizontal plane (x–z) (*b*) Vertical plane (x–y)

Figure 12-13 Pinion Shaft Bending Moments

These values are shown on Figure 12-12.

To solve for the reactions, we need to consider the horizontal (x-z) and the vertical (x-y) planes separately. It may help you to look also at Figure 12-13, which breaks out the forces on the pinion shaft in these two planes. Then each plane can be analyzed, using the fundamental equations of equilibrium.

1. Bearing reactions—pinion shaft: bearings A and B
   a. To find B_z and A_z: In the x-z plane, only W_{tP} acts. Summing moments about A:

$$0 = W_{tP}(a) - B_z(L_P) = 302(1.10) - B_z(3.25)$$
$$B_z = 102 \text{ lb}$$

   Summing moments about B:

$$0 = W_{tP}(b) - A_z(L_P) = 302(2.15) - A_z(3.25)$$
$$A_z = 200 \text{ lb}$$

   b. To find B_y and A_y: In the x-y plane, both W_{rP} and W_{xP} act. Summing moments about A:

$$0 = W_{rP}(a) + W_{xP}(r_m) - B_y(L_P)$$
$$0 = 104(1.10) + 35(0.87) - B_y(3.25)$$
$$B_y = 44.6 \text{ lb}$$

   Summing moments about B:

$$0 = W_{rP}(b) - W_{xP}(r_m) - A_y(L_P)$$
$$0 = 104(2.15) - 35(0.87) - A_y(3.25)$$
$$A_y = 59.4 \text{ lb}$$

c. To find B_x: Summing forces in the x direction:

$$B_x = W_{xP} = 35 \text{ lb}$$

This is the thrust force on bearing B.

d. To find the total radial force on each bearing, compute the resultant of the y and z components:

$$A = \sqrt{A_y^2 + A_z^2} = \sqrt{59.4^2 + 200^2} = 209 \text{ lb}$$
$$B = \sqrt{B_y^2 + B_z^2} = \sqrt{44.6^2 + 102^2} = 111 \text{ lb}$$

2. Bearing reactions—gear shaft: bearings C and D. Using similar methods, the following forces can be found (see Figure 12-14).

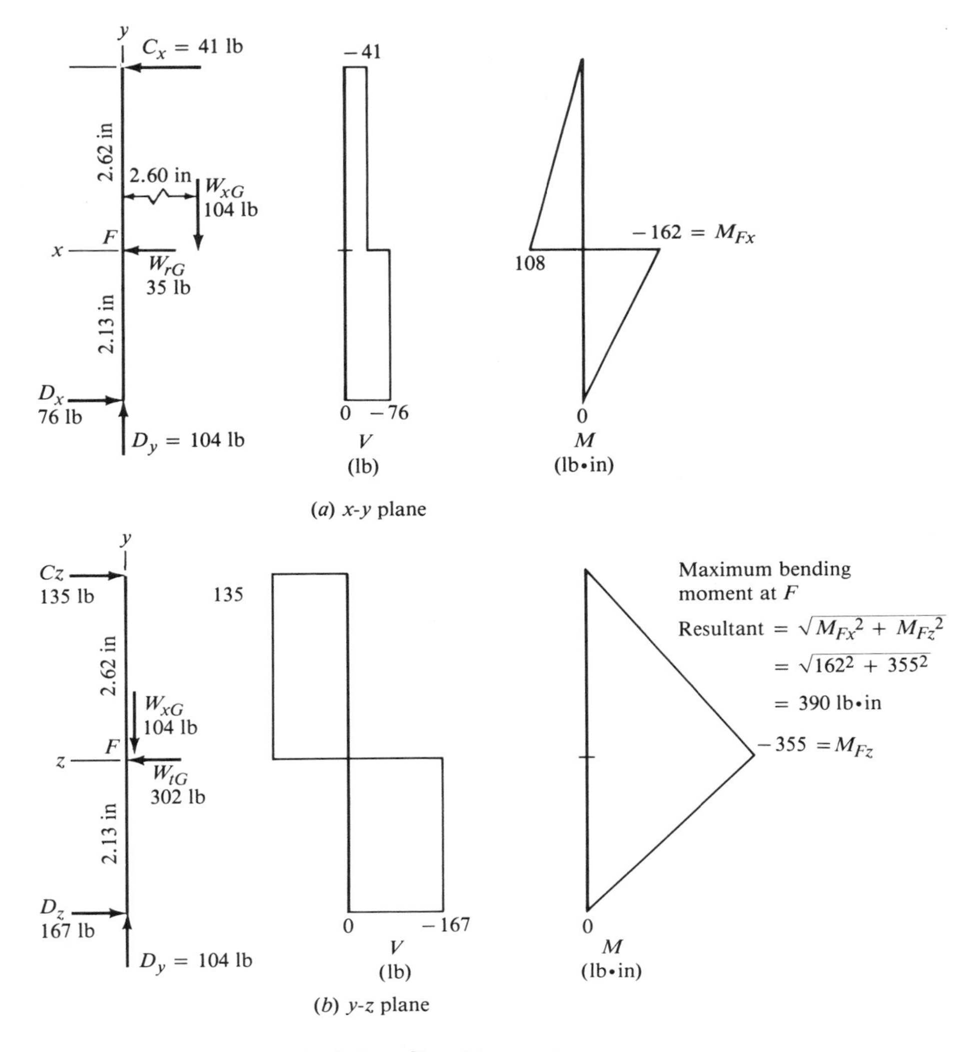

Figure 12-14 Gear Shaft Bending Moments

$$\left.\begin{aligned} C_z &= 135 \text{ lb} \\ C_x &= 41.2 \text{ lb} \end{aligned}\right] \quad C = 141 \text{ lb (radial force on } C\text{)}$$

$$\left.\begin{aligned} D_z &= 167 \text{ lb} \\ D_x &= 76.2 \text{ lb} \end{aligned}\right] \quad D = 184 \text{ lb (radial force on } D\text{)}$$

$$D_y = W_{xG} = 104 \text{ lb (thrust force on } D\text{)}$$

In summary, when selecting the bearings for these shafts, the following capacities are required:

Bearing A: 209-lb radial
Bearing B: 111-lb radial; 35-lb thrust
Bearing C: 141-lb radial
Bearing D: 184-lb radial; 104-lb thrust

12-9 BENDING MOMENTS ON SHAFTS CARRYING BEVEL GEARS

Because there are forces acting in two planes on bevel gears, as discussed in the preceding section, there is also bending in two planes. The analysis of the shearing force and bending moment diagrams for the shafts must take this into account.

Figures 12-13 and 12-14 show the resulting diagrams for the pinion and the gear shafts, respectively, for the gear pair used for example problems 12-4, 12-5, and 12-6. Notice that the axial thrust load on each shaft provides a concentrated moment to the shaft equal to the axial load times the distance it is offset from the axis of the shaft. Also notice that the maximum bending moment for each shaft is the resultant of the moments in the two planes. On the pinion shaft, the maximum moment is 240 lb · in at E, where the lines of action for the radial and tangential forces intersect the shaft. Similarly, on the gear shaft, the maximum moment is 390 lb · in at F. These data are used in the shaft design (as discussed in Chapter 9).

12-10 STRESSES IN STRAIGHT BEVEL GEAR TEETH

The stress analysis for bevel gear teeth is similar to that already presented for spur and helical gear teeth. The maximum bending stress occurs at the root of the tooth in the fillet. This stress can be computed from

$$\sigma_t = \frac{W_t K_a}{K_v} \frac{P_d}{F} \frac{K_s K_m}{J} \tag{12-25}$$

The terms have all been used before, except K_s, the size factor. But there are minor differences in the manner of evaluating the factors, so they will be reviewed here.

Tangential Load, W_t. Contrary to the way W_t was computed in the preceding section, it is computed here by using the diameter of the gear at its large end, rather than the diameter at the middle of the tooth. This is more convenient, and the adjustment for the actual force distribution on the teeth is made in the value of the geometry factor, J. Then

$$W_t = \frac{T}{r} = \frac{63\,000(P)}{n_P} \frac{1}{d/2} \tag{12-26}$$

where

$$T = \text{Torque transmitted (lb} \cdot \text{in)}$$
$$r = \text{Pitch radius of the pinion (in)}$$
$$P = \text{Power transmitted (hp)}$$
$$n_P = \text{Rotational speed of the pinion (rpm)}$$
$$d = \text{Pitch diameter of the pinion at its large end (in)}$$

Velocity Factor, K_v. Use the same equation as for spur gears of average quality (quality number 5).

$$K_v = 50/(50 + \sqrt{v_t})$$

where

$$v_t = \text{Pitch line speed (ft/min)} = \pi d n_P/12$$

Application Factor, K_a. Use the values from Table 11-13, Chapter 11.

Size Factor, K_s. Size factor is related to the size of the teeth. Use $K_s = P_d^{-0.25}$ for coarse pitch gears up to $P_d = 16$. Use $K_s = 0.5$ for teeth finer than $P_d = 16$.

Alignment Factor, K_m. Also called the *load distribution factor,* K_m is dependent on the rigidity of the mounting of the gears. Assuming that straddle mounting is the most rigid, general industrial practice calls for the following values:

$K_m = 1.10$ Pinion and gear straddle-mounted

$K_m = 1.25$ Pinion or gear straddle-mounted

$K_m = 1.40$ Neither straddle-mounted

Geometry Factor, J. Use Figure 12-15 if the pressure angle is 20° and the shaft angle is 90°.

The computed value for stress from equation (12-25) is then compared to the allowable stress, s_{at}, from Figure 12-16. From these data, the required material can be specified.

Example Problem 12-7. Compute the bending stress in the teeth of the bevel pinion shown in Figure 12-11. The data from example problems 12-4 and 12-5 apply: $N_P = 16$; $N_G = 48$; $n_P = 600$ rpm; $P = 2.50$ hp; $P_d = 8$; $d = 2.000$ in; $F = 0.84$ in. Assume the pinion is driven by an electric motor and the load provides moderate shock.

Solution.

$$W_t = \frac{T}{r} = \frac{63\,000(P)}{n_P}\frac{1}{d/2} = \frac{63\,000(2.50)}{600}\frac{1}{2.000/2} = 263 \text{ lb}$$

$$v_t = \pi d n_P/12 = \pi(2.000)(600)/12 = 314 \text{ ft/min}$$

$$K_v = 50/(50 + \sqrt{v_t}) = 50/(50 + \sqrt{314}) = 0.74$$

$$K_a = 1.25 \quad \text{(from Table 11-13)}$$

$$K_s = P_d^{-0.25} = (8)^{-0.25} = 0.595$$

$$K_m = 1.10 \quad \text{(both gears straddle-mounted)}$$

$$J = 0.230 \quad \text{(from Figure 12-15)}$$

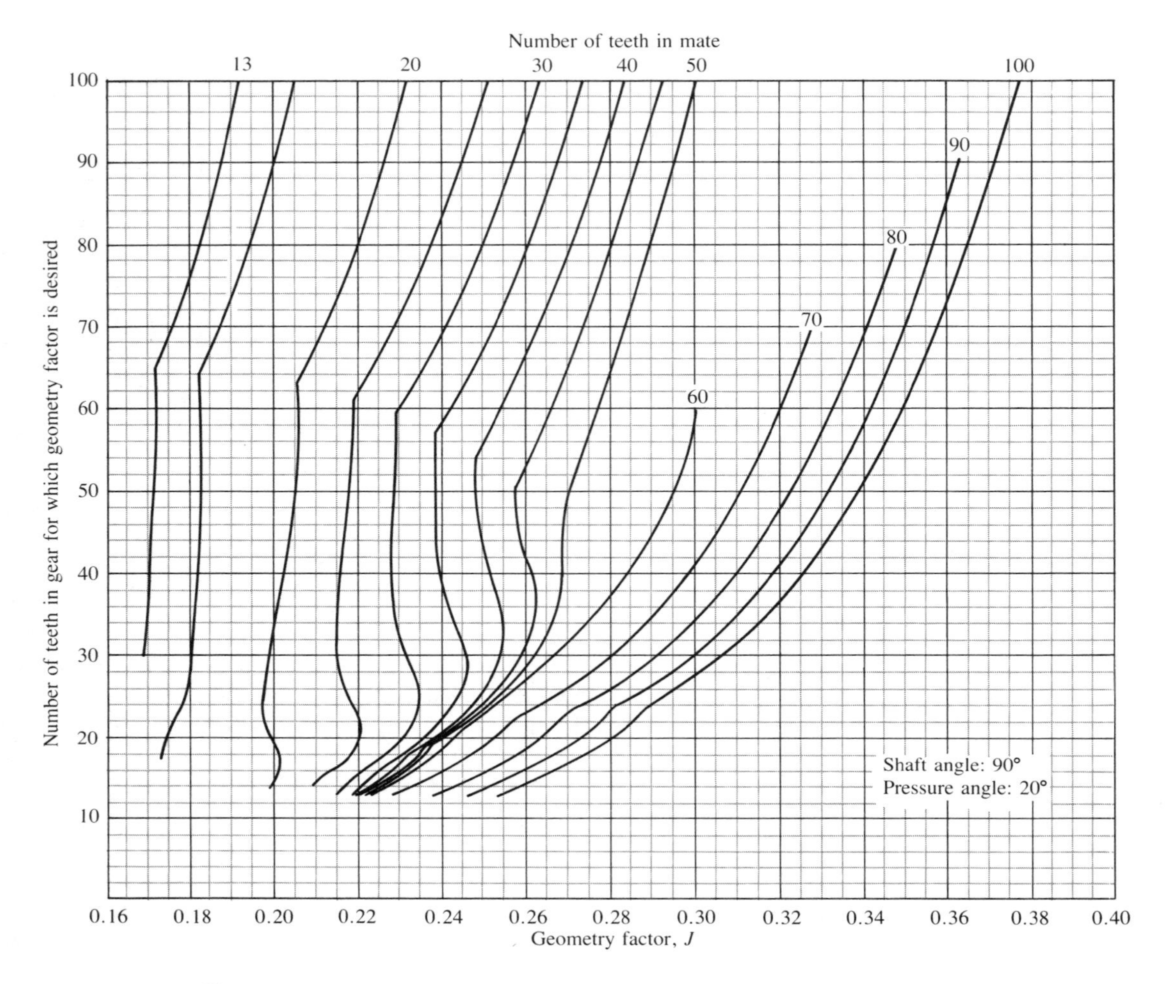

Figure 12-15 Geometry Factors for Straight Bevel Gears (Extracted from AGMA with the permission of the publisher, the American Gear Manufacturers Association, 1901 North Fort Myer Drive, Suite 1000, Arlington, Virginia 22209)

Then, from equation (12-25),

$$\sigma_t = \frac{W_t K_a}{K_v} \frac{P_d}{F} \frac{K_s K_m}{J} = \frac{(263)(1.25)}{0.74} \frac{8}{0.84} \frac{(0.595)(1.10)}{0.230} = 12\,200 \text{ psi}$$

Referring to Figure 12-16, it can be seen that this is a very modest stress level for steel gears, requiring only about HB 160 for the hardness. If stress were the only consideration, a redesign might be attempted to achieve a more compact system. However, normally the wear resistance or surface durability of the teeth require a harder material. The next section discusses wear.

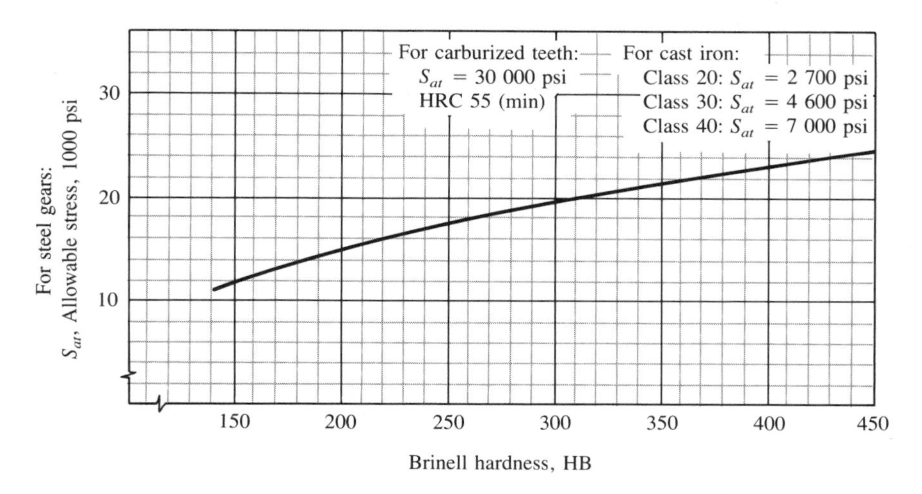

Figure 12-16 Allowable Fatigue Stress for Bevel Gears, s_{at}

12-11 DESIGN OF BEVEL GEARS FOR WEAR RESISTANCE

The approach to the design of bevel gears for wear resistance is similar to that for spur gears. The failure mode is fatigue of the surface of the teeth under the influence of the contact stress between the mating gears.

The contact stress, called the *Hertz stress*, σ_h, can be computed from

$$\sigma_h = C_p \sqrt{\frac{W_t C_a}{C_v} \frac{1}{Fd} \frac{C_m}{I}} \tag{12-27}$$

The factors C_a, C_v, and C_m are the same as K_a, K_v, and K_m, used for computing the bending stress in the preceding section. The terms W_t, F, and d have the same meanings, also. C_p is the elastic coefficient and is related to the modulus of elasticity and Poisson's ratio for the material of the pinion and the gear.

$$C_p = \sqrt{\frac{3}{2\pi} \frac{1}{[(1 - \nu_P^2)/E_P][(1 - \nu_G^2)/E_G]}} \tag{12-28}$$

For steel or cast iron gears,

$$C_p = 2\,800 \text{ for two steel gears } \quad (E = 30 \times 10^6 \text{ psi})$$

$$C_p = 2\,250 \text{ for two cast iron gears } (E = 19 \times 10^6 \text{ psi})$$

$$C_p = 2\,450 \text{ for a steel pinion and a cast iron gear}$$

The factor I is the geometry factor for surface durability and can be found from Figure 12-17.

The Hertz contact stress computed from equation (12-27) should be compared with the allowable contact stress number, s_{ac}, from Figure 11-30 in Chapter 11 if the material is steel. For cast iron, use the following values:

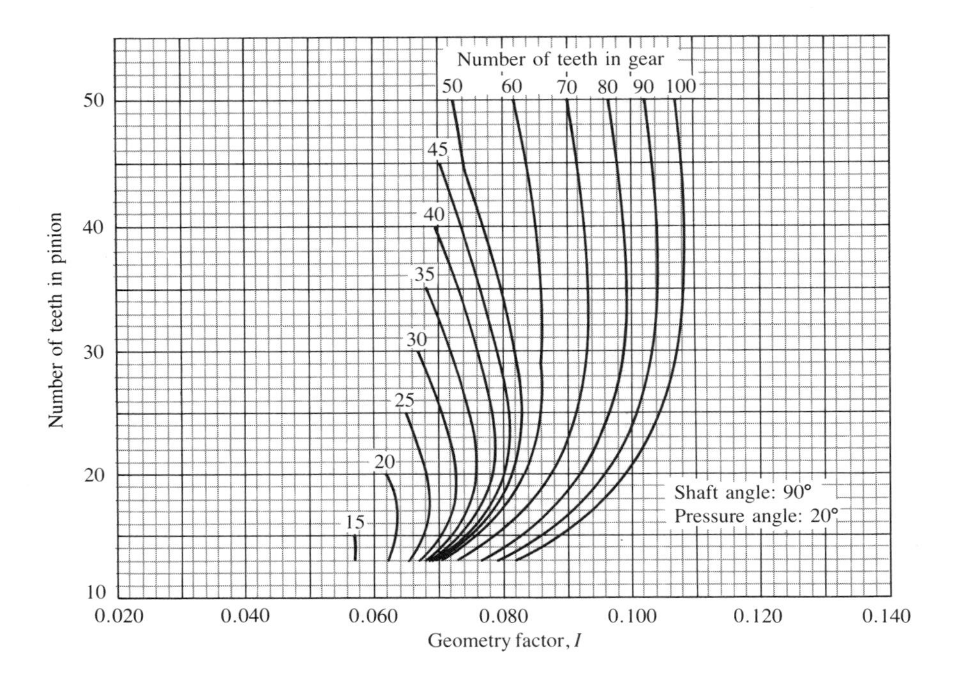

Figure 12-17 Geometry Factors for Straight and Zerol Bevel Gears (Extracted from AGMA with the permission of the publisher, the American Gear Manufacturers Association, 1901 North Fort Myer Drive, Suite 1000, Arlington, Virginia 22209)

$$\text{Class 20:} \quad s_{ac} = 30\,000 \text{ psi}$$
$$\text{Class 30:} \quad s_{ac} = 50\,000 \text{ psi}$$
$$\text{Class 40:} \quad s_{ac} = 65\,000 \text{ psi}$$

Example Problem 12-8. Compute the Hertz stress for the gear pair in Figure 12-11 for the conditions used in example problems 12-4 through 12-7: $N_P = 16$; $N_G = 48$; $n_P = 600$ rpm; $P_d = 8$; $F = 0.84$ in; $d = 2.000$ in. Both gears are to be steel. Specify a suitable steel for the gears, along with its heat treatment.

Solution. From example problem 12-7: $W_t = 263$ lb; $C_a = 1.25$; $C_v = 0.74$; and $C_m = 1.10$. For two steel gears, $C_p = 2\,800$. From Figure 12-17, $I = 0.077$. Then

$$\sigma_h = C_p \sqrt{\frac{W_t C_a}{C_v} \frac{1}{Fd} \frac{C_m}{I}} = 2\,800 \sqrt{\frac{(263)(1.25)}{0.74} \frac{1}{(0.84)(2.000)} \frac{1.10}{0.077}}$$
$$= 173\,300 \text{ psi}$$

Comparing this value with the allowable contact stress number from Figure 11-30 shows that approximately HB 450 is required for surface durability. Note that only about HB 160 was required for strength in example problem 12-7.

Material selection is based on satisfying both the strength and the wear requirements. In this case, the wear considerations govern the design if through-hardening is to be done.

From Appendix A-4-4, we could choose AISI 4140 OQT 700 for the pinion, which would provide a hardness of HB 461. A slightly lower hardness for the gear is usually specified, perhaps AISI 4140 OQT 800 with HB 429.

An alternative material specification would be to use a surface-hardening treatment, such as flame or induction hardening, on the same alloy but with an initial heat treatment that would leave the core of the teeth in a much more ductile form. For example, an initial treatment of OQT 1100 would produce a through-hardness of HB 311 in AISI 4140, which is sufficient for the strength requirement. Performing the subsequent surface hardening to bring the case up to a hardness of HRC 50 will satisfy the wear requirement. The induction heating process must be controlled to produce a sufficient case thickness, as suggested by Figure 11-16. For the 8-pitch teeth, a case thickness of approximately 0.022 in is recommended.

12-12 PRACTICAL CONSIDERATIONS FOR BEVEL GEARING

Similar factors to those discussed for spur and helical gears should be considered when designing systems using bevel gears. The accuracy of alignment and the accommodation of thrust loads discussed in the example problems are critical. Figures 12-18 and 12-19 show commercial applications.

12-13 TYPES OF WORMGEARING

Wormgearing is used to transmit motion and power between nonintersecting shafts, usually at 90° to each other. The drive consists of a worm on the high-speed shaft that has the general appearance of a power screw thread: a cylindrical, helical thread. The worm drives a wormgear, which has an appearance similar to that of a helical gear. Figure 12-20 shows a typical worm and wormgear set. Sometimes the wormgear is referred to as a *worm wheel* or simply a *wheel* or *gear*.

Several variations of the geometry of wormgear drives are available. The most common one, shown in Figures 12-20 and 12-21, employs a cylindrical worm mating with a wormgear having teeth that are throated, wrapping partially around the worm. This is called a *single enveloping type* of wormgear drive. The contact between the threads of the

Figure 12-18 Spiral Bevel Right Angle Reducer (Sumitomo Machinery Corporation of America, Teterboro, N.J.)

Figure 12-19 Final Drive for a Tractor (Ford Tractor Operations, Troy, Michigan)

Figure 12-20 Worms and Wormgears (Browning Mfg. Division, Emerson Electric Co., Maysville, Ky.)

worm and wormgear teeth is along a line, and the power transmission capacity is quite good. Many manufacturers offer this type of wormgear set as a stocked item. Installation of the worm is relatively easy because axial alignment is not very critical. However, the wormgear must be carefully aligned axially in order to achieve the benefit of the enveloping action. Figure 12-22 shows a cutaway of a commercial wormgear reducer.

A simpler form of wormgear drive allows a special cylindrical worm to be used with a standard spur gear or helical gear. Neither the worm nor the gear must be aligned with great accuracy, and the center distance is not critical. However, the contact between the worm threads and the wormgear teeth is theoretically a point, drastically reducing the

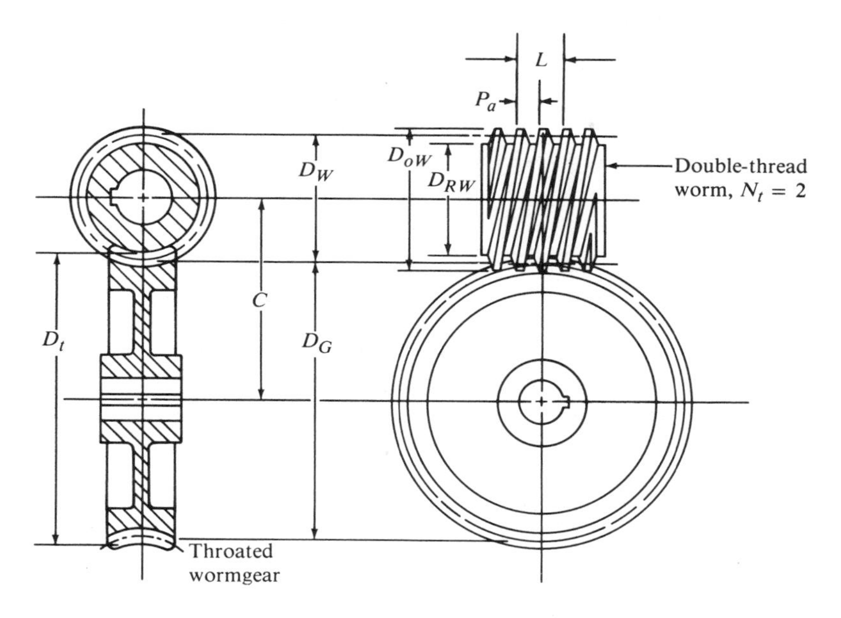

Figure 12-21 Single-Enveloping Wormgear Set

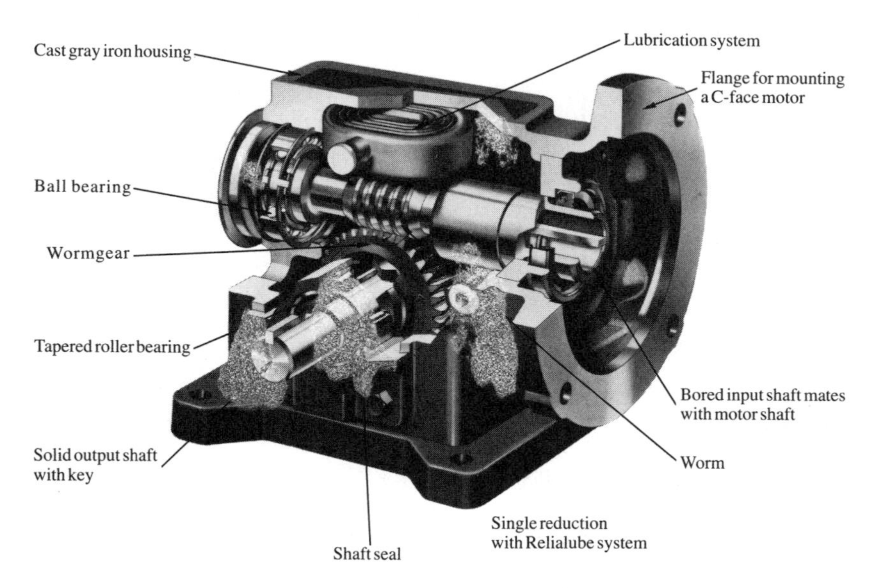

Figure 12-22 Wormgear Reducer (Reliance Electric Company, Greeneville, South Carolina)

power transmission capacity of the set. Thus this type is used mostly for nonprecision positioning applications at low speeds and low power levels.

A third type of wormgear set is the *double enveloping type* in which the worm is made in an hourglass shape and mates with an enveloping type of wormgear. This results in area contact rather than line or point contact and allows a much smaller system to transmit a given power at a given reduction ratio. However, the worm is more difficult to manufacture, and the alignment of both the worm and the wormgear is very critical.

12-14 GEOMETRY OF WORMS AND WORMGEARS

Pitches

A basic requirement of the worm and wormgear set is that the *axial pitch* of the worm must be equal to the *circular pitch* of the wormgear in order for them to mesh. Figure 12-21 shows the basic geometric features of a single enveloping worm and wormgear set. *Axial pitch, P_a*, is defined as the distance from a point on the worm thread to the corresponding point on the next adjacent thread, measured axially on the pitch cylinder. As before, the circular pitch is defined for the wormgear as the distance from a point on a tooth on the pitch circle of the gear to the corresponding point on the next adjacent tooth, measured along the pitch circle. Thus the circular pitch is an arc distance that can be calculated from

$$P_c = \pi D_G / N_G \tag{12-29}$$

where

D_G = Pitch diameter of the gear

N_G = Number of teeth in the gear

Some wormgears are made according to the circular pitch convention. But, as noted with spur gears, commercially available wormgear sets are usually made to a diametral pitch convention with the following pitches readily available: 48, 32, 24, 16, 12, 10, 8, 6, 5, 4, and 3. The diametral pitch is defined for the gear as

$$P_d = N_G / D_G \tag{12-30}$$

The conversion from diametral pitch to circular pitch can be made from the following equation:

$$P_d P_c = \pi \tag{12-31}$$

Number of Worm Threads

Worms can have a single thread, as in a typical screw, or multiple threads, usually 2 or 4 but sometimes 3, 5, 6, 8, or more. It is common to refer to the number of threads as N_t and then to treat that number as if it were the number of teeth in the worm. The number of threads in the worm is frequently referred to as the number of *starts*; this is convenient because, if you look at the end of a worm, you can count the number of threads that start at the end and wind down the cylindrical worm.

Lead, *L*

The *lead* of a worm is the axial distance that a point on the worm would move as the worm is rotated one revolution. Lead is related to the axial pitch by

$$L = N_t P_a \tag{12-32}$$

Lead Angle, λ

The *lead angle* is the angle between the tangent to the worm thread and the line perpendicular to the axis of the worm. The method of calculating the lead angle can be visualized by referring to Figure 12-23, which shows a simple triangle that would be formed if one thread of the worm were unwrapped from the pitch cylinder and laid flat on the paper. The length of the hypotenuse is the length of the thread itself. The vertical side is the lead, L. The horizontal side is the circumference of the pitch cylinder, πD_W, where D_W is the pitch diameter of the worm. Then,

$$\tan \lambda = L/\pi D_W \tag{12-33}$$

Velocity Ratio, *VR*

It is most convenient to calculate the velocity ratio of a worm and wormgear set from the ratio of the input rotational speed to the output rotational speed.

$$VR = \frac{\text{Speed of worm}}{\text{Speed of gear}} = \frac{n_W}{n_G} = \frac{N_G}{N_t} \tag{12-34}$$

Example Problem 12-9. A wormgear has 52 teeth and a diametral pitch of 6. It mates with a triple-threaded worm that rotates at 1 750 rpm. The pitch diameter of the worm is 2.000 in. Compute the circular pitch, the axial pitch, the lead, the lead angle, the pitch diameter of the wormgear, the center distance, the velocity ratio, and the rotational speed of the wormgear.

Solution.

Circular pitch	$P_c = \pi/P_d = \pi/6 = 0.5236$ in
Axial pitch	$P_a = P_c = 0.5236$ in
Lead	$L = N_t P_a = (3)(0.5236) = 1.5708$ in
Lead angle	$\lambda = \tan^{-1}(L/\pi D_w) = \tan^{-1}(1.5708/\pi 2.000)$
	$\lambda = 14.04°$
Pitch diameter	$D_G = N_G/P_d = 52/6 = 8.667$ in
Center distance	$C = (D_W + D_G)/2 = (2.000 + 8.667)/2 = 5.333$ in
Velocity ratio	$VR = N_G/N_t = 52/3 = 17.333$
Gear rpm	$n_G = n_W/VR = 1750/17.333 = 101$ rpm

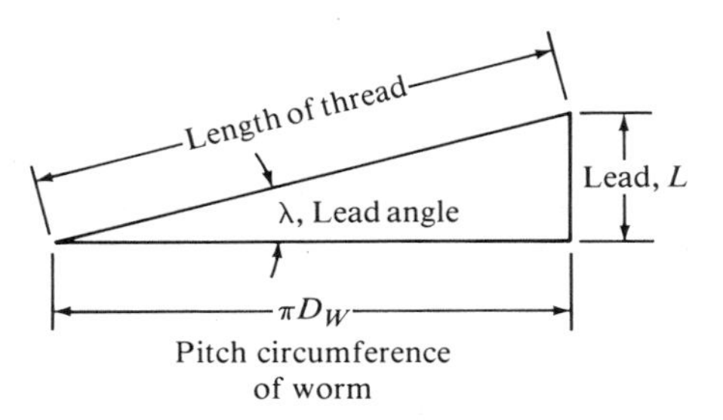

Figure 12-23 Lead Angle

Pressure Angle

Most commercially available wormgears are made with pressure angles of 14½°, 20°, 25°, or 30°. The low pressure angles are used with worms having a low lead angle and/or a low diametral pitch. For example, a 14½° pressure angle may be used for lead angles up to about 17°. For higher lead angles and with higher diametral pitches (smaller teeth), the 20° or 25° pressure angle is used to eliminate interference without excessive undercutting. The 20° pressure angle is the preferred value for lead angles up to 30°. From 30° to 45° of lead angle, the 25° pressure angle is recommended. Either the normal pressure angle, ϕ_n, or the transverse pressure angle, ϕ_t, may be specified. These are related by

$$\tan \phi_n = \tan \phi_t \cos \lambda \tag{12-35}$$

12-15 FORCES ON WORMGEAR SETS

The force system acting on the worm/wormgear set is usually considered to be made of three perpendicular components as was done for helical and bevel gears. There are a tangential load, a radial load, and an axial load acting on the worm and the wormgear. We will use the same notation here as in the bevel gear system.

Figure 12-24 shows two orthogonal views (front and side) of a worm/wormgear pair, showing only the pitch diameters of the gears. The figure shows the separate worm and wormgear with the forces acting on each. Note that, because of the 90° orientation of the two shafts,

$$\begin{aligned} W_{tG} &= W_{xW} \\ W_{xG} &= W_{tW} \\ W_{rG} &= W_{rW} \end{aligned} \tag{12-36}$$

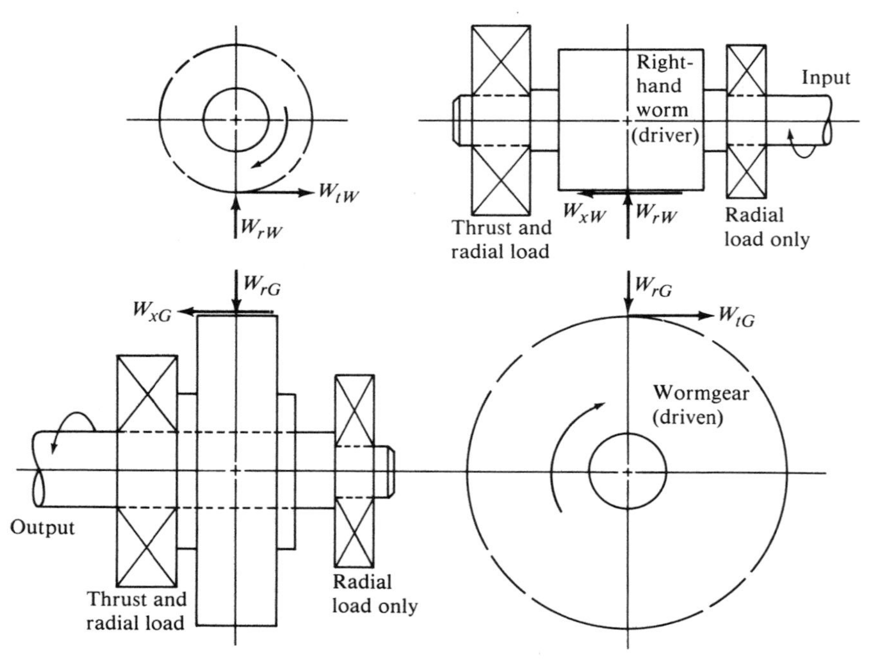

Figure 12-24 Forces on a Worm and Wormgear

Of course, the directions of the paired forces are opposite because of the action/reaction principle.

The tangential force on the wormgear is computed first and is based on the required operating conditions of torque, power, and speed at the output shaft. If a given system is being evaluated with a known motor power, it is conservative to assume that the motor power also appears at the output shaft. In reality, the output power is lower because of the power loss due to friction. More accurate calculations can be made after the efficiency is computed, as discussed later.

Coefficient of Friction

Friction plays a major part in the operation of a wormgear set because there is inherently sliding contact between the worm threads and the wormgear teeth. The coefficient of friction is dependent on the materials used, the lubricant, and the sliding velocity. Based on the pitch line speed of the gear, the sliding velocity is

$$v_s = v_{tG}/\sin\lambda \tag{12-37}$$

Based on the pitch line speed of the worm,

$$v_s = v_{tW}/\cos\lambda \tag{12-38}$$

Then, for a steel worm driving a bronze gear, with good oil lubrication, an estimate for the coefficient of friction is

$$f = 0.32/v_s^{0.36} \tag{12-39}$$

For a steel worm and a cast iron gear,

$$f = 0.40/v_s^{0.36} \tag{12-40}$$

Procedure for Calculating Forces on Worm and Wormgear Set

Given: Rotational speed of wormgear, n_G, in rpm
Pitch diameter of wormgear, D_G, in inches
Output power, P_o, in horsepower
Lead angle, λ
Pressure angle, ϕ_n

Compute:

$$v_{tG} = \pi D_G n_G/12 = \text{Pitch line speed of gear}$$

$$W_{tG} = 33\,000\, P_o/v_{tG}$$

$$W_{xG} = W_{tG}\frac{\cos\phi_n \sin\lambda + f\cos\lambda}{[\cos\phi_n \cos\lambda - f\sin\lambda]} \tag{12-41}$$

$$W_{rG} = \frac{W_{tG}\sin\phi_n}{\cos\phi_n\cos\lambda - f\sin\lambda} \tag{12-42}$$

The forces on the worm can be obtained by observation, using the equation set (12-36).

Efficiency of a Worm/Wormgear Set

Because of the sliding contact between the worm and the wormgear, the efficiency of a worm/wormgear set is typically much lower than for spur, helical, and bevel gears.

An extension of the force analysis summarized previously can be used to compute the friction force acting along the face of the gear teeth. The work done by this friction force must be added to the required output work to obtain the input work. Then the ratio of the output work to the input work is the efficiency of the drive. The resulting equation for efficiency is

$$\text{eff} = \frac{P_o}{P_i} = \tan\lambda\left[\frac{\cos\phi_n - f\tan\lambda}{\cos\phi_n\tan\lambda + f}\right] \quad (12\text{-}43)$$

Example Problem 12-10. The wormgear set described in example problem 12-9 is transmitting 6.68 hp at the output shaft where the rotational speed is 101 rpm. The transverse pressure angle is 20°. Compute the forces on the wormgear, the efficiency, and the input power to the worm. The worm is steel and the wormgear is bronze.

Solution. Recall from example problem 12-9:

$$\lambda = 14.04° \qquad D_G = 8.667 \text{ in}$$
$$n_W = 1\,750 \text{ rpm} \qquad D_W = 2.000 \text{ in}$$

The normal pressure angle is required. From equation (12-34),

$$\phi_n = \tan^{-1}(\tan\phi_t\cos\lambda) = \tan^{-1}(\tan 20° \cos 14.04°) = 19.45°$$

Because they recur in several formulas, let's compute

$$\cos\phi_n = \cos 19.45° = 0.943$$
$$\cos\lambda = \cos 14.04° = 0.970$$
$$\sin\lambda = \sin 14.04° = 0.243$$
$$\tan\lambda = \tan 14.04° = 0.250$$

The pitch line speed of the gear:

$$v_{tG} = \pi D_G n_G/12 = \pi(8.667)(101)/12 = 229 \text{ ft/min}$$

The tangential load on the gear:

$$W_{tG} = 33\,000\, P_o/v_{tG} = 33\,000(6.68)/229 = 962 \text{ lb}$$

The coefficient of friction, which requires the sliding velocity, v_s, is now required (see equations [12-37] and [12-39]).

$$v_s = v_{tG}/\sin\lambda = 229/\sin 14.04° = 944 \text{ ft/min}$$
$$f = 0.32/v_s^{0.36} = 0.32/(944)^{0.36} = 0.027$$

The axial load on the gear (from equation [12-41]):

$$W_{xG} = 962 \text{ lb}\left[\frac{(0.943)(0.243) + (0.027)(0.970)}{(0.943)(0.970) - (0.027)(0.243)}\right] = 270 \text{ lb}$$

The radial load on the gear (from equation [12-42]):

$$W_{rG} = \left[\frac{(962)(0.333)}{(0.943)(0.970) - (0.027)(0.243)}\right] = 353 \text{ lb}$$

The efficiency (from equation [12-43]):

$$\text{eff} = 0.250\left[\frac{0.943 - (0.027)(0.250)}{(0.943)(0.250) + 0.027}\right] = 0.891$$

This represents 89.1 percent efficiency.

Input power:

$$\text{Because eff} = P_o/P_i$$

$$P_i = P_o/\text{eff} = (6.68 \text{ hp})/0.891 = 7.50 \text{ hp}$$

Factors Affecting Efficiency

As can be seen in equation (12-43), the lead angle, the normal pressure angle, and the coefficient of friction all affect the efficiency. The one that has the largest affect, and over which the designer has the most control, is the lead angle, λ. The larger the lead angle, the higher the efficiency, up to approximately $\lambda = 45°$. Now, looking back to the definition of the lead angle, it is seen that the number of threads in the worm has a major effect on the lead angle. Therefore, to obtain a high efficiency, use multiple-threaded worms. But there is a disadvantage to this conclusion. More worm threads require more gear teeth to achieve the same ratio, resulting in a larger system overall. The designer is often forced to compromise.

Self-locking Wormgear Sets

Self-locking is the condition in which the worm drives the wormgear but, if torque is applied to the gear shaft, the worm does not turn. It is locked! The locking action is produced by the friction force between the worm threads and the wormgear teeth, and this is highly dependent on the lead angle. It is recommended that a lead angle no higher than about 5.0° be used in order to ensure that self-locking will occur. This lead angle usually requires the use of a single-threaded worm; the low lead angle results in a low efficiency, possibly as low as 60 to 70 percent.

12-16 STRESS AND WEAR IN WORMGEAR SETS

A modified form of the Lewis equation is used to compute the stress in the teeth of the wormgear. The threads of the worm are inherently stronger than the wormgear teeth, so only the latter need be analyzed. The calculation is only approximate because the shape of the teeth is not uniform across the width. However, sufficient accuracy is obtained to achieve a satisfactory design. In addition, the final rating of the gear set is more likely to depend on wear than on strength. In fact, because of the energy losses due to friction, the heat gain of the system may be the limiting factor in determining the power rating.

Table 12-3 Approximate Lewis Form Factor for Wormgear Teeth

ϕ_n	y
14½°	0.100
20°	0.125
25°	0.150
30°	0.175

The stress in the gear teeth can be computed from

$$\sigma = \frac{W_d}{yFP_{cn}} \tag{12-44}$$

where

W_d = Dynamic load on the gear teeth

y = Lewis form factor (see Table 12-3)

F = Face width of the gear

$$P_{cn} = \text{Normal circular pitch} = P_c \cos \lambda = \pi \cos \lambda / P_d \tag{12-45}$$

The dynamic load can be estimated from

$$W_d = W_{tG}/K_v \tag{12-46}$$

and

$$K_v = 1\,200/(1\,200 + v_{tG}) \tag{12-47}$$

$$v_{tG} = \pi D_G n_G/12 = \text{Pitch line speed of the gear} \tag{12-48}$$

Only one value is given for the Lewis form factor for a given pressure angle because the actual value is very difficult to calculate precisely and does not vary much with the number of teeth. The actual face width should be used, up to the limit of two-thirds of the pitch diameter of the worm.

The computed value of tooth bending stress from equation (12-44) can be compared with the fatigue strength of the material of the gear. For manganese gear bronze, use a fatigue strength of 17 000 psi; for phosphor gear bronze, use 24 000 psi. For cast iron, use approximately 0.35 times the ultimate strength, unless specific data are available for fatigue strength.

Wear Analysis

We will use the limiting load for wear to evaluate a wormgear set design for wear. We can compute

$$W_w = D_G F K_w \tag{12-49}$$

Table 12-4 Wear Load Factors for Worms

MATERIALS		K_w AT LISTED ϕ_n			
Worm	*Gear*	14½°	20°	25°	30°
Hardened steel	Chilled bronze	90	125	150	180
Hardened steel	Bronze	60	80	100	120
Steel	Bronze	36	50	60	72
Steel	Gray cast iron	55	75	90	110
Steel	Laminated phenolic	47	64	80	95

Note: Hardened steel: HB 500 or higher. Usually case-hardened. Other steel worms in "as wrought" condition, approximately HB 250 minimum. Chilled bronze produced in a special manner to obtain a hard surface. Good lubrication assumed.

Source: Faires, V. M. *Design of Machine Elements,* 4th ed. (New York: Macmillan Publishing Company, 1965).

where

W_w = Limiting load for wear

K_w = Wear load factor for wormgears (see Table 12-4)

The other terms have their usual meanings.

Example Problem 12-11. Is the wormgear set described in example problem 12-9 satisfactory with regard to strength and wear when operating under the conditions of example problem 12-10? The wormgear has a face width of 1.00 in.

Solution. From previous problems and solutions:

$$W_{tG} = 962 \text{ lb}$$

$$v_{tG} = 229 \text{ ft/min}$$

$$D_G = 8.667 \text{ in}$$

Steel worm: Assume HB 250 minimum.

Bronze gear: Assume regular "as-cast."

Stress:

$$K_v = 1\,200/(1\,200 + v_{tG}) = 1\,200/(1\,200 + 229) = 0.84$$

$$W_d = W_{tG}/K_v = 962/0.84 = 1\,145 \text{ lb}$$

$$F = 1.00 \text{ in}$$

$$y = 0.125 \quad \text{(from Table 12-3)}$$

$$P_{cn} = P_c \cos\lambda = 0.523\,6 \cos 14.04° = 0.508 \text{ in}$$

Then

$$\sigma = \frac{W_d}{yFP_{cn}} = \frac{1\,145}{(0.125)(1.00)(0.508)} = 18\,050 \text{ psi}$$

This seems satisfactory for the phosphor gear bronze, which has a fatigue strength of approximately 24 000 psi.

Wear: From Table 12-4, $K_w = 50$.

Then

$$W_w = D_G F K_w = (8.667)(1.00)(50) = 433 \text{ lb}$$

Because W_w is much less than W_d, the set would probably wear out in a short time. If the worm were to be surface-hardened above HB 500, and if the gear were to be made from chilled bronze, then the value of K_w would be 125. Then,

$$W_w = (8.667)(1.00)(125) = 1\,083 \text{ lb}$$

This is still lower than the dynamic load, but it is closer. Because these calculations are approximate, this design could give a satisfactorily long life. Performance testing would ensure acceptable life. Of course, a redesign can be done, probably increasing the size of the gear, and/or its face width, to improve the limiting load for wear.

If a guaranteed rating is required, the standard of the AGMA (4) should be consulted. Using the procedures described in this section should be conservative.

12-17 TYPICAL GEOMETRY OF WORMGEAR SETS

Considerable latitude is permissible in the design of wormgear sets because the worm and wormgear combination is designed as a unit. However, there are some guidelines.

Typical Tooth Dimensions

Table 12-5 shows typical values used for the dimensions of worm threads and gear teeth when both are the same.

Worm Diameter

The diameter of the worm affects the lead angle, which in turn affects the efficiency of the set. For this reason, small diameters are desirable. But for practical reasons and proper proportion with respect to the wormgear, it is recommended that the worm diameter be approximately $C^{0.875}/2.2$, where C is the center distance between the worm and the wormgear. Variation of 30 percent is allowed (4). Thus the worm diameter should fall in the range

$$1.7 < \frac{C^{0.875}}{D_W} < 3.0 \qquad (12\text{-}50)$$

But some commercially available wormgear sets fall outside this range, especially in the smaller sizes. Also, those worms designed to have a through hole bored in them for installation on a shaft are typically larger than you would find from equation (12-50). Proper proportion and efficient use of material should be the guide. The worm shaft must also be checked for deflection under operating loads. For worms machined integral with the shaft, the root of the worm threads determines the minimum shaft diameter. For worms having bored holes, sometimes called *shell worms,* care must be exercised to leave

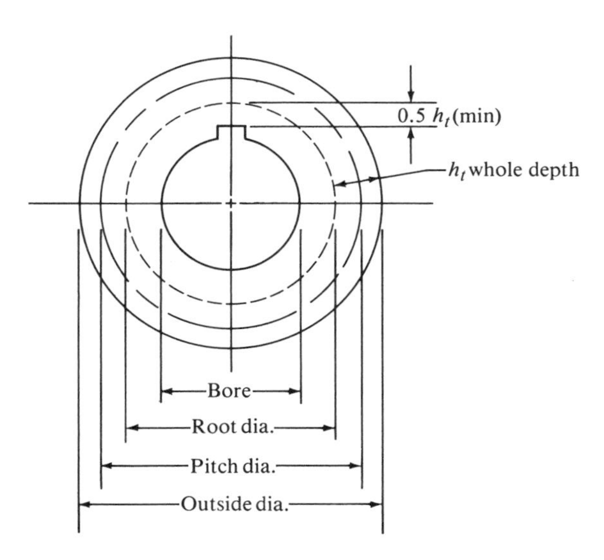

Figure 12-25 Shell Worm

sufficient material between the thread root and the keyway in the bore. Figure 12-25 shows the recommended thickness above the keyway to be a minimum of one-half the whole depth of the threads.

Wormgear Dimensions

We are concerned here with the single-enveloping type of wormgear, as shown in Figure 12-26. Its addendum, dedendum, and depth dimensions are assumed to be the same as those listed in Table 12-5, measured at the throat of the wormgear teeth. The throat is in line with the vertical centerline of the worm. The recommended face width for the wormgear is

$$F_G = (D_{oW}^2 - D_W^2)^{1/2} \tag{12-51}$$

Table 12-5 Typical Tooth Dimensions for Worms and Wormgears

Dimension	*Formula*
Addendum	$a = 0.3183\ P_a = 1/P_d$
Whole depth	$h_t = 0.6866\ P_a = 2.157/P_d$
Working depth	$h_k = 2a = 0.6366\ P_a = 2/P_d$
Dedendum	$b = h_t - a = 0.3683\ P_a = 1.157/P_d$
Root diameter of worm	$D_{rW} = D_W - 2b$
Outside diameter of worm	$D_{oW} = D_W + 2a = D_W + h_k$
Root diameter of gear	$D_{rG} = D_G - 2b$
Throat diameter of gear	$D_t = D_G + 2a$

Source: American Gear Manufacturers Association. *Design of General Industrial Coarse-Pitch Cylindrical Wormgearing*. Standard AGMA 341.02-1965 (R1970). Arlington, Va.; 1970.

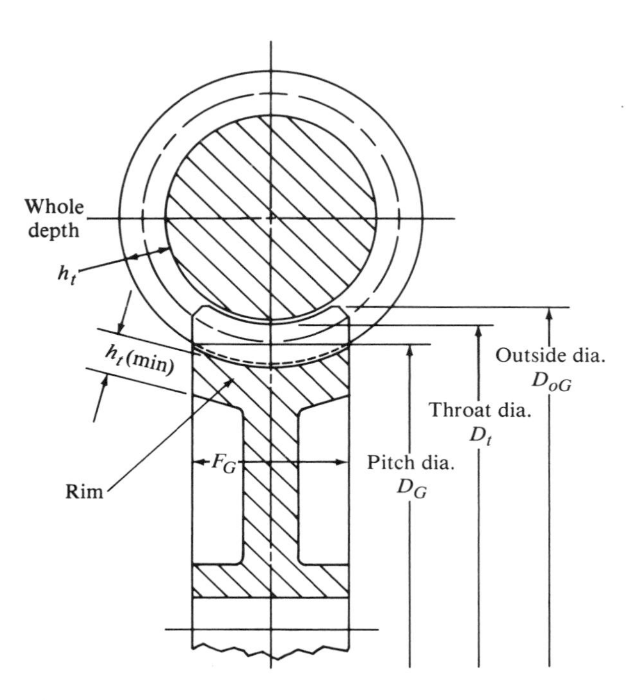

Figure 12-26 Wormgear Details

This corresponds to the length of the line tangent to the pitch circle of the worm and limited by the outside diameter of the worm. Any face width beyond this value would not be effective in resisting stress or wear, but a convenient value slightly greater than the minimum should be used. The outer edges of the wormgear teeth should be chamfered approximately as shown in Figure 12-26.

Another recommendation, which is convenient for initial design, is that the face width of the gear should be approximately 2.0 times the circular pitch. Because we are working in the diametral pitch system, we will use

$$F_G = 2P_c = 2\pi/P_d \tag{12-52}$$

But since this is only approximate and 2π is approximately 6, we will use

$$F_G = 6/P_d \tag{12-53}$$

If the gear web is thinned, a rim thickness at least equal to the whole depth of the teeth should be left.

Face Length of the Worm

For maximum load sharing, the worm face length should extend to at least the point where the outside diameter of the worm intersects the throat diameter of the wormgear. This length is

$$F_W = 2[(D_t/2)^2 - (D_G/2 - a)^2]^{1/2} \tag{12-54}$$

12-18 DESIGN OF WORMGEARING

The design of wormgearing is by no means standardized. Manufacturers sometimes use modified techniques to achieve a particular goal. And because the worm and wormgear are made as a matched set, there is little need for interchangeability. The design procedure presented here should enable you to produce a reasonable design and, in the process, learn the effects of different design decisions. With this experience, the selection of commercially available wormgearing can be approached more confidently. It is recommended that any tentative design be checked by performance testing or by comparing the predicted performance with the standards of the AGMA, listed at the end of this chapter.

General Statement of the Design Problem

It will be assumed that the following information is given:

Required torque at the output shaft, T_o
or
Output power, P_o

Output rotational speed, n_G

Input rotational speed, n_W
or
Velocity ratio, $VR = n_W/n_G$

Then the following tentative decisions must be made:

Diametral pitch

Pressure angle (usually 20°, normal pressure angle).

Type of material for the worm and the wormgear (see Table 12-4).

Number of threads in the worm (Remember: A small number of threads will allow a small number of wormgear teeth but may have low efficiency.).

Approximate face width of the gear: $F_G = 6/P_d$.

The design then works toward a satisfactory wear load limit, as compared with the dynamic load on the wormgear teeth.

Compute:

$$N_G = N_t(VR)$$

$$D_G = N_G/P_d$$

$$W_w = D_G F_G K_w$$

$$T_o = 63\,000(P_o)/n_G$$

$$W_{tG} = T_o/(D_G/2)$$

$$v_{tG} = \pi D_G n_G/12$$

$$W_d = W_{tG}\left[\frac{1\,200 + v_{tG}}{1\,200}\right]$$

Compare: W_w must be greater than W_d. If it is not, select a smaller value for P_d and repeat the process. If it is, continue with the stress analysis.

Select a trial value for D_W.
Compute: $C = (D_W + D_G)/2$ = Center distance
Compute: $C^{0.875}/D_W$
Compare: Is this value between 1.7 and 3.0? If so, D_W is satisfactory. If not, adjust D_W.
Compute:

$$P_a = P_c = \pi/P_d$$
$$L = N_t P_a$$
$$\lambda = \tan^{-1}[L/(\pi D_W)]$$
$$P_{cn} = P_c \cos\lambda = \pi \cos\lambda/P_d$$

Find y from Table 12-3.
Compute:

$$\sigma = \frac{W_d}{yF_G P_{cn}}$$

Compare: The value for σ must be greater than the fatigue strength of the material for the wormgear. If it is not, choose a smaller value for P_d and repeat the process. If it is, the design is satisfactory for strength and wear. The resulting design can be evaluated for efficiency, forces, and other geometric features.

Example Problem 12-12. Design a wormgear set to transmit 1.5 hp at the output shaft at a rotational speed of approximately 55 rpm. The worm will rotate at 1 750 rpm. Check both strength and wear.

Solution. Tentative design decisions:

$P_d = 10$

20° normal pressure angle

Hardened steel worm and chilled bronze gear ($K_w = 125$)

Double-threaded worm, $N_t = 2$

Approximate $F_G = 6/P_d = 6/10 = 0.600$ in

Use $F_G = 5/8$ in $= 0.625$ in

Computations:

$$VR = n_W/n_G = 1\,750/55 = 31.8 \quad \text{(ideal)}$$
$$\text{Let } VR = 32$$
$$n_G = n_W/(VR) = 1\,750/32 = 54.7 \text{ rpm} \quad \text{(OK)}$$
$$N_G = N_t(VR) = 2(32) = 64 \text{ teeth}$$
$$D_G = N_G/P_d = 64/10 = 6.400 \text{ in}$$
$$W_w = D_G F_G K_w = (6.400)(0.625)(125) = 500 \text{ lb}$$
$$T_o = 63\,000(P_o)/n_G = 63\,000(1.5)/54.7 = 1\,728 \text{ lb} \cdot \text{in}$$
$$W_{tG} = T_o/(D_G/2) = 1\,728/(6.400/2) = 540 \text{ lb}$$

The transmitted load on the gear teeth is itself larger than the limiting load for wear. The dynamic load will be even larger. Therefore, let's immediately reduce P_d to 8 and repeat the calculations. The formulas will not be repeated. We will continue with $N_t = 2$, $N_G = 64$. Then,

$$\approx F_G = 6/P_d = 6/8 = 0.75 \text{ in} \quad \text{(OK)}$$
$$D_G = 64/8 = 8.000 \text{ in}$$
$$W_w = (8.000)(0.75)(125) = 750 \text{ lb}$$
$$T_o = 1\,728 \text{ lb} \cdot \text{in} \quad \text{(no change)}$$
$$W_{tG} = 1\,728/(8.000/2) = 432 \text{ lb}$$
$$v_{tG} = \pi D_G n_G/12 = (\pi)(8.000)(54.7)/12 = 115 \text{ ft/min}$$
$$W_d = W_{tG}\left[\frac{1\,200 + v_{tG}}{1\,200}\right] = 432\left[\frac{1\,200 + 115}{1\,200}\right] = 473 \text{ lb}$$

Now W_d is less than the limiting load for wear, W_w, so the design is satisfactory for wear. We can continue with the stress analysis. Let's try a worm diameter $D_W = 1.500$ in. Then,

$$C = (D_W + D_G)/2 = (1.500 + 8.000)/2 = 4.750 \text{ in}$$
$$C^{0.875}/D_W = (4.750)^{0.875}/1.500 = 2.61$$

This value is between 1.7 and 3.0. The proportion is OK.

$$P_a = P_c = \pi/P_d = \pi/8 = 0.3927 \text{ in}$$
$$L = N_t P_a = (2)(0.392\,7) = 0.785\,4 \text{ in}$$
$$\lambda = \tan^{-1}[L/(\pi D_W)] = \tan^{-1}[0.785\,4/(\pi 1.500)] = 9.46°$$
$$P_{cn} = P_c \cos\lambda = 0.392\,7 \cos 9.46° = 0.387 \text{ in}$$
$$y = 0.125 \quad \text{(from Table 12-3)}$$
$$\sigma = \frac{W_d}{yF_G P_{cn}} = \frac{473}{(0.125)(0.75)(0.387)} = 13\,000 \text{ psi}$$

This should provide a satisfactory fatigue life, using bronze for the gear.

Other geometric features are

$$a = 1/P_d = 1/8 = 0.125 \text{ in}$$
$$D_t = D_G + 2a = 8.000 + 2(0.125) = 8.250 \text{ in}$$

The face length of the worm should be approximately [equation (12-54)]

$$F_W = 2[(8.25/2)^2 - (8.000/2 - 0.125)^2]^{1/2} = 2.828 \text{ in}$$

Let's round this off to 3.00 in.

Summary of the Design

Given:	$P_o = 1.5$ hp	$n_W = 1\,750$ rpm	$n_G = 55$ rpm (approx.)	
Results:	$N_t = 2$	$N_G = 64$	$VR = 32$	$n_G = 54.7$ rpm
	$P_d = 8$	$D_G = 8.000$ in	$D_W = 1.500$ in	$C = 4.750$ in
	$a = 0.125$ in	$D_t = 8.250$ in	$D_{oW} = 1.750$ in	
	$\lambda = 9.46°$	$F_G = 0.75$ in	$F_W = 3.00$ in	

12-19 COMPUTER PROGRAM FOR DESIGN AND ANALYSIS OF WORMGEARING

As the several example problems presented in this chapter show, numerous calculations are performed in designing wormgearing and analyzing a proposed design for geometry, strength, wear, and forces. The following is a computer program, written in BASIC, that aids in the design process. A sample of the output from the program is also given. The data used for the sample output are the same as those used in the example problem 12-12. You should follow along with the program and the output to see how each value is computed. The design procedure used in the program is that described in section 12-18. The flowchart is shown in Figure 12-27. The program is limited to designs using hardened steel worms and chilled bronze wormgears for wear analysis. Other material combinations could be added.

The variable names used in the program are listed next.

A = Normal pressure angle ϕ_n | A1 = Lead angle λ
A3 = Addendum
B = Dedendum | C = Center distance
C1 = cos(A) = cos(ϕ_n) | C2 = cos(A1) = cos(λ)
D1 = Worm pitch diameter | D2 = Gear pitch diameter
D3 = Gear throat diameter | D4 = Worm outside diameter
D5 = Worm root diameter | D6 = Gear root diameter
E = Efficiency | F = Coefficient of friction
F1 = Length of worm face | F2 = Width of gear face
K = Wear load factor | K1 = Velocity factor
L = Lead
N1 = Input speed | N2 = Output speed
P = Diametral pitch
P1 = Input power | P2 = Output power
P3 = π = 3.141 592 6 | P4 = Normal circular pitch
R = Gear ratio = T2/T1 | S = Bending stress on gear teeth
S1 = sin(A) = sin(ϕ_n) | S2 = sin(A1) = sin(λ)
T1 = No. of worm threads | T2 = No. of gear teeth
T3 = Output torque | T4 = tan(A1) = tan(λ)
T5 = Input torque
V1 = Worm pitch line speed | V2 = Gear pitch line speed
V3 = Sliding velocity
W1 = Wear load limit | W2 = Tangential load on gear
W3 = Dynamic load on gear | W4 = Axial force on gear
W5 = Radial force on gear | Y = Form factor for gear teeth
X1, Z, Z1 are dummy variables

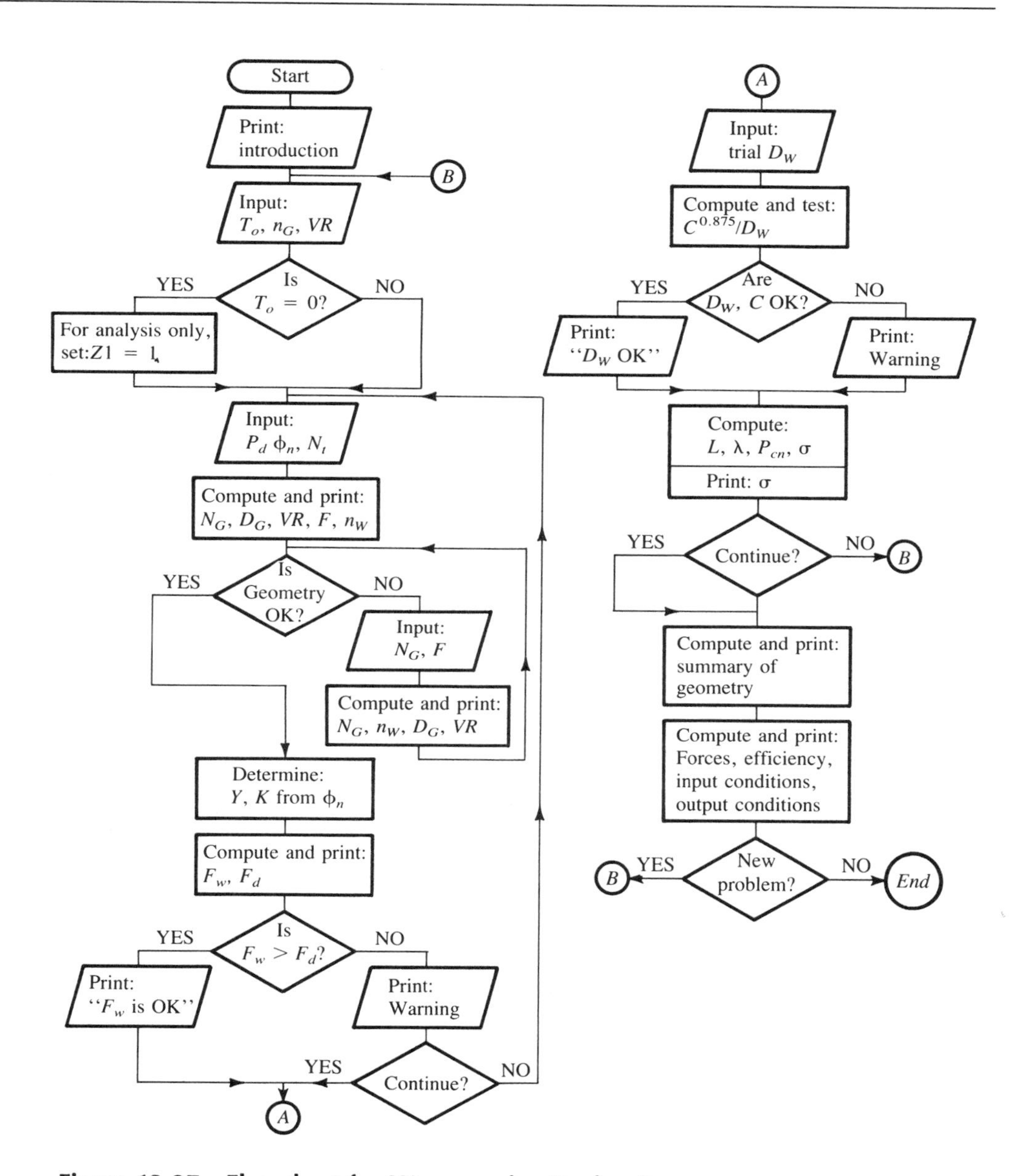

Figure 12-27 Flowchart for Wormgearing Design Program

Some of the features of the program follow.

Lines 100 to 320: After a brief introduction, the basic problem data are requested. For design, a specific value of the required output torque would be typed in. For an analysis of a given gear set, typing *T3* = *0* instructs the program to use the wear load limit to determine the torque rating of the set.

Lines 330 to 710: From the given diametral pitch, normal pressure angle, number of worm threads, and desired gear ratio, the number of gear teeth (an integer), the actual gear ratio, the gear face width, and the gear pitch diameter are computed. The designer is permitted to evaluate the geometry and input specific values or accept the computed results.

Lines 720 to 880: The wear load factor, K, and the form factor, Y, are specified for the given normal pressure angle (using the data of Tables 12-3 and 12-4) for a hardened steel worm and a chilled bronze wormgear.

Lines 890 to 1250: The wear load limit is compared with the dynamic load on the gear teeth, and the results of this trial design are printed. If the wear load limit is lower than the dynamic load, the designer is informed. The designer can then decide whether to continue with the design or try new values.

Lines 1260 to 1560: The worm diameter is specified, and the lead, lead angle, and normal circular pitch are computed. After a worm diameter is tried, the center distance is computed and the check of worm/wormgear proportion [equation (12-50)] is made. Finally, the bending stress is computed. The designer is permitted to accept the design at this point or to try new data.

Lines 1570 to 1820: The geometry data for the worm and the wormgear are completed and printed.

Lines 1830 to 2330: The forces on the gear, the efficiency, the general data for the set, a summary of the output conditions, and a summary of the input conditions complete the design.

Lines 2340 to 2390: The designer is given two choices of how to proceed.

Wormgear Design: Program Listing

```
100 PRINT "THIS PROGRAM ASSISTS IN THE DESIGN OF WORMGEARING"
110 PRINT "  CONSIDERING GEOMETRY, STRENGTH, WEAR, AND "
120 PRINT "  THE FORCES EXERTED ON THE WORMGEAR. "
130 PRINT
140 PRINT "HARDENED STEEL CYLINDRICAL WORMS WITH"
150 PRINT "  CHILLED BRONZE SINGLE-ENVELOPING WORMGEARS"
160 PRINT "  ARE USED. PROCEDURE AND DATA FROM"
170 PRINT "  MACHINE ELEMENTS IN MECHANICAL DESIGN,"
180 PRINT "  BY R. L. MOTT, CHARLES E. MERRILL PUBLISHING CO."
190 PRINT
200 PRINT "********* PROBLEM DATA *********"
210 PRINT
220 PRINT "TO BEGIN, TYPE IN THE FOLLOWING DATA"
230 PRINT "  T3 = OUTPUT TORQUE IN LB-IN"
240 PRINT "  (FOR ANALYSIS ONLY, LET T3 = 0)
250 PRINT "  N2 = OUTPUT SHAFT SPEED IN RPM"
260 PRINT "  VELOCITY RATIO = VR = INPUT/OUTPUT"
```

Introduction

```
270 PRINT
280 Z1 = 0
290 PRINT "TYPE IN T3, N2, VR"
300 INPUT T3, N2, R
310 IF T3 > 0 THEN 330
320 Z1 = 1
```

Input problem data

```
330 PRINT
340 PRINT "    **** DESIGN DATA ****"
350 PRINT
360 PRINT "NOW MAKE THE FOLLOWING DESIGN DECISIONS"
370 PRINT
```

Design decisions

```
380 PRINT "   P = DIAMETRAL PITCH"
390 PRINT "   A = NORMAL PRESSURE ANGLE: 14.5, 20, 25, 30"
400 PRINT "   T1 = NUMBER OF THREADS IN THE WORM"
410 PRINT
420 PRINT "TYPE IN P, A, T1"
430 INPUT P,A,T1
440 T2 = INT(T1*R + .5)
450 D2 = T2/P
460 R = T2/T1
470 F2 = 6/P
480 N1 = N2*R
490 PRINT "      **** TRIAL GEOMETRY ****"
500 PRINT
510 PRINT "T2 = NUMBER OF TEETH IN GEAR =";T2
520 PRINT "D2 = GEAR PITCH DIAMETER =";D2;"INCHES"
530 PRINT "F2 = NOMINAL FACE WIDTH = 6/P =";F2;"INCHES"
540 PRINT "ACTUAL GEAR RATIO = T2/T1 =";R
550 PRINT "INPUT SPEED =";N1;"RPM"
560 PRINT
570 PRINT "IS GEOMETRY OK? 1 = YES; 2 = NO"
580 INPUT Z
590 IF Z = 1 THEN 720
600 PRINT "TYPE IN DESIRED T2, F2"
610 INPUT T2,F2
620 R = T2/T1
630 N1 = N2*R
640 D2 = T2/P
650 PRINT
660 PRINT "T2 = NUMBER OF TEETH IN GEAR =";T2
670 PRINT "D2 = GEAR PITCH DIAMETER =";D2;"INCHES"
680 PRINT "F2 = ACTUAL FACE WIDTH =";F2;"INCHES"
690 PRINT "ACTUAL VELOCITY RATIO = T2/T1 =";R
700 PRINT "INPUT SPEED =";N1;"RPM"
710 GOTO 570
720 IF A > 14.5 THEN 760
730 K = 90
740 Y = .1
750 GOTO 890
760 IF A > 20 THEN 800
770 K = 125
780 Y = .125
790 GOTO 890
800 IF A > 25 THEN 840
810 K = 150
820 Y = .15
830 GOTO 890
840 IF A <= 30 THEN 870
850 PRINT "PRESSURE ANGLE LARGER THAN 30 DEGREES"
860 GOTO 360
870 K = 180
880 Y = .175
890 W1 = D2*F2*K
900 P3 = 3.1415926
910 V2 = P3*D2*N2/12
920 K1 = (1200 + V2)/1200
930 IF Z1 <> 1 THEN 980
940 W2 = W1/K1
950 T3 = W2*D2/2
960 W3 = W1
970 GOTO 1000
980 W2 = T3/(D2/2)
990 W3 = W2*K1
1000 PRINT
1010 P2 = T3*N2/63000
```

Trial geometry (lines 490–550)

Evaluate geometry (lines 570–710)

Determine *K* and *Y* (lines 720–880)

Evaluate wear load and dynamic load (lines 890–1010)

```
1020 PRINT " DATA: PITCH =";P;"      PRESSURE ANGLE=";
  A;"DEGREES"
1030 PRINT
1040 PRINT "RESULTS:"
1050 PRINT "WEAR LOAD FACTOR, KW =";K
1060 PRINT "LIMITING LOAD FOR WEAR =";W1;"POUNDS"
1070 PRINT
1080 PRINT "OUTPUT TORQUE =";T3;"LB-IN"
1090 PRINT "TANGENTIAL LOAD ON GEAR =";W2;"POUNDS"
1100 PRINT "PITCH LINE SPEED OF GEAR =";V2;"FT/MIN"
1110 PRINT "DYNAMIC LOAD =";W3;"POUNDS"
1120 IF W1 >= W3 THEN 1230
1130 PRINT
1140 PRINT "WEAR LOAD LESS THAN DYNAMIC LOAD"
1150 PRINT "INCREASE PITCH DIAMETER, DECREASE DIAMETRAL
  PITCH,"
1160 PRINT "      OR CHANGE MATERIAL"
1170 PRINT
1180 PRINT "TYPE 1 IF YOU WANT TO PROCEED TO STRESS ANALYSIS"
1190 PRINT "TYPE 2 TO TRY A NEW DESIGN NOW"
1200 INPUT Z
1210 IF Z = 1 GOTO 1250
1220 GOTO 330
1230 PRINT
1240 PRINT "WEAR LOAD SATISFACTORY"
1250 PRINT
1260 PRINT
1270 PRINT "BEGIN STRESS ANALYSIS"
1280 PRINT
1290 PRINT "INPUT TRIAL WORM DIAMETER, D1"
1300 INPUT D1
1310 C = (D1 + D2)/2
1320 PRINT "CENTER DISTANCE =";C;"IN."
1330 X1 = C^.875/D1
1340 PRINT "X1 = C^0.875/D1 =";X1
1350 IF X1 > 3 THEN 1390
1360 IF X1 < 1.7 THEN 1410
1370 PRINT "WORM DIAMETER SATISFACTORY; 1.7 < X1 < 3.0"
1380 GOTO 1430
1390 PRINT "WORM DIAMETER TOO SMALL; MAX X1 = 3.0
  RECOMMENDED"
1400 GOTO 1430
1410 PRINT "WORM DIAMETER TOO LARGE; MIN X1 = 1.7
  RECOMMENDED"
1430 PRINT
1440 L = T1*P3/P
1450 A1 = ATN(L/(P3*D1))
1460 P4 = P3*COS(A1)/P
1470 A1 = A1*180/P3
1480 S = W3/(Y*F2*P4)
1490 PRINT "BENDING STRESS =";S;"PSI"
1500 PRINT
1510 PRINT "TO TRY A NEW DESIGN, TYPE 1"
1520 PRINT "TO CONTINUE WITH THIS DESIGN, TYPE 2"
1530 INPUT Z
1540 PRINT
1550 ON Z GOTO 230,1560
1560 PRINT
1570 PRINT "****** SUMMARY OF DESIGN ******"
1580 PRINT
1590 PRINT "      WORM"
1600 A3 = 1/P
1610 B = 1.157/P
1620 D3 = D2 + 2*A3
1630 F1 = 2*SQR((D3/2)^2 - (D2/2 - A3)^2)
1640 PRINT
```

Decision on how to continue

Worm geometry

Stress analysis

Decision on how to continue

Summary of design

```
1650 PRINT "WORM DIAMETER =";D1;"IN."
1660 PRINT "NUMBER OF THREADS =";T1
1670 PRINT "LEAD ANGLE =";A1;"DEGREES"
1680 D4 = D1 + 2*A3
1690 D5 = D1 - 2*B
1700 PRINT "OUTSIDE DIAMETER =";D4;"IN."
1710 PRINT "ROOT DIAMETER =";D5;"IN."
1720 PRINT "FACE LENGTH =";F1;"IN."
1730 PRINT
1740 PRINT "     WORMGEAR"
1750 PRINT
1760 PRINT "PITCH DIAMETER =";D2;"IN."
1770 PRINT "NUMBER OF TEETH =";T2
1780 PRINT "FACE WIDTH =";F2;"IN."
1790 PRINT "THROAT DIAMETER =";D3;"IN."
1800 D6 = D2 - 2*B
1810 PRINT "ROOT DIAMETER =";D6;"IN."
1820 PRINT
1830 PRINT "      FORCES ON THE WORM AND WORMGEAR"
1840 PRINT
1850 V3 = V2/SIN(A1*P3/180)
1860 F = .32/(V3^.36)
1870 C1 = COS(A*P3/180)
1880 C2 = COS(A1*P3/180)
1890 S1 = SIN(A*P3/180)
1900 S2 = SIN(A1*P3/180)
1910 T4 = TAN(A1*P3/180)
1920 W4 = W2*(C1*S2 + F*C2)/(C1*C2 - F*S2)
1930 W5 = W2*S1/(C1*C2 - F*S2)
1940 E = T4*(C1 - F*T4)/(C1*T4 + F)
1950 P1 = P2/E
1960 T5 = 63000*P1/N1
1970 V1 = P3*D1*N1/12
1980 PRINT "TANGENTIAL FORCE ON THE WORMGEAR - AND -"
1990 PRINT "  AXIAL FORCE ON THE WORM:"
2000 PRINT W2;"POUNDS"
2010 PRINT
2020 PRINT "AXIAL FORCE ON THE WORMGEAR - AND -"
2030 PRINT "  TANGENTIAL FORCE ON THE WORM:"
2040 PRINT W4;"POUNDS"
2050 PRINT
2060 PRINT "RADIAL FORCE ON THE WORMGEAR - AND -"
2070 PRINT "  RADIAL FORCE ON THE WORM:"
2080 PRINT W5;"POUNDS"
2090 PRINT
2100 PRINT
2110 PRINT "        EFFICIENCY =";E*100;"PERCENT"
2120 PRINT
2130 PRINT
2140 PRINT "      DATA FOR THE SET"
2150 PRINT
2160 PRINT "VELOCITY RATIO = ";R
2170 PRINT "DIAMETRAL PITCH =";P
2180 PRINT "NORMAL PRESSURE ANGLE =";A;"DEGREES"
2190 PRINT "CENTER DISTANCE =";C;"IN."
2200 PRINT "ADDENDUM =";A3;"IN."
2210 PRINT "DEDENDUM =";B;"IN."
2220 PRINT
2230 PRINT "OUTPUT SPEED =";N2;"RPM (GIVEN)"
2240 PRINT "OUTPUT TORQUE =";T3;"LB-IN"
2250 PRINT "OUTPUT POWER RATING =";P2;"HORSEPOWER"
2260 PRINT "PITCH LINE SPEED OF THE GEAR =";V2;"FT/MIN"
2270 PRINT
2280 PRINT "INPUT SPEED =";N1;"RPM"
2290 PRINT "TORQUE ON INPUT SHAFT =";T5;"LB-IN"
```

Forces and efficiency

Final summary

```
2300 PRINT "POWER REQUIRED AT INPUT SHAFT=";
  P1;"HORSEPOWER"
2310 PRINT "PITCH LINE SPEED OF THE WORM =";V1;"FT/MIN"
2320 PRINT
2330 PRINT "      **** DESIGN COMPLETED ****"
2340 PRINT
2350 PRINT "TO BEGIN A NEW PROBLEM, TYPE 1"
2360 PRINT "TO END, TYPE 2"
2370 INPUT Z
2380 ON Z GOTO 190,2390
2390 END
```

Decision on how to continue

Wormgear Design: Sample Output[1]

```
THIS PROGRAM ASSISTS IN THE DESIGN OF WORMGEARING
  CONSIDERING GEOMETRY, STRENGTH, WEAR, AND
  THE FORCES EXERTED ON THE WORMGEAR.

HARDENED STEEL CYLINDRICAL WORMS WITH
  CHILLED BRONZE SINGLE-ENVELOPING WORMGEARS
  ARE USED. PROCEDURE AND DATA FROM
  MACHINE ELEMENTS IN MECHANICAL DESIGN,
  BY R. L. MOTT, CHARLES E. MERRILL PUBLISHING CO.

********* PROBLEM DATA *********

TO BEGIN, TYPE IN THE FOLLOWING DATA
  T3 = OUTPUT TORQUE IN LB-IN
    (FOR ANALYSIS ONLY, LET T3 = 0)
  N2 = OUTPUT SHAFT SPEED IN RPM
  VELOCITY RATIO = VR = INPUT/OUTPUT

TYPE IN T3, N2, VR
? 1718 ,55 ,31.8                                    Statement 300

        **** DESIGN DATA ****

NOW MAKE THE FOLLOWING DESIGN DECISIONS

  P = DIAMETRAL PITCH
  A = NORMAL PRESSURE ANGLE: 14.5, 20, 25, 30
  T1 = NUMBER OF THREADS IN THE WORM

TYPE IN P, A, T1
? 10 ,20 ,2                                         Statement 430

        **** TRIAL GEOMETRY ****

T2 = NUMBER OF TEETH IN GEAR = 64
D2 = GEAR PITCH DIAMETER = 6.4 INCHES
F2 = NOMINAL FACE WIDTH = 6/P = .6 INCHES
ACTUAL GEAR RATIO = T2/T1 = 32
INPUT SPEED = 1760 RPM

IS GEOMETRY OK? 1 = YES; 2 = NO
? 2                                                 Statement 580
TYPE IN DESIRED T2, F2
? 64 , .625                                         Statement 610

T2 = NUMBER OF TEETH IN GEAR = 64
D2 = GEAR PITCH DIAMETER = 6.4 INCHES
F2 = NOMINAL FACE WIDTH = .625 INCHES
ACTUAL VELOCITY RATIO = T2/T1 = 32
```

[1]Data from Example Problem 12-12

```
  INPUT SPEED = 1760 RPM
  IS GEOMETRY OK? 1 = YES; 2 = NO
▶ ? 1                                                    Statement 580

  DATA: PITCH = 10        PRESSURE ANGLE = 20 DEGREES

  RESULTS:
  WEAR LOAD FACTOR, KW = 125
  LIMITING LOAD FOR WEAR = 500 POUNDS

  OUTPUT TORQUE = 1718 LB-IN
  TANGENTIAL LOAD ON GEAR = 536.875 POUNDS
  PITCH LINE SPEED OF GEAR = 92.1534 FT/MIN
  DYNAMIC LOAD = 578.104 POUNDS

  WEAR LOAD LESS THAN DYNAMIC LOAD
    INCREASE PITCH DIAMETER, DECREASE DIAMETRAL PITCH,
      OR CHANGE MATERIAL

  TYPE 1 IF YOU WANT TO PROCEED TO STRESS ANALYSIS
  TYPE 2 TO TRY A NEW DESIGN NOW
▶ ? 2                                                    Statement 1200

          **** DESIGN DATA ****

  NOW MAKE THE FOLLOWING DECISIONS

    P = DIAMETRAL PITCH
    A = NORMAL PRESSURE ANGLE; 14.5, 20, 25, 30
    T1 = NUMBER OF THREADS IN THE WORM

  TYPE IN P, A, T1
▶ ? 8 , 20 , 2                                           Statement 430

          **** TRIAL GEOMETRY ****

  T2 = NUMBER OF TEETH IN GEAR = 64
  D2 = GEAR PITCH DIAMETER = 8 INCHES
  F2 = NOMINAL FACE WIDTH = 6/P = .75 INCHES
  ACTUAL GEAR RATIO = T2/T1 = 32
  INPUT SPEED = 1760 RPM

  IS GEOMETRY OK? 1 = YES; 2 = NO
▶ ? 1                                                    Statement 580

  DATA: PITCH = 8      PRESSURE ANGLE = 20 DEGREES

  RESULTS:
  WEAR LOAD FACTOR, KW = 125
  LIMITING LOAD FOR WEAR = 750 POUNDS

  OUTPUT TORQUE = 1718 LB-IN
  TANGENTIAL LOAD ON GEAR = 429.5 POUNDS
  PITCH LINE SPEED OF GEAR = 115.192 FT/MIN
  DYNAMIC LOAD = 470.729 POUNDS

  WEAR LOAD SATISFACTORY


  BEGIN STRESS ANALYSIS

  INPUT TRIAL WORM DIAMETER, D1
▶ ? 1.5                                                  Statement 1300
  CENTER DISTANCE = 4.75 IN.
  X1 = C^0.875/D1 = 2.60625
  WORM DIAMETER SATISFACTORY; 1.7 < X1 < 3.0
```

```
BENDING STRESS = 12962.5 PSI

TO TRY A NEW DESIGN, TYPE 1
TO CONTINUE WITH THIS DESIGN, TYPE 2
► ? 2                                                   Statement 1530

        **** SUMMARY OF DESIGN ****

      WORM

WORM DIAMETER = 1.5 IN.
NUMBER OF THREADS = 2
LEAD ANGLE = 9.46232 DEGREES
OUTSIDE DIAMETER = 1.75 IN.
ROOT DIAMETER = 1.21075 IN.
FACE LENGTH = 2.82843 IN.

      WORMGEAR

PITCH DIAMETER = 8 IN.
NUMBER OF TEETH = 64
FACE WIDTH = .75 IN.
THROAT DIAMETER = 8.25 IN.
ROOT DIAMETER = 7.71075 IN.

      FORCES ON THE WORM AND WORMGEAR

TANGENTIAL FORCE ON THE WORMGEAR - AND -
  AXIAL FORCE ON THE WORM:
  429.5 POUNDS

AXIAL FORCE ON THE WORMGEAR - AND -
  TANGENTIAL FORCE ON THE WORM:
  85.8714 POUNDS

RADIAL FORCE ON THE WORMGEAR - AND -
  RADIAL FORCE ON THE WORM:
  159.336 POUNDS

      EFFICIENCY = 83.3611 PERCENT

      DATA FOR THE SET

VELOCITY RATIO = 32
DIAMETRAL PITCH = 8
NORMAL PRESSURE ANGLE = 20 DEGREES
CENTER DISTANCE = 4.75 IN.
ADDENDUM = .125 IN.
DEDENDUM = .144625 IN.

OUTPUT SPEED = 55 RPM (GIVEN)
OUTPUT TORQUE = 1718 LB-IN
OUTPUT POWER RATING = 1.49984 HORSEPOWER
PITCH LINE SPEED OF THE GEAR = 115.192 FT/MIN

INPUT SPEED = 1760 RPM
TORQUE ON INPUT SHAFT = 64.4036 LB-IN
POWER REQUIRED AT INPUT SHAFT = 1.79921 HORSEPOWER
PITCH LINE SPEED OF THE WORM = 691.15 FT/MIN


        **** DESIGN COMPLETED ****

TO BEGIN A NEW PROBLEM, TYPE 1
TO END, TYPE 2
► ? 2                                                   Statement 2370
```

REFERENCES

1. American Gear Manufacturers Association. *Surface Durability (Pitting) Formulas for Straight Bevel and Zerol Bevel Gear Teeth.* Standard AGMA 212.02-1964 (R1974). Arlington, Va.: 1974.
2. American Gear Manufacturers Association. *Rating the Pitting Resistance and Bending Strength of Spur and Helical Involute Gear Teeth.* Standard AGMA 218.01-1982. Arlington, Va.: 1982.
3. American Gear Manufacturers Association. *Rating the Strength of Straight Bevel and Zerol Bevel Gear Teeth.* Standard AGMA 222.02-1964 (R1974). Arlington, Va: 1974.
4. American Gear Manufacturers Association. *Design of General Industrial Coarse-Pitch Cylindrical Wormgearing.* Standard AGMA 341.02-1965 (R1970). Arlington, Va.: 1970.
5. American Gear Manufacturers Association. *Standard Practice for Single and Double-Reduction Cylindrical-Worm and Helical-Worm Speed Reducers.* Standard AGMA 440.04-1971. Arlington, Va.: 1971.
6. Buckingham, Earl, and Ryffel, Henry H. *Design of Worm and Spiral Gears.* New York: Industrial Press, n.d.
7. Deutschman, A. D., Michels, W. J., and Wilson, C. E. *Machine Design.* New York: Macmillan Publishing Company, 1975.
8. Dudley, Darle W., ed. *Gear Handbook.* New York: McGraw-Hill Book Company, 1962.
9. Dudley, D. W. *Practical Gear Design.* New York: McGraw-Hill Book Company, 1954.
10. Faires, V. M. *Design of Machine Elements,* 4th ed. New York: Macmillan Publishing Company, 1965.
11. Shigley, J. E. *Mechanical Engineering Design,* 4th ed. New York: McGraw-Hill Book Company, 1983.

PROBLEMS

Helical Gearing

1. A helical gear has a transverse diametral pitch of 8, a transverse pressure angle of 14½°, 44 teeth, a face width of 2.00 in, and a helix angle of 30°. Compute the circular pitch, normal circular pitch, normal diametral pitch, axial pitch, pitch diameter, and normal pressure angle. Then compute the number of axial pitches in the face width.
2. A helical gear has a normal diametral pitch of 12, a normal pressure angle of 20°, 48 teeth, a face width of 1.50 in, and a helix angle of 45°. Compute the transverse diametral pitch, circular pitch, normal circular pitch, axial pitch, number of axial pitches in the face width, transverse pressure angle, and pitch diameter.
3. For the gear of problem 1, compute the transmitted load, the axial load, and the radial load if it transmits 5.00 hp at a speed of 1 250 rpm.
4. For the gear of problem 2, compute the transmitted load, the axial load, and the radial load if it transmits 2.50 hp at 1 750 rpm.
5. A helical gear has a transverse diametral pitch of 6, a transverse pressure angle of 14½°, 20 teeth, a face width of 1.00 in, and a helix angle of 45°. Compute the circular pitch, normal circular pitch, normal diametral pitch, axial pitch, pitch diameter, and normal pressure angle. Then compute the number of axial pitches in the face width.
6. A helical gear has a normal diametral pitch of 24, a normal pressure angle of 14½°, 72 teeth, a face width of 0.25 in, and a helix angle of 45°. Compute the transverse diametral pitch, circular pitch, normal circular pitch, axial pitch, number of axial pitches in the face width, transverse pressure angle, and pitch diameter.
7. For the gear of problem 5, compute the transmitted load, the axial load, and the radial load if it transmits 15.0 hp at a speed of 2 200 rpm.
8. For the gear of problem 6, compute the transmitted load, the axial load, and the radial load if it transmits 0.50 hp at 3 450 rpm.
9. For the gear of problem 3, compute the bending stress in the teeth if the power comes from an electric motor and the drive is to a reciprocating pump.
10. For the gear of problem 4, compute the bending stress in the teeth if the power comes from a steam turbine and the drive is to a centrifugal blower.
11. For the gear of problem 7, compute the bending stress in the teeth if the power comes from a six-cylinder gasoline engine and the drive is to a concrete mixer.
12. For the gear of problem 8, compute the bending stress in the teeth if the power comes from an electric motor and the drive is to a winch that will experience moderate shock.

Problems 13–16: For the given data, evaluate the required wear load factor, K. Each gear is part of a pair with a velocity ratio of 3.00. Then specify a suitable material if possible.

13. Use the data of problem 9.
14. Use the data of problem 10.
15. Use the data of problem 11.
16. Use the data of problem 12.

For the following problems, complete the design of a pair of helical gears to operate under the stated conditions. Specify the geometry of the gears and the material and its heat treatment. Assume that the drive is from an electric motor unless otherwise specified. Consider both strength and wear.

17. A pair of helical gears is to be designed to transmit 5.0 hp while the pinion rotates at 1 200 rpm. The gear drives a reciprocating compressor and must rotate between 385 and 390 rpm.
18. A helical gear pair is to be a part of the drive for a milling machine requiring 20.0 hp with the pinion speed at 550 rpm and the gear speed to be between 180 and 190 rpm.
19. A helical gear drive for a punch press requires 50.0 hp with the pinion rotating at 900 rpm and the gear speed at 225 to 230 rpm.
20. A single cylinder gasoline engine has the pinion of a helical gear pair on its output shaft. The gear is attached to the shaft of a small cement mixer. The mixer requires 2.5 hp while rotating at approximately 75 rpm. The engine is governed to run at approximately 900 rpm.
21. A four-cylinder industrial engine runs at 2 200 rpm and delivers 75 hp to the input gear of a helical gear drive for a large wood chipper used to prepare pulpwood chips for paper making. The output gear must run between 4 500 and 4 600 rpm.
22. A small commercial tractor is being designed for chores such as lawn mowing and snow removal. The wheel drive system is to be through a helical gear pair in which the pinion runs at 450 rpm while the gear, mounted on the hub of the wheel, runs at 75 to 80 rpm. The wheel has an 18-in diameter. The two-cylinder gasoline engine delivers 20.0 hp to the wheels.
23. A water turbine transmits 15.0 hp to a pair of helical gears at 4 500 rpm. The output of the gear pair must drive an electric power generator at 3 600 rpm. The center distance for the gear pair must not exceed 4.00 in.
24. Determine the power-transmitting capacity of a pair of helical gears having a normal pressure angle of 20°, a helix angle of 15°, a normal diametral pitch of 10, 20 teeth in the pinion, 75 teeth in the gear, and a face width of 2.50 in, if they are made from AISI 4140 OQT 1000 steel. They are of typical commercial quality. The pinion will rotate at 1 725 rpm on the shaft of an electric motor. The gear will drive a centrifugal pump.
25. Repeat problem 24 with the gears made from AISI 4620 DOQT 300 carburized, case-hardened steel. Then compute the axial and radial forces on the gears.

Bevel Gears

26. A straight bevel gear pair has the following data: $N_P = 15$; $N_G = 45$; $P_d = 6$; 20° pressure angle. Compute all the geometric features from Table 12-2.
27. If the gear pair of problem 26 is transmitting 3.0 hp, compute the forces on both the pinion and the gear. The pinion speed is 300 rpm. The face width is 1.25 in.
28. If the gear pair of problem 26 is transmitting 3.0 hp at a pinion speed of 300 rpm, compute the bending stress and the Hertz stress for the teeth and specify a suitable material and heat treatment. The gears are driven by a gasoline engine, and the load is a concrete mixer providing moderate shock. Assume that neither gear is straddle-mounted.
29. Draw the gear pair of problem 26 to scale. The following additional dimensions are given (refer to Figure 12-9). Mounting distance (MD) for the pinion = 5.250 in; MD for the gear = 3.000 in; face width = 1.250 in. Supply any other needed dimensions.
30. For the gear pair of problem 29, design the shaft and bearing system to support the gears, using the techniques of Chapters 9 and 15. Use approximately 1.0 in for the diameter of the bore of the gears. Use two bearings for each shaft. The gears may be straddle-mounted or overhung.
31. A straight bevel gear pair has the following data: $N_P = 25$; $N_G = 50$; $P_d = 10$; 20° pressure angle. Compute all the geometric features from Table 12-2.
32. If the gear pair of problem 31 is transmitting 3.5 hp, compute the forces on both the pinion and the gear. The pinion speed is 1 250 rpm. The face width is 0.70 in.
33. If the gear pair of problem 31 is transmitting 3.5 hp at a pinion speed of 1 250 rpm, compute the bending stress and the Hertz stress for the teeth and specify a suitable material and heat treatment. The gears are driven by a gasoline engine, and the load is a concrete mixer providing moderate shock. Assume that neither gear is straddle-mounted.
34. Draw the gear pair of problem 31 to scale. The following additional dimensions are given (refer to Figure 12-9). Mounting distance (MD) for the pinion = 3.375 in; MD for the gear = 2.625 in; face width = 0.700 in. Supply any other needed dimensions.
35. For the gear pair of problem 34, design the shaft and bearing system to support the gears, using the techniques of Chapters 9 and 15. Use approximately

1.0 in for the diameter of the bore of the gears. Use two bearings for each shaft. The gears may be straddle-mounted or overhung.

36. Design a pair of straight bevel gears to transmit 5.0 hp at a pinion speed of 850 rpm. The gear speed should be approximately 300 rpm. Consider both strength and wear. The driver is a gasoline engine, and the driven machine is a heavy-duty conveyor.
37. Design a pair of straight bevel gears to transmit 0.75 hp at a pinion speed of 1 800 rpm. The gear speed should be approximately 475 rpm. Consider both strength and wear. The driver is an electric motor, and the driven machine is a reciprocating saw.

Wormgearing

38. A wormgear set has a single-thread worm with a pitch diameter of 1.250 in, a diametral pitch of 10, and a normal pressure angle of 14.5°. If the worm meshes with a wormgear having 40 teeth and a face width of 0.625 in, compute the lead, axial pitch, circular pitch, lead angle, addendum, dedendum, worm outside diameter, worm root diameter, gear pitch diameter, center distance, and velocity ratio.
39. If the wormgear set described in problem 38 is transmitting 924 lb · in of torque at its output shaft, which is rotating at 30 rpm, compute the forces on the gear, efficiency, input speed, input power, and stress on the gear teeth. If the worm is hardened steel and the gear is chilled bronze, evaluate the wear load limit and determine whether the design is satisfactory for wear.
40. Three designs are being considered for a wormgear set to produce a velocity ratio of 20 when the wormgear rotates at 90 rpm. All three have a diametral pitch of 12, a worm pitch diameter of 1.000 in, a gear face width of 0.500 in, and a normal pressure angle of 14.5°. One has a single-thread worm and 20 wormgear teeth; the second has a double-thread worm and 40 wormgear teeth; the third has a four-thread worm and 80 wormgear teeth. For each design, compute the rated output torque, considering both strength and wear. The worms are hardened steel, and the wormgears are chilled bronze.
41. For each of the three designs proposed in problem 40, compute the efficiency.

Problems 42, 43, and 44: Design a wormgear set to produce the desired velocity ratio when transmitting the given torque at the output shaft for the given output rotational speed.

Problem	*VR*	*Torque (lb · in)*	*Output Speed (rpm)*
42.	7.5	984	80
43.	3	52.5	600
44.	40	4 200	45

45. Compare the two designs described next when each is transmitting 1 200 lb · in of torque at its output shaft, which rotates at 20 rpm. Compute the forces on the worm and the wormgear, the efficiency, and the input power required.

Design	P_d	N_t	N_G	D_W	F_G	*Pressure Angle*
A	6	1	30	2.000	1.000	14.5°
B	10	2	60	1.250	0.625	14.5°

13 Belts and Chains

13-1 OVERVIEW

Belts and chains represent the major types of flexible power transmission elements. In contrast with gear drives, which require relatively closely spaced and precise center distances, belt and chain drives can transmit power between shafts that are widely separated. The center distance, furthermore, is inherently adjustable and need not be nearly as precise as for gear drives.

In general, belt drives are applied where the rotational speeds are relatively high, as on the first stage of speed reduction from a motor or engine. The linear speed of a belt is usually 2 500 to 7 000 feet per minute (ft/min). At lower speeds, the tension in the belt becomes too large for typical belt cross sections. At higher speeds, dynamic effects such as centrifugal forces, belt whip, and vibration reduce the effectiveness of the drive and its life. A speed of 4 000 ft/min is generally ideal.

Chain drives are typically applied at lower speeds, with the consequent higher torques. The steel chain links have a very high tensile strength to withstand the high forces resulting from high torque. But at higher speeds, the noise, the impact between the chain links and the sprocket teeth, and the difficulty in providing lubrication become severe problems.

Thus belts and chains are complementary. In fact, it is not unusual to find a system like that sketched in Figure 13-1, in which a belt drive provides the first stage of speed reduction from a motor to the input of a gear type speed reducer. Then a chain drive provides the final reduction at the slower speed to the driven machine.

A wide variety of belt types is available. The simplest is the flat belt made from leather or various fabrics, sometimes treated with rubber to improve friction. But flat belts have a tendency to slip on the flat, smooth pulley surfaces, limiting the power transmission capacity. *Grooved belts,* sometimes called *timing belts,* are frequently applied to eliminate the slipping. The grooves (or raised cogs) engage the mating pulley to provide positive

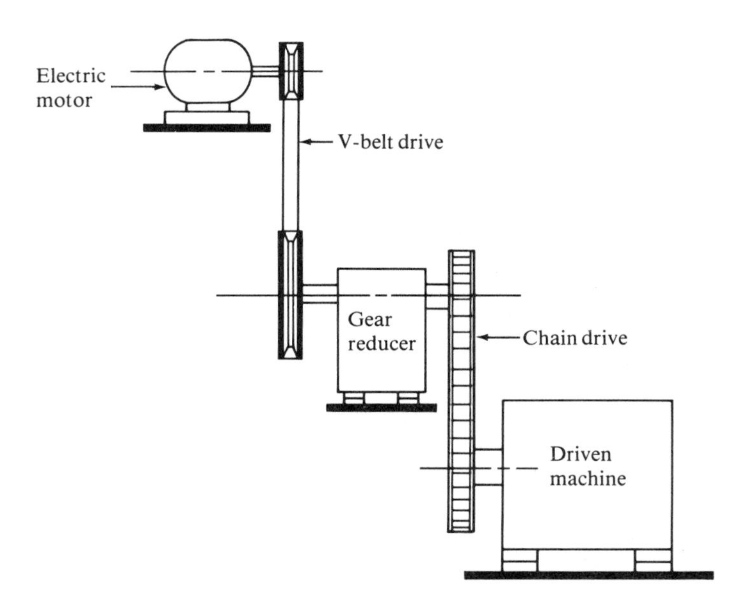

Figure 13-1 Combination Drive Employing V-belts, a Gear Reducer, and a Chain Drive

drive. But the most widely used type of belt drive is the V-belt drive, discussed in the following section.

13-2 V-BELT DRIVES

The typical arrangement of the elements of a V-belt drive is shown in Figure 13-2. The important observations to be derived from this arrangement are summarized here.

1. The pulley, with a circumferential groove carrying the belt, is called a *sheave* (usually pronounced *shiv*).
2. The size of a sheave is indicated by its pitch diameter, slightly smaller than the outside diameter of the sheave.
3. The speed ratio between the driving and the driven sheaves is inversely proportional to the ratio of the sheave pitch diameters. This follows from the observation that there is no slipping (under normal loads). Thus the linear speed of the pitch line of both sheaves is the same and equal to the belt speed, v_b. Then,

$$v_b = R_1 \omega_1 = R_2 \omega_2 \tag{13-1}$$

But $R_1 = D_1/2$ and $R_2 = D_2/2$. Then,

$$v_b = D_1 \omega_1/2 = D_2 \omega_2/2 \tag{13-1A}$$

The angular velocity ratio is

$$\frac{\omega_1}{\omega_2} = \frac{D_2}{D_1} \tag{13-2}$$

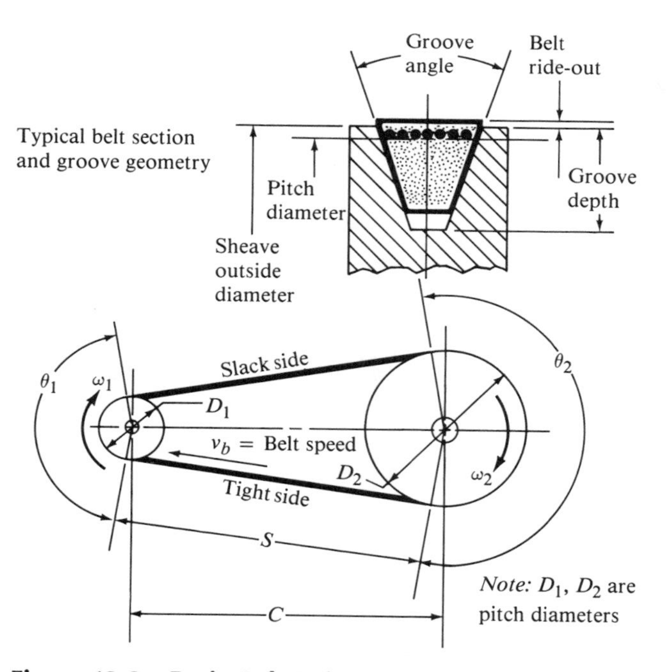

Figure 13-2 Basic Belt Drive Geometry

4. The relationships between pitch length, L, center distance, C, and the sheave diameters are

$$L = 2C + 1.57(D_2 + D_1) + \frac{(D_2 - D_1)^2}{4C} \tag{13-3}$$

$$C = \frac{B + \sqrt{B^2 - 32(D_2 - D_1)^2}}{16} \tag{13-4}$$

where $B = 4L - 6.28(D_2 + D_1)$.

5. The angle of contact of the belt on each sheave is

$$\theta_1 = 180° - 2 \sin^{-1}\left[\frac{D_2 - D_1}{2C}\right] \tag{13-5}$$

$$\theta_2 = 180° + 2 \sin^{-1}\left[\frac{D_2 - D_1}{2C}\right] \tag{13-6}$$

These angles are important because commercially available belts are rated with an assumed contact angle of 180°. This will occur only if the drive ratio is 1 (no speed change). The angle of contact on the smaller of the two sheaves will always be less than 180°, requiring a lower power rating.

6. The length of the span between the two sheaves, over which the belt is unsupported, is

$$S = \sqrt{C^2 - \left[\frac{D_2 - D_1}{2}\right]^2} \tag{13-7}$$

This is important for two reasons. The proper belt tension can be checked by measuring the amount of force required to deflect the belt at the middle of the span by a given amount. Also, the tendency for the belt to vibrate or whip is dependent on this length.

7. The contributors to the stress in the belt are
   a. The tensile force in the belt, maximum on the tight side of the belt.
   b. The bending of the belt around the sheaves, maximum as the tight side of the belt bends around the smaller sheave.
   c. Centrifugal forces created as the belt moves around the sheaves.

   The maximum total stress occurs where the belt enters the smaller sheave and the bending stress is a major part. Thus there are recommended minimum sheave diameters for standard belts. Using smaller sheaves drastically reduces belt life.

8. The design value of the ratio of the tight side tension to the slack side tension is 5.0 for V-belt drives. The actual value may range as high as 10.0.

Standard Belt Cross Sections

Commercially available belts are made to one of the following standards:

Industrial heavy-duty (Figure 13-3)

Inch sizes:	A	B	C	D	E
Metric sizes:	13C	16C	22C	32C	

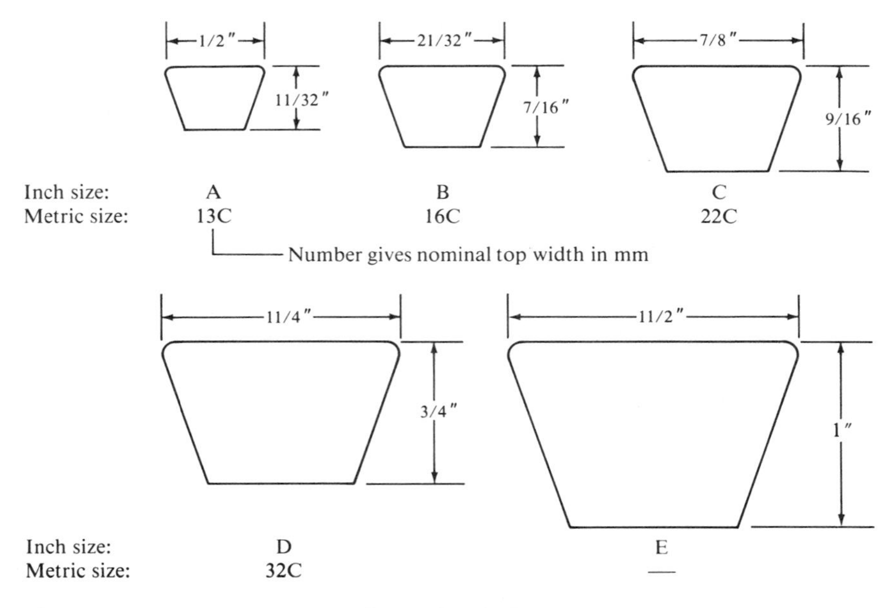

Figure 13-3 Heavy-Duty Industrial V-belts

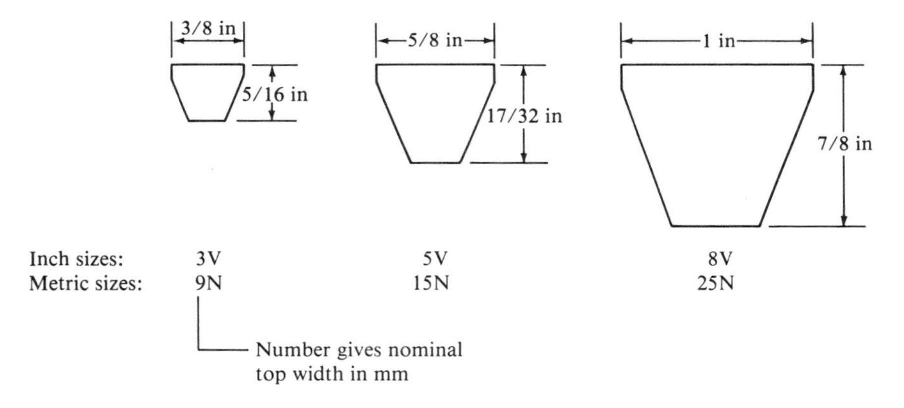

Figure 13-4 Industrial Narrow-section V-belts

Industrial narrow section (Figure 13-4)

Inch sizes:	3V	5V	8V
Metric sizes:	9N	15N	25N

Light duty (Figure 13-5)

Inch sizes:	2L	3L	4L	5L
Metric sizes:	6R	9R	12R	16R

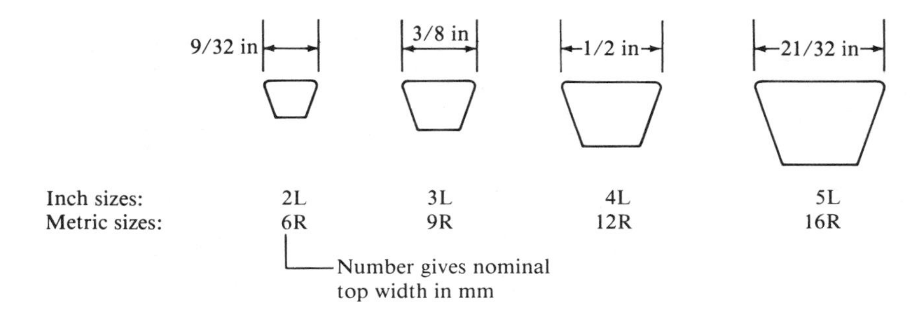

Figure 13-5 Light-duty, Fractional Horsepower (FHP) V-belts

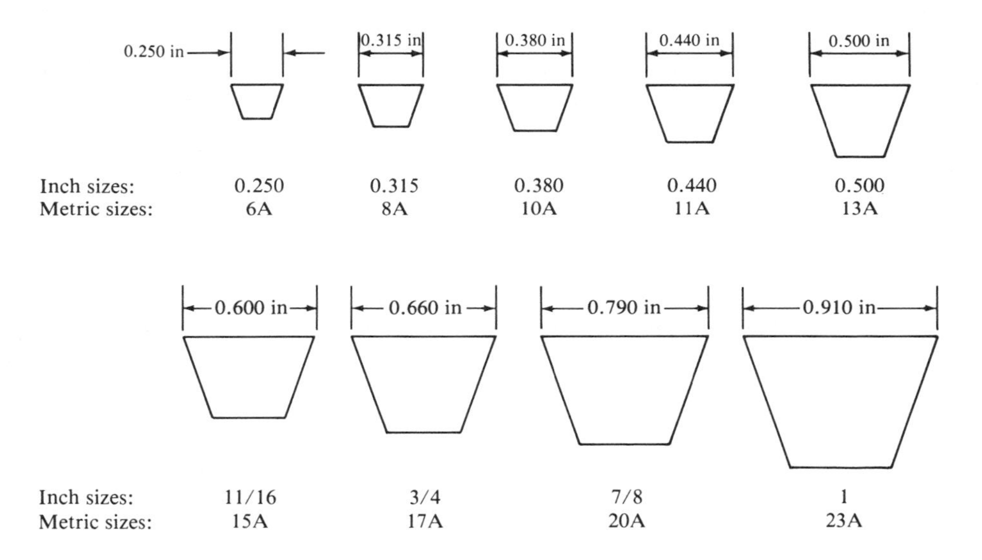

Figure 13-6 Automotive V-belts

Automotive (Figure 13-6)

Inch sizes:	0.250	0.315	0.380	0.440	0.500	11/16 (0.600)	3/4 (0.660)	7/8 (0.790)	1 (0.910)
Metric sizes:	6A	8A	10A	11A	13A	15A	17A	20A	23A

The alignment between the inch sizes and the metric sizes indicates that the paired sizes are really the same cross section. A "soft conversion" was used to rename the familiar inch sizes with the number for the metric sizes giving the nominal top width in millimeters.

The nominal value of the included angle between the sides of the V-groove ranges from 30° to 42°. The angle on the belt may be slightly different to achieve a tight fit in the groove. Some belts are designed to "ride out" of the groove somewhat.

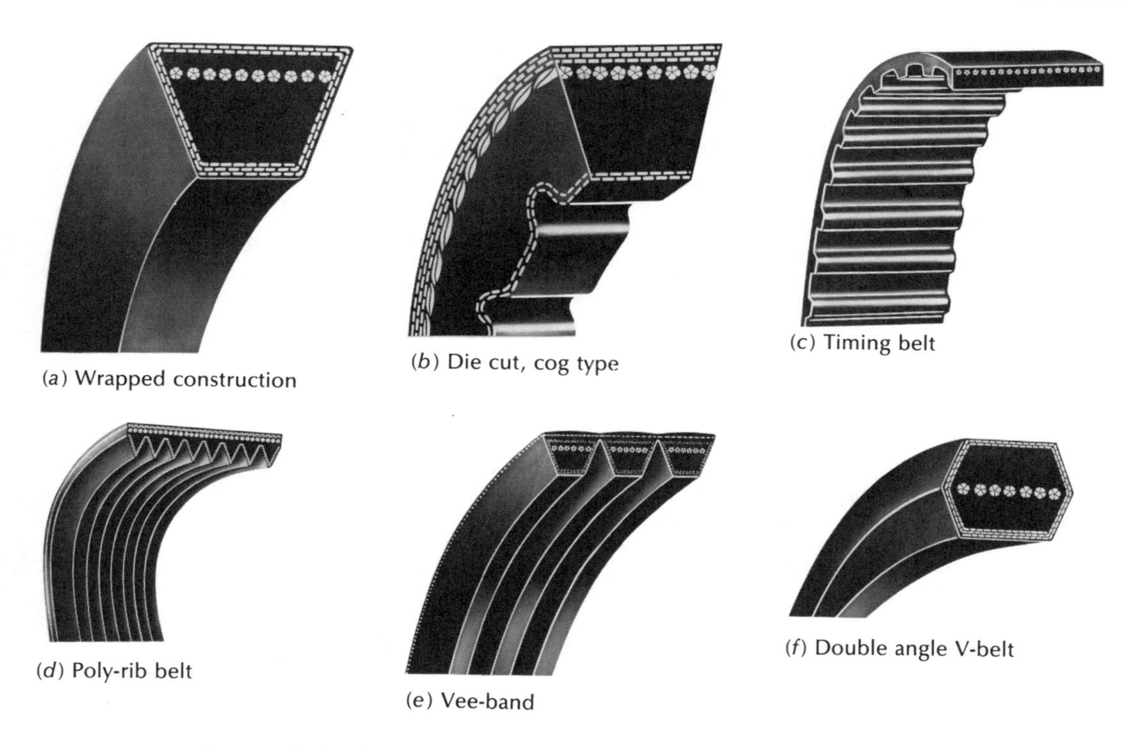

Figure 13-7 Examples of V-belt Construction (Dayco Corp., Dayton, Ohio)

Belt Construction

Belts from the various manufacturers differ. Typical belt constructions are shown in Figure 13-7. The *cords,* or woven fabric near the pitch line of the belt, carry the main tensile load. The lower portion is usually made from synthetic rubber, which provides support for the cords and undergoes compression as the belt passes over the sheaves. The upper portion is the tension section. Some belts are wrapped with a fabric cover. Others are die-cut from wide stock and are left with the cut edge exposed.

The cog-belt construction is designed to provide more flexibility than the standard belts, allowing smaller minimum diameter sheaves to be used. These belts are also more efficient. Most of the sections shown in Figures 13-3 to 13-6 are available in the cog-belt design; in those cases, an X is then added to the designation (AX, 13XC, BX, 16XC, 3VX, 9XN, and so on).

13-3 V-BELT DRIVE DESIGN

The factors involved in selecting a V-belt and the driving and driven sheaves and proper installation of the drive are summarized in this section. Abbreviated examples of the data available from suppliers are given for illustration. Catalogs contain extensive data and step-by-step instructions are given for their use.

The basic data required for drive selection are the following.

The rated power of the driving motor or other prime mover

The service factor based on the type of driver and driven load

The center distance

The power rating for one belt as a function of the size and speed of the smaller sheave

The belt length

The size of the driving and driven sheaves

The correction factor for belt length

The correction factor for the angle of wrap on the smaller sheave

The number of belts

The initial tension on the belt

Many design decisions depend on the application and space limitations. A few guidelines are given here.

Adjustment for the center distance must be provided in both directions from the nominal value. The center distance must be shortened at the time of installation to enable the belt to be placed in the grooves of the sheaves without force. Provision for increasing the center distance must be made to permit the initial tensioning of the drive and to take up for belt stretch. Manufacturers' catalogs give the data. One convenient way to accomplish the adjustment is the use of a take-up unit, as shown in Figure 15-16 in Chapter 15.

If fixed centers are required, idler pulleys should be used. It is best to use a grooved idler on the inside of the belt, close to the large sheave. Adjustable tensioners are commercially available to carry the idler.

The nominal range of center distances should be

$$D_2 < C < 3(D_2 + D_1) \tag{13-8}$$

The angle of wrap on the smaller sheave should be greater than 120°.

Most commercially available sheaves are cast iron, which should be limited to 6 500-ft/min belt speed.

Consider an alternative type of drive, such as a gear type or chain, if the belt speed is less than 1 000 ft/min.

Avoid elevated temperatures around belts.

Ensure that the shafts carrying mating sheaves are parallel and that the sheaves are in alignment so that the belts track smoothly into the grooves.

In multibelt installations, matched belts are required. Match numbers are printed on industrial belts, with *50* indicating a belt length very close to nominal. Longer belts carry match numbers above 50; shorter belts below 50.

Belts must be installed with the initial tension recommended by the manufacturer. Tension should be checked after the first few hours of operation because seating and initial stretch occur.

Design Data

Catalogs typically give several dozen pages of design data for the various sizes of belts and sheave combinations to ease the job of drive design. The following data are available in tabular form (1). Graphical form is used here so you can get a feel for the variation in performance with design choices. Any design made from the data in this book should be checked against a particular manufacturer's ratings before use.

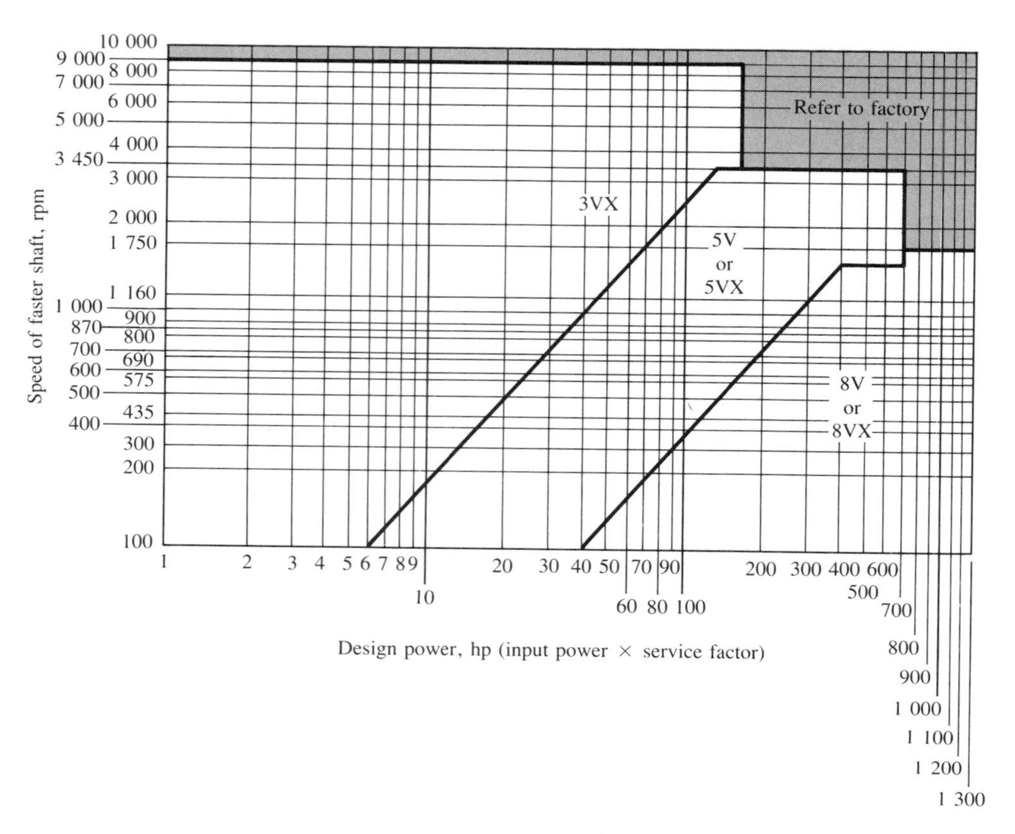

Figure 13-8 Selection Chart for Narrow-section Industrial V-belts (Dayco Corp., Dayton, Ohio)

The data given here are for the narrow-section belts: 3V, 5V, and 8V. These three sizes cover a wide range of power transmission capacities. Figure 13-8 can be used to choose the basic size for the belt cross section. Note that the power axis is *design power,* the rated power of the prime mover times the service factor from Table 13-1.

Figures 13-9, 13-10, and 13-11 give the rated power per belt for the three cross sections as a function of the pitch diameter of the smaller sheave and its speed of rotation. The labeled vertical lines in each figure give the standard sheave pitch diameters available.

The basic power rating for a speed ratio of 1.00 is given as the solid curve. A given belt can carry a greater power as the speed ratio increases, up to a ratio of approximately 3.38. Further increases have little effect and may lead to trouble with the angle of wrap on the smaller sheave, as well. Figure 13-12 is a plot of the data for power to be added to the basic rating as a function of speed ratio for the 5V belt size. The catalog data are given in a stepwise fashion. The maximum power added, for ratios of above 3.38, was used to draw the dashed curves in Figures 13-9, 13-10, and 13-11. In most cases, a rough interpolation between the two curves is satisfactory.

Figure 13-13 gives the value of a correction factor, C_θ, as a function of the angle of wrap of the belt on the small sheave.

Figure 13-14 gives the value of the correction factor, C_L, for belt length. A longer belt is desirable because it reduces the frequency that a given part of the belt encounters the

Table 13-1 V-belt Service Factors

	DRIVER TYPE					
	AC Motors: Normal-Torque[a] *DC Motors: Shunt-wound* *Engines: Multiple-cylinder*			*AC Motors: High-torque*[b] *DC Motors: Series-wound, Compound-wound* *Engines: 4-cylinder or Less*		
DRIVEN MACHINE TYPE	*<6 h per day*	*6–15 h per day*	*>15 h per day*	*<6 h per day*	*6–15 h per day*	*>15 h per day*
Agitators Blowers, fans Centrifugal pumps Conveyors, light	1.0	1.1	1.2	1.1	1.2	1.3
Generators Machine tools Mixers Conveyors, gravel	1.1	1.2	1.3	1.2	1.3	1.4
Bucket elevators Textile machines Hammer mills Conveyors, heavy	1.2	1.3	1.4	1.4	1.5	1.6
Crushers Ball mills Hoists Rubber extruders	1.3	1.4	1.5	1.5	1.6	1.8
Any machine that can choke	2.0	2.0	2.0	2.0	2.0	2.0

[a]Synchronous, split-phase, three-phase with starting torque or breakdown torque less than 250 percent of full load torque.

[b]Single-phase, three-phase with starting torque or breakdown torque greater than 250 percent of full load torque.

stress peak as it enters the small sheave. Only certain standard belt lengths are available (Table 13-2).

Example problem 13-1 illustrates the use of the design data.

Example Problem 13-1. Design a V-belt drive that has the input sheave on the shaft of an electric motor (normal torque) rated at 50.0 hp at 1 160 rpm full load speed. The drive is to a bucket elevator in a potash plant that is to be used 16 hours (h) daily at approximately 675 rpm.

Solution.

1. Compute the design power. From Table 13-1, for a normal torque electric motor running 16 h daily driving a bucket elevator, the service factor is 1.40. Then, the design power is 1.40(50.0 hp) = 70.0 hp.

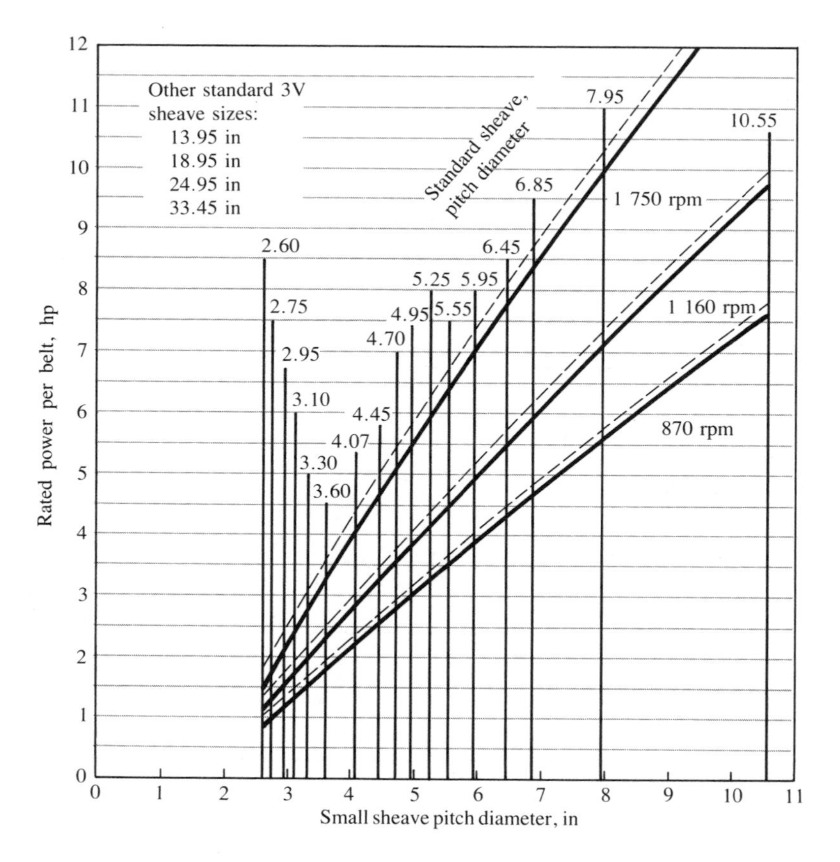

Figure 13-9 Power Rating: 3V Belts

2. Select the belt section. From Figure 13-8, a 5V belt is recommended for 70.0 hp at 1 160-rpm input speed.
3. Compute the nominal speed ratio.

$$\text{Ratio} = 1\,160/675 = 1.72$$

4. Compute the driving sheave size that would produce a belt speed of 4 000 ft/min, as a guide to select a standard sheave.

$$\text{Belt speed} = v_b = \pi D_1 n_1/(12 \text{ ft/min})$$

Then the required diameter to give $v_b = 4\,000$ ft/min is

$$D_1 = \frac{12v_b}{\pi n_1} = \frac{12(4\,000)}{\pi n_1} = \frac{15\,279}{n_1} = \frac{15\,279}{1\,160} = 13.17 \text{ in}$$

5. Select trial sizes for the input sheave and compute the desired size of the output sheave. Select a standard size for the output sheave and compute the actual ratio and output speed.

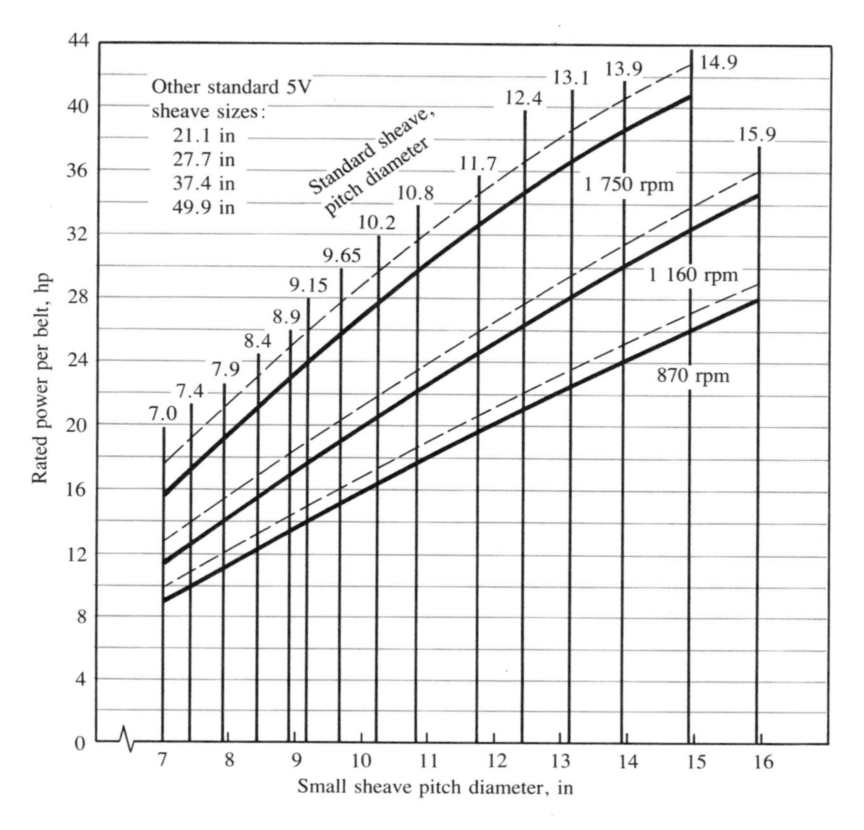

Figure 13-10 Power Rating: 5V Belts

For this problem, the trials are given in the following table (diameters are in inches).

Standard Driving Sheave Size, D_1	*Approximate Driven Sheave Size ($1.72D_1$)*	*Nearest Standard Sheave, D_2*	*Actual Output Speed (rpm)*
13.10	22.5	21.1	720
12.4	*21.3*	*21.1*	*682*
11.7	20.1	21.1	643
10.8	18.6	21.1	594
10.2	17.5	15.9	744
9.65	16.6	15.9	704
9.15	*15.7*	*15.9*	*668*
8.9	15.3	14.9	693

The two trials in italics give only about 1 percent variation from the desired output speed of 675 rpm, and the speed of a bucket elevator is not critical. Because no space limitations were given, let's choose the larger size.

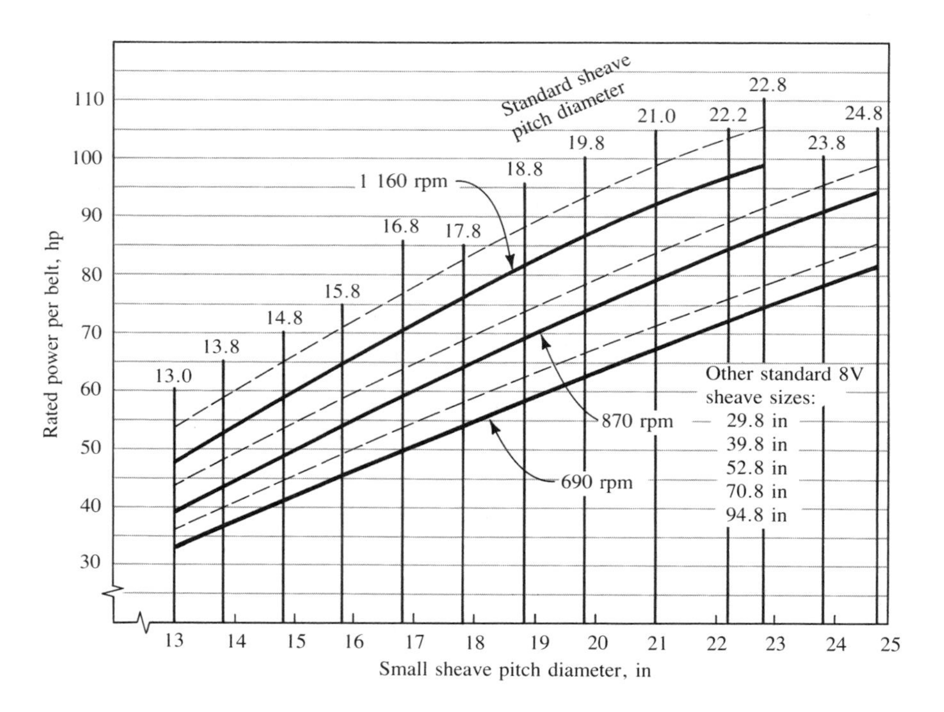

Figure 13-11 Power Rating: 8V Belts

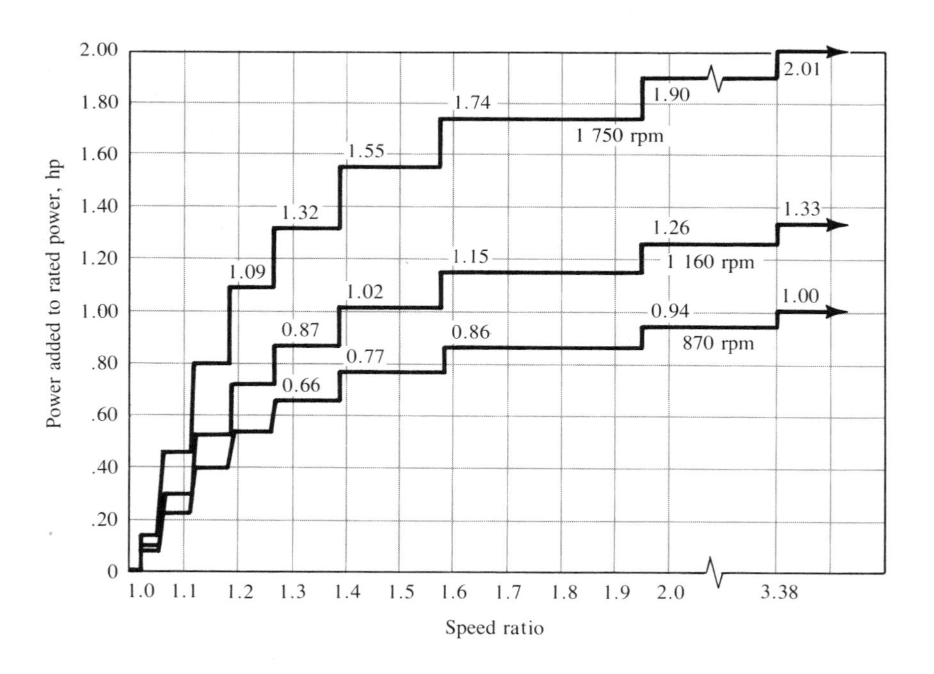

Figure 13-12 Power Added versus Speed Ratio: 5V Belts

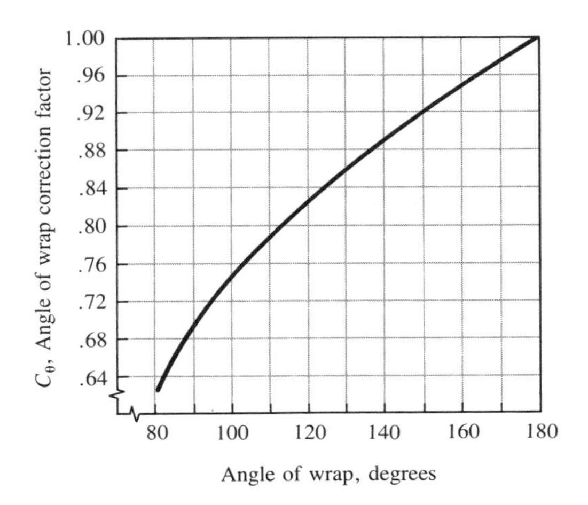

Figure 13-13 Angle of Wrap Correction Factor, C_θ

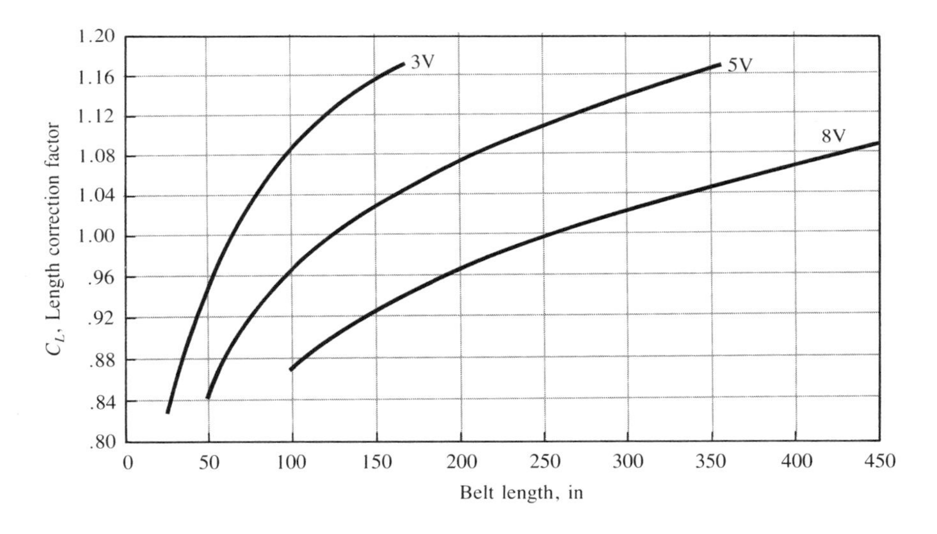

Figure 13-14 Belt Length Correction Factor, C_L

6. Determine the rated power from Figure 13-9, 13-10, or 13-11.

   For the 5V belt we have selected, Figure 13-10 is appropriate. For a 12.4-in sheave at 1 160 rpm, the basic rated power is 26.4 hp. The ratio is relatively high, indicating that some added power rating can be used. This can be estimated from Figure 13-10 or taken directly from Figure 13-12 for the 5V belt. Power added is 1.15 hp. Then the actual rated power is 26.4 + 1.15 = 27.55 hp.

7. Specify a trial center distance.

   Equation 13-8 can be used to determine a nominal acceptable range for C.

Table 13-2 Standard Belt Lengths for 3V, 5V, and 8V Belts (Inches)

3V Only	*3V and 5V*	*3V, 5V, and 8V*	*5V and 8V*	*8V Only*
25	50	100	150	375
26.5	53	106	160	400
28	56	112	170	425
30	60	118	180	450
31.5	63	125	190	475
33.5	67	132	200	500
35.5	71	140	212	
37.5	75		224	
40	80		236	
42.5	85		250	
45	90		265	
47.5	95		280	
			300	
165			315	
			335	
			355	

$$D_2 < C < 3(D_2 + D_1)$$
$$21.1 < C < 3(21.1 + 12.4)$$
$$21.1 < C < 100.5 \text{ in}$$

In the interest of conserving space, let's try $C = 24.0$ in.

8. Compute the required belt length from equation (13-3).

$$L = 2C + 1.57(D_2 + D_1) + \frac{(D_2 - D_1)^2}{4C}$$

$$L = 2(24.0) + 1.57(21.1 + 12.4) + \frac{(21.1 - 12.4)^2}{4(24.0)} = 101.4 \text{ in}$$

9. Select a standard belt length from Table 13-2 and compute the resulting actual center distance from equation (13-4).

   In this problem, the nearest standard length is 100.0 in. Then, from equation (13-4),

$$B = 4L - 6.28(D_2 + D_1) = 4(100) - 6.28(21.1 + 12.4) = 189.6$$

$$C = \frac{189.6 + \sqrt{(189.6)^2 - 32(21.1 - 12.4)^2}}{16} = 23.30 \text{ in}$$

10. Compute the angle of wrap of the belt on the small sheave from equation (13-5).

$$\theta_1 = 180° - 2\sin^{-1}\left[\frac{D_2 - D_1}{2C}\right] = 180° - 2\sin^{-1}\left[\frac{21.1 - 12.4}{2(23.30)}\right] = 158°$$

11. Determine the correction factors from Figures 13-13 and 13-14. For $\theta = 158°$, $C_\theta = 0.94$. For $L = 100$ in, $C_L = 0.96$.
12. Compute the corrected rated power per belt and the number of belts required to carry the design power.

$$\text{Corrected power} = C_\theta C_L P = (0.94)(0.96)(27.55 \text{ hp}) = 24.86 \text{ hp}$$

$$\text{No. of belts} = 70.0/24.86 = 2.82 \text{ belts} \quad (\text{Use 3 belts})$$

Summary of Design

Input: Electric motor: 50.0 hp at 1 160 rpm

Service factor: 1.4

Design power: 70.0 hp

Belt: 5V cross section, 100-in length, 3 belts

Sheaves: Driver: 12.4-in pitch diameter, 3 grooves, 5V; Driven: 21.1-in pitch diameter, 3 grooves, 5V

Actual output speed: 682 rpm

Center Distance: 23.30 in

Belt Tension

The initial tension given to a belt is critical because it ensures that the belt will not slip under the design load. At rest, the two sides of the belt have the same tension. As power is being transmitted, the tension in the tight side increases while the tension in the slack side decreases. Without the initial tension, the slack side would go totally loose and the belt would not seat in the groove. Thus it would slip. Manufacturers' catalogs give data for the proper belt-tensioning procedures.

13-4 CHAIN DRIVES

A chain is a power transmission element made as a series of pin-connected links. The design provides for flexibility while enabling the chain to transmit large tensile forces.

When transmitting power between rotating shafts, the chain engages mating toothed wheels, called *sprockets*. Figure 13-15 shows a typical chain drive.

The most common type of chain is the *roller chain,* in which the roller on each pin provides exceptionally low friction between the chain and the sprockets. Other types include a variety of extended link designs used mostly in conveyor applications (see Figure 13-16).

Roller chain is classified by its *pitch,* the distance between corresponding parts of adjacent links. The pitch is usually illustrated as the distance between the centers of adjacent pins. Standard roller chain carries a size designation from 40 to 240, as listed in Table 13-3. The digits (other than the final zero) indicate the pitch of the chain in eighths of an inch, as in the table. For example, the no. 100 chain has a pitch of 10/8 or 1¼ in. A series of heavy-duty sizes, with the suffix *H* on the designation (60H–240H), has the same basic dimensions as the standard chain of the same number except for thicker side plates. In addition, there are the smaller and lighter sizes: 25, 35, and 41.

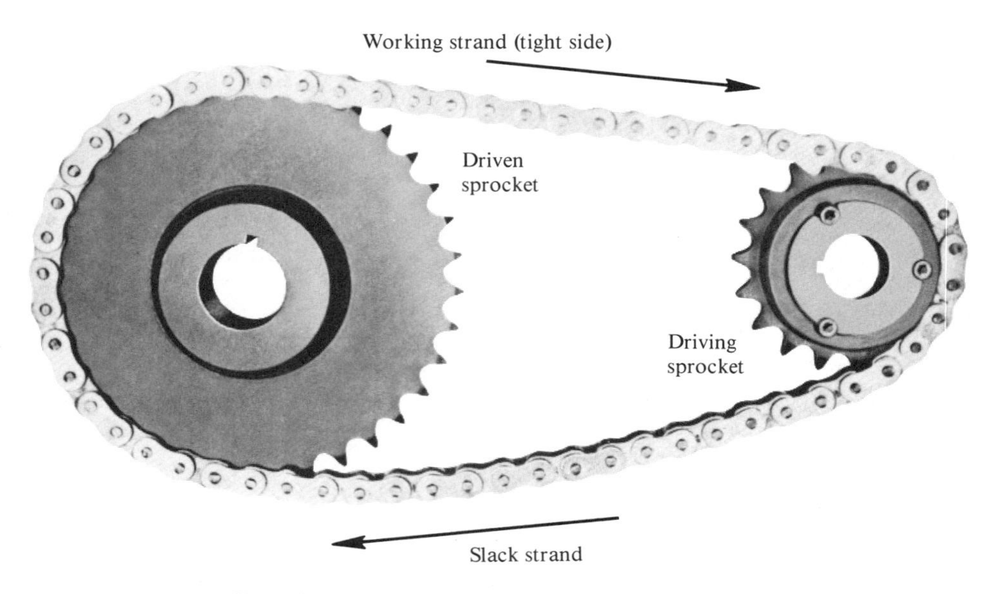

Figure 13-15 Roller Chain Drive (Rexnord, Inc., Milwaukee, Wis.)

Figure 13-16 Some Roller Chain Styles (Rexnord, Inc., Milwaukee, Wis.)

The average tensile strengths of the various chain sizes are also listed in Table 13-3. These data can be used for very-low-speed drives or for applications in which the function of the chain is to apply a tensile force or to support a load. It is recommended that only

Table 13-3 Roller Chain Sizes

Chain Number	*Pitch (in)*	*Average Tensile Strength (lb)*
25	1/4	925
35	3/8	2 100
41	1/2	2 000
40	1/2	3 700
50	5/8	6 100
60	3/4	8 500
80	1	14 500
100	1 1/4	24 000
120	1 1/2	34 000
140	1 3/4	46 000
160	2	58 000
180	2 1/4	80 000
200	2 1/2	95 000
240	3	130 000

10 percent of the average tensile strength be used in such applications. For power transmission, the rating of a given chain size as a function of the speed of rotation must be determined, as explained later in this chapter.

A wide variety of attachments is available to facilitate the application of roller chain to conveying or other material-handling uses. Usually in the form of extended plates or tabs with holes provided, the attachments make it easy to connect rods, buckets, parts pushers, part support devices, or conveyor slats to the chain. Figure 13-17 shows some attachment styles.

Figure 13-18 shows a variety of chain types used especially for conveying and similar applications. Such chain typically has a longer pitch than standard roller chain (usually twice the pitch), and the link plates are heavier. The larger sizes have cast link plates.

13-5 DESIGN OF CHAIN DRIVES

The rating of chain for its power transmission capacity considers three modes of failure: fatigue of the link plates due to the repeated application of the tension in the tight side of the chain, impact of the rollers as they engage the sprocket teeth, and galling between the pins of each link and the bushings on the pins.

The ratings are based on empirical data with a smooth driver and a smooth load (service factor = 1.0) and with a rated life of approximately 15 000 h. The important variables are the pitch of the chain and the size and rotational speed of the smaller sprocket. Lubrication is critical to the satisfactory operation of a chain drive. Manufacturers recommend the type of lubrication method for given combinations of chain size, sprocket size, and speed. Details are discussed later.

Tables 13-4, 13-5, and 13-6 list the rated power for three sizes of standard chain: no. 40 (1/2 in), no. 60 (3/4 in), and no. 80 (1.00 in). These are typical of the type of data available for all chain sizes in manufacturers' catalogs. Notice these features of the data:

(*a*) Slats assembled to attachments to form a flat conveying surface

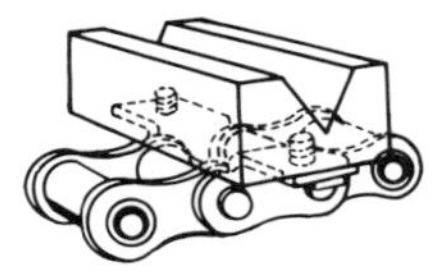

(*b*) *V* block assembled to attachments to convey round objects of varying diameters

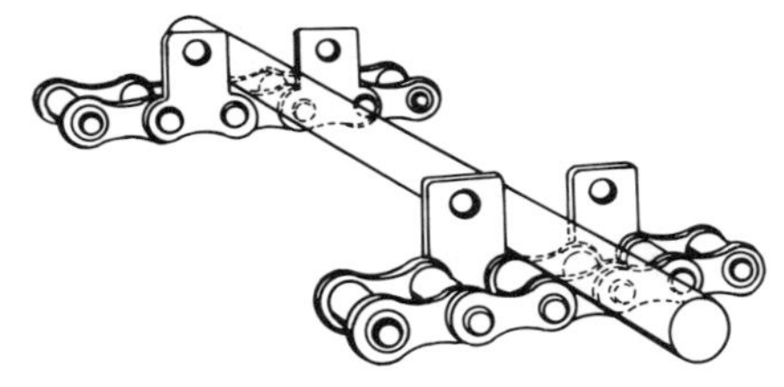

(*c*) Attachments used as spacers to convey and position long objects

Figure 13-17 Chain Attachments (Rexnord, Inc., Milwaukee, Wis.)

1. The ratings are based on the speed of the smaller sprocket.
2. For a given speed, the power capacity increases with the number of teeth on the sprocket. Of course, the larger the number of teeth, the larger the diameter of the sprocket. It should be noted that the use of a chain with a small pitch on a large sprocket produces the quieter drive.
3. For a given sprocket size (a given number of teeth), the power capacity increases with increasing speed up to a point; then it decreases. Fatigue due to the tension in the chain governs at the low to moderate speeds; impact on the sprockets governs at the higher speeds. Each sprocket size has an absolute upper limit speed due to the onset of galling between the pins and the bushings of the chain. This explains the abrupt drop in power capacity to zero at the limiting speed.
4. The ratings are for a single strand of chain. Although multiple strands do increase the power capacity, they do not provide a direct multiple of the single-strand capacity. Multiply the capacity in the tables by the following factors:

   Two strands: Factor = 1.7

   Three strands: Factor = 2.5

   Four strands: Factor = 3.3

5. The ratings are for a service factor of 1.0. Specify a service factor for a given application according to Table 13-7.

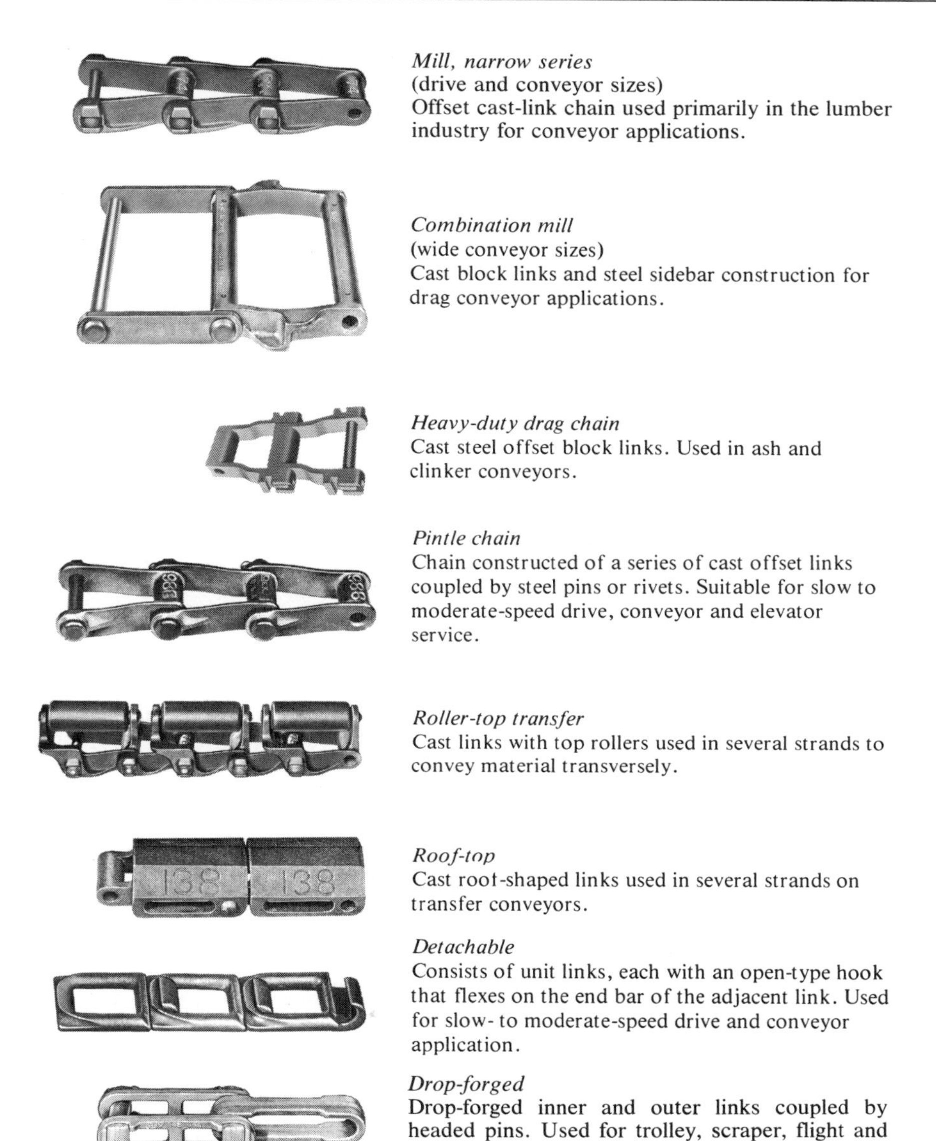

Figure 13-18 Conveyor Chains (Rexnord, Inc., Milwaukee, Wis.)

Design Guidelines for Chain Drives

The following are general recommendations for designing chain drives.

1. The minimum number of teeth in a sprocket should be 17 unless the drive is operating at a very low speed, under 100 rpm.
2. The maximum speed ratio should be 7.0, although higher ratios are feasible. Two or more stages of reduction can be used to achieve higher ratios.

Table 13-4 Horsepower Ratings, Standard Single-Strand Roller Chain, No. 40, ½-in. Pitch

NO. OF TEETH SMALL SPKT.	REVOLUTIONS PER MINUTE—SMALL SPROCKET																								
	10	25	50	100	200	300	400	500	700	900	1 000	1 200	1 400	1 600	1 800	2 100	2 400	2 700	3 000	3 500	4 000	5 000	6 000	7 000	8 000
9	0.04	0.10	0.19	0.35	0.65	0.93	1.21	1.48	2.00	2.51	2.75	3.25	3.73	4.12	3.45	2.74	2.24	1.88	1.60	1.27	1.04	0.75	0.57	0.45	0.37
10	0.05	0.11	0.21	0.39	0.73	1.04	1.35	1.65	2.24	2.81	3.09	3.64	4.18	4.71	4.04	3.21	2.63	2.20	1.88	1.49	1.22	0.87	0.66	0.53	0.43
11	0.05	0.12	0.23	0.43	0.80	1.16	1.50	1.83	2.48	3.11	3.42	4.03	4.63	5.22	4.66	3.70	3.03	2.54	2.17	1.72	1.41	1.01	0.77	0.61	0.50
12	0.06	0.14	0.25	0.47	0.88	1.27	1.65	2.01	2.73	3.42	3.76	4.43	5.09	5.74	5.31	4.22	3.45	2.89	2.47	1.96	1.60	1.15	0.87	0.69	0.57
13	0.06	0.15	0.28	0.52	0.96	1.39	1.80	2.20	2.97	3.73	4.10	4.83	5.55	6.26	5.99	4.76	3.89	3.26	2.79	2.21	1.81	1.29	0.98	0.78	0.64
14	0.07	0.16	0.30	0.56	1.04	1.50	1.95	2.38	3.22	4.04	4.44	5.23	6.01	6.78	6.70	5.31	4.35	3.65	3.11	2.47	2.02	1.45	1.10	0.87	0.71
15	0.07	0.17	0.32	0.60	1.12	1.62	2.10	2.56	3.47	4.35	4.78	5.64	6.47	7.30	7.43	5.89	4.82	4.04	3.45	2.74	2.24	1.60	1.22	0.97	0.79
16	0.08	0.19	0.35	0.65	1.20	1.74	2.25	2.75	3.72	4.66	5.13	6.04	6.94	7.83	8.18	6.49	5.31	4.45	3.80	3.02	2.47	1.77	1.34	1.07	0.87
17	0.08	0.20	0.37	0.69	1.29	1.85	2.40	2.93	3.97	4.98	5.48	6.45	7.41	8.36	8.96	7.11	5.82	4.88	4.17	3.31	2.71	1.94	1.47	1.17	0.96
18	0.09	0.21	0.39	0.73	1.37	1.97	2.55	3.12	4.22	5.30	5.82	6.86	7.88	8.89	9.76	7.75	6.34	5.31	4.54	3.60	2.95	2.11	1.60	1.27	0
19	0.09	0.22	0.42	0.78	1.45	2.09	2.71	3.31	4.48	5.62	6.17	7.27	8.36	9.42	10.5	8.40	6.88	5.76	4.92	3.91	3.20	2.29	1.74	1.38	0
20	0.10	0.24	0.44	0.82	1.53	2.21	2.86	3.50	4.73	5.94	6.53	7.69	8.83	9.96	11.1	9.07	7.43	6.22	5.31	4.22	3.45	2.47	1.88	1.49	0
21	0.11	0.25	0.46	0.87	1.62	2.33	3.02	3.69	4.99	6.26	6.88	8.11	9.31	10.5	11.7	9.76	7.99	6.70	5.72	4.54	3.71	2.66	2.02	1.60	0
22	0.11	0.26	0.49	0.91	1.70	2.45	3.17	3.88	5.25	6.58	7.23	8.52	9.79	11.0	12.3	10.5	8.57	7.18	6.13	4.87	3.98	2.85	2.17	1.72	0
23	0.12	0.27	0.51	0.96	1.78	2.57	3.33	4.07	5.51	6.90	7.59	8.94	10.3	11.6	12.9	11.2	9.16	7.68	6.55	5.20	4.26	3.05	2.32	1.84	0
24	0.13	0.29	0.54	1.00	1.87	2.69	3.48	4.26	5.76	7.23	7.95	9.36	10.8	12.1	13.5	11.9	9.76	8.18	6.99	5.54	4.54	3.25	2.47	1.96	0
25	0.13	0.30	0.56	1.05	1.95	2.81	3.64	4.45	6.02	7.55	8.30	9.78	11.2	12.7	14.1	12.7	10.4	8.70	7.43	5.89	4.82	3.45	2.63	0	
26	0.14	0.31	0.58	1.09	2.04	2.93	3.80	4.64	6.28	7.88	8.66	10.2	11.7	13.2	14.7	13.5	11.0	9.23	7.88	6.25	5.12	3.66	2.79	0	
28	0.15	0.34	0.63	1.18	2.20	3.18	4.11	5.03	6.81	8.54	9.39	11.1	12.7	14.3	15.9	15.0	12.3	10.3	8.80	6.99	5.72	4.09	3.11	0	
30	0.16	0.37	0.68	1.27	2.38	3.42	4.43	5.42	7.33	9.20	10.1	11.9	13.7	15.4	17.2	16.7	13.6	11.4	9.76	7.75	6.34	4.54	3.45	0	
32	0.17	0.39	0.73	1.36	2.55	3.67	4.75	5.81	7.86	9.86	10.8	12.8	14.7	16.5	18.4	18.4	15.0	12.6	10.8	8.64	6.99	5.00	0		
35	0.19	0.43	0.81	1.50	2.81	4.04	5.24	6.40	8.66	10.9	11.9	14.1	16.2	18.2	20.3	21.0	17.2	14.4	12.3	9.76	7.99	5.72	0		
40	0.22	0.50	0.93	1.74	3.24	4.67	6.05	7.39	10.0	12.5	13.8	16.3	18.7	21.1	23.4	25.7	21.0	17.6	15.0	11.9	9.76	6.99	0		
45	0.25	0.57	1.06	1.97	3.68	5.30	6.87	8.40	11.4	14.2	15.7	18.5	21.2	23.9	26.6	30.5	25.1	21.0	17.9	14.2	11.7	0			

TYPE I — TYPE II — TYPE III

Source: Reprinted from *Chains for Power Transmission and Material Handling*, p. 147, by courtesy of Marcel Dekker, Inc.

TYPE I: Manual or Drip Lubrication
TYPE II: Bath or Disc Lubrication
TYPE III: Oil Stream Lubrication

The limiting RPM for each lubrication type is read from the column to the left of the boundary line shown.

Table 13-5 Horsepower Ratings, Standard Single-Strand Roller Chain, No. 60, ¾-in Pitch

NO. OF TEETH SMALL SPKT.	REVOLUTIONS PER MINUTE—SMALL SPROCKET																							
	10	25	50	100	150	200	300	400	500	600	700	800	900	1 000	1 100	1 200	1 400	1 600	1 800	2 000	2 500	3 000	3 500	4 000	4 500
9	0.15	0.33	0.62	1.16	1.67	2.16	3.12	4.04	4.94	5.82	6.68	7.54	8.38	9.21	9.99	8.77	6.96	5.70	4.77	4.08	2.92	2.22	1.76	1.44	1.21
10	0.16	0.37	0.70	1.30	1.87	2.43	3.49	4.53	5.53	6.52	7.49	8.44	9.39	10.3	11.2	10.3	8.15	6.67	5.59	4.77	3.42	2.60	2.06	1.69	1.41
11	0.18	0.41	0.77	1.44	2.07	2.69	3.87	5.02	6.13	7.23	8.30	9.36	10.4	11.4	12.5	11.9	9.41	7.70	6.45	5.51	3.94	3.00	2.38	1.95	1.63
12	0.20	0.45	0.85	1.58	2.28	2.95	4.25	5.51	6.74	7.94	9.12	10.3	11.4	12.6	13.7	13.5	10.7	8.77	7.35	6.28	4.49	3.42	2.71	2.22	1.86
13	0.22	0.50	0.92	1.73	2.49	3.22	4.64	6.01	7.34	8.65	9.94	11.2	12.5	13.7	14.9	15.2	12.1	9.89	8.29	7.08	5.06	3.85	3.06	2.50	0
14	0.24	0.54	1.00	1.87	2.69	3.49	5.02	6.51	7.96	9.37	10.8	12.1	13.5	14.8	16.2	17.0	13.5	11.1	9.26	7.91	5.66	4.31	3.42	2.80	0
15	0.25	0.58	1.08	2.01	2.90	3.76	5.41	7.01	8.57	10.1	11.6	13.1	14.5	16.0	17.4	18.8	15.0	12.3	10.3	8.77	6.28	4.77	3.79	3.10	0
16	0.27	0.62	1.16	2.16	3.11	4.03	5.80	7.52	9.19	10.8	12.4	14.0	15.6	17.1	18.7	20.2	16.5	13.5	11.3	9.66	6.91	5.26	4.17	3.42	0
17	0.29	0.66	1.24	2.31	3.32	4.30	6.20	8.03	9.81	11.6	13.3	15.0	16.7	18.3	19.9	21.6	18.1	14.8	12.4	10.6	7.57	5.76	4.57	3.74	0
18	0.31	0.70	1.31	2.45	3.53	4.58	6.59	8.54	10.4	12.3	14.1	15.9	17.7	19.5	21.2	22.9	19.7	16.1	13.5	11.5	8.25	6.28	4.98	4.08	0
19	0.33	0.75	1.39	2.60	3.74	4.85	6.99	9.05	11.1	13.0	15.0	16.9	18.8	20.6	22.5	24.3	21.4	17.5	14.6	12.5	8.95	6.81	5.40	4.42	0
20	0.35	0.79	1.47	2.75	3.96	5.13	7.38	9.57	11.7	13.8	15.8	17.9	19.8	21.8	23.8	25.7	23.1	18.9	15.8	13.5	9.66	7.35	5.83	0	
21	0.36	0.83	1.55	2.90	4.17	5.40	7.78	10.1	12.3	14.5	16.7	18.8	20.9	23.0	25.1	27.1	24.8	20.3	17.0	14.5	10.4	7.91	6.28	0	
22	0.38	0.87	1.63	3.05	4.39	5.68	8.19	10.6	13.0	15.3	17.5	19.8	22.0	24.2	26.4	28.5	26.6	21.8	18.2	15.6	11.1	8.48	6.73	0	
23	0.40	0.92	1.71	3.19	4.60	5.96	8.59	11.1	13.6	16.0	18.4	20.8	23.1	25.4	27.7	29.9	28.4	23.3	19.5	16.7	11.9	9.07	7.19	0	
24	0.42	0.96	1.79	3.35	4.82	6.24	8.99	11.6	14.2	16.8	19.3	21.7	24.2	26.6	29.0	31.3	30.3	24.8	20.8	17.8	12.7	9.66	7.67	0	
25	0.44	1.00	1.87	3.50	5.04	6.52	9.40	12.2	14.9	17.5	20.1	22.7	25.3	27.8	30.3	32.7	32.2	26.4	22.1	18.9	13.5	10.3	8.15	0	
26	0.46	1.05	1.95	3.65	5.25	6.81	9.80	12.7	15.5	18.3	21.0	23.7	26.4	29.0	31.6	34.1	34.2	28.0	23.4	20.0	14.3	10.9	8.65	0	
28	0.50	1.13	2.12	3.95	5.69	7.37	10.6	13.8	16.8	19.8	22.8	25.7	28.5	31.4	34.2	37.0	38.2	31.3	26.2	22.4	16.0	12.2	0		
30	0.54	1.22	2.28	4.26	6.13	7.94	11.4	14.8	18.1	21.4	24.5	27.7	30.8	33.8	36.8	39.8	42.4	34.7	29.1	24.8	17.8	13.5	0		
32	0.57	1.31	2.45	4.56	6.57	8.52	12.3	15.9	19.4	22.9	26.3	29.7	33.0	36.3	39.5	42.7	46.7	38.2	32.0	27.3	19.6	14.9	0		
35	0.63	1.44	2.69	5.03	7.24	9.38	13.5	17.5	21.4	25.2	29.0	32.7	36.3	39.9	43.5	47.1	53.4	43.7	36.6	31.3	22.4	17.0	0		
40	0.73	1.67	3.11	5.81	8.37	10.8	15.6	20.2	24.7	29.1	33.5	37.7	42.0	46.1	50.3	54.4	62.5	53.4	44.7	38.2	27.3	0			
45	0.83	1.89	3.53	6.60	9.50	12.3	17.7	23.0	28.1	33.1	38.0	42.0	47.7	52.4	57.1	61.7	70.9	63.7	53.4	45.6	32.6	0			

TYPE I TYPE II TYPE III

Source: Reprinted from *Chains for Power Transmission and Material Handling,* p. 149, by courtesy of Marcel Dekker, Inc.

TYPE I: Manual or Drip Lubrication
TYPE II: Bath or Disc Lubrication
TYPE III: Oil Stream Lubrication

The limiting RPM for each lubrication type is read from the column to the left of the boundary line shown.

Table 13-6 Horsepower Ratings, Standard Single-Strand Roller Chain, No. 80, 1-in. Pitch

NO. OF TEETH SMALL SPKT.	REVOLUTIONS PER MINUTE—SMALL SPROCKET																							
	10	25	50	100	150	200	300	400	500	600	700	800	900	1 000	1 100	1200	1 400	1 600	1 800	2 000	2 200	2 400	2 700	3 000	3 400
9	0.34	0.78	1.45	2.71	3.90	5.05	7.28	9.43	11.5	13.6	15.6	17.6	17.0	14.5	12.6	11.0	8.76	7.17	6.01	5.13	4.45	3.90	3.27	2.79	2.32
10	0.38	0.87	1.63	3.03	4.37	5.66	8.16	10.6	12.9	15.2	17.5	19.7	19.9	17.0	14.7	12.9	10.3	8.40	7.04	6.01	5.21	4.57	3.83	3.27	2.71
11	0.42	0.97	1.80	3.36	4.84	6.28	9.04	11.7	14.3	16.9	19.4	21.9	23.0	19.6	17.0	14.9	11.8	9.69	8.12	6.93	6.01	5.27	4.42	3.77	1.70
12	0.47	1.06	1.98	3.69	5.32	6.89	9.93	12.9	15.7	18.5	21.3	24.0	26.2	22.3	19.4	17.0	13.5	11.0	9.25	7.90	6.85	6.01	5.04	4.30	0
13	0.51	1.16	2.16	4.03	5.80	7.52	10.8	14.0	17.1	20.2	23.2	26.2	29.1	25.2	21.8	19.2	15.2	12.5	10.4	8.91	7.72	6.78	5.68	4.85	0
14	0.55	1.25	2.34	4.36	6.29	8.14	11.7	15.2	18.6	21.9	25.1	28.4	31.5	28.2	24.4	21.4	17.0	13.9	11.7	9.96	8.63	7.57	6.35	5.42	0
15	0.59	1.35	2.52	4.70	6.77	8.77	12.6	16.4	20.0	23.6	27.1	30.6	34.0	31.2	27.1	23.8	18.9	15.4	12.9	11.0	9.57	8.40	7.04	6.01	0
16	0.63	1.45	2.70	5.04	7.26	9.41	13.5	17.6	21.5	25.3	29.0	32.8	36.4	34.4	29.8	26.2	20.8	17.0	14.2	12.2	10.5	9.25	7.76	6.62	0
17	0.68	1.55	2.88	5.38	7.75	10.0	14.5	18.7	22.9	27.0	31.0	35.0	38.9	37.7	32.7	28.7	22.7	18.6	15.6	13.3	11.5	10.1	8.49	7.25	0
18	0.72	1.64	3.07	5.72	8.25	10.7	15.4	19.9	24.4	28.7	33.0	37.2	41.4	41.1	35.6	31.2	24.8	20.3	17.0	14.5	12.6	11.0	9.25	7.90	0
19	0.76	1.74	3.25	6.07	8.74	11.3	16.3	21.1	25.8	30.4	35.0	39.4	43.8	44.5	38.6	33.9	26.9	22.0	18.4	15.7	13.6	12.0	10.0	8.57	0
20	0.81	1.84	3.44	6.41	9.24	12.0	17.2	22.3	27.3	32.2	37.0	41.7	46.3	48.1	41.7	36.6	29.0	23.8	19.9	17.0	14.7	12.9	10.8	0	
21	0.85	1.94	3.62	6.76	9.74	12.6	18.2	23.5	28.8	33.9	39.0	43.9	48.9	51.7	44.8	39.4	31.2	25.6	21.4	18.3	15.9	13.9	11.7	0	
22	0.90	2.04	3.81	7.11	10.2	13.3	19.1	24.8	30.3	35.7	41.0	46.2	51.4	55.5	48.1	42.2	33.5	27.4	23.0	19.6	17.0	14.9	12.5	0	
23	0.94	2.14	4.00	7.46	10.7	13.9	20.1	26.0	31.8	37.4	43.0	48.5	53.9	59.3	51.4	45.1	35.8	29.3	24.6	21.0	18.2	15.9	13.4	0	
24	0.98	2.24	4.19	7.81	11.3	14.6	21.0	27.2	33.2	39.2	45.0	50.8	56.4	62.0	54.8	48.1	38.2	31.2	26.2	22.3	19.4	17.0	14.2	0	
25	1.03	2.34	4.37	8.16	11.8	15.2	21.9	28.4	34.7	40.9	47.0	53.0	59.0	64.8	58.2	51.1	40.6	33.2	27.8	23.8	20.6	18.1	15.1	0	
26	1.07	2.45	4.56	8.52	12.3	15.9	22.9	29.7	36.2	42.7	49.1	55.3	61.5	67.6	61.8	54.2	43.0	35.2	29.5	25.2	21.8	19.2	16.1	0	
28	1.16	2.65	4.94	9.23	13.3	17.2	24.8	32.1	39.3	46.3	53.2	59.9	66.7	73.3	69.0	60.6	48.1	39.4	33.0	28.2	24.4	21.4	0		
30	1.25	2.85	5.33	9.94	14.3	18.5	26.7	34.6	42.3	49.9	57.3	64.6	71.8	78.9	76.6	67.2	53.3	43.6	36.6	31.2	27.1	23.8	0		
32	1.34	3.06	5.71	10.7	15.3	19.9	28.6	37.1	45.4	53.5	61.4	69.2	77.0	84.6	84.3	74.0	58.7	48.1	40.3	34.4	29.8	26.2	0		
35	1.48	3.37	6.29	11.7	16.9	21.9	31.6	40.9	50.0	58.9	67.6	76.3	84.8	93.3	96.5	84.7	67.2	55.0	46.1	39.4	34.1	0			
40	1.71	3.89	7.27	13.6	19.5	25.3	36.4	47.2	57.7	68.0	78.1	88.1	98.0	108	117	103	82.1	67.2	56.3	48.1	20.0	0			
45	1.94	4.42	8.25	15.4	22.2	28.7	41.4	53.6	65.6	77.2	88.7	100	111	122	133	123	98.0	80.2	67.2	54.1	0				

TYPE I | TYPE II | TYPE III

Source: Reprinted from *Chains for Power Transmission and Material Handling*, p. 149, by courtesy of Marcel Dekker, Inc.

TYPE I: Manual or Drip Lubrication
TYPE II: Bath or Disc Lubrication
TYPE III: Oil Stream Lubrication

The limiting RPM for each lubrication type is read from the column to the left of the boundary line shown

Table 13-7 Service Factors for Chain Drives

	TYPE OF DRIVER		
LOAD TYPE	*Hydraulic Drive*	*Electric Motor or Turbine*	*Internal Combustion Engine with Mechanical Drive*
Smooth (agitators, fans, light uniformly loaded conveyors)	1.0	1.0	1.2
Moderate shock (machine tools, cranes, heavy conveyors, food mixers and grinders)	1.2	1.3	1.4
Heavy shock (punch presses, hammer mills, reciprocating conveyors, rolling mill drive)	1.4	1.5	1.7

3. The center distance between the sprocket axes should be approximately 30 to 50 pitches (30 to 50 times the pitch of the chain).
4. The arc of contact of the chain on the smaller sprocket should be no smaller than 120°.
5. The larger sprocket should normally have no more than 120 teeth.
6. The preferred arrangement for a chain drive is with the centerline of the sprockets horizontal and with the tight side on top.
7. The chain length must be an integral multiple of the pitch, and an even number of pitches is recommended. The center distance should be made adjustable to accommodate the chain length and to take up for tolerances and wear. Excessive sag on the slack side should be avoided, especially on drives that are not horizontal. A convenient relation between the center distance (C), chain length (L), number of teeth in the small sprocket (N_1), and number of teeth in the large sprocket (N_2), expressed in pitches, is

$$L = 2C + \frac{N_2 + N_1}{2} + \frac{(N_2 - N_1)^2}{4\pi^2 C} \tag{13-9}$$

The exact theoretical center distance for a given chain length, again in pitches, is

$$C = \frac{1}{4}\left[L - \frac{N_2 + N_1}{2} + \sqrt{\left[L - \frac{N_2 + N_1}{2}\right]^2 - \frac{8(N_2 - N_1)^2}{4\pi^2}}\right] \tag{13-10}$$

The theoretical center distance assumes no sag in either the tight or slack side of the chain and is thus a *maximum*. Negative tolerances or adjustment must be provided.
8. The pitch diameter of a sprocket with N teeth for a chain with a pitch of p is:

$$D = \frac{p}{\sin(180°/N)} \tag{13-11}$$

9. The minimum sprocket diameter and therefore the minimum number of teeth in a sprocket is often limited by the size of the shaft on which it is mounted. Check the sprocket catalog.

Lubrication

Chain manufacturers recommend three different methods of applying lubrication, depending on the linear speed of the chain. A constant supply of clean oil is essential to smooth operation and satisfactory life of the chain drive. Although there may be modest differences between manufacturers, the following are the general guidelines for the limits of speed. Refer to Figure 13-19 for illustrations of the methods.

Type I (170 to 650 ft/min). Manual or drip lubrication: For manual lubrication, oil is applied with a brush or a spout can, preferably at least once every 8 h of operation. For drip feed lubrication, oil is fed directly onto the link plates of each chain strand.

Type II (650 to 1 500 ft/min). Bath or disc lubrication: The chain cover provides a sump of oil into which the chain dips continuously. Alternatively, a disc or slinger can be attached to one of the shafts to lift oil to a trough above the lower strand of chain. The trough then delivers a stream of oil to the chain. The chain itself, then, does not need to dip into the oil.

Type III (Above 1 500 ft/min). Oil stream lubrication: An oil pump delivers a continuous stream of oil on the lower part of the chain.

Example problem 13-2 presents a chain drive design.

Example Problem 13-2. Design a chain drive for a uniformly loaded conveyor to be driven by a gasoline engine through a mechanical drive. The input speed will be 900 rpm, and the desired output speed is 230 to 240 rpm. The conveyor requires 15.0 hp.

Solution.

1. Specify a service factor and compute the design power. From Table 13-7, for moderate shock and a gasoline engine drive through a mechanical drive, $SF = 1.4$. A lower value can be used because of the uniform loading.

$$\text{Design power} = 1.4(15.0) = 21.0 \text{ hp}$$

2. Compute the desired ratio. Using the middle of the required range of output speeds,

$$\text{Ratio} = (900 \text{ rpm})/(235 \text{ rpm}) = 3.83$$

3. Refer to the tables for power capacity (Tables 13-4, 13-5, and 13-6) and select the chain pitch. For a single strand, the no. 60 chain with $p = 3/4$ in seems best. A 21-tooth sprocket is rated at 20.9 hp at 900 rpm. At this speed, type II lubrication (oil bath) is required.
4. Compute the required number of teeth on the large sprocket.

$$N_2 = N_1 \times \text{ratio} = 21(3.83) = 80.43$$

Let's use the integer: 80 teeth.

5. Compute the actual expected output speed.

$$n_2 = n_1(N_1/N_2) = 900 \text{ rpm}(21/80) = 236 \text{ rpm} \quad \text{(OK!)}$$

6. Compute the pitch diameters of the sprockets.

$$D_1 = p/\sin(180°/N_1) = 0.75/\sin(180°/21) = 5.032 \text{ in}$$
$$D_2 = p/\sin(180°/N_2) = 0.75/\sin(180°/80) = 19.103 \text{ in}$$

7. Specify the nominal center distance. Let's use the middle of the recommended range, 40 pitches.

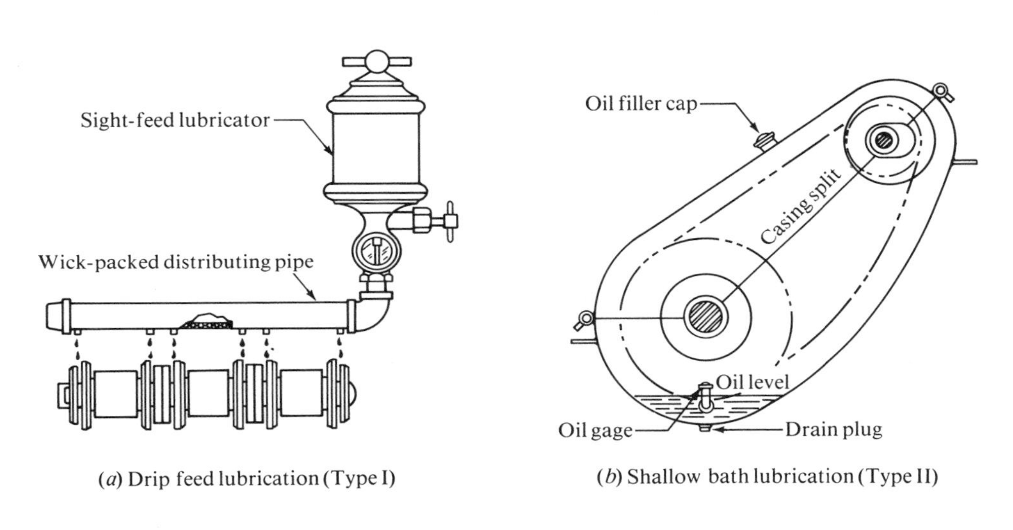

(*a*) Drip feed lubrication (Type I)

(*b*) Shallow bath lubrication (Type II)

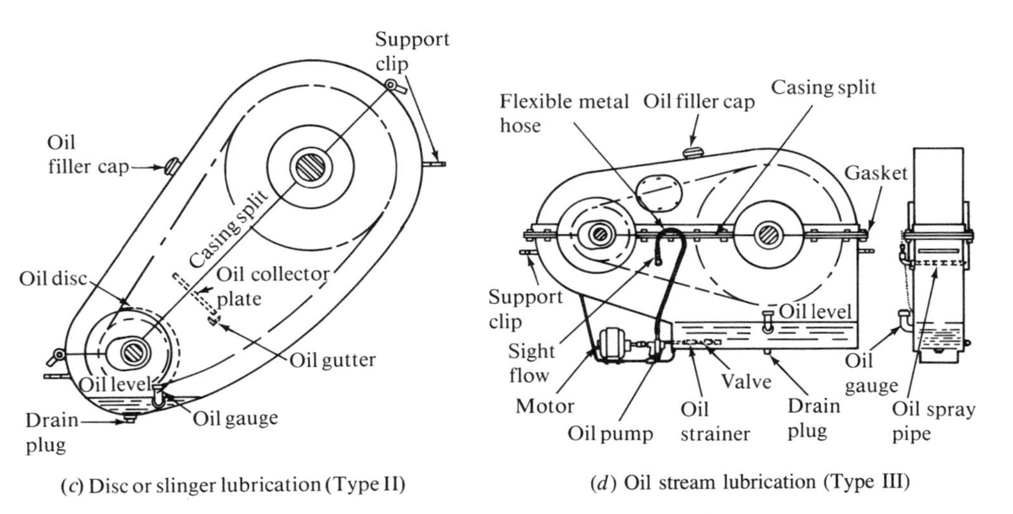

(*c*) Disc or slinger lubrication (Type II)

(*d*) Oil stream lubrication (Type III)

Figure 13-19 Lubrication Methods (American Chain Association, Washington, D.C.)

8. Compute the required chain length in pitches from equation (13-9).

$$L = 2C + \frac{N_2 + N_1}{2} + \frac{(N_2 - N_1)^2}{4\pi^2 C} \tag{13-9}$$

$$L = 2(40) + \frac{80 + 21}{2} + \frac{(80 - 21)^2}{4\pi^2(40)} = 132.7 \text{ pitches}$$

9. Specify an integral number of pitches for the chain length and compute the actual theoretical center distance. Let's use 132 pitches, an even number. Then, from equation (13-10),

$$C = \frac{1}{4}\left[L - \frac{N_2 + N_1}{2} + \sqrt{\left[L - \frac{N_2 + N_1}{2}\right]^2 - \frac{8(N_2 - N_1)^2}{4\pi^2}}\right] \tag{13-10}$$

$$C = \frac{1}{4}\left[132 - \frac{80 + 21}{2} + \sqrt{\left[132 - \frac{80 + 21}{2}\right]^2 - \frac{8(80 - 21)^2}{4\pi^2}}\right]$$

$$C = 39.638 \text{ pitches} = 39.638(0.75 \text{ in}) = 29.73 \text{ in}$$

10. Summarize the design. Figure 13-20 shows a sketch of the design to scale.

Pitch: no. 60 chain, ¾-in pitch
Length: 132 pitches = 132(0.75) = 99.0 in
Center distance: C = 29.73 in (maximum)
Sprockets: Single-strand, no. 60, ¾-in pitch
  Small: 21 teeth, D = 5.033 in
  Large: 80 teeth, D = 19.103 in
Type II lubrication required. Large sprocket can dip into an oil bath.

Example Problem 13-3. Create an alternate design for the conditions of example problem 13-2 to produce a smaller drive.

Solution. In order to permit a smaller drive, a smaller pitch is desirable. To handle the same design power (21.0 hp) at the same speed (900 rpm), with a smaller chain, let's consider a multistrand design.

Let's try a four-strand chain for which the power capacity factor is 3.3. Then the required power per strand is

$$P = 21.0/3.3 = 6.36 \text{ hp}$$

From Table 13-4, we find that a no. 40 chain (½-in pitch) with a 22-tooth sprocket will be satisfactory. Type II lubrication, oil bath, can be used.

Required large sprocket:

$$N_2 = N_1 \times \text{ratio} = 22(3.83) = 84.26$$

Let's use N_2 = 84 teeth.

Sprocket diameters:

$$D_1 = p/\sin(180°/22) = 3.513 \text{ in}$$

$$D_2 = p/\sin(180°/84) = 13.372 \text{ in}$$

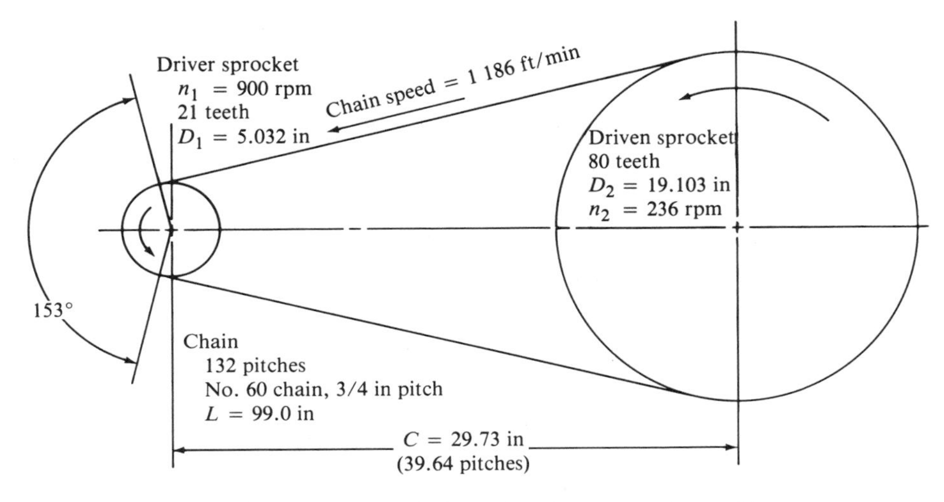

Figure 13-20 Scale Layout of Chain Drive for Example Problem 13-2

Center distance: Let's try the minimum recommended: $C = 30$ pitches.

$$30(0.50 \text{ in}) = 15.0 \text{ in}$$

Chain length:

$$L = 2(30) + \frac{84 + 22}{2} + \frac{(84 - 22)^2}{4\pi^2(30)} = 116.3 \text{ pitches}$$

Specify the integer length, $L = 116$ pitches $= 116(0.50) = 58.0$ in. Actual maximum center distance:

$$C = \frac{1}{4}\left[116 - \frac{84 + 22}{2} + \sqrt{\left[116 - \frac{84 + 22}{2}\right]^2 - \frac{8(84 - 22)^2}{4\pi^2}}\right]$$

$$C = 29.87 \text{ pitches} = 29.87(0.50) = 14.94 \text{ in}$$

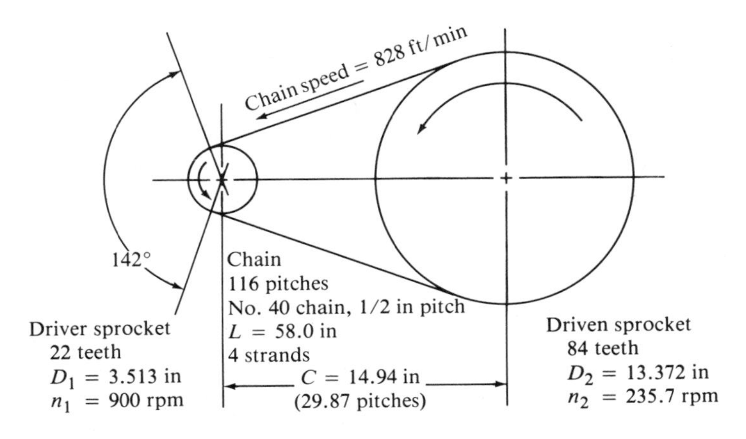

Figure 13-21 Scale Layout of Chain Drive for Example Problem 13-3

Summary. Figure 13-21 shows the new design to the same scale as the first design. The space reduction is significant.

Chain: No. 40, ½-in pitch, four-strand, 116 pitches
Sprockets: No. 40-4 (four strands), ½-in pitch
Small: 22 teeth, $D = 3.513$ in
Large: 84 teeth, $D = 13.372$ in
Maximum center distance: 14.94 in
Type II lubrication (oil bath)

REFERENCES

1. American Chain Association. *Chains for Power Transmission and Material Handling*. New York: Marcel Dekker, 1982.
2. Dayco Corporation. *Dayco Engineering Guide for V-Belt Drives*. Dayton, Ohio: Rubber Products Company, 1981.
3. Dayco Corporation. *Engineering Handbook of Automotive V-Belt Drives*. Automotive O.E.M. Division. Melrose Park, Ill.
4. Emerson Electric Company. *Power Transmission Equipment Catalog No. 9*. Browning Manufacturing Division. Maysville, Ky.: 1980.
5. The Gates Rubber Company. *V-Belt Drive Design Manual*. Denver.
6. Motherway, J. E. "Designing V-Belt Drives With a Microcomputer." *Computers in Mechanical Engineering*, 2, no. 1 (July 1983).
7. Reliance Electric Company. *V-Belt Drives Bulletin 30E*. Dodge Division. Mishawaka, Ind.
8. Rexnord, Incorporated, *Catalog of Power Transmission and Conveying Components*. Milwaukee.
9. Rubber Manufacturers Association. *V-Belt Drives with Twist and Non-Alignment Including Quarter Turn*. Power Transmission Belt Technical Information Bulletin No. 10.
10. Society of Automotive Engineers. "Standard SAE J636c V-Belts and Pulleys" and "Standard SAE J637b Automotive V-Belt Drives," *SAE Handbook*. Warrendale, Pa.: 1984.
11. T. B. Wood's Sons Company. *V-Belt Drive Manual*. Chambersburg, Pa.

PROBLEMS

V-belt Drives

1. Specify the standard 3V belt length (from Table 13-2) that would be applied to two sheaves with pitch diameters of 5.25 in and 13.95 in with a center distance of no more than 24.0 in.
2. For the standard belt specified in problem 1, compute the actual center distance that would result.
3. For the standard belt specified in problem 1, compute the angle of wrap on both the sheaves.
4. Specify the standard 5V belt length (from Table 13-2) that would be applied to two sheaves with pitch diameters of 8.4 in and 27.7 in with a center distance of no more than 60.0 in.
5. For the standard belt specified in problem 4, compute the actual center distance that would result.
6. For the standard belt specified in problem 4, compute the angle of wrap on both the sheaves.
7. Specify the standard 8V belt length (from Table 13-2) that would be applied to two sheaves with pitch diameters of 13.8 in and 94.8 in with a center distance of no more than 144 in.
8. For the standard belt specified in problem 7, compute the actual center distance that would result.
9. For the standard belt specified in problem 7, compute the angle of wrap on both the sheaves.
10. If the small sheave of problem 1 is rotating at 1 750 rpm, compute the linear speed of the belt.
11. If the small sheave of problem 4 is rotating at 1 160 rpm, compute the linear speed of the belt.
12. If the small sheave of problem 7 is rotating at 870 rpm, compute the linear speed of the belt.
13. For the belt drive from problems 1 and 10, compute the rated power, considering corrections for speed ratio, belt length, and angle of wrap.
14. For the belt drive from problems 4 and 11, compute the rated power, considering corrections for speed ratio, belt length, and angle of wrap.
15. For the belt drive from problems 7 and 12, compute the rated power, considering corrections for speed ratio, belt length, and angle of wrap.

Problem Number	*Driver Type*	*Driven Machine*	*Service (h/day)*	*Input Speed (rpm)*	*Input Power (hp)*	*Nominal Output Speed (rpm)*
18.	AC motor (HT)	Hammer mill	8	870	25	310
19.	AC motor (NT)	Fan	22	1 750	5	725
20.	6-cylinder engine	Conveyor, heavy	16	1 500	40	550
21.	DC motor (compound)	Milling machine	16	1 250	20	695
22.	AC motor (HT)	Rock crusher	8	870	100	625

Note: *NT* indicates a normal-torque electric motor. *HT* indicates a high-torque electric motor.

16. Describe a standard 15N belt cross section. To what size belt (inches) would it be closest?
17. Describe a standard 17A belt cross section. To what size belt (inches) would it be closest?

For problems 18–22, design a V-belt drive. Specify the belt size, sheave sizes, number of belts, actual output speed, and center distance.

Roller Chain

23. Describe a standard roller chain no. 140.
24. Describe a standard roller chain no. 60.
25. Specify a suitable standard chain to exert a static pulling force of 1 250 lb.
26. Roller chain is used in a hydraulic fork lift truck to elevate the forks. If two strands support the load equally, which size would you specify for a design load of 5 000 lb?
27. List three typical failure modes of roller chain.
28. Determine the power rating of a no. 60 chain, single-strand, operating on a 20-tooth sprocket at 750 rpm. Describe the preferred method of lubrication. The chain connects a hydraulic drive with a meat grinder.
29. For the data of problem 28, what would be the rating for three strands?
30. Determine the power rating of a no. 40 chain, single-strand, operating on a 12-tooth sprocket at 860 rpm. Describe the preferred method of lubrication. The small sprocket is applied to the shaft of an electric motor. The output is to a coal conveyor.
31. For the data of problem 30, what would be the rating for four strands?
32. Determine the power rating of a no. 80 chain, single-strand, operating on a 32-tooth sprocket at 1 160 rpm. Describe the preferred method of lubrication. The input is an internal combustion engine, and the output is to a fluid agitator.
33. For the data of problem 32, what would be the rating for two strands?
34. Specify the required length of no. 60 chain to mount on sprockets having 15 and 50 teeth with a center distance of no more than 36 in.
35. For the chain specified in problem 34, compute the exact theoretical center distance.
36. Specify the required length of no. 40 chain to mount on sprockets having 11 and 45 teeth with a center distance of no more than 24 in.
37. For the chain specified in problem 36, compute the exact theoretical center distance.

For problems 38–42, design a roller chain drive. Specify the chain size, sizes and number of teeth in the sprockets, number of chain pitches, and center distance.

Problem Number	*Driver Type*	*Driven Machine*	*Input Speed (rpm)*	*Input Power (hp)*	*Nominal Output Speed (rpm)*
38.	AC motor	Hammer mill	310	25	160
39.	AC motor	Agitator	750	5	325
40.	6-cylinder engine	Conveyor, heavy	550	40	250
41.	Steam turbine	Centrifugal pump	2 200	20	775
42.	Hydraulic drive	Rock crusher	625	100	225

14 Plain Surface Bearings

14-1 OVERVIEW

The term *plain surface bearing* refers to that kind of bearing in which two surfaces move relative to each other without the benefit of rolling contact. Thus there is sliding contact. The actual shape of the surfaces can be anything which permits the relative motion. The most common shapes are flat surfaces and concentric cylinders.

The name *journal bearing* is sometimes used for plain surface bearings. This is derived from the terminology of the components of the complete bearing system. For the case of a bearing on a rotating shaft, the portion of the rotating shaft at the bearing is called the *journal*. The stationary part that supports the load is the *bearing*. Figure 14-1 shows the basic geometry of a journal bearing.

A given bearing system can operate with any of three types of lubrication:

Boundary lubrication: There is actual contact between the solid surfaces of the moving and the stationary parts of the bearing system, although a film of lubricant is present.

Mixed-film lubrication: A transition zone between boundary and full-film lubrication.

Full-film lubrication: The moving and stationary parts of the bearing system are separated by a complete film of lubricant that supports the load. The term *Hydrodynamic lubrication* is often used to describe this type.

All of these types of lubrication can be encountered in a bearing without external pressurization of the bearing. If lubricant under pressure is supplied to the bearing, it is called a *hydrostatic bearing,* which is discussed separately. Running dry surfaces together is not recommended unless there is inherently good lubricity between the mating materials. Some plastics are used dry, as discussed in sections 14-3 and 14-4.

14-2 BEARING PARAMETER, *Zn/p*

The performance of a bearing differs radically, depending on which type of lubrication occurs. There is a marked decrease in the coefficient of friction when the operation changes from boundary to full-film lubrication. Wear also decreases with full-film lubrication. Thus it is desirable to understand the conditions under which one or the other type of lubrication occurs.

The creation of full-film lubrication, the most desirable type, is encouraged by light loads, high relative speed between the moving and stationary parts, and presence of a high-viscosity lubricant in the bearing in copious supply. For a rotating journal bearing, the combined effect of these three factors, as it relates to the friction in the bearing, can be evaluated by computing the *bearing parameter, Zn/p*. The viscosity of the lubricant is indicated by Z; the rotational speed by n; and the bearing load by the pressure, p. The pressure is computed by dividing the applied radial load on the bearing by the *projected area* of the bearing: that is, the product of the length times the diameter.

Any units for the terms in the bearing parameter can be used. One common set expresses the viscosity Z in centipoise, the speed n in rpm, and the pressure P in pounds per square inch (psi). This is by no means a coherent unit system, but it represents one traditional set of units for the three terms. Besides, the bearing parameter is used mostly for comparing the relative performance of different bearing designs, so the actual units are usually irrelevant.

The effect of the bearing parameter can be seen in Figure 14-2, which is a plot of the coefficient of friction, f, for the bearing versus the value of Zn/p. At low values of Zn/p,

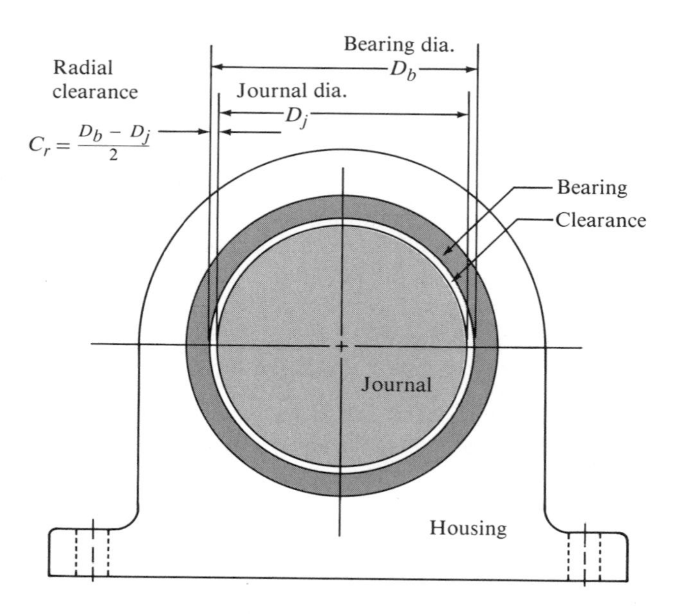

Figure 14-1 Bearing Geometry

boundary lubrication occurs and the coefficient of friction is high. For example, with a steel shaft sliding slowly in a lubricated bronze bearing (boundary lubrication), the value of f would be approximately 0.08 to 0.14. At high values of Zn/p, the full hydrodynamic film is created, and the value of f is normally in the range of 0.001 to 0.005. Note that this compares favorably with precision rolling contact bearings. Between boundary and full-film lubrication, the *mixed-film* type, which is some combination of the other two, occurs.

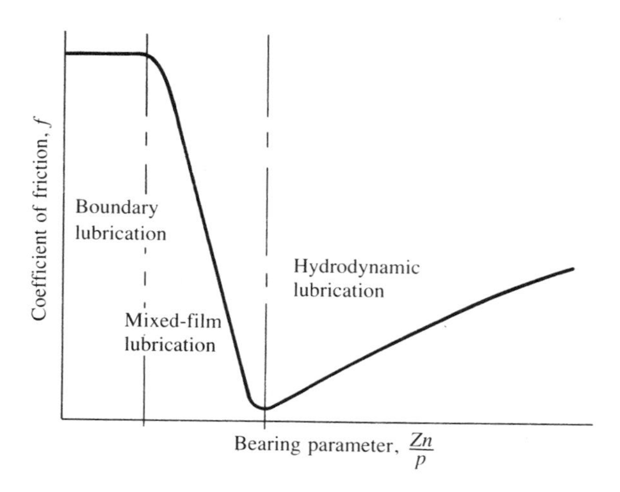

Figure 14-2 Bearing Performance and Types of Lubrication Related to the Bearing Parameter Zn/p

It is recommended that designers avoid the mixed-film zone because it is virtually impossible to predict the performance of the bearing system. Also notice that the curve is very steep in this zone. Thus, a small change in any of the three factors, Z, n, or p, produces a large change in f, resulting in erratic performance of the machine.

The value of Zn/p at which full-film lubrication is produced is very difficult to predict. Besides the individual factors of speed, pressure (load), and viscosity (a function of the type of lubricant and its temperature), the variables affecting the production of the film include the quantity of lubricant supplied, the adhesion of the lubricant to the surfaces, the materials of the journal and the bearing, the structural rigidity of the journal and the bearing, and the surface roughness of the journal and the bearing. After completing the design process presented later in this chapter, the designer is advised to test the design. As a very rough estimate, a value of $Zn/p = 50$ or higher with the previously listed units will produce full-film lubrication with typical bearing designs.

In general, boundary lubrication should be expected for slow-speed operation with a surface speed less than about 10 ft/min (0.05 m/s). Reciprocating or oscillating motion or a combination of a light lubricant and high pressure would also produce boundary lubrication.

The design of bearings to produce full-film lubrication is described in section 14-5; in general, it requires a surface speed of greater than 25 ft/min (0.13 m/s) continuously in one direction with an adequate supply of oil at a proper viscosity.

14-3 BEARING MATERIALS

In rotational applications, the journal on the shaft is frequently steel. The stationary bearing may be made of any of a wide variety of materials, including the following:

Bronze   Babbitt
Aluminum   Porous Metals
Plastics (Nylon, TFE, PTFE, phenolic, acetal, polycarbonate, filled polyimide, and others)

The properties desirable for materials used for plain bearings are somewhat unique and compromises must frequently be made. The following is a discussion of these properties.

Strength

The function of the bearing is to carry the applied load and deliver it to the supporting structure. At times, the loads are varied, requiring fatigue strength as well as static strength.

Embeddability

The property of embeddability relates to the ability of the material to hold contaminants in the bearing without causing damage to the rotating journal. Thus, a relatively soft material is desirable.

Corrosion Resistance

The total environment of the bearing must be considered, including the journal material, the lubricant, temperature, air-borne particulates, and corrosive gases or vapors.

Cost

Always an important factor, cost includes not only the material cost but also processing and installation costs.

A brief discussion of the performance of some of the bearing materials follows.

Cast Bronze

The name *bronze* refers to several alloys of copper with tin, lead, zinc, or aluminum, either singly or in combination. The leaded bronzes contain 25 to 35 percent lead, which gives them good embeddability and resistance to seizing under conditions of boundary lubrication. However, their strength is relatively low. The cast bearing bronze, SAE CA932, has 83 percent copper, 7 percent tin, 7 percent lead, and 3 percent zinc. It possesses a good combination of properties for such uses as pumps, machinery, and appliances. Tin and aluminum bronzes have higher strength and hardness and can carry greater loads, particularly in impact situations. But they rate lower on embeddability.

Babbitt

Babbitts may be lead-based or tin-based, having nominally 80 percent of the parent metal. Various alloy compositions of copper and antimony (as well as lead and tin) can tailor the properties to suit a particular application. Because of the softness of babbitts they have outstanding embeddability and resistance to seizure, important in applications in which boundary lubrication occurs. They have rather low strength, however, and are often applied as liners in steel or cast iron housings.

Aluminum

With the highest strength of the commonly used bearing materials, aluminum is used in severe applications in engines, pumps, and aircraft. The high hardness of the aluminum bearings results in poor embeddability, requiring clean lubricants.

Porous Metals

Products of the powder metal industry, porous metals are sintered from powders of bronze, iron, and aluminum; some mixed with lead or copper. The sintering leaves a large number of voids in the bearing material into which lubricating oil is forced. Then, during operation, the oil migrates out of the pores, supplying the bearing. Such bearings are particularly good for slow-speed, reciprocating, or oscillating motions.

Plastics

Generally referred to as *self-lubricating materials,* plastics used in bearing applications have inherently low friction characteristics. They can be operated dry, but most improve in performance with a lubricant present. Embeddability is usually good, as is resistance to seizure. But many have low strength, limiting their load-carrying capacity. Backing with metal sleeves is frequently done to improve load-carrying capacity. Major advantages are corrosion resistance and, when operated dry, freedom from contamination. These properties are particularly important in the processing of foodstuffs and chemical products.

Because of the complex chemical names for plastic materials and the virtually infinite combinations of base materials, reinforcements, and fillers used, it is difficult to characterize bearing plastics. Most are composites of several components. The group known as *fluoropolymers* are popular because of the very low coefficient of friction (0.05 to 0.15) and good wear resistance. Phenolics, polycarbonates, acetals, nylons, and many other plastics are also used for bearings. Among the chemical names and abbreviations found in this field are the following:

PTFE: Polytetrafluoroethylene
PA: Polyamide
PPS: Polyphenylene sulfide
PVDF: Polyvinylidene fluoride
PEEK: Polyetheretherketone
PEI: Polyetherimide
PES: Polyethersulfone
PFA: Perfluoroalkoxy-modified tetrafluoroethylene

Reinforcements and fillers used with plastic bearing materials include glass fibers, milled glass, carbon fibers, bronze powders, PTFE, PPS, and some solid lubricants, such as graphite and molybdenum disulfide.

14-4 DESIGN OF BOUNDARY-LUBRICATED BEARINGS

The factors to be considered when selecting materials for bearings and specifying the design details include the following:

Coefficient of friction: Both static and dynamic conditions should be considered.

Load capacity, p: Radial load divided by the projected area of the bearing.

Speed of operation, V: The relative speed between the moving and stationary components, usually in feet per minute.

Temperature at operating conditions.

Wear limitations.

Producibility: Machining, molding, fastening, assembly, and service.

pV Factor

In addition to the individual consideration of load capacity (p) and speed of operation (V), the product pV is an important performance parameter for bearing design when boundary lubrication occurs. The pV value is a measure of the ability of the bearing material to accommodate the frictional energy generated in the bearing. At the limiting pV value, the bearing will not achieve a stable temperature limit, and rapid failure will occur. A practical design value for pV is one-half of the limiting pV value, given in Table 14-1.

Wear Factor

The wear factor, K, is measured under fixed conditions with the material loaded as a thrust washer. When equilibrium is reached, wear is measured as volumetric loss of material as a function of time. The load and the velocity also affect wear; thus, K is defined as

$$K = W/FVT \tag{14-1}$$

Table 14-1 Performance Parameters for Bearing Materials in Boundary Lubrication at Room Temperature

Material	*pV* (psi-fpm)	*Wear Factor, K* (10^{-10} in^3 · min)/(ft · lb · h)	*Coefficient of Friction*[a]
Porous bronze	50 000	NA	NA
PTFE fabric	25 000	NA	NA
PTFE-Bronze filled	21 000	5	0.13
PTFE-Glass filled	18 000	7	0.09
PTFE-PPS filled	15 000	1	0.13
PA-Filled with PTFE and glass fiber	20 000	16	0.26
PPS-Filled with PTFE and glass fiber	30 000	110	0.17

[a]Dynamic; static coefficient generally lower.

Source: *Machine Design Magazine Mechanical Drives Reference Issue*. Cleveland: Penton/IPC, 1981; Wolverton, M. P., et al. "How Plastic Composites Wear at High Temperatures." *Machine Design Magazine,* Feb. 10, 1983.

where

$$W = \text{Wear, volume of material lost (in}^3)$$
$$F = \text{Applied load (lb)}$$
$$V = \text{Linear velocity (ft/min)}$$
$$T = \text{Time (h)}$$

Then the units for K are

$$K = \frac{\text{in}^3}{\text{lb}}\frac{1}{\text{ft/min}}\frac{1}{\text{h}} = \frac{\text{in}^3 \cdot \text{min}}{\text{lb} \cdot \text{ft} \cdot \text{h}}$$

But because wear is a very slow phenomenon, the values are very small. Note in Table 14-1 that the values are to be multiplied by 10^{-10}

Example Problem 14-1. A bearing is to be designed to carry a radial load of 150 lb from a shaft having a diameter of 1.50 in and rotating at 500 rpm. The bearing is to be 2.00 in long. Compute the pV value and compare it with the values for the materials listed in Table 14-1.

Solution. First compute the pressure.

$$p = F/LD = 150/(2.00)(1.50) = 50 \text{ psi}$$

Now compute the linear speed. With D in inches and n in rpm,

$$V = \pi Dn/12 = (\pi)(1.50)(500)/12 = 196 \text{ ft/min}$$

Then

$$pV = (50)(196) = 9\,800 \text{ psi-fpm}$$

It is recommended that the operating pV value be no more than about one-half of the limiting value from Table 14-1. Therefore the required limiting value is approximately 19 600 psi-fpm. The bronze-filled PTFE bearing material would be satisfactory.

If a shorter bearing is desired, we could compute the minimum acceptable length with the porous bronze material.

Let the operating $pV = 50\,000/2 = 25\,000$ psi-fpm

$$V = 196 \text{ ft/min} \quad \text{(as before)}$$

Then the required maximum p is

$$p = pV/V = 25\,000/196 = 128 \text{ psi}$$

But, $p = F/LD$. Then the minimum acceptable length is

$$L = F/pD = 150/(128)(1.50) = 0.78 \text{ in}$$

This results in an L/D ratio of just over 0.50, a recommended minimum. A typical upper end of the range of L/D ratio is 2.0. With careful design, these limits can be exceeded. But with long bearings, sloping of the shaft relative to the bearing tends to reduce the clearance and may result in nonuniform loading.

14-5 FULL-FILM HYDRODYNAMIC BEARINGS

In the *full-film hydrodynamic bearing* the load on the bearing is supported on a continuous film of lubricant, usually oil, so that no contact between the bearing and the rotating journal occurs. There must be a pressure developed in the oil in order to carry the load. With proper design, the motion of the journal inside the bearing creates the necessary pressure.

Figure 14-3 shows the progressive action in a plain surface bearing from start-up to the steady-state hydrodynamic operation. It should be observed that boundary lubrication and mixed-film lubrication preceed the establishment of full-film hydrodynamic lubrication. At start-up, the radial load applied through the journal to the bearing forces the journal off center in the direction of the load, taking up all clearance [Figure 14-3(a)]. At the initial slow rotational speeds, the friction between the journal and the bearing causes the journal to climb up the bearing wall, somewhat as shown in Figure 14-3(b). Because of the viscous shear stresses developed in the oil, the moving journal draws oil into the converging wedge-shaped area above the region of contact. The resulting pumping action produces a pressure in the oil film; when the pressure is sufficiently high, the journal is lifted from the bearing. The frictional forces are greatly reduced under this operating condition, and the journal moves eventually to the steady state position shown in Figure 14-3(c). Note that the journal is offset from the direction of the load; that there is a certain eccentricity, e, between the geometric center of the bearing and the center of the journal; and that there is a point of minimum film thickness, h_o, at the nose of the wedge-shaped pressurized zone.

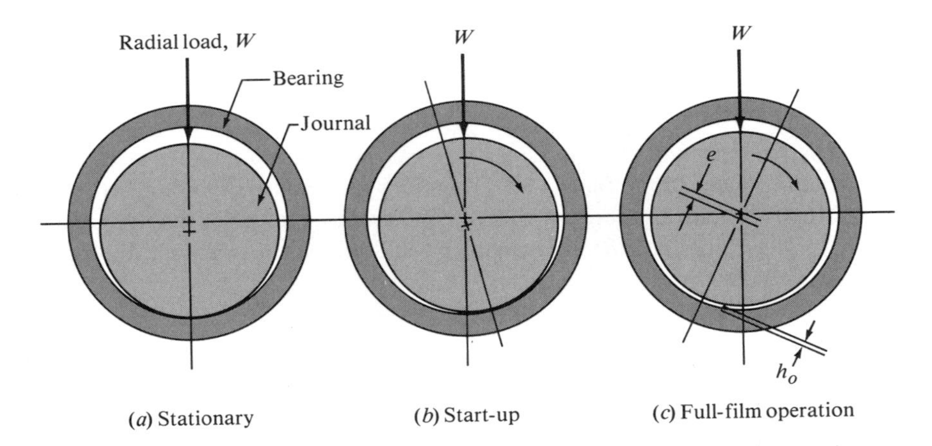

Figure 14-3 Position of the Journal Relative to the Bearing as a Function of Mode of Operation

The following discussion presents several guidelines for bearing design for typical industrial applications.

Surface Roughness

A ground journal with a surface roughness of 16 to 32 microinches (μin) root mean square (RMS) is recommended for bearings of good quality. The bearing should be equally smooth or made from one of the softer materials so that "wearing in" can smooth the high spots, creating a good fit between the journal and the bearing. In high-precision equipment, polishing or lapping can be used to produce a surface finish of the order of 8 to 16 μin RMS.

Minimum Film Thickness

The limiting acceptable value of the minimum film thickness depends on the surface roughness of the journal and the bearing because the film must be thick enough to eliminate any solid contact during expected operating conditions. The suggested design value also depends on the size of the journal. For ground journals, the following relationship can be used to estimate the design value.

$$h_o = 0.000\,25D \qquad (14\text{-}2)$$

where D is the diameter of the bearing.

Diametral Clearance

The clearance between the journal and the bearing depends on the nominal diameter of the bearing, the precision of the machine for which the bearing is being designed, and the surface roughness of the journal. The coefficient of thermal expansion for both the journal and the bearing must also be considered to ensure a satisfactory clearance under all

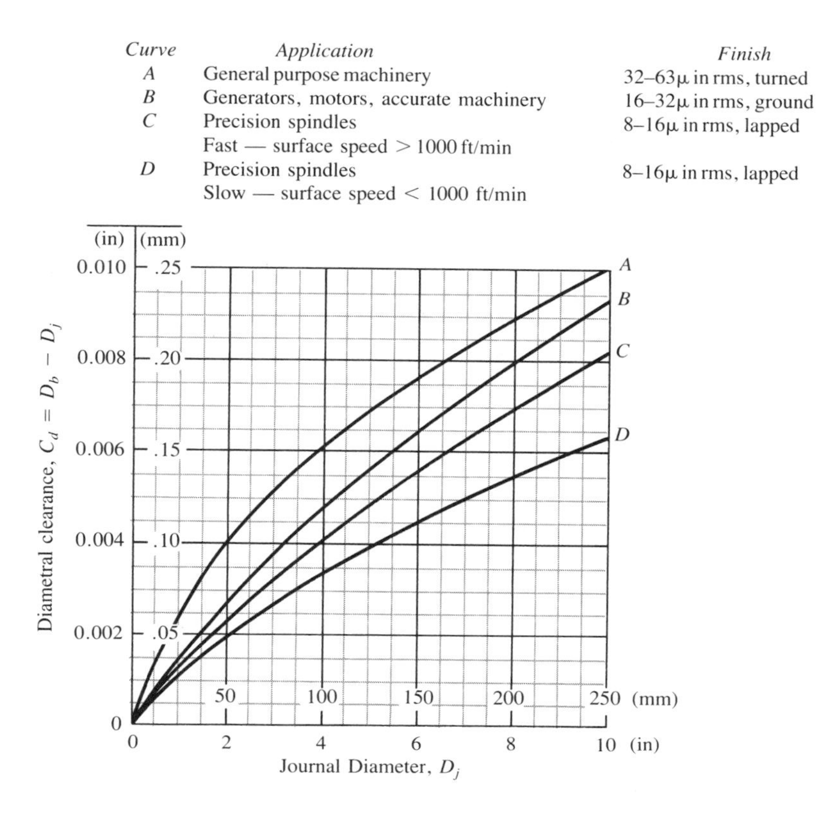

Figure 14-4 Recommended Diametral Clearance versus Journal Diameter for Different Applications

expected operating conditions. An overall guideline of making the clearance in the range of 0.001 to 0.002 times the bearing diameter can be used. Figure 14-4 shows a graph of the recommended diametral clearance as a function of the bearing diameter and the type of service. Some variation above and below the values from the curve is permissible. For journals approximately 5.0 in (125 mm) and smaller, ± 0.0005 in may be allowed from the curves. For the larger shafts, a deviation of ±0.001 in is acceptable.

Bearing Length to Diameter Ratio

The diameter of the journal, because it is a part of the shaft itself, is usually limited by stress and deflection considerations of the type discussed in Chapter 9. Then the length of the bearing is specified to provide for a suitable level of bearing pressure. General-purpose industrial machinery bearings typically operate at approximately 200 to 500 psi (1.4 to 3.4 MPa) bearing pressure, based on the projected area of the bearing [$p = \text{load}/(LD)$]. The pressure may range from as low as 50 psi (0.34 MPa) for light-duty equipment to 2 000 psi (13.4 MPa) for heavy machinery under varying loads, such as in internal combustion engines. The leakage of oil from the bearing is also affected by the bearing length. The typical range of length to diameter ratio (L/D) for full-film hydrodynamic bearings is from 0.50 to 1.5. But many successful bearings operate outside this range.

Lubricant Temperature

The viscosity of the oil is a critical parameter in the performance of a bearing. Figure 14-5 shows the great variation of viscosity with temperature, which indicates that temperature control is advisable. Also, most petroleum lubricating oils should be limited to approximately 160°F (70°C) in order to retard oxidation. The temperature of interest is, of course, that inside the bearing. Frictional energy or thermal energy from the equipment itself can increase the oil temperature above that in the supply reservoir. In design examples we will select the lubricant that will ensure a satisfactory viscosity at 160°F, unless otherwise noted. If the actual operating temperature is lower, the resulting film thickness will be greater than the design value: a conservative result. It is incumbent on the designer to ensure that the limiting temperature is not exceeded, using forced cooling if necessary.

Lubricant Viscosity

The specification of the lubricant for the bearing is one of the final decisions to be made in the design procedure which follows. In the calculations, it is the dynamic viscosity, μ,

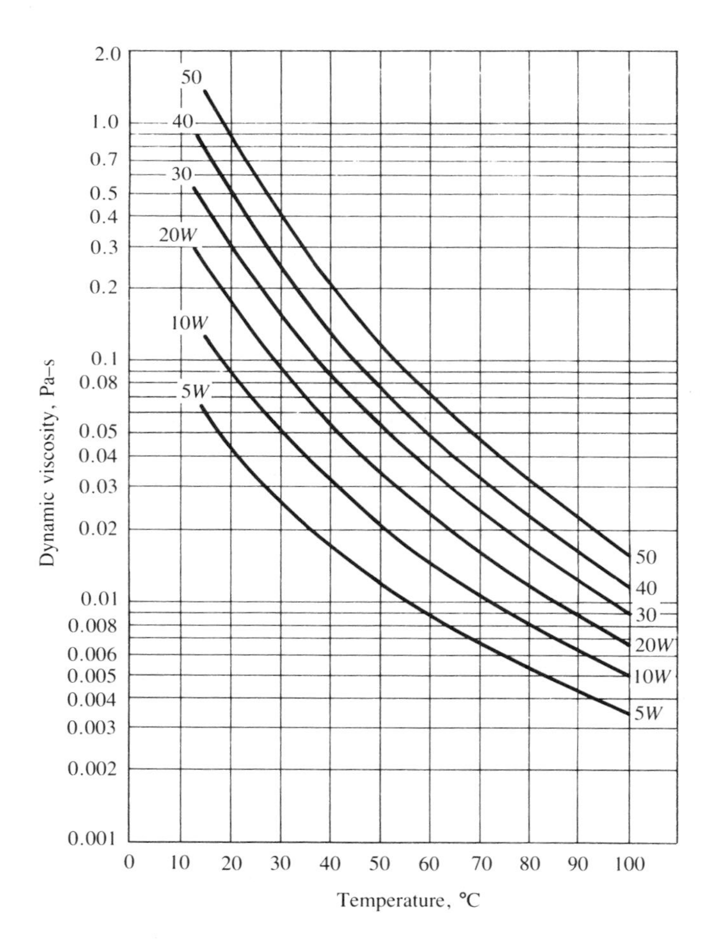

Figure 14-5 Viscosity versus Temperature for SAE Oils

which is used. In the U.S. Customary unit (lb-in-s) system, the dynamic viscosity is expressed in lb · s/in², which is given the name *reyn* in honor of Osbourne Reynolds, who performed much significant work in fluid flow. In the SI units, the standard unit is N · s/m² or Pa · s. Past work frequently used the now obsolete unit of poise or centipoise, derived from the dyne-cm-s metric system. Some useful conversions are

$$1.0 \text{ reyn} = 6\,895 \text{ Pa} \cdot \text{s}$$
$$1.0 \text{ Pa} \cdot \text{s} = 1\,000 \text{ centipoise}$$

Other viscosity conversions may also be useful (8).

Sommerfeld Number

The combined effect of many of the variables involved in the operation of a bearing under hydrodynamic lubrication can be characterized by the dimensionless number, S, called the *Sommerfeld number.* In fact, some refer to this as the *bearing characteristic number,* defined as follows:

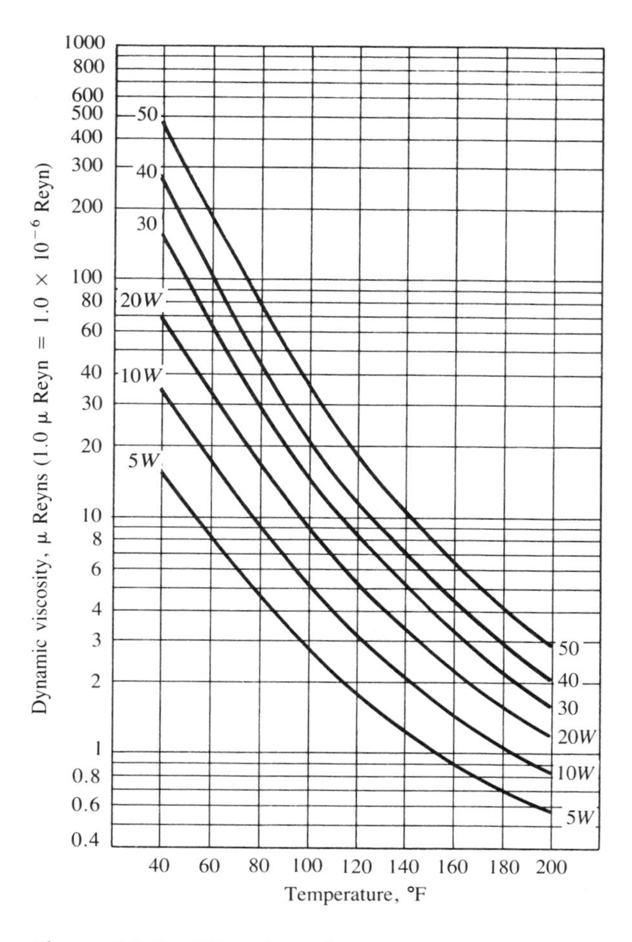

Figure 14-5 (Continued)

$$S = \frac{\mu n_s (R/C_r)^2}{p} \tag{14-3}$$

Note that S is similar to the bearing parameter, Zn/p, discussed in section 14-2, in that it involves the combined effect of viscosity, rotational speed, and bearing pressure. In order for S to be dimensionless, the following units should be used for the factors:

	U.S. Customary Units	*SI Units*
μ	lb · s/in^2 (reyns)	Pa · s (N · s/m^2)
n_s	rev/s	rev/s
p	lb/in^2 (psi)	Pa (N/m^2)
R, C_r	in	m or mm

Any consistent units can be used. Figure 14-6 shows the relationship between the Sommerfeld number and the film thickness ratio, h_o/C_r. Figure 14-7 shows the relationship between S and the coefficient of friction variable, $f(R/C_r)$. These values are used in the design procedure that follows. Because many design decisions are required, several acceptable solutions are possible.

Design Procedure

Because the bearing design is usually done after the stress analysis of the shaft is completed, the following items are typically known:

Radial load on the bearing, W, usually in lb or N

Rotational speed, n, usually in rpm;

Nominal shaft diameter at the journal, sometimes specified as a minimum acceptable diameter based on strength and stiffness.

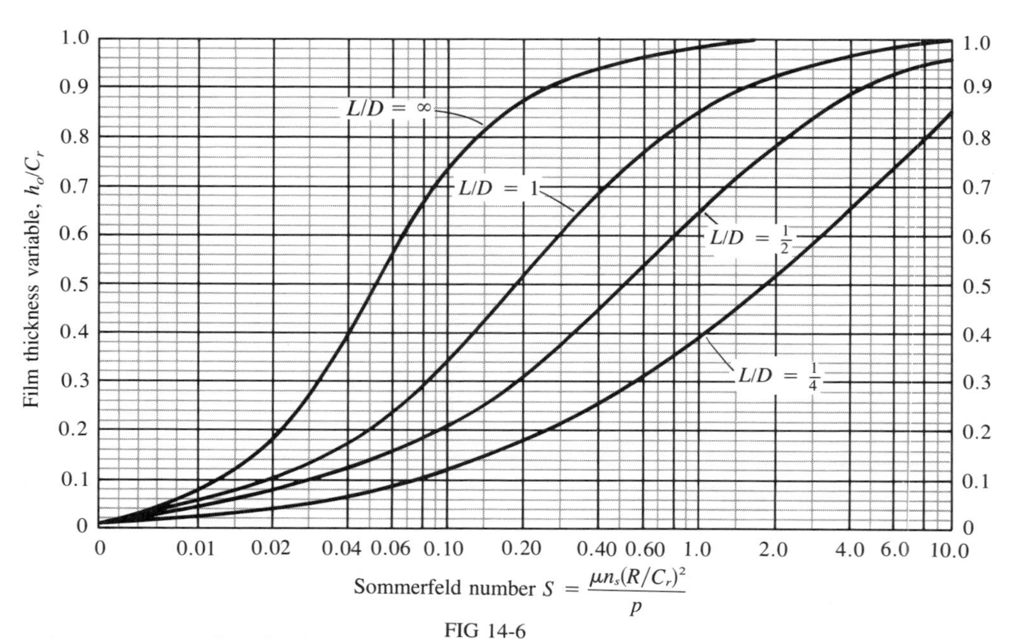

Figure 14-6 Film Thickness Variable, h_o/C_r, versus Sommerfeld Number, S

The results of the design procedure produce values for the actual journal diameter, the bearing length, the diametral clearance, the minimum film thickness of the lubricant during operation, the surface finish for the journal, the lubricant and its maximum operating temperature, the coefficient of friction, the friction torque, and the power dissipated because of friction.

1. Specify a trial value for the journal diameter, D, and the radius, $R = D/2$.
2. Specify a nominal operating bearing pressure, usually 200 to 500 psi (1.4 to 3.4 MPa), where $p = W/LD$. Solve for L:

$$L = W/pD$$

   Then compute L/D. It might be desirable to redefine L/D to be a convenient value from 0.50 to 1.5 in order to use the design charts available. Finally, specify the actual design value of L/D and L and compute $p = W/LD$.
3. Refer to the chart in Figure 14-4 and specify the diametral clearance, C_d, based on the application. Then compute $C_r = C_d/2$ and the ratio R/C_r.
4. Specify the desired surface finish for the journal and bearing, also based on the application. Refer to Figure 14-4.
5. Compute the nominal minimum film thickness from

$$h_o = 0.000\,25\ D$$

6. Compute h_o/C_r, the film thickness ratio.
7. From Figure 14-6, determine the value of the Sommerfeld number for the selected film thickness ratio and L/D ratio. Exercise care in interpolating in this chart because

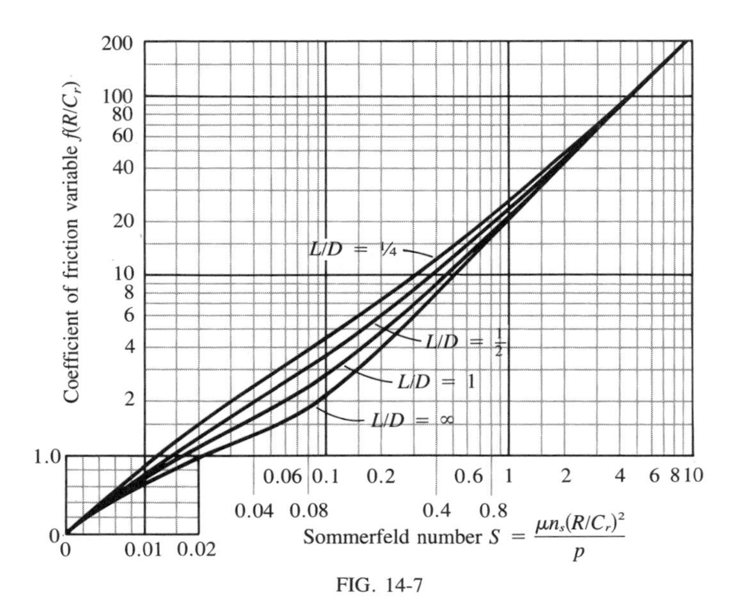

Figure 14-7 Coefficient of Friction Variable, $f(R/C_r)$, versus Sommerfeld Number, S

of the logarithmic axis and the non-linear spread between curves. For $L/D > 1$, only approximate data can be obtained. For $L/D = 1.5$, interpolate approximately one-fourth of the way between the curves for $L/D = 1$ and $L/D =$ infinity. For $L/D = 2$, go about half way (4).

8. Compute the rotational speed n_s in revolutions per second.

$$n_s = n/60 \quad \text{where } n \text{ is in rpm}$$

9. Because each factor of the Sommerfeld number is known except the lubricant viscosity, μ, solve for the required minimum viscosity which will produce the desired minimum film thickness.

$$\mu = \frac{Sp}{n_s(R/C_r)^2} \tag{14-3}$$

10. Specify a maximum acceptable lubricant temperature, usually 160°F or 70°C. Select a lubricant from Figure 14-5 that will have at least the required viscosity at the operating temperature. If the selected lubricant has a viscosity greater than that computed in step 9, recompute S for the new value of viscosity. The resulting value for the minimum film thickness will now be somewhat greater than the design value, a generally desirable result. Figure 14-6 can be consulted again to determine the new value of the minimum film thickness, if desired.
11. From Figure 14-7, obtain the coefficient of friction variable, $f(R/C_r)$.
12. Compute $f = f(R/C_r)/(R/C_r) =$ coefficient of friction.
13. Compute the friction torque. The product of the coefficient of friction and the load W gives the frictional force at the surface of the journal. That times the radius gives torque:

$$T_f = F_f R = fWR$$

14. Compute the power dissipated in the bearing, from the relationship between power, torque, and speed that we have used many times.

$$P_f = T_f n/63\,000 \text{ hp}$$

Example problem 14-2 will illustrate the procedure.

Example Problem 14-2. Design a plain surface journal bearing to carry a steady radial load of 1 500 lb while the shaft rotates at 850 rpm. The shaft stress analysis determines that the minimum acceptable diameter at the journal is 2.10 in. The shaft is part of a machine requiring good precision.

Solution.

1. Let's select $D = 2.50$ in. Then $R = 1.25$ in.
2. For $p = 200$ psi, L must be

$$L = W/pD = 1\,500/(200)(2.50) = 3.00 \text{ in}$$

For this value of L, $L/D = 3.00/2.50 = 1.20$. In order to use one of the standard design charts, let's change L to 2.50 in so that $L/D = 1.0$. This is not essential but it eliminates interpolation. The actual pressure is then

$$p = W/LD = 1\,500/(2.50)(2.50) = 240 \text{ psi}$$

This is an acceptable pressure.

3. From Figure 14-4, $C_d = 0.003$ in is suitable for the diametral clearance, being between curves B and C, which represent accurate machinery. And, $C_r = C_d/2 = 0.001\,5$ in. Also,

$$R/C_r = 1.25/0.001\,5 = 833$$

This value is used in later calculations.

4. For the precision desired in this machine, use a surface finish of 16 to 32 μin, RMS, requiring a ground journal.
5. Minimum film thickness (design value):

$$h_o = 0.000\,25D = 0.000\,25(2.50) = 0.000\,6 \text{ in (approx.)}$$

6. Film thickness variable:

$$h_o/C_r = 0.000\,6/0.001\,5 = 0.40$$

7. From Figure 14-6, for $h_o/C_r = 0.40$ and $L/D = 1$, we can read $S = 0.13$.
8. Rotational speed in revolutions per second:

$$n_s = n/60 = 850/60 = 14.2 \text{ rev/s}$$

9. Solving for the viscosity from the Sommerfeld number, S:

$$\mu = \frac{Sp}{n_s(R/C_r)^2} = \frac{(0.13)(240)}{(14.2)(833)^2} = 3.17 \times 10^{-6} \text{ reyn}$$

10. From the viscosity chart, Figure 14-5, SAE 40 oil is required to ensure a sufficient viscosity at 160°F. The actual expected viscosity of SAE 40 at 160°F is approximately 3.3×10^{-6} reyn.
11. For the actual viscosity, the Sommerfeld number would be

$$S = \frac{\mu n_s(R/C_r)^2}{p} = \frac{(3.3 \times 10^{-6})(14.2)(833)^2}{240} = 0.135$$

12. Coefficient of friction (from Figure 14-7): $f(R/C_r) = 3.5$ for $S = 0.135$ and $L/D = 1$. Now, because $R/C_r = 833$,

$$f = 3.5/833 = 0.004\,2$$

13. Friction torque:

$$T_f = fWR = (0.004\,2)(1\,500)(1.25) = 7.9 \text{ lb} \cdot \text{in}$$

14. Frictional power:

$$P = T_f n/63\,000 = (7.9)(850)/63\,000 = 0.11 \text{ hp}$$

A qualitative evaluation of the result would require more knowledge about the application. But it should be noted that a coefficient of friction of 0.004 2 is quite low. It is likely that a machine requiring such a large shaft and with such high bearing forces also requires a large power to drive it. Then the 0.11-hp friction power would appear small.

14-6 PRACTICAL CONSIDERATIONS FOR PLAIN SURFACE BEARINGS

The design of the bearing system must consider the method of delivery of the lubricant to the bearing, the distribution of the lubricant within the bearing, the quantity of lubricant required, the amount of heat generated in the bearing and its effect on the temperature of the lubricant, the dissipation of heat from the bearing, the maintenance of the lubricant in a clean condition, and the performance of the bearing over the complete range of operating conditions the bearing is likely to experience.

Many of these factors are simply design details that must be worked out along with the other aspects of the machine design. But some guidelines and general recommendations will be presented here.

The lubricant can be delivered to the bearing by a pump, perhaps driven from the same source which drives the entire machine. In some gear drives, one of the gears is designed to dip into an oil sump and carry oil up to the gear mesh and to the bearings. An external oil cup can be used to supply oil by gravity if the lubricant quantity required is small.

Methods of estimating the quantity of oil required, considering the leakage of oil from the ends of the bearing, are available (1, 5, 6, 9).

The delivery of oil to the bearing should always be in an area opposite the location of the hydrodynamic pressure which supports the load. Otherwise, the oil delivery hole would destroy the pressurization of the film.

Grooving is frequently used to distribute the oil along the length of the bearing. Oil would be delivered through a radial hole in the bearing at the mid-length point. The groove would extend axially in both directions from the hole but would terminate somewhat before the end of the bearing in order to keep the oil from leaking to the side. The rotation of the journal then carries the oil around to the area where the hydrodynamic film is generated.

Cooling of the bearing itself, or of the oil in the sump which supplies the oil, must always be considered. Natural convection may be sufficient to transfer heat away and maintain an acceptable bearing temperature. If not, forced convection can be used. In severe cases of heat generation, especially where the bearing system operates in a hot area such as a furnace, liquid coolant can be pumped through a jacket around the bearing. Some commercially available bearings provide this feature. Placing a heat exchanger in the oil sump or pumping the oil through an external heat exchanger can also be done.

The lubricant can be cleaned by passing it through filters as it is pumped to the bearing. Magnetic plugs within the sump can be effective in attracting and holding metallic particles that can score the bearing if allowed to enter the clearance space between the journal and the bearing. Of course, frequent changing of the oil is also desirable.

The design procedure used in the preceding section was completed for one set of conditions: a given temperature, diametral clearance, load, and rotational speed. If any of these factors varies during the operation of the machine, the performance of the bearing must be evaluated under the new conditions. The testing of a prototype under a variety of conditions is also desirable.

14-7 HYDROSTATIC BEARINGS

Recall that hydrodynamic lubrication resulted from the creation of a pressurized film of oil sufficient to carry the load on the bearing, with the film being generated by the motion of the journal itself within the bearing. It was noted that a steady relative motion between the journal and the bearing is required to generate and maintain the film.

In some types of equipment, the conditions are such that a hydrodynamic film cannot be developed. Reciprocating or oscillating devices or very-slow-moving machines are examples. If the load on the bearing is very high, it may not be possible to generate a sufficiently high pressure in the film to support the load. Even in cases in which hydrodynamic lubrication can be developed during normal operation of the machine, there is still mixed-film or boundary lubrication during start-up and shut-down cycles. This may be unacceptable.

Consider the design of the mount for a telescope or antenna system in which rotation of the base is required at very slow speed and with smooth motion. Also, low friction is desired in order to keep the drive system small and to provide fast response and accurate positioning. The mount is essentially a thrust bearing supporting the weight of the system.

In this type of application, *hydrostatic lubrication* is desirable. Lubricant is supplied to the bearing under high pressure, several hundred psi or higher, and the pressure acting on the area of the bearing literally lifts the load off the bearing even with the equipment stationary.

Figure 14-8 shows the main elements of a hydrostatic bearing system. A positive displacement pump draws oil from a reservoir and delivers it under pressure to a supply

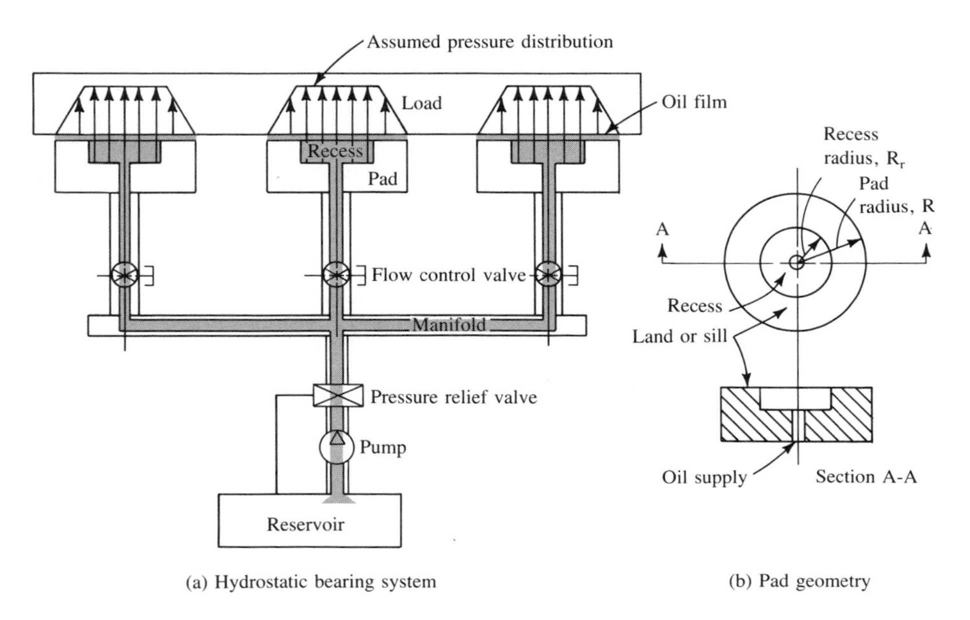

(a) Hydrostatic bearing system

(b) Pad geometry

Figure 14-8 Main Elements of a Hydrostatic Bearing System

manifold from which several bearing pads can be supplied. At each bearing pad, the oil passes through a control element which permits the balancing of the system. The control element may be a flow-control valve, a length of small diameter tubing, or an orifice, any of which offers a resistance to the flow of oil and permits the several bearing pads to operate at a sufficiently high pressure to lift the load on that pad. When in operation, the oil enters a recess within the bearing pad. For example, Figure 14-8(b) shows a circular pad with a circular recess in its center supplied with oil through a central hole. The load initially rests on the land area, sealing the recess. As the pressure in the recess reaches the level where the product of the pressure times the recess area equals the applied load, the load is lifted from the pad. Immediately there is a flow of oil across the land area under the lifted load and the pressure decreases to atmospheric pressure at the outside of the pad. The flow of oil must be maintained at a level which matches the outflow from the pad. When equilibrium occurs, the integrated product of the local pressure times the area lifts the load a certain distance, h, usually in the range of 0.001 to 0.010 in (0.025 to 0.25 mm). The film thickness, h, must be large enough to ensure no solid contact over the range of operating conditions, but it should be kept as small as possible to minimize the flow of oil through each bearing and the pump power required to drive the system.

Hydrostatic Bearing Performance

Three factors characterizing the performance of a hydrostatic bearing are its load-carrying capacity, the flow of oil required, and the pumping power required, as indicated by the dimensionless coefficients a_f, q_f, and H_f. The magnitudes of the coefficients depend on the design of the pad.

$$W = a_f A_p p_r \tag{14-4}$$

$$Q = q_f \frac{W}{A_p} \frac{h^3}{\mu} \tag{14-5}$$

$$P = p_r Q = H_f \left(\frac{W}{A_p}\right)^2 \frac{h^3}{\mu} \tag{14-6}$$

where

W = Load on the bearing, lb or N

Q = Volume flow rate of oil, in^3/s or m^3/s

P = Pumping power, lb · in/s or N · m/s (watts)

a_f = Pad load coefficient, dimensionless

q_f = Pad flow coefficient, dimensionless

H_f = Pad power coefficient, dimensionless (Note: $H_f = q_f/a_f$)

A_p = Pad area, in^2 or m^2

p_r = Oil pressure in the recess of the pad, psi or Pa

h = Film thickness, in or m

μ = Dynamic viscosity of the oil, lb · s/in^2 (reyn) or Pa · s

Figure 14-9 shows the typical variation of the dimensionless coefficients as a function of the pad geometry for a circular pad with a circular recess. As the size of the recess

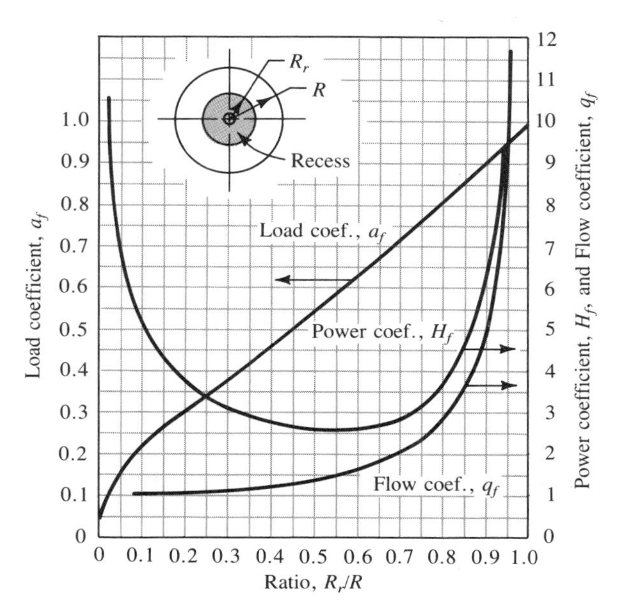

Figure 14-9 Dimensionless Performance Coefficients for Circular Pad Hydrostatic Bearing (Cast Bronze Institute. *Cast Bronze Hydrostatic Bearing Design Manual.* Chicago: 1975)

(R_r/R) increases, the load-carrying capacity increases, as indicated by a_f. But at the same time, the flow through the bearing increases, as indicated by q_f. The increase is gradual up to a value of R_r/R of approximately 0.7, and then rapid for higher ratios. This higher flow rate requires much higher pumping power, as indicated by the rapidly increasing power coefficient. At very low ratios of R_r/R, the load coefficient decreases rapidly. The pressure in the recess would have to increase to compensate in order to lift the load. The higher pressure requires more pumping power. Therefore, the power coefficient is high at either very small ratios of R_r/R or at high ratios. The minimum power is required for ratios between 0.4 and 0.6.

These general characteristics of hydrostatic bearing pad performance are typical for many different geometries of pads. Extensive data for the performance of different pad shapes are published (2).

Example problem 14-3 will illustrate the basic design procedure for hydrostatic bearings.

Example Problem 14-3. A large antenna mount weighing 12 000 lb is to be supported on three hydrostatic bearings such that each bearing pad carries 4 000 lb. A positive displacement pump will be used to deliver oil at a pressure up to 500 psi. Design the hydrostatic bearings.

Solution. We will choose the circular pad design for which the performance coefficients are available in Figure 14-9. The results of the design will specify the dimensions of the pads, the required oil pressure in the recess of each pad, the type of oil required and its temperature, the thickness of the film of oil when the bearings are supporting the load, the flow rate of oil required, and the pumping power required.

1. From Figure 14-9, the minimum power required for a circular pad bearing would occur with the ratio R_r/R of approximately 0.50. For that ratio, the value of the load coefficient, $a_f = 0.55$. The pressure at the bearing recess will be somewhat below the maximum available of 500 psi because of the pressure drop in the restrictor placed between the supply manifold and the pad. Let's design for a recess pressure of approximately 400 psi. Then, from equation (14-4),

$$A_p = \frac{W}{a_f p_r} = \frac{4\,000\text{ lb}}{0.55(400\text{ lb/in}^2)} = 18.2\text{ in}^2$$

But $A_p = \pi D^2/4$. Then, the required pad diameter is

$$D = \sqrt{4A_p/\pi} = \sqrt{4(18.2)/\pi} = 4.81\text{ in}$$

For convenience, let's specify $D = 5.00$ in. Then the actual pad area will be

$$A_p = \pi D^2/4 = (\pi)(5.00\text{ in})^2/4 = 19.6\text{ in}^2$$

The required recess pressure is then

$$p_r = \frac{W}{a_f A_p} = \frac{4\,000\text{ lb}}{0.55(19.6\text{ in}^2)} = 370\text{ lb/in}^2$$

Also,

$$R = D/2 = 5.00\text{ in}/2 = 2.50\text{ in}$$
$$R_r = 0.50R = 0.50(2.50\text{ in}) = 1.25\text{ in}$$

2. Specify the design value of the film thickness, h. It is recommended that h be between 0.001 and 0.010 in. Let's use $h = 0.005$ in.
3. Specify the lubricant and the operating temperature. Let's select SAE 30 oil and assume that the maximum oil temperature in the oil film will be 120°F. A method of estimating the actual film temperature during operation may be consulted (2). From the viscosity/temperature curves, Figure 14-5, the viscosity is approximately 8.3×10^{-6} reyn (lb · s/in²).
4. Compute the oil flow through the bearing from equation 14-5. The value of $q_f = 1.4$ can be found from Figure 14-9.

$$Q = q_f \frac{W}{A_p}\frac{h^3}{\mu} = (1.4)\frac{4\,000\text{ lb}}{19.6\text{ in}^2}\frac{(0.005\text{ in})^3}{8.3 \times 10^{-6}\text{ lb}\cdot\text{s/in}^2}$$
$$Q = 4.30\text{ in}^3\text{/s}$$

5. Compute the pumping power required from equation (14-6). The value of $H_f = 2.6$ can be found from Figure 14-9.

$$P = p_r Q = H_f\left(\frac{W}{A_p}\right)^2\frac{h^3}{\mu} = 2.6\left(\frac{4\,000}{19.6}\right)^2\frac{(0.005)^3}{8.3 \times 10^{-6}} = 1\,631\text{ lb}\cdot\text{in/s}$$

For convenience, we can convert this to horsepower.

$$P = \frac{1\,631\text{ lb}\cdot\text{in}}{\text{s}}\frac{1.0\text{ ft}}{12\text{ in}}\frac{1.0\text{ hp}}{550\text{ lb}\cdot\text{ft/s}} = 0.247\text{ hp}$$

REFERENCES

1. Cast Bronze Institute. *Cast Bronze Bearing Design Manual*. Chicago: 1979.
2. ———. *Cast Bronze Hydrostatic Bearing Design Manual*. Chicago: 1975.
3. Clauss, F. J. *Solid Lubricants and Self-Lubricating Solids*. New York: Academic Press, 1972.
4. Faires, V. M. *Design of Machine Elements*. New York: Macmillan Publishing Company, 1965.
5. Fuller, D. D. *Theory and Practice of Lubrication for Engineers*. New York: John Wiley & Sons, 1956.
6. Juvinall, R. C. *Fundamentals of Machine Component Design*. New York: John Wiley & Sons, 1983.
7. *Machine Design Magazine Mechanical Drives Reference Issue*. Cleveland: Penton/IPC, 1981.
8. Mott, R. L. *Applied Fluid Mechanics*, 2d ed. Columbus: Charles E. Merrill Publishing, 1980.
9. Trumpler, P. R. *Design of Film Bearings*. New York: Macmillan Publishing Company, 1966.
10. Wills, J. George. *Lubrication Fundamentals*. New York: Marcel Dekker, 1980.
11. Wolverton, M. P., et al. "How Plastic Composites Wear at High Temperatures." *Machine Design Magazine*, Feb. 10, 1983.

PROBLEMS

For problems 1–8, design a plain surface bearing, using the boundary lubricated approach of section 14-4. Use an L/D ratio for the bearing in the range of 0.50 to 1.50. Compute the pV factor and specify a material from Table 14-1.

Problem Number	*Radial Load (lb)*	*Shaft Diameter (in)*	*Shaft Speed (rpm)*
1	225	3.00	1 750
2	100	1.50	1 150
3	200	1.25	850
4	75	0.50	600
5	850	4.50	625
6	500	3.75	450
7	800	3.00	350
8	60	0.75	750

For problems 9–18, design a hydrodynamically lubricated bearing, using the method outlined in section 14-5. Specify the journal nominal diameter, the bearing length, the diametral clearance, the minimum film thickness of the lubricant during operation, the surface finish of the journal and bearing, the lubricant, and its maximum operating temperature. For your design, compute the coefficient of friction, the friction torque, and the power dissipated as the result of friction.

Problem Number	*Radial Load*	*Min. Shaft Diameter*	*Shaft Speed (rpm)*	*Application (Fig. 14-4)*
9	1 250 lb	2.60 in	1 750	Electric motor (B)
10	2 250 lb	3.50 in	850	Conveyor (A)
11	1 875 lb	2.25 in	1 150	Air compressor (B)
12	1 250 lb	1.75 in	600	Precision spindle (D)
13	500 lb	1.15 in	2 500	Precision spindle (C)
14	850 lb	1.45 in	1 200	Idler sheave (A)
15	4 200 lb	4.30 in	450	Shaft for chain drive (A)
16	18.7 kN	100 mm	500	Conveyor (A)
17	2.25 kN	25 mm	2 200	Machine tool (B)
18	5.75 kN	65 mm	1 750	Printer (B)

For problems 19–28, design a hydrostatic bearing of circular shape. Specify the pad diameter, the recess diameter, the recess pressure, the film thickness, the lubricant and its temperature, the oil flow rate, and the pumping power. The load specified is for a single bearing. You may choose to use multiple bearings. (The supply pressure is the maximum available at the pump.)

Problem Number	*Load*	*Supply Pressure*
19	1 250 lb	300 psi
20	5 000 lb	300 psi
21	3 500 lb	500 psi
22	750 lb	500 psi
23	250 lb	150 psi
24	500 lb	150 psi
25	22.5 kN	2.0 MPa
26	1.20 kN	750 kPa
27	8.25 kN	1.5 MPa
28	12.5 kN	1.5 MPa

15 Rolling Contact Bearings

15-1 OVERVIEW

The purpose of a bearing is to support a load while permitting relative motion between two elements of a machine. The term *rolling contact bearings* refers to the wide variety of bearings that use spherical balls or some type of roller between the stationary and the moving elements. The most common type of bearing supports a rotating shaft, resisting purely radial loads or a combination of radial and axial (thrust) loads. Some bearings are designed to carry only thrust loads. Most bearings are used in applications involving rotation, but some are used in linear motion applications.

The components of a typical rolling contact bearing are the inner race, the outer race, and the rolling elements. Figure 15-1 shows the common single row, deep groove ball bearing. Usually the outer race is stationary and is held by the housing of the machine. The inner race is pressed onto the rotating shaft and thus rotates with it. Then the balls roll between the outer and inner races. The load path is from the shaft, to the inner race, to the balls, to the outer race, and finally to the housing. The presence of the balls allows a very smooth, low friction rotation of the shaft. The typical coefficient of friction for a rolling contact bearing is approximately 0.001 to 0.005. These values reflect only the rolling elements themselves and the means of retaining them in the bearing. The presence of seals, excessive lubricant, or unusual loading increases these values.

The types of rolling contact bearings usually applied to machine design for supporting shafts are listed next. Figures 15-1 through 15-7 show the construction of these bearings.

Single row, deep groove ball bearing	(Figure 15-1)
Double row, deep groove ball bearing	(Figure 15-2)
Angular contact bearing	(Figure 15-3)
Cylindrical roller bearing	(Figure 15-4)
Needle bearing	(Figure 15-5)
Spherical roller bearing	(Figure 15-6)
Tapered roller bearing	(Figure 15-7)

Selection of the type of bearing to use in a given application can be aided by referring to the comparison chart in Table 15-1. The bearings differ widely in their ability to withstand radial loads and thrust loads and to accommodate misalignment. In general, the ratings for misalignment capability imply the ability to operate satisfactorily up to 4.0° for an excellent rating, less than 0.05° for a poor rating, and up to approximately 0.15° for a fair rating. Manufacturers' data should be consulted.

The comparative ratings shown in Table 15-1 can be justified by the following discussion. With regard to radial load-carrying capacity, roller bearings are better than ball bearings because of the shape and size of the area over which the load is spread. A spherical ball rolling on the curved races of a bearing would have only a point contact if the elasticity of the materials were not considered. Of course, there is some deformation of both the ball and the races, resulting in the load's acting on a small circular area. Contrast this with a cylindrical roller, on which there is nominally line contact instead of point contact. Then, with elasticity considered, the contact area is a rectangle. The resulting contact stresses are lower for the cylindrical roller than for the ball. Thus, a roller bearing of a given size withstands a higher radial load than an equivalent ball bearing. Similar characterizations can be made about the other types of bearings.

Thrust load capacity varies dramatically with bearing design. The grooves in the races of the deep groove ball bearing permit the transfer of moderate thrust loads in combination

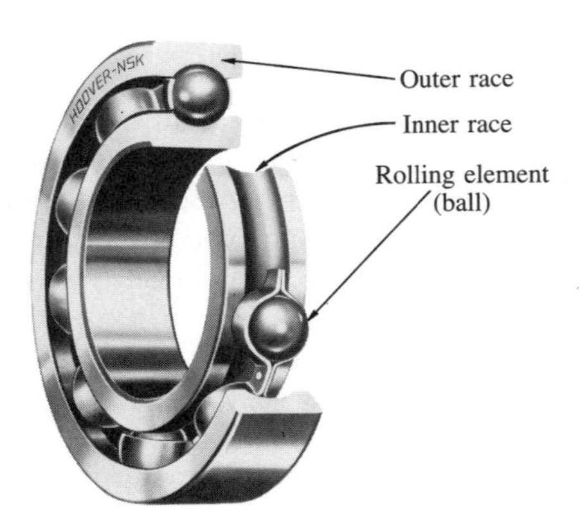

Figure 15-1 Single Row, Deep Groove Ball Bearing (Hoover-NSK Bearing Company, Ann Arbor, Mich.)

Figure 15-2 Double Row, Deep Groove Ball Bearing (Hoover-NSK Bearing Company, Ann Arbor, Mich.)

Figure 15-3 Angular Contact Ball Bearing (Hoover-NSK Bearing Company, Ann Arbor, Mich.)

Figure 15-4 Cylindrical Roller Bearing (Hoover-NSK Bearing Company, Ann Arbor, Mich.)

with radial loads. But the angular contact bearing is better than the plain ball bearing because the races are higher on one side, providing a more favorable load path for the resultant of the radial and the thrust load which would act on the bearing at an angle. Cylindrical and needle roller bearings should not be subjected to any thrust load at all because such loads would have to be transferred to the rollers through their ends, resulting in rubbing and high friction. Tapered roller bearings are virtually ideal for high thrust loads combined with moderate radial loads.

An attempt to apply a needle bearing where there is significant angular misalignment would result in nonuniform loading of the rollers, dramatically reducing its capacity. This is the reason for its poor rating for misalignment. Conversely, the spherical roller bearing adjusts easily to angular misalignment. In fact, some refer to this type of bearing as a "self-aligning" bearing.

(*a*) Single- and double-row needle bearings

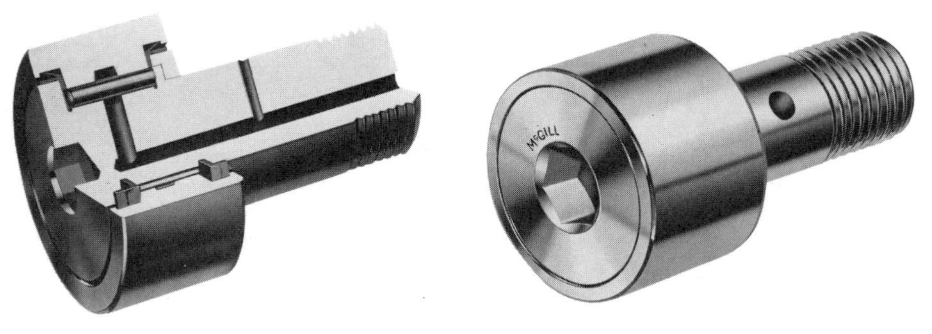

(*b*) Needle bearings adapted to cam followers

Figure 15-5 Needle Bearings (McGill Manufacturing Co., Inc., Bearing Division, Valparaiso, Ind.)

Figure 15-6 Spherical Roller Bearing (Hoover-NSK Bearing Company, Ann Arbor, Mich.)

Figure 15-7 Tapered Roller Bearing (Hoover-NSK Bearing Company, Ann Arbor, Mich.)

15-2 LOAD/LIFE RELATIONSHIP

As mentioned, the load on a rolling contact bearing is exerted on a small area. The resulting contact stresses are quite high, regardless of the type of bearing. Contact stresses

Table 15-1 Comparison of Bearing Types

Bearing Type	*Radial Load Capacity*	*Thrust Load Capacity*	*Misalignment Capability*
Single-row, deep groove ball	Good	Fair	Fair
Double-row, deep groove ball	Excellent	Good	Fair
Angular contact	Good	Excellent	Poor
Cylindrical roller	Excellent	Poor	Fair
Needle	Excellent	Poor	Poor
Spherical roller	Excellent	Fair/good	Excellent
Tapered roller	Excellent	Excellent	Poor

of approximately 300 000 psi are not uncommon in commercially available bearings. To withstand such high stresses, the balls, rollers, and races are made from a very hard, high-strength steel. A typical bearing steel is AISI 52100, which has 1.00 percent carbon and a high chromium content. The heat treatment produces a hardness in the range of HRC 60.

Despite using very high strength steels, all bearings have a finite life and will eventually fail due to fatigue because of the high contact stresses. But, obviously, the lighter the load the longer the life, and vice versa. The relationship between load, P, and life, L, for rolling contact bearings can be stated as

$$\frac{L_2}{L_1} = \left(\frac{P_1}{P_2}\right)^k \tag{15-1}$$

where $k = 3.00$ for ball bearings and $k = 3.33$ for roller bearings.

15-3 BEARING MANUFACTURERS' DATA

The selection of a rolling contact bearing from a manufacturer's catalog involves considerations of load-carrying capacity and the geometry of the bearing. Table 15-2 shows a portion of the data from a catalog for two sizes of single row, deep groove ball bearings.

Considering load-carrying capacity first, the data reported for each bearing design will include a basic dynamic load rating, C, and a basic static load rating, C_o.

The *basic static load rating* is the load that the bearing can withstand without permanent deformation of any component. If this load is exceeded, the most probable result would be the indention of one of the bearing races by the rolling elements. The deformation would be similar to that produced in the Brinell hardness test and the failure is sometimes referred to as *Brinelling*. Operation of the bearing after brinelling would be very noisy, and the impact loads on the indented area would produce rapid wear and progressive failure of the bearing.

To understand the basic dynamic load rating, it is necessary first to discuss the concept of the rated life of a bearing. Fatigue occurs over a large number of cycles of loading; for a bearing, that would be a large number of revolutions. Also, fatigue is a statistical phenomenon with considerable spread of the actual life of a group of bearings of a given design. The rated life is the standard means of reporting the results of many tests

Table 15-2 Bearing Selection Data for Single Row, Deep Groove, Conrad Type Ball Bearings

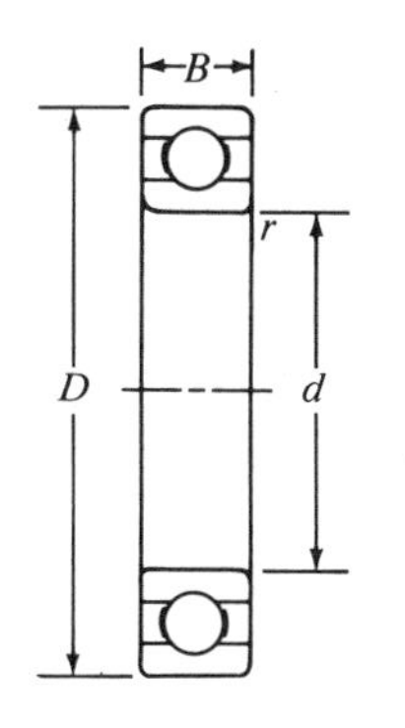

SERIES 6200

BEARING NUMBER	NOMINAL BEARING DIMENSIONS							PREFERRED SHOULDER DIAMETER		BEARING WEIGHT	BASIC STATIC LOAD RATING C_o	BASIC DYNAMIC LOAD RATING C
	d		D		B		r^*	*Shaft*	*Housing*			
	mm	in	mm	in	mm	in	in	in	in	lb	lb	lb
6200	10	0.3937	30	1.1811	9	0.3543	0.024	0.500	0.984	0.07	520	885
6201	12	0.4724	32	1.2598	10	0.3937	0.024	0.578	1.063	0.08	675	1180
6202	15	0.5906	35	1.3780	11	0.4331	0.024	0.703	1.181	0.10	790	1320
6203	17	0.6693	40	1.5748	12	0.4724	0.024	0.787	1.380	0.14	1010	1660
6204	20	0.7874	47	1.8504	14	0.5512	0.039	0.969	1.614	0.23	1400	2210
6205	25	0.9843	52	2.0472	15	0.5906	0.039	1.172	1.811	0.29	1610	2430
6206	30	1.1811	62	2.4409	16	0.6299	0.039	1.406	2.205	0.44	2320	3350
6207	35	1.3780	72	2.8346	17	0.6693	0.039	1.614	2.559	0.64	3150	4450
6208	40	1.5748	80	3.1496	18	0.7087	0.039	1.811	2.874	0.82	3650	5050
6209	45	1.7717	85	3.3465	19	0.7480	0.039	2.008	3.071	0.89	4150	5650
6210	50	1.9685	90	3.5433	20	0.7874	0.039	2.205	3.268	1.02	4650	6050
6211	55	2.1654	100	3.9370	21	0.8268	0.059	2.441	3.602	1.36	5850	7500
6212	60	2.3622	110	4.3307	22	0.8661	0.059	2.717	3.996	1.73	7250	9050
6213	65	2.5591	120	4.7244	23	0.9055	0.059	2.913	4.390	2.18	8000	9900
6214	70	2.7559	125	4.9213	24	0.9449	0.059	3.110	4.587	2.31	8800	10800
6215	75	2.9528	130	5.1181	25	0.9843	0.059	3.307	4.783	2.64	9700	11400
6216	80	3.1496	140	5.5118	26	1.0236	0.079	3.504	5.118	3.09	10500	12600
6217	85	3.3465	150	5.9055	28	1.1024	0.079	3.740	5.512	3.97	12300	14600
6218	90	3.5433	160	6.2992	30	1.1811	0.079	3.937	5.906	4.74	14200	16600
6219	95	3.7402	170	6.6929	32	1.2598	0.079	4.213	6.220	5.73	16300	18800
6220	100	3.9370	180	7.0866	34	1.3386	0.079	4.409	6.614	6.94	18600	21100
6221	105	4.1339	190	7.4803	36	1.4173	0.079	4.606	7.008	8.15	20900	23000
6222	110	4.3307	200	7.8740	38	1.4961	0.079	4.803	7.402	9.59	23400	24900
6224	120	4.7244	215	8.4646	40	1.5748	0.079	5.197	7.992	11.4	26200	26900

Table 15-2 Bearing Selection Data for Single Row, Deep Groove, Conrad Type Ball Bearings (continued)

SERIES 6200

BEARING NUMBER	NOMINAL BEARING DIMENSIONS							PREFERRED SHOULDER DIAMETER		BEARING WEIGHT	BASIC STATIC LOAD RATING C_o	BASIC DYNAMIC LOAD RATING C
	d		*D*		*B*		*r**	*Shaft*	*Housing*			
	mm	in	mm	in	mm	in	in	in	in	lb	lb	lb
6226	130	5.1181	230	9.0551	40	1.5748	0.098	5.669	8.504	12.7	29 100	28 700
6228	140	5.5118	250	9.8425	42	1.6535	0.098	6.063	9.291	19.6	29 300	28 700
6230	150	5.9055	270	10.6299	45	1.7717	0.098	6.457	10.079	25.3	32 500	30 000
6232	160	6.2992	290	11.4173	48	1.8898	0.098	6.850	10.886	32.0	35 500	32 000
6234	170	6.6929	310	12.2047	52	2.0472	0.118	7.362	11.535	38.5	43 000	36 500
6236	180	7.0866	320	12.5984	52	2.0472	0.118	7.758	11.929	41.0	46 500	39 000
6238	190	7.4803	340	13.3858	55	2.1654	0.118	8.150	12.717	50.5	54 500	44 000
6240	200	7.8740	360	14.1732	58	2.2835	0.118	8.543	13.504	61.5	60 000	46 500

SERIES 6300

BEARING NUMBER	*d* mm	*d* in	*D* mm	*D* in	*B* mm	*B* in	*r** in	*Shaft* in	*Housing* in	BEARING WEIGHT lb	C_o lb	C lb
6300	10	0.3937	35	1.3780	11	0.4331	0.024	0.563	1.181	0.12	805	1 400
6301	12	0.4724	37	1.4567	12	0.4724	0.039	0.656	1.220	0.13	990	1 680
6302	15	0.5906	42	1.6535	13	0.5118	0.039	0.781	1.417	0.18	1 200	1 980
6303	17	0.6693	47	1.8504	14	0.5512	0.039	0.875	1.614	0.25	1 460	2 360
6304	20	0.7874	52	2.0472	15	0.5906	0.039	1.016	1.772	0.32	1 730	2 760
6305	25	0.9843	62	2.4409	17	0.6693	0.039	1.220	2.165	0.52	2 370	3 550
6306	30	1.1811	72	2.8346	19	0.7480	0.039	1.469	2.559	0.76	3 150	4 600
6307	35	1.3780	80	3.1496	21	0.8268	0.059	1.688	2.795	1.01	4 050	5 800
6308	40	1.5748	90	3.5433	23	0.9055	0.059	1.929	3.189	1.40	5 050	7 050
6309	45	1.7717	100	3.9370	25	0.9843	0.059	2.126	3.583	1.84	6 800	9 150
6310	50	1.9685	110	4.3307	27	1.0630	0.079	2.362	3.937	2.42	8 100	10 700
6311	55	2.1654	120	4.7244	29	1.1417	0.079	2.559	4.331	2.98	9 450	12 300
6312	60	2.3622	130	5.1181	31	1.2205	0.079	2.835	4.646	3.75	11 000	14 100
6313	65	2.5591	140	5.5118	33	1.2992	0.079	3.031	5.039	4.63	12 600	16 000
6314	70	2.7559	150	5.9055	35	1.3780	0.079	3.228	5.433	5.51	14 400	18 000
6315	75	2.9528	160	6.2992	37	1.4567	0.079	3.425	5.827	6.61	16 300	19 600

Table 15-2 Bearing Selection Data for Single Row, Deep Groove, Conrad Type Ball Bearings (continued)

SERIES 6300

BEARING NUMBER	NOMINAL BEARING DIMENSIONS							PREFERRED SHOULDER DIAMETER		BEARING WEIGHT	BASIC STATIC LOAD RATING C_o	BASIC DYNAMIC LOAD RATING C
	d		*D*		*B*		*r**	*Shaft*	*Housing*			
	mm	in	mm	in	mm	in	in	in	in	lb	lb	lb
6316	80	3.1496	170	6.6929	39	1.5354	0.079	3.622	6.220	7.93	18300	21300
6317	85	3.3465	180	7.0866	41	1.6142	0.098	3.898	6.535	9.37	20400	22900
6318	90	3.5433	190	7.4803	43	1.6929	0.098	4.094	6.929	10.8	22500	24700
6319	95	3.7402	200	7.8740	45	1.7717	0.098	4.291	7.323	12.5	24900	26400
6320	100	3.9370	215	8.4646	47	1.8504	0.098	4.488	7.913	15.3	29800	30000
6321	105	4.1339	225	8.8583	49	1.9291	0.098	4.685	8.307	17.9	32500	31700
6322	110	4.3307	240	9.4488	50	1.9685	0.098	4.882	8.898	21.0	38000	35500
6324	120	4.7244	260	10.2362	55	2.1654	0.098	5.276	9.685	27.6	38500	36000
6326	130	5.1181	280	11.0236	58	2.2835	0.118	5.827	10.315	40.8	44500	39500
6328	140	5.5118	300	11.8110	62	2.4409	0.118	6.220	11.102	48.5	51000	43500
6330	150	5.9055	320	12.5984	65	2.5591	0.118	6.614	11.890	57.3	58000	47500
6332	160	6.2992	340	13.3858	68	2.6772	0.118	7.008	12.677	58	58500	48000
6334	170	6.6929	360	14.1732	72	2.8346	0.118	7.402	13.465	84	73500	56500
6336	180	7.0866	380	14.9606	75	2.9528	0.118	7.795	14.252	98	84000	61500
6338	190	7.4803	400	15.7480	78	3.0709	0.157	8.346	14.882	112	84000	61500
6340	200	7.8740	420	16.5354	80	3.1496	0.157	8.740	15.669	127	91500	65500

*Max fillet that corner radius will clear

Source: Hoover-NSK Bearing Company, Ann Arbor, Mich.

of bearings of a given design. It represents the life that 90 percent of the bearings would achieve successfully at a rated load. Note that it also represents the life that 10 percent of the bearings would not achieve. The rated life is thus typically referred to as the *L_{10} life* at the rated load.

Now the *basic dynamic load* rating can be defined as that load to which the bearings can be subjected while achieving a rated life (L_{10}) of 1 million revolutions (rev). Thus, the manufacturer supplies you with one set of data relating load and life. Equation (15-1) can be used to compute the expected life at any other load.

You should be aware that different manufacturers use other bases for the rated life. For example, some use 90 million cycles as the rated life and determine the rated load for that life. Care should be exercised to understand the basis for the ratings in any given catalog.

Example Problem 15-1. A catalog lists the basic dynamic load rating for a ball bearing to be 7 050 lb for a rated life of 1 million rev. What would be the expected L_{10} life of the bearing if it were subjected to a load of 3 500 lb?

Solution. In equation (15-1),

$$P_1 = C = 7\,050 \text{ lb} \quad \text{(Basic dynamic load rating)}$$

$$P_2 = P_d = 3\,500 \text{ lb} \quad \text{(Design load)}$$

$$L_1 = 10^6 \text{ rev} \quad (L_{10} \text{ life at load } C)$$

$$k = 3 \quad \text{(Ball bearing)}$$

Then letting the life, L_2, be called the *design life*, L_d, at the design load,

$$L_2 = L_d = L_1\left(\frac{P_1}{P_2}\right)^k = 10^6\left(\frac{7\,050}{3\,500}\right)^{3.00} = 8.17 \times 10^6 \text{ rev}$$

This must be interpreted as the L_{10} life at a load of 3 500 lb.

From the preceding problem, it was shown that, if the rated life is 1 million rev, equation (15-1) can be written

$$L_d = (C/P_d)^k\,(10^6) \tag{15-2}$$

The required C for a given design load and life would be

$$C = P_d(L_d/10^6)^{1/k} \tag{15-3}$$

Most people do not think in terms of the number of revolutions that a shaft makes. Rather, they consider the speed of rotation of the shaft, usually in rpm, and the design life of the machine, usually in hours of operation. The design life is specified by the designer, considering the application. As a guide, Table 15-3 can be used. Now, for a specified design life in hours, and a known speed of rotation in rpm, the number of design revolutions for the bearing would be

$$L_d = (\text{h})\,(\text{rpm})\,(60 \text{ min/h})$$

Example Problem 15-2. Compute the required basic dynamic load rating, C, for a ball bearing to carry a radial load of 650 lb from a shaft rotating at 600 rpm that is part of an assembly conveyor in a manufacturing plant.

Table 15-3 Recommended Design Life for Bearings

Application	*Design Life* L_{10}, *h*
Domestic appliances	1 000–2 000
Aircraft engines	1 000–4 000
Automotive	1 500–5 000
Agricultural equipment	3 000–6 000
Elevators, industrial fans, multipurpose gearing	8 000–15 000
Electric motors, industrial blowers, general industrial machines	20 000–30 000
Pumps and compressors	40 000–60 000
Critical equipment in continuous 24-h operation	100 000–200 000

Source: Baumeister, Theodore, editor-in-chief. *Marks' Standard Handbook for Mechanical Engineers*, 8th ed. New York: McGraw-Hill Book Company, 1978.

Solution. From Table 15-3, let's select a design life of 30 000 h. Then L_d is

$$L_d = (30\,000\text{ h})(600\text{ rpm})(60\text{ min/h}) = 1.08 \times 10^9\text{ rev}$$

From equation (15-3),

$$C = 650(1.08 \times 10^9/10^6)^{1/3} = 6\,670\text{ lb}$$

To facilitate calculations, some manufacturers provide charts or tables of life factors and speed factors that make computing the number of revolutions unnecessary. Note that the rated life of 1 million rev would be achieved by a shaft rotating at 33⅓ rpm for 500 h. If the actual speed or desired life is different from these two values, a speed factor, f_N, and a life factor, f_L, can be determined from charts like those shown in Figure 15-8. The factors account for the load/life relationship of equation (15-1). The required basic dynamic load rating, C, for a bearing to carry a design load, P_d, would then be

$$C = P_d f_L / f_N \tag{15-4}$$

Other catalogs use different approaches, but they are all based on the load/ life relationship of equation (15-1).

If example problem 15-2 were solved by using the charts of Figure 15-8, the following would result.

$$f_N = 0.381 \quad \text{(For 600 rpm)}$$
$$f_L = 3.90 \quad \text{(For 30 000-h life)}$$
$$C = 650(3.90)/(0.381) = 6\,654\text{ lb}$$

This compares closely with the value of 6 670 lb found previously.

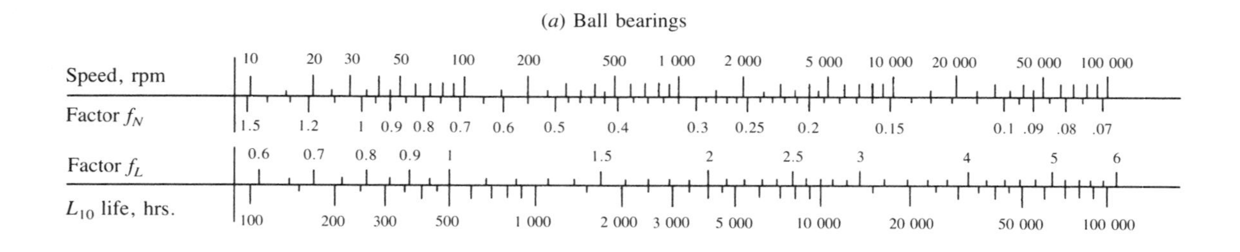

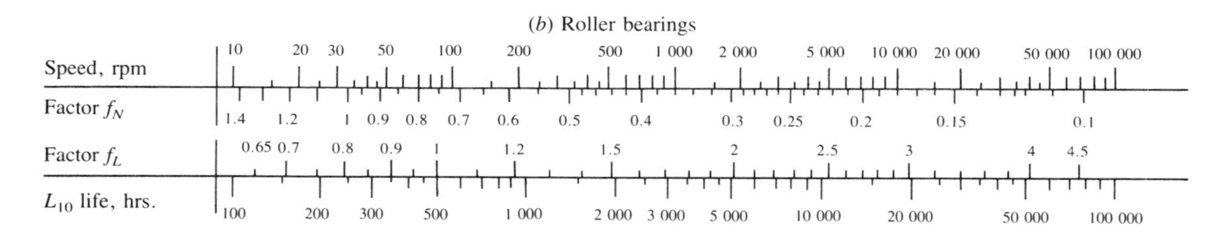

Figure 15-8 Life and Speed Factors for Ball and Roller Bearings (FAG Bearing Corporation, Stamford, Conn.)

15-4 BEARING SELECTION

The selection of a bearing takes into consideration the load capacity, as discussed, and the geometry of the bearing that will ensure that it can be installed conveniently in the machine. We will first consider unmounted bearings carrying radial loads only. Then we will consider unmounted bearings carrying a combination of radial and thrust loads. The term *unmounted* refers to the case in which the designer must provide for the proper application of the bearing onto the shaft and into a housing. Mounted bearings, sometimes called *pillow blocks,* will be discussed in a later section.

The bearing is normally selected after the shaft design has progressed to the point where the required minimum diameter of the shaft has been determined, using the techniques presented in Chapter 9. The radial and thrust loads are also known, along with the orientation of the bearings with respect to other elements in the machine.

The selection process would proceed as follows:

1. Specify the design load on the bearing, usually called *equivalent load,* as discussed later.
2. Determine the minimum acceptable diameter of the shaft that will limit the bore size of the bearing.
3. Select the type of bearing, using Table 15-1 as a guide.
4. Specify the design life of the bearing, using Table 15-3.
5. Determine the speed factor and life factor if such tables are available for the selected type of bearing.
6. Compute the required basic dynamic load rating, C, from equation (15-1), (15-3), or (15-4).
7. Identify a set of candidate bearings that have the required basic dynamic load rating.

8. Select the bearing having the most convenient geometry, also considering its cost and availability.
9. Determine mounting conditions, such as shaft seat diameter and tolerance, housing bore diameter and tolerance, means of locating the bearing axially, and special needs such as seals or shields.

Equivalent Load: Radial Load Only

The *equivalent load, P,* should be used as the design load in the calculation of the required basic dynamic load rating, C. The method of determining the equivalent load when only a radial load, R, is applied takes into account whether the inner or the outer race rotates.

$$P = VR \tag{15-5}$$

The factor V is called a *rotation factor* and takes the value of 1.0 if the inner race of the bearing rotates, the usual case. Use $V = 1.2$ if the outer race rotates.

Standard bearings are available in several classes, typically extra-light, light, medium, and heavy classes. The designs differ in the size and number of load-carrying elements (balls or rollers) in the bearing. The bearing number usually indicates the class and the size of the bore of the bearing. Most bearings are manufactured with nominal dimensions in metric units, and the last two digits of the bearing number indicate the nominal bore size. The bore size convention can be seen from the data in Table 15-2. Note that for bore sizes 04 and above, the nominal bore dimension in millimeters is five times the last two digits in the bearing number.

The number preceding the last two digits indicates the class. For example, several manufacturers use series 100 to indicate extra-light, 200 for light, 300 for medium, and 400 for heavy-duty classes. The three digits may be preceded by others to indicate a special design code of the manufacturer, as is the case in Table 15-2. Figure 15-9 shows the relative size of the different classes of bearings.

Inch-type bearings are available with bores ranging from 0.125 0 through 15.000 in.

Example Problem 15-3. Select a single row, deep groove ball bearing to carry 650 lb of pure radial load from a shaft that rotates at 600 rpm. The design life is to be 30 000 h. The bearing is to be mounted on a shaft with a minimum acceptable diameter of 1.48 in.

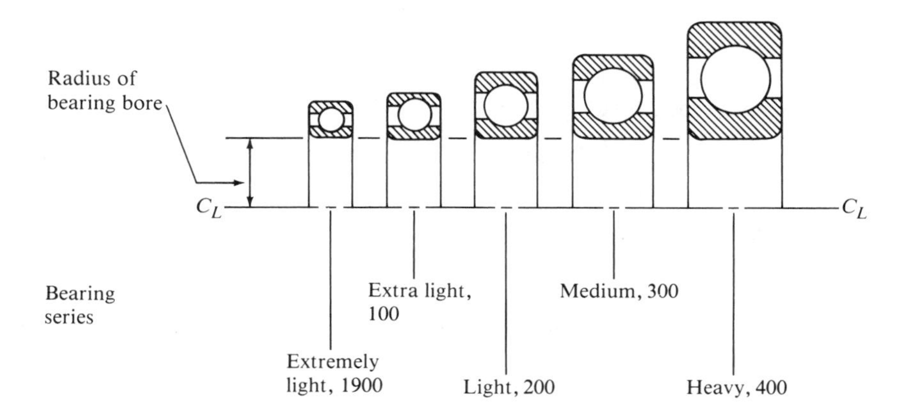

Figure 15-9 Relative Sizes of Bearing Series

Solution. Note that this is a pure radial load and the inner race is to be pressed onto the shaft and rotate with it. Therefore, the factor $V = 1.0$ in equation (15-5) and the design load is equal to the radial load. These are the same data used in example problem 15-2, where we found the required basic dynamic load rating, C, to be 6 670 lb. From Table 15-2, giving design data for two classes of bearings, we find we could use a bearing 6211 or a bearing 6308. Either has a rated C of just over 6 670 lb. But note that the 6211 has a bore of 55 mm (2.165 4 in), and the 6308 has a bore of 40 mm (1.574 8 in). The 6308 is more nearly in line with the desired shaft size.

Equivalent Load: Radial and Thrust Load

When both radial and thrust loads are exerted on a bearing, the equivalent load is the constant radial load that would produce the same rated life for the bearing as the combined loading. The method of computing the equivalent load, P, for such cases is presented in the manufacturer's catalog and takes the form

$$P = VXR + YT \tag{15-6}$$

where

P = Equivalent load

V = Rotation factor (as defined)

R = Applied radial load

T = Applied thrust load

X = Radial factor

Y = Thrust factor

The values of X and Y vary with the specific design of the bearing and with the magnitude of the thrust load relative to the radial load. For relatively small thrust loads, $X = 1$ and $Y = 0$, so the equivalent load equation reverts to the form of equation (15-5) for pure radial loads. To indicate the limiting thrust load for which this is the case, manufacturers list a factor called e. If the ratio $T/R > e$, equation (15-6) must be used to compute P. If $T/R < e$, equation (15-5) is used. Table 15-4 shows one set of data for a single row, deep groove ball bearing. Note that both e and Y depend on the ratio T/C_o, where C_o is the static load rating of a particular bearing. This presents a difficulty in bearing selection because the value of C_o is not known until the bearing has been selected. Therefore, a simple trial-and-error method is applied. If a significant thrust load is applied to a bearing along with a radial load, perform the following steps:

1. Assume a value of Y from Table 15-4. The value $Y = 1.50$ is reasonable, being at about the middle of the range of possible values.
2. Compute $P = VXR + YT$
3. Compute the required basic dynamic load rating C from equation (15-1), (15-3), or (15-4).
4. Select a candidate bearing having a value of C at least equal to the required value.
5. For the selected bearing, determine C_o.
6. Compute T/C_o.
7. From Table 15-4, determine e.

Table 15-4 Radial and Thrust Factors for Single Row, Deep Groove Ball Bearings

e	T/C_o	Y	e	T/C_o	Y
0.19	0.014	2.30	0.34	0.170	1.31
0.22	0.028	1.99	0.38	0.280	1.15
0.26	0.056	1.71	0.42	0.420	1.04
0.28	0.084	1.55	0.44	0.560	1.00
0.30	0.110	1.45			

Note: $X = 0.56$ for all values of Y.

8. If $T/R > e$, then determine Y from Table 15-4.
9. If the new value of Y is different from that assumed in step 1, repeat the process.
10. If $T/R < e$, use equation (15-5) to compute P and proceed as for a pure radial load.

Example Problem 15-4. Select a single row, deep groove ball bearing from Table 15-2 to carry a radial load of 1 850 lb and a thrust load of 675 lb. The shaft is to rotate at 1 150 rpm, and a design life of 20 000 h is desired. The minimum acceptable diameter for the shaft is 3.10 in.

Solution. Using the procedure outlined above:

1. Assume $Y = 1.50$.
2. $P = VXR + YT = (1.0)(0.56)(1\,850) + (1.50)(675) = 2\,049$ lb.
3. From Figure 15-8, the speed factor $f_N = 0.30$ and the life factor $f_L = 3.41$. Then the required basic dynamic load rating C is

$$C = Pf_L/f_N = 2\,049(3.41)/(0.30) = 23\,300 \text{ lb}$$

4. From Table 15-2, we could use either bearing number 6222 or 6318. The 6318 has a bore of 3.543 3 in and is well suited to this application.
5. For bearing number 6318, $C_o = 22\,500$ lb.
6. $T/C_o = 675/22\,500 = 0.03$
7. From Table 15-4, $e = 0.22$ (approximately).
8. $T/R = 675/1\,850 = 0.36$. Because $T/R > e$, we can find $Y = 1.97$ from Table 15-4 by interpolation based on $T/C_o = 0.03$.
9. Recompute $P = (1.0)(0.56)(1\,850) + (1.97)(675) = 2\,366$ lb:

$$C = 2\,366(3.41)/(0.30) = 26\,900 \text{ lb}$$

The bearing number 6318 is not satisfactory at this load. Let's choose bearing number 6320 and repeat from step 5.

5. $C_o = 29\,800$ lb.
6. $T/C_o = 675/29\,800 = 0.023$.
7. $e = 0.20$.

8. $T/R > e$. Then $Y = 2.10$.
9. $P = (1.0)(0.56)(1\,850) + (2.10)(675) = 2\,454$ lb.

$$C = 2\,454(3.41)/(0.30) = 27\,900 \text{ lb}$$

Because bearing number 6320 has a value of $C = 30\,000$ lb, it is satisfactory.

15-5 MOUNTING OF BEARINGS

Up to this point, we have considered the load-carrying capacity of the bearings and the bore size in selecting a bearing for a given application. Although these are the most critical parameters, the successful application of a bearing must consider its proper mounting. Bearings are precision machine elements. Great care must be exercised in their handling, mounting, installation, and lubrication.

The primary considerations in the mounting of a bearing are as follows:

The shaft seat diameter and tolerances

The housing internal bore and tolerances

The shaft shoulder diameter against which the inner race of the bearing will be located

The housing shoulder diameter provided for locating the outer race

The radius of the fillets at the base of the shaft and housing shoulders

The means of retaining the bearing in position

In a typical installation, the bore of the bearing makes a light interference fit on the shaft, and the outside diameter of the outer race makes a close clearance fit in the housing bore. To ensure proper operation and life, the mounting dimensions must be controlled to a total tolerance of only *a few ten-thousandths of an inch*. Most catalogs specify the limit dimensions for both the shaft seat diameter and the housing bore diameter.

Likewise, the catalog will specify the desirable shoulder diameters for the shaft and the housing that will provide a secure surface against which to locate the bearing while ensuring that the shaft shoulder contacts only the inner race and the housing shoulder contacts only the outer race. Table 15-2 includes these values.

The fillet radius specified in the catalog (see r in Table 15-2) is the maximum permissible radius *on the shaft and in the housing* that will clear the external radius on the bearing races. Using too large a radius would not permit the bearing to seat tightly against the shoulder. Of course, the actual fillet radius should be made as large as possible up to the maximum to minimize the stress concentration at the shoulder.

Bearings can be retained in the axial direction by many of the means described in Chapter 10. Three popular methods are retaining rings, end caps, and locknuts. Figure 15-10 shows one possible arrangement. Note that for the left bearing the shaft diameter is slightly smaller to the left of the bearing seat. This allows the bearing to be slid easily over the shaft up to the place where it must be pressed on. The retaining ring for the outer race could be supplied as a part of the outer race instead of as a separate piece.

The right bearing is held on the shaft with a locknut threaded on the end of the shaft. See Figure 15-11 for the design of standard locknuts. The internal tab on the lockwasher engages a groove in the shaft, and one of the external tabs is bent into a groove on the nut after it is seated to keep the nut from backing off. The external cap not only protects the bearing but also retains the outer race in position.

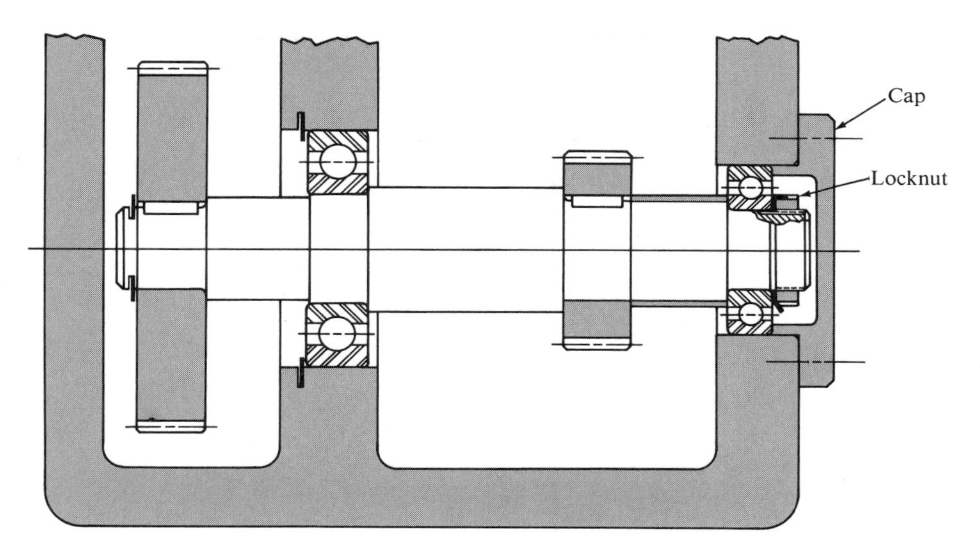

Figure 15-10 Bearing Mounting Illustration

15-6 TAPERED ROLLER BEARINGS

The taper on the rollers of tapered roller bearings, evident in Figure 15-7, results in a different load path than for the bearings discussed thus far. Figure 15-12 shows two tapered roller bearings supporting a shaft with a combination of a radial load and a thrust load. The design of the shaft is such that the thrust load is resisted by the left bearing. But a peculiar feature of this type of bearing is that a radial load on one of the bearings creates a thrust on the opposing bearing, also; this feature must be considered in analyzing the bearing.

The location of the radial reaction must also be determined with care. Part (b) of Figure 15-12 shows a dimension a that is found by the intersection of a line perpendicular to the axis of the roller and the centerline of the shaft. The radial reaction at the bearing acts through this point. The distance a is reported in the tables of data for the bearings.

The Anti-Friction Bearings Manufacturers' Association (AFBMA) recommends the following approach in computing the equivalent loads on a tapered roller bearing.

$$P_A = 0.4F_{rA} + 0.5\frac{Y_A}{Y_B}F_{rB} + Y_A T_A \tag{15-7}$$

$$P_B = F_{rB} \tag{15-8}$$

where

P_A = Equivalent radial load on bearing A

P_B = Equivalent radial load on bearing B

F_{rA} = Applied radial load on bearing A

F_{rB} = Applied radial load on bearing B

T_A = Thrust load on bearing A

Y_A = Thrust factor for bearing A from tables

Y_B = Thrust factor for bearing B from tables

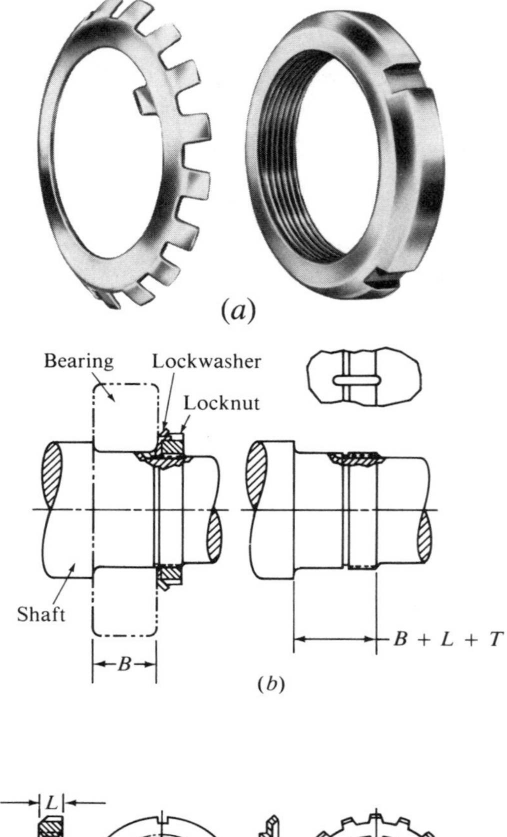

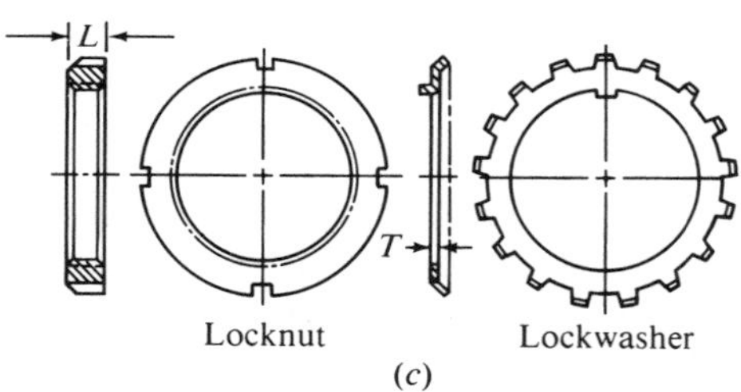

Figure 15-11 Locknut and Lockwasher for Retaining Bearings (SKF Industries, Inc., King of Prussia, Penn.)

Table 15-5 shows an abbreviated set of data from a catalog to illustrate the method of computing equivalent loads.

For the several hundred designs of standard tapered roller bearings available commercially, the value of the thrust factor varies from as small as 1.07 to as high as 2.26. In design problems a trial-and-error procedure is usually necessary. Example problem 15-4 illustrates one approach.

Example Problem 15-4. The shaft shown in Figure 15-12 carries a transverse load of 6 800 lb and a thrust load of 2 500 lb. The thrust is resisted by bearing A. The shaft rotates at 350 rpm and is to be used in a piece of agricultural equipment. Specify suitable tapered roller bearings for the shaft.

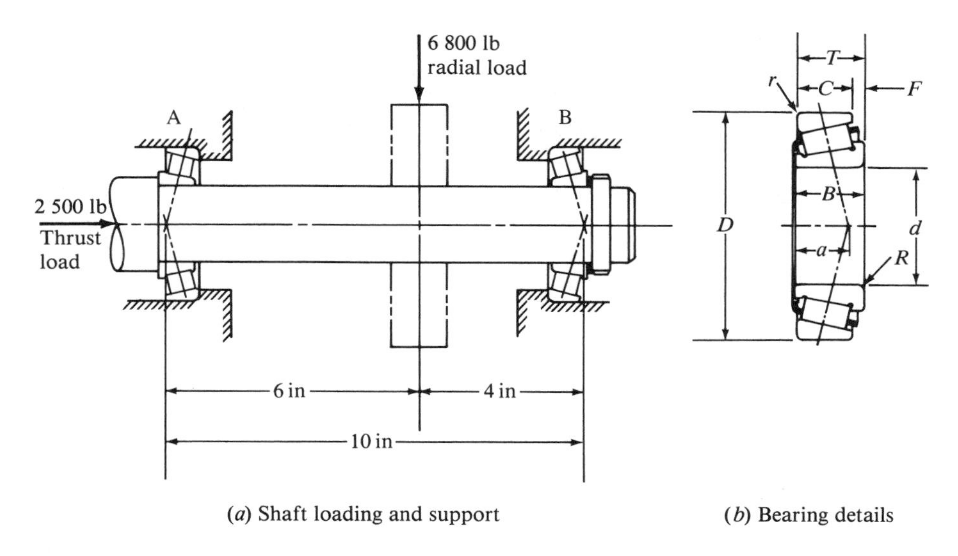

(*a*) Shaft loading and support

(*b*) Bearing details

Figure 15-12 Example of Tapered Roller Bearing Installation

Solution. The radial loads on the bearings are

$$F_{rA} = 6\,800\ (4\ \text{in}/10\ \text{in}) = 2\,720\ \text{lb}$$
$$F_{rB} = 6\,800\ (6\ \text{in}/10\ \text{in}) = 4\,080\ \text{lb}$$
$$T_A = 2\,500\ \text{lb}$$

To use equation (15-6), values of Y_A and Y_B must be assumed. Let's use $Y_A = Y_B = 1.75$. Then,

$$P_A = 0.40(2\,720) + 0.5\frac{1.75}{1.75}4\,080 + 1.75(2\,500) = 7\,503\ \text{lb}$$
$$P_B = F_{rB} = 4\,080\ \text{lb}$$

Table 15-5 Tapered Roller Bearing Data

Bore	*Outside Diameter*	*Width*	*a*	*Thrust Factor, Y*	*Basic Dynamic Load Rating, C*
1.0000	2.5000	0.8125	0.583	1.71	8 370
1.5000	3.0000	0.9375	0.690	1.98	12 800
1.7500	4.0000	1.2500	0.970	1.50	21 400
2.0000	4.3750	1.5000	0.975	2.02	26 200
2.5000	5.0000	1.4375	1.100	1.65	29 300
3.0000	6.0000	1.6250	1.320	1.47	39 700
3.5000	6.3750	1.8750	1.430	1.76	47 700

Note: Dimensions in inches. Load C in pounds for an L_{10} life of 1 million rev.

Using Table 15-3 as a guide, let's select 4 000 h as a design life. Then the number of revolutions would be

$$L_d = (4\,000 \text{ h})(350 \text{ rpm})(60 \text{ min/h}) = 8.4 \times 10^7 \text{ rev}$$

The required basic dynamic load rating can now be calculated from equation (15-3), using $k = 3.33$.

$$C_A = P_A(L_d/10^6)^{1/k}$$

$$C_A = 7\,503(8.4 \times 10^7/10^6)^{0.30} = 28\,400 \text{ lb}$$

Similarly,

$$C_B = 4\,080(8.4 \times 10^7/10^6)^{0.30} = 15\,400 \text{ lb}$$

From Table 15-4, we can choose the following bearings:

Bearing A:

$$d = 2.5000 \text{ in} \qquad D = 5.0000 \text{ in}$$

$$C = 29\,300 \text{ lb} \qquad Y_A = 1.65$$

Bearing B:

$$d = 1.7500 \text{ in} \qquad D = 4.0000 \text{ in}$$

$$C = 21\,400 \text{ lb} \qquad Y_B = 1.50$$

We can now recompute the equivalent loads.

$$P_A = 0.40(2\,720) + 0.5\frac{1.65}{1.50}4\,080 + 1.65(2\,500) = 7\,457 \text{ lb}$$

$$P_B = F_{rB} = 4\,080 \text{ lb}$$

From these, the new values of $C_A = 28\,200$ lb and $C_B = 15\,400$ lb are still satisfactory for the selected bearings.

One caution must be observed in using the equations for equivalent loads for tapered roller bearings. If, from equation (15-7), the equivalent load on bearing A is less than the applied radial load, the following equations should be used.

If $P_A < F_{rA}$, then let $P_A = F_{rA}$ and compute P_B.

$$P_B = 0.4F_{rB} + 0.5\frac{Y_B}{Y_A}F_{rA} - Y_B T_A \tag{15-9}$$

A similar analysis is used for angular contact ball bearings in which the design of the races results in a similar load path as that for tapered roller bearings. Figure 15-13 shows an angular contact bearing and the angle through the pressure center. This is equivalent to the line perpendicular to the axis of the tapered roller bearing. The radial reaction on the bearing acts through the intersection of this line and the axis of the shaft. Also, a radial load on one bearing induces a thrust load on the opposing bearing, requiring the application of the equivalent load formulas of the type used in equations (15-7) and (15-9). The angle of the load line in commercially available angular contact bearings ranges from 15° to 40°.

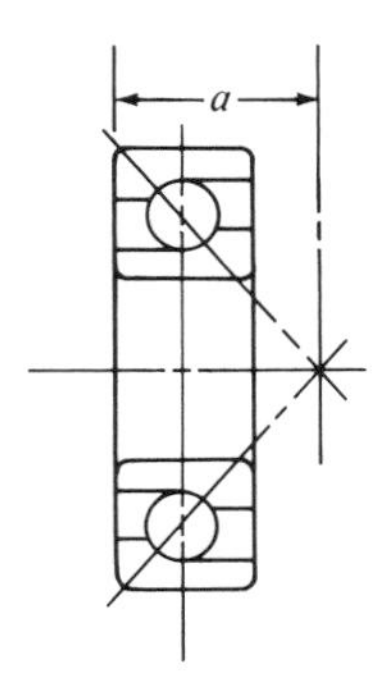

Figure 15-13 Angular Contact Bearing

(*a*) Example of ball thrust bearings

(*c*) Examples of roller thrust bearing

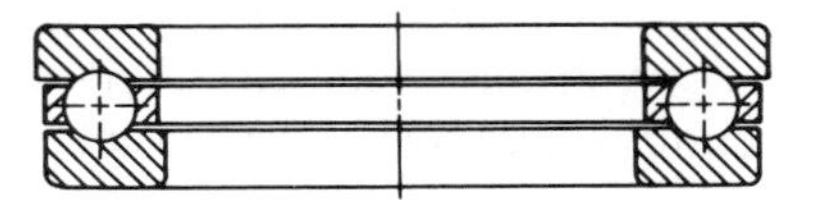

(*b*) Typical cross section of ball thrust bearing

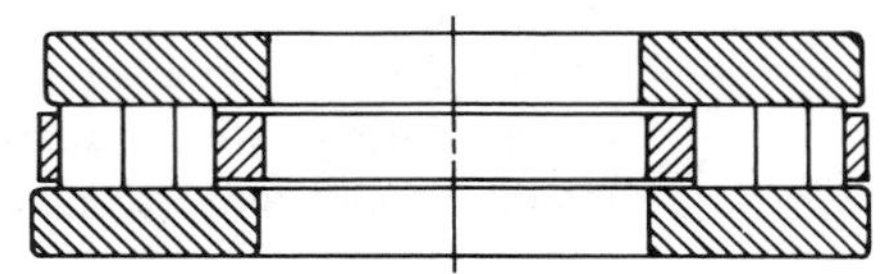

Standard cylindrical roller-thrust bearing

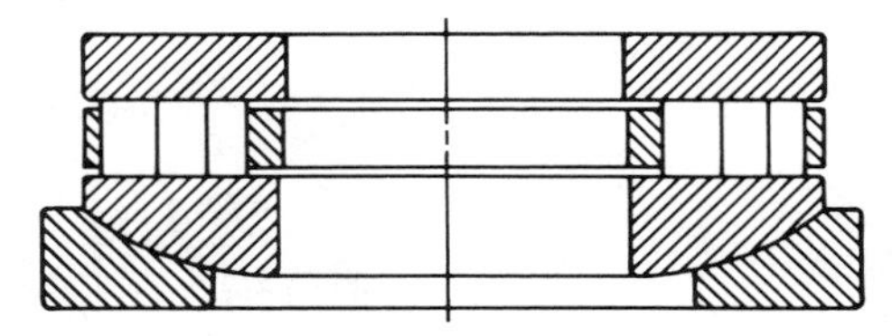

Self-aligning thrust bearing

Figure 15-14 Thrust Bearings (Andrews Bearing Corp., Spartanburg, S.C.)

15-7 THRUST BEARINGS

The bearings discussed so far in this chapter have been designed to carry radial loads or a combination of radial and thrust loads. Many machine design projects demand a bearing that resists only thrust loads, and several types of standard thrust bearings are commercially available. The same types of rolling elements are used: spherical balls, cylindrical rollers, and tapered rollers (see Figure 15-14).

Most thrust bearings can take little or no radial load. Then the design and selection of such bearings is dependent only on the magnitude of the thrust load and the design life. The data for basic dynamic load rating and basic static load rating are reported in manufacturers' catalogs in the same way as they are for radial bearings. Then equations (15-1) through (15-4) are used to select a suitable bearing.

15-8 MOUNTED BEARINGS

In many types of heavy machines and special machines produced in small quantities, mounted bearings are selected rather than unmounted bearings. The mounted bearings provide means to attach the bearing unit directly to the frame of the machine with bolts rather than inserting it into a machined recess in a housing as is required in unmounted bearings.

Figure 15-15 shows the most common configuration for a mounted bearing, the *pillow block*. The housing is made from formed steel, cast iron, or cast steel with holes or slots provided for attachment during assembly of the machine at which time alignment of the bearing unit is adjusted. The bearings themselves can be of virtually any of the types discussed in the preceding sections, ball, tapered roller, or spherical roller being preferred. Misalignment capability is an important application consideration because of the conditions of use of such bearings. This capability is provided either in the bearing construction itself or in the housing.

Because the bearing itself is similar to those already discussed, the selection process is also similar. Most catalogs provide extensive charts of data listing the load-carrying capacity at specified rated life values (L_{10}). This precludes the need to compute life or load as done in this chapter.

Other forms of mounted bearings are shown in Figure 15-16. The *flange units* are designed to be mounted on the vertical side frames of machines, holding horizontal shafts. Again several bearing types and sizes are available. The term *take-up unit* refers to a bearing mounted in a housing, which in turn is mounted in a frame that allows movement of the bearing with the shaft in place. Used on conveyors, chain drives, belt drives, and similar applications, the take-up unit permits adjustment of the center distance of the drive

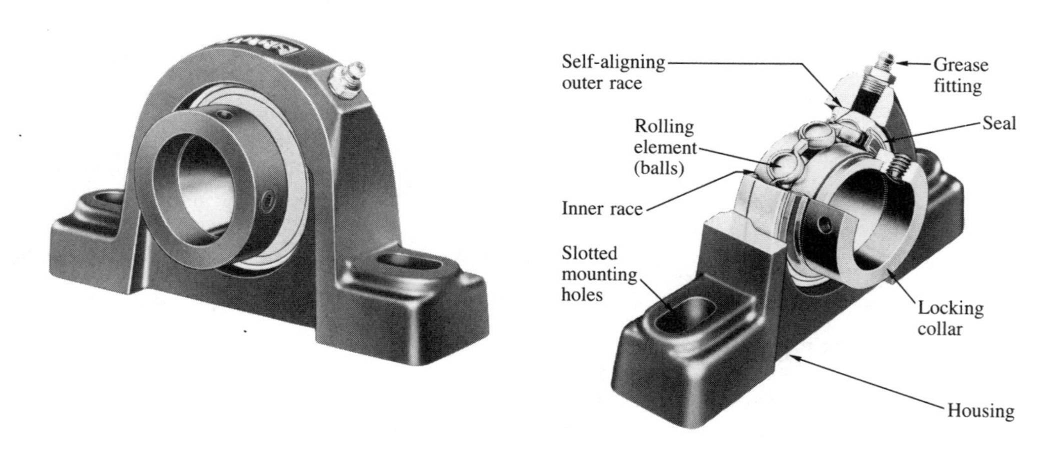

Figure 15-15 Ball Bearing Pillow Block (Dodge Division, Reliance Electric Co.)

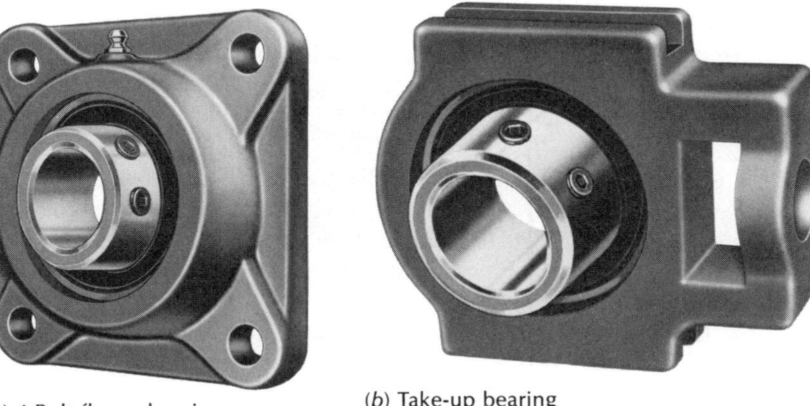

(a) 4-Bolt flange bearing

(b) Take-up bearing

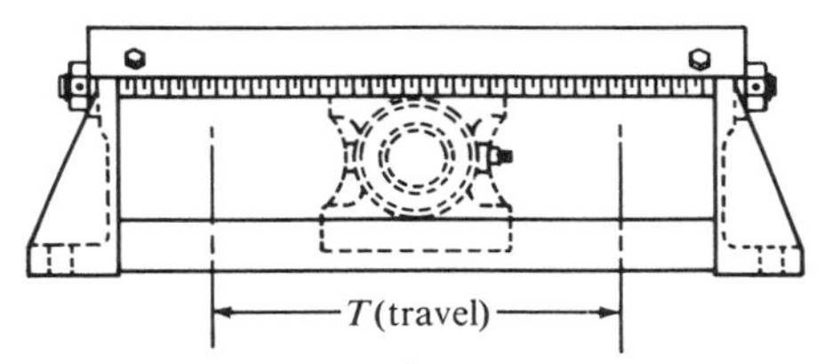

(c) Top angle take-up frames

Figure 15-16 Forms of Mounted Bearings (Dodge Division, Reliance Electric Co.)

components at the time of installation and during operation to accommodate wear or stretch of parts of the assembly.

15-9 PRACTICAL CONSIDERATIONS IN THE APPLICATION OF BEARINGS

This section will discuss lubrication of bearings, installation, preloading, stiffness, operation under varying loads, sealing, limiting speeds, bearing tolerance classes, and standards related to the manufacture and application of bearings.

Lubrication

The functions of lubrication in a bearing unit are as follows.

1. To provide a low-friction film between the rolling elements and the races of the bearing and at points of contact with cages, guiding surfaces, retainers, and so on

2. To protect the bearing components from corrosion
3. To help to dissipate heat from the bearing unit
4. To carry heat away from the bearing unit
5. To help dispel contaminants and moisture from the bearing

Rolling contact bearings are usually lubricated with either grease or oil. Under normal ambient temperatures (approximately 70°F) and relatively slow speeds (under 500 rpm), grease is satisfactory. At higher speeds or higher ambient temperatures, oil lubrication applied in a continuous flow is required, possibly with external cooling of the oil.

Oils used in bearing lubrication are usually clean, stable mineral oils. Under lighter loads and lower speeds, light oil is used. Heavier loads and/or higher speeds require grease or heavier oils up to SAE 30. A recommended upper limit for lubricant temperature is 160°F. The choice of the correct oil or grease depends on many factors, so each application should be discussed with the bearing manufacturer. In general, a viscosity of 70 to 100 Saybolt Universal Seconds (SUS) should be maintained at the operating temperature of the lubricant in the bearing.

In some critical applications such as bearings in jet engines and very high speed devices, lubricating oil is pumped under pressure to an enclosed housing for the bearing where the oil is directed at the rolling elements themselves. A controlled return path is also provided. The temperature of the oil in the sump is monitored and controlled with heat exchangers or refrigeration to maintain oil viscosity within acceptable limits. Such systems provide reliable lubrication and ensure the removal of heat from the bearing.

Greases used in bearings are mixtures of lubricating oils and thickening agents, usually soaps such as lithium or barium. The soaps act as carriers for the oil that is drawn out at the point of need within the bearing. Additives to resist corrosion or oxidation of the oil itself are sometimes added. Classifications of greases specify the operating temperatures to which the greases will be exposed, as defined by the Anti-Friction Bearing Manufacturers' Association (AFBMA), and outlined next.

Group	*Type of Grease*	*Operation Temperature Range (°F)*
I	General-purpose	−40–250
II	High-temperature	0–300
III	Medium-temperature	32–200
IV	Low-temperature	−67–225
V	Extreme-high-temperature	up to 450

Installation

It has already been stated that most bearings should be installed with a light interference fit between the bore of the bearing and the shaft to preclude the possibility of rotation of the inner race of the bearing with respect to the shaft. Such a condition would result in uneven wear of the bearing elements and early failure. To install the bearing then requires rather heavy forces applied axially. Care must be exercised so that the bearing is not damaged during installation. The installation force should be applied directly to the inner race of the bearing.

If the force were applied through the outer race, the load would be transferred through the rolling elements to the inner race. Because of the small contact area, it is likely that such transfer of forces would overstress some element, exceeding the static load capacity. Brinelling would result, along with the noise and rapid wear that accompany this condition. For large bearings it may be necessary to heat the bearing to expand its diameter in order to keep the installation forces within reason. Removal of bearings intended for reuse requires similar precautions. Bearing pullers are available to facilitate this task.

Preloading

Some bearings are made with internal clearances that must be taken up in a particular direction to ensure satisfactory operation. In such cases, preloading must be provided, usually in the axial direction. On horizontal shafts, springs are typically used with axial adjustment of the spring deflection sometimes provided to adjust the amount of preload. When space is limited, the use of Belleville washers is desirable because they provide high forces with small deflections. Shims can be used to adjust the actual deflection and preload obtained (see Chapter 7). On vertical shafts the weight of the shaft assembly itself may be sufficient to provide the required preload.

Bearing Stiffness

Stiffness is the deflection that a given bearing undergoes when carrying a given load. Usually the radial stiffness is most important because the dynamic behavior of the rotating shaft system is affected. The critical speed and the mode of vibration are both functions of the bearing stiffness. Generally speaking, the softer the bearing (low stiffness), the lower the critical speed of the shaft assembly. Stiffness is measured in the units used for springs, such as pounds per inch or newtons per millimeter. Of course the stiffness values are quite high, with values of 500 000 to 1 000 000 lb/in reasonable. The manufacturer should be consulted when such information is needed, because it is rarely included in standard catalogs.

Varying Loads

The load/life relationships used thus far assume that the load is reasonably constant in magnitude and direction. If the load varies considerably, an effective mean load must be used for determining the expected life of the bearing (1, 2). Oscillating loads also require special analysis because only a few of the rolling elements share the load.

Sealing

When the bearing is to operate in dirty or moist environments, special shields and seals are usually specified. They can be provided on either or both sides of the rolling elements. Shields are typically metal and are fixed to the stationary race but remain clear of the rotating race. Seals are made of elastomeric materials and do contact the rotating race. Bearings fitted with both seals and shields and precharged at the factory with grease are sometimes called *permanently lubricated*. Although such bearings are likely to give many years of satisfactory service, extreme conditions can produce a degradation of the lubricating properties of the grease. The presence of seals also increases the friction in a bearing. Sealing can be provided outside the bearing in the housing or at the shaft/housing interface. On high-speed shafts, a *labyrinth seal,* consisting of a noncontacting ring around the shaft with a few thousandths of an inch radial clearance, is frequently used.

Grooves, sometimes in the form of a thread, are machined in the ring; the relative motion of the shaft with respect to the ring creates the sealing action.

Limiting Speeds

Most catalogs list limiting speeds for each bearing. Exceeding these limits may result in excessively high operating temperatures due to friction between the cages supporting the rolling elements. Generally the limiting speed is lower for larger bearings than for smaller bearings. Also, a given bearing will have a lower limiting speed as loads increase. With special care, either in the fabrication of the bearing cage or in the lubrication of the bearing, bearings can be operated at higher speeds than those listed in the catalog. The manufacturer should be consulted in such applications.

Standards

Several groups are involved in standard setting for the bearing industry. A partial list is given here.

Anti-Friction Bearing Manufacturers' Association	(AFBMA)
Annular Bearing Engineers Committee	(ABEC)
Roller Bearing Engineers Committee	(RBEC)
Ball Manufacturers Engineers Committee	(BMEC)
American National Standards Institute	(ANSI)
International Standards Organization	(ISO)

Many standards are co-listed by the ANSI and AFBMA organizations. The catalog of ANSI lists 22 different standards related to bearings. A few of them are listed here:

Terminology and Definitions for Ball and Roller Bearings and Parts, ANSI B3.7.

Load Ratings and Fatigue Life for Ball Bearings, ANSI/AFBMA 9.

Load Ratings and Fatigue Life for Roller Bearings, ANSI/AFBMA 11.

Tolerances

Several different tolerance classes are recognized in the bearing industry to accommodate the needs of the wide variety of equipment using rolling contact bearings. In general, of course, all bearings are precision machine elements and should be treated as such. As noted before, the general range of tolerances is of the order of a few ten-thousandths of an inch. The standard tolerance classes are defined by ABEC, as identified below.

ABEC 1: Standard radial ball and roller bearings
ABEC 3: Semi-precision instrument ball bearings
ABEC 5: Precision radial ball and roller bearings
ABEC 5P: Precision instrument ball bearings
ABEC 7: High precision radial ball bearings
ABEC 7P: High precision instrument ball bearings

Most machine applications would use ABEC 1 tolerances, the data for which are usually provided in the catalogs. Machine tool spindles requiring extra smooth and accurate running would use ABEC 5 or ABEC 7 classes.

REFERENCES

1. SKF Industries, Inc. *SKF Engineering Data*. Philadelphia, Pa.
2. FMC Corporation. *Bearing Technical Journal*. FMC Bearing Division, Indianapolis.
3. Dodge Division of Reliance Electric Company. *Dodge Mounted Bearings Bulletin*. Mishawaka, Ind.
4. Hoover NSK Bearing Company. *Catalog No. AM-1, Ball, Tapered Roller, Cylindrical Roller & Spherical Roller Bearings*. Ann Arbor, Mich.
5. FAG Bearings Corporation. *Precision Bearings Catalog*. Stamford, Conn.
6. The Timken Company. *Tapered Roller Bearing Engineering Journal*. Canton, Ohio.
7. Baumeister, Theodore, editor-in-chief. *Marks' Standard Handbook for Mechanical Engineers*, 8th ed. New York: McGraw-Hill Book Company, 1978.

PROBLEMS

1. A radial ball bearing has a basic dynamic load rating of 2 350 lb for a rated (L_{10}) life of 1 million rev. What would be its L_{10} life when operating at a load of 1 675 lb?
2. Determine the required basic dynamic load rating for a bearing to carry 1 250 lb from a shaft rotating at 880 rpm if the design life is to be 20 000 h?
3. A catalog lists the basic dynamic load rating for a ball bearing to be 3 150 lb for a rated life of 1 million rev. What would be the L_{10} life of the bearing if it were subjected to a load of (a) 2 200 lb? (b) 4 500 lb?
4. Compute the required basic dynamic load rating, C, for a ball bearing to carry a radial load of 1 450 lb at a shaft speed of 1 150 rpm for an industrial fan.
5. Specify suitable bearings for the shaft of example problem 9-1. Note the data contained in Figures 9-1, 9-2, 9-8, and 9-9.
6. Specify suitable bearings for the shaft of example problem 9-3. Note the data contained in Figures 9-10, 9-11, and 9-12.
7. Specify suitable bearings for the shaft of example problem 9-4. Note the data contained in Figures 9-13 and 9-14.
8. For any of the bearings specified in problems 2 through 7, make a scale drawing of the shaft, bearings, and that part of the housing to support the outer races of the bearings. Be sure to consider fillet radii and axial location of the bearings.
9. A bearing is to carry a radial load of 455 lb and no thrust load. Specify a suitable bearing from Table 15-2 if the shaft rotates at 1 150 rpm and the design life is 20 000 h.

   For each of the following problems, repeat problem 9 with the new data.

Problem Number	*Radial Load*	*Thrust Load*	*rpm*	*Design Life, h*
10	875 lb	0	450	30 000
11	1 265 lb	645 lb	210	5 000
12	235 lb	88 lb	1 750	20 000
13	2 875 lb	1 350 lb	600	15 000
14	3.8 kN	0	3 450	15 000
15	5.6 kN	2.8 kN	450	2 000
16	10.5 kN	0	1 150	20 000
17	1.2 kN	0.85 kN	860	20 000

10. In Chapter 9, Figures 9-15 through 9-27 showed shaft design exercises related to the problems at the end of the chapter. For each bearing of each shaft, specify a suitable bearing from Table 15-2. If the shaft design has already been completed to the point that the minimum acceptable diameter of the shaft at the bearing seat is known, consider that diameter when specifying the bearing. Note the Chapter 9 problem statements for the speed of the shaft and the loading data.

16 Motion Control: Clutches and Brakes

16-1 OVERVIEW

Machine systems require control whenever the speed or direction of the motion of one or more components is to be changed. When a device is started initially, it must be accelerated from rest to the operating speed. As a function is completed, the system must frequently be brought to rest. In continuously operating systems, changing speeds to adjust to different operating conditions is often necessary. Safety sometimes dictates motion control, as with a load being lowered by a hoist or an elevator.

In this chapter we will be concerned mostly with control of rotary motion in systems driven by electric motors, engines, turbines, and the like. Ultimately, linear motion may be involved through linkages, conveyors, or other mechanisms.

The principles of physics tell us that whenever the speed or direction of the motion of a body is changed, there must be a force exerted on the body. If the body is in rotation, a torque must be applied to the system to speed it up or to slow it down. When a change in speed occurs, a corresponding change in the kinetic energy of the system occurs. Thus, motion control inherently involves the control of energy, either adding energy to accelerate a system or absorbing energy to decelerate it.

The machine elements most frequently used for motion control are the clutch and the brake, defined as follows.

A *clutch* is a device used to connect or disconnect a driven component from the prime mover of the system. For example, in a machine that must cycle frequently, the driving motor is left running continuously and a clutch is interposed between the motor and the driven machine. Then the clutch is cycled on and off to connect and disconnect the load. This permits the motor to operate at an efficient speed and also permits the system to cycle more rapidly because there is no need to accelerate the heavy motor rotor with each cycle.

A *brake* is a device used to bring a moving system to rest, to slow its speed, or to control its speed to a certain value under varying conditions.

Several arrangements of clutches and brakes are shown in Figure 16-1. By convention, the term *clutch* is reserved for the application in which the connection is made to a shaft parallel to the motor shaft as illustrated in Figure 16-1(a). If the connection is to a shaft in-line with the motor, the term *clutch coupling* is used, as shown in Figure 16-1(b).

Also by convention, a brake (Figure 16-1c) is actuated by some overt action: the application of a fluid pressure, the switching on of an electric current, or the moving of a lever by hand. A brake that is spring-applied automatically in the absence of an overt action is called a *fail-safe brake* (Figure 16-1d). When the power goes off, the brake goes on.

When the functions of both a clutch and a brake are required in a system, they are frequently arranged in the same unit, the *clutch-brake module*. When the clutch is activated, the brake is deactivated, and vice versa (see Figure 16-1e).

A slip clutch is a clutch that, by design, transmits only a limited torque; thus it will slip at any higher torque. It is used to provide a controlled acceleration to a load that is smooth and which requires a smaller motor power. It is also used as a safety device, protecting expensive or sensitive components in the event a system is jammed.

Most of the discussion in this chapter will concern clutches and brakes that transmit motion through friction at the interface between two rotating parts moving at different speeds. Other types are discussed briefly in the last section.

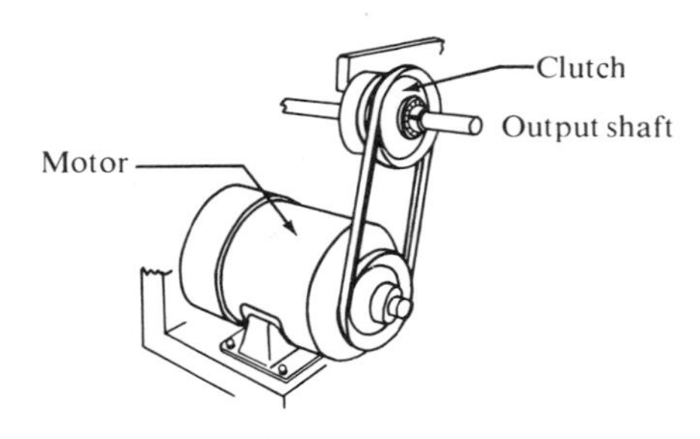

(*a*) Clutch: Transmits rotary motion to a parallel shaft only when coil is energized, by using sheaves, sprockets, gears, or timing pulleys.

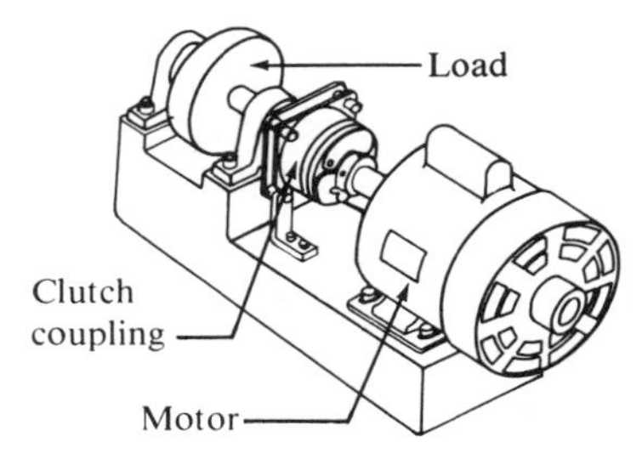

(*b*) Clutch coupling: Transmits rotary motion to an in-line shaft only when coil is energized. Split shaft applications.

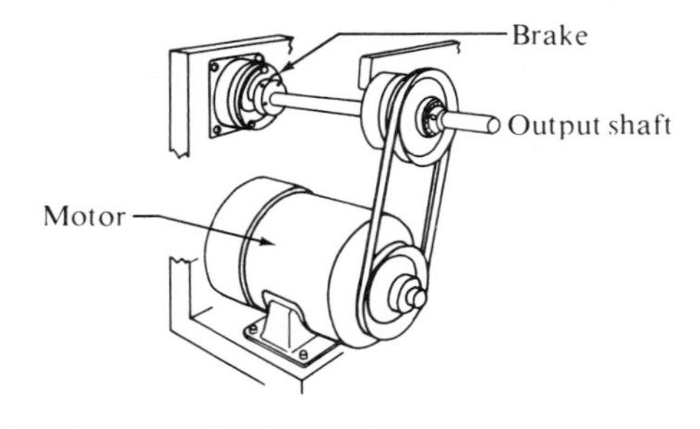

(*c*) Brake: Stops (brakes) load when coil is energized. Panel shows clutch imparting rotary motion to the output (load) shaft while brake is de-energized. Conversely, de-energizing clutch coil and energizing brake coil causes load to stop.

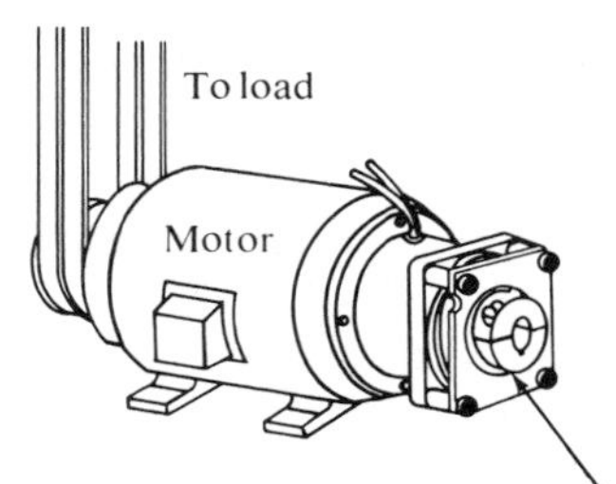

(*d*) Fail-safe brake: Stops load by de-energization of coil; power off—brake on.

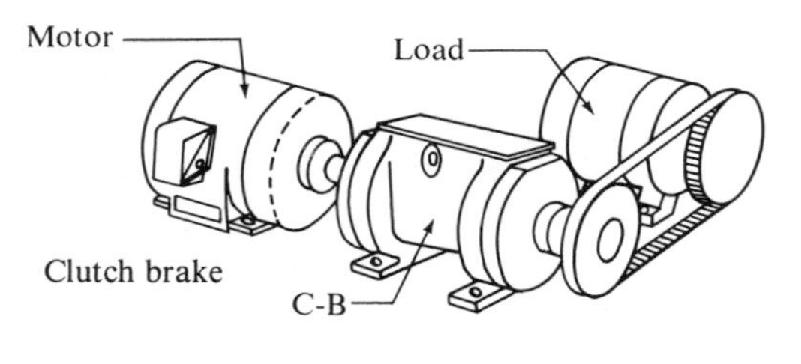

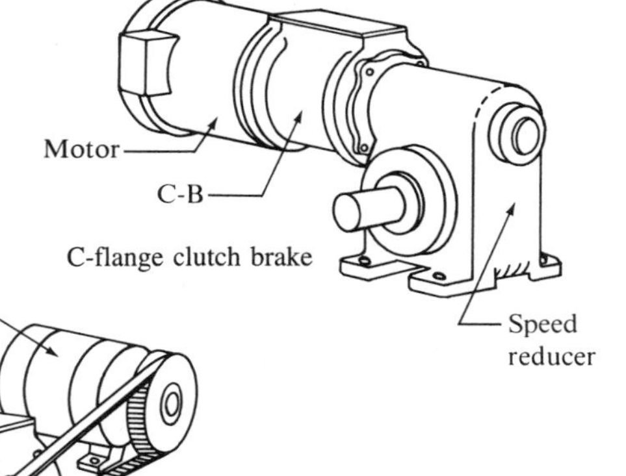

(*e*) Three types of mountings for clutch-brake modules: *Clutch brake* combines functions of clutch and brake in a complete preassembled package with input and output shafts. *C-flange clutch brake* performs same function but for use between NEMA "C" flange motor and speed reducer. *Motor clutch-brake* preassembled module for mounting to a NEMA "C" flange motor and presenting an output shaft for connection to load.

Figure 16-1 Typical Applications of Clutches and Brakes (Electroid Corp., Springfield, N.J.)

16-2 PERFORMANCE PARAMETERS

The parameters involved in the rating of clutches and brakes are described in the following list.

1. Torque required to accelerate or decelerate the system
2. Time required to accomplish the speed change
3. The cycling rate: number of on/off cycles per unit time
4. The inertia of the rotating or translating parts
5. The environment of the system: temperature, cooling effects, and so on
6. Energy dissipation capability of the clutch or brake
7. Physical size and configuration
8. Actuation means
9. Life and reliability of the system
10. Cost and availability

Two basic methods are used to determine the torque capacity required of a clutch or brake. One relates the capacity to the power of the motor driving the system. Recall that, in general, power = torque × rotational speed ($P = Tn$). The required torque capacity is then usually expressed in the form

$$T = \frac{CPK}{n} \tag{16-1}$$

where C is a conversion factor for units and K is a service factor based on the application. More will be said about these later.

Note that the torque required is inversely proportional to the rotational speed. For this reason, it is advisable to locate the clutch or brake on the highest speed shaft in the system so that the required torque is a minimum. The size, cost, and response time are all typically lower when the torque is lower. One disadvantage is that the shaft to be accelerated or decelerated must undergo a larger change in speed and the amount of slipping may be greater. This effect may generate more frictional heat, leading to thermal problems. However, it is offset by the increased cooling effect because of the faster motion of the clutch or brake parts.

The value for the K factor in the torque equation is largely a design decision. Some of the typical guidelines follow.

1. For brakes under average conditions, use $K = 1.0$.
2. For clutches in light duty where the output shaft does not assume its normal load until after it is up to speed, use $K = 1.5$.
3. For clutches in heavy duty where large attached loads must be accelerated, use $K = 3.0$.
4. For clutches in systems having varying loads, use a K factor at least equal to the factor by which the motor breakdown torque exceeds the full load torque. This is discussed in Chapter 17, but for a typical industrial motor (design B), use $K = 2.75$. For a high starting torque motor (design C or capacitor start motor), $K = 4.0$ may be required.

This ensures that the clutch will be able to transmit at least as much torque as the motor and that it will not slip after getting up to speed.

5. For clutches in systems driven by gasoline engines, diesel engines, or other prime movers, consider the peak torque capability of the driver. $K = 5.0$ may be required.

The following table relates the value for C to typically used units for torque, power, and rotational speed. For example, if power is in hp, and speed is in rpm, then, to obtain torque in lb · ft, use $T = 5\,252(P/n)$.

Torque	*Power*	*Speed*	*C*
lb · ft	hp	rpm	5 252
lb · in	hp	rpm	63 025
N · m	W	rad/s	1
N · m	W	rpm	9.549
N · m	kW	rpm	9 549

Although the torque computation method of equation (16-1) will produce generally satisfactory performance in typical applications, it does not provide a means of estimating the actual time required to accelerate the load with a clutch or to decelerate the load with a brake. The second method, described next, does.

Time Required To Accelerate a Load

The basic principle involved is taken from dynamics:

$$T = I\alpha$$

where I is the mass moment of inertia of the components being accelerated and α (alpha) is the angular acceleration, that is, the rate of change of angular velocity. The usual objective of such an analysis is to determine the torque required to produce a change in rotational speed, Δn, of a system in a given amount of time, t. But $\Delta n/t = \alpha$. Also, it is more convenient to express the mass moment of inertia in terms of the *radius of gyration, k*. By definition,

$$k = \sqrt{I/m} \qquad \text{or} \qquad k^2 = I/m$$

where m is the mass and $m = W/g$. Then,

$$I = mk^2 = Wk^2/g$$

The equation for torque then becomes

$$T = I\alpha = \frac{Wk^2}{g}\frac{(\Delta n)}{t} \tag{16-2}$$

The term Wk^2 is frequently called simply the *inertia* of the load, although that designation is not strictly correct. A large proportion of the components in a machine system to be accelerated are in the form of cylinders or discs. Figure 16-2 gives the relationships for the radius of gyration and Wk^2 for hollow discs. Solid discs are simply a special case with an inside radius of zero. More complex objects can be analyzed by considering them to be made from a set of simpler discs. Example problem 16-1 illustrates the process.

Example Problem 16-1. Compute the value of Wk^2 for the steel flat-belt pulley shown in Figure 16-3.

Solution. The pulley can be considered to be made up of three components, each of which is a hollow disc. The Wk^2 for the total pulley is the sum of that for each component.

Part 1. Using the formula from Figure 16-2 for a steel disc,

$$Wk^2 = \frac{(R_1^4 - R_2^4)(L)}{323.9}\ \text{lb}\cdot\text{ft}^2 = \frac{[(10.0)^4 - (9.0)^4](6.0)}{323.9}$$

$$Wk^2 = 63.70\ \text{lb}\cdot\text{ft}^2$$

Part 2.

$$Wk^2 = \frac{[(9.0)^4 - (3.0)^4](0.75)}{323.9} = 15.00\ \text{lb}\cdot\text{ft}^2$$

Part 3.

$$Wk^2 = \frac{[(3.0)^4 - (1.5)^4](4.0)}{323.9} = 0.94\ \text{lb}\cdot\text{ft}^2$$

$$\text{Total } Wk^2 = 63.70 + 15.00 + 0.94 = 79.64\ \text{lb}\cdot\text{ft}^2$$

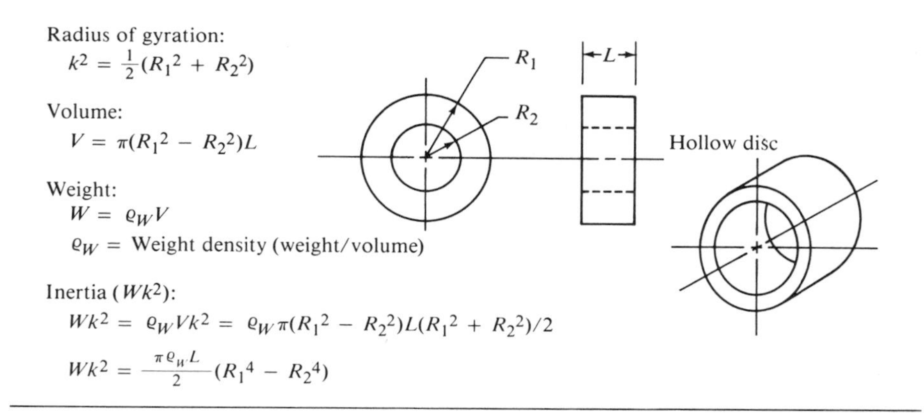

Typical units: L, R_1, R_2 in inches

ϱ_W in lb/in³

Wk^2 in lb•ft²

$$Wk^2 = \frac{\pi}{2} \times \varrho_W \frac{\text{lb}}{\text{in}^3} \times L(\text{in}) \times (R_1{}^4 - R_2{}^4)\ \text{in}^4 \times \frac{1\ \text{ft}^2}{144\ \text{in}^2}$$

$$Wk^2 = \frac{\varrho_W L(R_1{}^4 - R_2{}^4)}{91.67}\ \text{lb}\cdot\text{ft}^2$$

Special case for steel: $\varrho_W = 0.283$ lb/in³

$$Wk^2 = \frac{L(R_1{}^4 - R_2{}^4)}{323.9}\ \text{lb}\cdot\text{ft}^2$$

Figure 16-2 Properties of a Hollow Disc

Now the torque required to accelerate the pulley can be computed. Equation (16-2) can be put into a more convenient form by noting that T is usually expressed in lb · ft, Wk^2 in lb · ft^2, n in rpm, and t in sec. Using $g = 32.2$ ft/s^2 and converting units gives

$$T = \frac{Wk^2(\Delta n)}{308t} \text{ lb} \cdot \text{ft} \tag{16-3}$$

Example Problem 16-2. Compute the torque that a clutch must transmit to accelerate the pulley of Figure 16-3 from rest to 550 rpm in 2.50 sec. From example problem 16-1, $Wk^2 = 79.64$ lb · ft^2.

Solution. Using equation (16-3),

$$T = \frac{(79.64)(550)}{308(2.5)} = 56.9 \text{ lb} \cdot \text{ft}$$

Inertia of a System Referred to the Clutch Shaft Speed

In many practical machine systems, there are several elements on several shafts, operating at differing speeds. It is required that the effective inertia of the entire system *as it affects the clutch* be determined. The effective inertia of a connected load operating at rotational speed different from that of the clutch is proportional to the square of the ratio of the speeds. That is:

$$Wk_e^2 = Wk^2\left(\frac{n}{n_c}\right)^2 \tag{16-4}$$

where n is the speed of the load of interest and n_c is the speed of the clutch.

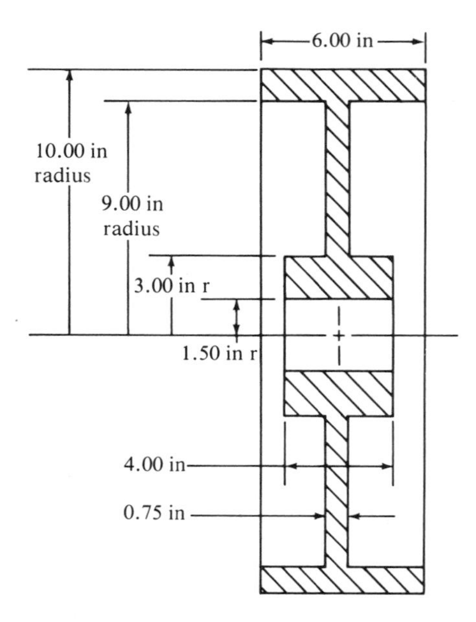

Figure 16-3 Pulley for Example Problem 16-1 and 16-2

Example Problem 16-3. Compute the total inertia of the system in Figure 16-4, as seen by the clutch. Then compute the time required to accelerate the system from rest to a motor speed of 550 rpm if the clutch exerts a torque of 24.0 lb · ft. The Wk^2 for the armature of the clutch, which also must be accelerated, is 0.22 lb · ft².

Solution. The clutch and gear A will be rotating at 550 rpm, but, because of the gear reduction, gear B, its shaft, and the pulley rotate at

$$n_2 = 550\ \text{rpm}(24/66) = 200\ \text{rpm}$$

Now compute the inertia for each element referred to the clutch speed. Assume the gears are discs having outside diameters equal to the pitch diameter of the gear and inside diameters equal to the shaft diameter. The equation in Figure 16-2 for a steel disc will be used to compute Wk^2.

Gear A:

$$Wk^2 = [(2.00)^4 - (0.625)^4](2.50)/323.9 = 0.122\ \text{lb} \cdot \text{ft}^2$$

Gear B:

$$Wk^2 = [(5.50)^4 - (1.50)^4](2.50)/323.9 = 7.02\ \text{lb} \cdot \text{ft}^2$$

But, because of the speed difference, the effective inertia is

$$Wk_e^2 = 7.02(200/550)^2 = 0.93\ \text{lb} \cdot \text{ft}^2$$

Pulley: From example problem 16-1, $Wk^2 = 79.64$ lb · ft². The effective inertia is

$$Wk_e^2 = 79.64(200/550)^2 = 10.53\ \text{lb} \cdot \text{ft}^2$$

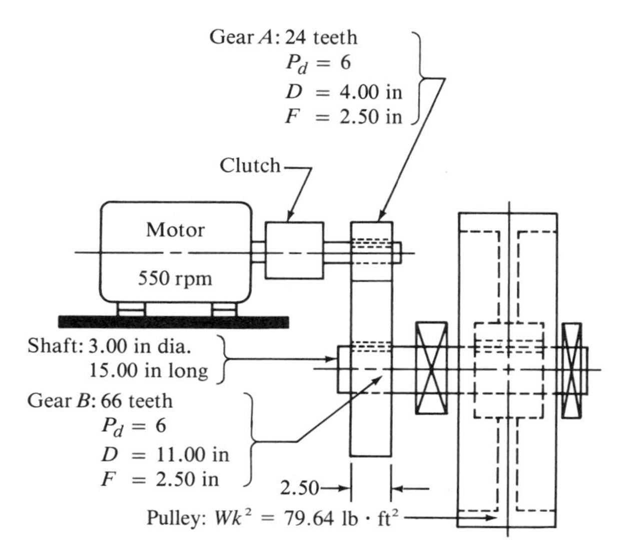

Figure 16-4 System for Example Problem 16-3

Shaft:

$$Wk^2 = (1.50)^4(15.0)/323.9 = 0.234 \text{ lb} \cdot \text{ft}^2$$

The effective inertia is

$$Wk_e^2 = 0.234(200/550)^2 = 0.03 \text{ lb} \cdot \text{ft}^2$$

The total effective inertia as seen by the clutch is

$$Wk_e^2 = 0.22 + 0.12 + 0.93 + 10.53 + 0.03 = 11.83 \text{ lb} \cdot \text{ft}^2$$

Solving equation (16-3) for the time gives

$$t = \frac{Wk_e^2(\Delta n)}{308T} = \frac{(11.83)(550)}{308(24.0)} = 0.88 \text{ sec}$$

Effective Inertia for Bodies Moving Linearly

To this point, we have dealt only with components that rotate. Many systems include linear devices such as conveyors, hoist cables and their loads, or reciprocating racks driven by gears that also have inertia and must be accelerated. It would be convenient to represent these devices with an effective inertia measured by Wk^2 as we have for rotating bodies. This can be accomplished by relating the equations for kinetic energy for linear and rotary motion. The actual kinetic energy of a translating body is

$$KE = \frac{1}{2}mV^2 = \frac{1}{2}\frac{W}{g}V^2 = \frac{WV^2}{2g}$$

where V is the linear velocity of the body. We will use ft/min for the units of velocity. For a rotating body

$$KE = \frac{1}{2}I\omega^2 = \frac{1}{2}\frac{Wk^2}{g}\omega^2 = \frac{Wk^2\omega^2}{2g}$$

Letting Wk^2 be the effective inertia and equating these two formulas gives

$$Wk_e^2 = W\left(\frac{V}{\omega}\right)^2$$

where ω must be in rad/min to be consistent. Using n rpm rather than ω rad/min, we must substitute $\omega = 2\pi n$. Then

$$Wk_e^2 = W\left(\frac{V}{2\pi n}\right)^2 \qquad (16\text{-}5)$$

Example Problem 16-4. The conveyor in Figure 16-5 moves at 80 ft/min. The combined weight of the belt and the parts on it is 140 lb. Compute the equivalent Wk^2 inertia for the conveyor referred to the shaft driving the belt.

Solution. The rotational speed of the shaft is

$$\omega = \frac{V}{R} = \frac{80\text{ ft}}{\text{min}}\,\frac{1}{5.0\text{ in}}\,\frac{12\text{ in}}{\text{ft}} = 192\text{ rad/min}$$

Then, the equivalent Wk^2 would be

$$Wk_e^2 = W\left(\frac{V}{\omega}\right)^2 = (140\text{ lb})\left(\frac{80\text{ ft/min}}{192\text{ rad/min}}\right)^2 = 24.3\text{ lb}\cdot\text{ft}^2$$

Energy Absorption: Heat Dissipation Requirements

When using a brake to stop a rotating object or using a clutch to accelerate it, the clutch or brake must transmit energy through the friction surfaces as they slip in relation to each other. Heat is generated at these surfaces, tending to increase the temperature of the unit. Of course, heat is then dissipated from the unit, and for a given set of operating conditions an equilibrium temperature is achieved. That temperature must be sufficiently low to ensure a long life of the friction elements and other operating parts of the unit, such as electric coils, springs, and bearings.

The energy to be absorbed or dissipated by the unit per cycle is equal to the change in kinetic energy of the components being accelerated or stopped. That is,

$$E = \Delta KE = \frac{1}{2}I\omega^2 = \frac{1}{2}mk^2\omega^2 = \frac{Wk^2\omega^2}{2g}$$

For typical units in the U.S. Customary system ($\omega = n$ rpm; Wk^2 in lb · ft; and $g = 32.2\text{ ft/s}^2$), we get

$$E = \frac{Wk^2(\text{lb}\cdot\text{ft})}{2(32.2\text{ ft/s}^2)}\,\frac{n^2\text{ rev}^2}{\text{min}^2}\,\frac{(2\pi)^2\text{ rad}}{\text{rev}^2}\,\frac{1\text{ min}^2}{60^2\text{ s}^2}$$

$$E = 1.7\times 10^{-4}\,Wk^2n^2\text{ lb}\cdot\text{ft} \qquad (16\text{-}6)$$

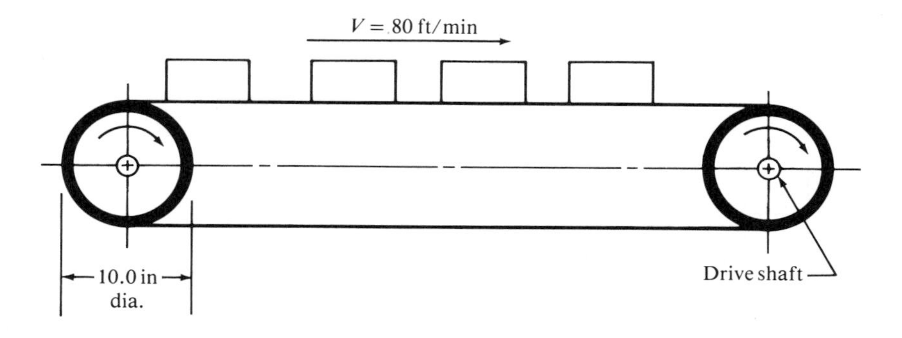

Figure 16-5 Conveyor for Example Problem 16-4

In SI units, mass is in kilograms (kg), radius of gyration is in meters (m), and angular velocity is in radians per second (rad/s). Then

$$E = \frac{1}{2} I\omega^2 = \frac{1}{2} mk^2 \omega^2 \text{ (kg} \cdot \text{m}^2/\text{s}^2)$$

But the newton unit is equal to the kg · m/s². Then,

$$E = \frac{1}{2} mk^2 \omega^2 \text{ N} \cdot \text{m} \tag{16-7}$$

No further conversion factors are required.

If there are repetitive cycles of operation, the energy from equation (16-6) or (16-7) must be multiplied by the cycling rate, usually in cycles/min in U.S. Customary units and cycles/s for SI units. The result would be the energy generation per unit time, which must be compared with the heat dissipation capacity of the clutch or brake being considered for the application.

When the clutch-brake cycles on and off, part of its operation is at the full operating speed of the system and part is at rest. The combined heat dissipation capacity is the average of the capacity at each speed weighted by the proportion of the cycle at each speed (see example problem 16-5).

Response Time

The term *response time* refers to the time required for the unit (clutch or brake) to accomplish its function after action is initiated by the application of an electric current, air pressure, spring force, or manual force. Figure 16-6 shows a complete cycle using a clutch-brake module. The straight line curve is idealized while the curved line gives the general form of system motion. The actual response time will vary, even for a given unit, with variations in the load, environment, or other operating conditions.

Commercially available clutches and brakes for typical machine applications have response times from a few milliseconds (1/1000 sec) for a small device, such as a paper

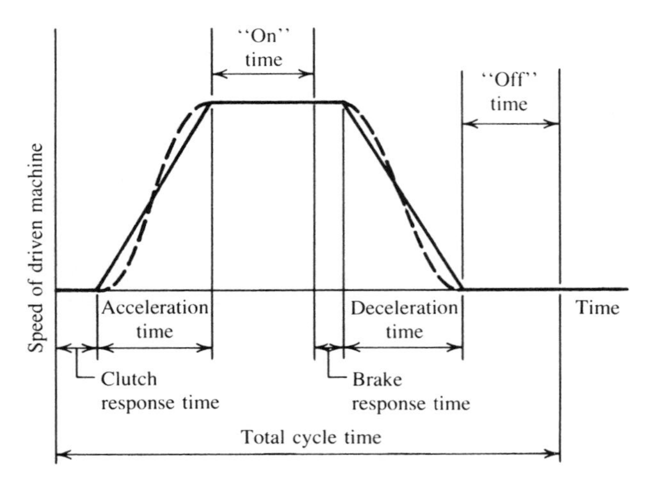

Figure 16-6 Typical Cycle of Engagement and Disengagement for a Clutch

Table 16-1 Clutch-brake Performance Data

UNIT SIZE	TORQUE CAPACITY (lb · ft)	INERTIA Wk^2 (lb · ft²)	HEAT DISSIPATION (ft · lb/min)		RESPONSE TIME (sec)	
			At rest	*1 800 rpm*	*Clutch*	*Brake*
A	0.42	0.000 17	750	800	0.022	0.019
B	1.25	0.001 4	800	1 200	0.032	0.024
C	6.25	0.021	1 050	2 250	0.042	0.040
D	20.0	0.108	2 000	6 000	0.090	0.089
E	50.0	0.420	3 000	13 000	0.110	0.105
F	150.	1.17	9 000	62 000	0.250	0.243
G	240.	2.29	18 000	52 000	0.235	0.235
H	465.	5.54	20 000	90 000	0.350	0.350
I	700.	13.82	26 000	190 000	0.512	0.512

Note: Torque ratings are static. The torque capacity decreases as the speed difference between the parts being engaged increases. Interpolation may be used on heat dissipation data.

transport, to approximately 1.0 sec for larger machines, such as an assembly conveyor. Manufacturers' literature should be consulted.

Sample Clutch-brake Performance Data

To give you a feel for the capabilities of commercially available clutches and brakes, Table 16-1 gives sample data for electrically powered units.

Example Problem 16-5. For the system in Figure 16-4, and using the data from example problem 16-3, estimate the total cycle time if the system is controlled by unit G from Table 16-1 and the system must stay on (at steady speed) for 1.50 sec, and off (at rest) for 0.75 sec; also estimate the response time of the clutch-brake and the acceleration and deceleration times. If the system cycles continuously, compute the heat dissipation rate and compare it with the capacity of the unit.

Solution. Figure 16-7 shows the estimated total cycle time for the system to be 2.896 sec. From Table 16-1, we find that the clutch-brake exerts 240 lb · ft of torque and has a response time of 0.235 sec for both the clutch and the brake.

Acceleration and deceleration time (equation [16-3]):

$$t = \frac{Wk_e^2(\Delta n)}{308T} = \frac{(11.83)(550)}{308(240)} = 0.088 \text{ s}$$

Cycling Rate and Heat Dissipation:

For a total cycle time of 2.896 sec, the number of cycles per minute would be

$$C = \frac{1.0 \text{ cycle}}{2.896 \text{ s}} \frac{60 \text{ s}}{\text{min}} = 20.7 \text{ cycles/min}$$

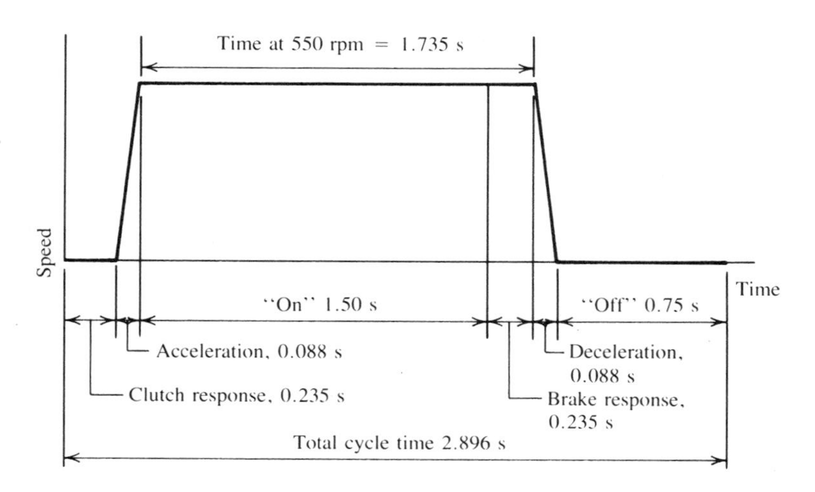

Figure 16-7 Cycle Time for Example Problem 16-5

The energy generated with each engagement of either the clutch or the brake is

$$E = 1.7 \times 10^{-4}\, Wk^2 n^2 = 1.7 \times 10^{-4}(11.83)(550)^2 = 608 \text{ ft} \cdot \text{lb}$$

The energy generation per minute is

$$E_t = 2EC = (2)(608 \text{ ft} \cdot \text{lb/cycle})(20.7 \text{ cycles/min}) = 25\,200 \text{ ft} \cdot \text{lb/min}$$

This is greater than the heat dissipation capacity of unit G at rest (18 000 ft · lb/min). Then let's compute a weighted average capacity for this cycle. First, referring to Figure 16-7, approximately 1.735 sec is "at speed", 550 rpm. The balance of the cycle, 1.161 sec, is at rest. From Table 16-1, and interpolating between zero speed and 1 800 rpm, the heat dissipation rate at 550 rpm is approximately 28 400 ft · lb/min. Then the weighted average capacity for unit G is

$$E_{avg} = \frac{t_0}{t_t} E_0 + \frac{t_{550}}{t_t} E_{550}$$

where

$$t_t = \text{Total cycle time}$$
$$t_0 = \text{Time at rest} \quad (0 \text{ rpm})$$
$$t_{550} = \text{Time at 550 rpm}$$
$$E_0 = \text{Heat dissipation capacity at rest}$$
$$E_{550} = \text{Heat dissipation capacity at 550 rpm}$$

Then

$$E_{avg} = \frac{1.161}{2.896}(18\,000) + \frac{1.735}{2.896}(28\,400) = 24\,230 \text{ ft} \cdot \text{lb/min}$$

This is a little lower than required and the design would be marginal. Fewer cycles per minute should be specified.

16-3 TYPES OF FRICTION CLUTCHES AND BRAKES

Clutches and brakes that use friction surfaces as the means of transmitting the torque to start or stop a mechanism can be classified according to the general geometry of the friction surfaces and by the means used to actuate them. In some cases, the same basic geometry can be used either as a clutch or a brake by selectively attaching the friction elements to the driver, the driven machine, or the stationary frame of the machine.

The following types of clutches and brakes are sketched in Figure 16-8.

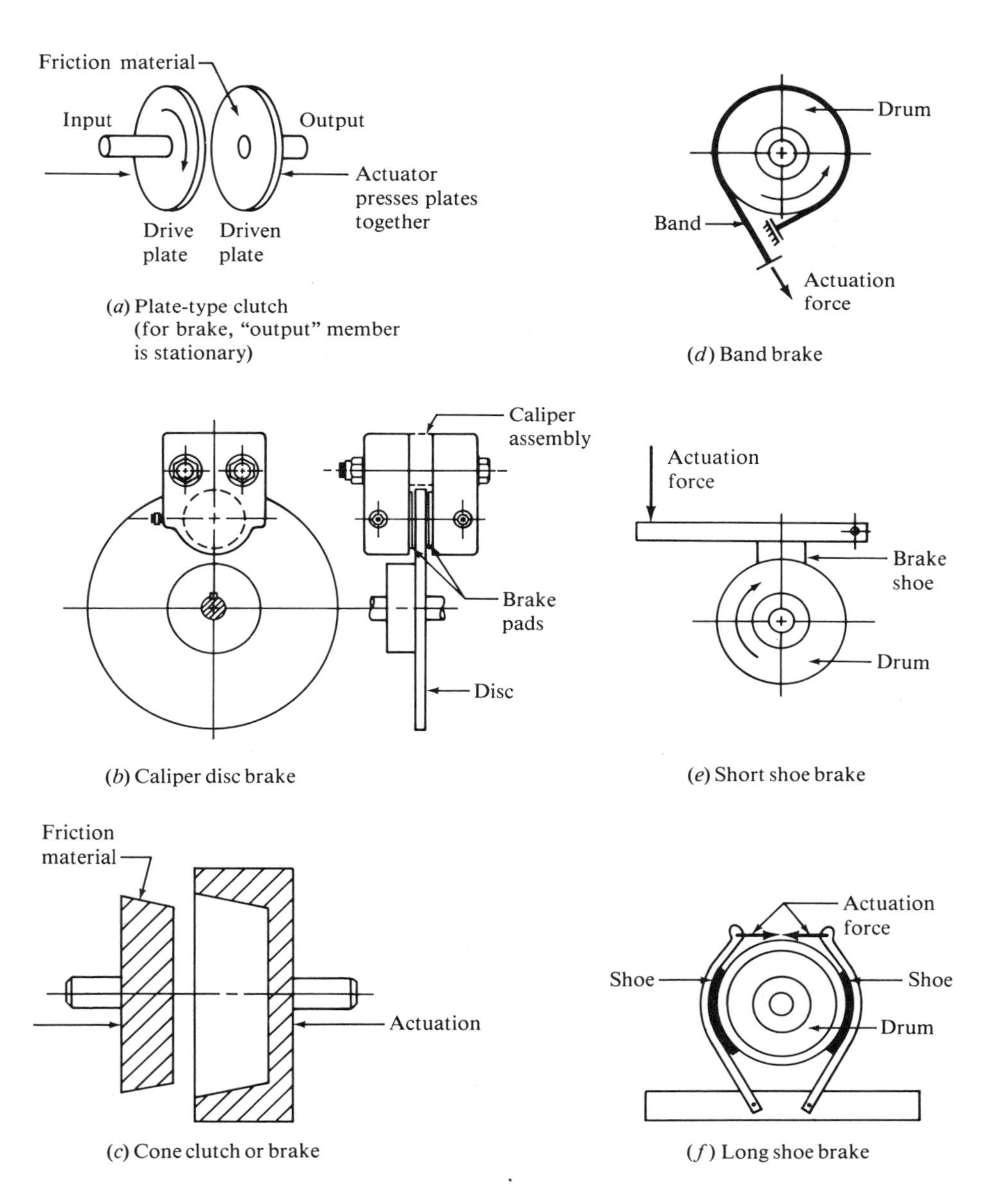

Figure 16-8 Types of Friction Clutches and Brakes (b, Tol-O-Matic, Minneapolis, Minn.)

Plate Clutch or Brake

Each friction surface is in the shape of an annulus on a flat plate. One or more friction plates moves axially to contact a mating smooth plate, usually made of steel, to which the friction torque is transmitted.

Caliper Disc Brake

A disc-shaped rotor is attached to the machine to be controlled. Friction pads covering only a small portion of the disc are contained in a fixed assembly called a *caliper* and are forced against the disc by air pressure or hydraulic pressure.

Cone Clutch or Brake

A cone clutch or brake is similar to a plate clutch or brake except that the mating surfaces are on a portion of a cone instead of on a flat plate.

Band Brake

Used only as a brake, the friction material is on a flexible band that nearly surrounds a cylindrical drum attached to the machine to be controlled. When braking is desired, the band is tightened on the drum, exerting a tangential force to stop the load.

Block or Shoe Brake

Curved rigid pads faced with the friction material are forced against the surface of a drum, either from the outside or the inside, exerting a tangential force to stop the load.

Actuation

The following are the means used to actuate clutches or brakes. Each may be applied to several of the types described. Figures 16-9 through 16-15 show a variety of commercially available designs.

Manual. The operator provides the force, usually through a lever arrangement to achieve force multiplication.

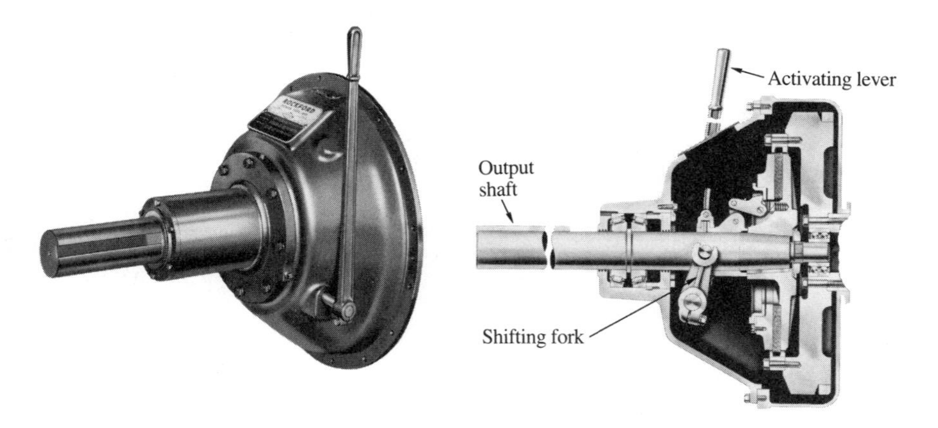

Figure 16-9 Manually Operated Clutch (Rockford Division, Borg-Warner Corp., Rockford, Ill.)

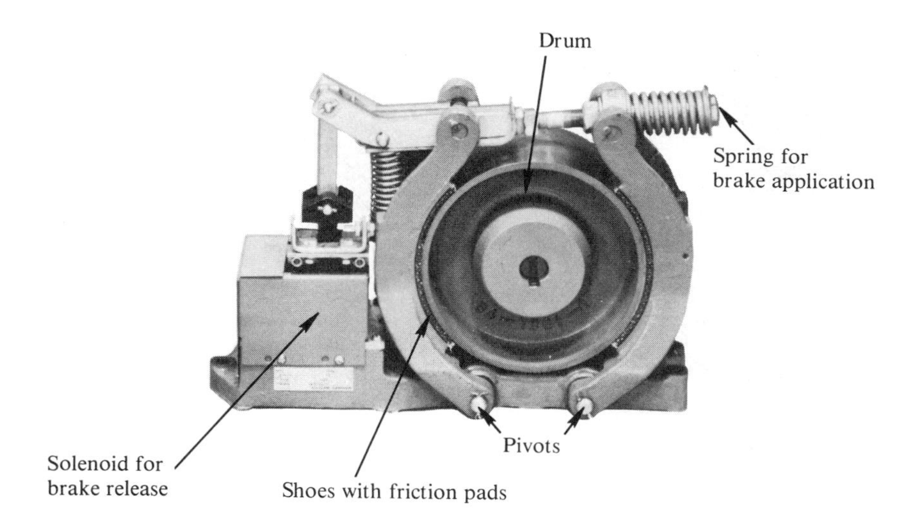

Figure 16-10 Spring Applied, Electrically Released, Long Shoe Brake (Eaton Corp., Cutler-Hammer Products, Milwaukee, Wis.)

Spring-applied. Sometimes called a *fail-safe* design when applied to a brake, the brake is applied automatically by springs unless there is some opposing force present. Thus if power fails, or if air pressure or hydraulic pressure is lost, or if the operator is unable to

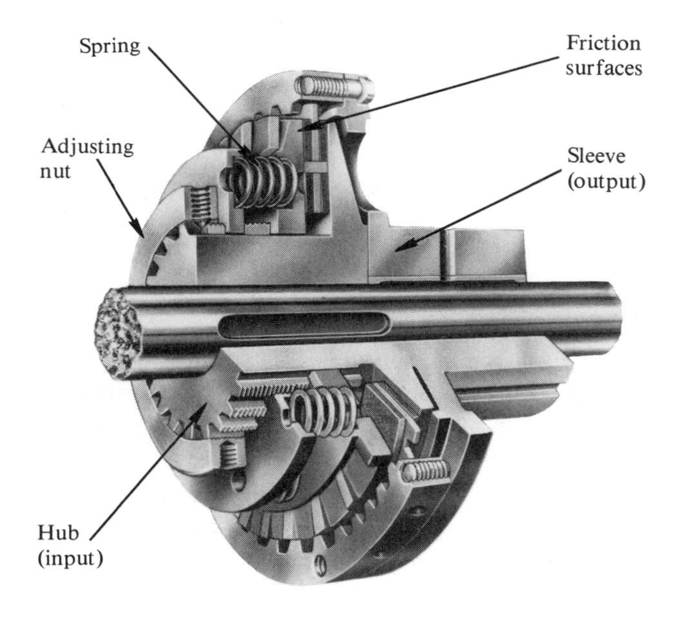

Figure 16-11 Slip Clutch. Springs apply normal pressure on friction plates; spring force adjustable to vary torque level at which clutch will slip. (The Hilliard Corp., Elmira, N.Y.)

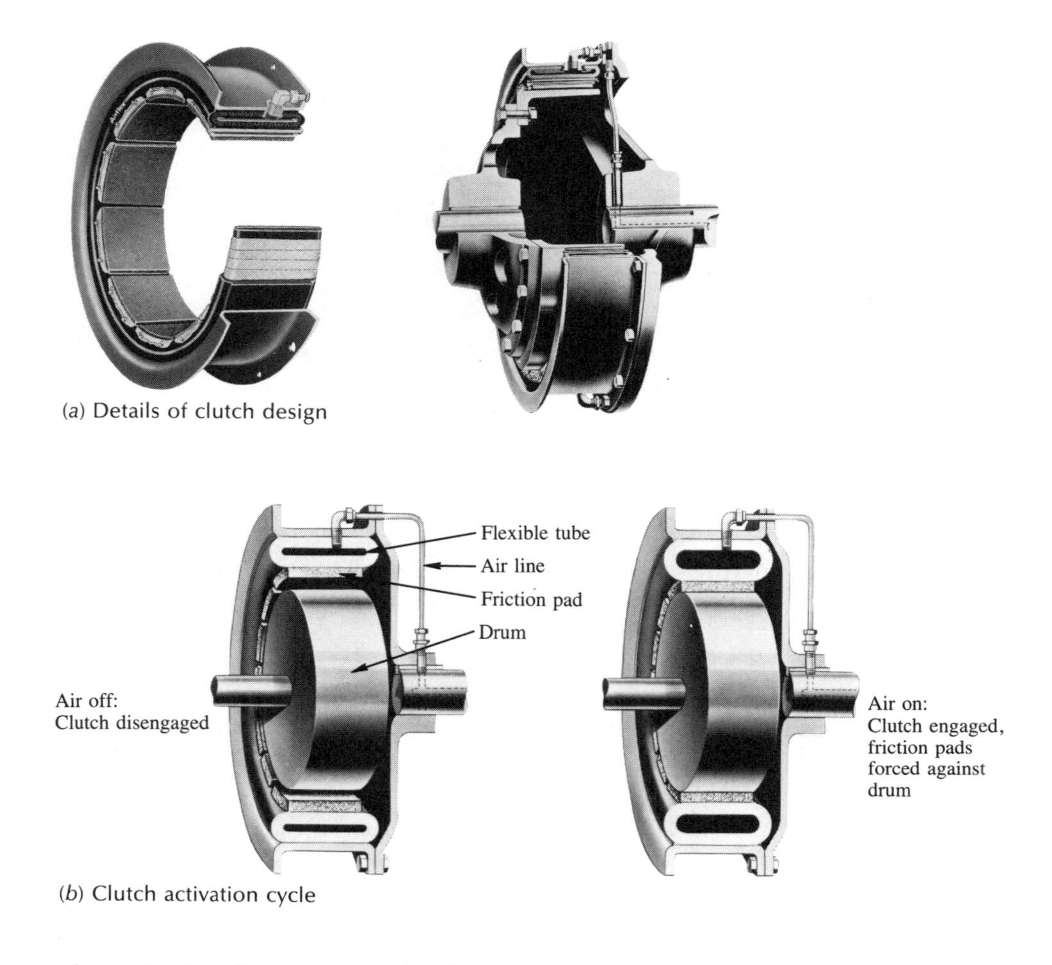

Figure 16-12 Air-actuated Clutch or Brake (Eaton Corp., Airflex Division, Cleveland, Ohio)

perform the function, the springs apply the brake and stop the load. The concept may also be used to engage or to disengage a clutch.

Centrifugal. A centrifugal clutch is sometimes employed to permit the driving system to accelerate without a connected load. Then, at a preselected speed, centrifugal force moves the clutch elements into contact to connect the load. As the system slows, the load would be automatically disconnected.

Pneumatic. Compressed air is introduced into a cylinder or some other chamber. The force produced by the pressure on a piston or a diaphragm brings the friction surfaces into contact with the members connected to the load.

Hydraulic. Similar to the pneumatic type except that it uses oil hydraulic fluids instead of air, the hydraulic actuator is usually applied where high actuation forces are required.

Electromagnetic. An electric current is applied to a coil, creating an electromagnetic flux. The magnetic force then attracts an armature attached to the machine that is to be controlled. The armature is usually of the plate type.

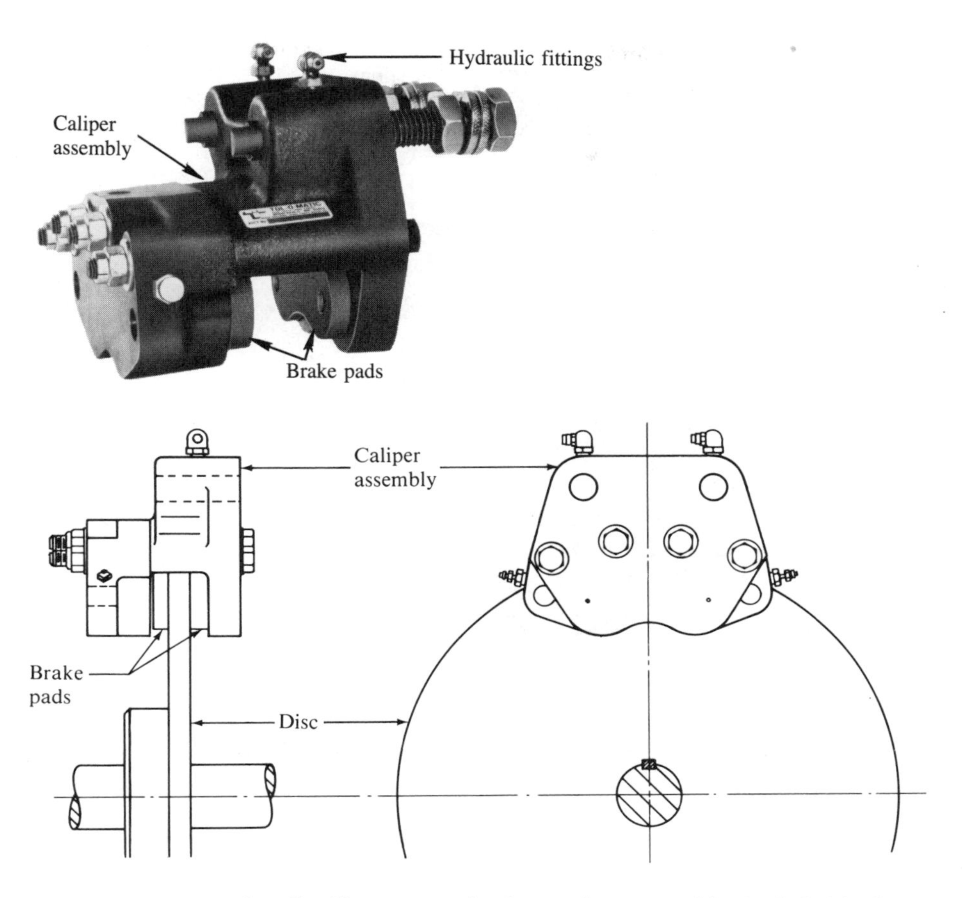

Figure 16-13 Hydraulically Actuated Disc Brake Assembly (Tol-O-Matic, Minneapolis, Minn.)

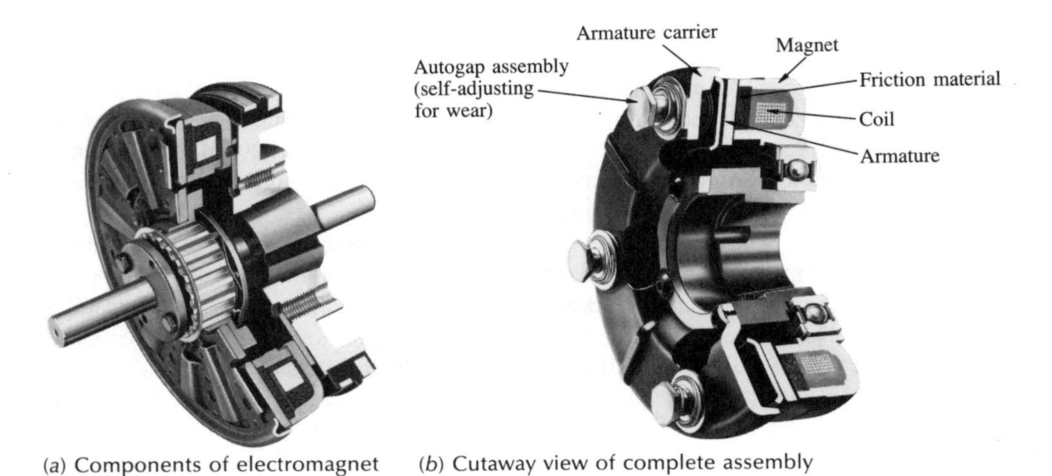

(*a*) Components of electromagnet (*b*) Cutaway view of complete assembly

Figure 16-14 Electrically Actuated Plate Type Clutch or Brake (Warner Electric Brake & Clutch Co., South Beloit, Ill.)

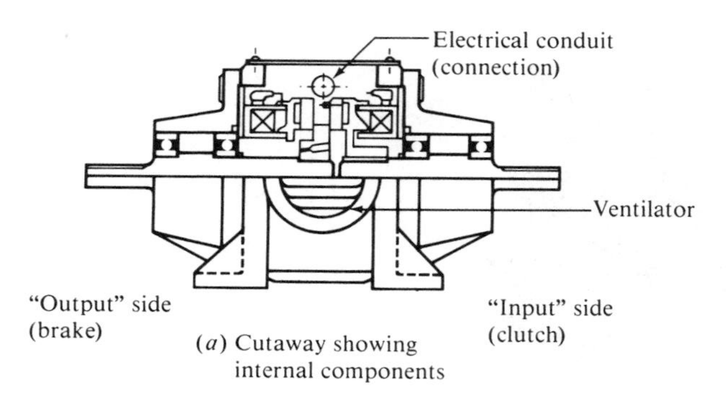

(*a*) Cutaway showing internal components

Figure 16-15 Electrically Operated Clutch/Brake Module (Electroid Corp., Springfield, N.J.)

16-4 PLATE TYPE CLUTCH OR BRAKE

When two bodies are brought into contact with a normal force between them, a friction force that tends to resist relative motion is created. This is the principle on which the plate type clutch or brake is based.

As the friction plate rotates in relation to the mating plate with an axial force pressing them together, the friction force acts in a tangential direction, producing the brake or clutch torque. At any point, the local pressure times the differential area at the point is the *normal force*. The normal force times the coefficient of friction is the *friction force*. The friction force times the radius of the point is the *torque* produced at this point. The *total torque* is the sum of all the torques over the entire area of the plate. The sum can be found by integrating over the area.

There is usually some variation of pressure over the surface of the friction plate, and some assumption about the nature of the variation must be made before the total torque can be computed. A conservative assumption that yields a useful result is that the friction surface will wear uniformly over the entire area as the brake or clutch operates. This assumption implies that the product of the local pressure (p) times the linear relative speed (V) between the plates is constant. Wear has been found to be approximately proportional to the product of p times V.

Taking all these factors into account and completing the analysis produces the following result for friction torque:

$$T_f = fN(R_o + R_i)/2$$

But the last part of this relation is the mean radius, R_m, of the annular plate. Then,

$$T_f = fNR_m \tag{16-8}$$

As stated before, this is a conservative result, meaning that the actual torque produced would be slightly larger than predicted.

Note that the torque is proportional to the mean radius but that no area relationship is involved in equation (16-8). Thus the completion of the design for final dimensions requires some other parameter. The missing factor in equation (16-8) is the expected wear rate of the friction material. It should be obvious that even with the same mean radius, a brake with a larger area would wear less than one with a smaller area.

Manufacturers of friction materials can assist in the final determination of the relationship between wear and the area of the friction surface. However, the following guidelines allow the estimation of the physical size of brakes and will be used for problem solutions in this book.

The wear rating (WR) will be based on the frictional power (P_f) absorbed by the brake per unit area (A), where

$$P_f = T_f\omega \tag{16-9}$$

and ω is the angular velocity of the disc. In SI units, with torque in N · m and ω in rad/s, the frictional power is in N · m/s or watts. In the U.S. Customary system, with torque in lb · in and angular velocity expressed as n rpm, the frictional power is in hp, computed from

$$P_f = \frac{T_f n}{63\,000}\ \text{hp} \tag{16-10}$$

For industrial applications, we will use

$$WR = P_f/A \tag{16-11}$$

$WR = 0.04$ hp/in^2 for frequent applications, a conservative rating

$WR = 0.10$ hp/in^2 for average service

$WR = 0.40$ hp/in^2 for infrequently used brakes allowed to cool somewhat between applications

Coefficient of Friction

Values for the coefficient of friction of commercially available friction materials vary widely with composition, speed, pressure, cleanliness, and type of mating material. We will use $f = 0.25$ as a design value that is conservative on the low side for most materials.

Example Problem 16-6. Compute the dimensions of an annular plate type brake to produce a braking torque of 300 lb · in. Springs will provide a normal force of 320 lb between the friction surfaces. The coefficient of friction is 0.25. The brake will be used in average industrial service, stopping a load from 750 rpm.

Solution.

1. Compute the required mean radius. From equation (16-8),

$$R_m = \frac{T_f}{fN} = \frac{300 \text{ lb} \cdot \text{in}}{(0.25)(320 \text{ lb})} = 3.75 \text{ in}$$

2. Specify a desired ratio of R_o/R_i and solve for the dimensions. A reasonable value for the ratio is approximately 1.50. The range can be from 1.2 to about 2.5, at the designer's choice. Using 1.50, $R_o = 1.50R_i$, and,

$$R_m = (R_o + R_i)/2 = (1.5R_i + R_i)/2 = 1.25R_i$$

Then

$$R_i = R_m/1.25 = (3.75 \text{ in})/1.25 = 3.00 \text{ in}$$
$$R_o = 1.50R_i = 1.50(3.00) = 4.50 \text{ in}$$

3. Compute the area of the friction surface.

$$A = \pi(R_o^2 - R_i^2) = \pi[(4.50)^2 - (3.00)^2] = 35.3 \text{ in}^2$$

4. Compute the frictional power absorbed.

$$P_f = \frac{T_f n}{63\,000} = \frac{(300)(750)}{63\,000} = 3.57 \text{ hp}$$

5. Compute the wear ratio.

$$WR = \frac{P_f}{A} = \frac{3.57 \text{ hp}}{35.3 \text{ in}^2} = 0.101 \text{ hp/in}^2$$

6. Judge the suitability of *WR*. If *WR* is too high, return to step 2 and increase the ratio. If *WR* is too low, decrease the ratio. In this example, *WR* is acceptable.

A more compact unit can be designed if more than one friction plate is used. The friction torque for one plate is multiplied by the number of plates to determine the total friction torque. One disadvantage of this approach is that the heat dissipation is relatively poorer than for the single plate.

16-5 CALIPER DISC BRAKES

The disc brake pads are brought into contact with the rotating disc by fluid pressure acting on a piston in the caliper. The pads are either round or in a short crescent shape to cover more of the disc surface (see Figure 16-8b and Figure 16-13). However, one advantage of the disc brake is that the disc is exposed to the atmosphere, promoting the dissipation of heat. Because the disc rotates with the machine to be controlled, heat dissipation is further enhanced. The cooling effect improves the fade resistance of this type of brake relative to the shoe type brake.

Design for the friction torque and wear rating are similar to that explained for plate type brakes.

16-6 CONE CLUTCH OR BRAKE

The angle of inclination of the conical surface of the cone clutch or brake is typically 12°. A lower angle could be used with care, but there is a tendency for the friction surfaces to engage suddenly with a jerk. As the angle increases, the amount of axial force required to produce a given friction torque increases. Thus, 12° is a reasonable compromise.

Referring to Figure 16-16, we see that, as an axial force F_a is applied by a spring, manually or by fluid pressure, a normal force N is created between the mating friction surfaces, all around the periphery of the cone. The desired friction force F_f is produced in the tangential direction, where $F_f = fN$. It is assumed that the friction force acts at the mean radius of the cone so that the friction torque is

$$T_f = F_f R_m = fNR_m \tag{16-8}$$

In addition to the tangentially directed friction force, a friction force develops along the surface of the cone and opposes the tendency for the member having the internal cone surface to move axially away from the external cone. We will call this force F_f' and compute it also from

$$F_f' = fN$$

For the equilibrium condition of the external cone, the sum of the horizontal forces must be zero. Then,

$$F_a = N \sin \alpha + F_f' \cos \alpha = N \sin \alpha + fN \cos \alpha = N(\sin \alpha + f \cos \alpha)$$

or

$$N = \frac{F_a}{\sin \alpha + f \cos \alpha} \tag{16-12}$$

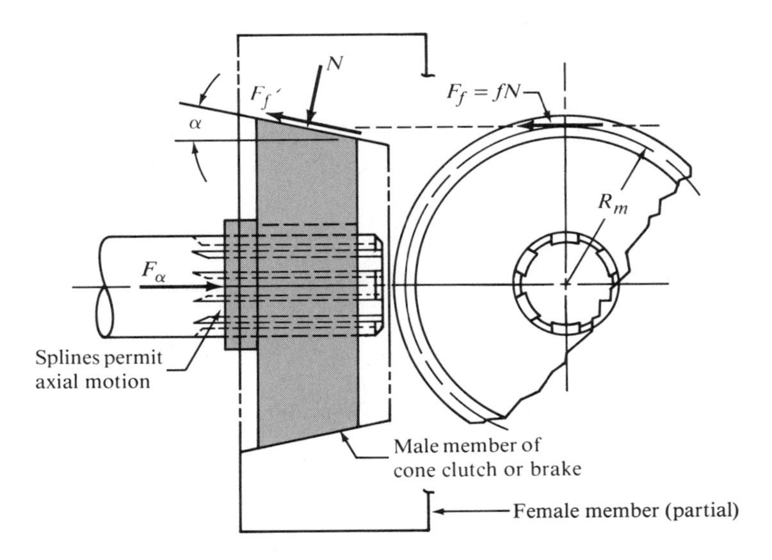

Figure 16-16 Cone Clutch or Brake

Substituting this into equation (16-8) gives

$$T_f = \frac{fR_m F_a}{\sin \alpha + f \cos \alpha} \tag{16-13}$$

Example Problem 16-7. Compute the axial force required for a cone brake if it is to exert a braking torque of 50 lb · ft. The mean radius of the cone is 5.0 in. Use $f = 0.25$. Try cone angles of 10°, 12°, and 15°.

Solution. We can solve equation (16-13) for the axial force F_a.

$$F_a = \frac{T_f(\sin \alpha + f \cos \alpha)}{fR_m} = \frac{(50 \text{ lb} \cdot \text{ft})(\sin \alpha + 0.25 \cos \alpha)}{(0.25)(5.0/12) \text{ ft}}$$

$$F_a = 480(\sin \alpha + 0.25 \cos \alpha) \text{ lb}$$

Then the values of F_a as a function of the cone angle are

For $\alpha = 10°$, $F_a = 202$ lb

For $\alpha = 12°$, $F_a = 217$ lb

For $\alpha = 15°$, $F_a = 240$ lb

16-7 DRUM BRAKES

Short Shoe Drum Brake

Figure 16-17 shows a sketch of a drum brake in which the actuating force W acts on the lever which pivots on pin A. This creates a normal force between the shoe and the rotating drum. The resulting friction force is assumed to act tangential to the drum if the shoe is short. The friction force times the radius of the drum is the friction torque which slows the drum.

The objectives of the analysis are to determine the relationship between the applied load and the friction force and to be able to evaluate the effect of design decisions such as the size of the drum, the lever dimensions, and the placement of the pivot A. The free body diagrams in Figure 16-17(a) support this analysis. For the lever, we can sum moments about the pivot A.

$$\sum M_a = 0 = WL - Na + F_f b \tag{16-14}$$

But note that $F_f = fN$ or $N = F_f/f$. Then,

$$0 = WL - F_f a/f + F_f b = WL - F_f(a/f - b)$$

Solving for W gives

$$W = \frac{F_f(a/f - b)}{L} \tag{16-15}$$

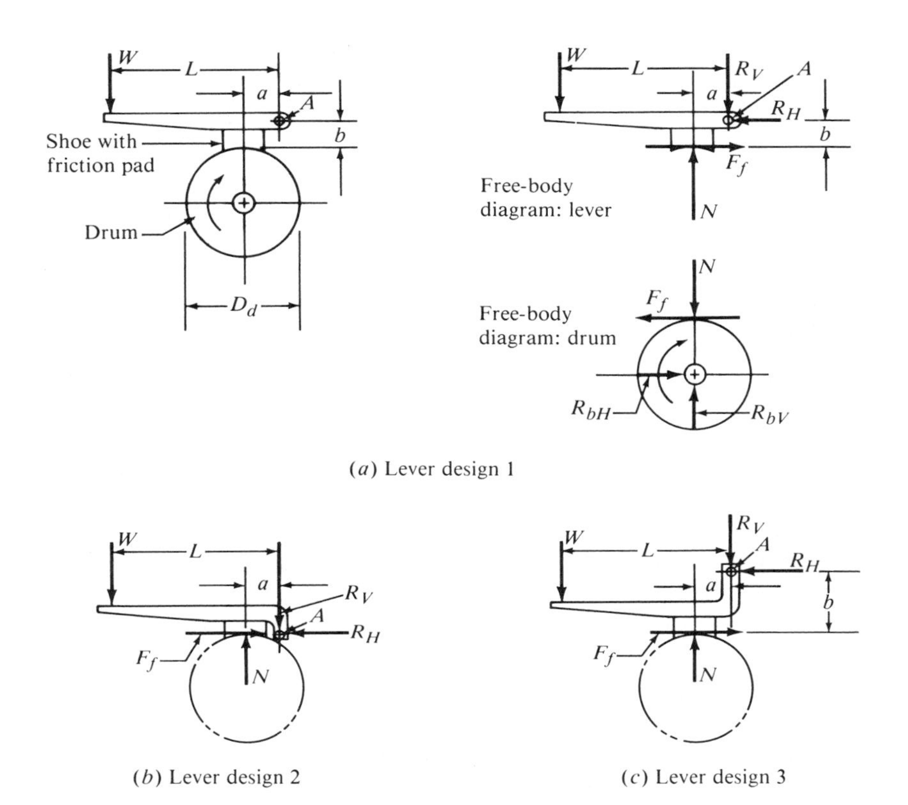

Figure 16-17 Short Shoe Drum Brake

Solving for F_f gives

$$F_f = \frac{WL}{(a/f - b)} \tag{16-16}$$

These equations can be used for the friction torque by noting

$$T_f = F_f D_d/2 \tag{16-17}$$

where D_d is the diameter of the drum.

Note the alternate positions of the pivot in parts (b) and (c) of Figure 16-17. In (b), the dimension $b = 0$.

Example Problem 16-8. Compute the actuation force required for the short shoe drum brake of Figure 16-17 to produce a friction torque of 50 lb · ft. Use a drum diameter of 10.0 in, $a = 3.0$ in, and $L = 15.0$ in. Consider values of f of 0.25, 0.50, and 0.75 and different points of location of the pivot A such that b ranges from 0 to 6.0 in.

Solution. The required friction force can be found from equation (16-17).

$$F_f = 2T_f/D_d = (2)(50 \text{ lb} \cdot \text{ft})/(10/12 \text{ ft}) = 120 \text{ lb}$$

In equation (16-15), we can substitute for a, L, and F_f.

$$W = \frac{F_f(a/f - b)}{L} = \frac{120 \text{ lb}[(3.0 \text{ in})/f - b]}{15.0 \text{ in}} = 8(3/f - b) \text{ lb}$$

The varying values of f and b can be substituted into this last equation to compute the data for the curves of Figure 16-18 showing the actuating load versus the distance b for different values of f. Note that for some combinations, the value of W is *negative*. This means that the brake is *self-actuating* and that an upward force on the lever would be required to release the brake.

Long Shoe Drum Brakes

The assumption used for short shoe brakes, that the resultant friction force acted at the middle of the shoe, cannot be used in the case of shoes covering more than about 45° of the drum. In such cases, the pressure between the friction lining and the drum is very nonuniform, as is the moment of the friction force and of the normal force with respect to the pivot of the shoe.

The following equations govern the performance of a long shoe brake, using the terminology from Figure 16-19 (3).

1. Friction torque on drum

$$T_f = r^2 f w p_{\text{max}}(\cos \theta_1 - \cos \theta_2) \tag{16-18}$$

2. Actuation force

$$W = (M_N + M_f)/L \tag{16-19}$$

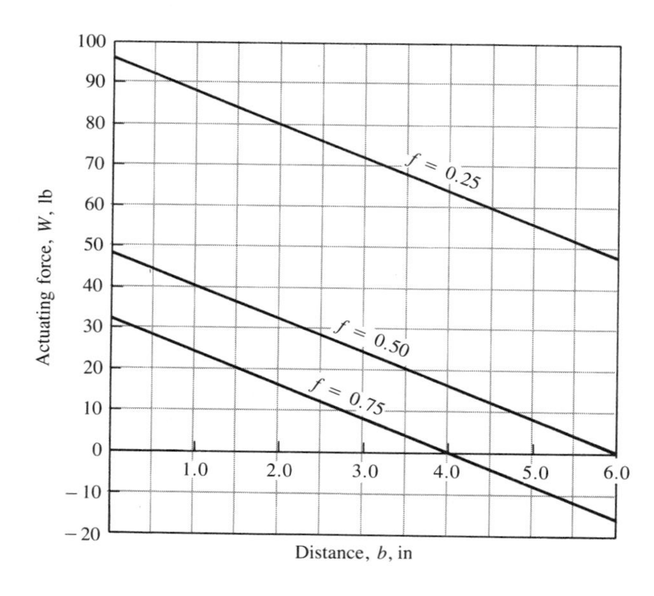

Figure 16-18 Results From Example Problem 16-8

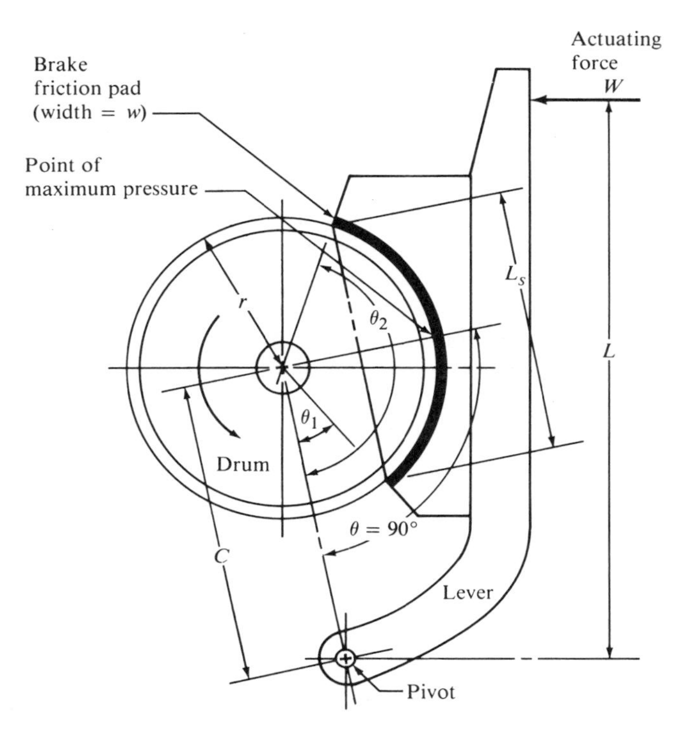

Figure 16-19 Terminology For Long Shoe Drum Brake

where

M_N = Moment of normal force with respect to the hinge pin

$$M_N = 0.25p_{max}wrC[2(\theta_2 - \theta_1) - \sin 2\theta_2 + \sin 2\theta_1] \quad (16\text{-}20)$$

M_f = Moment of friction force with respect to the hinge pin

$$M_f = fp_{max}wr[r(\cos\theta_1 - \cos\theta_2) + 0.25C(\cos 2\theta_2 - \cos 2\theta_1)] \quad (16\text{-}21)$$

The sign of M_f is negative (−) if the drum surface is moving away from the pivot and positive (+) if it is moving toward the pivot.

3. Friction power

$$P_f = T_f n/63\,000 \text{ hp} \quad (16\text{-}22)$$

where n = rotational speed in rpm.

4. Brake shoe area (Note: Projected area is used.)

$$A = L_s w = 2wr \sin[(\theta_2 - \theta_1)/2] \quad (16\text{-}23)$$

5. Wear ratio

$$WR = P_f/A \quad (16\text{-}24)$$

The use of these relationships in the design and analysis of a long shoe brake is shown in example problem 16-9.

Example Problem 16-9. Design a long shoe drum brake to produce a friction torque of 750 lb · in to stop a drum from 120 rpm.

Solution.

1. Select a brake friction material and specify the desired maximum pressure and the design value for the coefficient of friction. Table 16-2 lists some general properties for friction materials. Actual test values or specific manufacturer's data should be used where possible. The design value for p_{max} should be far less than the permissible pressure listed in Table 16-2 in order to improve wear life.

   For this problem, let's choose a molded asbestos material and design for approximately 75 psi maximum pressure. Note, as shown in Figure 16-19, that the maximum pressure occurs at the section 90° from the pivot. If the shoe does not extend at least 90°, the equations used here are not valid (3). Also, let's use $f = 0.25$ for design.
2. Propose trial values for the geometry of the brake drum and the brake pad. Several design decisions must be made here. The general arrangement shown in Figure 16-19 can be used as a guide. But your specific application and creativity may lead you to modify the arrangement.

   Trial values: $r = 4.0$ in; $C = 8.0$ in; $L = 15$ in; $\theta_1 = 30°$; $\theta_2 = 150°$.
3. Solve for the required width of the shoe from equation (16-18).

$$w = \frac{T_f}{r^2 f p_{max}(\cos\theta_1 - \cos\theta_2)}$$

   for this problem.

$$w = \frac{750 \text{ lb} \cdot \text{in}}{(4.0 \text{ in})^2(0.25)(75 \text{ lb/in}^2)(\cos 30° - \cos 150°)} = 1.44 \text{ in}$$

   For convenience, let $w = 1.50$ in. Because the maximum pressure is proportional to the width, the actual maximum pressure will be

$$p_{max} = 75 \text{ psi}(1.44/1.50) = 72 \text{ psi}$$

4. Compute M_N from equation (16-20). The value of $\theta_2 - \theta_1$ must be in radians, with π radians = 180°. Then

$$\theta_2 - \theta_1 = 120°(\pi \text{ rad}/180°) = 2.09 \text{ rad}$$

Table 16-2 Properties of Friction Materials

MATERIAL	f (DRY)	PERMISSIBLE PRESSURE		MAXIMUM TEMPERATURE	
		psi	*kPa*	°*F*	°*C*
Woven asbestos	0.25–0.45	100	690	500	260
Molded asbestos	0.25–0.45	300	2 070	500	260
Sintered metal	0.15–0.45	300	2 070	1 250	675

The moment of the normal force on the shoe is

$$M_N = 0.25(72\ \text{lb/in}^2)(1.50\ \text{in})(4.0\ \text{in})(8.0\ \text{in})$$
$$[2(2.09) - \sin(300°) + \sin(60°)]$$
$$M_N = 5\,108\ \text{lb} \cdot \text{in}$$

5. Compute the moment of the friction force on the shoe, M_f, from equation (16-21).

$$M_f = 0.25(72\ \text{lb/in}^2)(1.50\ \text{in})(4.0\ \text{in})$$
$$[(4.0\ \text{in})(\cos 30° - \cos 150°)$$
$$+ 0.25(8.0\ \text{in})(\cos 300° - \cos 60°)]$$
$$M_f = 748\ \text{lb} \cdot \text{in}$$

6. Compute the required actuation force, W, from equation (16-19).

$$W = (M_N - M_f)/L = (5\,108 - 748)/(15) = 291\ \text{lb}$$

Note the minus sign because the drum surface is moving away from the pivot.

7. Compute the frictional power from equation (16-22).

$$P_f = T_f n/(63\,000) = (750)(120)/(63\,000) = 1.43\ \text{hp}$$

8. Compute the projected area of the brake shoe from equation (16-23).

$$A = L_s w = 2wr \sin[(\theta_2 - \theta_1)/2]$$
$$A = 2(1.50\ \text{in})(4.0\ \text{in}) \sin(120°/2) = 10.4\ \text{in}^2$$

9. Compute the wear ratio, WR.

$$WR = P_f/A = 1.43\ \text{hp}/10.4\ \text{in}^2 = 0.14\ \text{hp/in}^2$$

10. Evaluate the suitability of the results. In this problem, more information about the application would be needed to evaluate the results. However, the wear ratio seems reasonable for average service (see section 16-4), and the geometry seems acceptable.

16-8 BAND BRAKES

Figure 16-20 shows the typical configuration of a band brake. The flexible band, usually made of steel, is faced with a friction material that can conform to the curvature of the drum. The application of a force to the lever puts tension in the band and forces the friction material against the drum. The normal force, thus created, causes the friction force tangential to the drum surface to be created, retarding the drum.

The tension in the band decreases from the value P_1 at the pivot side of the band to P_2 at the lever side. The net torque on the drum is then

$$T_f = (P_1 - P_2)r \tag{16-25}$$

where r is the radius of the drum.

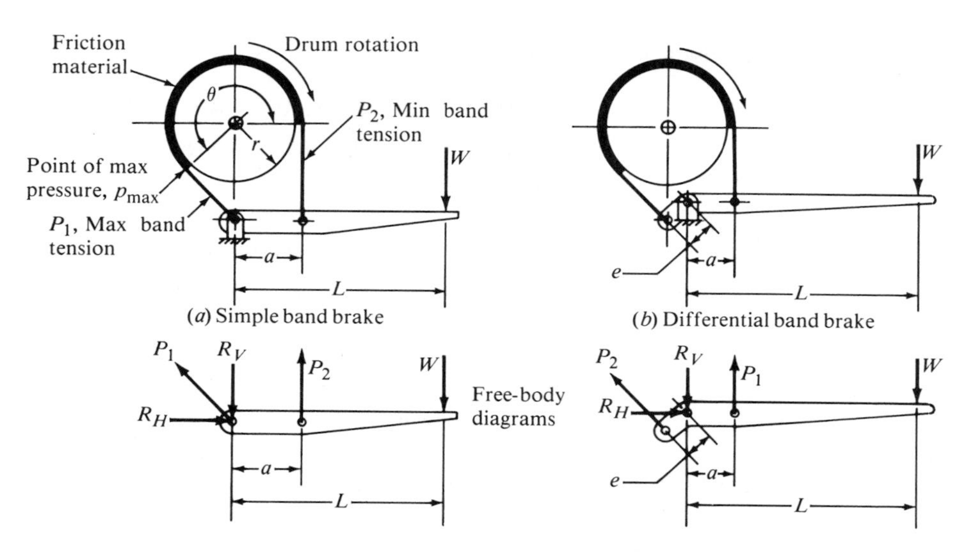

Figure 16-20 Band Brake Design

The relationship between P_1 and P_2 can be shown (3) to be the logarithmic function

$$P_2 = P_1/e^{f\theta} \tag{16-26}$$

where θ is the total angle of coverage of the band in radians.

The point of maximum pressure on the friction material occurs at the end nearest the highest tension, P_1, where

$$P_1 = p_{max}rw \tag{16-27}$$

and w is the width of the band.

For the two types of band brakes shown in Figure 16-20, the free body diagrams of the levers can be used to show the following relationships for the actuating force, W, as a function of the tensions in the band. For the simple band brake of Figure 16-20(a),

$$W = P_2a/L \tag{16-28}$$

The style shown in Figure 16-20(b) is called a *differential band brake,* where the actuation force is

$$W = (P_2a - P_1e)/L \tag{16-29}$$

The design procedure is presented in the context of example problem 16-10.

Example Problem 16-10. Design a band brake to exert a braking torque of 720 lb · in while slowing the drum from 120 rpm.

Solution.

1. Select a material and specify a design value for the maximum pressure. A woven friction material is desirable to facilitate the conformance to the cylindrical drum shape. Let's use $p_{max} = 25$ psi and a design value of $f = 0.25$.

2. Specify a trial geometry: r, θ, w. For this problem, let's try $r = 6.0$ in, $\theta = 225°$, $w = 2.0$ in. Note that $225° = 3.93$ radians.
3. Compute the maximum band tension, P_1, from equation (16-27).

$$P_1 = p_{max}rw = (25 \text{ lb/in}^2)(6.0 \text{ in})(2.0 \text{ in}) = 300 \text{ lb}$$

4. Compute tension P_2 from Equation (16-26).

$$P_2 = \frac{P_1}{e^{f\theta}} = \frac{300 \text{ lb}}{e^{(0.25)(3.93)}} = 112 \text{ lb}$$

5. Compute the friction torque, T_f.

$$T_f = (P_1 - P_2)r = (300 - 112)(6.0) = 1\,128 \text{ lb} \cdot \text{in}$$

Note: Repeat steps 2 to 5 until you achieve a satisfactory geometry and friction torque. Let's try a smaller design, say $r = 5.0$ in.

$$P_1 = (25)(5.0)(2.0) = 250 \text{ lb}$$

$$P_2 = \frac{250 \text{ lb}}{e^{(0.25)(3.93)}} = 93.7 \text{ lb}$$

$$T_f = (250 - 93.7)(5.0) = 782 \text{ lb} \cdot \text{in} \quad \text{(OK)}$$

6. Specify the geometry of the lever and compute the required actuation force. Let's use $a = 5.0$ in; $L = 15.0$ in. Then,

$$W = P_2(a/L) = 93.7 \text{ lb } (5.0/15.0) = 31.2 \text{ lb}$$

7. Compute the average wear ratio from $WR = P_f/A$.

$$A = 2\pi rw(\theta/360) = 2(\pi)(5.0 \text{ in})(2.0 \text{ in})(225/360) = 39.3 \text{ in}^2$$

$$P_f = T_f n/63\,000 = (782)(120)/(63\,000) = 1.50 \text{ hp}$$

$$WR = P_f/A = (1.50 \text{ hp})/(39.3 \text{ in}^2) = 0.038 \text{ hp/in}^2$$

This should be conservative for average service.

16-9 OTHER TYPES OF CLUTCHES AND BRAKES

The chapter thus far has concentrated on clutches and brakes using friction materials to transmit the torque between rotating members. Many other types are available. Brief descriptions are given below but specific design information is not given. Most are unique to the manufacturer and application data are available through catalogs.

Jaw Clutch

The teeth of the mating sets of jaws are brought into engagement by sliding one or both members axially. The teeth may be straight-sided or triangular, or may incorporate some smooth curve to facilitate engagement. Once engaged, there is a positive transmission of torque. The jaw clutch is normally engaged while the system is stopped or running very slowly.

Ratchet

Although not strictly a clutch, the familiar ratchet and pawl permits alternate engagement and disengagement of moving members and thus can be used in similar applications. Typically, the ratchet moves only a small fraction of a revolution per cycle.

Sprag, Roller, and Cam Clutches

There are differences in the specific geometry of sprag, roller, and cam clutches, but they all perform similar functions. When the input shaft is rotating in the driving direction, the internal elements (sprags, rollers, or cams) are wedged between the driving and driven members and thus transmit torque. But when the input member rotates in the opposite direction, the internal elements move out of engagement and no torque is transmitted. Thus they can be used for applications similar to ratchets but with much smoother operation and with a virtually infinite amount of incremental motion. Another application is *back-stopping,* in which the clutch runs free when the machine is being driven in the intended direction. But if the drive is disengaged and the load starts to reverse direction, the clutch locks up and prevents motion. This type of clutch is also used for *overrunning:* a positive drive as long as the load rotates no faster than the driver. If the load tends to run faster (overrun) than the driver, the clutch elements disengage. This protects equipment which might be damaged by overspeeding.

Fiber Clutch

A fiber clutch operates in a manner similar to the overrunning clutches described previously. But instead of driving through solid elements, the torque is transmitted through stiff fibers that have a preferred orientation. When rotated in a direction opposite the preferred direction, the fibers "lie down" and no torque is transmitted.

Wrap Spring Clutch

Again used in cases similar to the overrunning clutches, the wrap spring is made from a rectangular wire and normally has an inside diameter slightly larger than the shaft on which it is installed. Thus no torque is transmitted. But when one end of the spring is restrained, the spring "wraps down" tightly on the surface of the shaft and torque is transmitted positively through the spring.

Single-revolution Clutches

It is frequently desired to have a machine cycle one complete revolution and then come to a stop. The single-revolution clutch provides this feature. After it is tripped, it drives the output shaft until reaching a positive stop at the end of one revolution. Some types can be engaged for more than one revolution but they will return to a fixed position, say at the top of the stroke of a press.

Fluid Clutches

The fluid clutch is made up of two separate parts with no mechanical connection between them. A fluid fills a cavity between the parts, and, as one member rotates, it tends to shear the fluid, causing torque to be transmitted to the mating element. The resulting drive is very smooth and soft because load peaks will simply cause one member to move relative to the other. In this situation, it is similar to the slip clutch described earlier.

Eddy Current Drive

When a conducting disc moves through a magnetic field, eddy currents are induced in the disc which cause a force to be exerted on the disc in a direction opposite to the direction of rotation. The force can be used to brake the disc or to transmit torque to a mating part as a clutch. An advantage of this type of unit is that there is no mechanical connection between the elements. The torque can be controlled by varying the current to the electromagnets.

Overload Clutches

The drive is positive, provided the torque is below some set value. At higher torques, some element is disengaged automatically. One type uses a series of spherical balls positioned in detents and held under a spring force. When the tripping torque level is reached, the balls are forced out of the detents and disengage the drive.

REFERENCES

1. Electroid Corporation. *Engineering Catalog No. 900E*. Springfield, N.J.
2. Faires, V. M. *Design of Machine Elements,* 4th ed. New York: Macmillan Publishing Company, 1965.
3. Juvinall, Robert C. *Fundamentals of Machine Component Design*. New York: John Wiley & Sons, 1983.
4. *Machine Design Magazine, Mechanical Drives Reference Issue*. Cleveland: Penton/IPC, 56, no. 15, June 28, 1984.
5. Shigley, J. E., and Mitchell, L. D. *Mechanical Engineering Design,* 4th ed. New York: McGraw-Hill Book Company, 1983.
6. Warner Electric Brake & Clutch Company. *Catalog P-137-15*. Beloit, Wisc.

PROBLEMS

1. Specify the required torque rating for a clutch to be attached to a motor shaft running at 1 750 rpm. The motor is rated at 5.0 hp and is of the design *B* type.
2. Specify the required torque rating for a clutch to be attached to a diesel engine shaft running at 2 500 rpm. The engine is rated at 75.0 hp.
3. Specify the required torque rating for a clutch to be attached to an electric motor shaft running at 1 150 rpm. The motor is rated at 0.50 hp and drives a light fan.
4. An alternate design for the system described in problem 1 is being considered. Instead of putting the clutch on the motor shaft, it is desired to place it on the output shaft of a speed reducer that rotates at 180 rpm. The power transmission is still approximately 5.0 hp. Specify the required torque rating for the clutch.
5. Specify the required brake torque rating for each of the conditions in problems 1 to 4 for average industrial conditions.
6. Specify the required torque rating for a clutch in N · m if it is attached to a design *B* electric motor shaft rated at 20.0 kW and rotating at 3 450 rpm.
7. A clutch/brake module is to be connected between a design *C* electric motor and a speed reducer. The motor is rated at 50.0 kW at 900 rpm. Specify the required torque ratings for the clutch and the brake portions of the module for average industrial service. The drive is to a large conveyor.
8. Compute the torque required to accelerate a solid steel disc from rest to 550 rpm in 2.0 sec. The disc has a diameter of 24.0 in and is 2.5 in thick.
9. The assembly shown in Figure 16-21 is to be stopped by a brake from 775 rpm to zero in 0.50 sec or less. Compute the required brake torque.
10. Compute the required clutch torque to accelerate the system shown in Figure 16-22 from rest to a motor speed of 1 750 rpm in 1.50 sec. Neglect the inertia of the clutch.

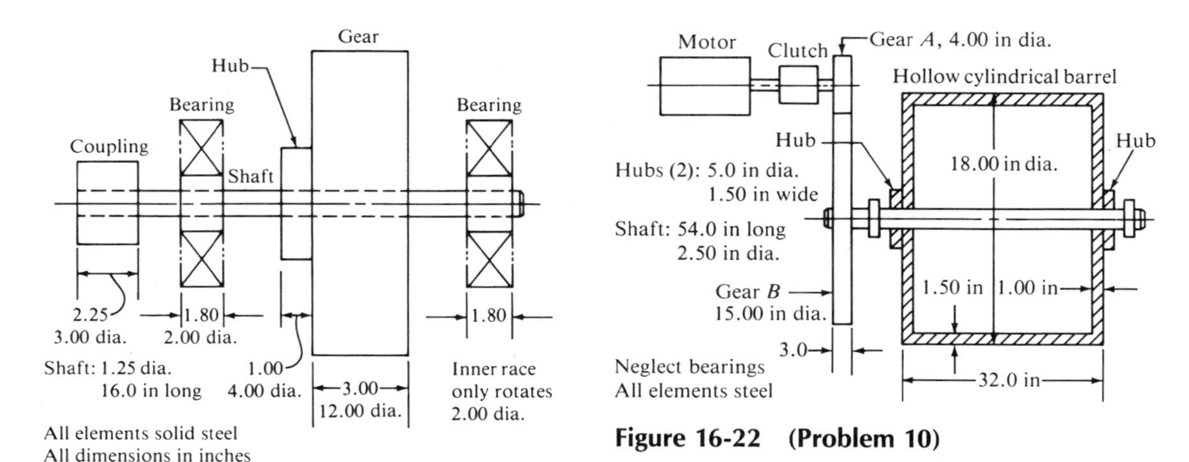

Figure 16-21 (Problem 9)

Figure 16-22 (Problem 10)

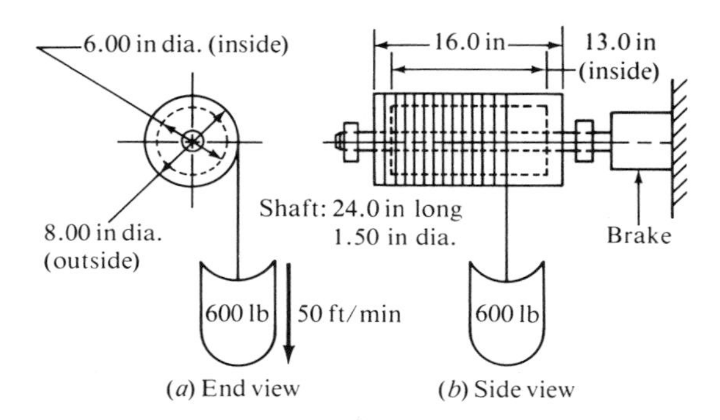

Figure 16-23 (Problem 11)

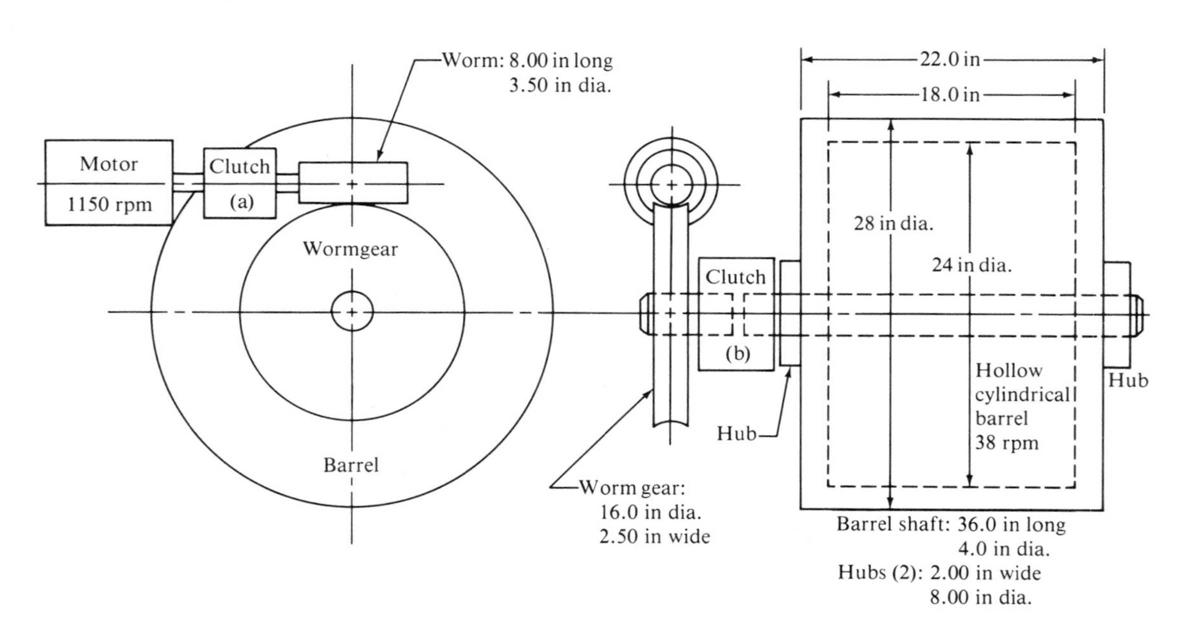

Figure 16-24 (Problem 12)

11. A winch, sketched in Figure 16-23, is lowering a load at the speed of 50 ft/min. Compute the required torque rating for the brake on the winch shaft to stop the system in 0.25 sec.
12. Figure 16-24 shows a tumbling barrel being driven through a wormgear reduction unit. Evaluate the torque rating required for a clutch to accelerate the barrel to 38.0 rpm from rest in 2.0 sec, (a) if the clutch is placed on the motor shaft, and (b) if it is placed at the output of the reducer. Neglect the inertia of the gear shafts, the bearing races, and the clutch. Consider the worm and wormgear to be solid cylinders.
13. Compute the dimensions of an annular plate type brake to produce a braking torque of 75 lb · in. Air pressure will develop a normal force of 150 lb between the friction surfaces. Use a coefficient of friction of 0.25. The brake will be used in average industrial service, stopping a load from 1 150 rpm.
14. Design a plate type brake for the application described in problem 9. Specify the design coefficient of friction, the dimensions of the plate, and the axial force required.
15. Compute the axial force required for a cone clutch if it is to exert a driving torque of 15 lb · ft. The cone surface has a mean diameter of 6.0 in and an angle of 12°. Use $f = 0.25$.
16. Design a cone brake for the application described in problem 9. Specify the design coefficient of friction, the mean diameter of the cone surface, and the axial force required.
17. Compute the actuation force required for the short shoe drum brake of Figure 16-17 to produce a friction torque of 150 lb · ft. Use a drum diameter of 12.0 in, $a = 4.0$ in, and $L = 24.0$ in. Use $f = 0.25$ and $b = 5.0$ in.
18. For all other data from problem 17 being the same, determine the required dimension b for the brake to be self-actuating.
19. Design a short shoe drum brake to produce a torque of 100 lb · ft. Specify the drum diameter, the configuration of the actuation lever, and the actuation force.
20. Design a long shoe drum brake to produce a friction torque of 100 lb · ft to stop a load from 480 rpm. Specify the friction material, the drum size, the shoe configuration, pivot locations, and actuation force.
21. Design a band brake to exert a braking torque of 75 lb · ft while slowing a drum from 350 rpm to rest. Specify a material, the drum diameter, the band width, the angle of coverage of the friction material, the actuation lever configuration, and the actuation force.

17 Electric Motors

17-1 OVERVIEW

The electric motor is very widely used for providing the prime power to industrial machinery, consumer products, and business equipment. This chapter will describe the various types of motors and discuss their performance characteristics. The objective is to provide you with the background needed to specify motors and to communicate with vendors to acquire the proper motor for a given application.

As a minimum, the following items must be specified for motors:

Motor type: AC, DC, single-phase, three-phase, and so on

Power rating and speed

Operating voltage and frequency

Type of enclosure

Frame size

Mounting details

In addition, there may be many special requirements, which must be communicated to the vendor.

The primary factors to be considered in selecting a motor include the following:

Operating torque, operating speed, and power rating. Note that these are related by the equation, power = torque × speed.

Starting torque

Load variations expected and corresponding speed variations which can be tolerated.

Current limitations during the running and starting phases of operation.

Duty cycle: how frequently the motor is to be started and stopped.

Environmental factors: temperature, presence of corrosive or explosive atmospheres, exposure to weather or to liquids, availability of cooling air, and so on.

Voltage variations expected: Most motors will tolerate up to ± 10 percent variation from the rated voltage. Beyond this, special designs are required.

Shaft loading: particularly side loads and thrust loads that can affect the life of shaft bearings.

Motor Size

A rough classification of motors by size is used to group motors of similar design. The horsepower (hp) is currently most frequently used, with the metric unit of watts or kilowatts also used at times. The conversion is

$$1.0 \text{ hp} = 0.746 \text{ kW} = 746 \text{ W}$$

The classifications are as follows:

Subfractional horsepower: 1 to 40 millihorsepower (mhp), where 1 mhp = 0.001 hp. Thus, this range includes 0.001 to 0.040 hp (0.75 W to 30 W approximately).

Fractional horsepower: $\frac{1}{20}$ to 1.0 hp (37 W to 746 W approximately).

Integral horsepower: 1.0 hp (0.75 kW) and larger.

The following sections will discuss the types of AC power available, AC motors, DC power, DC motors, other motor types, and motor controls.

17-2 AC POWER AND GENERAL INFORMATION ABOUT AC MOTORS

Alternating current (AC) power is produced by the electric utility and delivered to the industrial, commercial, or residential consumer in a variety of forms. In the United States, AC power has a frequency of 60 hertz (Hz) or 60 cycles/sec. In many other countries, 50 Hz is used. Some aircraft use 400-Hz power from an on-board generator.

AC power is also classified as single-phase or three-phase. Most residential units and light commercial installations have only single phase power, carried by two conductors plus ground. The wave form of the power would appear like that in Figure 17-1, a single continuous sine wave at the system frequency whose amplitude is the rated voltage of the power. Three-phase power is carried on a three-wire system and is composed of three distinct waves of the same amplitude and frequency, with each phase offset from the next by 120°, as illustrated in Figure 17-2. Industrial and large commercial installations use three phase power for the larger electrical loads because smaller motors are possible and there are economies of operation.

AC Voltages

Some of the more popular voltage ratings available in AC power are listed in Table 17-1. Given are the nominal system voltage and the typical motor voltage rating for that system in both single-phase and three-phase. In most cases, the highest voltage available should be used because the current flow for a given power is smaller. This allows smaller conductors to be used.

Speeds of AC Motors

An AC motor at zero load would tend to operate at or near its *synchronous speed,* which is related to the frequency, f, of the AC power and to the number of electrical poles, p, wound into the motor, according to the equation

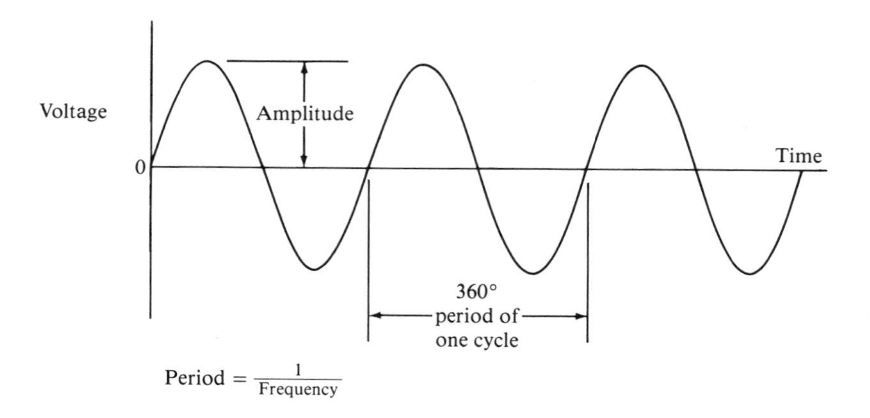

Figure 17-1 Single-phase AC Power

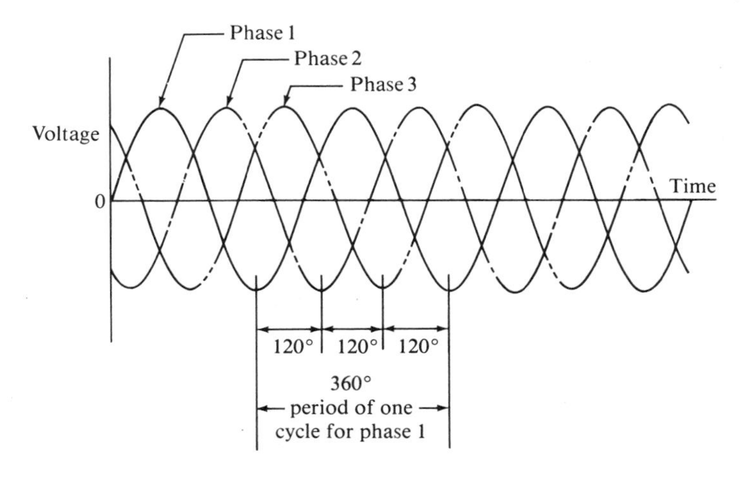

Figure 17-2 Three-phase AC Power

$$n_s = \frac{120f}{p} \tag{17-1}$$

Motors have an even number of poles, usually from 2 to 12, resulting in the synchronous speeds listed in Table 17-2 for 60-Hz power. But the induction motor, the most widely used type, operates at a speed progressively slower than its synchronous speed as the load (torque) demand increases. When the motor is delivering its rated torque it would be operating near its rated or full-load speed, also listed in Table 17-2. Note that the full-load speed is not precise and that those listed are for motors with normal slip of approximately 5 percent. Some motors described later are "high slip" motors having lower full-load speeds. Some 4-pole motors are rated at 1 750 rpm at full load, indicating only about 3 percent slip. *Synchronous motors* operate precisely at the synchronous speed with no slip.

Table 17-1 AC Motor Voltages

	MOTOR VOLTAGE RATINGS	
SYSTEM VOLTAGE	*Single Phase*	*Three Phase*
120	115	115
120/208	115	200
240	230	230
480	—	460
600	—	575

Table 17-2 AC Motor Speeds

Number of Poles	*Synchronous Speed (rpm)*	*Full Load[a] Speed (rpm)*
2	3 600	3 450
4	1 800	1 725
6	1 200	1 140
8	900	850
10	720	690
12	600	575

[a]Approximately 95 percent of synchronous speed (normal slip).

Speed Control

AC motors are not normally speed controlled but operate at a constant speed for a given load as determined by the speed/torque curve discussed later. However there are solid state electronic controls available that vary the frequency of the AC power delivered to the motor. As can be seen from equation (17-1), the speed varies directly with the frequency. With such a control, speeds can be matched to machine requirements, acceleration and deceleration can be controlled for "soft" starts and stops, and automatic reversing or dynamic braking can be provided.

If a set of discrete speeds is acceptable rather than continuously varying speeds, a motor can be designed with switchable electrical circuits which change the number of electrical poles in the motor. Equation (17-1) shows that the speed is inversely proportional to the number of poles. Thus if a motor is connected as a 2-pole/4-pole motor, its two synchronous speeds would be 3 600 rpm and 1 800 rpm.

Some AC motors, notably the wound rotor type, can have adjustable speeds to a modest extent as described later. Of course, speed control can be provided by mechanical means using variable speed belt sheaves, speed change gear boxes, or other such devices.

Frame Types

The design of the equipment in which the motor is to be mounted determines the type of frame required. Some types are described below.

Foot-mounted

The most widely used type for industrial machinery, the foot-mounted frame has integral feet with a standard hole pattern for bolting the motor to the machine (see Figure 17-3).

Cushion Base

A foot mounting is provided with resilient isolation of the motor from the frame of the machine to reduce vibration and noise (see Figure 17-4).

C-face Mounting

A machined face is provided on the shaft end of the motor which has a standard pattern of tapped holes. Driven equipment is then bolted directly to the motor. The design of the face is standardized by the National Electrical Manufacturers Association (NEMA). See Figures 17-5 and 17-6.

D-flange Mounting

A machined flange is provided on the shaft end of the motor with a standard pattern of through clearance holes for bolts for attaching the motor to the driven equipment. The flange design is controlled by NEMA.

Vertical Mounting

Vertical mounting is a special design because of the effects of the vertical orientation on the bearings of the motor. Attachment to the driven equipment is usually through C-face or D-flange bolt holes as described above (see Figure 17-7).

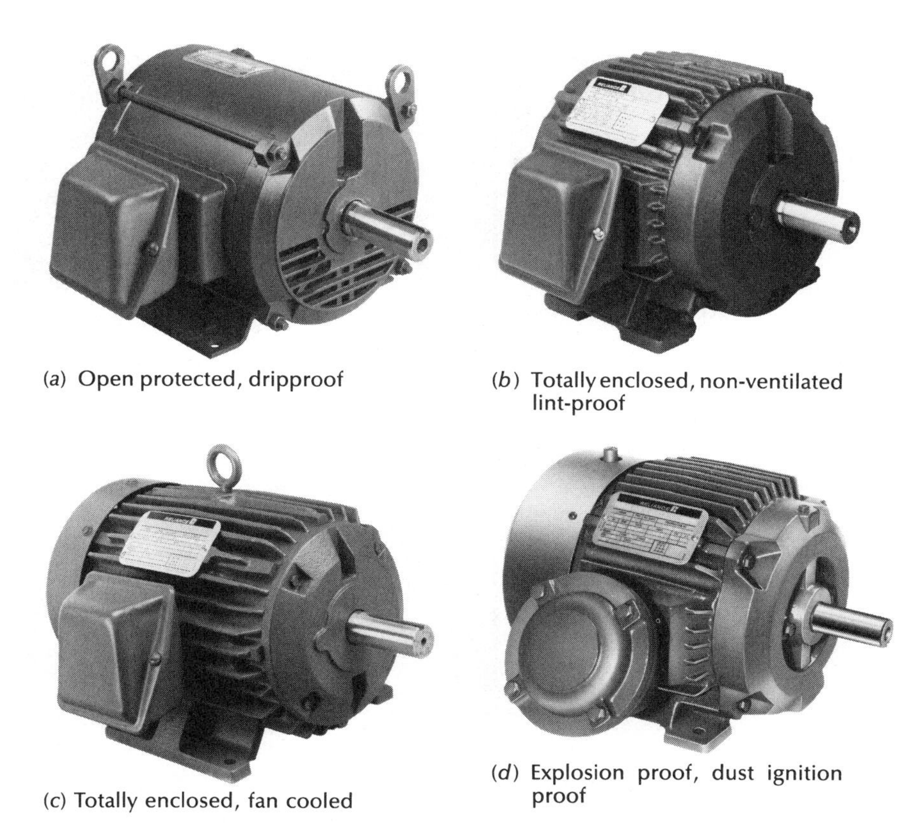

(*a*) Open protected, dripproof

(*b*) Totally enclosed, non-ventilated lint-proof

(*c*) Totally enclosed, fan cooled

(*d*) Explosion proof, dust ignition proof

Figure 17-3 Foot-mounted Motors with Various Enclosure Types
(Reliance Electric Co., Cleveland, Ohio)

Unmounted

Some equipment manufacturers purchase only the bare rotor and stator from the motor manufacturer and then build them into their machine. Compressors for refrigeration equipment are usually built in this manner.

Special-purpose Mountings

Many special designs are made for fans, pumps, oil burners, and so on.

Enclosures

The housings around the motor which support the active parts and protect them vary with the degree of protection required. Some of the enclosure types are shown in Figures 17-3 through 17-8 and are described next.

Open

Typically a light gage sheet metal housing is provided around the stator with end plates to support the shaft bearings. The housing contains several holes or slots which permit

Figure 17-4 Enclosed, Cushion Base Motor with Double End Shaft (Century Electric, Inc., St. Louis, Mo.)

Figure 17-5 C-face Motor. See Figure 12-22 for an example of a reducer designed to mate with a C-face motor. (Reliance Electric Co., Cleveland, Ohio)

Figure 17-6 Rigid Base Dripproof NEMA C-face Motor for Application to a Close Coupled Pump (Century Electric, Inc., St. Louis, Mo.)

Figure 17-7 Totally Enclosed, Nonventilated, C-face Motor with Dripcover for Vertical Operation (Century Electric, Inc., St. Louis, Mo.)

cooling air to enter the motor. Such a motor must be protected by the housing of the machine itself (see Figure 17-8).

Protected

Sometimes called *dripproof,* ventilating openings are provided only on the lower part of the housing so that liquids dripping on the motor from above cannot enter the motor. This is probably the most widely used type (see Figure 17-3a).

Totally Enclosed Nonventilated (TENV)

No openings at all are provided in the housing and no special provisions are made for cooling the motor except for fins cast into the frame to promote convective cooling. The design protects the motor from harmful atmospheres (see Figure 17-3b).

Figure 17-8 Open Frame Motor, Shown with Cushion Base. Mounting may also be by through bolts, belly-band, for cushion ring. (Century Electric, Inc., St. Louis, Mo.)

Totally Enclosed Fan-cooled (TEFC)

TEFC design is similar to TENV except a fan is mounted to one end of the shaft to draw air over the finned housing (see Figure 17-3c).

TEFC-XP (Explosion Proof)

TEFC-XP design is similar to the TEFC housing, except special protection is provided for electrical connections to prohibit fire or explosion in hazardous environments (see Figure 17-3d).

Frame Sizes

The critical dimensions of motor frames are controlled by NEMA frame sizes. Included are the overall height and width; the height from the base to the shaft centerline; the shaft

Table 17-3 Motor Frame Sizes

		DIMENSIONS (IN)								
hp	*Frame Size*	*A*	*C*	*D*	*E*	F	*O*	*U*	*V*	*Keyway*
1/4	48	5.63	9.44	3.00	2.13	1.38	5.88	0.500	1.50	0.05 flat
1/2	56	6.50	10.07	3.50	2.44	1.50	6.75	0.625	1.88	3/16 × 3/32
1	143T	7.00	10.69	3.50	2.75	2.00	7.00	0.875	2.00	3/16 × 3/32
2	145T	7.00	11.69	3.50	2.75	2.50	7.00	0.875	2.00	3/16 × 3/32
5	184T	9.00	13.69	4.50	3.75	2.50	9.00	1.125	2.50	1/4 × 1/8
10	215T	10.50	17.25	5.25	4.25	3.50	10.56	1.375	3.13	5/16 × 5/32
15	254T	12.50	22.25	6.25	5.00	4.13	12.50	1.625	3.75	3/8 × 3/16
20	256T	12.50	22.25	6.25	5.00	5.00	12.50	1.625	3.75	3/8 × 3/16
25	284T	14.00	23.38	7.00	5.50	4.75	14.00	1.875	4.38	1/2 × 1/4
30	286T	14.00	24.88	7.00	5.50	5.50	14.00	1.875	4.38	1/2 × 1/4
40	324T	16.00	26.00	8.00	6.25	5.25	16.00	2.125	5.00	1/2 × 1/4
50	326T	16.00	27.50	8.00	6.25	6.00	16.00	2.125	5.00	1/2 × 1/4

Note: All motors are 4-pole, 3-phase, 60-Hz AC induction motors.
Refer to Figure 17-9 for description of dimensions.

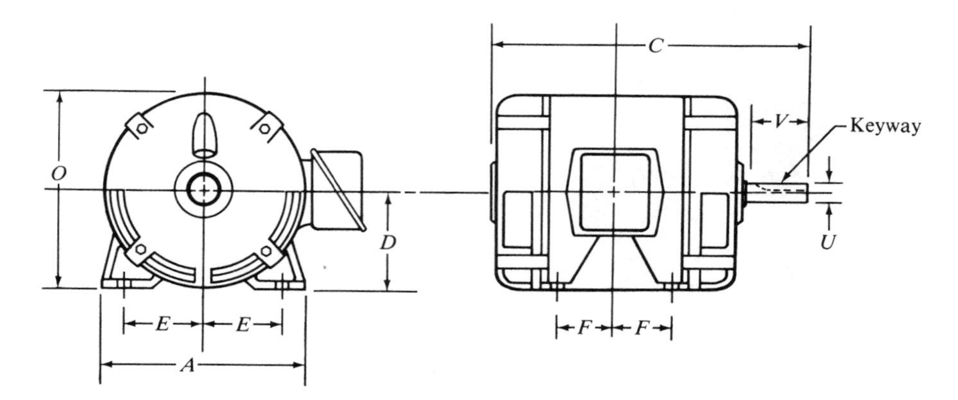

Figure 17-9 Key for NEMA Standard Motor Frame Dimensions Listed in Table 17-3

diameter, length, and keyway size; and mounting hole pattern dimensions. A few selected motor frame sizes for 1 725-rpm, three-phase, induction, foot-mounted dripproof motors are listed in Table 17-3. For the descriptions of the dimensions, refer to Figure 17-9.

17-3 AC MOTOR TYPES AND PERFORMANCE

This section will describe the several types of AC motors and discuss their typical performance characteristics and uses. There are four basic classes of AC motors: single-phase, three-phase, universal, and synchronous. Each class includes several specific designs.

The performance of electric motors is usually displayed on a graph of speed versus torque, as shown in Figure 17-10. The *vertical axis* is the rotational speed of the motor as a percentage of synchronous speed. The *horizontal axis* is the torque developed by the motor as a percentage of the full-load or rated torque. When exerting its full-load torque, the motor operates at its full-load speed and delivers the rated power. See Table 17-2 for a listing of synchronous speeds and full-load speeds.

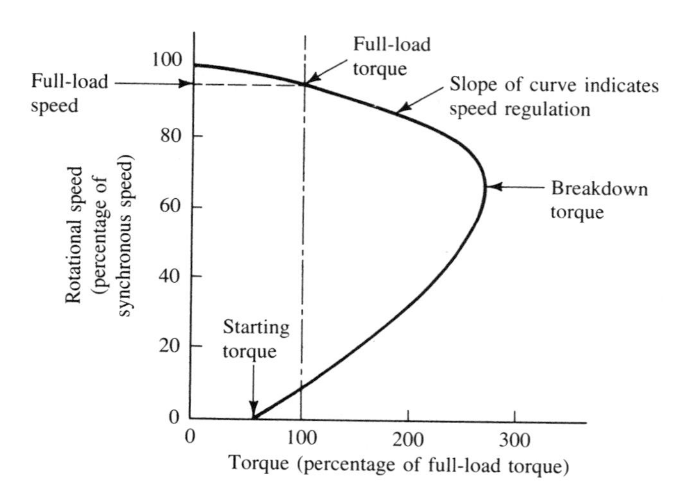

Figure 17-10 General Form of Motor Performance Curve

The torque at the bottom of the curve where the speed is zero is called the *starting torque* or *locked rotor torque*. It is the torque available to initially get the load moving and begin its acceleration. This is one of the most important selection parameters for motors, as will be discussed in the descriptions of the individual types of motors.

The "knee" of the curve is called the *breakdown torque* and is the maximum torque developed by the motor during acceleration. The slope of the speed/torque curve in the vicinity of the full-load operating point is an indication of *speed regulation*. A flat curve (a low slope) indicates good speed regulation with little variation in speed as load varies. Conversely, a steep curve (a high slope) indicates poor speed regulation and the motor would exhibit wide swings in speed as load varies. Such motors produce a "soft" acceleration of a load that may be an advantage in some applications. But where fairly constant speed is desired, a motor with good speed regulation should be selected.

Single-phase Motors

The four most common types of single-phase motors are the *split-phase, capacitor-start, permanent-split capacitor,* and *shaded pole*. Each is unique in its physical construction and in the manner in which the electrical components are connected to provide for starting and running of the motor. The emphasis here is not on how to design the motors, but rather on the performance, so a suitable motor can be selected.

Figure 17-11 shows the performance characteristics of these four types of motors so they can be compared. The special features of the performance curves for the four types of motors are discussed in later sections.

In general, the single-phase induction motor incorporates a fixed *stator* around the outside, formed by several windings of electrical wires. The single-phase power flows through the stator and acts similar to the primary of a transformer to induce currents in the rotor, which acts like the secondary of a transformer. The *rotor* is solid and consists of soft iron laminations in a cylindrical form with a number of metal bars inserted in parallel and connected at the ends of the rotor with caps or rings which short-circuit the bars. Because of the appearance of the bars and the end caps, the rotor is often called a *squirrel*

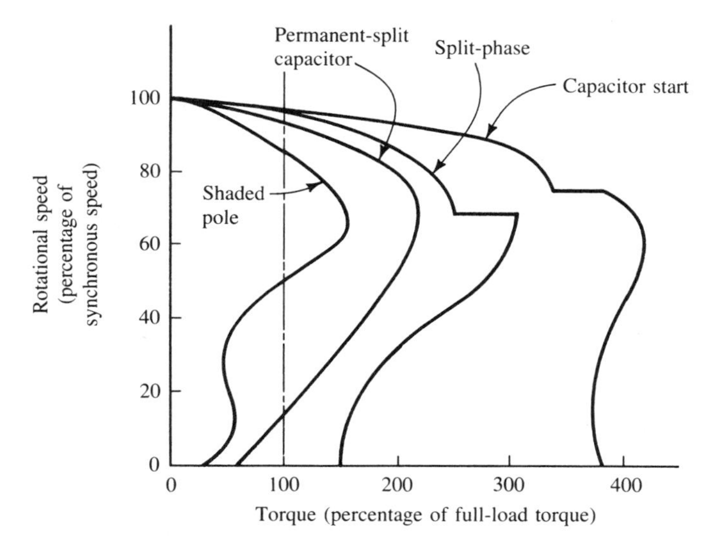

Figure 17-11 Peformance Curves of Four Types of Single-Phase Electric Motors

cage. There are no electrical connections to the rotor. As the rotor rotates, slightly slower than the rotating field developed by the stator, magnetic lines of force are cut, creating the torque to rotate the rotor.

Single-phase motors are usually in the subfractional or fractional horsepower range from 1/50 hp (15 W) to 1.0 hp (750 W), although some are available up to 10 hp (7.5 kW).

Split-phase Motors

The stator of the split-phase motor has two windings: the main winding, which is continuously connected to the power line, and the starting winding, which is connected only during the starting of the motor. The starting winding creates a slight phase shift that creates the initial torque to start and accelerate the rotor. After the rotor reaches approximately 75 percent of its synchronous speed, the starting winding is cut out by a centrifugal switch and the rotor continues to run on the main winding.

The performance curve for the split-phase motor is shown in Figure 17-11. It has moderate starting torque, approximately 150 percent of full-load torque. It has good efficiency and is designed for continuous operation. Speed regulation is good. One of the disadvantages is that it requires a centrifugal switch to cut out the starting winding. The step in the speed/torque curve indicates this cutout.

These characteristics make the split-phase motor one of the most popular types, used in business machines, machine tools, centrifugal pumps, electric lawn mowers, and similar applications.

Capacitor-start Motors

Like the split-phase motor, the capacitor-start motor also has two windings, a main or running winding and a starting winding. But in it, a capacitor is connected in series with the starting winding, giving a much higher starting torque than the split-phase motor. A starting torque of 250 percent of full load or higher is common. Again a centrifugal switch is used to cut out the starting winding and the capacitor. The running characteristics of the motor are then very similar to the split-phase motor: good speed regulation, and good efficiency for continuous operation.

Disadvantages include the switch and the relatively bulky capacitor. Frequently the capacitor is mounted right on top of the motor and is very evident (see Figure 17-8). It may also be integrated into a package containing the starting switch, a relay, or other control elements.

Uses for the capacitor-start motor include the many types of machines which need the high starting torque. Examples include heavily loaded conveyors, refrigeration compressors, and pumps and agitators for heavy fluids.

Permanent-split Capacitor Motors

A capacitor is connected in series with the starting winding at all times. The starting torque of the permanent-split capacitor motor is typically quite low, approximately 40 percent or less of full-load torque. Thus only low-inertia loads such as fans and blowers are usually used. An advantage is that the running performance and speed regulation can be tailored to match the load by selecting the appropriate capacitor value. Also, no centrifugal switch is required.

Shaded Pole Motors

The shaded pole motor has only one winding, the main or running winding. The starting reaction is created by the presence of a copper band around one side of each pole. The low-resistance band "shades" the pole to produce a rotating magnetic field to start the motor.

The shaded pole motor is simple and inexpensive but it has a low efficiency and a very low starting torque. Speed regulation is poor and it must be fan-cooled during normal operation. Thus its primary use is in shaft mounted fans and blowers where the air is drawn over the motor. Some small pumps, toys, and intermittently used household items also employ shaded pole motors because of their low cost.

Three-phase Squirrel Cage Motors

Three of the most commonly used three-phase alternating current motors are simply designated as designs B, C, and D by NEMA. They differ primarily in the value of starting torque and in the speed regulation near full load. Figure 17-12 shows the performance curves for these three designs for comparison. Each of these designs employs the solid "squirrel cage" type of rotor, and thus there is no electrical connection to the rotor.

The 4-pole design with a synchronous speed of 1 800 rpm is the most common and is available in virtually all power ratings from ¼ hp to 500 hp. Certain sizes are available in 2-pole (3 600 rpm), 6-pole (1 200 rpm), 8-pole (900 rpm), 10-pole (720 rpm), and 12-pole (600 rpm) designs.

NEMA Design B

The performance of the three-phase design B motor is similar to that of the single-phase split-phase motor described previously. It has a moderate starting torque (about 150 percent of full-load torque) and good speed regulation. The breakdown torque is high, usually 200 percent of full-load torque or more. Starting current is fairly high, at approximately six times full-load current. The starting circuit must be selected to be able to handle this current for the short time required to bring the motor up to speed.

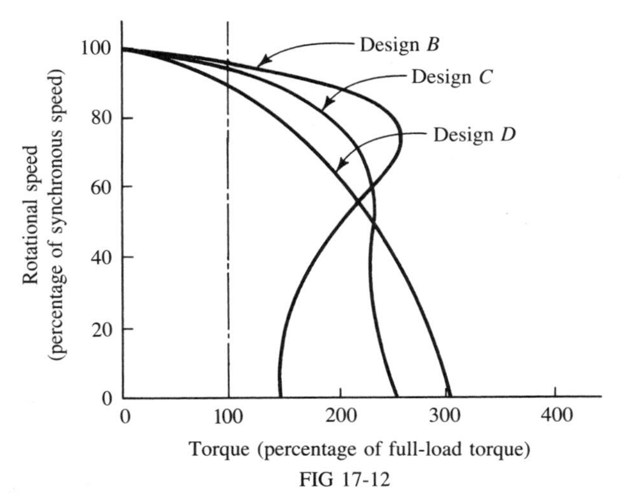

Figure 17-12 Performance Curves for Three-Phase Motors: Designs B, C, D

Typical uses for the design B motor are centrifugal pumps, fans, blowers, and machine tools such as grinders and lathes.

NEMA Design C

High starting torque is the main advantage of the design C motor. Loads requiring 200 to 300 percent of full-load torque to start can be driven. Starting current is typically lower than for the design B motor for the same starting torque. Speed regulation is good and is about the same as for the design B motor. Reciprocating compressors, refrigeration systems, heavily loaded conveyors, and ball and rod mills are typical uses.

NEMA Design D

The design D motor has a high starting torque, about 300 percent of full-load torque. But it also has poor speed regulation, which results in large speed changes with varying loads. Sometimes called a *high slip motor,* it operates at 5 to 13 percent slip at full load, whereas the designs B and C operate at 3 to 5 percent slip. Thus, the full-load speed will be lower for the design D motor.

The poor speed regulation is considered an advantage in some applications and is the main reason for selecting the design D motor for such uses as punch presses, shears, sheet metal press brakes, cranes, elevators, and oil well pumps. Allowing the motor to slow down significantly when loads increase gives the system a "soft" response, reducing shock and jerk felt by the drive system and the driven machine. Consider an elevator: When a heavily loaded elevator cab is started, the acceleration should be smooth and soft and the cruising speed should be approached without excessive jerk. This comment also applies to a crane. If a large jerk occurs when the crane hook is heavily loaded, the peak acceleration would be high. The resulting high inertia force may break the cable.

Wound Rotor Motors

As the name implies, the rotor of the wound rotor motor has electrical windings that are connected through slip rings to the external power circuit. The selective insertion of resistance in the rotor circuit allows the performance of the motor to be tailored to the needs of the system and to be changed with relative ease to accommodate system changes or to actually vary the speed of the motor.

Figure 17-13 shows the results obtained by changing the resistance in the rotor circuit. Note that the four curves are all for the same motor, with curve 0 giving the performance with zero external resistance. This is similar to the design B. Curves 1, 2, and 3 show the performance with progressively higher levels of resistance in the rotor circuit. Thus, the starting torque and the speed regulation (softness) can be tuned to the load. Speed adjustment under a given load up to approximately 50 percent of full-load speed can be obtained.

The wound rotor design is used in such applications as printing presses, crushing equipment, conveyors, and hoists.

Synchronous Motors

Entirely different from the squirrel cage induction motor or the wound rotor motor, the synchronous motor operates precisely at the synchronous speed with no slip. Such motors are available in sizes from subfractional, used for timers and instruments, to several hundred horsepower to drive large air compressors, pumps, or blowers.

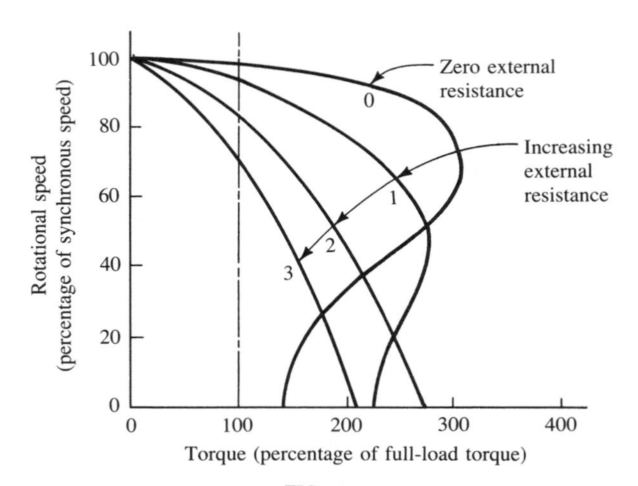

Figure 17-13 Performance Curves for a Three-Phase Wound Rotor Motor with Varying External Resistance in the Rotor Circuit

The synchronous motor must be started and accelerated by a means separate from the synchronous motor components themselves because they provide very little torque at zero speed. Typically there will be a separate squirrel cage type winding within the normal rotor which initially accelerates the motor shaft. When the speed of the rotor is within a few percent of the synchronous speed, the field poles of the motor are excited and the rotor is pulled into synchronism. At that point the squirrel cage becomes ineffective and the motor continues to run at speed regardless of load variations, up to a limit called the *pull-out torque*. A load above the pull-out torque will pull the motor out of synchronism and cause it to stop.

Universal Motors

Universal motors operate on either AC or direct current (DC) power. Their construction is similar to a series-wound DC motor which is described later in section 17-5. The rotor has electrical coils in it which are connected to the external circuit through a commutator on the shaft, a kind of slip ring assembly made of several copper segments on which stationary carbon brushes ride. Contact is maintained with light spring pressure.

Universal motors usually run at high speeds, 3 500 rpm to 20 000 rpm. This results in a high power-to-weight and high power-to-size ratio for this type of motor, making it desirable for hand held tools such as drills, saws, food mixers, and so on. Vacuum cleaners and sewing machines also frequently use universal motors. Figure 17-14 shows a typical set of speed/torque curves for a high-speed version of the universal motor, showing the performance for 60-Hz and 25-Hz AC power and DC power. It can be seen that performance at near the rated load is similar, regardless of the nature of the input power, and that these motors have poor speed regulation. That is, the speed varies greatly with load.

17-4 DC POWER

DC motors have several inherent advantages over AC motors, as is discussed in the next section. A disadvantage of DC motors is that a source of DC power must be available.

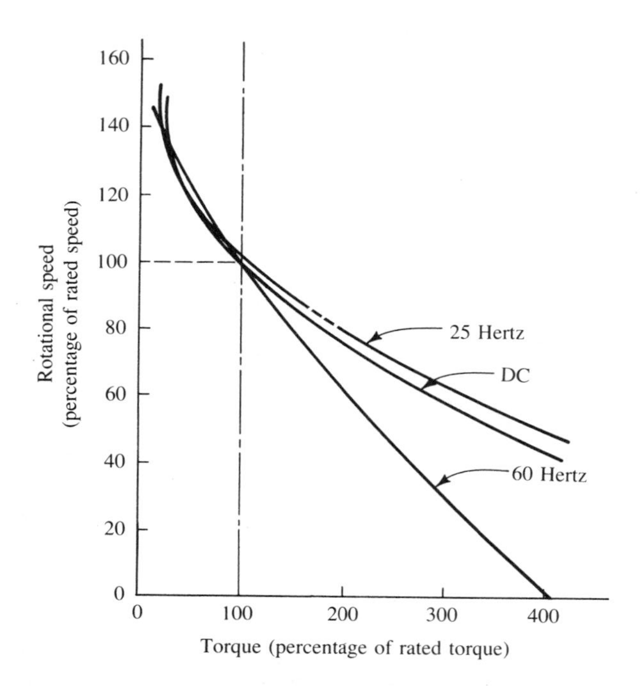

Figure 17-14 Performance Curves for a Universal Motor

Most residential, commercial, and industrial locations have only AC power provided by the local utility. Three approaches are used to provide DC power.

1. Batteries: Typically batteries are available in voltages of 1.5, 6.0, 12.0, and 24.0 volts (V). They are used for portable devices or for mobile applications. The power is pure DC but voltage varies with time as the battery discharges. The bulkiness, weight, and finite life are disadvantages.
2. Generators: Powered by AC electric motors, internal combustion engines, turbine engines, wind devices, water turbines or the like, DC generators produce pure DC. The usual voltages are 115 V and 230 V. Some industries maintain such generators to provide DC power throughout the plant.
3. Rectifiers: *Rectification* is the process of converting AC power with its sinusoidal variation of voltage with time to DC power, which ideally is nonvarying. A readily available device is the *silicon-controlled rectifier (SCR)*. One difficulty with rectification of AC power to produce DC power is that there is always some amount of "ripple", a small variation of voltage as a function of time. Excessive ripple can cause overheating of the DC motor. Most commercially available SCR devices produce DC power with an acceptably low ripple. Table 17-4 lists the commonly used DC voltage ratings for motors powered by rectified AC power as defined by NEMA.

17-5 DC MOTORS

The advantages of direct current motors are summarized below.

The speed is adjustable by using a simple rheostat to adjust the voltage applied to the motor.

Table 17-4 DC Motor Voltage Ratings

Input AC Voltage	*DC Motor Rating*	*NEMA Code*
115 V AC, 1-phase	90 V DC	K
230 V AC, 1-phase	180 V DC	K
230 V AC, 3-phase	240 V DC	C or D
460 V AC, 3-phase	500 V DC or 550 V DC	C or D
460 V AC, 3-phase	240 V DC	E

The direction of rotation is reversible by switching the polarity of the voltage applied to the motor.

Automatic control of speed is simple to provide for matching of the speeds of two or more motors or to program a variation of speed as a function of time.

Acceleration and deceleration can be controlled to provide the desired response time or to decrease jerk.

Torque can be controlled by varying the current applied to the motor. This is desirable in tension control applications such as the winding of film on a spool.

Dynamic braking can be obtained by reversing the polarity of the power while the motor is rotating. The reversed effective torque slows the motor without the need for mechanical braking.

DC motors typically have quick response, accelerating quickly when voltage is changed, because they have a small rotor diameter giving them a high ratio of torque to inertia.

DC motors have electric windings in the rotor and each coil has two connections to the commutator on the shaft. The commutator is a series of copper segments through which the electric power is transferred to the rotor. The current path from the stationary part of the motor to the commutator is through a pair of brushes, usually made of carbon, which are held against the commutator by light coil or leaf springs. Maintenance of the brushes is one of the disadvantages of DC motors.

DC Motor Types

Four commonly used DC motor types are the *shunt-wound, series-wound, compound-wound,* and *permanent magnet* motors. They are described in terms of their speed/torque curves in a manner similar to that used for AC motors. One difference here is that the speed axis is expressed in percentage of *full-load rated speed,* rather than a percentage of synchronous speed, as that term does not apply to DC motors.

Shunt-wound DC Motor

The electromagnetic field is connected in parallel with the rotating armature, as sketched in Figure 17-15. The speed/torque curve shows relatively good speed regulation up to approximately two times full-load torque with a rapid drop in speed after that point. The speed at no load is only slightly higher than the full-load speed. Contrast this with the series-wound motor described next. Shunt-wound motors are used mainly for small fans and blowers.

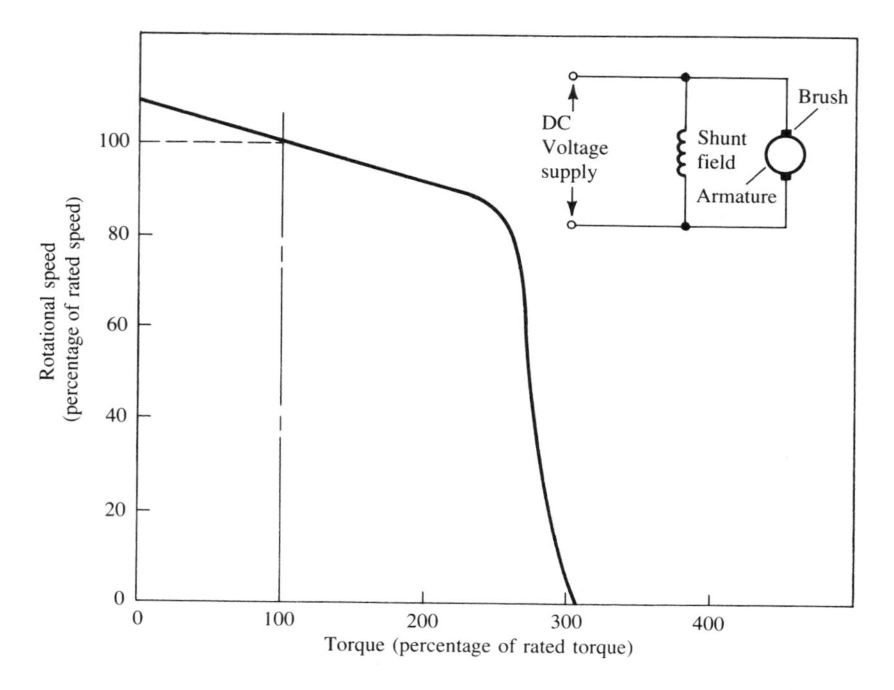

Figure 17-15 Shunt-wound DC Motor Performance Curve

Series-wound DC Motor

The electromagnetic field is connected in series with the rotating armature, as shown in Figure 17-16. The speed/torque curve is very steep, giving the motor a soft performance that is desirable in cranes, hoists, and traction drives for vehicles. The starting torque is very high, as much as 800 percent of full-load rated torque. A major difficulty, however, with series-wound motors is that the no-load speed is theoretically unlimited. The motor could reach a dangerous speed if the load were to be accidentally disconnected. Safety devices, such as overspeed detectors that shut off the power, should be used.

Compound-wound DC Motors

The compound-wound DC motor employs both a series field and a shunt field, as sketched in Figure 17-17. It has a performance somewhat between that of the series-wound and shunt-wound motors. It has fairly high starting torque and a soft speed characteristic, but it has an inherently controlled no-load speed. This makes it good for cranes, which may suddenly lose their loads. The motor would run slowly when heavily loaded for safety and control and fast when lightly loaded to improve productivity.

Permanent Magnet DC Motors

Instead of using electromagnets, the permanent magnet DC motor uses permanent magnets to provide the field for the armature. The direct current passes through the armature, as shown in Figure 17-18. The field is nearly constant at all times and results in a linear speed/torque curve. Current draw also varies linearly with torque. Applications include fans and blowers to cool electronics packages in aircraft, small actuators for controls in

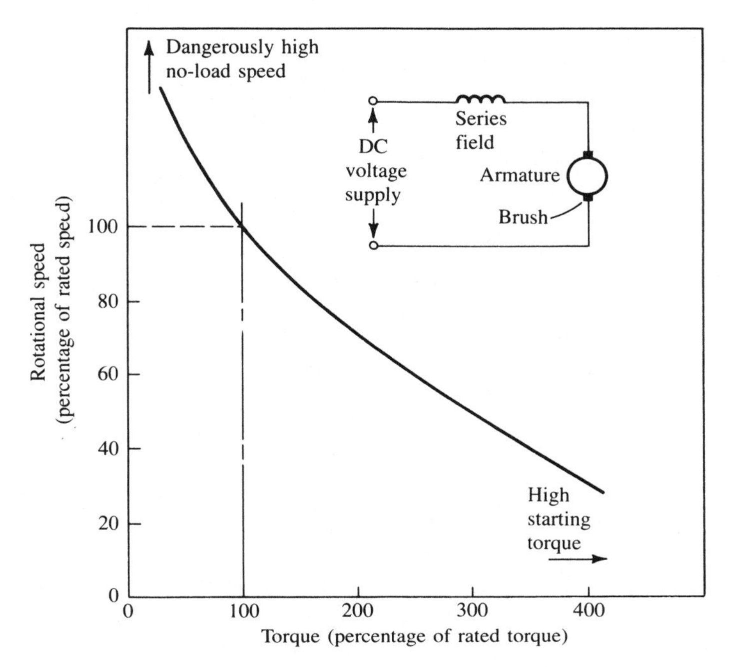

Figure 17-16 Series-wound DC Motor Performance Curve

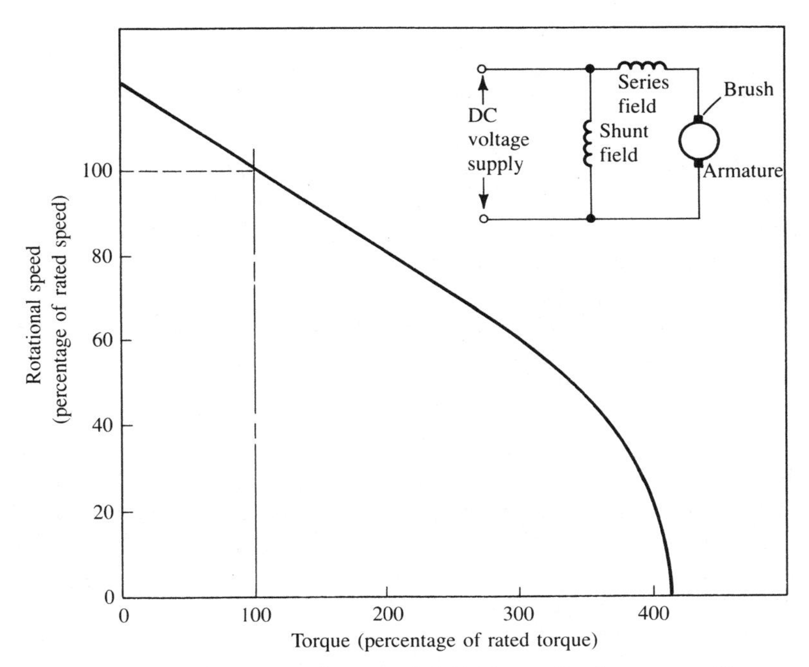

Figure 17-17 Compound-wound DC Motor Performance Curve

aircraft, automotive power assists for windows and seats, and fans in automobiles for heating and air conditioning. These motors frequently have built-in gear type speed reducers to produce a low-speed, high-torque output.

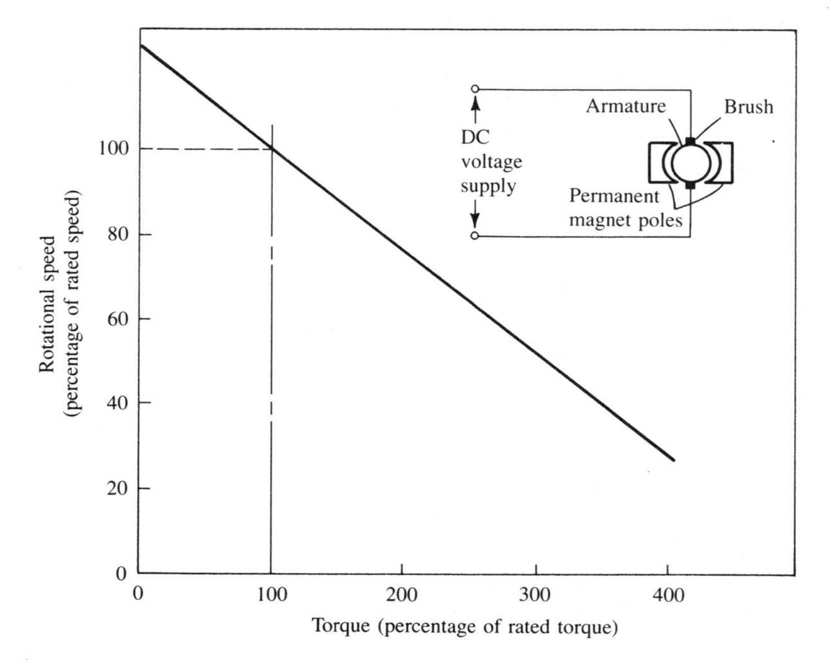

Figure 17-18 Permanent Magnet DC Motor Performance Curve

17-6 OTHER TYPES OF MOTORS

Torque Motors

As the name implies, *torque motors* are selected for their ability to exert a certain torque rather than for a rated power. Frequently this type of motor is operated at a stalled condition to maintain a set tension on a load. The continuous operation at slow speed or at zero speed causes heat generation to be a potential problem. In severe cases, external cooling fans may be required.

By special design, several of the AC and DC motor types discussed elsewhere in this chapter can be used as torque motors.

Servomotors

Either AC or DC servomotors are available to provide automatic control of position or speed of a mechanism in response to a control signal. Such motors are used in aircraft actuators, instruments, computer printers, and machine tools. Most have rapid response characteristics because of the low inertia of the rotating components and the relatively high torque exerted by the motor.

Stepper Motors

A stream of electronic pulses is delivered to a stepper motor, which then responds with a fixed rotation (step) for each pulse. Thus a very precise angular position can be obtained by counting and controlling the number of pulses delivered to the motor. Several step angles are available in commercially provided motors, such as 1.8°, 3.6°, 7.5°, 15°, 30°,

45°, and 90°. When the pulses are stopped, the motor stops automatically and is held in position. Because many of these motors are connected through a gear type speed reducer to the load, very precise positioning is possible to a small fraction of a step. Also, the reducer provides a torque increase.

Brushless Motors

The typical DC motor requires brushes to make contact with the rotating commutator on the shaft of the motor. This is a major failure mode of such motors. In the brushless DC motor, the switching of the rotor coils is accomplished by solid state electronic devices, resulting in very long life. The emission of electromagnetic interference is likewise reduced as compared with brush type DC motors.

Printed Circuit Motors

The rotor of the printed circuit motor is a flat disc that operates between two permanent magnets. The resulting design has a relatively large diameter and small axial length; sometimes it is called a *pancake motor*. The rotor has a very low inertia, so high acceleration rates are possible.

17-7 MOTOR CONTROLS

There are several functions that motor controls must perform, as outlined in Figure 17-19. The complexity of the control depends on the size and type of motor involved. Small fractional or subfractional motors may sometimes be started with a simple switch that connects the motor directly to the full line voltage. Larger motors, and some smaller motors on critical equipment, require greater protection.

The functions of motor controls are as follows:

1. To start and stop the motor
2. To protect the motor from overloads that would cause the motor to draw dangerously high current levels
3. To protect the motor from overheating
4. To protect personnel from contact with hazardous parts of the electrical system
5. To protect the controls from the environment
6. To prohibit the controls from causing a fire or explosion
7. To provide controlled torque, acceleration, speed, or deceleration of the motor
8. To provide for the sequential starting of a series of motors or other devices
9. To provide for the coordinated operation of different parts of a system, possibly including several motors
10. To protect the conductors of the branch circuit in which the motor is connected

The proper selection of a motor control system requires knowledge of at least the following factors.

1. The type of electrical service
   a. AC: Voltage and frequency; single- or three-phase; current limitations
   b. DC: Voltage; percent ripple; current limitations

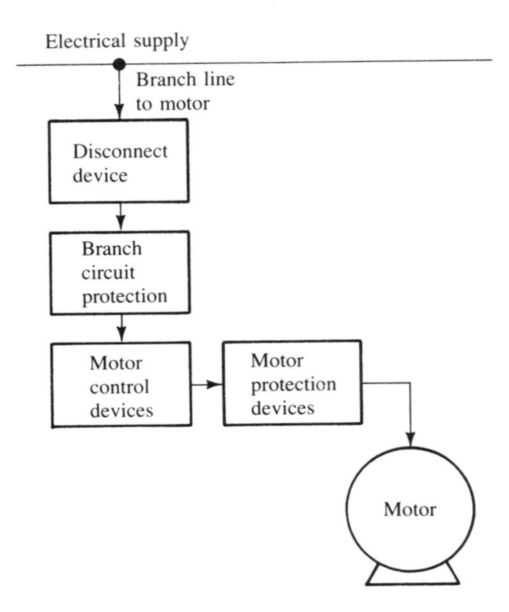

Figure 17-19 Motor Control Block Diagram

2. The type and size of motor: Power and speed ratings; full-load current rating; locked rotor current rating
3. Operation desired: Duty cycle (continuous, start/stop, or intermittent); single or multiple discrete speeds, or variable speed operation; one-direction or reversing
4. Environment: Temperature; water (rain, snow, sleet, sprayed or splashed water); dust and dirt; corrosive gases or liquids; explosive vapors or dusts; oils or lubricants
5. Space limitations
6. Accessibility of controls
7. Noise or appearance factors

The following discussion is for the control of AC induction motors. Control of DC motors is presented separately.

Starters

There are several classifications of motor starters: manual or magnetic; one-direction or reversing; two-wire or three-wire control; full-voltage or reduced-voltage starting; single speed or multiple speed; normal stopping, braking, or plug stopping. All of these typically include some form of overload protection which will be discussed later.

Manual and Magnetic, Full-voltage, One-direction Starting

Figure 17-20 shows the schematic connection diagram for manual starters for single-phase and three-phase motors. The symbol M indicates a normally open contactor (switch) that is actuated manually, for example, by throwing a lever. The contactors are rated according

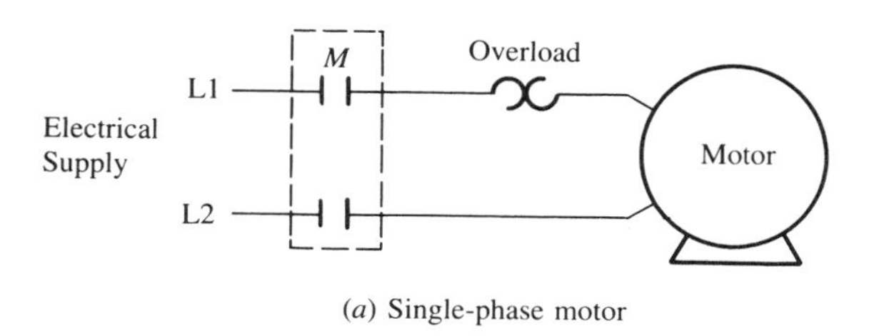

(*a*) Single-phase motor

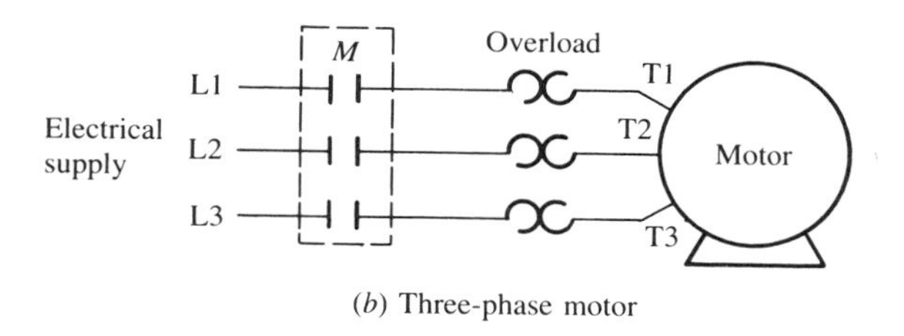

(*b*) Three-phase motor

Figure 17-20 Manual Starters. *M* = Normally open contactors. All actuate together.

to the motor power that they can safely handle. The power rating indirectly relates to the current drawn by the motor and the contactor design must safely make contact during the startup of the motor, considering the high starting current; carry the expected range of operating current without overheating; and break contact without excessive arcing that could burn the contacts. The ratings are established by NEMA. Tables 17-5 and 17-6 show the ratings for selected NEMA starter sizes.

Note in Figure 17-20 that overload protection is required in all three lines for three-phase motors but in only one line of the single-phase motors.

Figure 17-21 shows the schematic connection diagrams for magnetic starters using two-wire and three-wire controls. The "start" button in the three-wire control is a momentary contact type. As it is actuated manually, the coil in parallel with the switch is

Table 17-5 Ratings of AC Full-Voltage Starters for Single-phase Power

NEMA SIZE NUMBER	CURRENT RATING (AMPERES)	POWER RATING AT GIVEN VOLTAGES					
		110 V		*220 V*		*440 and 550 V*	
		hp	*kW*	*hp*	*kW*	*hp*	*kW*
00	—	½	0.37	¾	0.56	—	—
0	15	1	0.75	1½	1.12	1½	1.12
1	25	1½	1.12	3	2.24	5	3.73
2[a]	50	3	2.24	7½	5.60	10	7.46
3[a]	100	7½	5.60	15	11.19	25	18.65

[a]Applies to magnetically operated starters only.

Table 17-6 Ratings of AC Full-Voltage Starters for Three-phase Power

NEMA SIZE NUMBER	CURRENT RATING (AMPERES)	POWER RATING AT GIVEN VOLTAGES					
		110 V		*220 V*		*440 and 550 V*	
		hp	*kW*	*hp*	*kW*	*hp*	*kW*
00	—	¾	0.56	1	0.75	1	0.75
0	15	1½	1.12	2	1.49	2	1.49
1	25	3	2.24	5	3.73	7½	5.60
2	50	7½	5.60	15	11.19	25	18.65
3	100	15	11.19	30	22.38	50	37.30

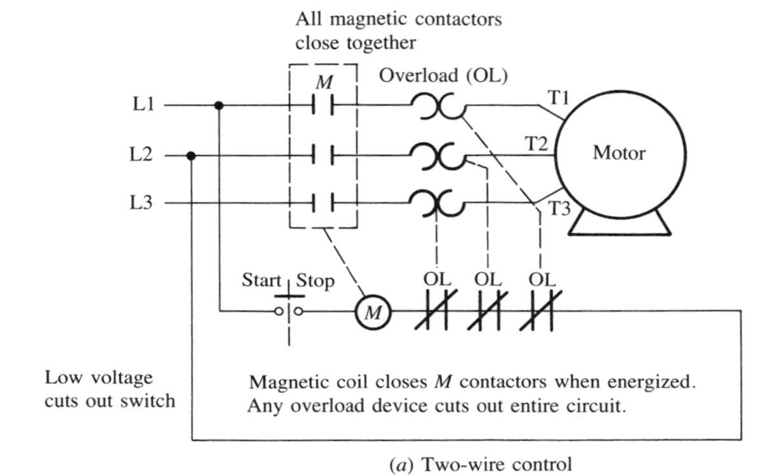

(*a*) Two-wire control

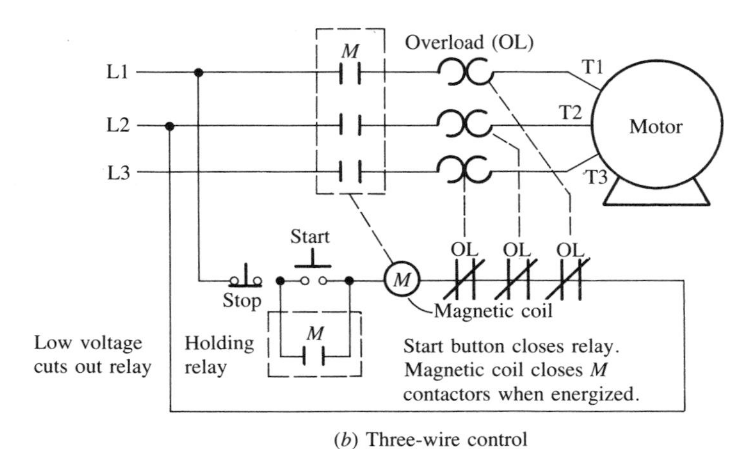

(*b*) Three-wire control

Figure 17-21 Magnetic Starters for Three-phase Motors

energized and it magnetically closes the line contactors marked *M*. The contacts remain closed until the stop button is pushed or until the line voltage drops to a set low value. (Remember, a low line voltage causes the motor to draw excessive current.) Either case causes the magnetic contactors to open, stopping the motor. The start button must be manually pushed again to restart the motor.

The two-wire control has a manually operated start button which stays engaged after the motor starts. As a safety feature, the switch will open when a low-voltage condition occurs. But when the voltage rises again to an acceptable level, the contacts close, restarting the motor. You must ensure that this is a safe operating mode.

Reversing Starters

Figure 17-22 shows the connection for a reversing starter for a three-phase motor. The direction of rotation of a three-phase motor can be reversed by interchanging any two of the three power lines. The *F* contactors are used for the forward direction. The *R* contactors would interchange L1 and L3 to reverse the direction. The *FOR* and *REV* pushbuttons actuate only one of the sets of contactors.

Reduced Voltage Starting

The motors discussed in the previous sections and the circuits shown in Figures 17-20 through 17-22 employ full-voltage starting. That is, when the system is actuated, the full line voltage is applied to the motor terminals. This would give the maximum starting

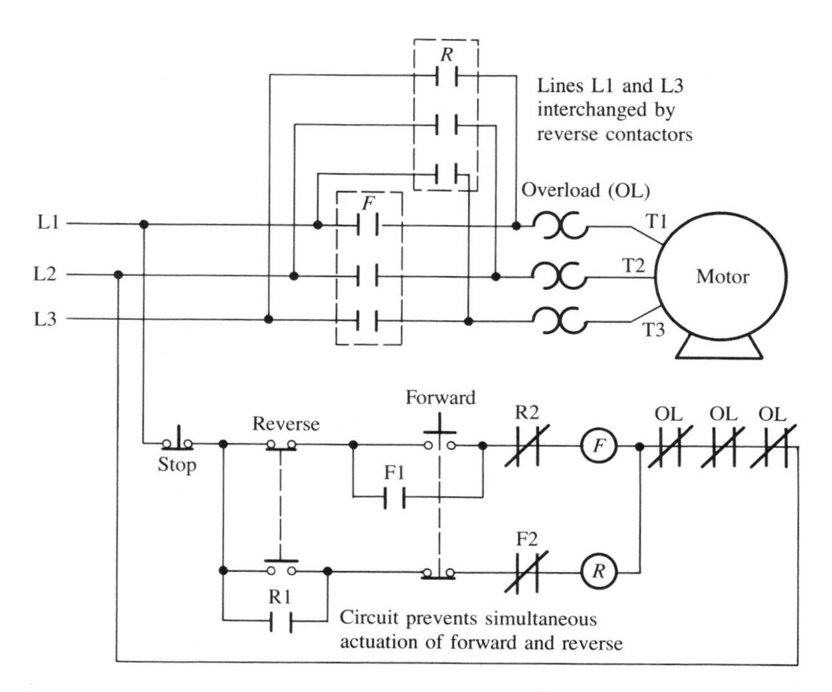

Figure 17-22 Reversing Control for a Three-phase Motor

effort, but in some cases it is not desirable. To limit jerk, to control the acceleration of a load, and to limit the starting current, reduced voltage starting is sometimes used. This gentle start is used on some conveyors, hoists, pumps, and similar loads.

Figure 17-23 shows one method of providing a reduced voltage to the motor when starting. The first action is the closing of the contactors marked *A*. Thus the power to the motor passes through a set of resistors which reduces the voltage at each motor terminal. A typical reduction would be to approximately 65 percent of normal line voltage; the peak line current would be reduced to 65 percent of normal locked rotor current, and the starting torque would be 42 percent of normal locked rotor torque (4). After the motor is accelerated, the main contactors *M* are closed and full voltage is applied to the motor. A timer is typically used to control the sequencing of the *A* and *M* contactors.

Dual-speed Motor Starting

A dual-speed motor with two separate windings to produce the different speeds can be started with the circuit shown in Figure 17-24. The operator selectively closes either the *F* (fast) or *S* (slow) contacts to obtain the desired speed. The other features of starting circuits discussed earlier can also be applied to this circuit.

Stopping the Motor

Where no special conditions exist when the system is shut down, the motor can be permitted to coast to a stop after the power is interrupted. The time required to stop will

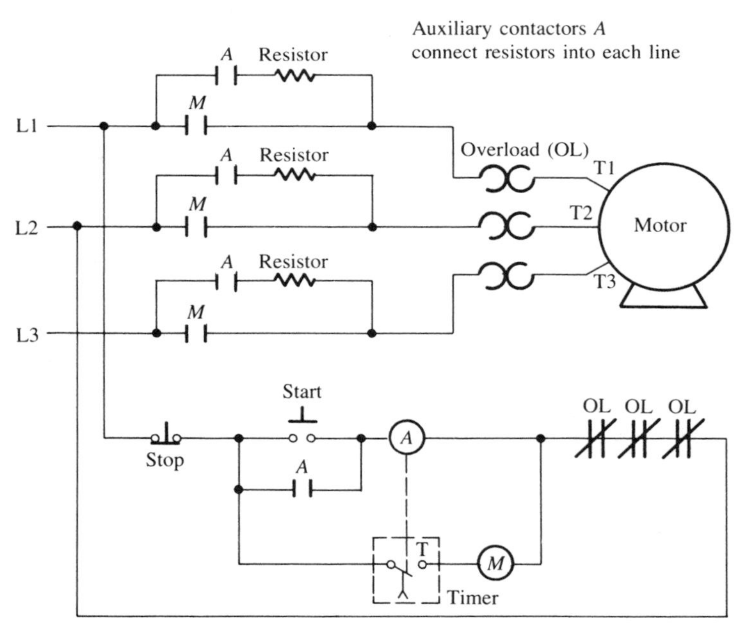

Figure 17-23 Reduced-voltage Starting by Primary Resistor Method

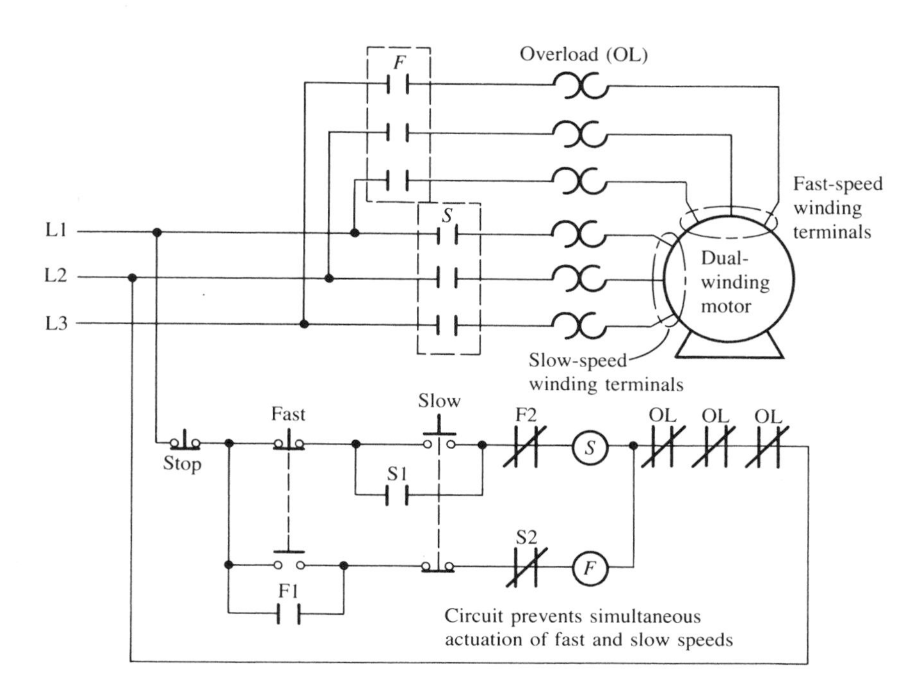

Figure 17-24 Speed Control for a Dual-winding Three-phase Motor

depend on the inertia and friction in the system. If controlled, rapid stopping is required, external brakes can be used. *Brake motors,* which have a brake integral with the motor, are available. Typically the design is of the "fail-safe" nature, in which the brake is disengaged by an electromagnetic coil when the motor is energized. When the motor is de-energized, either on purpose or because of power failure, the brake is actuated by mechanical spring force.

On circuits with reversing starters, *plug stopping* can be used. When it is desired to stop the motor running in the forward direction, the control can be switched immediately to reverse. There would then be a decelerating torque applied to the rotor, stopping it quickly. Care must be exercised to cut out the reversing circuit when the motor is at rest to prevent it from continuing in the reverse direction.

Overload Protection

The chief cause of failure in electric motors is overheating of the wound coils due to excessive current. The current is dependent on the load on the motor. A short circuit, of course, would cause a virtually instantaneously high current of a damaging level.

The protection against a short circuit can be provided by *fuses,* but careful application of fuses to motors is essential. A fuse contains an element which literally melts when a particular level of current flows through it. Thus the circuit is opened. Reactivating the circuit would require replacing the fuse. Time-delay fuses, or "slow-blowing" fuses, are needed for motor circuits to prevent the fuses from blowing when the motor starts, drawing the relatively high starting current which is normal and not damaging. After the motor starts, the fuse would blow at a set value of overcurrent.

Fuses are inadequate for larger or more critical motors because they provide protection at only one level of overcurrent. Each motor design has a characteristic *overheating curve,* as shown in Figure 17-25. This indicates that the motor could withstand different levels of overcurrent for different periods of time. For example, for the motor heating curve of Figure 17-25, a current twice as high as the full-load current (200 percent) could exist for up to 9 min before a damaging temperature is produced in the windings. But a 400 percent overload would cause damage in less than 2 min. An ideal overload protection device would parallel the overheating curve of the given motor, always cutting out the motor at a safe current level, as shown in Figure 17-25. Devices are available commercially to provide this protection. Some use special melting alloys, bimetallic strips similar to a thermostat, or magnetic coils that are sensitive to the current flowing in them. Most large motor starters include overload protection integral with the starter.

Another type of overload protection uses a temperature sensitive device inserted in the windings of the motor when it is made. Then it opens the motor circuit when the windings reach a dangerous temperature, regardless of the reason.

Enclosures for Motor Controls

As stated before, one of the functions of a motor control system is to protect personnel from contact with dangerous parts of the electrical system. Also, the protection of the system from the environment must be provided. These functions are accomplished by the enclosure.

NEMA has established standards for enclosures for the variety of environments encountered by motor controls. The most frequent types are described in Table 17-7.

AC Variable Speed Drives

Standard AC motors operate at a fixed speed for a given load if powered by AC power at a fixed frequency, say 60 Hz. Variable speed operation can be obtained with a control system which produces a variable frequency power. Two such control types are in common

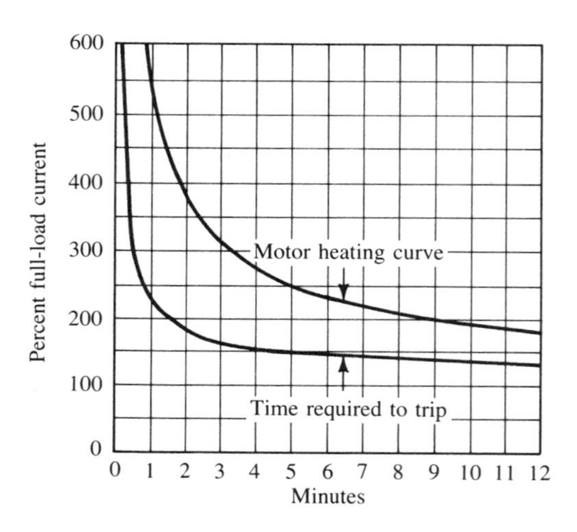

Figure 17-25 Motor Heating Curve and Response Curve of a Typical Overload Protector (Anderson, E. P., and Miller, Rex. *Electric Motors,* 3d ed. Indianapolis, Ind.: Theodore Audel & Co., 1977)

Table 17-7 Motor Control Enclosures

NEMA Design Number	*Description*
1	General-purpose: indoor use; not dusttight
3	Dusttight, raintight: outdoor weather resistant
3R	Dusttight, rainproof, sleet resistant
4	Watertight: can withstand a direct spray of water from a hose; used on ships and in food processing plants where washdown is required
4X	Watertight, corrosion resistant
7	Hazardous locations, class I: can operate in areas where flammable gases or vapors are present
9	Hazardous locations, class II: combustible dust areas
12	Industrial use: resistant to dust, lint, oil, coolants
13	Oiltight, dusttight

use, the *six-step method* and the *pulse width modulation (PWM) method*. Either system takes the 60-Hz line voltage and first rectifies it to a DC voltage. The six-step method then uses an inverter to produce a series of square waves that provides a voltage to the motor

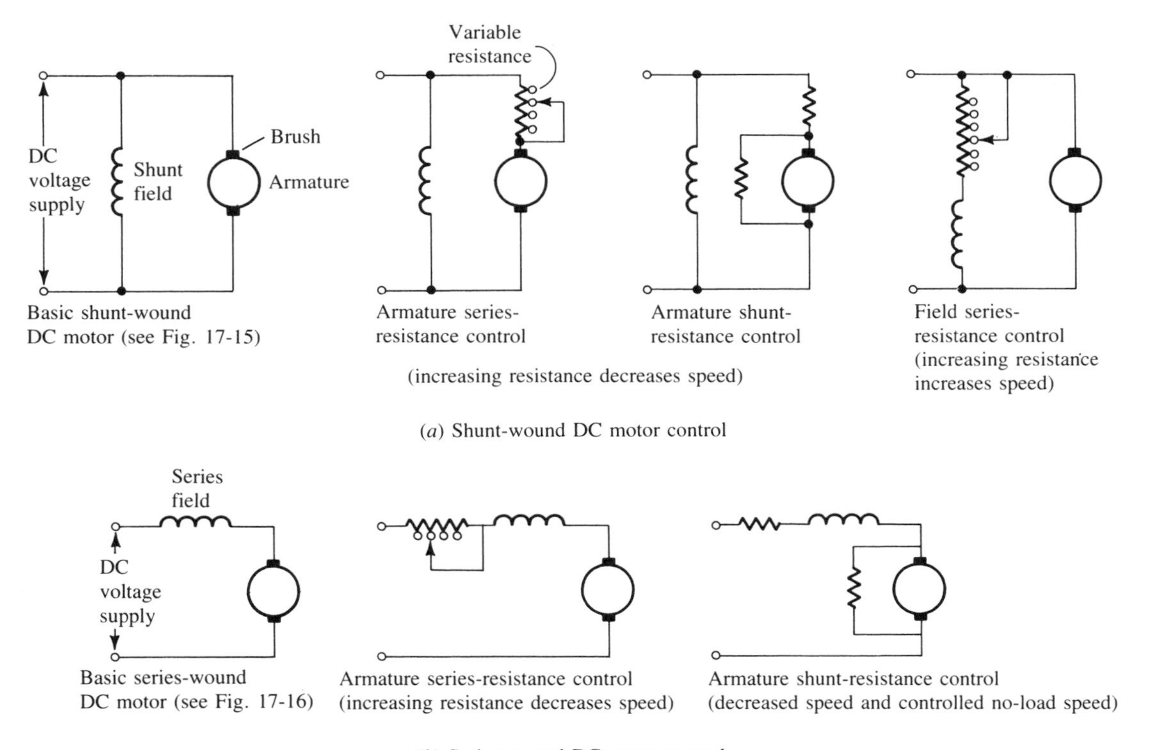

(*a*) Shunt-wound DC motor control

(*b*) Series-wound DC motor control

Figure 17-26 DC Motor Control

winding that varies in six steps per cycle. The frequency of the resulting six-step voltage is variable. In the PWM system, the DC voltage is input to an inverter that produces a series of pulses of variable width. The rate of polarity reversals determines the frequency applied to the motor.

DC Motor Control

Starting DC motors presents essentially the same problems as discussed for AC motors in terms of limiting the starting current and the provision of switching devices and holding relays of sufficient capacity to handle the operating loads. The situation is made somewhat more severe, however, by the presence of the commutators in the rotor circuit which are more sensitive to overcurrent.

Speed control is provided by variation of the resistance in the lines containing the armature or the field of the motor. The details depend on whether the motor is a series, shunt, or compound type. The variable resistance device is sometimes called a *rheostat* and can provide either step-wise variation in resistance or continuously varying resistance. Figure 17-26 shows the schematic diagrams for several types of DC motor speed controls.

REFERENCES

1. Anderson, E. P., and Miller, R. *Electric Motors,* 3d ed. Indianapolis: Theodore Audel & Company, 1977.
2. Baumeister, T., Avalonne, E. A., and Baumeister, T., eds. *Marks' Standard Handbook for Mechanical Engineers,* 8th ed. New York: McGraw-Hill Book Company, 1978.
3. Bose, B. K. "A Review of AC Drives Technology." Report No. 81CRD127. Schenectady: General Electric Company, June 1981.
4. Penton/IPC. "Electrical & Electronics Reference Issue." Machine Design Magazine. May 1981.

PROBLEMS

1. List six items that must be specified for electric motors.
2. List eight factors to be considered when selecting an electric motor.
3. Define *duty cycle.*
4. How much variation in voltage will most AC motors tolerate?
5. State the relationship between torque, power, and speed.
6. What does the abbreviation *AC* stand for?
7. Describe and sketch the form of single-phase AC power.
8. Describe and sketch the form of three-phase AC power.
9. What is the standard frequency of AC power in the United States?
10. What is the standard frequency of AC power in Europe?
11. What type of electrical power would be available in a typical American residence?
12. How many conductors are required to carry single-phase power? How many for three-phase power?
13. Assume that you are selecting an electric motor for a machine to be used in an industrial plant. The following types of AC power are available: 120-V, single-phase; 240-V, single-phase; 240-V, three-phase; 480-V, three-phase. In general, for which type of power would you specify your motor?
14. Define *synchronous speed* for an AC motor.
15. Define *full-load speed* for an AC motor.
16. What is the synchronous speed of a 4-pole AC motor when running in the United States? In France?
17. A motor name plate lists the full-load speed to be 3 450 rpm. How many poles does the motor have? What would be its approximate speed at zero load?
18. If a 4-pole motor operates on 400-Hz AC power, what would its synchronous speed be?
19. If an AC motor is listed as a normal slip 4-pole/6-pole motor, what would be its approximate full-load speeds?
20. What type of control would you use to make an AC motor run at variable speeds?
21. Describe a C-face motor.
22. Describe a D-flange motor.

23. What does the abbreviation *NEMA* stand for?
24. Describe a protected motor.
25. Describe a *TEFC* motor.
26. Describe a *TENV* motor.
27. What type of motor enclosure would you specify to be used in a plant that manufactures baking flour?
28. What type of motor would you specify for a meat grinder if the motor is to be exposed?
29. Figure 17-27 shows a machine that is to be driven by a 5-hp, protected, foot-mounted AC motor having a 184T frame. The motor is to be aligned with the shaft of the driven machine. Make a complete dimensioned drawing, showing standard side and top views of the machine and the motor. Design a suitable mounting base for the motor showing the motor mounting holes.
30. Define *locked rotor torque*. What is another term used for this parameter?
31. What is meant if one motor has a poorer speed regulation than another?
32. Define *breakdown torque*.
33. Name the four most common types of single-phase AC motors.
34. Refer to the AC motor performance curve in Figure 17-28.
    a. What type of motor is the curve likely to represent?
    b. If the motor is a 6-pole type, rated at 0.75 hp, how much torque can it exert at the rated load?
    c. How much torque can the motor develop to start a load?
    d. What is the breakdown torque for the motor?

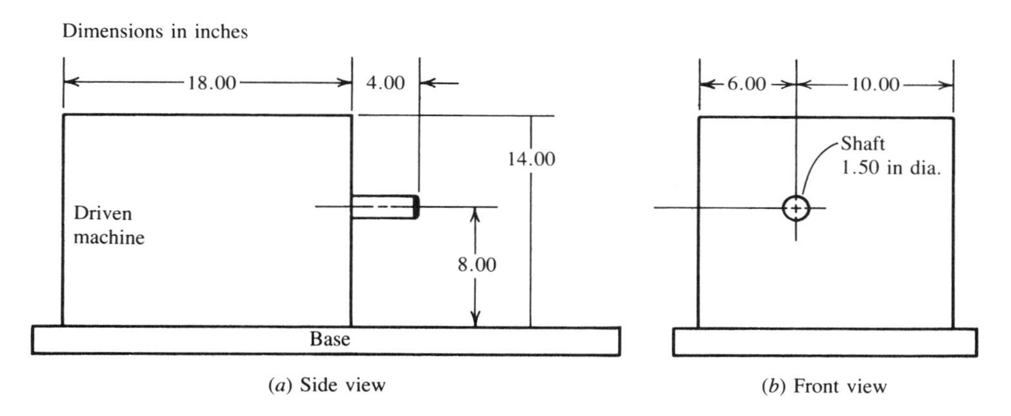

(*a*) Side view (*b*) Front view

Figure 17-27 (Problem 29)

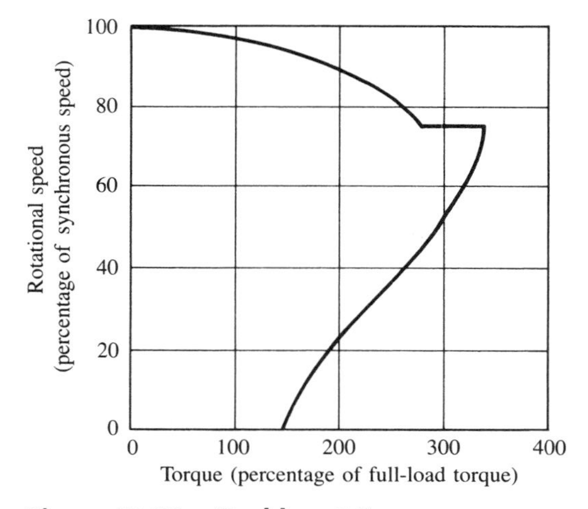

Figure 17-28 (Problem 34)

35. Repeat parts b, c, and d of problem 34 if the motor is a 2-pole type, rated at 1.50 kW.
36. A cooling fan for a computer is to operate at 1 725 rpm, direct-driven by an electric motor. The speed/torque curve for the fan is shown in Figure 17-29. Specify a suitable motor, giving the type of motor, horsepower, and number of poles.
37. Figure 17-30 shows the speed/torque curve for a household refrigeration compressor, designed to operate at 3 450 rpm. Specify a suitable motor, giving the type, power rating in watts, and number of poles.
38. How is speed adjusted for a wound rotor motor?
39. What is the full-load speed of a 10-pole synchronous motor?
40. What is meant by the term *pull-out torque,* as applied to a syncronous motor?
41. Discuss the reasons that universal motors are frequently used for hand-held tools and small appliances.
42. Why is the adjective *universal* used to describe a universal motor?
43. Name three ways to produce DC power.
44. List 12 common DC voltages.
45. What is an *SCR* control? For what is it used?
46. If a DC motor drive advertises that it produces *low ripple,* what does the term mean?
47. If you want to use a DC motor in your home, and your home has only standard 115 VAC, single-phase power, what would you need? What type of motor should you get?
48. List seven advantages of DC motors with respect to AC motors.
49. Discuss two disadvantages of DC motors.
50. Name four types of DC motors.
51. What happens to a series-wound DC motor if the load on the motor falls to near zero?
52. Assume that a permanent magnet DC motor can exert a torque of 15.0 N · m when operating at 3 000 rpm. What torque could it exert at 2 200 rpm?
53. List ten functions of a motor control.
54. What size motor starter is required for a 10-hp, three-phase motor, operating on 220 V?
55. A single-phase, 110-V AC motor has a name-plate rating of 1.00 kW. What size motor starter is required?
56. What does the term *plug stopping* mean and how is it accomplished?
57. Why is a fuse an inadequate protection device for an industrial motor?
58. What type of motor control enclosure would you specify for use in a car wash?
59. What could you do to the control circuit for a standard series DC motor to give it a controlled no-load speed?
60. What happens if you connect a resistance in series with the armature of a shunt-wound DC motor?
61. What happens if you connect a resistance in series with the shunt field of a shunt-wound DC motor?

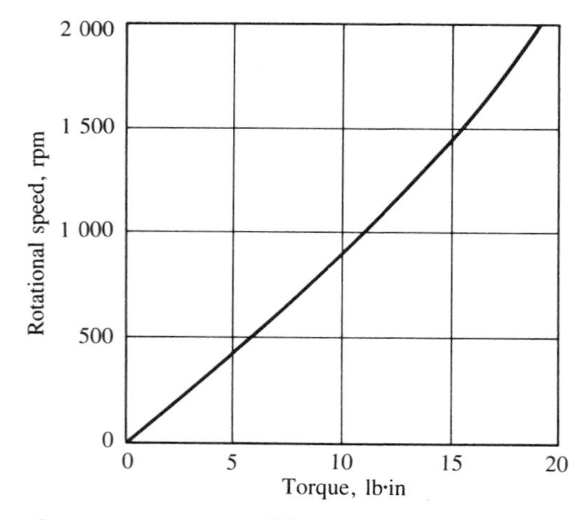

Figure 17-29 (Problem 36)

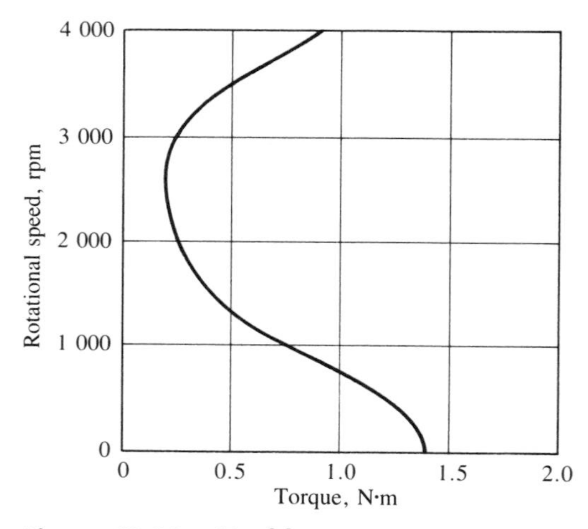

Figure 17-30 (Problem 37)

18 Power Screws, Ball Screws, and Fasteners

18-1 OVERVIEW

Power screws are designed to transmit force through the classic inclined plane concept inherent in the screw thread. Several configurations are possible. If the screw is rotated while being held against translation, then the member having the internal threads, analogous to a nut on a bolt, will be translated. If the nut is held, the screw is translated.

Applications of power screws are seen in machine tools such as lathes and milling machines. The carriage of the lathe that carries the cutting tool is moved along the bed of the lathe by a power screw. The rotational speed of the screw is controlled to give a particular speed of travel and to coordinate the motion of the carriage with the rotation of the spindle. In a milling machine, the X and Y motions of the table are usually controlled by power screws. The Z axis is also moved by a screw in some mills.

Some types of mechanical jacks use the power screw to lift large loads with a modest torque applied to rotate the screw. Materials testing devices, such as the tensile test machine, use power screws to exert the force required to pull the tensile specimen to failure. Others use hydraulic cylinders to apply the force but use power screws to permit adjustment of the crosshead of the machine. Then the full force applied to the specimen is transferred through the screw thread.

Figure 18-1 shows three types of power screw threads: the square thread, the Acme thread, and the buttress thread. Of these, the square and buttress threads are the most efficient. That is, they require the least torque to move a given load along the screw. However, the Acme thread is not greatly less efficient and it is easier to machine. The buttress thread is desirable when force is to be transmitted in only one direction.

Table 18-1 gives the preferred combinations of basic major diameter, D, and number of threads per inch, n, for Acme screw threads. The pitch, p, is the distance from a point on one thread to the corresponding point on the adjacent thread, and $p = 1/n$.

Other pertinent dimensions listed in Table 18-1 include the minimum minor diameter and the minimum pitch diameter of a screw with an external thread. In performing stress analyses on the screw, the safest approach is to compute the area corresponding to the minor diameter for tensile or compressive stresses. However, a more accurate stress computation results from using the *tensile stress area* (listed in the table), found from

$$A_t = \frac{\pi}{4}\left[\frac{D_r + D_p}{2}\right]^2 \tag{18-1}$$

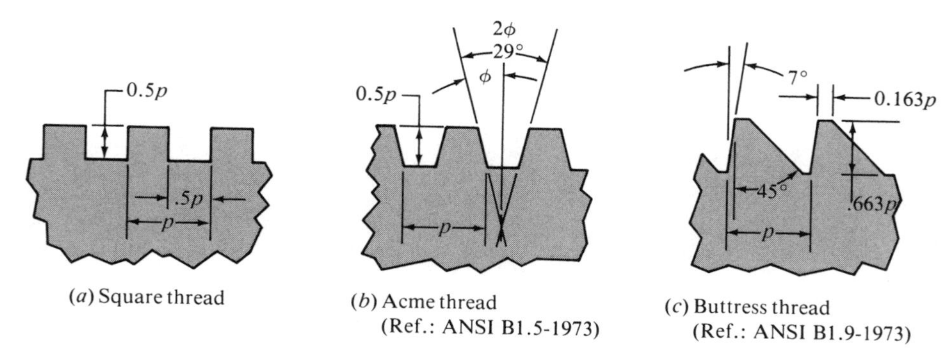

Figure 18-1 Forms of Power Screw Threads (b, ANSI B1.5—1973; c, ANSI B1.9—1973)

Table 18-1 Preferred Acme Screw Threads

Nominal Major Diameter (in)	*Threads per in n*	*Pitch, $p = 1/n$ (in)*	*Minimum Minor Diameter (in)*	*Minimum Pitch Diameter (in)*	*Tensile Stress Area (in²)*	*Shear Stress Area (in²)[a]*
1/4	16	0.062 5	0.161 8	0.204 3	0.026 32	0.335 5
5/16	14	0.071 4	0.214 0	0.261 4	0.044 38	0.434 4
3/8	12	0.083 3	0.263 2	0.316 1	0.065 89	0.527 6
7/16	12	0.083 3	0.325 3	0.378 3	0.097 20	0.639 6
1/2	10	0.100 0	0.359 4	0.430 6	0.122 5	0.727 8
5/8	8	0.125 0	0.457 0	0.540 8	0.195 5	0.918 0
3/4	6	0.166 7	0.537 1	0.642 4	0.273 2	1.084
7/8	6	0.166 7	0.661 5	0.766 3	0.400 3	1.313
1	5	0.200 0	0.750 9	0.872 6	0.517 5	1.493
1 1/8	5	0.200 0	0.875 3	0.996 7	0.688 1	1.722
1 1/4	5	0.200 0	0.999 8	1.121 0	0.883 1	1.952
1 3/8	4	0.250 0	1.071 9	1.218 8	1.030	2.110
1 1/2	4	0.250 0	1.196 5	1.342 9	1.266	2.341
1 3/4	4	0.250 0	1.445 6	1.591 6	1.811	2.803
2	4	0.250 0	1.694 8	1.840 2	2.454	3.262
2 1/4	3	0.333 3	1.857 2	2.045 0	2.982	3.610
2 1/2	3	0.333 3	2.106 5	2.293 9	3.802	4.075
2 3/4	3	0.333 3	2.355 8	2.542 7	4.711	4.538
3	2	0.500 0	2.432 6	2.704 4	5.181	4.757
3 1/2	2	0.500 0	2.931 4	3.202 6	7.388	5.700
4	2	0.500 0	3.430 2	3.700 8	9.985	6.640
4 1/2	2	0.500 0	3.929 1	4.199 1	12.972	7.577
5	2	0.500 0	4.428 1	4.697 3	16.351	8.511

[a]Per inch of length of engagement.

This is the area corresponding to the average of the minor (or root) diameter, D_r, and the pitch diameter, D_p. The data reflect the minimums for commercially available screws according to recommended tolerances (3).

The shear stress area, A_s, listed in the table is also found in published data (3) and represents the area in shear approximately at the pitch line of the threads for a 1.0-in length of engagement. Other lengths would require that the area be modified by the ratio of the actual length to 1.0 in.

Torque Required to Move a Load

When using a power screw to exert a force, as with a jack raising a load, there is a need to know how much torque must be applied to the nut of the screw to move the load. The parameters involved are the force to be moved, F; the size of the screw, as indicated by its pitch diameter, D_p; the lead of the screw, L; and the coefficient of friction, f. Note that the *lead* is defined as the axial distance that the screw would move in one complete revolution. For the usual case of a single-threaded screw, the lead is equal to the pitch and can be read from Table 18-1 or computed from $L = p = 1/n$.

In the development of equation (18-2) for the torque required to turn the screw, Figure 18-2(a), which depicts a load being pushed up an inclined plane against a friction

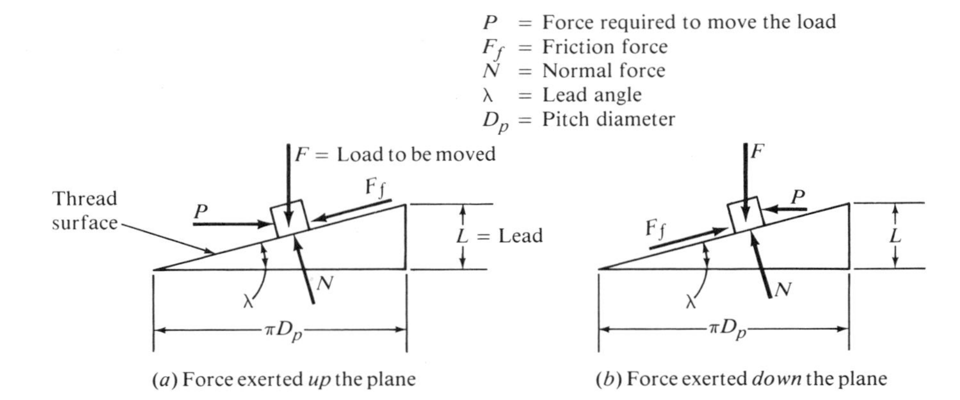

(a) Force exerted *up* the plane (b) Force exerted *down* the plane

Figure 18-2 Screw Thread Force Analysis

force, is used. This is a reasonable representation for a square thread if you think of the thread as being unwrapped from the screw and laid flat. The torque for an Acme thread is slightly different from this due to the thread angle. The revised equation for the Acme thread will be shown later.

The torque computed from equation (18-2) is called T_u, implying that the force is applied to move a load up the plane, that is, to raise the load. This observation is completely appropriate if the load is raised vertically, as with a jack. If, however, the load is horizontal or at some angle, equation (18-2) is still valid if the load is to be advanced along the screw "up the thread." Equation (18-4) shows the required torque, T_d, to lower a load or move a load "down the thread."

The torque to move a load up the thread is

$$T_u = \frac{FD_p}{2}\left[\frac{L + \pi f D_p}{\pi D_p - fL}\right] \tag{18-2}$$

This equation accounts for the force required to overcome friction between the screw and the nut in addition to the force just required to move the load. If the screw or the nut bears against a stationary surface while rotating, there would be an additional friction torque developed at that surface. For this reason, many jacks and similar devices incorporate antifriction bearings at such points.

The coefficient of friction for use in equation (18-2) depends on the materials used and the manner of lubricating the screw. For well-lubricated steel screws acting in steel nuts, $f = 0.15$ should be conservative.

An important factor in the analysis for torque is the angle of inclination of the plane. In a screw thread, the angle of inclination is referred to as the *lead angle*, λ. It is the angle between the tangent to the helix of the thread and the plane transverse to the axis of the screw. It can be seen from the figure that

$$\tan \lambda = L/(\pi D_p) \tag{18-3}$$

where πD_p is the circumference of the pitch line of the screw. Then, if the rotation of the screw tends to raise the load (move it up the incline), the friction force opposes the motion and acts down the plane.

Conversely, if the rotation of the screw tends to lower the load, the friction force would act up the plane, as shown in Figure 18-2(b). The torque analysis changes, producing equation (18-4):

$$T_d = \frac{FD_p}{2}\left[\frac{\pi f D_p - L}{\pi D_p + fL}\right] \tag{18-4}$$

If the screw thread is very steep (that is, it has a high lead angle), the friction force may not be able to overcome the tendency for the load to "slide" down the plane and the load would fall due to gravity. In most cases for power screws with single threads, however, the lead angle is rather small and the friction force is large enough to oppose the load and keep it from sliding down the plane. Such a screw is called *self-locking,* a desirable characteristic for jacks and similar devices. Quantitatively, the condition that must be met for self-locking is

$$f > \tan \lambda \tag{18-5}$$

The coefficient of friction must be greater than the tangent of the lead angle. For $f = 0.15$, the corresponding value of the lead angle is 8.5°. For $f = 0.1$, for very smooth, well-lubricated surfaces, the lead angle for self-locking is 5.7°. The lead angles for the screw designs listed in Table 18-1 range from 1.94° to 5.57°. Thus it is expected that all would be self-locking. However, operation under vibration should be avoided as this may still cause movement of the screw.

Efficiency of a Power Screw

Efficiency for the transmission of a force by a power screw can be expressed as the ratio of the torque required to move the load without friction to that with friction. Equation (18-2) gives the torque required with friction, T_u. Letting $f = 0$, the torque required without friction, T', is

$$T' = \frac{FD_p}{2}\frac{L}{\pi D_p} = \frac{FL}{2\pi} \tag{18-6}$$

Then the efficiency, e, is

$$e = \frac{T'}{T_u} = \frac{FL}{2\pi T_u} \tag{18-7}$$

Alternate Forms of the Torque Equations

Equations (18-2) and (18-4) can be expressed in terms of the lead angle, rather than the lead and the pitch diameter, by noting the relationship in equation (18-3). With this substitution, the torque required to move the load would be

$$T_u = \frac{FD_p}{2}\left[\frac{(\tan \lambda + f)}{(1 - f \tan \lambda)}\right] \tag{18-8}$$

And the torque required to lower the load is

$$T_d = \frac{FD_p}{2}\left[\frac{(f - \tan \lambda)}{(1 + f \tan \lambda)}\right] \tag{18-9}$$

Adjustment for Acme Threads

The difference between Acme threads and the square threads is the presence of the thread angle, ϕ. It can be seen from Figure 18-1 that $2\phi = 29°$, and therefore $\phi = 14.5°$. This changes the direction of action of the forces on the thread from that depicted in Figure 18-2. Figure 18-3 shows that F would have to be replaced by $F/\cos\phi$. Carrying this through the analysis for torque would give modified forms of equations (18-8) and (18-9):

The torque to move the load up the thread would be

$$T_u = \frac{FD_p}{2}\left[\frac{(\cos\phi \tan\lambda + f)}{(\cos\phi - f\tan\lambda)}\right] \tag{18-10}$$

And the torque to move the load down the thread is

$$T_d = \frac{FD_p}{2}\left[\frac{(f - \cos\phi \tan\lambda)}{(\cos\phi + f\tan\lambda)}\right] \tag{18-11}$$

Example Problem 18-1. Two Acme threaded power screws are to be used to raise a heavy access hatch, as sketched in Figure 18-4. The total weight of the hatch is 25 000 lb, divided equally between the two screws. Select a satisfactory screw from Table 18-1 on the basis of tensile strength, limiting the tensile stress to 10 000 psi. Then determine the required thickness of the yoke that acts as the nut on the screw to limit the shear stress in the threads to 5 000 psi. For the screw thus designed, compute the lead angle, the torque required to raise the load, the efficiency of the screw, and the torque required to lower the load. Use a coefficient of friction of 0.15.

Solution. The load to be lifted places each screw in direct tension. Therefore, the required tensile stress area is

$$A_t = \frac{F}{\sigma_a} = \frac{12\,500 \text{ lb}}{10\,000 \text{ lb/in}^2} = 1.25 \text{ in}^2$$

From Table 18-1, a 1½-in diameter Acme thread screw with four threads per inch would provide a tensile stress area of 1.266 in^2.

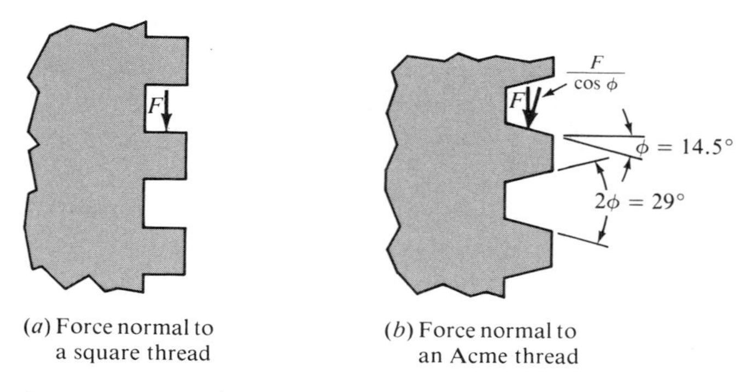

Figure 18-3 Force on an Acme Thread

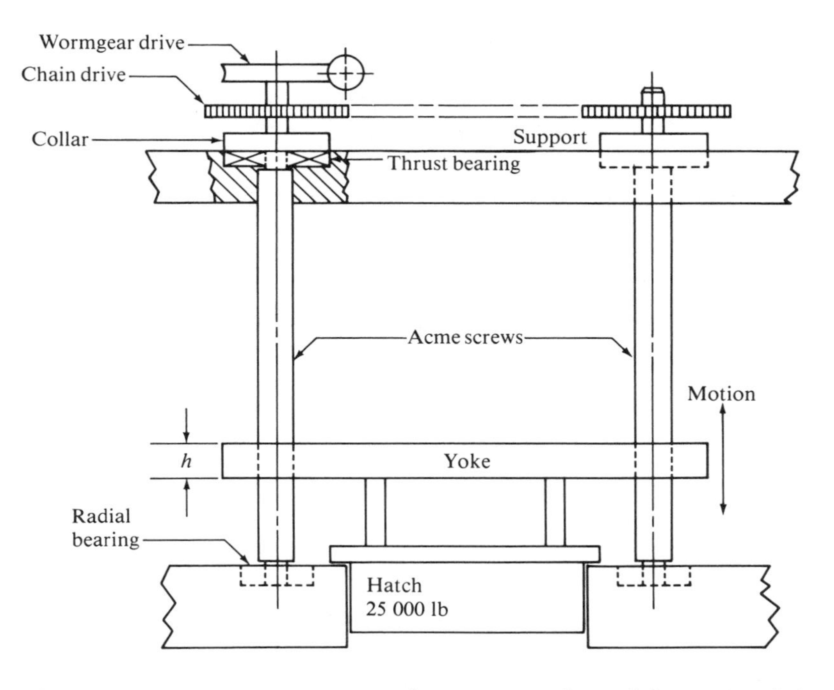

Figure 18-4 An Acme Screw Driven System for Raising a Hatch (Example Problem 18-1)

For this screw, each inch of length of a nut would provide 2.341 in^2 of shear stress area in the threads. The required shear area is then

$$A_s = \frac{F}{\tau_a} = \frac{12\,500 \text{ lb}}{5\,000 \text{ lb/in}^2} = 2.50 \text{ in}^2$$

Then the required length of the yoke would be

$$h = 2.5 \text{ in}^2 \left[\frac{1.0 \text{ in}}{2.341 \text{ in}^2}\right] = 1.07 \text{ in}$$

For convenience, let's specify $h = 1.25$ in

The lead angle is (remember $L = p = 1/n = 1/4 = 0.250$ in)

$$\lambda = \tan^{-1}\frac{L}{\pi D_p} = \tan^{-1}\frac{0.250}{\pi(1.342\,9)} = 3.39°$$

The torque required to raise the load can be computed from equation (18-10).

$$T_u = \frac{FD_p}{2}\left[\frac{(\cos\phi \tan\lambda + f)}{(\cos\phi - f\tan\lambda)}\right] \tag{18-10}$$

Using $\cos\phi = \cos(14.5°) = 0.968$, and $\tan\lambda = \tan(3.39°) = 0.059\,2$:

$$T_u = \frac{(12\,500 \text{ lb})(1.342\,9 \text{ in})}{2}\frac{[(0.968)(0.059\,2) + 0.15]}{[0.968 - (0.15)(0.059\,2)]} = 1\,809 \text{ lb} \cdot \text{in}$$

The efficiency can be computed from equation (18-7).

$$e = \frac{FL}{2\pi T_u} = \frac{(12\,500 \text{ lb})(0.250 \text{ in})}{2(\pi)(1\,809 \text{ lb} \cdot \text{in})} = 0.275 \quad \text{or 27.5 percent}$$

The torque required to lower the load can be computed from equation (18-11).

$$T_d = \frac{FD_p}{2}\left[\frac{(f - \cos\phi \tan\lambda)}{(\cos\phi + f\tan\lambda)}\right] \tag{18-11}$$

$$T_d = \frac{(12\,500 \text{ lb})(1.342\,9 \text{ in})}{2} \frac{[0.15 - (0.968)(0.059\,2)]}{[0.968 + (0.15)(0.059\,2)]} = 796 \text{ lb} \cdot \text{in}$$

Power Required to Drive a Power Screw

If the torque required to rotate the screw is applied at a constant rotational speed, n, then the power in horsepower to drive the screw is

$$P = \frac{Tn}{63\,000}$$

Example Problem 18-2. It is desired to raise the hatch in Figure 18-4 a total of 15.0 in in no more than 12.0 sec. Compute the required rotational speed for the screws and the power required.

Solution. The screw selected in the solution for example problem 18-1 was a 1½-in Acme threaded screw with four threads per inch. Thus the load would be moved ¼ in with each revolution. The linear speed required is

$$V = \frac{15.0 \text{ in}}{12.0 \text{ sec}} = 1.25 \text{ in/s}$$

The required rotational speed would be

$$n = \frac{1.25 \text{ in}}{\text{s}} \frac{1 \text{ rev}}{0.25 \text{ in}} \frac{60 \text{ s}}{\text{min}} = 300 \text{ rpm}$$

Then the power required to drive each screw would be

$$P = \frac{Tn}{63\,000} = \frac{(1\,809 \text{ lb} \cdot \text{in})(300 \text{ rpm})}{63\,000} = 8.61 \text{ hp}$$

18-2 BALL SCREWS

The basic action of using screws to produce linear motion from rotation has been described in section 18-1 on power screws. A special adaptation of this action which minimizes the friction between the screw threads and the mating nut is the ball screw.

Figure 18-5 shows a cutaway view of a commercially available ball screw. It replaces the sliding friction of the conventional power screw with the rolling friction of bearing balls. The bearing balls circulate in hardened steel races formed by concave helical

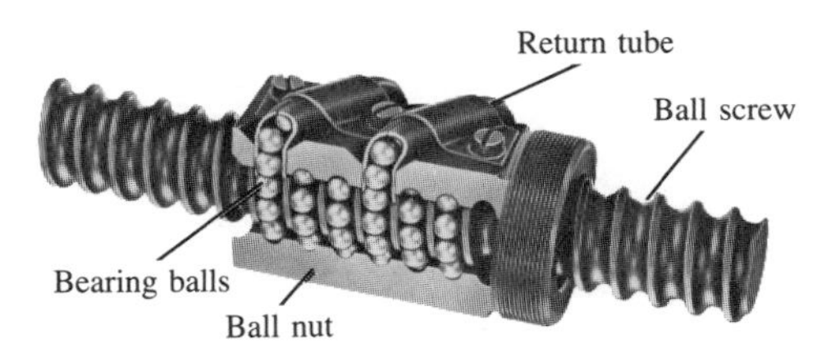

Figure 18-5 Ball Bearing Screw (Saginaw Steering Gear Division, General Motors Corp., Saginaw, Mich.)

grooves in the screw and nut. All reactive loads between the screw and the nut are carried by the bearing balls that provide the only physical contact between these members. As the screw and nut rotate relative to each other, the bearing balls are diverted from one end and carried by the ball guide return tubes to the opposite end of the ball nut. This recirculation permits unrestricted travel of the nut in relation to the screw (6).

Applications of ball screws occur in automotive steering systems, machine tool tables and linear actuators, jacking and positioning mechanisms, aircraft controls such as flap actuating devices, packaging equipment, instruments, and many similar systems. Figure 18-6 shows a machine with a ball screw installed to move a component along the ways of the bed.

The application parameters to be considered in selecting a ball screw include the following:

The axial load to be exerted by the screw during rotation

The rotational speed of the screw

Figure 18-6 Application of a Ball Screw (Saginaw Steering Gear Division, General Motors Corp., Saginaw, Mich.)

The maximum static load on the screw
The direction of the load
The manner of supporting the ends of the screw
The length of the screw
The expected life
The environmental conditions

When transmitting a load, a ball screw experiences stresses similar to those on a ball bearing, as discussed in Chapter 15. The load is transferred from the screw to the balls, from the balls to the nut, and from the nut to the driven device. The contact stress between the balls and the races in which they roll eventually causes fatigue failure, indicated by pitting of the balls or the races.

Thus, the rating of ball screws gives the load capacity of the screw for a given life which 90 percent of the screws of a given design will survive. This is similar to the L_{10} life of ball bearings. Because ball screws are typically used as linear actuators, the most pertinent life parameter is the distance travelled by the nut relative to the screw.

Manufacturers usually report the rated load that a given screw can exert for 1 million in (25.4 km) of cumulative travel. The relationship between load, P, and life, L, is also similar to that for a ball bearing.

$$\frac{L_2}{L_1} = \left(\frac{P_1}{P_2}\right)^3 \tag{18-12}$$

Thus if the load on a ball screw is doubled, the life is reduced to one-eighth of the original life. If the load is cut in half, the life is increased by eight times. Figure 18-7 shows the nominal performance of ball screws of a small variety of sizes. Many more sizes and styles are available.

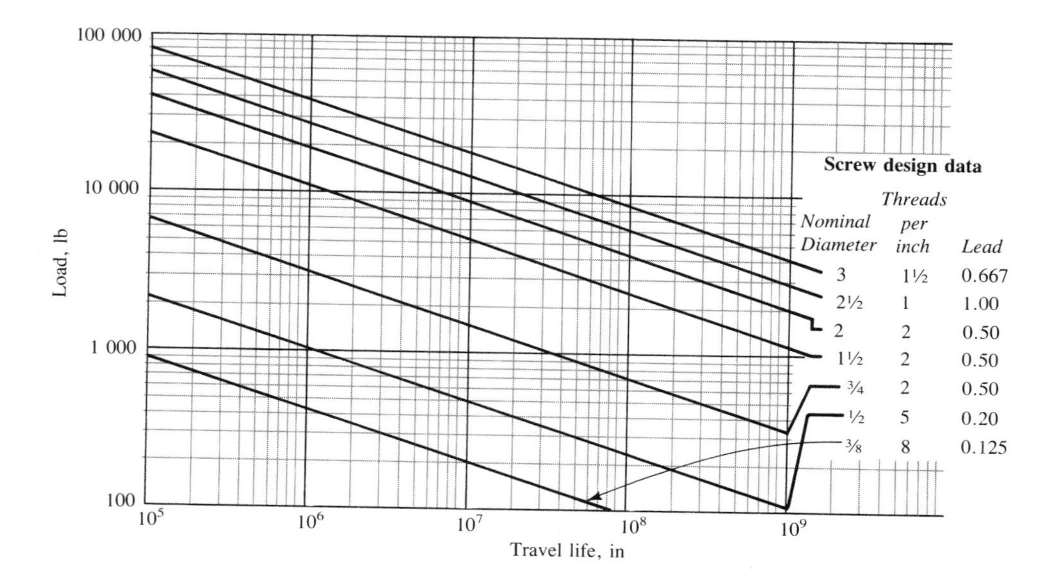

Figure 18-7 Ball Screw Performance

In addition to the load/life considerations, the proper application of ball screws must take into account the vibration tendencies and the tendency for the screw to buckle. Because of the low friction inherent in a ball screw, many are applied at a high rotational speed. Care must be exercised that the operational speed is not near a critical speed of the screw. The critical speed is dependent on the length (actually the length squared!), the screw diameter, and the manner of supporting the ends. Thus one of the worst cases is a long, small-diameter screw, supported in one simple bearing at each end. Such a system would have a relatively low critical speed. One of the best designs, resulting in a high critical speed, is a short, large-diameter screw, mounted rigidly in a pair of widely spaced bearings on each end. Catalogs give specific design data for critical speeds. As long as the operating speed is below 0.80 times the critical speed, the operation should be satisfactory.

Column buckling should be checked whenever a ball screw is loaded in compression. The parameters are the same as those discussed in Chapter 14. The good and bad design approaches for column loading are similar to those described for critical speed. Of course, whenever possible, the load on the screw should be tensile. Then column buckling is not a factor.

Torque and Efficiency

The efficiency of a ball bearing screw is typically taken to be 90 percent. This far exceeds the efficiency for power screws without rolling contact which are typically in the range of 20 to 30 percent. Thus far less torque is required to exert a given load with a given size of screw. Power is correspondingly reduced. The computation of the torque to turn is adapted from equation (18-7), relating efficiency to torque.

$$e = \frac{FL}{2\pi T_u} \qquad (18\text{-}7)$$

Then, using $e = 0.90$,

$$T_u = \frac{FL}{2\pi e} = 0.177FL \qquad (18\text{-}13)$$

Because of the low friction, ball screws are virtually never self-locking. In fact, this is also used to advantage by purposely using the applied load on the nut to rotate the screw. This is called *backdriving;* backdriving torque can be computed from

$$T_b = \frac{FLe}{2\pi} = 0.143FL \qquad (18\text{-}14)$$

Example Problem 18-3. Select suitable ball screws for the application described in example problem 18-1 and illustrated in Figure 18-4. The hatch must be lifted 15.0 in to open it eight times per day and then closed. The design life is 10 years. The lifting or lowering is to be completed in no more than 12.0 sec.

For the screw selected, compute the torque to turn the screw, the power required, and the actual expected life.

Solution. The data required to select a screw from Figure 18-7 are the load and the travel of the nut on the screw over the desired life. The load is 12 500 lb on each screw.

$$\text{Travel} = \frac{15.0 \text{ in}}{\text{stroke}} \frac{2 \text{ strokes}}{\text{cycle}} \frac{8 \text{ cycles}}{\text{day}} \frac{365 \text{ days}}{\text{year}} \frac{10 \text{ years}}{} = 8.76 \times 10^5 \text{ in}$$

From Figure 18-7, the 2-in screw with two threads per inch and a lead of 0.50 in is satisfactory.

The torque required to turn the screw is

$$T_u = 0.177FL = 0.177(12\,500)(0.50) = 1\,106 \text{ lb} \cdot \text{in}$$

The rotational speed required is

$$n = \frac{1 \text{ rev}}{0.50 \text{ in}} \frac{15.0 \text{ in}}{12.0 \text{ s}} \frac{60 \text{ s}}{\text{min}} = 150 \text{ rpm}$$

The power required for each screw is

$$P = \frac{Tn}{63\,000} = \frac{(1\,106)(150)}{63\,000} = 2.63 \text{ hp}$$

Compare this with the 8.61 hp required for the Acme screw in example problem 18-1.

The actual travel life expected for this screw at 12 500 lb load would be approximately 3.2×10^6 in, using Figure 18-7. This is 3.65 times longer than required.

18-3 FASTENERS

A *fastener* is any device used to connect or join two or more components. Literally hundreds of fastener types and variations are available. The most common are threaded fasteners referred to by many names, among them: bolts, screws, nuts, studs, lag screws, and set screws.

A *bolt* is a threaded fastener designed to pass through holes in the mating members and to be secured by tightening a nut from the end opposite the head of the bolt. See Figure 18-8(a). Several types of bolts are shown in Figure 18-9.

A *screw* is a threaded fastener designed to be inserted through a hole in one member to be joined and into a threaded hole in the mating member. See Figure 18-8(b). The threaded hole may have been preformed, say by tapping, or it may be formed by the screw itself as it is forced into the material. *Machine screws,* also called *cap screws,* are precision fasteners with straight threaded bodies that are turned into tapped holes (see

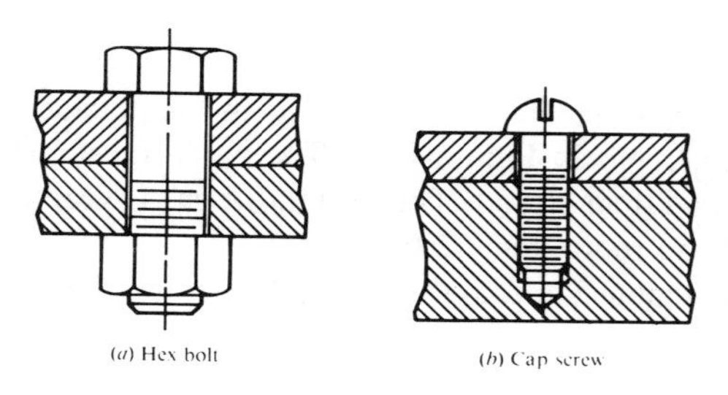

Figure 18-8 Comparison of a Bolt with a Screw (Hoelscher, et al., *Graphics for Engineers,* New York: John Wiley & Sons, 1968)

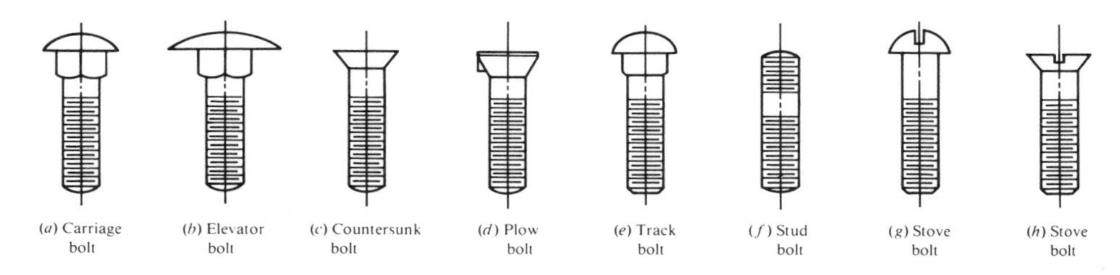

Figure 18-9 Bolt Styles (Hoelscher, et al., *Graphics for Engineers,* New York: John Wiley & Sons, Inc., 1968)

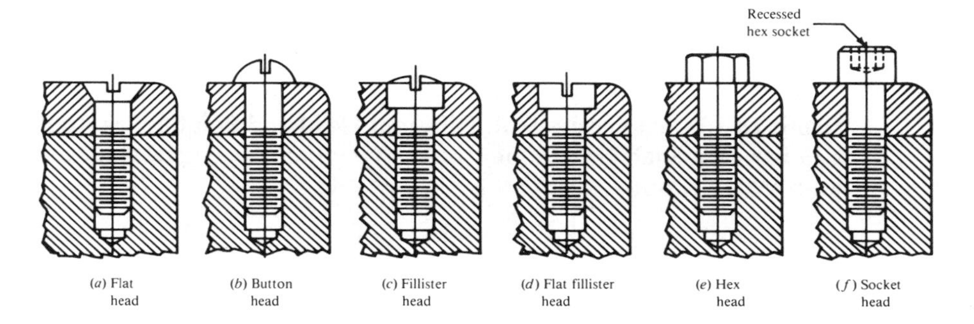

Figure 18-10 Cap Screws or Machine Screws (Hoelscher, et al., *Graphics for Engineers,* New York: John Wiley & Sons, 1968)

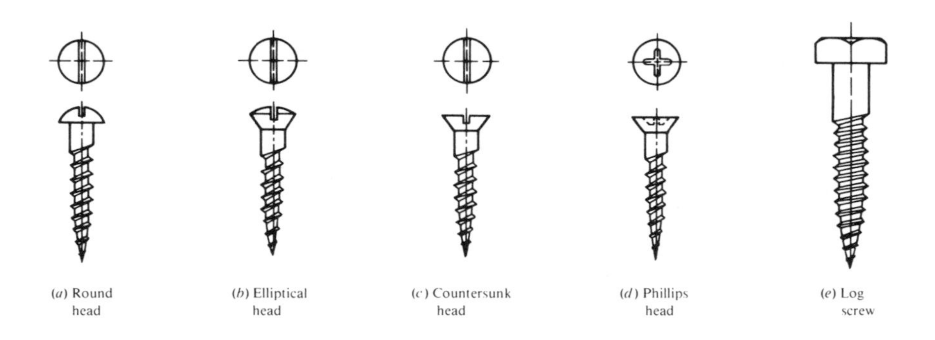

Figure 18-11 Screws (Hoelscher, et al., *Graphics for Engineers,* New York: John Wiley & Sons, 1968)

Figure 18-10). *Sheet metal screws, lag screws, self-tapping screws,* and *wood screws* usually form their own threads. Figure 18-11 shows a few styles.

Most bolts and screws have enlarged heads that bear down on the part to be clamped and thus exert the clamping force. *Set screws* are headless, are inserted into tapped holes, and are designed to bear directly on the mating part, locking it into place. Figure 18-12 shows several styles of points and drive means for set screws. Caution must be used with set screws, as with any threaded fastener, so that vibration does not loosen the screw.

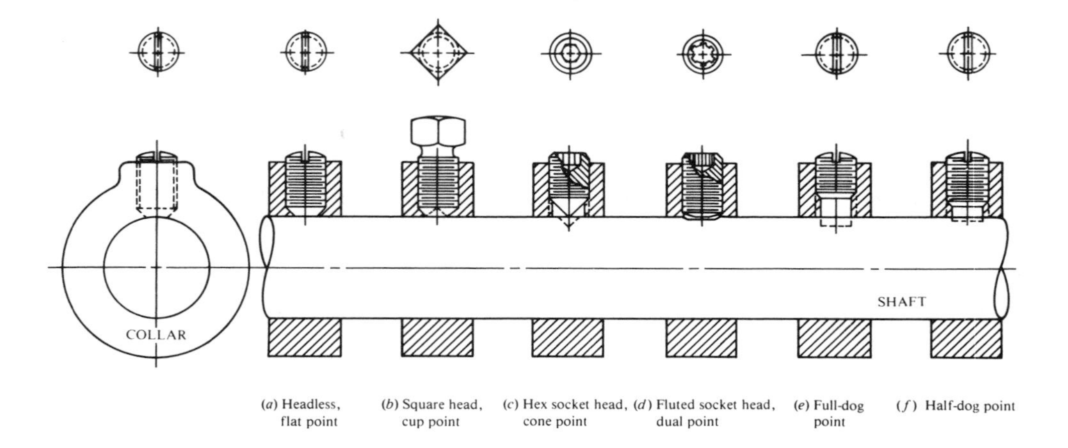

Figure 18-12 Set Screws with Different Head and Point Styles Applied to Hold a Collar on a Shaft (Hoelscher, et al., *Graphics for Engineers,* New York: John Wiley & Sons, 1968)

A *washer* may be used under either or both the bolt head and the nut to distribute the clamping load over a wide area and to provide a bearing surface for the relative rotation of the nut. The basic type of washer is the plain flat washer, a flat disc with a hole in it through which the bolt or screw passes. Other styles, called *lock washers,* have axial deformations or projections which produce axial forces on the fastener when compressed. These forces keep the threads of the mating parts in intimate contact and decrease the probability of the fastener loosening in service. Figure 18-13 shows several means of using

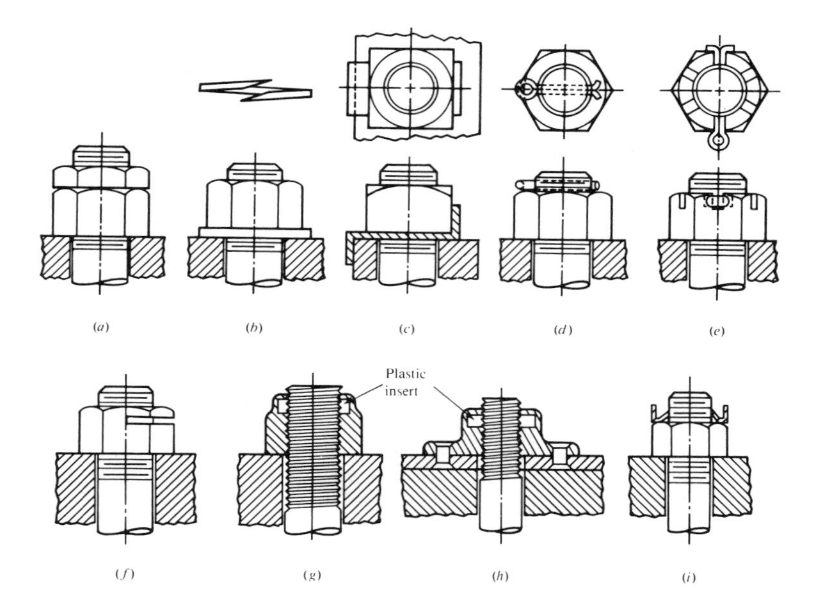

Figure 18-13 Locking Devices (Hoelscher, et al., *Graphics for Engineers,* New York: John Wiley & Sons, 1968)

washers and other types of locking devices. Part (a) is a jam nut tightened against the regular nut. Part (b) is the standard lock washer. Part (c) is a locking tab which keeps the nut from turning. Part (d) is a cotter inserted through a hole drilled through the bolt. Part (e) uses a cotter, but it also passes through slots in the nut. Part (f) is one of several types of thread deformation techniques used. Part (g), an *elastic stop nut,* uses a plastic insert to keep the threads of the nut in tight contact with the bolt. This may be used on machine screws as well. In part (h), the elastic stop nut is riveted to a thin plate allowing a mating part to be bolted from the opposite side. The thin metal device in (i) bears against the top of the nut and grips the threads preventing axial motion of the nut.

A *stud* is like a stationary bolt attached permanently to a part of one member to be joined. The mating member is then placed over the stud and a nut is tightened to clamp the parts together.

Additional variations occur when these types of fasteners are combined with different head styles. Several of these are shown in the figures already discussed. Others are listed next.

Square	Hex	Heavy hex	Hex jam
Hex castle	Hex flat	Hex slotted	12-point
High crown	Low crown	Round	T-head
Pan	Truss	Hex washer	Flat countersunk
Plow	Cross recess	Fillister	Oval countersunk
Hex socket	Spline socket	Button	Binding

Additional combinations are created by considering American National Standards or British Standard (Metric); material grades; finishes; thread sizes; lengths; class (tolerance grade); manner of forming heads (machining, forging, cold heading); and the manner of forming threads (machining, die cutting, tapping, rolling, plastic molding).

Thus it can be seen that comprehensive treatment of threaded fasteners encompasses extensive data (3, 4, 5, 8). This section will give some basic concepts related to the application of threaded fasteners.

Bolt Materials and Strength

In machine design, most fasteners are made from steel because of its high strength, good ductility, and good machinability and formability. But varying compositions and conditions of steel are used. The strength of steels used for bolts and screws is used to determine its *grade,* according to one of several standards. Three strength ratings are frequently available, the familiar tensile strength and yield strength plus the proof strength. The *proof strength,* similar to the elastic limit, is defined as the stress at which the bolt or screw would undergo permanent deformation. It usually ranges between 0.90 and 0.95 times the yield strength.

The SAE uses grade numbers ranging from 1 to 8, with increasing numbers indicating greater strength. Table 18-2 lists some aspects of this grading system taken from SAE Standard J429. The markings shown are embossed into the head of the bolt.

The ASTM publishes five standards relating to bolt steel strength, as listed in Table 18-3.

Metric bolts and screws use a numerical code system ranging from 4.6 to 12.9, with higher numbers indicating higher strengths. The numbers before the decimal point are approximately 0.01 times the tensile strength of the material in MPa. The last digit with the decimal point is the approximate ratio of the yield strength of the material to the tensile strength. Table 18-4 shows pertinent data from SAE Standard J1199.

Table 18-2 SAE Grades of Steels for Fasteners

Grade Number	*Bolt Size (in)*	*Tensile Strength (Ksi)*	*Yield Strength (Ksi)*	*Proof Strength (Ksi)*	*Head Marking*
1	1/4–1 1/2	60	36	33	None
2	1/4–3/4	74	57	55	None
	>3/4–1 1/2	60	36	33	
4	1/4–1 1/2	115	100	65	None
5	1/4–1	120	92	85	
	>1–1 1/2	105	81	74	
7	1/4–1 1/2	133	115	105	
8	1/4–1 1/2	150	130	120	

Table 18-3 ASTM Standards for Bolt Steels

ASTM Grade	*Bolt Size (in)*	*Tensile Strength (Ksi)*	*Yield Strength (Ksi)*	*Proof Strength (Ksi)*	*Head Marking*
A307	1/4–4	60	(Not reported)		None
A325	1/2–1	120	92	85	A 325
	>1–1 1/2	105	81	74	
A354-BC	1/4–2 1/2	125	109	105	BC
A354-BD	1/4–2 1/2	150	130	120	
A449	1/4–1	120	92	85	
	>1–1 1/2	105	81	74	
	>1 1/2–3	90	58	55	
A574	0.060–1/2	180	—	140	(Socket Head Capscrews)
	5/8–4	170	—	135	

Thread Designations

Table 18-5 shows pertinent dimensions for threads in the American Standard and SI Metric styles. For consideration of strength and size, the designer must know the basic major diameter, the pitch of the threads, and the area available to resist tensile loads. Note that the pitch is equal to $1/n$, where n is the number of threads per inch in the American Standard system. In the SI system, the pitch in millimeters is designated directly. The tensile stress area listed in the table takes into account the actual area cut by a transverse plane. Because of the helical path of the thread on the screw, such a plane will cut near the root on one side of the screw but will cut near the major diameter on the other. The equation for the tensile stress area for American Standard threads is

Table 18-4 Metric Grades of Steels for Bolts

Grade	*Bolt Size*	*Tensile Strength (MPa)*	*Yield Strength (MPa)*	*Proof Strength (MPa)*
4.6	M5–M36	400	240	225
4.8	M1.6–M16	420	340[a]	310
5.8	M5–M24	520	415[a]	380
8.8	M17–M36	830	660	600
9.8	M1.6–M16	900	720[a]	650
10.9	M6–M36	1 040	940	830
12.9	M1.6–M36	1 220	1 100	970

[a]Yield strengths approximate and not included in standard.

$$A_t = (0.7854)(D - 0.9743p)^2$$

where D is the major diameter and p is the pitch of the thread. For metric threads, the tensile stress area is

$$A_t = (0.7854)(D - 0.9382p)^2$$

For most standard screw thread sizes, at least two pitches are available, the *coarse* and *fine thread* series. Both are included in Table 18-5.

The smaller American Standard threads use a number designation from 0 to 12. The corresponding major diameter is listed in Table 18-5(a). The larger sizes use fractional inch designations. The decimal equivalent for the major diameter is shown in Table 18-5(b). Metric threads list the major diameter and the pitch in millimeters, as shown in Table 18-5(c). Samples of the standard designations for a thread are given next.

American Standard: Basic size followed by number of threads per inch and the thread series designation.

10–24 UNC	10–32 UNF
½–13 UNC	½–20 UNF
1½–6 UNC	1½–12 UNF

Metric: M (for "metric"), followed by the basic major diameter and then the pitch in millimeters.

M3×0.5 M3×0.35 M10×1.5

Socket Head Cap Screws

A very popular type of machine screw is the socket head cap screw. The usual configuration, shown in Figure 18-10, has a cylindrical head with a recessed hex-socket. Also readily available are flat head styles for countersinking to produce a flush surface, button head styles for a low profile appearance, and shoulder screws providing a precision

Table 18-5(a) American Standard Thread Dimensions, Numbered Sizes

SIZE	BASIC MAJOR DIAMETER (IN)	COARSE THREADS: UNC		FINE THREADS: UNF	
		Threads per in	*Tensile Stress Area (in^2)*	*Threads per in*	*Tensile Stress Area (in^2)*
0	0.060 0	—	—	80	0.001 80
1	0.073 0	64	0.002 63	72	0.002 78
2	0.086 0	56	0.003 70	64	0.003 94
3	0.099 0	48	0.004 87	56	0.005 23
4	0.112 0	40	0.006 04	48	0.006 61
5	0.125 0	40	0.007 96	44	0.008 30
6	0.138 0	32	0.009 09	40	0.010 15
8	0.164 0	32	0.014 0	36	0.014 74
10	0.190 0	24	0.017 5	32	0.020 0
12	0.216 0	24	0.024 2	28	0.025 8

Table 18-5(b) American Standard Thread Dimensions, Fractional Sizes

SIZE	BASIC MAJOR DIAMETER (IN)	COARSE THREADS: UNC		FINE THREADS: UNF	
		Threads per in	*Tensile Stress Area (in^2)*	*Threads per in*	*Tensile Stress Area (in^2)*
¼	0.250 0	20	0.031 8	28	0.036 4
5/16	0.312 5	18	0.052 4	24	0.058 0
⅜	0.375 0	16	0.077 5	24	0.087 8
7/16	0.437 5	14	0.106 3	20	0.118 7
½	0.500 0	13	0.141 9	20	0.159 9
9/16	0.562 5	12	0.182	18	0.203
⅝	0.625 0	11	0.226	18	0.256
¾	0.750 0	10	0.334	16	0.373
⅞	0.875 0	9	0.462	14	0.509
1	1.000	8	0.606	12	0.663
1⅛	1.125	7	0.763	12	0.856
1¼	1.250	7	0.969	12	1.073
1⅜	1.375	6	1.155	12	1.315
1½	1.500	6	1.405	12	1.581
1¾	1.750	5	1.90	—	—
2	2.000	4½	2.50	—	—

Table 18-5(c) Metric Thread Dimensions

	COARSE THREADS		FINE THREADS	
BASIC MAJOR DIAMETER *(mm)*	*Pitch (mm)*	*Tensile Stress Area (mm^2)*	*Pitch (mm)*	*Tensile Stress Area (mm^2)*
1	0.25	0.460	—	—
1.6	0.35	1.27	0.20	1.57
2	0.4	2.07	0.25	2.45
2.5	0.45	3.39	0.35	3.70
3	0.5	5.03	0.35	5.61
4	0.7	8.78	0.5	9.79
5	0.8	14.2	0.5	16.1
6	1	20.1	0.75	22.0
8	1.25	36.6	1	39.2
10	1.5	58.0	1.25	61.2
12	1.75	84.3	1.25	92.1
16	2	157	1.5	167
20	2.5	245	1.5	272
24	3	353	2	384
30	3.5	561	2	621
36	4	817	3	865
42	4.5	1 121	—	—
48	5	1 473	—	—

bearing surface for location or pivoting. The *1960 Series* of socket head cap screws is made from a heat-treated alloy steel having the following strengths:

Size Range	*Tensile Strength (Ksi)*	*Yield Strength (Ksi)*
0–⅝	190	170
¾–3	180	155

Roughly equivalent performance is obtained from metric socket head cap screws made to the metric strength grade 12.9. The same geometry is available in corrosion resistant stainless steel, typically 18-8, at somewhat lower strength levels. Consult the manufacturers.

Clamping Load

When a bolt or screw is used to clamp two parts, the force exerted between the parts is the *clamping load*. The designer is responsible for specifying the clamping load and for ensuring that the fastener is capable of withstanding the load. The maximum clamping load is often taken to be 0.75 times the proof load, where the proof load is the product of the proof stress times the tensile stress area of the bolt or screw.

Tightening Torque

The clamping load is created in the bolt or screw by exerting a tightening torque on the nut or on the head of the screw. An approximate relationship between the torque and the axial tensile force in the bolt or screw (the clamping force) is

$$T = KDP \tag{18-15}$$

where

T = Torque, lb · in

D = Nominal outside diameter of threads, in

P = Clamping load, lb

K = Constant dependent on the lubrication present

For average commercial conditions, use $K = 0.15$ if any lubrication at all is present. Even cutting fluids or other residual deposits on the threads will produce conditions consistent with $K = 0.15$. If the threads are well cleaned and dried, $K = 0.20$ is better. Of course, these values are approximate and variations among seemingly identical assemblies should be expected. Testing and statistical analysis of the results are recommended.

Example Problem 18-4. A set of three bolts is to be used to provide a clamping force of 12 000 lb between two components of a machine. The load is shared equally among the three bolts. Specify suitable bolts, including the grade of the material, if each is to be stressed to 75 percent of its proof strength. Then compute the required tightening torque.

Solution. The load on each screw is to be 4 000 pounds. Let's specify a bolt made from SAE grade 5 steel, having a proof strength of 85 000 psi. Then the allowable stress is

$$\sigma_a = 0.75(85\,000 \text{ psi}) = 63\,750 \text{ psi}$$

The required tensile stress area for the bolt is then

$$A_t = \frac{\text{Load}}{\sigma_a} = \frac{4\,000 \text{ lb}}{63\,750 \text{ lb/in}^2} = 0.062\,7 \text{ in}^2$$

From Table 18-5, we find that the 3/8–16 UNC thread has the required tensile stress area. The required tightening torque will be

$$T = KDP = 0.15(0.375\,0 \text{ in})(4\,000 \text{ lb}) = 225 \text{ lb} \cdot \text{in}$$

Externally Applied Force on a Bolted Joint

The analysis shown in example problem 18-4 only considers the stress in the bolt under static conditions and for only the clamping load. It was recommended that the tension on the bolt be very high, approximately 75 percent of the proof load for the bolt. Such a load will use the available strength of the bolt efficiently and will prevent the separation of the connected members.

When a load is applied to a bolted joint over and above the clamping load, special consideration must be given to the behavior of the joint. Initially, the force on the bolt (in tension) is equal to the force on the clamped members (in compression). Then, some of the additional load will act to stretch the bolt beyond its length assumed after the clamping load was applied. Another portion will result in a *decrease* in the compressive force in the clamped member. Thus, only a part of the applied force is carried by the bolt. The amount is dependent on the relative stiffness of the bolt and the clamped members.

If a very stiff bolt is clamping a flexible member, such as a resilient gasket, most of the added force will be taken by the bolt because it takes little force to change the compression in the gasket. In this case, the bolt design must take into account not only the initial clamping force but also the added force.

Conversely, if the bolt is relatively flexible compared with the clamped members, virtually all of the externally applied load will initially go to decreasing the clamping force until the members actually separate, a condition usually interpreted as failure of the joint. Then the bolt will carry the full amount of the external load.

In practical joint design, a situation between the extremes described above would normally occur. Typical "hard" joints (without a soft gasket) have the stiffness of the clamped members approximately three times as stiff as the bolt. The externally applied load then is shared by the bolt and the clamped members according to their relative stiffnesses as follows:

$$F_b = P + \frac{k_b}{k_b + k_c} F_e \qquad (18\text{-}16)$$

$$F_c = P - \frac{k_c}{k_b + k_c} F_e \qquad (18\text{-}17)$$

where

F_e = Externally applied load

P = Initial clamping load (as used in equation [18-15])

F_b = Final force in bolt

F_c = Final force on clamped members

k_b = Stiffness of bolt

k_c = Stiffness of clamped members

Example Problem 18-5. Assume that the joint described in example problem 18-4 was subjected to an additional external load of 3 000 lb after the initial clamping load of 4 000 lb was applied. Also assume that the stiffness of the clamped members is three times that of the bolt. Compute the force in the bolt, the force in the clamped members, and the final stress in the bolt after the external load is applied.

Solution. We will first use equation (18-16) and (18-17) with $P = 4\,000$ lb, $F_e = 3\,000$ lb, and $k_c = 3k_b$.

$$F_b = P + \frac{k_b}{k_b + k_c} F_e = P + \frac{k_b}{k_b + 3k_b} F_e = P + \frac{k_b}{4k_b} F_e$$

$$F_b = P + F_e/4 = 4\,000 + 3\,000/4 = 4\,750 \text{ lb}$$

$$F_c = P - \frac{k_c}{k_b + k_c} F_e = P - \frac{3k_b}{k_b + 3k_b} F_e = P - \frac{3k_b}{4k_b} F_e$$

$$F_c = P - 3F_e/4 = 4\,000 - 3(3\,000)/4 = 1\,750 \text{ lb}$$

Because F_c is still greater than zero, the joint is still tight. Now the stress in the bolt can be found. For the ⅜–16 bolt, the tensile stress area is 0.077 5 in^2.

$$\sigma = \frac{P}{A_t} = \frac{4\,750 \text{ lb}}{0.077\,5 \text{ in}^2} = 61\,300 \text{ psi}$$

The proof strength of the grade 5 material is 85 000 psi, and this stress is approximately 72 percent of the proof strength. Therefore, the selected bolt is still safe. But consider what would happen with a relatively "soft" joint.

Example Problem 18-6. Solve the problem stated in example problem 18-5 again but assume that the joint has a flexible elastomeric gasket separating the clamping members and that the stiffness of the bolt is then ten times that of the joint.

Solution. The procedure will be the same as that used previously, but now $k_b = 10k_c$.

$$F_b = P + \frac{k_b}{k_b + k_c}F_e = P + \frac{10k_c}{10k_c + k_c}F_e = P + \frac{10k_c}{11k_c}F_e$$

$$F_b = P + 10F_e/11 = 4\,000 + 10(3\,000)/11 = 6\,727 \text{ lb}$$

$$F_c = P - \frac{k_c}{k_b + k_c}F_e = P - \frac{k_c}{10k_c + k_c}F_e = P - \frac{k_c}{11k_c}F_e$$

$$F_c = P - F_e/11 = 4\,000 - 3\,000/11 = 4\,273 \text{ lb}$$

The stress in the bolt would be

$$\sigma = \frac{6\,727 \text{ lb}}{0.077\,5 \text{ in}^2} = 86\,800 \text{ psi}$$

This exceeds the proof strength of the grade 5 material and is dangerously close to the yield strength.

18-4 OTHER MEANS OF FASTENING

The chapter thus far has focused on steel screws and bolts because of their wide applications. Other materials for screws and bolts and other types of fastening means will now be discussed.

Aluminum is used for its corrosion resistance, light weight, and fair strength level. Its good thermal and electrical conductivity may also be desirable. The most widely used alloys are 2024-T4, 2011-T3, and 6061-T6. Properties of these materials are listed in Appendix A-10.

Brass, copper, and *bronze* are also used for their corrosion resistance. Ease of machining and an attractive appearance are also advantages. Certain alloys are particularly good for resistance to corrosion in marine applications.

Nickel and its alloys such as *Monel* and *Inconel* (from the International Nickel Company) provide good performance at elevated temperatures while also having good corrosion resistance, toughness at low temperatures, and an attractive appearance.

Stainless steels are used primarily for their corrosion resistance. Alloys used for fasteners include 18-8, 410, 416, 430, and 431. In addition, stainless steels in the 300 series are nonmagnetic. See Appendix A-6 for properties.

A high strength-to-weight ratio is the chief advantage of *titanium* alloys used for fasteners in aerospace applications. Appendix A-11 gives a list of properties of several alloys.

Coatings and *finishes* are provided for metallic fasteners to improve appearance or corrosion resistance. Some also lower the coefficient of friction for more consistent results relating tightening torque to clamping force. Steel fasteners can be finished with black oxide, blueing, bright nickel, phosphate, and hot-dip zinc. Plating can be used to deposit cadmium, copper, chromium, nickel, silver, tin, and zinc. Various paints, lacquers, and chromate finishes are also used. Aluminum is usually anodized.

Plastics are used widely because of their light weight, corrosion resistance, insulating ability, and ease of manufacture. Nylon 6/6 is the most frequently used material but others include ABS, Acetal, TFE fluorocarbons, polycarbonate, polyethylene, polypropylene, and polyvinylchloride. Appendix A-13 lists several plastics and their properties. In addition to screws and bolts, plastics are used extensively where the fastener is designed specially for the particular application.

Rivets are nonthreaded fasteners, usually made of steel or aluminum. Originally made with one head, the opposite end is formed after the rivet is inserted through holes in the parts to be joined. Steel rivets are formed hot, whereas aluminum can be formed at room temperatures. Of course, riveted joints are not designed to be assembled more than once.

A large variety of *quick-operating fasteners* is available. Many are of the quarter-turn type requiring just a 90° rotation to connect or disconnect the fastener. Access panels, hatches, covers, and brackets for removable equipment are attached with such fasteners. Similarly, many varieties of *latches* are available to provide quick action with, perhaps, added holding power.

Welding involves the metallurgical bonding of metals, usually by the application of heat with an electric arc, gas flame, or electrical resistance heating under heavy pressure. Welding is discussed in Chapter 19.

Brazing and *soldering* use heat to melt a bonding agent which flows into the space between parts to be joined, adhering to both parts, and then solidifying as it cools. *Brazing* uses relatively high temperatures, above 840°F (450°C), using alloys of copper, silver, aluminum, silicon, or zinc. Of course the metals to be joined must have a significantly higher melting temperature. Metals successfully brazed include plain carbon and alloy steels, stainless steels, nickel alloys, copper, aluminum, and magnesium. *Soldering* is similar to brazing, except that it is performed at lower temperatures, less than 840°F. Several soldering alloys of lead-tin, tin-zinc, tin-silver, lead-silver, zinc-cadmium, zinc-aluminum, and others are used. Brazed joints are generally stronger than soldered joints due to the inherently higher strength of the brazing alloys. Most soldered joints are fabricated with interlocking lap joints to provide mechanical strength and then the solder is used to hold the assembly together and possibly to provide sealing. Joints in piping and tubing are frequently soldered.

Adhesives are seeing wide use. Versatility and ease of application are strong advantages of adhesives used in an array of products from toys and household appliances to automotive and aerospace structures. Some types include the following:

Acrylics: Used for many metals and plastics.

Cyanoacrylates: Very fast curing; flows easily between well-mated surfaces.

Epoxies: Good structural strength; joint is usually rigid; some require two-part formulations; a large variety of formulations and properties is available.

Anaerobics: Used for securing nuts and bolts and other joints with small clearances; cures in the absence of oxygen.

Silicones: Flexible adhesive with good high-temperature performance (400°F, 200°C).

Polyester hot melt: Good structural adhesive; easy to apply with special equipment.

Polyurethane: Good bonding; provides a flexible joint.

REFERENCES

1. Armco Steel Corporation. *Armco Standard Fasteners*. Kansas City, Mo.
2. Faires, V. M. *Design of Machine Elements,* 5th ed. New York: Macmillan Publishing Company, 1965.
3. Oberg, E., Jones, F. D., and Horton, H. L. *Machinery's Handbook,* 22d ed. New York: Industrial Press, 1984.
4. Parmley, R. O., editor-in-chief. *Standard Handbook of Fastening and Joining*. New York: McGraw-Hill Book Company, 1977.
5. Penton/IPC. *Machine Design Fastening and Joining Reference Issue*. 55, no. 26, Nov. 17, 1983.
6. Saginaw Steering Gear Division, General Motors Corporation. *Industrial Ball Bearing Screw Catalog*. Saginaw, Mich.
7. Shigley, J. E., and Mitchell, L. D. *Mechanical Engineering Design,* 4th ed. New York: McGraw-Hill Book Company, 1983.
8. Society of Automotive Engineers. *SAE Handbook*. Warrendale, Pa., 1982.
9. SPS Technology, Inc. *Unbrako Catalog—Socket Head Cap Screws*.
10. Warner Electric Brake & Clutch Company. *Ball Bearing Screws Catalog*. Beloit, Wis.

PROBLEMS

1. Name three types of threads used for power screws.
2. Make a scale drawing of an acme thread having a major diameter of 1½ in and four threads per inch. Draw a section 2.0 in long.
3. Repeat problem 2 for a buttress thread.
4. Repeat problem 2 for a square thread.
5. If an Acme thread power screw is loaded in tension with a force of 30 000 lb, what size screw from Table 18-1 should be used to maintain a tensile stress below 10 000 psi?
6. For the screw chosen in problem 5, what would be the required axial length of the nut on the screw that transfers the load to the frame of the machine if the shear stress in the threads must be less than 6 000 psi?
7. Compute the torque required to raise the load of 30 000 lb with the Acme screw selected in problem 5. Use a coefficient of friction of 0.15.
8. Compute the torque required to lower the load with the screw from problem 5.
9. If a square thread screw having a major diameter of ¾ in and six threads per inch is used to lift a load of 4 000 lb, compute the torque required to rotate the screw. Use a coefficient of friction of 0.15.
10. For the screw of problem 9, compute the torque required to rotate the screw when lowering the load.
11. Compute the lead angle for the screw of problem 9. Is it self locking?
12. Compute the efficiency for the screw of problem 9.
13. If the load of 4 000 lb is lifted by the screw described in problem 9 at the rate of 0.5 in/sec, compute the required rotational speed of the screw and the power required to drive it.
14. A ball screw for a machine table drive is to be selected. The axial force to be transmitted by the screw is 600 lb. The table moves 24 in per cycle and it is expected to cycle ten times per hour for a design life of 10 years. Select an appropriate screw.
15. For the screw selected in problem 14, compute the torque required to drive the screw.
16. For the screw selected in problem 14, the normal travel speed of the table is 10.0 in/min. Compute the power required to drive the screw.
17. If the cycle time for the machine in problem 14 is decreased to obtain 20 cycles/h instead of 10, what would be the expected life in years of the screw originally selected?
18. Describe the difference between a screw and a bolt.
19. Define the term *proof strength*.
20. Define the term *clamping load*.
21. Specify suitable machine screws to be installed in a pattern of four, equally spaced around a flange, if the clamping force between the flange and the mating structure is to be 6 000 lb. Then recommend a suitable tightening torque for each screw.
22. What would be the tensile force in a machine screw having an 8–32 thread if it is made from SAE grade 5 steel and is stressed to its proof strength?

23. What would be the tensile proof force in newtons (N) in a machine screw having a major diameter of 4 mm with standard fine threads if it is made from steel having a metric strength grade of 8.6?
24. What would be the nearest standard metric screw thread size to the American Standard 7/8–14 thread? By how much do their major diameters differ?
25. A machine screw is found with no information given as to its size. The following data are found by using a standard micrometer caliper. Major diameter = 0.196 in; the axial length for 20 full threads is 0.630 in. Identify the thread.
26. A threaded fastener is made from nylon 6/6 with an M10×1.5 thread. Compute the maximum tensile force that can be permitted in the fastener if it is to be stressed to 75 percent of the tensile strength of the nylon. See Appendix Table A-13.
27. Compare the tensile force that can be carried by a 1/4–20 screw if it is to be stressed to 50 percent of its tensile strength and if it is made from each of the following materials:
    a. Steel, SAE grade 2
    b. Steel, SAE grade 5
    c. Steel, SAE grade 8
    d. Steel, ASTM grade A307
    e. Steel, ASTM grade A574
    f. Steel, metric grade 8.8
    g. Aluminum 2024-T4
    h. AISI 430 annealed
    i. Ti-6Al-4V annealed
    j. Nylon 6/6
    k. Polycarbonate
    l. High-impact ABS
28. Describe the differences among welding, brazing, and soldering.
29. What types of metals are typically brazed?
30. What are some common brazing alloys?
31. What materials make up commonly used solders?
32. Name five common adhesives and give the typical properties of each.
33. The label of a common household adhesive describes it as a *cyanoacrylate*. What would you expect its properties to be?
34. Find three commercially available adhesives from your home, a laboratory, a machine shop, or your workplace. Try to identify the generic nature of the adhesive and compare it with the list presented in this chapter.

19 Bolted Connections, Welded Joints, and Machine Frames

19-1 OVERVIEW

This chapter will focus on the joining of machine parts and the design and fabrication of machine frames. Most of the other parts of the book have dealt with individual machine members. But, in keeping with the theme of machine elements in mechanical design, it is necessary to consider how the elements would be assembled together and how they would be housed and supported.

Chapter 18 covered part of the story, that dealing with bolts loaded in pure tension, as in a clamping function. This chapter extends this to consider eccentrically loaded joints, those that must resist a combination of direct shear and a bending moment on a bolt pattern.

The ability of a welded joint to carry a variety of loads is discussed with the objective of designing the weld. Here both uniformly loaded and eccentrically loaded joints are treated.

The subject of machine frames and structures is quite complex. It is discussed from the standpoint of general principles and guidelines, rather than specific design techniques. Critical frames are typically designed with computerized finite element analysis. Also, experimental stress analysis techniques are often used to verify designs.

19-2 ECCENTRICALLY LOADED BOLTED JOINTS

Figure 19-1 shows an example of an eccentrically loaded bolted joint. The motor on the extended bracket places the bolts in shear because its weight acts directly downward. But there also exists a moment equal to $P \times a$ that must be resisted. The moment tends to rotate the bracket and thus to shear the bolts.

The basic approach to the analysis and design of eccentrically loaded joints is to determine the forces that act on each bolt because of all the applied loads. Then, by a

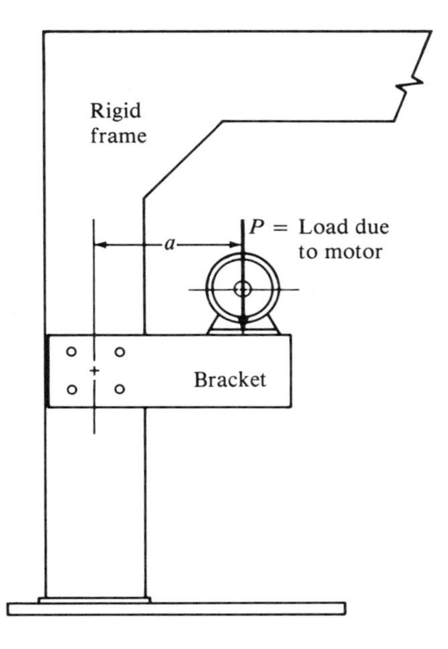

Figure 19-1 Eccentrically Loaded Bolted Joint

process of superposition, the loads are combined vectorially to determine which bolt carries the greatest load. That bolt is then sized. The method will be illustrated in the context of an example problem.

The American Institute of Steel Construction (AISC) lists allowable stresses for bolts made from ASTM grade steels, as shown in Table 19-1. These data are for bolts used in standard sized holes, 1/16 in larger than the bolt. Also, a *friction type connection,* in which the clamping force is sufficiently large that the friction between the mating parts helps to carry some of the shear load, is assumed.

In the design of bolted joints, you should ensure that there are no threads in the plane where shear occurs. The body of the bolt will then have a diameter equal to the major diameter of the thread. The tables in Chapter 18 can be used to select the standard size for a bolt.

Example Problem 19-1. For the bracket in Figure 19-1, assume that the total force P is 3 500 lb and the distance a is 12 in. Design the bolted joint, including the location and number of bolts, the material, and the diameter.

Solution. The solution shown is an outline of a procedure that can be used to analyze similar joints. The data of this problem illustrate the procedure.

1. Propose the number of bolts and the pattern. This is a design decision, based on your judgment and the geometry of the connected parts.

   In this problem, let's try a pattern of four bolts placed as shown in Figure 19-2.

2. Determine the direct shear force on the bolt pattern and on each individual bolt, assuming that all bolts share the shear load equally.

$$\text{Shear load} = P = 3\,500 \text{ lb}$$
$$\text{Load per bolt} = F_s = P/4 = 3\,500 \text{ lb}/4 = 875 \text{ lb/bolt}$$

   The shear force acts directly downward on each bolt.

3. Compute the *moment* to be resisted by the bolt pattern: the product of the overhanging load and the distance to the *centroid* of the bolt pattern.

   In this problem, $M = P \times a = (3\,500 \text{ lb})(12 \text{ in}) = 42\,000 \text{ lb} \cdot \text{in}$.

4. Compute the radial distance from the centroid of the bolt pattern to the center of each bolt. In this problem, each bolt has a radial distance of

$$r = \sqrt{(1.50 \text{ in})^2 + (2.00 \text{ in})^2} = 2.50 \text{ in}$$

Table 19-1 Allowable Stresses for Bolts

ASTM Grade	*Allowable Shear Stress*	*Allowable Tensile Stress*
A307	10 Ksi (69 MPa)	20 Ksi (138 MPa)
A325 and A449	17.5 Ksi (121 MPa)	44 Ksi (303 MPa)
A490	22 Ksi (152 MPa)	54 Ksi (372 MPa)

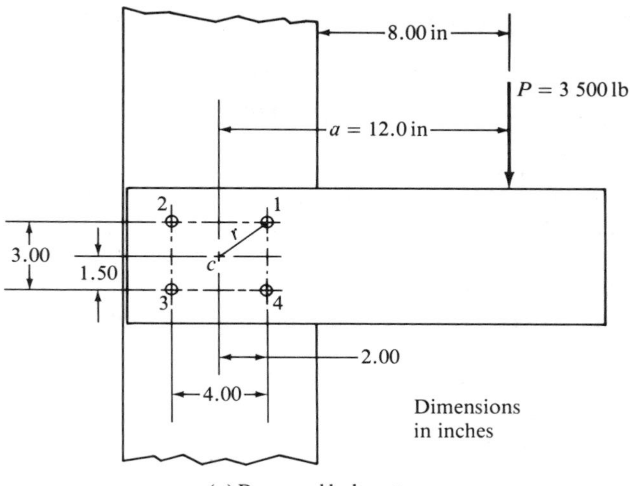

(*a*) Proposed bolt pattern

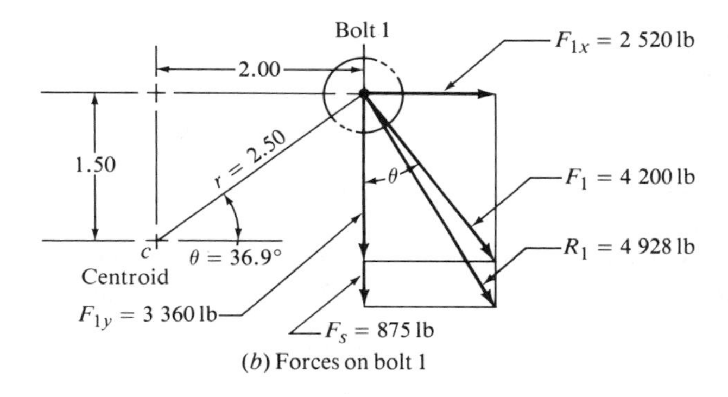

(*b*) Forces on bolt 1

Figure 19-2 Geometry of Bolted Joint for Example Problem 19-1 and Forces on Bolt 1

5. Compute the sum of the *squares* of all radial distances to all bolts. In this problem, all four bolts have the same *r*. Then

$$\sum r^2 = 4(2.50 \text{ in})^2 = 25.0 \text{ in}^2$$

6. Compute the force on each bolt required to resist the bending moment from the relation

$$F_i = \frac{Mr_i}{\sum r^2} \qquad (19\text{-}1)$$

where

r_i = Radial distance from the centroid of the bolt pattern to the *i*th bolt

F_i = Force on the *i*th bolt due to the moment.
The force acts perpendicular to the radius.

In this problem, all such forces are equal. For example, for bolt 1

$$F_1 = \frac{Mr_1}{\sum r^2} = \frac{(42\,000 \text{ lb} \cdot \text{in})(2.50 \text{ in})}{25.0 \text{ in}^2} = 4\,200 \text{ lb}$$

7. Determine the resultant of all forces acting on each bolt. A vector summation can be performed either analytically or graphically. Or each force can be resolved into horizontal and vertical components. The components can be summed and then the resultant can be computed.

   Let's use the latter approach for this problem. The shear force acts directly downward, in the y direction. The x and y components of F_1 are

$$F_{1x} = F_1 \sin \theta = (4\,200 \text{ lb}) \sin (36.9°) = 2\,520 \text{ lb}$$
$$F_{1y} = F_1 \cos \theta = (4\,200 \text{ lb}) \cos (36.9°) = 3\,360 \text{ lb}$$

   The total force in the y direction is then

$$F_{1y} + F_s = 3\,360 + 875 = 4\,235 \text{ lb}$$

   Then the resultant force on bolt 1 is

$$R_1 = \sqrt{(2\,520)^2 + (4\,235)^2} = 4\,928 \text{ lb}$$

8. Specify the bolt material; compute the required area for the bolt; and select an appropriate size.

   For this problem, let's specify ASTM A325 steel for the bolts having an allowable shear stress of 17 500 psi from Table 19-1. Then the required area for the bolt is

$$A_s = \frac{R_1}{\tau_a} = \frac{4\,928 \text{ lb}}{17\,500 \text{ lb/in}^2} = 0.282 \text{ in}^2$$

   The required diameter would be

$$D = \sqrt{\frac{4A_s}{\pi}} = \sqrt{\frac{4(0.282 \text{ in}^2)}{\pi}} = 0.599 \text{ in}$$

   Let's specify a ⅝-in bolt having a diameter of 0.625 in.

19-3 WELDED JOINTS

The design of welded joints requires consideration of the manner of loading on the joint, the types of materials in the weld and in the members to be joined, and the geometry of the joint itself. The load may be either uniformly distributed over the weld such that all parts of the weld are stressed to the same level; or the load may be eccentrically applied. Both are discussed in this section.

The materials of the weld and the parent members determine the allowable stresses. Table 19-2 lists several examples for steel and aluminum. The allowables listed are for shear on fillet welds. For steel, welded by the electric arc method, the type of electrode is an indication of the tensile strength of the filler metal. For example, the E70 electrode has a minimum tensile strength of 70 Ksi (483 MPa). Additional data are available in publications of the American Welding Society (AWS), the American Institute for Steel Construction (AISC), and the Aluminum Association (AA).

Table 19-2 Allowable Shear Stresses on Fillet Welds

STEEL

Electrode Type	*Typical Metals Joined (ASTM Grade)*	*Allowable Shear Stress*
E60	A36, A500	18 Ksi (124 MPa)
E70	A242, A441	21 Ksi (145 MPa)
E80	A572 Grade 65	24 Ksi (165 MPa)
E90	—	27 Ksi (186 MPa)
E100	—	30 Ksi (207 MPa)
E110	—	33 Ksi (228 MPa)

ALUMINUM

	Filler Alloy							
	1100		*4043*		*5356*		*5556*	
	Allowable Shear Stress							
Metal Joined	*Ksi*	*MPa*	*Ksi*	*MPa*	*Ksi*	*MPa*	*Ksi*	*MPa*
1100	3.2	22	4.8	33	—	—	—	—
3003	3.2	22	5.0	34	—	—	—	—
6061	—	—	5.0	34	7.0	48	8.5	59
6063	—	—	5.0	34	6.5	45	6.5	45

Types of Joints

Joint type refers to the relationship between mating parts, as illustrated in Figure 19-3. The butt weld allows a joint to be the same nominal thickness as the mating parts and is usually loaded in tension. If the joint is properly made with the appropriate weld metal, the joint will be stronger than the parent metal. Thus, no special analysis of the joint is required if the joined members themselves are shown to be safe. Caution is advised, however, when the materials to be joined are adversely affected by the heat of the welding process. Heat-treated steels and many aluminum alloys are examples. The other types of joints in Figure 19-3 are assumed to place the weld in shear.

Types of Welds

Figure 19-4 shows several types of welds named for the geometry of the edges of the parts to be joined. Note the special edge preparation required, especially for thick plates, to permit the welding rod to enter the joint and build a continuous weld bead.

Size of Weld

The five types of groove-type welds in Figure 19-4 are made as complete penetration welds. Then, as indicated before for butt welds, the weld is stronger than the parent metals and no further analysis is required.

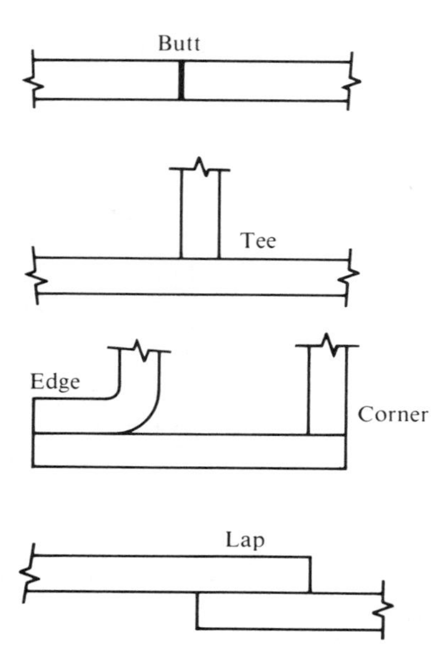

Figure 19-3 Types of Weld Joints

Fillet welds are typically made as equal leg right triangles with the size of the weld indicated by the length of the leg. A fillet weld loaded in shear would tend to fail along the shortest dimension of the weld which is the line from the root of the weld to the

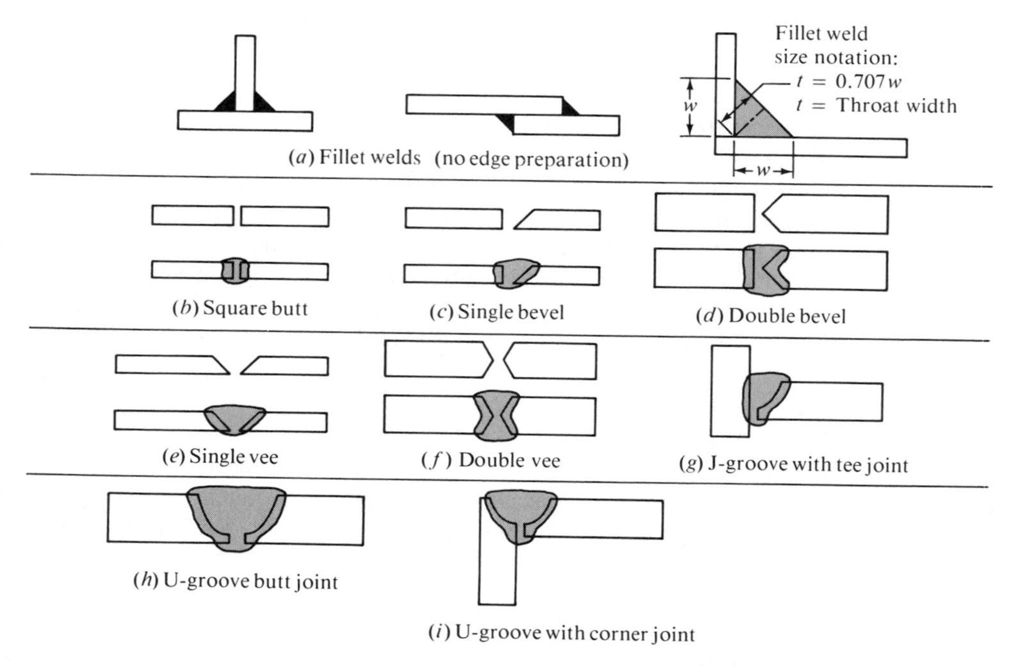

Figure 19-4 Some Types of Welds, Showing Edge Preparation

Table 19-3 Allowable Shear Stresses and Forces on Welds

BASE METAL ASTM GRADE	ELECTRODE	ALLOWABLE SHEAR STRESS	ALLOWABLE FORCE PER INCH OF LEG
Building Type Structures			
A36, A441	E60	13 600 psi	9 600 lb/in
A36, A441	E70	15 800 psi	11 200 lb/in
Bridge Type Structures			
A36	E60	12 400 psi	8 800 lb/in
A441, A242	E70	14 700 psi	10 400 lb/in

theoretical face of the weld and normal to the face. The length of this line is found from simple trigonometry to be $0.707w$, where w is the leg dimension.

The objectives of the design of a fillet welded joint are to specify the length of the legs of the fillet; the pattern of the weld; and the length of the weld. Presented here is the method that treats the weld as a line having no thickness. The method involves determining the *maximum force per inch* of weld leg length. Comparing the actual force with an allowable force allows the calculation of the required leg length.

Table 19-3 gives data for the allowable shear stress and the allowable force per inch for some combinations of base metal and welding electrode. In general, the allowables for building type structures are for steady loads. The values for bridge type loading accounts for the cyclic effects. For true fatigue type repeated loading, refer to the literature (2, 3).

Method of Treating Weld as a Line

Four different types of loading are considered here: direct tension or compression, direct vertical shear, bending, and twisting. The method allows the designer to perform calculations in a manner very similar to that used to design the load carrying members themselves. In general, the weld is analyzed separately for each type of loading to determine the force per inch of weld size due to each load. The loads are then combined vectorially to determine the maximum force. This maximum force is compared with the allowables from Table 19-3 to determine the size of the weld required.

The relationships used are summarized next.

Type of Loading	*Formula for Force Per Inch of Weld*	
Direct tension or compression	$f = P/A_w$	(19-2)
Direct vertical shear	$f = V/A_w$	(19-3)
Bending	$f = M/Z_w$	(19-4)
Twisting	$f = Tc/J_w$	(19-5)

In these formulas, the geometry of the weld is used to evaluate the terms A_w, Z_w, and J_w, using the relationships shown in Figure 19-5. Note the similarity between these formulas and those used to perform the stress analysis. Note, also, the similarity between the geometry factors for welds and the properties of areas used for the stress analysis. Because the weld is treated as a line having no thickness, the units for the geometry factors are different from those of the area properties, as indicated in Figure 19-5.

The use of this method of weld analysis will be demonstrated with example problems. In general, the method requires the following steps:

1. Propose the geometry of the joint and the design of the members to be joined.
2. Identify the types of stresses to which the joint is subjected (bending, twisting, vertical shear, direct tension or compression).
3. Analyze the joint to determine the magnitude and direction of the force on the weld due to each type of load.
4. Combine the forces vectorially at the point or points of the weld where the forces appear to be maximum.
5. Divide the maximum force on the weld by the allowable force from Table 19-3 to determine the required leg size for the weld. Note that when thick plates are welded, there are minimum acceptable sizes for the welds as listed in Table 19-4.

Example Problem 19-2. Design a bracket similar to that in Figure 19-1 but use welding to attach the bracket to the column. The bracket is 6.00 in high and is made from ASTM A36 steel having a thickness of ½ in. The column is also made from A36 steel and is 8.00 in wide.

Solution.

Step 1. The proposed geometry is a design decision and may have to be subjected to some iteration to achieve an optimum design. For a first trial, let's use the C-shaped weld pattern shown in Figure 19-6.

Step 2. The weld will be subjected to direct vertical shear and twisting caused by the 3 500-lb load on the bracket.

Step 3. To compute the forces on the weld, the geometry factors A_w and J_w must be known. Also, the location of the centroid of the weld pattern must be computed (see Figure 19-6b).

$$A_w = 2b + d = 2(4) + 6 = 14 \text{ in}$$

$$J_w = \frac{(2b + d)^3}{12} - \frac{b^2(b + d)^2}{(2b + d)} = \frac{(14)^3}{12} - \frac{16(10)^2}{14} = 114.4 \text{ in}^3$$

$$x = \frac{b^2}{2b + d} = \frac{16}{14} = 1.14 \text{ in}$$

Force due to vertical shear:

$$V = P = 3\,500 \text{ lb}$$

$$f_s = P/A_w = (3\,500 \text{ lb})/14 \text{ in} = 250 \text{ lb/in}$$

This force acts vertically downward on all parts of the weld.

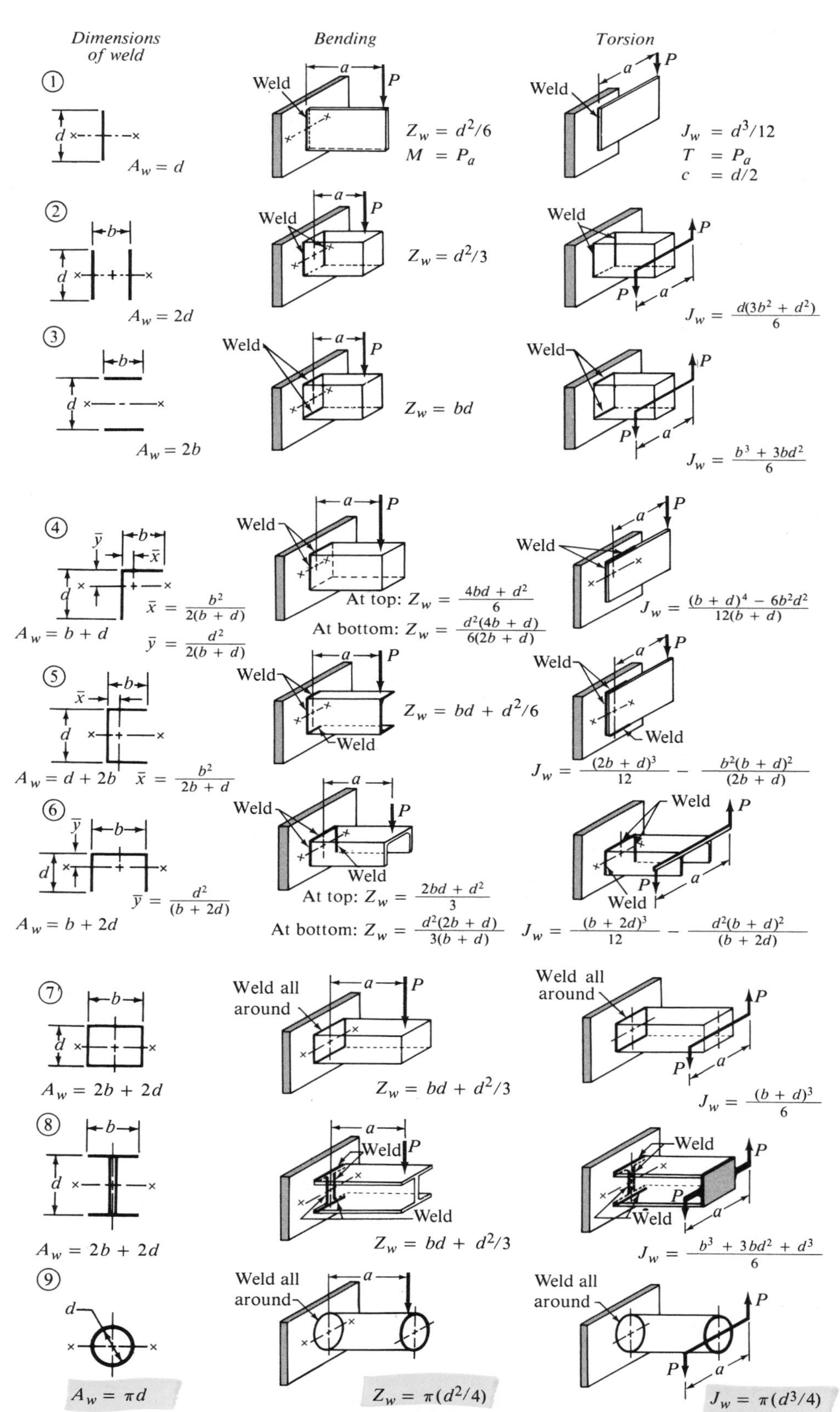

Figure 19-5 Geometry Factors for Weld Analysis

Table 19-4 Minimum Weld Sizes for Thick Plates

Plate Thickness (inch)	*Minimum Leg Size for Fillet Weld (inch)*
$\leq 1/2$	3/16
$>1/2$–3/4	1/4
$>3/4$–1 1/2	5/16
$>1\,1/2$–2 1/4	3/8
$>2\,1/4$–6	1/2
>6	5/8

Forces due to the twisting moment:

$$T = P[8.00 + (b - x)] = 3\,500[8.00 + (4.00 - 1.14)]$$

$$T = 3\,500(10.86) = 38\,010 \text{ lb} \cdot \text{in}$$

The twisting moment causes a force to be exerted on the weld which is perpendicular to a radial line from the centroid of the weld pattern to the point of interest. In this case, the end of the weld to the upper right experiences the greatest force. It is most convenient

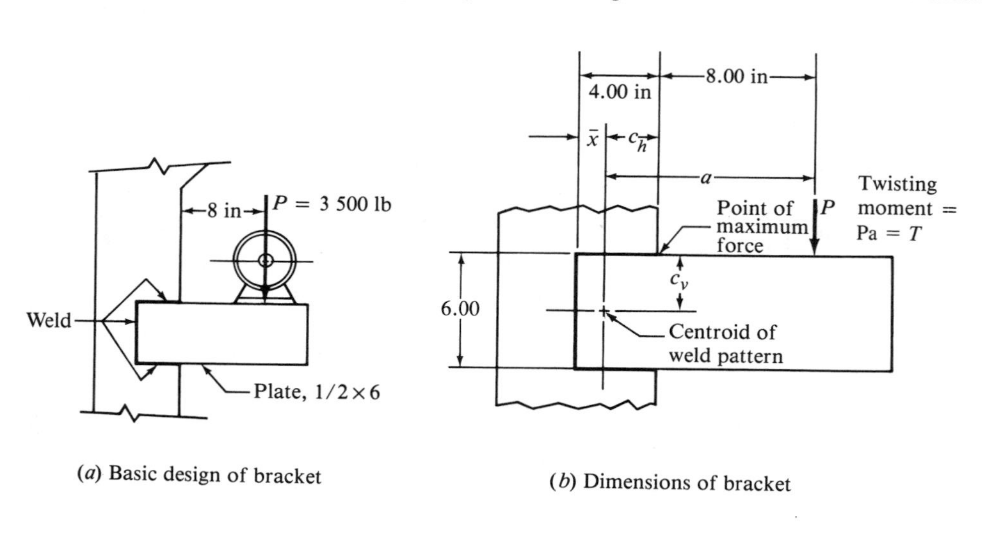

(*a*) Basic design of bracket

(*b*) Dimensions of bracket

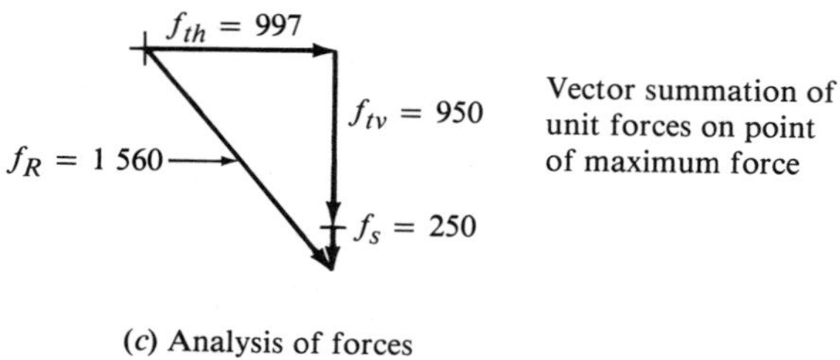

(*c*) Analysis of forces

Figure 19-6 Weld Bracket for Example Problem 19-2

to break the force down into horizontal and vertical components, then subsequently recombine all such components to compute the resultant force.

$$f_{th} = \frac{Tc_v}{J_w} = \frac{(38\,010)\,(3.00)}{114.4} = 997 \text{ lb/in}$$

$$f_{tv} = \frac{Tc_h}{J_w} = \frac{(38\,010)\,(2.86)}{114.4} = 950 \text{ lb/in}$$

Step 4. The vectorial combination of the forces on the weld is shown in Figure 19-6(c). Thus the maximum force is 1 560 lb/in.

Step 5. Selecting an E60 electrode for the welding, the allowable force per inch of weld leg size is 9 600 lb/in (Table 19-3). Then the required weld leg size is

$$w = \frac{1\,560 \text{ lb/in}}{9\,600 \text{ lb/in per inch of leg}} = 0.163 \text{ in}$$

Table 19-4 shows that the minimum size weld for a ½-in plate is 3/16 in (0.188 in). That size should be specified.

Example Problem 19-3. A steel strap, ¼-in thick, is to be welded to a rigid frame to carry a dead load of 12 500 lb, as shown in Figure 19-7. Design the strap and its weld.

Solution. Basically the objectives of the design are to specify a suitable material for the strap, the welding electrode, the size of the weld, and the dimensions W and h, as shown in the figure.

Let's specify that the strap is to be made from ASTM A441 structural steel and that it is to be welded with an E70 electrode, using the minimum size weld, 3/16 in. Appendix A-7 gives the yield strength of the A441 steel as 42 000 psi. Using a design factor of 2, we can compute an allowable stress of

$$\sigma_a = 42\,000/2 = 21\,000 \text{ psi}$$

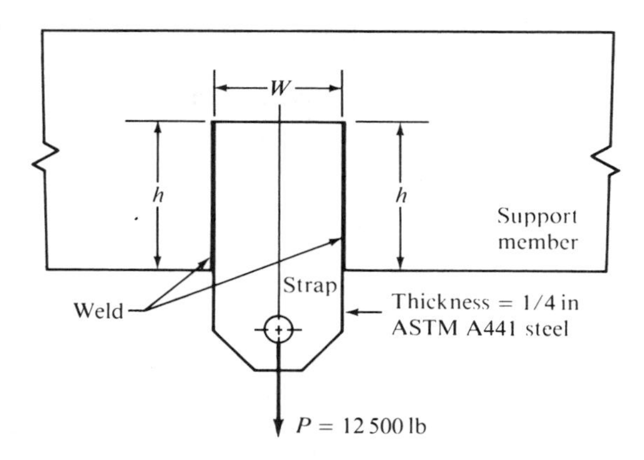

Figure 19-7 Strap for Example Problem 19-3

Then the required area of the strap is

$$A = \frac{P}{\sigma_a} = \frac{12\,500 \text{ lb}}{21\,000 \text{ lb/in}^2} = 0.595 \text{ in}^2$$

But the area is $W \times t$, where $t = 0.25$ in. Then the required width W is

$$W = A/t = 0.595/0.25 = 2.38 \text{ in}$$

Let's specify $W = 2.50$ in.

To compute the required length of the weld h, we need the allowable force on the $^3\!/_{16}$-in weld. Table 19-3 indicates the allowable force on the A441 steel welded with an E70 electrode to be 11 200 lb/in per inch of leg size. Then

$$f_a = \frac{11\,200 \text{ lb/in}}{1.0 \text{ in leg}} \times 0.188 \text{ in leg} = 2\,100 \text{ lb/in}$$

The actual force on the weld is

$$f_a = P/A_w = P/2h$$

Then solving for h gives

$$h = \frac{P}{2(f_a)} = \frac{12\,500 \text{ lb}}{2(2\,100 \text{ lb/in})} = 2.98 \text{ in}$$

Let's specify $h = 3.00$ in.

Example Problem 19-4. Evaluate the design shown in Figure 19-8 with regard to stress in the welds. All parts of the assembly are made of ASTM A36 structural steel and welded with an E60 electrode. The 2 500-lb load is a dead load.

Solution. The critical point would be the weld at the top of the tube where it is joined to the vertical surface. At this point, there is a three-dimensional force system acting on the weld as shown in Figure 19-9. The offset location of the load causes a twisting on the weld which produces a force f_t on the weld toward the left in the y direction. The bending produces a force f_b acting outward along the x axis. The vertical shear force f_s acts downward along the z axis.

From statics, the resultant of the three force components would be

$$f_R = \sqrt{f_t^2 + f_b^2 + f_s^2}$$

Now each component force on the weld will be computed.

Twisting force f_t:

$$f_t = \frac{Tc}{J_w}$$

$$T = (2\,500 \text{ lb})(8.00 \text{ in}) = 20\,000 \text{ lb} \cdot \text{in}$$

$$c = OD/2 = 4.500/2 = 2.25 \text{ in}$$

$$J_w = (\pi)(OD)^3/4 = (\pi)(4.500)^3/4 = 71.57 \text{ in}^3$$

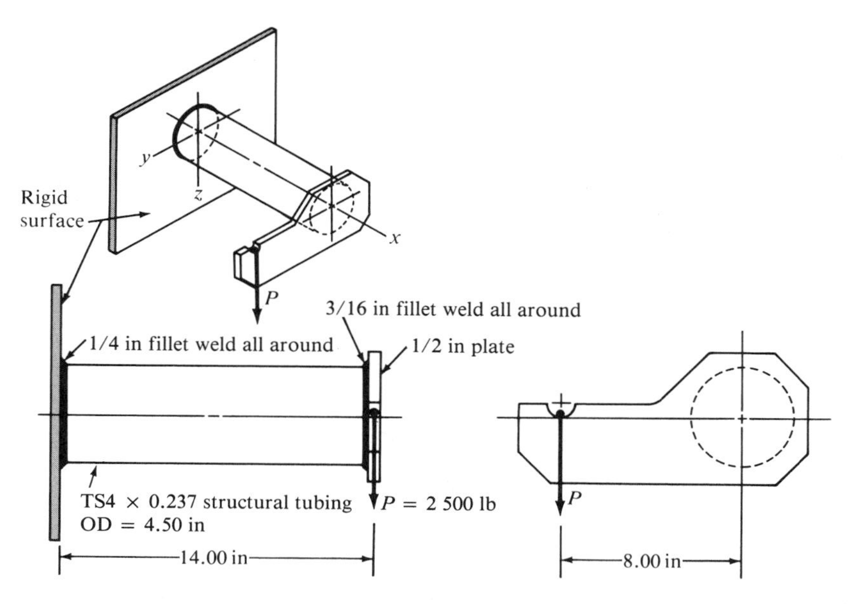

Figure 19-8 Bracket for Example Problem 19-4

Then

$$f_t = \frac{Tc}{J_w} = \frac{(20\,000)(2.25)}{71.57} = 629 \text{ lb/in}$$

Bending force f_b:

$$f_b = \frac{M}{Z_w}$$

$$M = (2\,500 \text{ lb})(14.00 \text{ in}) = 35\,000 \text{ lb} \cdot \text{in}$$

$$Z_w = (\pi)(OD)^2/4 = (\pi)(4.500)^2/4 = 15.90 \text{ in}^2$$

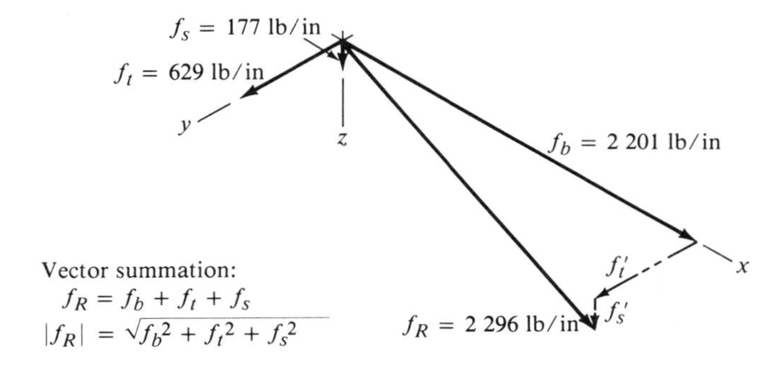

Figure 19-9 Force Vectors For Example Problem 19-4

Then

$$f_b = \frac{M}{Z_w} = \frac{35\,000}{15.90} = 2\,201 \text{ lb/in}$$

Vertical shear force f_s:

$$f_s = \frac{P}{A_w}$$

$$A_w = (\pi)(OD) = (\pi)(4.500 \text{ in}) = 14.14 \text{ in}$$

$$f_s = \frac{P}{A_w} = \frac{2\,500}{14.14} = 177 \text{ lb/in}$$

Now the resultant can be computed.

$$f_R = \sqrt{f_t^2 + f_b^2 + f_s^2}$$

$$f_R = \sqrt{629^2 + 2201^2 + 177^2} = 2\,296 \text{ lb/in}$$

Comparing this with the allowable force on a 1.0-in weld gives

$$w = \frac{2\,296 \text{ lb/in}}{9\,600 \text{ lb/in per inch of leg size}} = 0.239 \text{ in}$$

The ¼-in fillet specified in Figure 19-8 is satisfactory.

(has 1/4" weld though - so its ok

19-4 MACHINE FRAMES AND STRUCTURES

The design of machine frames and structures is largely art in that the components of the machine must be accommodated. The designer is often restricted in where supports can be placed in order not to interfere with the operation of the machine or in order to provide access for assembly or service.

But, of course, there are technical requirements which must be met as well for the structure itself. Some of the more important design parameters include the following:

Strength	Stiffness
Appearance	Cost to manufacture
Corrosion resistance	Weight
Size	Noise reduction
Vibration limitation	Life

Because of the virtually infinite possibilities for design details for frames and structures, this section will concentrate on general guidelines. The implementation of the guidelines would depend on the specific application. Factors to consider in starting a design project for a frame are now summarized.

Forces exerted by the components of the machine through mounting points such as bearings, pivots, brackets, and feet of other machine elements

Manner of support of the frame itself

Precision of the system: allowable deflection of components

Environment in which the unit will operate

Quantity of production and facilities available

Availability of analytical tools such as computerized stress analysis, past experience with similar products, and experimental stress analysis

Relationship to other machines, walls, and so on

Again, many of these factors require judgment by the designer. The parameters over which the designer has the most control are material selection, the geometry of load carrying parts of the frame, and manufacturing processes. A review of some possibilities is presented below.

Materials

As with machine elements discussed throughout this book, the material properties of strength and stiffness are of prime importance. Chapter 2 presented an extensive amount of information about materials and the appendixes contain much useful information. In general, steel ranks high in strength compared with competing materials for frames. But it is often better to consider more than just yield strength, ultimate tensile strength, or endurance strength alone. The complete design can be executed in several candidate materials to evaluate the overall performance. The *ratio of strength to density,* sometimes referred to as the *strength-to-weight ratio,* may yield a different material selection. Indeed, this is one reason for the use of aluminum, titanium, and composite materials in aircraft, aerospace vehicles, and transportation equipment.

Rigidity of a structure or frame is frequently the determining factor in the design, rather than strength. In these cases, the stiffness of the material, indicated by its modulus of elasticity, is the most important factor. Again, the ratio of stiffness to density may need to be evaluated. Table 19-5 gives some data for reference.

Recommended Deflection Limits

Actually, only intimate knowledge of the application of a machine member or frame can give a value for an acceptable deflection. But some guidelines are available to give you a place to start (3).

Deflection due to bending

General machine part: 0.000 5 to 0.003 in/in of beam length
Moderate precision: 0.000 01 to 0.000 5 in/in
High precision: 0.000 001 to 0.000 01 in/in

Deflection (rotation) due to torsion

General machine part: 0.001 to 0.01°/in of length
Moderate precision: 0.000 02 to 0.000 4°/in
High precision: 0.000 001 to 0.000 02°/in

Suggestions for Design to Resist Bending

Scrutiny of a table of deflection formulas for beams in bending would yield the following form for the deflection

$$\Delta = \frac{PL^3}{KEI} \tag{19-6}$$

Table 19-5 Comparison of Materials

Material	*Yield Strength (Ksi)*	*Modulus of Elasticity (Ksi)*	*Weight Density (lb/in^3)*	*Ratio s_y/ Density*	*Ratio E/Density*
Steel					
1020 HR	30	30 000	0.283	106	106 000
1050 HR	49	30 000	0.283	173	106 000
5160 OQT 1300	100	30 000	0.283	353	106 000
5160 OQT 700	237	30 000	0.283	837	106 000
Aluminum					
6061-T6	40	10 000	0.100	400	100 000
2014-T6	60	10 600	0.100	600	106 000
7075-T6	73	10 400	0.100	730	104 000
Titanium					
6Al-4V, Aged	150	16 500	0.160	938	103 000
3Al-13V-11Cr	175	16 000	0.160	1094	100 000
Ductile iron					
60-40-18	40	22 000	0.260	155	84 600
119-90-02	90	22 000	0.260	346	84 600
Magnesium					
AZ 63A-T6	19	6 500	0.066	288	98 500
Plastics					
Nylon 6/6	12	420	0.041	290	10 300
High Impact ABS	5	250	0.041	120	6 100
Phenolic	6.5	1 100	0.051	127	21 600

where

P = Load

L = Length between supports

E = Modulus of elasticity of the material in the beam

I = Moment of inertia of the cross section of the beam

K = A factor depending on the manner of loading and support

Some obvious conclusions from equation (19-6) are that the load and the length should be kept small and the values of E and I should be large. Note the cubic function of the length. This means, for example, that reducing the length by a factor of 2.0 would reduce the deflection by a factor of 8.0, obviously a desirable effect.

An appreciation for the factor K can be had by referring to Figure 19-10, which shows three beams carrying a single concentrated load.

In summary, the following suggestions are made for designing to resist bending:

1. Keep the length of the beam as short as possible and place loads close to the supports.
2. Maximize the moment of inertia of the cross section in the direction of bending. In general, this can be done by placing as much of the material as far away from the neutral axis of bending as possible, as in a wide flange beam or a hollow rectangular section.

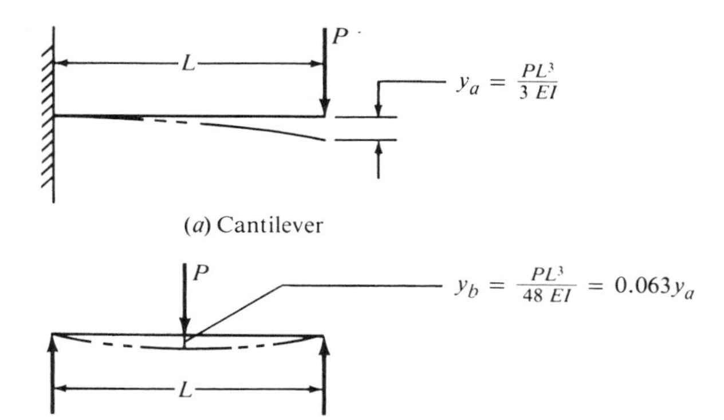

(b) Simply supported beam

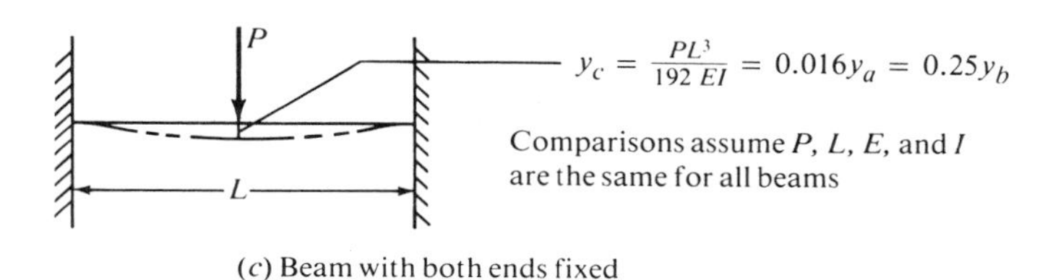

(c) Beam with both ends fixed

Figure 19-10 Beam Deflection Comparisons

3. Use a material with a high modulus of elasticity.
4. Use fixed ends for the beam where possible.
5. Consider lateral deflection in addition to deflection in the primary load direction. Such loads could be encountered during fabrication, handling, shipping, careless use, or casual bumping.
6. Be sure to evaluate the final design with regard to both strength and rigidity. Some approaches to improving rigidity (increasing I) can actually increase the stress in the beam because the section modulus is decreased.
7. Provide rigid corner bracing in open frames.
8. Cover an open frame section with a sheet material to resist distortion. This is sometimes called *panel stiffening*.
9. Consider a truss-type construction to obtain structural stiffness with light weight members.
10. When designing an open space frame, use diagonal bracing to break sections into triangular parts, an inherently rigid shape.
11. Consider stiffeners for large panels to reduce vibration and noise.
12. Add bracing and gussets to areas where loads are applied or at supports to help transfer the forces into adjoining members.
13. Beware of load-carrying members with thin extended flanges that may be placed in compression. Local buckling, sometimes called *crippling* or *wrinkling,* could occur.
14. Place connections at points of low stress if possible.

Suggestions for Design of Members to Resist Torsion

Torsion can be created in a machine frame member in a variety of ways: A support surface may be uneven; a machine or motor may transmit a reaction torque to the frame; a load acting to the side of the axis of the beam (or any place away from the flexural center of the beam) would produce twisting.

In general, the torsional deflection of a member is computed from

$$\theta = \frac{TL}{GR} \tag{19-7}$$

where

T = Applied torque or twisting moment

L = Length over which torque acts

G = Shear modulus of elasticity of the material

R = Torsional rigidity constant

The designer must choose the shape of the torsion member carefully to obtain a rigid structure. The following suggestions are made:

1. Use closed sections wherever possible. Examples are solid bars with large cross section, hollow pipe and tubing, closed rectangular or square tubing, special closed shapes which approximate a tube.
2. Conversely, avoid open sections made from thin materials. Figure 19-11 shows a dramatic illustration.
3. For wide frames, brackets, tables, bases, etc., use diagonal braces placed at 45° to the sides of the frame (see Figure 19-12).
4. Use rigid connections, such as by welding members together.

Most of the suggestions made in this section can be implemented regardless of the specific type of frame designed: castings made from cast iron, cast steel, aluminum, zinc,

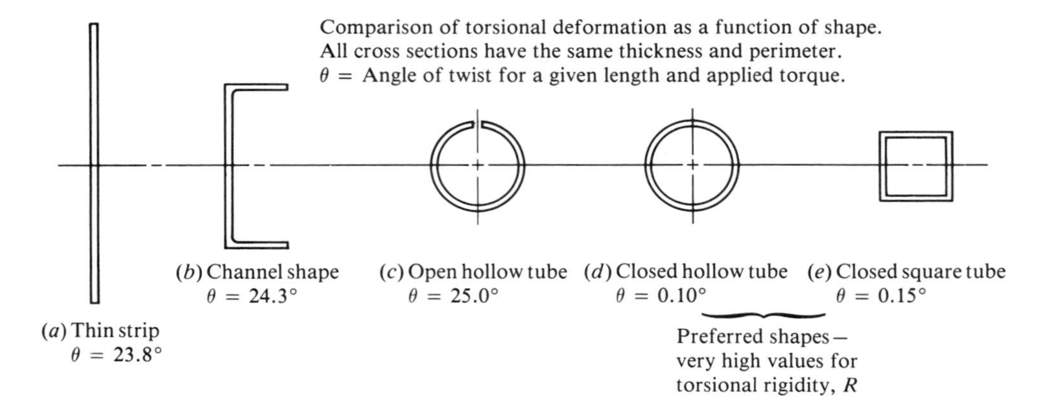

Figure 19-11 Comparison of Torsional Deformation as a Function of Shape. All cross sections have the same thickness and perimeter. θ = angle of twist for a given length and applied torque.

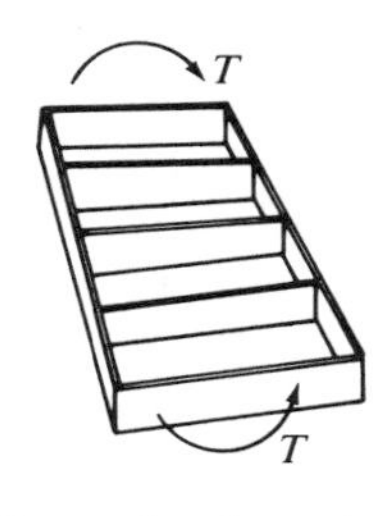

(*a*) Conventional cross bracing
$\theta = 10.8°$

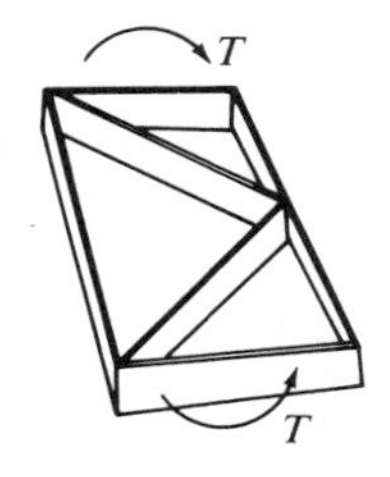

(*b*) Single diagonal bracing
$\theta = 0.30°$

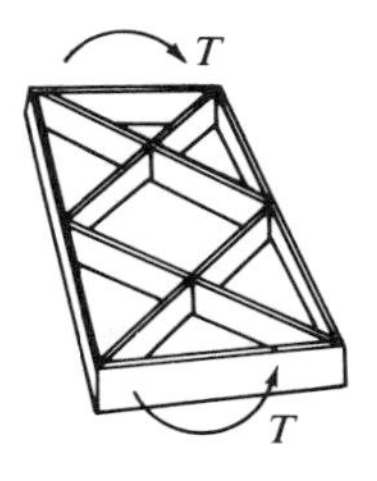

(*c*) Double diagonal bracing
$\theta = 0.10°$

Figure 19-12 Comparison of Torsional Angle of Twist, θ, for Boxlike Frames. Each has the same basic dimensions and applied torque.

or magnesium; weldments made from steel or aluminum plate; formed housings from sheet metal or plate; or plastic moldings.

REFERENCES

1. American Institute of Steel Construction. *AISC Handbook,* 8th ed. New York: 1980.
2. Blodgett, Omer W. *Design of Welded Structures.* Cleveland: James F. Lincoln Arc Welding Foundation, 1966.
3. Blodgett, Omer W. *Design of Weldments.* Cleveland: James F. Lincoln Arc Welding Foundation, 1963.
4. Weiser, Peter F., ed. *Steel Casting Handbook,* 5th ed. Rocky River, Ohio: Steel Founder's Society of America, 1980.

PROBLEMS

For Problems 1–6 design a bolted joint to join the two members shown in the appropriate figure. Specify the number of bolts, the pattern, the bolt grade, and the bolt size.

1. Figure 19-13
2. Figure 19-14
3. Figure 19-15
4. Figure 19-16
5. Figure 19-17
6. Figure 19-18

For Problems 7–12 design a welded joint to join the two members shown in the appropriate figure. Specify the weld pattern, the type of electrode to be used, and the size of weld. In problems 7, 8, and 9, the members are ASTM A36 steel. In problems 10, 11, and 12, the members are ASTM A441 steel. Use the method of treating the joint as a line and use the allowable forces per inch of leg for building type structures from Table 19-3.

7. Figure 19-13
8. Figure 19-16
9. Figure 19-17
10. Figure 19-18
11. Figure 19-19
12. Figure 19-19 (but $P_2 = 0$)

For Problems 13–16, design a welded joint to join the two aluminum members shown in the appropriate figure. Specify the weld pattern, the type of filler alloy, and the size of weld. The types of materials joined are listed in the problems.

13. Figure 19-13: 6061 alloy (but $P = 4\,000$ lb)
14. Figure 19-20: 6061 alloy
15. Figure 19-21: 6063 alloy
16. Figure 19-22: 3003 alloy
17. Compare the weight of a tensile rod carrying a dead load of 4 800 lb if it is made from (a) AISI 1020 HR steel, (b) AISI 5160 OQT 1300 steel, (c) Aluminum 2014-T6, (d) Aluminum 7075-T6, (e) Titanium 6Al-4V, and (f) Titanium 3Al-13V-11Cr. Use the data from Table 19-5. Use $N = 2$ based on yield strength.

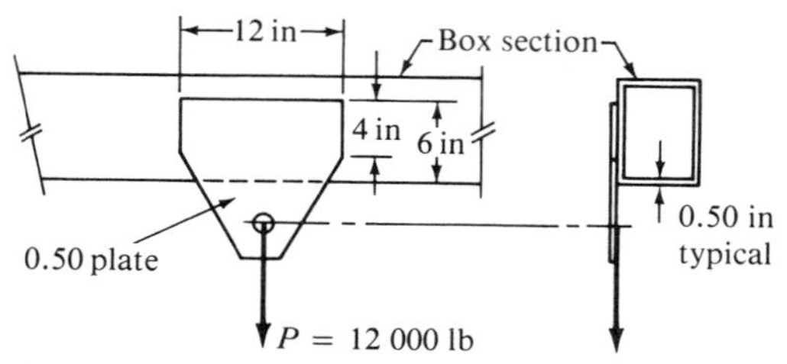

Figure 19-13 (Problems 1, 7, and 13)

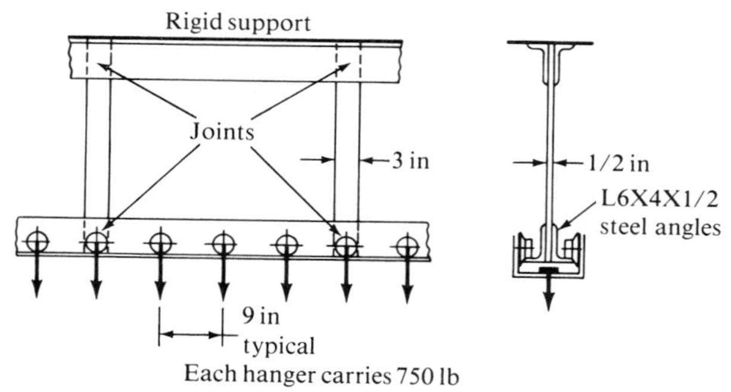

Figure 19-14 (Problem 2)

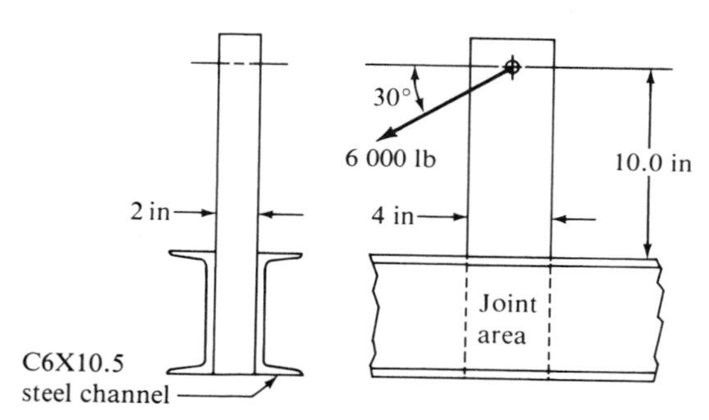

Figure 19-15 (Problem 3)

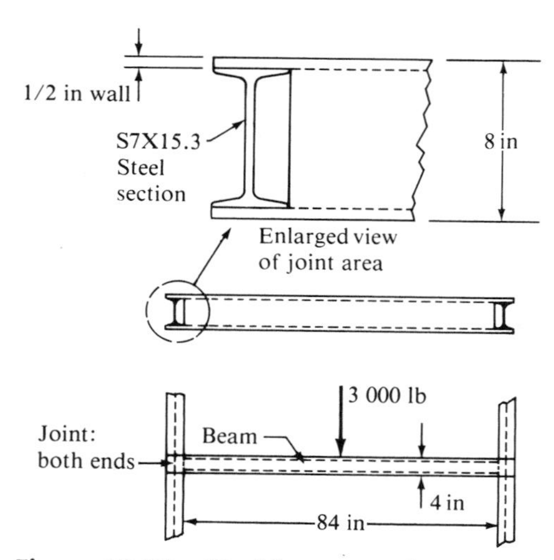

Figure 19-16 (Problems 4 and 8)

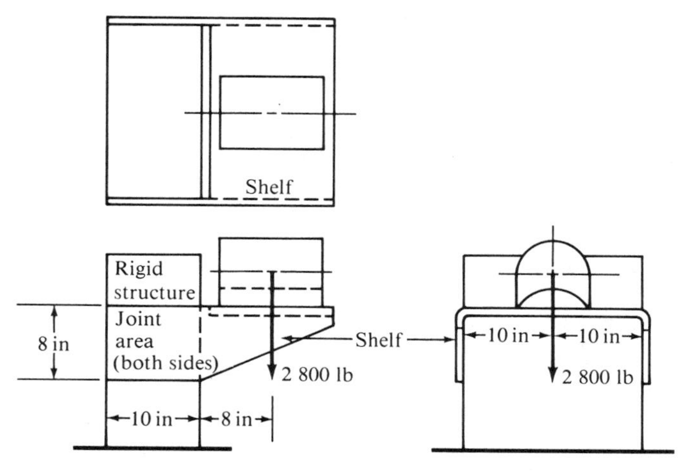

Figure 19-17 (Problems 5 and 9)

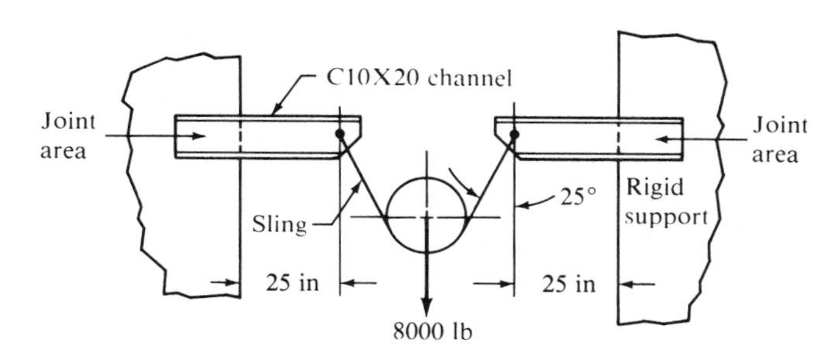

Figure 19-18 (Problems 6 and 10)

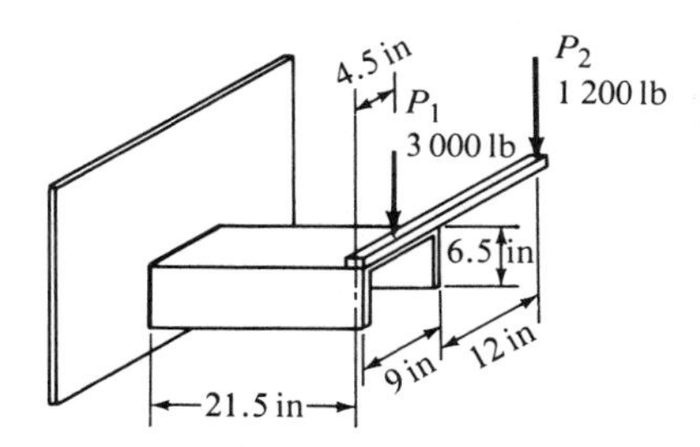

Figure 19-19 (Problems 11 and 12)

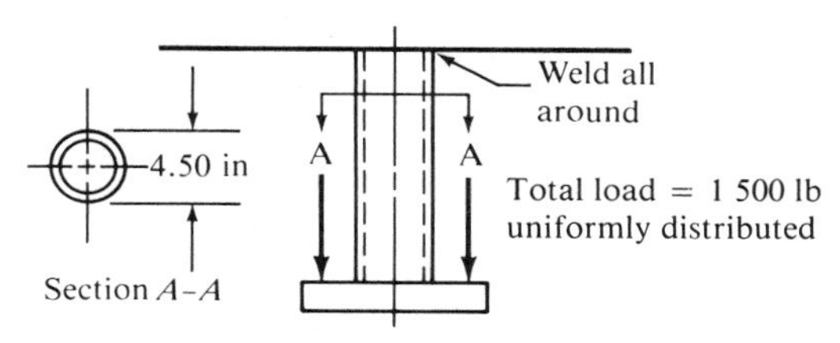

Figure 19-20 (Problem 14)

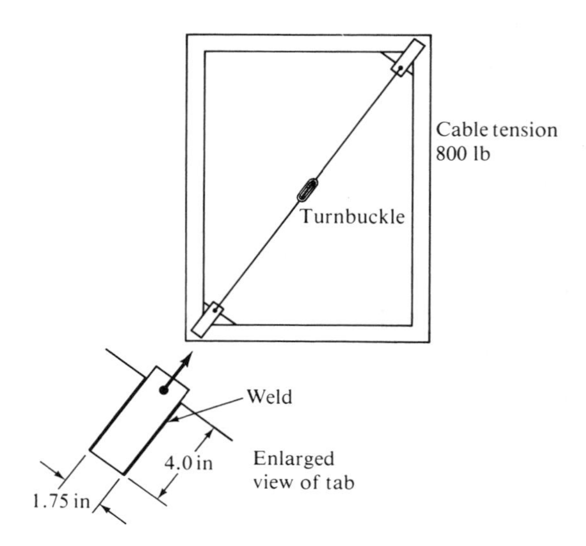

Figure 19-21 (Problem 15)

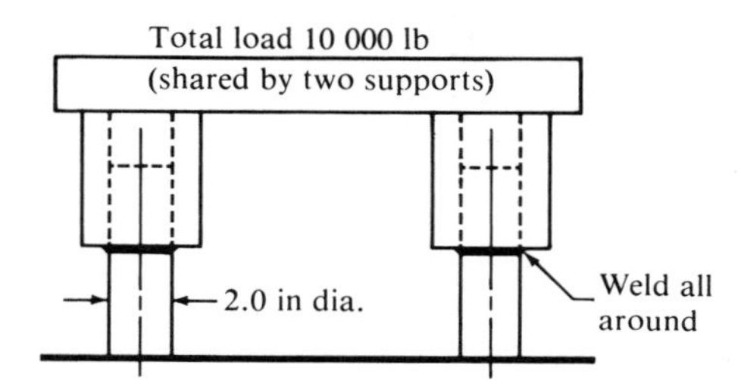

Figure 19-22 (Problem 16)

20 Design Projects

20-1 OVERVIEW

One of the primary focuses of this book has been to emphasize the integration of machine elements into complete mechanical designs. The interfaces between machine elements have been discussed for many examples. The forces exerted on one element by another have been computed. Commercially available components and complete devices have been shown in several figures throughout the book.

Although these discussions and examples are helpful, one of the better ways to learn mechanical design is to do mechanical design. You should decide the detailed functions and design requirements for the design. You should conceptualize several approaches to a design. You should decide which approach to complete. You should complete the design of each element in detail. You should make assembly and detail drawings to communicate your design to others who may use it or be responsible for its fabrication. You should specify completely the purchased components that are part of the design.

Following are several projects that call for you to do these operations. You or the instructor may modify or amplify the projects to suit individual needs or the available time or information, such as manufacturers' catalogs. As in most design, many solutions are possible. Different solutions from several members of a class could be compared and critiqued to enhance the learning achieved. It may be helpful at this time to review sections 1-4 and 1-5 in Chapter 1 about the functions and design requirements for design and a philosophy of design. Also, the questions at the end of Chapter 1 asked that you write a set of functions and design requirements for several of the same devices. If you have already done so, they can be used as a part of these exercises.

20-2 DESIGN PROJECTS

Automobile Hood Latch

Design a hood latch for an automobile. The latch must be able to hold the hood securely closed during operation of the vehicle. But it should be easy to open for servicing the contents of the engine compartment. Theft-proofing is an important design goal. Attachment of the latch to the frame of the car and to the hood should be defined. Mass production should be a requirement.

Hydraulic Lift

Design a hydraulic lift to be used for car repair. Obtain pertinent dimensions from representative cars for initial height, extended height, design of the pads which contact the car, and so on. The lift will raise the entire car.

Portable Crane

Design a portable crane to be used in homes, small industries, warehouses, and garages. It should have a capacity of at least 1 000 lb (4.45 kN). Typical uses would be to remove an engine from a car, lift machine components, or load trucks.

Can Crusher

Design a machine to crush soft drink or beer cans. The crusher would be used in homes or restaurants as an aid to recycling efforts. It could be operated either by hand or electrically. It should crush the cans to approximately 20 percent of their original volume.

Transfer Device

Design an automatic transfer device for a production line. The parts to be handled are steel castings having the following characteristics:

Weight: 42.0 lb (187 N).

Size: cylindrical; 6.75-in diameter and 10.0 in high. Exterior surface is free of projections or holes and has a reasonably smooth, as-cast finish.

Transfer rate: continuous flow, 2.00 sec between parts.

Parts enter at a 24.0-in elevation on a roller conveyor. They must be elevated to 48.0 in in a space of 60.0 in horizontally. They leave on a separate conveyor.

Drum Dumper

Design a drum dumper. The machine is to raise a 55-gallon drum of bulk material from floor level to a height of 60.0 in and dump the contents of the drum into a hopper.

Paper Feeder

Design a paper feed device for a copier. The paper must be fed at a rate of 120 sheets per min.

Gravel Conveyor

Design a conveyor to elevate gravel into a truck. The top edge of the truck bed is 8.0 ft (2.44 m) off the ground. The bed is 6.5 ft wide, 12.0 ft long, and 4.0 ft deep (1.98 m × 3.66 m × 1.22 m). It is desired to fill the truck in 5.0 min or less.

Construction Lift

Design a construction lift. The lift will raise building materials from ground level to any height up to 40.0 ft (12.2 m). The lift will be at the top of a rigid scaffold which is not a part of the design project. It will raise a load of up to 500 lb (2.22 kN) at the rate of 1.0 ft/sec (0.30 m/s). The load will be on a pallet, 3.0 ft by 4.0 ft (0.91 m × 1.22 m). At the top of the lift, means must be provided to bring the load onto a platform which supports the lift.

Packaging Machine

Design a packaging machine. Toothpaste tubes are to be taken from a continuous belt and inserted into cartons. Any standard tube size may be chosen. The device may include the means to close the cartons after the tube is in place.

Carton Packer

Design a machine to insert 24 cartons of toothpaste into a shipping case.

Robot Gripper

Design a gripper for a robot to grasp a spare tire assembly from a rack and insert it into the trunk of an automobile on an assembly line. Obtain dimensions from a particular car.

Weld Positioner

Design a weld positioner. A heavy frame is made of welded steel plate in the shape shown in Figure 20-1. The welding unit will be robot-guided but it is essential that the weld line

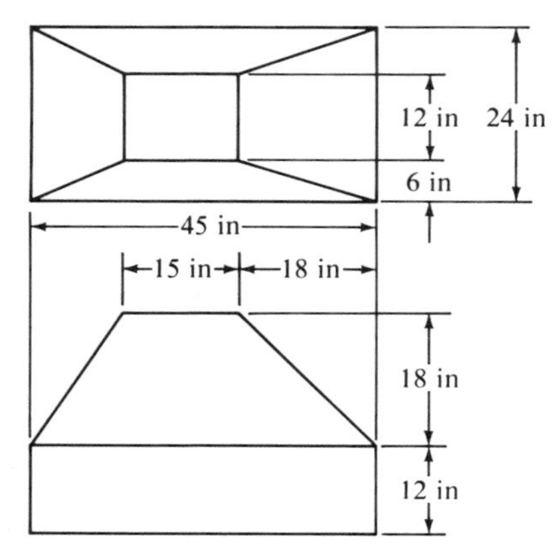

Weld all seams with weld line in horizontal plane

Figure 20-1 Frame for Weld Positioner

be horizontal as the weld proceeds. Design the device to hold the frame securely and move it to present the part to the robot. The plate has a thickness of ⅜ in (9.53 mm).

Garage Door Opener

Design a garage door opener.

Car Jack

Design a floor jack for a car to lift either the entire front end or the entire rear end. The jack may be powered by hand, using mechanical or hydraulic actuation. Or it may be powered by pneumatic pressure or electrical power.

Spur Gear Speed Reducer, Single Reduction

Design a complete single-reduction spur gear type speed reducer. Specify the two gears, two shafts, four bearings, and a housing. Use any of the data from Chapter 11, problems 40 through 46.

Spur Gear Speed Reducer, Double Reduction

Design a complete double-reduction spur gear type speed reducer. Specify the four gears, three shafts, six bearings, and a housing. Use any of the data from Chapter 11, problems 51, 52, or 53.

Helical Gear Speed Reducer, Single Reduction

Design a complete single-reduction helical gear type speed reducer. Use any of the data from Chapter 12, problems 17 through 23.

Bevel Gear Speed Reducer, Single Reduction

Design a complete single-reduction bevel gear type speed reducer. Use any of the data from Chapter 12, problems 26 through 37.

Wormgear Speed Reducer, Single Reduction

Design a complete single-reduction wormgear type speed reducer. Use any of the data from Chapter 12, problems 38 through 44.

Lift Device Using Acme Screws

Design a device similar to that sketched in Figure 18-4 in Chapter 18. An electric motor drives the worm at a speed of 1 750 rpm. The two Acme screws rotate and lift the yoke which in turn lifts the hatch. See example problem 18-1 for additional details. Complete the entire unit, including the wormgear set, the chain drive, the Acme screws, the bearings, and their mountings. The hatch is 60 in (1524 mm) in diameter at its top surface. The screws should be nominally 30 in (762 mm) long. The total motion of the yoke will be 24 in, to be completed in 15.0 sec or less.

Lift Device Using Ball Screws

Repeat design of the lift device, using ball screws instead of Acme screws.

Brake for a Drive Shaft

Design a brake. A rotating load (as sketched in Figure 16-21) is to be stopped from 775 rpm in 0.50 sec or less. Use any type of brake described in Chapter 16 and complete the design details, including the actuating means: springs, air pressure, manual lever, and so on. Show the brake attached to the shaft of Figure 16-21.

Brake for a Winch

Design a complete brake for the application shown in Figure 16-23 and described in problem 11 in Chapter 16.

Indexing Drive

Design an indexing drive for an automatic assembly system. The items to be moved are mounted on a square steel fixture plate, 6.0 in (152 mm) on a side and 0.50 in (12.7 mm) thick. The total weight of each assembly is 10.0 lb (44.5 N). The center of each fixture (intersection of its diagonals) is to move 12.0 in (305 mm) with each index. The index is to be completed in 1.0 sec or less and the fixture must be held stationary at each station for a minimum of 2.0 sec. Four assembly stations are required. The arrangement may be linear, rotary, or any other provided the fixtures move in a horizontal plane.

Child's Ferris Wheel

Design a child's ferris wheel. It should be capable of holding one to four children weighing up to 80 lb (356 N) each. The rotational speed should be one revolution in 6.0 sec. It should be driven by an electric motor.

Backyard Amusement Ride

Design a backyard amusement ride in which small coaster wagons are pulled along a circular path. The ride should be powered by an electric motor. Each wagon is 1.0 m long (39.4 in) and 0.50 m wide (19.7 in). The four wheels are 150 mm (6.0 in) in diameter. The wagon is to be attached to the drive rod at the point where the handle would normally be. The radial distance to the attachment point is to be 2.0 m (6.6 ft). The wagons are to make one revolution in 8.0 sec. Provide a means of starting and stopping the drive.

Transfer Device

Design a device to move automotive camshafts between processing stations. Each movement is to be 9.0 in (229 mm). The camshaft is to be supported on two unfinished bearing surfaces having a diameter of 3.80 in (96.5 mm) and an axial length of 0.75 in (19.0 mm). The spread between the bearing surfaces is 15.75 in (400.0 mm). Each camshaft weighs 16.3 lb (72.5 N). One motion cycle is to be completed each 2.50 sec. Design the complete mechanism, including the drive from an electric motor.

Appendixes

APPENDIX 1 PROPERTIES OF AREAS

Circle

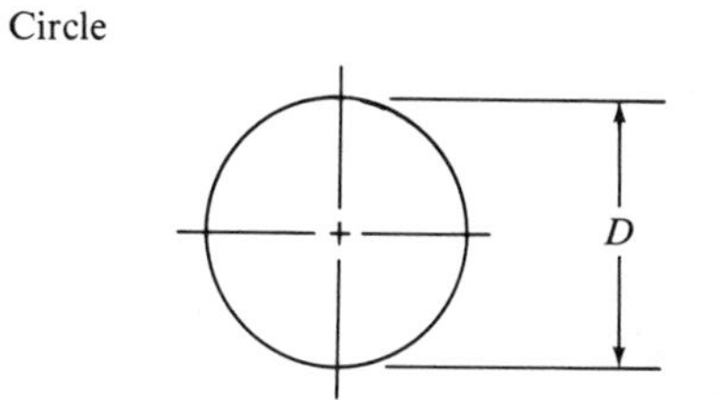

$A = \pi D^2/4$ $\quad r = D/4$

$I = \pi D^4/64$ $\quad J = \pi D^4/32$

$Z = \pi D^3/32$ $\quad Z_p = \pi D^3/16$

Hollow circle (tube)

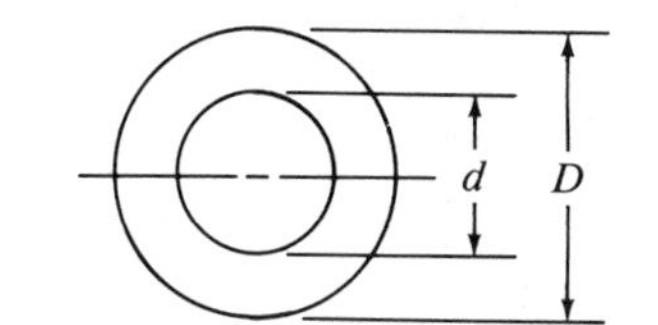

$A = \pi(D^2 - d^2)/4$ $\quad r = \sqrt{D^2 + d^2}/4$

$I = \pi(D^4 - d^4)/64$ $\quad J = \pi(D^4 - d^4)/32$

$Z = \pi(D^4 - d^4)/32D$ $\quad Z_p = \pi(D^4 - d^4)/16D$

Square

$A = S^2$ $\quad r = S/\sqrt{12}$

$I = S^4/12$

$Z = S^3/6$

Rectangle

$A = BH$ $\quad r_x = H/\sqrt{12}$

$I_x = BH^3/12$ $\quad r_y = B/\sqrt{12}$

$Z_x = BH^2/6$

Triangle

$A = BH/2$ $\quad r = H/\sqrt{18}$

$I = BH^3/36$

$Z = BH^2/24$

Semicircle

$A = \pi D^2/8$ $\quad r = 0.132D$

$I = 0.007D^4$

$Z = 0.024D^3$

Regular hexagon

$A = 0.866D^2$ $\quad r = 0.264D$

$I = 0.06D^4$

$Z = 0.12D^3$

A = area

I = moment of inertia

Z = section modulus

r = Radius of gyration = $\sqrt{I/A}$

J = Polar moment of inertia

Z_p = Polar section modulus

APPENDIX 2 CONVERSION FACTORS

U.S. Customary Unit System to SI Units

Quantity	*U.S. Customary Unit*		*SI Unit*		*Symbol*	*Equivalent Units*
Length	1 foot (ft)	=	0.304 8	meter	m	—
Mass	1 slug	=	14.59	kilogram	kg	—
Time	1 second	=	1.0	second	s	—
Force	1 pound (lb)	=	4.448	newton	N	$kg \cdot m/s^2$
Pressure	1 lb/in^2	=	6 895	pascal	Pa	N/m^2 or $kg/m \cdot s^2$
Energy	1 ft-lb	=	1.356	joule	J	$N \cdot m$ or $kg \cdot m^2/s^2$
Power	1 ft-lb/s	=	1.356	watt	W	J/s

Other Convenient Conversion Factors

Length

1 ft = 0.304 8 m
1 in = 25.4 mm
1 mi = 5 280 ft
1 mi = 1.609 km

1 km = 1000 m
1 cm = 10 mm
1 m = 1000 mm

Area

1 ft^2 = 0.092 9 m^2
1 in^2 = 645.2 mm^2

1 m^2 = 10.76 ft^2
1 m^2 = 10^6 mm^2

Volume

1 ft^3 = 7.48 gal
1 ft^3 = 1 728 in^3
1 ft^3 = 0.028 3 m^3

1 gal = 0.003 79 m^3
1 gal = 3.785 L
1 m^3 = 1000 L

Volume Flow Rate

1 ft^3/s = 449 gal/min
1 ft^3/s = 0.028 3 m^3/s
1 gal/min = 6.309×10^{-5} m^3/s

1 gal/min = 3.785 L/min
1 L/min = 16.67×10^{-6} m^3/s

Temperature

$T(°C) = [T(°F) - 32]\frac{5}{9}$
$T(°F) = \frac{9}{5}[T(°C)] + 32$

Power

1 hp = 550 $ft \cdot lb/s$
1 hp = 745.7 W

1 $ft \cdot lb/s$ = 1.356 W
1 Btu/h = 0.293 W

Density

1 $slug/ft^3$ = 515.4 kg/m^3

Specific Weight

1 lb/ft^3 = 157.1 N/m^3

Energy

1 $ft \cdot lb$ = 1.356 J
1 Btu = 1.055 kJ
1 $W \cdot h$ = 3.600 kJ

Torque or Moment

1 $lb \cdot in$ = 0.113 0 $N \cdot m$

Stress, Pressure, or Unit Loading

1 lb/in^2 = 6.895 kPa
1 lb/ft^2 = 0.047 9 kPa
1 kip/in^2 = 6.895 MPa

Section Modulus

1 in^3 = 1.639×10^4 mm^3

Moment of Inertia

1 in^4 = 4.162×10^5 mm^4

Hardness Conversion Table

BRINELL		ROCKWELL		TENSILE STRENGTH, (1 000 psi approx.)	BRINELL		ROCKWELL		TENSILE STRENGTH (1 000 psi approx.)
Indent. Diam. (mm)	*No.**	*B*	*C*		*Indent. Diam. (mm)*	*No.**	*B*	*C*	
2.25	745		65.3		3.75	262	(103.0)	26.6	127
2.30	712		—		3.80	255	(102.0)	25.4	123
2.35	682		61.7		3.85	248	(101.0)	24.2	120
2.40	653		60.0		3.90	241	100.0	22.8	116
2.45	627		58.7		3.95	235	99.0	21.7	114
2.50	601		57.3		4.00	229	98.2	20.5	111
2.55	578		56.0		4.05	223	97.3	(18.8)	—
2.60	555		54.7	298	4.10	217	96.4	(17.5)	105
2.65	534		53.5	288	4.15	212	95.5	(16.0)	102
2.70	514		52.1	274	4.20	207	94.6	(15.2)	100
2.75	495		51.6	269	4.25	201	93.8	(13.8)	98
2.80	477		50.3	258	4.30	197	92.8	(12.7)	95
2.85	461		48.8	244	4.35	192	91.9	(11.5)	93
2.90	444		47.2	231	4.40	187	90.7	(10.0)	90
2.95	429		45.7	219	4.45	183	90.0	(9.0)	89
3.00	415		44.5	212	4.50	179	89.0	(8.0)	87
3.05	401		43.1	202	4.55	174	87.8	(6.4)	85
3.10	388		41.8	193	4.60	170	86.8	(5.4)	83
3.15	375		40.4	184	4.65	167	86.0	(4.4)	81
3.20	363		39.1	177	4.70	163	85.0	(3.3)	79
3.25	352	(110.0)	37.9	171	4.80	156	82.9	(0.9)	76
3.30	341	(109.0)	36.6	164	4.90	149	80.8		73
3.35	331	(108.5)	35.5	159	5.00	143	78.7		71
3.40	321	(108.0)	34.3	154	5.10	137	76.4		67
3.45	311	(107.5)	33.1	149	5.20	131	74.0		65
3.50	302	(107.0)	32.1	146	5.30	126	72.0		63
3.55	293	(106.0)	30.9	141	5.40	121	69.8		60
3.60	285	(105.5)	29.9	138	5.50	116	67.6		58
3.65	277	(104.5)	28.8	134	5.60	111	65.7		56
3.70	269	(104.0)	27.6	130					

Note: This is a condensation of Table 2, Report J417b, SAE 1971 Handbook. Values in ( ) are beyond normal range, and are presented for information only.

*Values above 500 are for tungsten carbide ball; below 500 for standard ball.

Source: *Modern Steels and Their Properties,* Bethlehem Steel Co., Bethlehem, Pa.

APPENDIX 3 DESIGN PROPERTIES OF CARBON AND ALLOY STEELS

MATERIAL DESIGNATION *(AISI Number)*	CONDITION	TENSILE STRENGTH		YIELD STRENGTH		DUCTILITY *(Percent Elongation in 2 in)*	BRINELL HARDNESS *(HB)*
		Ksi	*MPa*	*Ksi*	*MPa*		
1020	Hot rolled	55	379	30	207	25	111
1020	Cold drawn	61	420	51	352	15	122
1020	Annealed	60	414	43	296	38	121
1040	Hot rolled	72	496	42	290	18	144
1040	Cold drawn	80	552	71	490	12	160
1040	OQT 1300	88	607	61	421	33	183
1040	OQT 400	113	779	87	600	19	262
1050	Hot rolled	90	620	49	338	15	180
1050	Cold drawn	100	690	84	579	10	200
1050	OQT 1300	96	662	61	421	30	192
1050	OQT 400	143	986	110	758	10	321
1117	Hot rolled	62	427	34	234	33	124
1117	Cold drawn	69	476	51	352	20	138
1117	WQT 350	89	614	50	345	22	178
1137	Hot rolled	88	607	48	331	15	176
1137	Cold drawn	98	676	82	565	10	196
1137	OQT 1300	87	600	60	414	28	174
1137	OQT 400	157	1 083	136	938	5	352
1144	Hot rolled	94	648	51	352	15	188
1144	Cold drawn	100	690	90	621	10	200
1144	OQT 1300	96	662	68	469	25	200
1144	OQT 400	127	876	91	627	16	277
1213	Hot rolled	55	379	33	228	25	110
1213	Cold drawn	75	517	58	340	10	150
12L13	Hot rolled	57	393	34	234	22	114
12L13	Cold drawn	70	483	60	414	10	140
1340	Annealed	102	703	63	434	26	207
1340	OQT 400	285	1 960	234	1 610	8	578
1340	OQT 700	221	1 520	197	1 360	10	444
1340	OQT 1000	144	993	132	910	17	363
1340	OQT 1300	100	690	75	517	25	235
3140	Annealed	95	655	67	462	25	187
3140	OQT 400	280	1 930	248	1 710	11	555
3140	OQT 700	220	1 520	200	1 380	13	461
3140	OQT 1000	152	1 050	133	920	17	311
3140	OQT 1300	115	792	94	648	23	233
4130	Annealed	81	558	52	359	28	156
4130	WQT 400	234	1 610	197	1 360	12	461
4130	WQT 700	208	1 430	180	1 240	13	415
4130	WQT 1000	143	986	132	910	16	302
4130	WQT 1300	98	676	89	614	28	202
4140	Annealed	95	655	60	414	26	197
4140	OQT 400	290	2 000	251	1 730	11	578
4140	OQT 700	231	1 590	212	1 460	13	461
4140	OQT 1000	168	1 160	152	1 050	17	341
4140	OQT 1300	117	807	100	690	23	235

MATERIAL DESIGNATION *(AISI Number)*	CONDITION	TENSILE STRENGTH		YIELD STRENGTH		DUCTILITY *(Percent Elongation in 2 in)*	BRINELL HARDNESS *(HB)*
		Ksi	*MPa*	*Ksi*	*MPa*		
4150	Annealed	106	731	55	379	20	197
4150	OQT 400	300	2 070	248	1 710	10	578
4150	OQT 700	247	1 700	229	1 580	10	495
4150	OQT 1000	197	1 360	181	1 250	11	401
4150	OQT 1300	127	880	116	800	20	262
4340	Annealed	108	745	68	469	22	217
4340	OQT 400	283	1 950	228	1 570	11	555
4340	OQT 700	230	1 590	206	1 420	12	461
4340	OQT 1000	171	1 180	158	1 090	16	363
4340	OQT 1300	140	965	120	827	23	280
5140	Annealed	83	572	42	290	29	167
5140	OQT 400	276	1 900	226	1 560	7	534
5140	OQT 700	220	1 520	200	1 380	11	429
5140	OQT 1000	145	1 000	130	896	18	302
5140	OQT 1300	104	717	83	572	27	207
5150	Annealed	98	676	52	359	22	197
5150	OQT 400	312	2 150	250	1 720	8	601
5150	OQT 700	240	1 650	220	1 520	10	461
5150	OQT 1000	160	1 100	149	1 030	15	321
5150	OQT 1300	116	800	102	700	22	241
5160	Annealed	105	724	40	276	17	197
5160	OQT 400	322	2 220	260	1 790	4	627
5160	OQT 700	263	1 810	237	1 630	9	514
5160	OQT 1000	170	1 170	151	1 040	14	341
5160	OQT 1300	115	793	100	690	23	229
6150	Annealed	96	662	59	407	23	197
6150	OQT 400	315	2 170	270	1 860	7	601
6150	OQT 700	247	1 700	223	1 540	10	495
6150	OQT 1000	183	1 260	173	1 190	12	375
6150	OQT 1300	118	814	107	738	21	241
8650	Annealed	104	717	56	386	22	212
8650	OQT 400	282	1 940	250	1 720	11	555
8650	OQT 700	240	1 650	222	1 530	12	495
8650	OQT 1000	176	1 210	155	1 070	14	363
8650	OQT 1300	122	841	113	779	21	255
8740	Annealed	100	690	60	414	22	201
8740	OQT 400	290	2 000	240	1 650	10	578
8740	OQT 700	228	1 570	212	1 460	12	461
8740	OQT 1000	175	1 210	167	1 150	15	363
8740	OQT 1300	119	820	100	690	25	241
9255	Annealed	113	780	71	490	22	229
9255	Q&T 400	310	2 140	287	1 980	2	601
9255	Q&T 700	260	1 790	240	1 650	5	534
9255	Q&T 1000	181	1 250	160	1 100	14	352
9255	Q&T 1300	130	896	102	703	21	262

Properties common to all carbon and alloy steels:
Density: 0.283 lb/in^3; 7 680 kg/m^3
Modulus of elasticity: 30×10^6 psi; 207 GPa
Poisson's ratio: 0.27
Shear modulus: 11.5×10^6 psi; 80 GPa
Coefficient of thermal expansion: 6.5×10^{-6} °F^{-1}

APPENDIX 4 PROPERTIES OF HEAT-TREATED STEELS

Appendix 4-1 Properties of Heat-treated AISI 1040

Treatment: Normalized at 1650 F; reheated to 1550 F; quenched in water.
1-in. Round Treated; .505-in. Round Tested. As-quenched HB 534.

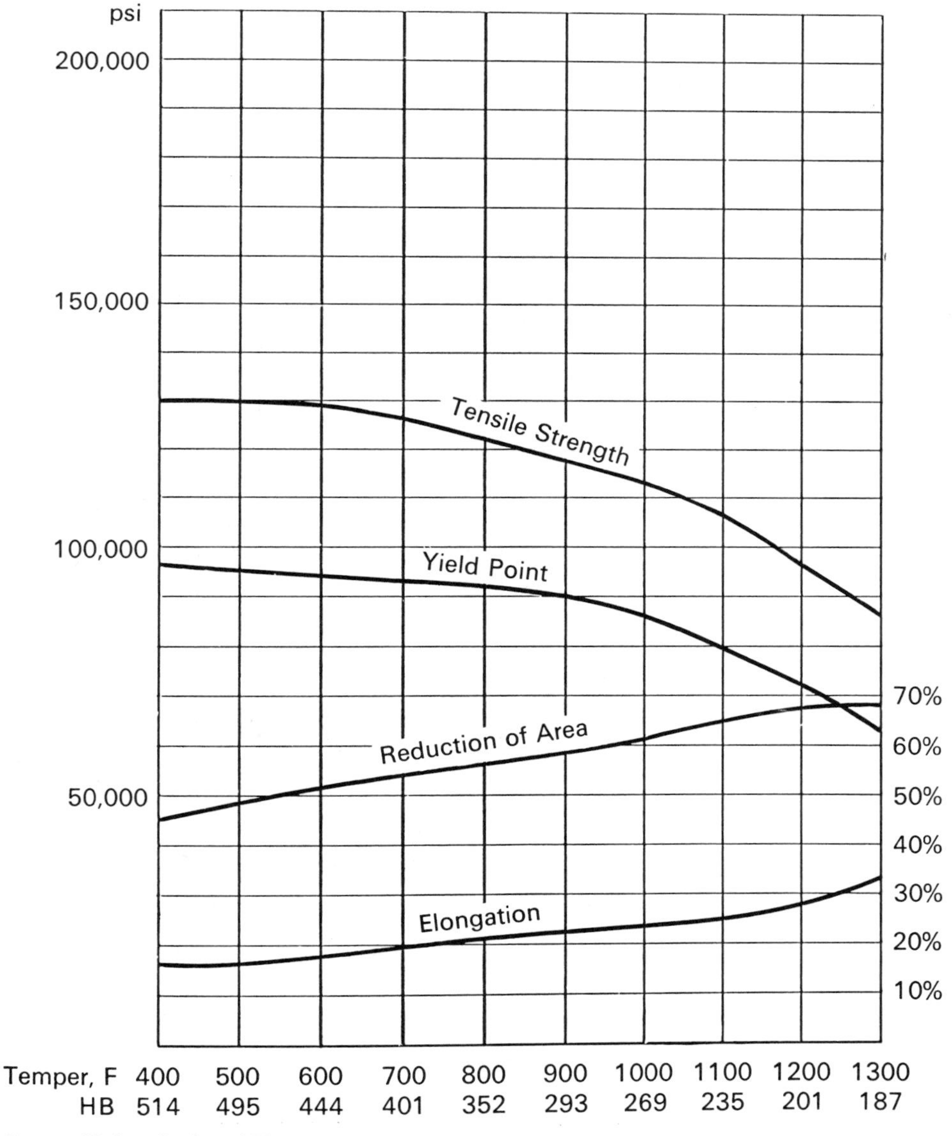

Source: *Modern Steels and Their Properties,* Bethlehem Steel Co., Bethlehem, Pa.

Appendix 4-2 Properties of Heat-treated AISI 1144

Treatment: Normalized at 1650 F; reheated to 1550 F; quenched in oil.
1-in. Round Treated; .505-in. Round Tested. As-quenched HB 285.

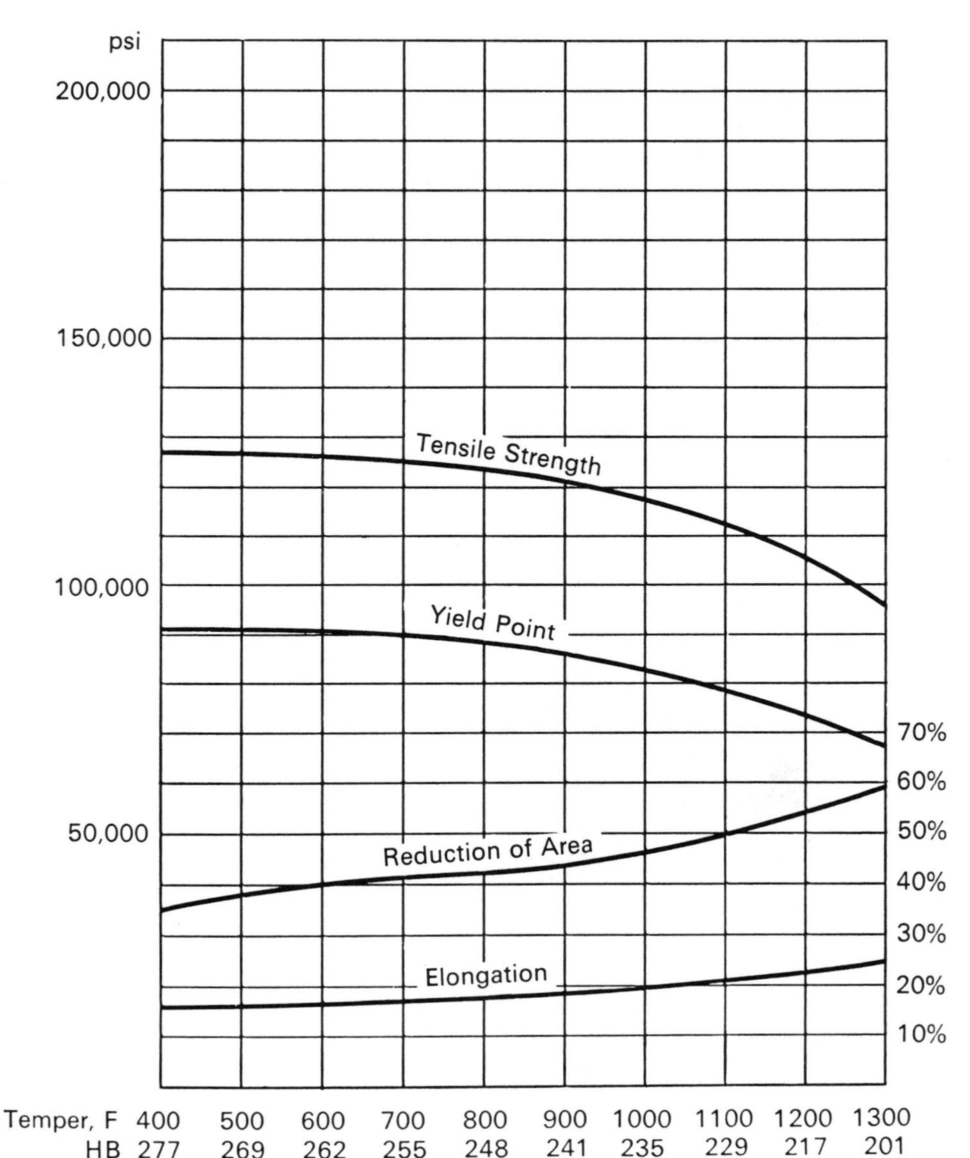

Source: *Modern Steels and Their Properties,* Bethlehem Steel Co., Bethlehem, Pa.

Appendix 4-3 Properties of Heat-treated AISI 1340

SINGLE HEAT RESULTS

	C	Mn	P	S	Si	Ni	Cr	Mo	Grain Size
Ladle	.43	1.70	.015	.039	.23	.03	.02	—	6-8

Critical Points, F: Ac_1 1340 Ac_3 1420 Ar_3 1195 Ar_1 1160

Treatment: Normalized at 1600 F; reheated to 1525 F; quenched in agitated oil.
.565-in. Round Treated; .505-in. Round Tested. As-quenched HB 601.

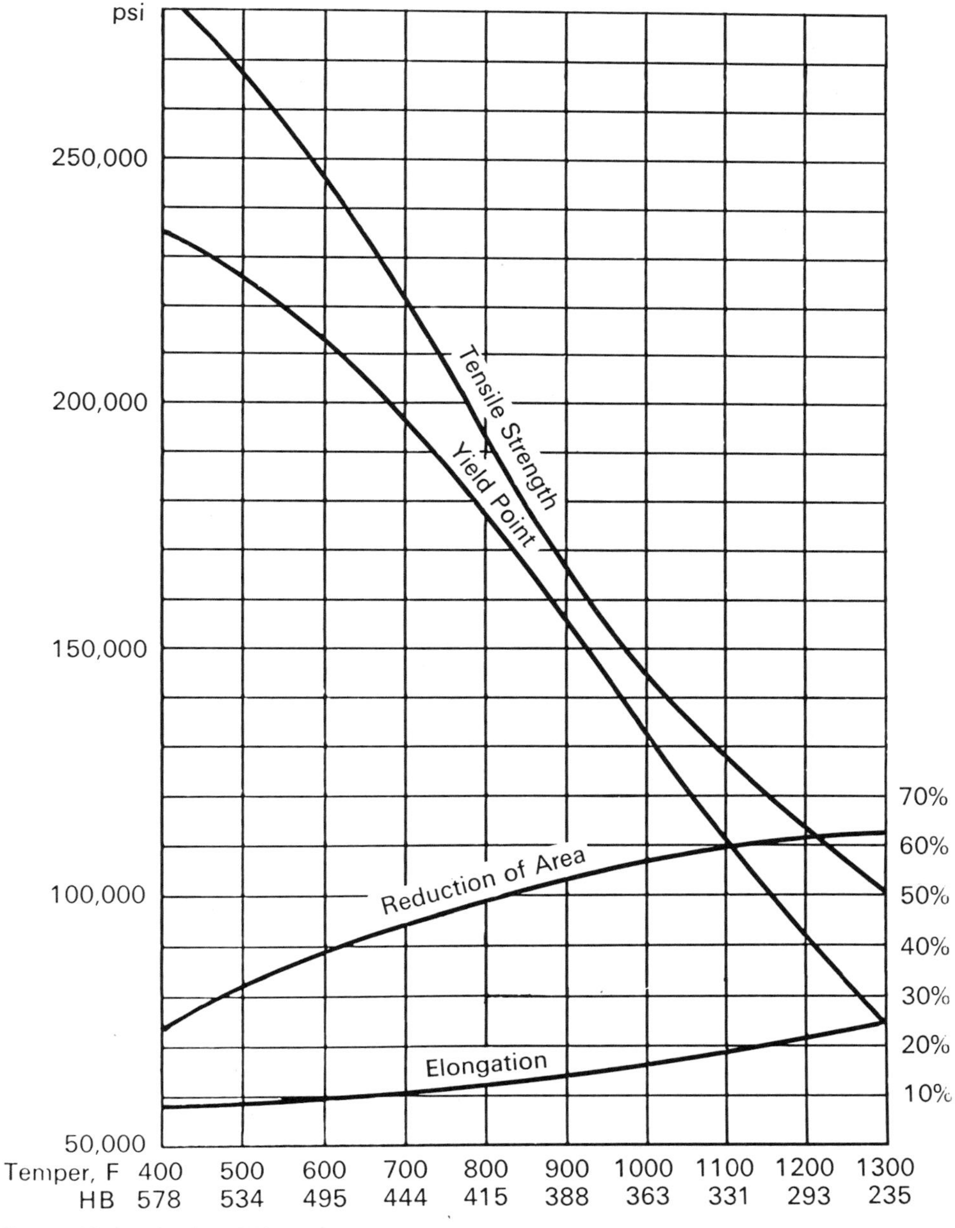

Source: *Modern Steels and Their Properties*, Bethlehem Steel Co., Bethlehem, Pa.

Appendix 4-4 Properties of Heat-treated AISI 4140

SINGLE HEAT RESULTS

	C	Mn	P	S	Si	Ni	Cr	Mo	Grain Size
Ladle	.41	.85	.024	.031	.20	.12	1.01	.24	6-8

Critical Points, F: Ac_1 1395 Ac_3 1450 Ar_3 1330 Ar_1 1280

Treatment: Normalized at 1600 F; reheated to 1550 F; quenched in agitated oil.
.530-in. Round Treated; .505-in. Round Tested. As-quenched HB 601.

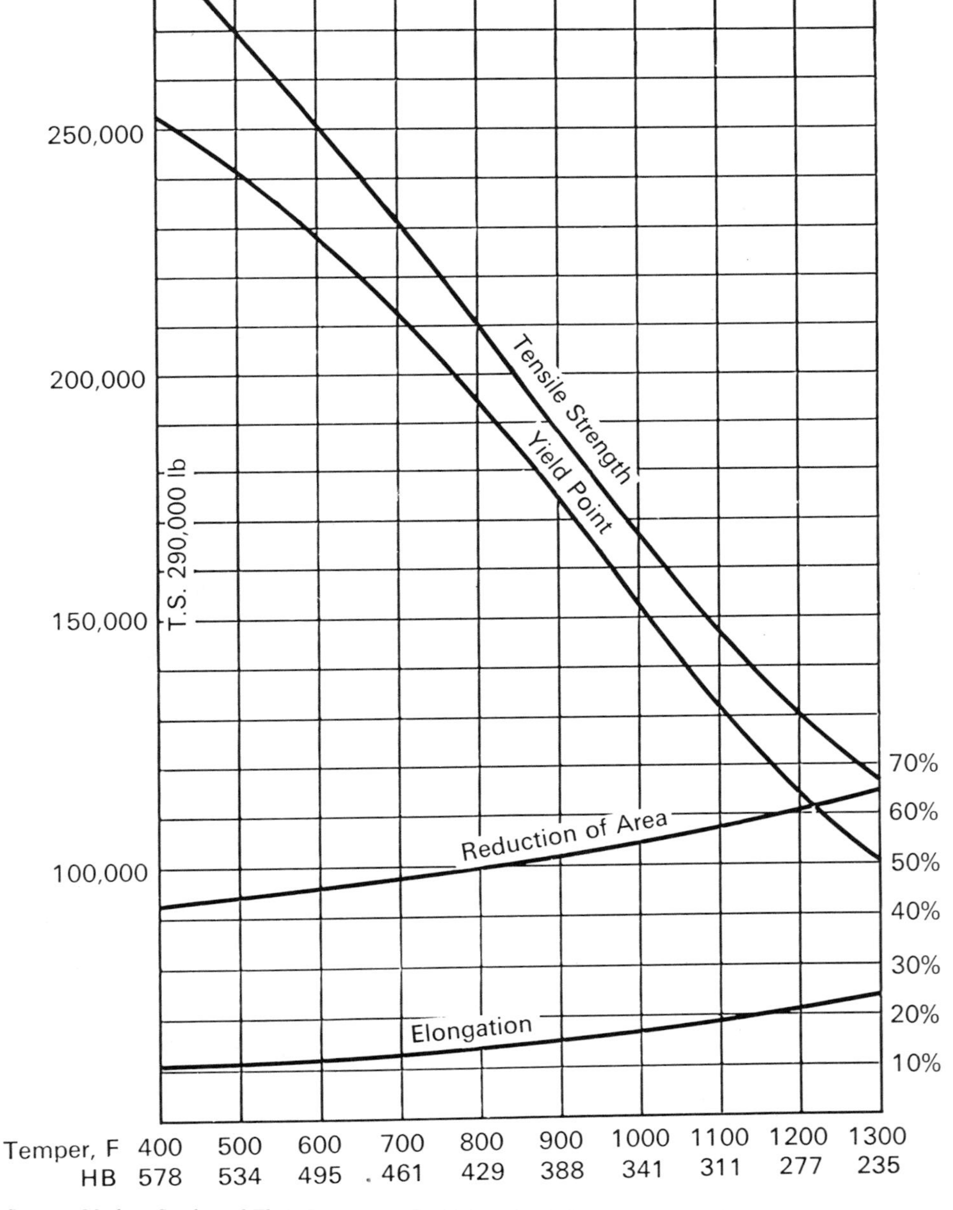

Source: *Modern Steels and Their Properties*, Bethlehem Steel Co., Bethlehem, Pa.

Appendix 4-5 Properties of Heat-treated AISI 4340

SINGLE HEAT RESULTS

	C	Mn	P	S	Si	Ni	Cr	Mo	Grain Size
Ladle	.41	.67	.023	.018	.26	1.77	.78	.26	6-8

Critical Points, F: Ac_1 1350 Ac_3 1415 Ar_3 890 Ar_1 720

Treatment: Normalized at 1600 F; reheated to 1475 F; quenched in agitated oil.
.530-in. Round Treated; .505-in. Round Tested. As-quenched HB 601.

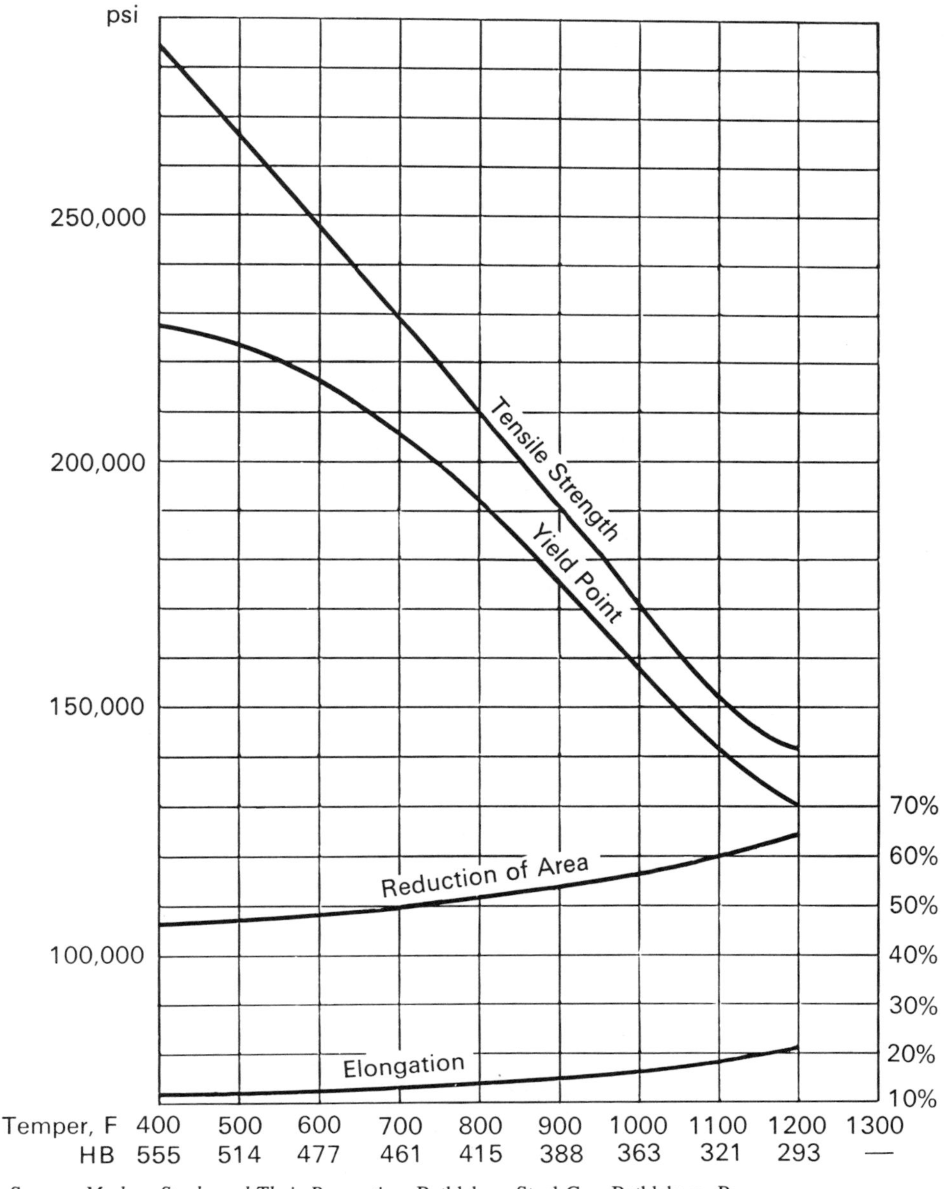

Source: *Modern Steels and Their Properties,* Bethlehem Steel Co., Bethlehem, Pa.

Appendix 4-6 Properties of Heat-treated AISI 6150

SINGLE HEAT RESULTS

	C	Mn	P	S	Si	Ni	Cr	Mo	V	Grain Size
Ladle	.49	.78	.012	.016	.29	.18	1.00	.05	.17	6-8

Critical Points, F: Ac_1 1395 Ac_3 1445 Ar_3 1315 Ar_1 1290

Treatment: Normalized at 1600 F; reheated to 1550 F; quenched in agitated oil.
.565-in. Round Treated; .505-in. Round Tested. As-quenched HB 627.

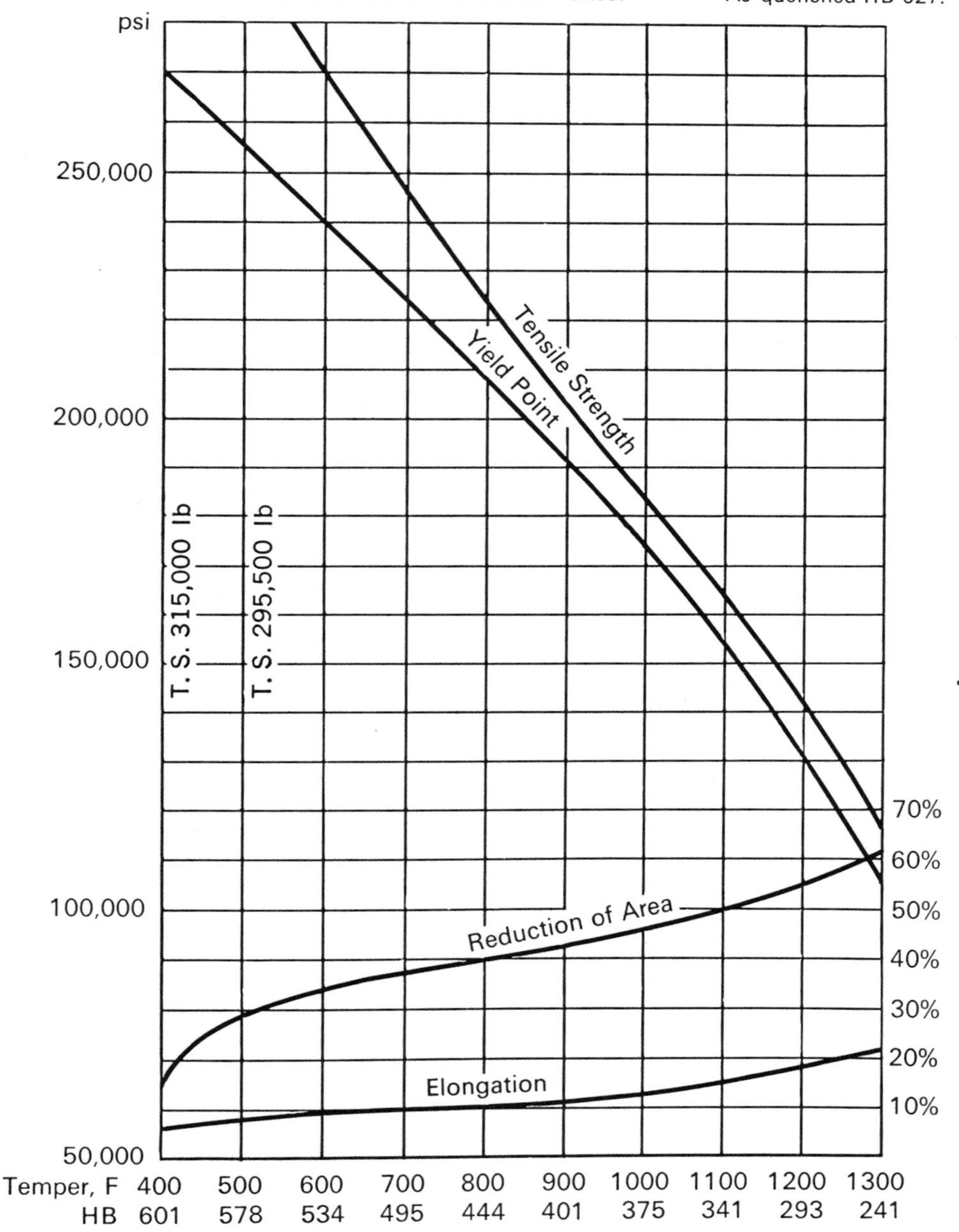

Source: *Modern Steels and Their Properties,* Bethlehem Steel Co., Bethlehem, Pa.

APPENDIX 5 PROPERTIES OF CARBURIZED STEELS

MATERIAL DESIGNATION (*AISI Number*)	CONDITION	CORE PROPERTIES							
		TENSILE STRENGTH		YIELD STRENGTH		DUCTILITY (*Percent Elongation in 2 in*)	BRINELL HARDNESS (*HB*)	CASE HARDNESS (*HRC*)
		Ksi	*MPa*	*Ksi*	*MPa*			
1015	SWQT 350	106	731	60	414	15	217	62
1020	SWQT 350	129	889	72	496	11	255	62
1022	SWQT 350	135	931	75	517	14	262	62
1117	SWQT 350	125	862	66	455	10	235	65
1118	SWQT 350	144	993	90	621	13	285	61
4118	SOQT 300	143	986	93	641	17	293	62
4118	DOQT 300	126	869	63	434	21	241	62
4118	SOQT 450	138	952	89	614	17	277	56
4118	DOQT 450	120	827	63	434	22	229	56
4320	SOQT 300	218	1 500	178	1 230	13	429	62
4320	DOQT 300	151	1 040	97	669	19	302	62
4320	SOQT 450	211	1 450	173	1 190	12	415	59
4320	DOQT 450	145	1 000	94	648	21	293	59
4620	SOQT 300	119	820	83	572	19	277	62
4620	DOQT 300	122	841	77	531	22	248	62
4620	SOQT 450	147	1 010	80	552	20	248	59
4620	DOQT 450	115	793	77	531	22	235	59
4820	SOQT 300	207	1 430	167	1 150	13	415	61
4820	DOQT 300	204	1 405	165	1 140	13	415	60
4820	SOQT 450	205	1 410	184	1 270	13	415	57
4820	DOQT 450	196	1 350	171	1 180	13	401	56
8620	SOQT 300	188	1 300	149	1 030	11	388	64
8620	DOQT 300	133	917	83	572	20	269	64
8620	SOQT 450	167	1 150	120	827	14	341	61
8620	DOQT 450	130	896	77	531	22	262	61
E9310	SOQT 300	173	1 190	135	931	15	363	62
E9310	DOQT 300	174	1 200	139	958	15	363	60
E9310	SOQT 450	168	1 160	137	945	15	341	59
E9310	DOQT 450	169	1 170	138	952	15	352	58

Note: Properties given are for a single set of tests on ½-in round bars.
SWQT: single water-quenched and tempered
SOQT: single oil-quenched and tempered
DOQT: double oil-quenched and tempered
300 and 450 are the tempering temperatures in °F. Carburized for 8 h. Case depth ranged from 0.045 to 0.075 in.

APPENDIX 6 PROPERTIES OF STAINLESS STEELS

MATERIAL DESIGNATION		CONDITION	TENSILE STRENGTH		YIELD STRENGTH		DUCTILITY (*Percent Elongation in 2 in*)
AISI Number	*UNS*		*Ksi*	*MPa*	*Ksi*	*MPa*	
Austenitic steels							
201	S20100	Annealed	115	793	55	379	55
		¼ hard	125	862	75	517	20
		½ hard	150	1 030	110	758	10
		¾ hard	175	1 210	135	931	5
		Full hard	185	1 280	140	966	4
301	S30100	Annealed	110	758	40	276	60
		¼ hard	125	862	75	517	25
		½ hard	150	1 030	110	758	15
		¾ hard	175	1 210	135	931	12
		Full hard	185	1 280	140	966	8
304	S30400	Annealed	85	586	35	241	60
310	S31000	Annealed	95	655	45	310	45
316	S31600	Annealed	80	552	30	207	60
Ferritic steels							
405	S40500	Annealed	70	483	40	276	30
430	S43000	Annealed	75	517	40	276	30
446	S44600	Annealed	80	552	50	345	25
Martensitic steels							
410	S41000	Annealed	75	517	40	276	30
416	S41600	Q&T 600	180	1 240	140	966	15
		Q&T 1000	145	1 000	115	793	20
		Q&T 1400	90	621	60	414	30
431	S43100	Q&T 600	195	1 344	150	1 034	15
440A	S44002	Q&T 600	280	1 930	270	1 860	3
Precipitation hardening steels							
17-4PH	S17400	H 900	200	1 380	185	1 280	14
		H 1150	145	1 000	125	862	19
17-7PH	S17700	RH 950	200	1 380	175	1 210	10
		TH 1050	175	1 210	155	1 070	12

APPENDIX 7 PROPERTIES OF STRUCTURAL STEELS

MATERIAL DESIGNATION *(ASTM Number)*	GRADE	TENSILE STRENGTH		YIELD STRENGTH		DUCTILITY *(Percent Elongation in 8 in)*
		Ksi	*MPa*	*Ksi*	*MPa*	
A36	—	58	400	36	248	20
A283	A	45	310	24	165	28
A283	D	60	414	33	228	20
A242	—	70	485	50	345	18
A440	—	67	460	46	315	18
A441	—	63	435	42	290	18
A514	Quenched and tempered	115	800	100	700	18 percent (in 2 in)
A572	42	60	414	42	290	—
A572	45	60	414	45	310	—
A572	50	65	448	50	345	—
A572	55	70	483	55	379	—
A572	60	75	517	60	414	—
A572	65	80	552	65	448	—
A588	—	70	485	50	345	18

Note: ASTM A572 is one of the high-strength, low-alloy steels (HSLA) and has similar properties to the SAE J410b steel specified by the SAE.

APPENDIX 8 DESIGN PROPERTIES OF CAST IRON

MATERIAL DESIGNATION *(ASTM Number)*	GRADE	TENSILE STRENGTH		YIELD STRENGTH		DUCTILITY *(Percent Elongation in 2 in)*	MODULUS OF ELASTICITY	
		Ksi	*MPa*	*Ksi*	*MPa*		*10^6 psi*	*GPa*
Gray iron								
A48-76	20	20	138	—	—	<1	12	83
	25	25	172	—	—	<1	13	90
	30	30	207	—	—	<1	15	103
	40	40	276	—	—	<1	17	117
	50	50	345	—	—	<1	19	131
	60	60	414	—	—	<1	20	138
Ductile iron								
A536-77	60-40-18	60	414	40	310	18	22	152
	80-55-06	80	552	55	379	6	22	152
	100-70-03	100	690	70	483	3	22	152
	120-90-02	120	827	90	621	2	22	152
Malleable iron								
A47-77	32510	50	345	32	221	10	25	172
	35018	53	365	35	241	18	25	172
A220-76	40010	60	414	40	276	10	26	179
	45006	65	448	45	310	6	26	179
	50005	70	483	50	345	5	26	179
	70003	85	586	70	483	3	26	179
	90001	105	724	90	621	1	26	179

Note: Strength values are typical. Casting variables and section size affect final values. Modulus of elasticity may also vary. Density of cast irons ranges from 0.25–0.27 lb/in^3 (6 920–7 480 kg/m^3).

APPENDIX 9 PROPERTIES OF POWDER METALS

DESIGNATION *(MPIF)*	TENSILE STRENGTH		YIELD STRENGTH		DUCTILITY *(Percent Elongation in 2 in)*	MODULUS OF ELASTICITY		IMPACT STRENGTH *(ft · lb)*
	Ksi	*MPa*	*Ksi*	*MPa*		10^6 *psi*	*GPa*	
Iron								
F-0000-N	16	110	11	76	2.0	10.5	72	3.0
F-0000-R	24	165	16	110	5.0	16.0	110	9.5
F-0000-T	40	276	26	179	15.0	23.0	156	25.0
F-0005-R	32	221	23	156	2.5	16.0	110	5.0
Medium carbon steel								
F-0005-R[a]	60	414	57	393	<0.5	—	—	—
F-0005-S	43	296	28	193	3.5	19.0	131	9.0
F-0005-S[a]	80	552	75	517	<0.5	—	—	—
Medium carbon, 2% nickel steel								
FN-0205-S[a]	110	758	88	607	1.0	21.0	145	16.0
FN-0205-T[a]	134	924	105	724	2.0	23.0	159	28.0
AISI 316 stainless steel								
SS-316-R	54	372	40	276	4.0	—	—	—

[a]Improved by quenching; others are as sintered.
Source: Materials Reference Issue, *Machine Design Magazine,* 1981.

APPENDIX 10 TYPICAL PROPERTIES OF ALUMINUM

ALLOY AND TEMPER	TENSILE STRENGTH		YIELD STRENGTH		DUCTILITY (*Percent Elongation in 2 in*)	SHEARING STRENGTH		ENDURANCE STRENGTH	
	Ksi	*MPa*	*Ksi*	*MPa*		*Ksi*	*MPa*	*Ksi*	*MPa*
1060-O	10	69	4	28	43	7	48	3	21
1060-H14	14	97	11	76	12	9	62	5	34
1060-H18	19	131	18	124	6	11	121	6	41
1350-O	12	83	4	28	28	8	55	—	—
1350-H14	16	110	14	97	—	10	69	—	—
1350-H19	27	186	24	165	—	15	103	7	48
2014-O	27	186	14	97	18	18	124	13	90
2014-T4	62	427	42	290	20	38	262	20	138
2014-T6	70	483	60	414	13	42	290	18	124
2024-O	27	186	11	76	22	18	124	13	90
2024-T4	68	469	47	324	19	41	283	20	138
2024-T361	72	496	57	393	12	42	290	18	124
2219-O	25	172	11	76	18	—	—	—	—
2219-T62	60	414	42	290	10	—	—	15	103
2219-T87	69	476	57	393	10	—	—	15	103
3003-O	16	110	6	41	40	11	121	7	48
3003-H14	22	152	21	145	16	14	97	9	62
3003-H18	29	200	27	186	10	16	110	10	69
5052-O	28	193	13	90	30	18	124	16	110
5052-H34	38	262	31	214	14	21	145	18	124
5052-H38	42	290	37	255	8	24	165	20	138
6061-O	18	124	8	55	30	12	83	9	62
6061-T4	35	241	21	145	25	24	165	14	97
6061-T6	45	310	40	276	17	30	207	14	97
6063-O	13	90	7	48	—	10	69	8	55
6063-T4	25	172	13	90	22	—	—	—	—
6063-T6	35	241	31	214	12	22	152	10	69
7001-O	37	255	22	152	14	—	—	—	—
7001-T6	98	676	91	627	9	—	—	22	152
7075-O	33	228	15	103	16	22	152	—	—
7075-T6	83	572	73	503	11	48	331	23	159

Note: Common properties:
Density: 0.095 to 0.102 lb/in^3 (2 635–2 829 kg/m^3).
Modulus of elasticity: 10 to 10.6 $\times$ 10^6 psi (69–73 GPa).

APPENDIX 11 PROPERTIES OF TITANIUM ALLOYS

MATERIAL DESIGNATION	CONDITION	TENSILE STRENGTH		YIELD STRENGTH		DUCTILITY (*Percent Elongation in 2 in*)	MODULUS OF ELASTICITY	
		Ksi	*MPa*	*Ksi*	*MPa*		10^6 *psi*	*GPa*
Commercially pure Alpha titanium (density = 0.163 lb/in^3; 4 515 kg/m^3)								
Ti-35A	Wrought	35	241	25	172	24	15.0	103
Ti-50A	Wrought	50	345	40	276	20	15.0	103
Ti-65A	Wrought	65	448	55	379	18	15.0	103
Alpha alloy (density = 0.163 lb/in^3; 4 515 kg/m^3)								
Ti-0.2Pd	Wrought	50	345	40	276	20	14.9	103
Beta alloy (density = 0.176 lb/in^3; 4 875 kg/m^3)								
Ti-3Al-13V-11Cr	Air-cooled from 1400°F	135	931	130	896	16	14.7	101
Ti-3Al-13V-11Cr	Air-cooled from 1400°F and aged	185	1 280	175	1 210	6	16.0	110
Alpha-beta alloy (density = 0.160 lb/in^3; 4 432 kg/m^3)								
Ti-6Al-4V	Annealed	130	896	120	827	10	16.5	114
Ti-6Al-4V	Quenched and aged at 1000°F	160	1 100	150	1 030	7	16.5	114

APPENDIX 12 PROPERTIES OF BRONZES

MATERIAL	UNS NUMBER DESIGNATION	TENSILE STRENGTH		YIELD STRENGTH		DUCTILITY (*Percent Elongation in 2 in*)	MODULUS OF ELASTICITY	
		Ksi	*MPa*	*Ksi*	*MPa*		*10^6 psi*	*MPa*
Leaded phosphor bronze	C54400	68	469	57	393	20	15	103
Silicon bronze	C65500	58	400	22	152	60	15	103
Manganese bronze	C67500	65	448	30	207	33	15	103
	C86200	95	655	48	331	20	15	103
Bearing bronze	C93200	35	241	18	124	20	14.5	100
Aluminum bronze	C95400	85	586	35	241	18	15.5	107
Copper-nickel alloy	C96200	45	310	25	172	20	18	124
Copper-nickel-zinc alloy (also called nickel silver)	C97300	35	241	17	117	20	16	110

APPENDIX 13 TYPICAL PROPERTIES OF SELECTED PLASTICS

MATERIAL	TYPE	TENSILE STRENGTH		TENSILE MODULUS		FLEXURAL STRENGTH		FLEXURAL MODULUS		IMPACT STRENGTH IZOD (*ft · lb/in of notch*)
		Ksi	*MPa*	*Ksi*	*MPa*	*Ksi*	*MPa*	*Ksi*	*MPa*	
Nylon	6/6	12.0	83	420	2 900	—	—	410	2 830	1.0
	11	8.5	59	180	1 240	—	—	150	1 030	3.3
ABS	Medium-impact grade	6.0	41	360	2 480	11.5	79	310	2 140	4.0
	High-impact	5.0	34	250	1 720	8.0	55	260	1 790	7.0
Polycarbonate	General-purpose	9.0	62	340	2 340	11.0	76	300	2 070	12.0
Acrylic	Standard	10.5	72	430	2 960	16.0	110	460	3 170	0.4
	High-impact	5.4	37	220	1 520	7.0	48	230	1 590	1.2
PVC	Rigid	6.0	41	350	2 410	—	—	300	2 070	0.4–20.0 (varies widely)
Polyimide	25 percent graphite powder filler	5.7	39	—	—	12.8	88	900	6 210	0.25
	Glass fiber filler	27.0	186	—	—	50.0	345	3 250	22 400	17.0
	Laminate	50.0	345	—	—	70.0	483	4 000	27 580	13.0
Acetal	Copolymer	8.0	55	410	2 830	13.0	90	375	2 590	1.3
Polyurethane	Elastomer	5.0	34	100	690	0.6	4	—	—	No break
Phenolic	General	6.5	45	1 100	7 580	9.0	62	1 100	7 580	0.3
Polyester with glass-fiber mat reinforcement (approx. 30% glass by weight)										
	Lay-up, contact mold	9.0	62	—	—	16.0	110	800	5 520	—
	Cold press molded	12.0	83	—	—	22.0	152	1 300	8 960	—
	Compression molded	25.0	172	—	—	10.0	69	1 300	8 960	—

APPENDIX 14 BEAM DEFLECTION FORMULAS

Type of Beam	*Deflections**	
	General Formula for Deflection at any Point	*Deflections at Critical Points*
Case 1: Supported at both ends, uniform load TOTAL LOAD W  $\frac{W}{2}$ x l $\frac{W}{2}$	$y = \frac{Wx(l-x)}{24\,EIl}\,[l^2 + x(l-x)]$	Maximum deflection, at center, $\frac{5}{384}\,\frac{Wl^3}{EI}$
Case 2: Supported at both ends, load at center 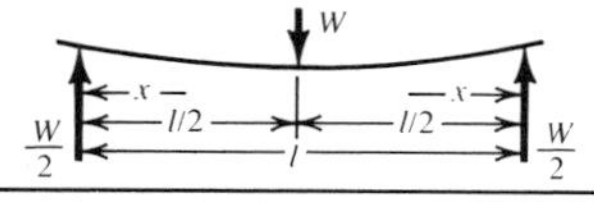 W x x $l/2$ $l/2$ l $\frac{W}{2}$ $\frac{W}{2}$	Between each support and load, $y = \frac{Wx}{48\,EI}\,(3\,l^2 - 4x^2)$	Maximum deflection at load, $\frac{Wl^3}{48\,EI}$
Case 3: Supported at both ends, load at any point 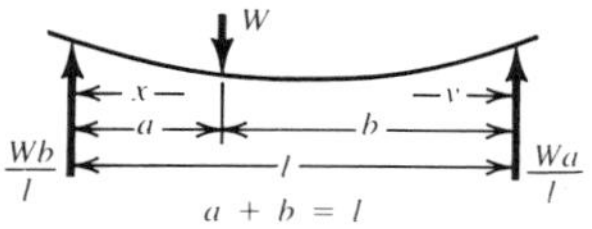 W x v a b l $\frac{Wb}{l}$ $\frac{Wa}{l}$ $a + b = l$	For segment of length a, $y = \frac{Wbx}{6\,EIl}\,(l^2 - x^2 - b^2)$ For segment of length b, $y = \frac{Wav}{6\,EIl}\,(l^2 - v^2 - a^2)$	Deflection at load, $\frac{Wa^2b^2}{3\,EIl}$ Let a be the length of the shorter segment and b of the longer one. The maximum deflection is in the longer segment, at $v = b\sqrt{\frac{1}{3} + \frac{2a}{3b}} = v_1$, and is $\frac{Wav_1^3}{3\,EIl}$
Case 4: Supported at both ends, two symmetrical loads 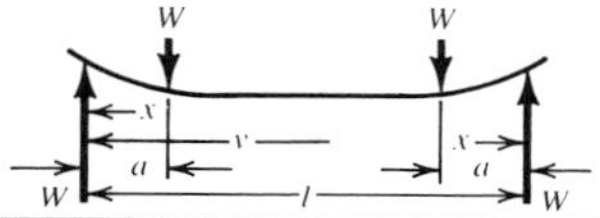 W W x v x a a l W W	Between each support and adjacent load, $y = \frac{Wx}{6\,EI}\,[3\,a(l-a) - x^2]$ Between loads, $y = \frac{Wa}{6\,EI}\,[3v(l-v) - a^2]$	Maximum deflection at center, $\frac{Wa}{24\,EI}\,(3l^2 - 4\,a^2)$ Deflection at loads $\frac{Wa^2}{6\,EI}\,(3l - 4a)$
Case 5: Both ends overhanging supports symmetrically, uniform load Total load W 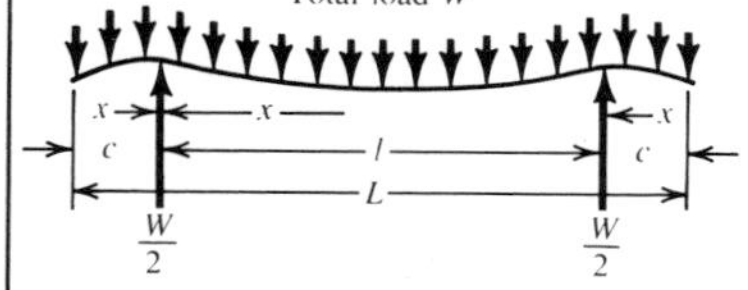 x c x l L x c $\frac{W}{2}$ $\frac{W}{2}$ $L = l + 2o$	Between each support and adjacent end, $y = \frac{Wu}{24\,EIL}[6\,c^2(l+u) - u^2(4c-u) - l^3]$ Between supports, $y = \frac{Wx(l-x)}{24\,EIL}\,[x(l-x) + l^2 - 6c^2]$	Deflection at ends, $\frac{Wc}{24\,EIL}\,[3\,c^2(c + 2l) - l^3]$ Deflection at center, $\frac{Wl^2}{384\,EIL}\,(5\,l^2 - 24c^2)$ If l is between $2\,c$ and $2.449\,c$, there are maximum upward deflections at points $\sqrt{3\,(\frac{1}{4}l^2 - c^2)}$ on both sides of the center, which are, $-\frac{W}{96\,EIL}\,(6c^2 - l^2)^2$

*The deflections apply only to cases where the cross-section of the beam is constant for its entire length

Source: Oberg, Erik, et al. *Machinery's Handbook,* 22nd ed. New York: Industrial Press, 1984.

Type of Beam	Deflections* — General Formula for Deflections at any Point	Deflections* — Deflections at Critical Points
Case 6: Both ends overhanging supports unsymmetrically, uniform load Total load W 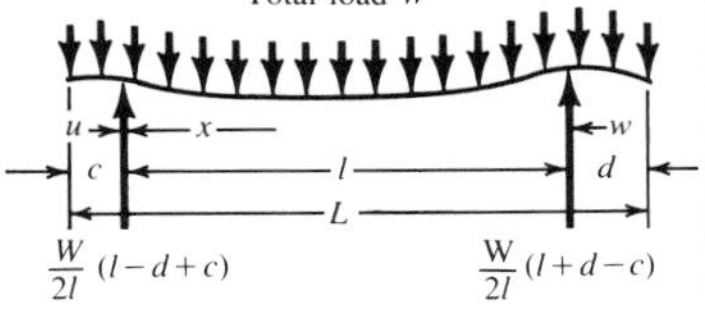 Reactions: $\frac{W}{2l}(l-d+c)$, $\frac{W}{2l}(l+d-c)$	For overhanging end of length c, $y = \frac{Wu}{24\,EIL}\,[2l(d^2+2c^2) + 6c^2u - u^2(4c-u) - l^3]$ Between supports, $y = \frac{Wx(l-x)}{24\,EIL}\left\{x(l-x) + l^2 - 2(d^2+c^2) - \frac{2}{l}[d^2x + c^2(l-x)]\right\}$ For overhanging end of length d, $y = \frac{Ww}{24\,EIL}\,[2l(c^2+2d^2) + 6\,d^2w - w^2(4\,d-w) - l^3]$	Deflection at end c, $\frac{Wc}{24\,EIL}\,[2l(d^2+2c^2)+3c^3-l^3]$ Deflection at end d, $\frac{Wd}{24\,EIL}\,[2l(c^2+2d^2)+3d^3-l^3]$ This case is so complicated that convenient general expressions for the critical deflections between supports cannot be obtained.
Case 7: Both ends overhanging supports, load at any point between 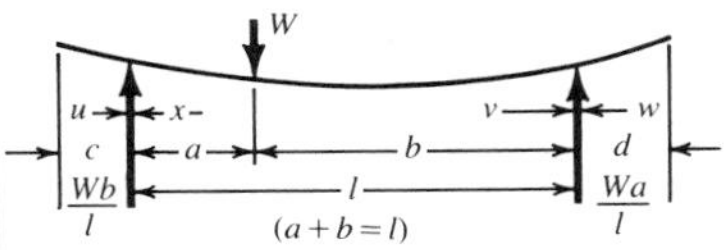 Reactions: $\frac{Wb}{l}$, $\frac{Wa}{l}$ $(a+b=l)$	Between supports, same as Case 3. For overhanging end of length c, $y = -\frac{Wabu}{6\,Ell}(l+b)$ For overhanging end of length d, $y = -\frac{Wabw}{6\,Ell}(l+a)$	Between supports, same as Case 3. Deflection at end c, $-\frac{Wabc}{6\,Ell}(l+b)$ Deflection at end d, $-\frac{Wabd}{6\,Ell}(l+a)$
Case 8: Both ends overhanging supports, single overhanging load 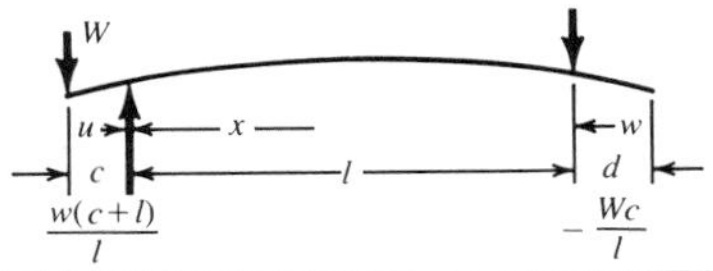 Reactions: $\frac{w(c+l)}{l}$, $-\frac{Wc}{l}$	Between load and adjacent support, $y = \frac{Wu}{6\,EI}(3cu - u^2 + 2cl)$ Between supports, $y = -\frac{Wcx}{6\,Ell}(l-x)(2l-x)$ Between unloaded end and adjacent support, $y = \frac{Wclw}{6\,EI}$	Deflection at load, $\frac{Wc^2}{3\,EI}(c+l)$ Maximum upward deflection is at $x = 0.42265\,l$, and is $-\frac{Wcl^2}{15.55\,EI}$ Deflection at unloaded end, $\frac{Wcld}{6\,EI}$
Case 9: Both ends overhanging supports, symmetrical overhanging loads 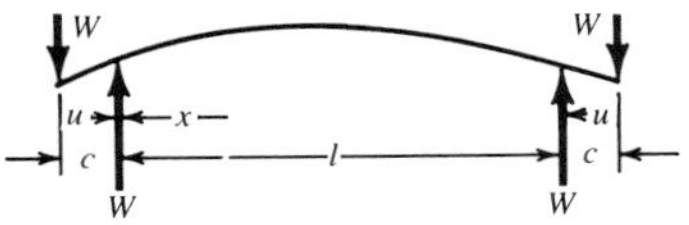	Between each load and adjacent support, $y = \frac{Wu}{6\,EI}[3c(l+u) - u^2]$ Between supports, $y = -\frac{Wcx}{2\,EI}(l-x)$	Deflections at loads, $\frac{Wc^2}{6EI}(2c+3l)$ Deflection at center, $-\frac{Wcl^2}{8\,EI}$
	The above expressions involve the usual approximations of the theory of flexure, and hold only for small deflections. Exact expressions for deflections of any magnitude are as follows: Between supports the curve is a circle of radius $r = \frac{EI}{Wc}$; $y = \sqrt{r^2 - \frac{1}{4}l^2} - \sqrt{r^2 - (\frac{1}{2}l - x)^2}$ Deflection at center, $\sqrt{r^2 - \frac{1}{4}l^2} - r$	

Source: Oberg, Erik, et al. *Machinery's Handbook,* 22nd ed. New York: Industrial Press, 1984.

Type of Beam	Deflections*	
	General Formula for Deflection at any Point	*Deflections at Critical Points*
Case 10: Fixed at one end, uniform load (Total load W; reactions $\frac{Wl}{2}$, W)	$y = \frac{Wx^2}{24\,EIl}\,[2l^2 + (2l - x)^2]$	Maximum deflection, at end, $\frac{Wl^3}{8\,EI}$
Case 11: Fixed at one end, load at other (reactions Wl, W)	$y = \frac{Wx^2}{6EI}\,(3l - x)$	Maximum deflection, at end, $\frac{Wl^3}{3\,EI}$
Case 12: Fixed at one end, intermediate load (reactions Wl, W)	Between support and load, $y = \frac{Wx^2}{6\,EI}\,(3l - x)$ Beyond load, $y = \frac{Wl^2}{6\,EI}\,(3v - l)$	Deflection at load, $\frac{Wl^3}{3\,EI}$ Maximum deflection, at end, $\frac{Wl^2}{6\,EI}\,(2l + 3b)$
Case 13: Fixed at one end, supported at the other, uniform load (Total load W; reactions $\frac{Wl}{3}$, $\frac{5}{8}W$, $\frac{3}{8}W$)	$y = \frac{Wx^2(l - x)}{48\,EIl}\,(3l - 2x)$	Maximum deflection is at $x = 0.5785\,l$, and is $\frac{Wl^3}{185\,EI}$ Deflection at center, $\frac{Wl^3}{192\,EI}$ Deflection at point of greatest negative stress, at $x = \frac{5}{8}l$ is $\frac{Wl^3}{187\,EI}$
Case 14: Fixed at one end, supported at the other, load at center (reactions $\frac{3}{16}Wl$, $\frac{11}{16}W$, $\frac{5}{16}W$)	Between point of fixture and load, $y = \frac{Wx^2}{96EI}\,(9\,l - 11\,x)$ Between support and load, $y = \frac{Wv}{96\,EI}\,(3l^2 - 5v^2)$	Maximum deflection is at $v = 0.4472\,l$, and is $\frac{Wl^3}{107.33\,EI}$ Deflection at load, $\frac{7}{768}\,\frac{Wl^3}{EI}$

Source: Oberg, Erik, et al. *Machinery's Handbook,* 22nd ed. New York: Industrial Press, 1984.

Type of Beam	Deflections*	
	General Formula for Deflections at any Point	*Deflections at Critical Points*
Case 15: Fixed at one end, supported at the other, load at any point $m = (l+a)(l+b)+al$ $n = al(l+b)$ 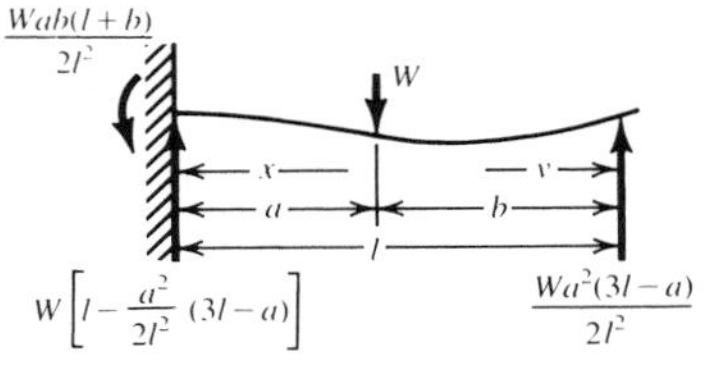 Reactions/moments: $\frac{Wab(l+b)}{2l^2}$; $W\left[l - \frac{a^2}{2l^2}(3l-a)\right]$; $\frac{Wa^2(3l-a)}{2l^2}$	Between point of fixture and load, $y = \frac{Wx^2b}{12\,EIl^3}(3n - mx)$ Between support and load, $y = \frac{Wa^2v}{12\,EIl^3}[3l^2b - v^2(3l-a)]$	Deflection at load, $\frac{Wa^3b^2}{12\,EIl^3}(3l+b)$ If $a < 0.5858\,l$, maximum deflection is between load and support, at $v = l\sqrt{\frac{b}{2l+b}}$ and is $\frac{Wa^2b}{6\,EI}\sqrt{\frac{b}{2l+b}}$ If $a = 0.5858\,l$, maximum deflection is at load and is $\frac{Wl^3}{101.9\,EI}$ If $a > 0.5858\,l$, maximum deflection is between load and point of fixture, at $x = \frac{2n}{m}$, and is $\frac{Wbn^3}{3\,EIm^{2/3}}$
Case 16: Fixed at one end, free but guided at the other, uniform load 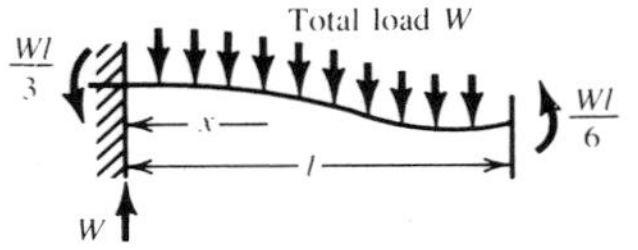 Total load W Reactions/moments: $\frac{Wl}{3}$; $\frac{Wl}{6}$; W	$y = \frac{Wx^2}{24\,EIl}(2l-x)^2$	Maximum deflection, at free end, $\frac{Wl^3}{24\,EI}$
Case 17: Fixed at one end, free but guided at the other, with load 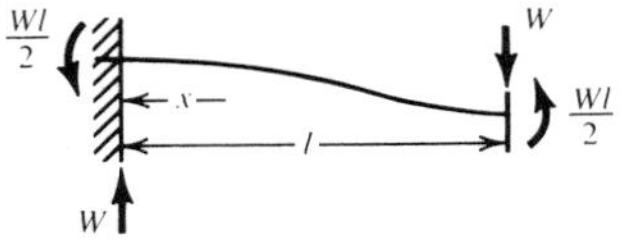 Reactions/moments: $\frac{Wl}{2}$; W; $\frac{Wl}{2}$; W	$y = \frac{Wx^2}{12\,EI}(3l-2x)$	Maximum deflection, at free end, $\frac{Wl^3}{12\,EI}$
Case 18: Fixed at both ends, uniform load 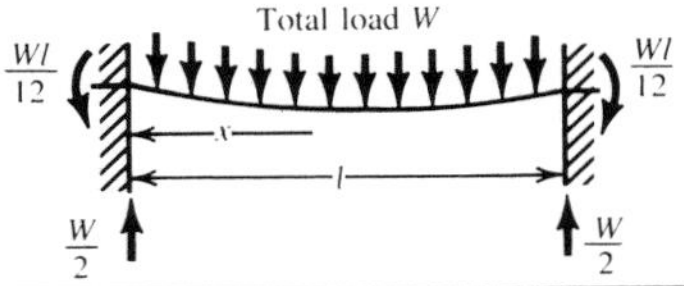 Total load W Reactions/moments: $\frac{Wl}{12}$; $\frac{Wl}{12}$; $\frac{W}{2}$; $\frac{W}{2}$	$y = \frac{Wx^2}{24\,EIl}(l-x)^2$	Maximum deflection, at center, $\frac{Wl^3}{384\,EI}$

Source: Oberg, Erik, et al. *Machinery's Handbook,* 22nd ed. New York: Industrial Press, 1984.

Type of Beam	Deflections*	
	General Formula for Deflections at any Point	*Deflections at Critical Points*
Case 19: Fixed at both ends, load at center 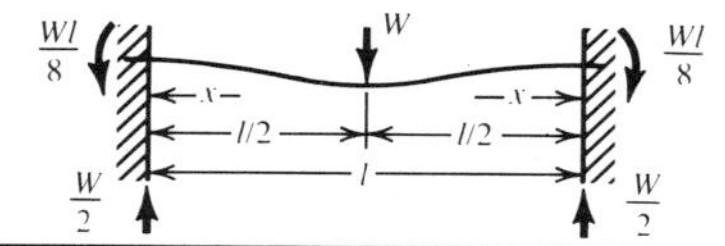 $\frac{Wl}{8}$, W, $\frac{Wl}{8}$, $\frac{W}{2}$, $\frac{W}{2}$	$y = \frac{Wx^2}{48\,EI}(3l - 4x)$	Maximum deflection, at load, $\frac{Wl^3}{192\,EI}$
Case 20: Fixed at both ends, load at any point 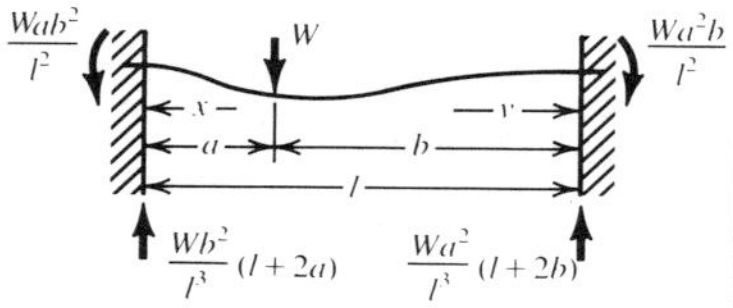 $\frac{Wab^2}{l^2}$, W, $\frac{Wa^2b}{l^2}$, $\frac{Wb^2}{l^3}(l+2a)$, $\frac{Wa^2}{l^3}(l+2b)$	For segment of length a, $y = \frac{Wx^2b^2}{6\,EIl^3}[2\,a\,(l - x) + l\,(a - x)]$ For segment of length b, $y = \frac{Wv^2a^2}{6\,EIl^3}[2\,b\,(l - v) + l\,(b - v)]$	Deflection at load, $\frac{Wa^3b^3}{3\,EIl^3}$ Let b be the length of the longer segment and a of the shorter one. The maximum deflection is the longer segment, at $v = \frac{2\,bl}{l + 2\,b}$, and is $\frac{2Wa^2b^3}{3\,EI\,(l + 2b)^2}$
Case 21: Continuous beam, with two equal spans, uniform load Total load on each span, W 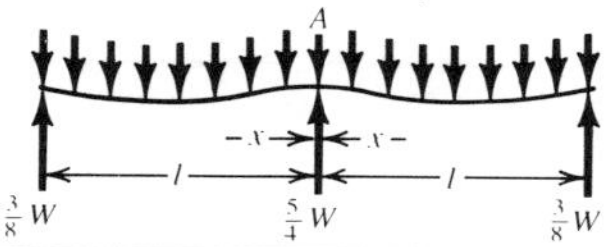 $\frac{3}{8}W$, $\frac{5}{4}W$, $\frac{3}{8}W$	$y = \frac{Wx^2\,(l - x)}{48\,EIl}(3l - 2x)$	Maximum deflection is at $x = 0.5785\,l$, and is $\frac{Wl^3}{185\,EI}$ Deflection at center of span, $\frac{Wl^3}{192\,EI}$ Deflection at point of greatest negative stress, at $x = \frac{5}{8}l$ is $\frac{Wl^3}{187\,EI}$
Case 22: Continuous beam, with two unequal spans, unequal, uniform loads Total load W_1 Total load W_2 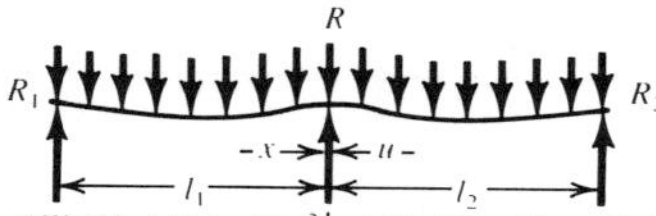 $R_1 = \frac{l_1W_1(3l_1+4l_2) - W_2l_2^2}{8l_1(l_1+l_2)}$ $R_2 = \frac{l_2W_2(3l_2+4l_1) - W_1l_1^2}{8l_2(l_1+l_2)}$ $R = \left(\frac{W_1+W_2}{2}\right) + \frac{1}{8}\left(\frac{W_1l_1}{l_2} + \frac{W_2l_2}{l_1}\right)$	Between R_1 and R, $y = \frac{x\,(l_1 - x)}{24\,EI}\left\{(2\,l_1 - x)\,(4\,r_1 - W_1) - \frac{W_1\,(l_1 - x)^2}{l_1}\right\}$ Between R_2 and R, $y = \frac{u\,(l_2 - u)}{24\,EI}\left\{(2\,l_2 - u)\,(4\,r_2 - W_2) - \frac{W_2(l_2 - u)^2}{l_2}\right\}$	This case is so complicated that convenient general expressions for the critical deflections cannot be obtained.

Source: Oberg, Erik, et al. *Machinery's Handbook,* 22nd ed. New York: Industrial Press, 1984.

APPENDIX 15 STRESS CONCENTRATION FACTORS

Appendix 15-1 Stepped Round Shaft

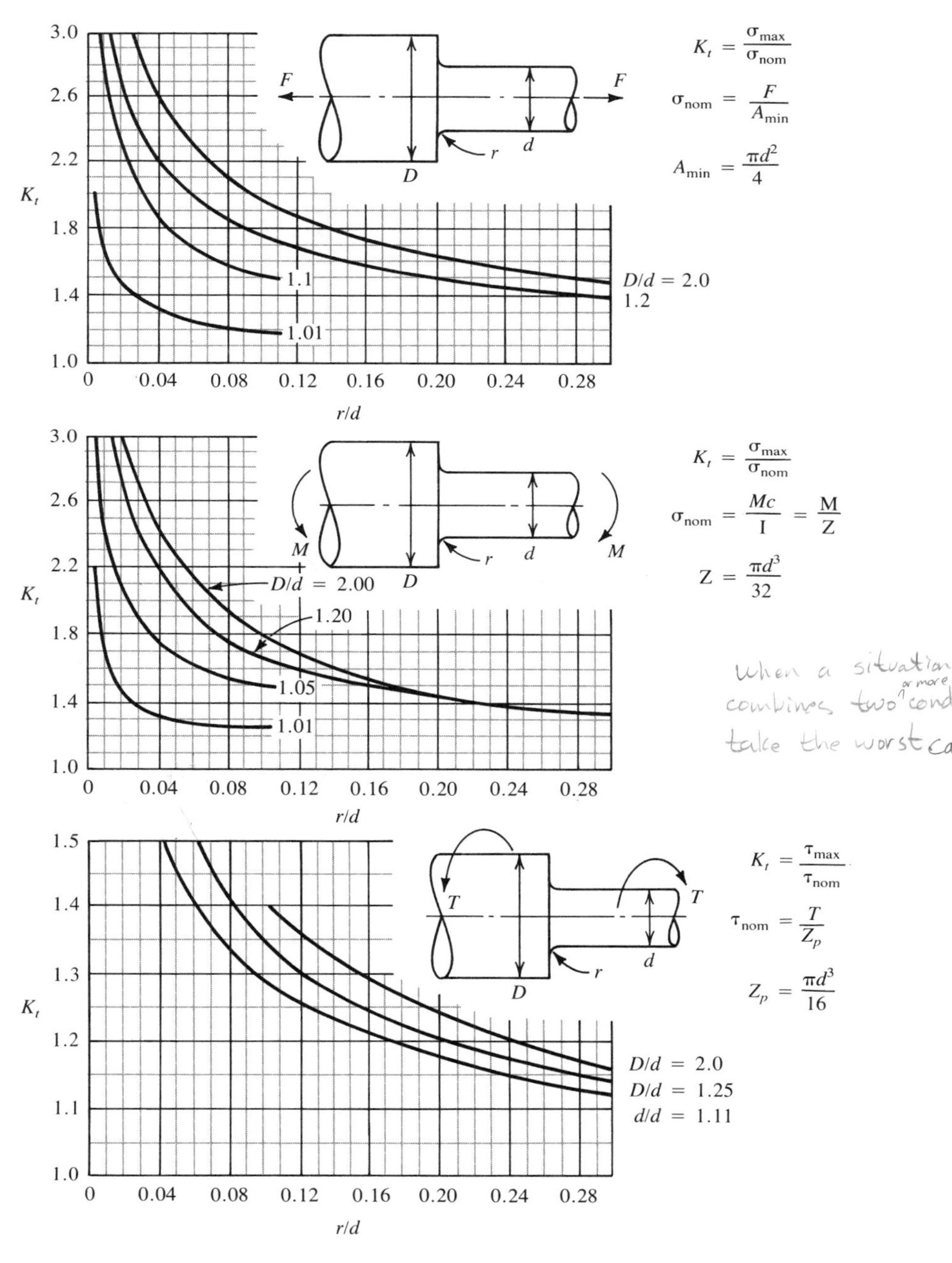

APPENDIX A15-1

Appendix 15-2 Stepped Flat Plate with Fillets

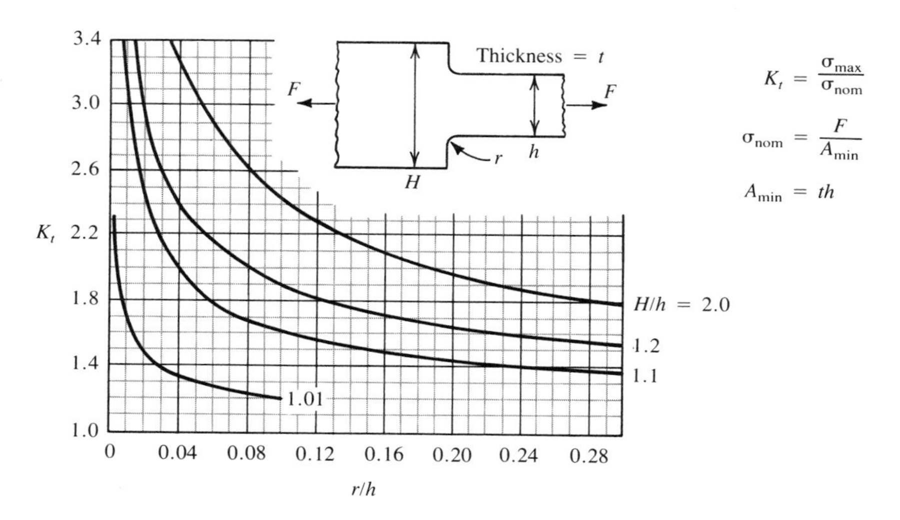

$$K_t = \frac{\sigma_{max}}{\sigma_{nom}}$$

$$\sigma_{nom} = \frac{F}{A_{min}}$$

$$A_{min} = th$$

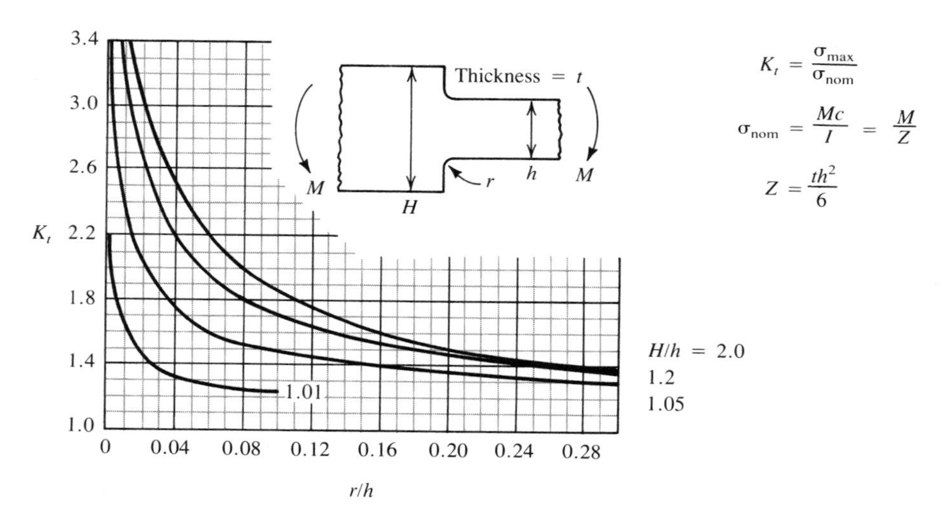

$$K_t = \frac{\sigma_{max}}{\sigma_{nom}}$$

$$\sigma_{nom} = \frac{Mc}{I} = \frac{M}{Z}$$

$$Z = \frac{th^2}{6}$$

APP A15-2

Appendix 15-3 Flat Plate with a Central Hole

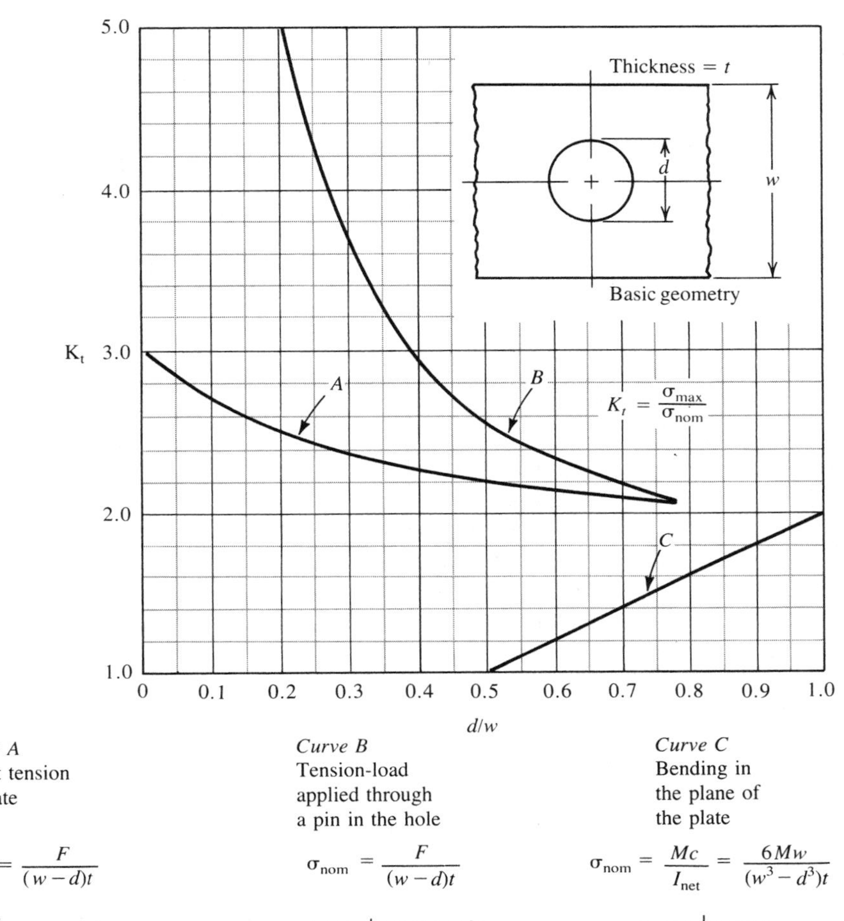

Curve A
Direct tension on plate

$$\sigma_{nom} = \frac{F}{(w-d)t}$$

F F = total load

Curve B
Tension-load applied through a pin in the hole

$$\sigma_{nom} = \frac{F}{(w-d)t}$$

F

Curve C
Bending in the plane of the plate

$$\sigma_{nom} = \frac{Mc}{I_{net}} = \frac{6Mw}{(w^3-d^3)t}$$

M M

NOTE: $K_t = 1.0$ for $d/w < 0.5$

APP A15-3

Appendix 15-4 Hole in Shaft

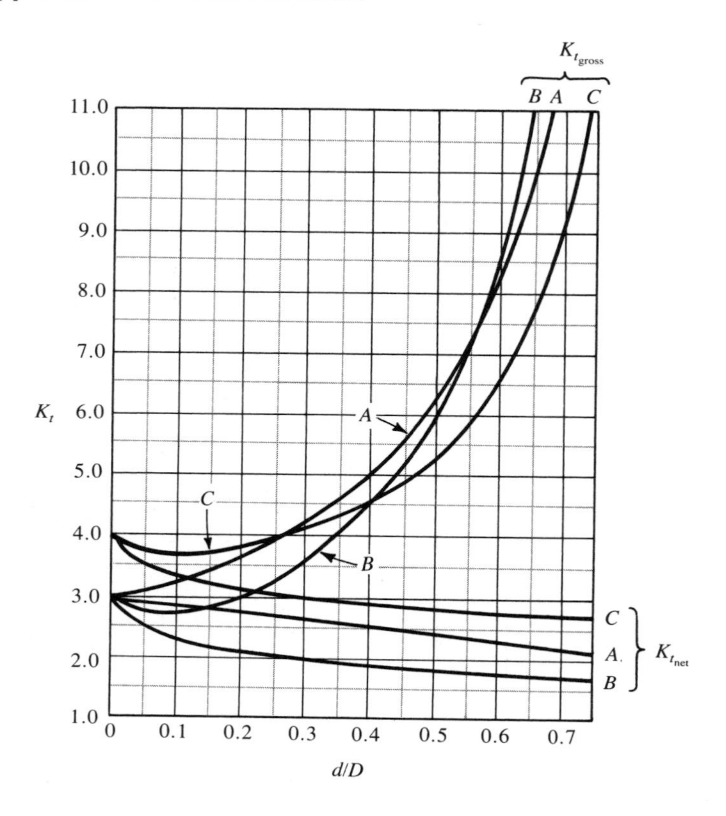

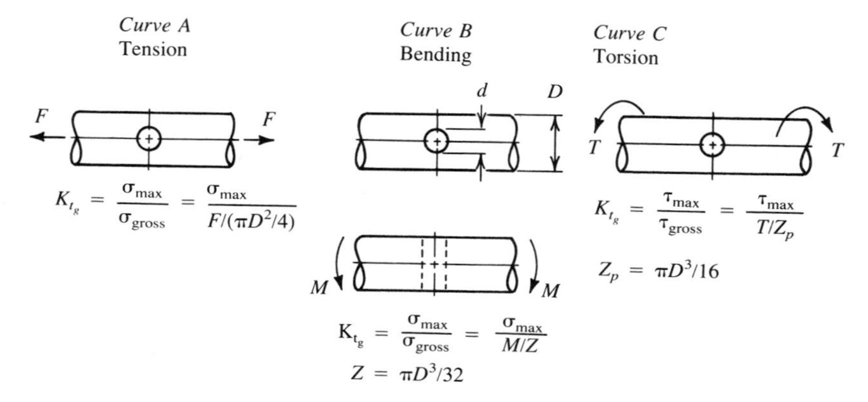

W

Wide Flange Shapes

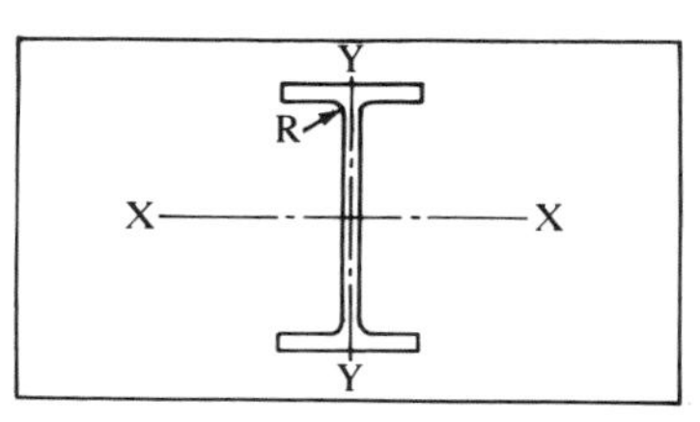

Properties for Designing

Designation and Nominal Size	Weight per Foot	Area	Depth	Flange		Web Thickness	Axis X-X			Axis Y-Y			Producing Mill Location
				Width	Thickness		I	Z	r	I	Z	r	
In.	Lbs.	In.^{2}	In.	In.	In.	In.	In.^{4}	In.^{3}	In.	In.^{4}	In.^{3}	In.	
W16 16 x 7	57	16.8	16.43	7.120	.715	.430	758	92.2	6.72	43.1	12.1	1.60	HSG
	50	14.7	16.26	7.070	.630	.380	659	81.0	6.68	37.2	10.5	1.59	
	45	13.3	16.13	7.035	.565	.345	586	72.7	6.65	32.8	9.34	1.57	
	40	11.8	16.01	6.995	.505	.305	518	64.7	6.63	28.9	8.25	1.57	
	36	10.6	15.86	6.985	.430	.295	448	56.5	6.51	24.5	7.00	1.52	
W16 16 x 5½	31	9.12	15.88	5.525	.440	.275	375	47.2	6.41	12.4	4.49	1.17	HG
	26	7.68	15.69	5.500	.345	.250	301	38.4	6.26	9.59	3.49	1.12	
W14 14 x 16	730†	215	22.42	17.890	4.910	3.070	14300	1280	8.17	4720	527	4.69	H
	665†	196	21.64	17.650	4.520	2.830	12400	1150	7.98	4170	472	4.62	
	605†	178	20.92	17.415	4.160	2.595	10800	1040	7.80	3680	423	4.55	
	550†	162	20.24	17.200	3.820	2.380	9430	931	7.63	3250	378	4.49	
	500†	147	19.60	17.010	3.500	2.190	8210	838	7.48	2880	339	4.43	
	455†	134	19.02	16.835	3.210	2.015	7190	756	7.33	2560	304	4.38	
	426†	125	18.67	16.695	3.035	1.875	6600	707	7.26	2360	283	4.34	HS
	398†	117	18.29	16.590	2.845	1.770	6000	656	7.16	2170	262	4.31	
	370†	109	17.92	16.475	2.660	1.655	5440	607	7.07	1990	241	4.27	
	342†	101	17.54	16.360	2.470	1.540	4900	559	6.98	1810	221	4.24	
	311†	91.4	17.12	16.230	2.260	1.410	4330	506	6.88	1610	199	4.20	
	283†	83.3	16.74	16.110	2.070	1.290	3840	459	6.79	1440	179	4.17	
	257†	75.6	16.38	15.995	1.890	1.175	3400	415	6.71	1290	161	4.13	
	233†	68.5	16.04	15.890	1.720	1.070	3010	375	6.63	1150	145	4.10	
	211†	62.0	15.72	15.800	1.560	.980	2660	338	6.55	1030	130	4.07	
	193	56.8	15.48	15.710	1.440	.890	2400	310	6.50	931	119	4.05	
	176	51.8	15.22	15.650	1.310	.830	2140	281	6.43	838	107	4.02	
	159	46.7	14.98	15.565	1.190	.745	1900	254	6.38	748	96.2	4.00	
	145	42.7	14.78	15.500	1.090	.680	1710	232	6.33	677	87.3	3.98	
W14 14 x 14½	132	38.8	14.66	14.725	1.030	.645	1530	209	6.28	548	74.5	3.76	HS
	120	35.3	14.48	14.670	.940	.590	1380	190	6.24	495	67.5	3.74	
	109	32.0	14.32	14.605	.860	.525	1240	173	6.22	447	61.2	3.73	
	99	29.1	14.16	14.565	.780	.485	1110	157	6.17	402	55.2	3.71	
	90	26.5	14.02	14.520	.710	.440	999	143	6.14	362	49.9	3.70	
W14 14 x 10	82	24.1	14.31	10.130	.855	.510	882	123	6.05	148	29.3	2.48	HS
	74	21.8	14.17	10.070	.785	.450	796	112	6.04	134	26.6	2.48	
	68	20.0	14.04	10.035	.720	.415	723	103	6.01	121	24.2	2.46	
	61	17.9	13.89	9.995	.645	.375	640	92.2	5.98	107	21.5	2.45	
W14 14 x 8	53	15.6	13.92	8.060	.660	.370	541	77.8	5.89	57.7	14.3	1.92	HSG
	48	14.1	13.79	8.030	.595	.340	485	70.3	5.85	51.4	12.8	1.91	
	43	12.6	13.66	7.995	.530	.305	428	62.7	5.82	45.2	11.3	1.89	
W14 14 x 6¾	38	11.2	14.10	6.770	.515	.310	385	54.6	5.87	26.7	7.88	1.55	HG
	34	10.0	13.98	6.745	.455	.285	340	48.6	5.83	23.3	6.91	1.53	
	30	8.85	13.84	6.730	.385	.270	291	42.0	5.73	19.6	5.82	1.49	

†These designated shapes are used principally as columns. When these shapes are used in other than column applications, existing specifications may require modification to include killed steel and/or other special metallurgical requirements. Buyers must determine if their applications require killed steel and/or other metallurgical requirements.

Source: *Structural Steel Shapes* (Pittsburgh, Pa.: U.S. Steel Corporation, 1982).

W

Wide Flange Shapes

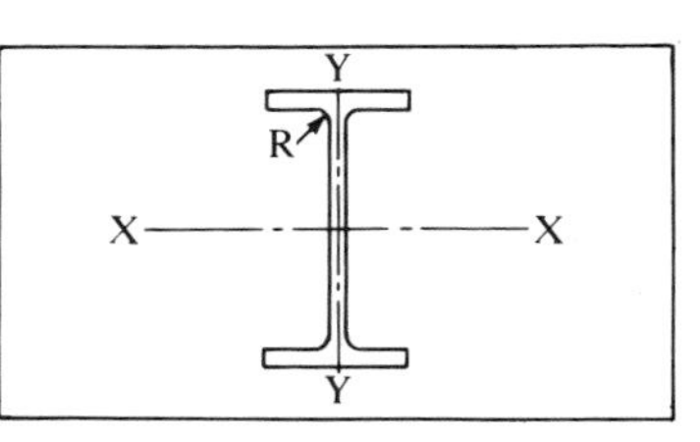

Properties for Designing

Designation and Nominal Size	Weight per Foot	Area	Depth	Flange		Web Thickness	Axis X-X			Axis Y-Y			Producing Mill Location
				Width	Thickness		I	Z	r	I	Z	r	
In.	Lbs.	In.²	In.	In.	In.	In.	In.⁴	In.³	In.	In.⁴	In.³	In.	
W14 14 x 5	26	7.69	13.91	5.025	.420	.255	245	35.3	5.65	8.91	3.54	1.08	HG
	22	6.49	13.74	5.000	.335	.230	199	29.0	5.54	7.00	2.80	1.04	
W12 12 x 12	190	55.8	14.38	12.670	1.735	1.060	1890	263	5.82	589	93.0	3.25	HS
	170	50.0	14.03	12.570	1.560	.960	1650	235	5.74	517	82.3	3.22	
	152	44.7	13.71	12.480	1.400	.870	1430	209	5.66	454	72.8	3.19	
	136	39.9	13.41	12.400	1.250	.790	1240	186	5.58	398	64.2	3.16	
	120	35.3	13.12	12.320	1.105	.710	1070	163	5.51	345	56.0	3.13	
	106	31.2	12.89	12.220	.990	.610	933	145	5.47	301	49.3	3.11	
	96	28.2	12.71	12.160	.900	.550	833	131	5.44	270	44.4	3.09	
	87	25.6	12.53	12.125	.810	.515	740	118	5.38	241	39.7	3.07	
	79	23.2	12.38	12.080	.735	.470	662	107	5.34	216	35.8	3.05	
	72	21.1	12.25	12.040	.670	.430	597	97.4	5.31	195	32.4	3.04	
	65	19.1	12.12	12.000	.605	.390	533	87.9	5.28	174	29.1	3.02	
W12 12 x 10	58	17.0	12.19	10.010	.640	.360	475	78.0	5.28	107	21.4	2.51	HS
	53	15.6	12.06	9.995	.575	.345	425	70.6	5.23	95.8	19.2	2.48	
W12 12 x 8	50	14.7	12.19	8.080	.640	.370	394	64.7	5.18	56.3	13.9	1.96	HSG
	45	13.2	12.06	8.045	.575	.335	350	58.1	5.15	50.0	12.4	1.94	
	40	11.8	11.94	8.005	.515	.295	310	51.9	5.13	44.1	11.0	1.93	
W12 12 x 6½	35	10.3	12.50	6.560	.520	.300	285	45.6	5.25	24.5	7.47	1.54	HG
	30	8.79	12.34	6.520	.440	.260	238	38.6	5.21	20.3	6.24	1.52	
	26	7.65	12.22	6.490	.380	.230	204	33.4	5.17	17.3	5.34	1.51	
W12 12 x 4	22	6.48	12.31	4.030	.425	.260	156	25.4	4.91	4.66	2.31	.847	G
	19	5.57	12.16	4.005	.350	.235	130	21.3	4.82	3.76	1.88	.822	
	16	4.71	11.99	3.990	.265	.220	103	17.1	4.67	2.82	1.41	.773	
	14	4.16	11.91	3.970	.225	.200	88.6	14.9	4.62	2.36	1.19	.753	
W10 10 x 10	112	32.9	11.36	10.415	1.250	.755	716	126	4.66	236	45.3	2.68	HS
	100	29.4	11.10	10.340	1.120	.680	623	112	4.60	207	40.0	2.65	
	88	25.9	10.84	10.265	.990	.605	534	98.5	4.54	179	34.8	2.63	
	77	22.6	10.60	10.190	.870	.530	455	85.9	4.49	154	30.1	2.60	
	68	20.0	10.40	10.130	.770	.470	394	75.7	4.44	134	26.4	2.59	
	60	17.6	10.22	10.080	.680	.420	341	66.7	4.39	116	23.0	2.57	
	54	15.8	10.09	10.030	.615	.370	303	60.0	4.37	103	20.6	2.56	
	49	14.4	9.98	10.000	.560	.340	272	54.6	4.35	93.4	18.7	2.54	
W10 10 x 8	45	13.3	10.10	8.020	.620	.350	248	49.1	4.32	53.4	13.3	2.01	HSG
	39	11.5	9.92	7.985	.530	.315	209	42.1	4.27	45.0	11.3	1.98	
	33	9.71	9.73	7.960	.435	.290	170	35.0	4.19	36.6	9.20	1.94	
W10 10 x 5¾	30	8.84	10.47	5.810	.510	.300	170	32.4	4.38	16.7	5.75	1.37	HG
	26	7.61	10.33	5.770	.440	.260	144	27.9	4.35	14.1	4.89	1.36	
	22	6.49	10.17	5.750	.360	.240	118	23.2	4.27	11.4	3.97	1.33	

Important: The sizes and limitations listed in the tables are subject to change without notice. The listing should not be considered as a commitment by United States Steel Corporation that the material will be available in the future. Please contact your nearest United States Steel Sales Office for the latest information.

Engineering and design data in these tables are conceptual in nature and are not intended to be utilized as the basis for a specific application. The accuracy, adequacy, and applicability of all data should be verified by the user's engineering staff or consultant.

Source: *Structural Steel Shapes* (Pittsburgh, Pa.: U.S. Steel Corporation, 1982).

W

Wide Flange Shapes

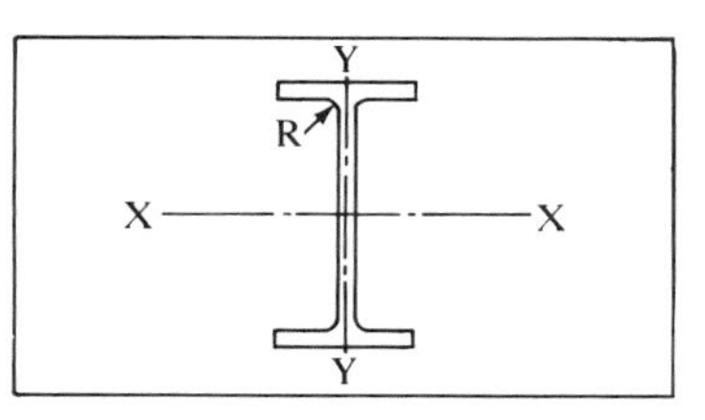

Properties for Designing

Designation and Nominal Size	Weight per Foot	Area	Depth	Flange		Web Thickness	Axis X-X			Axis Y-Y			Producing Mill Location
				Width	Thickness		I	Z	r	I	Z	r	
In.	Lbs.	In.²	In.	In.	In.	In.	In.⁴	In.³	In.	In.⁴	In.³	In.	
W10 10 x 4	19	5.62	10.24	4.020	.395	.250	96.3	18.8	4.14	4.29	2.14	.874	G
	17	4.99	10.11	4.010	.330	.240	81.9	16.2	4.05	3.56	1.78	.844	
	15	4.41	9.99	4.000	.270	.230	68.9	13.8	3.95	2.89	1.45	.810	
W8 8 x 8	67	19.7	9.00	8.280	.935	.570	272	60.4	3.72	88.6	21.4	2.12	HSG
	58	17.1	8.75	8.220	.810	.510	228	52.0	3.65	75.1	18.3	2.10	
	48	14.1	8.50	8.110	.685	.400	184	43.3	3.61	60.9	15.0	2.08	
	40	11.7	8.25	8.070	.560	.360	146	35.5	3.53	49.1	12.2	2.04	
	35	10.3	8.12	8.020	.495	.310	127	31.2	3.51	42.6	10.6	2.03	
	31	9.13	8.00	7.995	.435	.285	110	27.5	3.47	37.1	9.27	2.02	
W8 8 x 6½	28	8.25	8.06	6.535	.465	.285	98.0	24.3	3.45	21.7	6.63	1.62	HG
	24	7.08	7.93	6.495	.400	.245	82.8	20.9	3.42	18.3	5.63	1.61	
W8 8 x 5¼	21	6.16	8.28	5.270	.400	.250	75.3	18.2	3.49	9.77	3.71	1.26	G
	18	5.26	8.14	5.250	.330	.230	61.9	15.2	3.43	7.97	3.04	1.23	
W8 8 x 4	15	4.44	8.11	4.015	.315	.245	48.0	11.8	3.29	3.41	1.70	.876	G
	13	3.84	7.99	4.000	.255	.230	39.6	9.91	3.21	2.73	1.37	.843	
W6 6 x 6	25	7.34	6.38	6.080	.455	.320	53.4	16.7	2.70	17.1	5.61	1.52	G
	20	5.87	6.20	6.020	.365	.260	41.4	13.4	2.66	13.3	4.41	1.50	
	15	4.43	5.99	5.990	.260	.230	29.1	9.72	2.56	9.32	3.11	1.45	
W6 6 x 4	16	4.74	6.28	4.030	.405	.260	32.1	10.2	2.60	4.43	2.20	.966	G
	12	3.55	6.03	4.000	.280	.230	22.1	7.31	2.49	2.99	1.50	.918	

Source: *Structural Steel Shapes* (Pittsburgh, Pa.: U.S. Steel Corporation, 1982).

S

American Standard Beams

Properties for Designing

Designation and Nominal Size	Weight per Foot	Area	Depth	Width of Flange	Aver. Flange Thick-ness	Web Thick-ness	Axis X-X			Axis Y-Y			Producing Mill Location
							I	Z	r	I	Z	r	
In.	Lbs.	In.²	In.	In.	In.	In.	In.⁴	In.³	In.	In.⁴	In.³	In.	
S18 18 x 6	70	20.6	18.00	6.251	.691	.711	926	103	6.71	24.1	7.72	1.08	H
	54.7	16.1	18.00	6.001	.691	.461	804	89.4	7.07	20.8	6.94	1.14	
S15 15 x 5½	50	14.7	15.00	5.640	.622	.550	486	64.8	5.75	15.7	5.57	1.03	H
	42.9	12.6	15.00	5.501	.622	.411	447	59.6	5.95	14.4	5.23	1.07	
S12 12 x 5¼	50	14.7	12.00	5.477	.659	.687	305	50.8	4.55	15.7	5.74	1.03	H
	40.8	12.0	12.00	5.252	.659	.462	272	45.4	4.77	13.6	5.16	1.06	
S12 12 x 5	35	10.3	12.00	5.078	.544	.428	229	38.2	4.72	9.87	3.89	.980	HF
	31.8	9.35	12.00	5.000	.544	.350	218	36.4	4.83	9.36	3.74	1.00	
S10 10 x 4⅝	35	10.3	10.00	4.944	.491	.594	147	29.4	3.78	8.36	3.38	.901	HFG
	25.4	7.46	10.00	4.661	.491	.311	124	24.7	4.07	6.79	2.91	.954	
S8 8 x 4	23	6.77	8.00	4.171	.426	.441	64.9	16.2	3.10	4.31	2.07	.798	CFG
	18.4	5.41	8.00	4.001	.426	.271	57.6	14.4	3.26	3.73	1.86	.831	
S7 7 x 3⅝	15.3	4.50	7.00	3.662	.392	.252	36.7	10.5	2.86	2.64	1.44	.766	C
S6 6 x 3⅜	17.25	5.07	6.00	3.565	.359	.465	26.3	8.77	2.28	2.31	1.30	.675	CFG
	12.5	3.67	6.00	3.332	.359	.232	22.1	7.37	2.45	1.82	1.09	.705	
S5 5 x 3	10	2.94	5.00	3.004	.326	.214	12.3	4.92	2.05	1.22	.809	.643	C
S4 4 x 2⅝	9.5	2.79	4.00	2.796	.293	.326	6.79	3.39	1.56	.903	.646	.569	G
	7.7	2.26	4.00	2.663	.293	.193	6.08	3.04	1.64	.764	.574	.581	CG

Source: *Structural Steel Shapes* (Pittsburgh, Pa.: U.S. Steel Corporation, 1982).

C

American Standard Channels

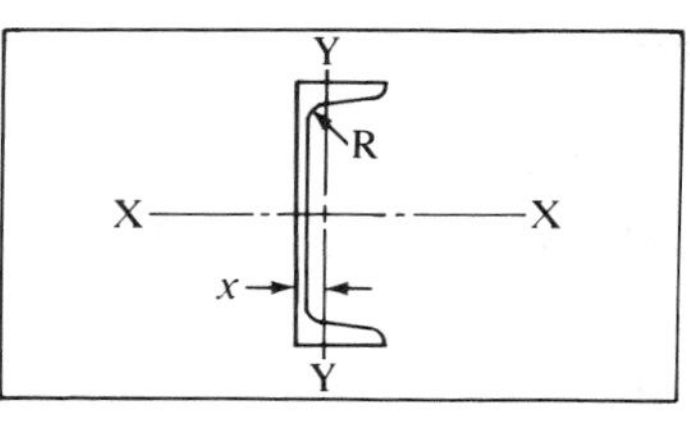

Properties for Designing

Designation and Nominal Size	Weight per Foot	Area	Depth	Width of Flange	Aver Flange Thick-ness	Web Thickness	Axis X-X			Axis Y-Y				Pro-ducing Mill Location
							I	Z	r	I	Z	r	x	
In.	Lbs.	In.^{2}	In.	In.	In.	In.	In.^{4}	In.^{3}	In.	In.^{4}	In.^{3}	In.	In.	
C15 15 x 3⅜	50	14.7	15.00	3.716	.650	.716	404	53.8	5.24	11.0	3.78	.867	.798	HFG
	40	11.8	15.00	3.520	.650	.520	349	46.5	5.44	9.23	3.37	.886	.777	
	33.9	9.96	15.00	3.400	.650	.400	315	42.0	5.62	8.13	3.11	.904	.787	
C12 12 x 3	30	8.82	12.00	3.170	.501	.510	162	27.0	4.29	5.14	2.06	.763	.674	HFG
	25	7.35	12.00	3.047	.501	.387	144	24.1	4.43	4.47	1.88	.780	.674	
	20.7	6.09	12.00	2.942	.501	.282	129	21.5	4.61	3.88	1.73	.799	.698	
C10 10 x 2⅝	30	8.82	10.00	3.033	.436	.673	103	20.7	3.42	3.94	1.65	.669	.649	CFG
	25	7.35	10.00	2.886	.436	.526	91.2	18.2	3.52	3.36	1.48	.676	.617	
	20	5.88	10.00	2.739	.436	.379	78.9	15.8	3.66	2.81	1.32	.692	.606	
	15.3	4.49	10.00	2.600	.436	.240	67.4	13.5	3.87	2.28	1.16	.713	.634	
C9 9 x 2½	15	4.41	9.00	2.485	.413	.285	51.0	11.3	3.40	1.93	1.01	.661	.586	CFG
	13.4	3.94	9.00	2.433	.413	.233	47.9	10.6	3.48	1.76	.962	.669	.601	
C8 8 x 2¼	18.75	5.51	8.00	2.527	.390	.487	44.0	11.0	2.82	1.98	1.01	.599	.565	CFG
	13.75	4.04	8.00	2.343	.390	.303	36.1	9.03	2.99	1.53	.854	.615	.553	
	11.5	3.38	8.00	2.260	.390	.220	32.6	8.14	3.11	1.32	.781	.625	.571	
C7 7 x 2⅛	12.25	3.60	7.00	2.194	.366	.314	24.2	6.93	2.60	1.17	.703	.571	.525	CFG
	9.8	2.87	7.00	2.090	.366	.210	21.3	6.08	2.72	.968	.625	.581	.540	
C6 6 x 2	13	3.83	6.00	2.157	.343	.437	17.4	5.80	2.13	1.05	.642	.525	.514	CFG
	10.5	3.09	6.00	2.034	.343	.314	15.2	5.06	2.22	.866	.564	.529	.499	
	8.2	2.40	6.00	1.920	.343	.200	13.1	4.38	2.34	.693	.492	.537	.511	
C5 5 x 1¾	9	2.64	5.00	1.885	.320	.325	8.90	3.56	1.83	.632	.450	.489	.478	CF
	6.7	1.97	5.00	1.750	.320	.190	7.49	3.00	1.95	.479	.378	.493	.484	

Source: *Structural Steel Shapes* (Pittsburgh, Pa.: U.S. Steel Corporation, 1982).

L

Angles
Equal Leg

Dimensions and Properties for Designing

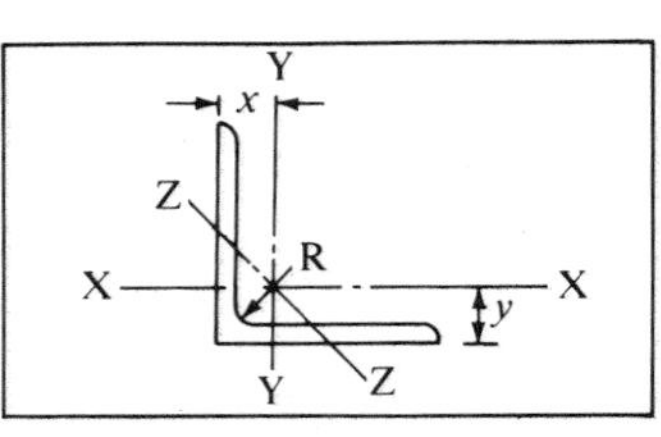

Designation and Nominal Size	Thick-ness	Weight per Foot	Area	Axis X-X and Axis Y-Y					Fillet Radius R	Pro-ducing Mill Location
				I	Z	r	x or y	r_{min}		
In.	In.	Lbs.	In.²	In.⁴	In.³	In.	In.	In.	In.	
L8x8	1⅛	56.9	16.7	98.0	17.5	2.42	2.41	1.56	⅝	HF
	1	51.0	15.0	89.0	15.8	2.44	2.37	1.56		
	⅞	45.0	13.2	79.6	14.0	2.45	2.32	1.57		
	¾	38.9	11.4	69.7	12.2	2.47	2.28	1.58		
	⅝	32.7	9.61	59.4	10.3	2.49	2.23	1.58		
	½	26.4	7.75	48.6	8.36	2.50	2.19	1.59		
L6x6	1	37.4	11.0	35.5	8.57	1.80	1.86	1.17	½	CFG
	⅞	33.1	9.73	31.9	7.63	1.81	1.82	1.17		
	¾	28.7	8.44	28.2	6.66	1.83	1.78	1.17		
	⅝	24.2	7.11	24.2	5.66	1.84	1.73	1.18		
	½	19.6	5.75	19.9	4.61	1.86	1.68	1.18		
	⅜	14.9	4.36	15.4	3.53	1.88	1.64	1.19		
L5x5	⅞	27.2	7.98	17.8	5.17	1.49	1.57	.973	½	C
	¾	23.6	6.94	15.7	4.53	1.51	1.52	.975		CF
	½	16.2	4.75	11.3	3.16	1.54	1.43	.983		
	⅜	12.3	3.61	8.74	2.42	1.56	1.39	.990		
	5/16	10.3	3.03	7.42	2.04	1.57	1.37	.994		
L4x4	¾	18.5	5.44	7.67	2.81	1.19	1.27	.778	⅜	CF
	⅝	15.7	4.61	6.66	2.40	1.20	1.23	.779		
	½	12.8	3.75	5.56	1.97	1.22	1.18	.782		
	⅜	9.8	2.86	4.36	1.52	1.23	1.14	.788		
	5/16	8.2	2.40	3.71	1.29	1.24	1.12	.791		
	¼	6.6	1.94	3.04	1.05	1.25	1.09	.795		
L3½x3½	⅜	8.5	2.48	2.87	1.15	1.07	1.01	.687	⅜	CF
	5/16	7.2	2.09	2.45	.976	1.08	.990	.690		
	¼	5.8	1.69	2.01	.794	1.09	.968	.694		
L3x3	½	9.4	2.75	2.22	1.07	.898	.932	.584	5/16	F
	⅜	7.2	2.11	1.76	.833	.913	.888	.587		
	5/16	6.1	1.78	1.51	.707	.922	.865	.589		
	¼	4.9	1.44	1.24	.577	.930	.842	.592		

Source: *Structural Steel Shapes* (Pittsburgh, Pa.: U.S. Steel Corporation, 1982).

L

Bar Size Angles Equal Legs

Dimensions and Properties for Designing

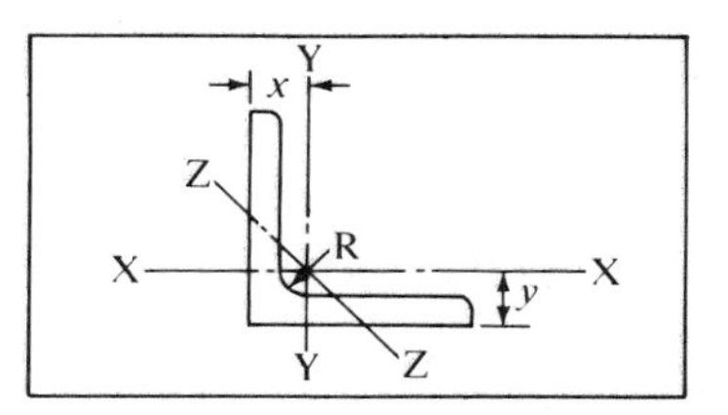

Designation and Nominal Size	Thickness	Weight per Foot	Area	Axis X-X and Axis Y-Y				Axis Z-Z	Fillet Radius R	Producing Mill Locations
				I	Z	r	x or y	$r_{min.}$		
In.	In.	Lbs.	In.²	In.⁴	In.³	In.	In.	In.	In.	
L2½x2½	⅜	5.9	1.73	.984	.566	.753	.762	.487	³⁄₁₆	†
	⁵⁄₁₆	5.0	1.46	.849	.482	.761	.740	.489		
	¼	4.1	1.19	.703	.394	.769	.717	.491		
	³⁄₁₆	3.07	.902	.547	.303	.778	.694	.495		
L2x2	⅜	4.7	1.36	.479	.351	.594	.636	.389	³⁄₁₆	†
	⁵⁄₁₆	3.92	1.15	.416	.300	.601	.614	.390		
	¼	3.19	.938	.348	.247	.609	.592	.391		
	.205	2.65	.778	.294	.207	.615	.575	.393		
	³⁄₁₆	2.44	.715	.272	.190	.617	.569	.394		
L1¾x1¾	¼	2.77	.813	.227	.186	.529	.529	.341	³⁄₁₆	†
	³⁄₁₆	2.12	.621	.179	.144	.537	.506	.343		
L1½x1½	¼	2.34	.688	.139	.134	.449	.466	.292	³⁄₁₆	†
	³⁄₁₆	1.80	.527	.110	.104	.457	.444	.293		
L1¼x1¼	¼	1.92	.563	.077	.091	.369	.403	.243	³⁄₁₆	†
	³⁄₁₆	1.48	.434	.061	.071	.377	.381	.244		

†Fairless Works (Fairless, Pa.)

For bar size shapes ordered to chemical composition rather than mechanical properties the chart on page 67 may be useful.

Source: *Structural Steel Shapes* (Pittsburgh, Pa.: U.S. Steel Corporation, 1982).

L

Angles Unequal Leg

Dimensions and Properties for Designing

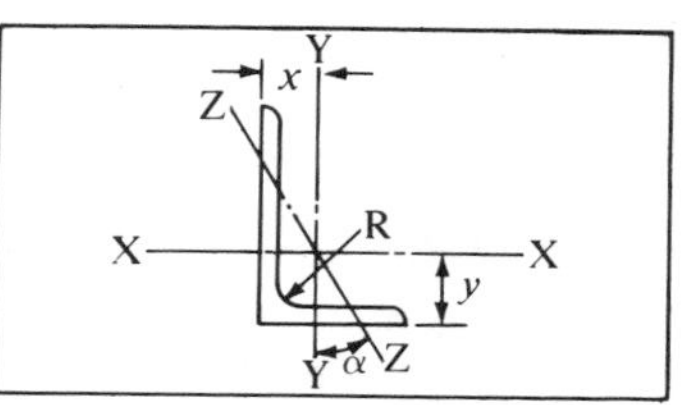

Designation and Nominal Size	Thickness	Weight per Foot	Area	Axis X-X I	Axis X-X Z	Axis X-X r	Axis X-X y	Axis Y-Y I	Axis Y-Y Z	Axis Y-Y r	Axis Y-Y x	Axis Z-Z r_{min}	Axis Z-Z Tan α	Fillet Radius R	Producing Mill Location
In.	In.	Lbs.	In.²	In.⁴	In.³	In.	In.	In.⁴	In.³	In.	In.	In.		In.	
L8x6	1	44.2	13.0	80.8	15.1	2.49	2.65	38.8	8.92	1.73	1.65	1.28	.543	½	H
	¾	33.8	9.94	63.4	11.7	2.53	2.56	30.7	6.92	1.76	1.56	1.29	.551		
	½	23.0	6.75	44.3	8.02	2.56	2.47	21.7	4.79	1.79	1.47	1.30	.558		
L8x4	1	37.4	11.0	69.6	14.1	2.52	3.05	11.6	3.94	1.03	1.05	.846	.247	½	H
	¾	28.7	8.44	54.9	10.9	2.55	2.95	9.36	3.07	1.05	.953	.852	.258		
	½	19.6	5.75	38.5	7.49	2.59	2.86	6.74	2.15	1.08	.859	.865	.267		
L7x4	¾	26.2	7.69	37.8	8.42	2.22	2.51	9.05	3.03	1.09	1.01	.860	.324	½	CG
	½	17.9	5.25	26.7	5.81	2.25	2.42	6.53	2.12	1.11	.917	.872	.335		
	⅜	13.6	3.98	20.6	4.44	2.27	2.37	5.10	1.63	1.13	.870	.880	.340		
L6x4	¾	23.6	6.94	24.5	6.25	1.88	2.08	8.68	2.97	1.12	1.08	.860	.428	½	CFG
	½	16.2	4.75	17.4	4.33	1.91	1.99	6.27	2.08	1.15	.987	.870	.440		
	⅜	12.3	3.61	13.5	3.32	1.93	1.94	4.90	1.60	1.17	.941	.877	.446		
L6x3½	⅜	11.7	3.42	12.9	3.24	1.94	2.04	3.34	1.23	.988	.787	.767	.350	½	CF
	5/16	9.8	2.87	10.9	2.73	1.95	2.01	2.85	1.04	.996	.763	.772	.352		
L5x3½	¾	19.8	5.81	13.9	4.28	1.55	1.75	5.55	2.22	.977	.996	.748	.464	7/16	CG
	½	13.6	4.00	9.99	2.99	1.58	1.66	4.05	1.56	1.01	.906	.755	.479		CFG
	⅜	10.4	3.05	7.78	2.29	1.60	1.61	3.18	1.21	1.02	.861	.762	.486		
	5/16	8.7	2.56	6.60	1.94	1.61	1.59	2.72	1.02	1.03	.838	.766	.489		
L5x3	½	12.8	3.75	9.45	2.91	1.59	1.75	2.58	1.15	.829	.750	.648	.357	⅜	CF
	⅜	9.8	2.86	7.37	2.24	1.61	1.70	2.04	.888	.845	.704	.654	.364		
	5/16	8.2	2.40	6.26	1.89	1.61	1.68	1.75	.753	.853	.681	.658	.368		
	¼	6.6	1.94	5.11	1.53	1.62	1.66	1.44	.614	.861	.657	.663	.371		
L4x3½	½	11.9	3.50	5.32	1.94	1.23	1.25	3.79	1.52	1.04	1.00	.722	.750	⅜	CF
	⅜	9.1	2.67	4.18	1.49	1.25	1.21	2.95	1.17	1.06	.955	.727	.755		
	5/16	7.7	2.25	3.56	1.26	1.26	1.18	2.55	.994	1.07	.932	.730	.757		
	¼	6.2	1.81	2.91	1.03	1.27	1.16	2.09	.808	1.07	.909	.734	.759		
L4x3	½	11.1	3.25	5.05	1.89	1.25	1.33	2.42	1.12	.864	.827	.639	.543	⅜	CF
	⅜	8.5	2.48	3.96	1.46	1.26	1.28	1.92	.866	.879	.782	.644	.551		
	5/16	7.2	2.09	3.38	1.23	1.27	1.26	1.65	.734	.887	.759	.647	.554		
	¼	5.8	1.69	2.77	1.00	1.28	1.24	1.36	.599	.896	.736	.651	.558		
L3½x3	⅜	7.9	2.30	2.72	1.13	1.09	1.08	1.85	.851	.897	.830	.625	.721	⅜	F
	5/16	6.6	1.93	2.33	.954	1.10	1.06	1.58	.722	.905	.808	.627	.724		
	¼	5.4	1.56	1.91	.776	1.11	1.04	1.30	.589	.914	.785	.631	.727		
L3½x2½	⅜	7.2	2.11	2.56	1.09	1.10	1.16	1.09	.592	.719	.660	.537	.496	5/16	F
	5/16	6.1	1.78	2.19	.927	1.11	1.14	.939	.504	.727	.637	.540	.501		
	¼	4.9	1.44	1.80	.755	1.12	1.11	.777	.412	.735	.614	.544	.506		
L3x2½	⅜	6.6	1.92	1.66	.810	.928	.956	1.04	.581	.736	.706	.522	.676	5/16	F
	5/16	5.6	1.62	1.42	.688	.937	.933	.898	.494	.744	.683	.525	.680		
	¼	4.5	1.31	1.17	.561	.945	.911	.743	.404	.753	.661	.528	.684		

Source: *Structural Steel Shapes* (Pittsburgh, Pa.: U.S. Steel Corporation, 1982).

L

Angles Unequal Leg

Dimensions and Properties for Designing

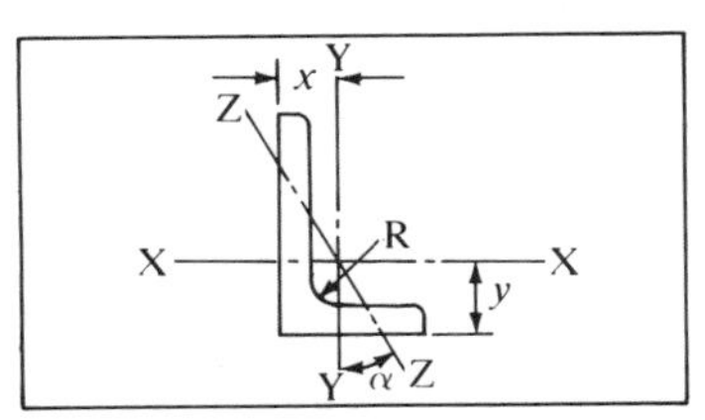

Designation and Nominal Size	Thick-ness	Weight per Foot	Area	Axis X-X				Axis Y-Y				Axis Z-Z		Fillet Radius R	Pro-ducing Mill Location
				I	Z	r	y	I	Z	r	x	r_{min}	Tan α		
In.	In.	Lbs.	In.²	In.⁴	In.³	In.	In.	In.⁴	In.³	In.	In.	In.		In.	
L8x6	1	44.2	13.0	80.8	15.1	2.49	2.65	38.8	8.92	1.73	1.65	1.28	.543	½	H
	¾	33.8	9.94	63.4	11.7	2.53	2.56	30.7	6.92	1.76	1.56	1.29	.551		
	½	23.0	6.75	44.3	8.02	2.56	2.47	21.7	4.79	1.79	1.47	1.30	.558		
L8x4	1	37.4	11.0	69.6	14.1	2.52	3.05	11.6	3.94	1.03	1.05	.846	.247	½	H
	¾	28.7	8.44	54.9	10.9	2.55	2.95	9.36	3.07	1.05	.953	.852	.258		
	½	19.6	5.75	38.5	7.49	2.59	2.86	6.74	2.15	1.08	.859	.865	.267		
L7x4	¾	26.2	7.69	37.8	8.42	2.22	2.51	9.05	3.03	1.09	1.01	.860	.324	½	CG
	½	17.9	5.25	26.7	5.81	2.25	2.42	6.53	2.12	1.11	.917	.872	.335		
	⅜	13.6	3.98	20.6	4.44	2.27	2.37	5.10	1.63	1.13	.870	.880	.340		
L6x4	¾	23.6	6.94	24.5	6.25	1.88	2.08	8.68	2.97	1.12	1.08	.860	.428	½	CFG
	½	16.2	4.75	17.4	4.33	1.91	1.99	6.27	2.08	1.15	.987	.870	.440		
	⅜	12.3	3.61	13.5	3.32	1.93	1.94	4.90	1.60	1.17	.941	.877	.446		
L6x3½	⅜	11.7	3.42	12.9	3.24	1.94	2.04	3.34	1.23	.988	.787	.767	.350	½	CF
	5/16	9.8	2.87	10.9	2.73	1.95	2.01	2.85	1.04	.996	.763	.772	.352		
L5x3½	¾	19.8	5.81	13.9	4.28	1.55	1.75	5.55	2.22	.977	.996	.748	.464	7/16	CG
	½	13.6	4.00	9.99	2.99	1.58	1.66	4.05	1.56	1.01	.906	.755	.479		CFG
	⅜	10.4	3.05	7.78	2.29	1.60	1.61	3.18	1.21	1.02	.861	.762	.486		
	5/16	8.7	2.56	6.60	1.94	1.61	1.59	2.72	1.02	1.03	.838	.766	.489		
L5x3	½	12.8	3.75	9.45	2.91	1.59	1.75	2.58	1.15	.829	.750	.648	.357	⅜	CF
	⅜	9.8	2.86	7.37	2.24	1.61	1.70	2.04	.888	.845	.704	.654	.364		
	5/16	8.2	2.40	6.26	1.89	1.61	1.68	1.75	.753	.853	.681	.658	.368		
	¼	6.6	1.94	5.11	1.53	1.62	1.66	1.44	.614	.861	.657	.663	.371		
L4x3½	½	11.9	3.50	5.32	1.94	1.23	1.25	3.79	1.52	1.04	1.00	.722	.750	⅜	CF
	⅜	9.1	2.67	4.18	1.49	1.25	1.21	2.95	1.17	1.06	.955	.727	.755		
	5/16	7.7	2.25	3.56	1.26	1.26	1.18	2.55	.994	1.07	.932	.730	.757		
	¼	6.2	1.81	2.91	1.03	1.27	1.16	2.09	.808	1.07	.909	.734	.759		
L4x3	½	11.1	3.25	5.05	1.89	1.25	1.33	2.42	1.12	.864	.827	.639	.543	⅜	CF
	⅜	8.5	2.48	3.96	1.46	1.26	1.28	1.92	.866	.879	.782	.644	.551		
	5/16	7.2	2.09	3.38	1.23	1.27	1.26	1.65	.734	.887	.759	.647	.554		
	¼	5.8	1.69	2.77	1.00	1.28	1.24	1.36	.599	.896	.736	.651	.558		
L3½x3	⅜	7.9	2.30	2.72	1.13	1.09	1.08	1.85	.851	.897	.830	.625	.721	⅜	F
	5/16	6.6	1.93	2.33	.954	1.10	1.06	1.58	.722	.905	.808	.627	.724		
	¼	5.4	1.56	1.91	.776	1.11	1.04	1.30	.589	.914	.785	.631	.727		
L3½x2½	⅜	7.2	2.11	2.56	1.09	1.10	1.16	1.09	.592	.719	.660	.537	.496	5/16	F
	5/16	6.1	1.78	2.19	.927	1.11	1.14	.939	.504	.727	.637	.540	.501		
	¼	4.9	1.44	1.80	.755	1.12	1.11	.777	.412	.735	.614	.544	.506		
L3x2½	⅜	6.6	1.92	1.66	.810	.928	.956	1.04	.581	.736	.706	.522	.676	5/16	F
	5/16	5.6	1.62	1.42	.688	.937	.933	.898	.494	.744	.683	.525	.680		
	¼	4.5	1.31	1.17	.561	.945	.911	.743	.404	.753	.661	.528	.684		

Source: *Structural Steel Shapes* (Pittsburgh, Pa.: U.S. Steel Corporation, 1982).

TS

Structural Tubing ASTM A501

Dimensions and Properties for Designing

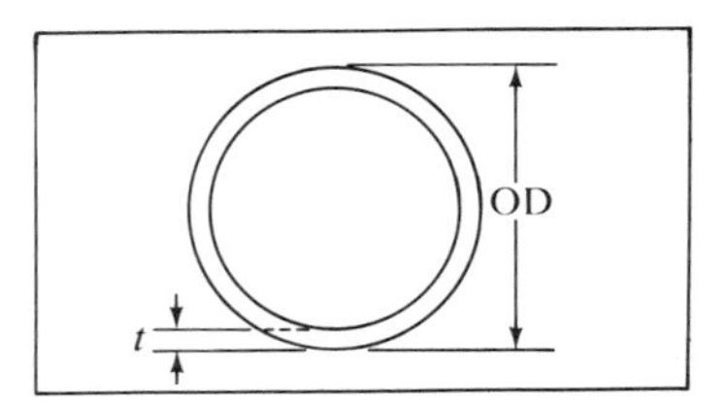

Nominal Size (OD) In.	Wall Thickness t In.	Weight per Foot Lbs.	Area In.²	I	Z	r	Producing Mill Location
TS½ (.840)	.109 .147	.85 1.09	.250 .320	.017 .020	.041 .048	.261 .250	*See Below
TS¾ (1.050)	.113 .154	1.13 1.47	.333 .433	.037 .045	.071 .085	.334 .321	*See Below
TS1 (1.315)	.133 .179	1.68 2.17	.494 .639	.087 .106	.133 .161	.421 .407	*See Below
TS1¼ (1.660)	.140 .191	2.27 3.00	.669 .881	.195 .242	.235 .291	.540 .524	*See Below
TS1½ (1.900)	.145 .200	2.72 3.63	.799 1.07	.310 .391	.326 .412	.623 .605	*See Below
TS2 (2.375)	.154 .218	3.65 5.02	1.07 1.48	.666 .868	.561 .731	.787 .766	*See Below
TS2½ (2.875)	.203 .276	5.79 7.66	1.70 2.25	1.53 1.92	1.06 1.34	.947 .924	*See Below
TS3 (3.500)	.216 .300	7.58 10.25	2.23 3.02	3.02 3.89	1.72 2.23	1.16 1.14	*See Below
TS3½ (4.000)	.226 .318	9.11 12.50	2.68 3.68	4.79 6.28	2.39 3.14	1.34 1.31	*See Below
TS4 (4.500)	.237 .337	10.79 14.98	3.17 4.41	7.23 9.61	3.21 4.27	1.51 1.48	*See Below

*Produced at Fairless, Pennsylvania.
It is available in carbon steel only. For larger sizes—please inquire.

ASTM A501 Structural Tubing may be ordered black or with hot-dipped galvanized coating. When specified with galvanized coating, the coating shall comply with the requirements of the latest edition of ASTM A120.

Unless otherwise agreed upon and so specified on the purchase order, all sizes and wall thicknesses shall be furnished with plain ends cut square. All burrs shall be removed from the outside ends. The finished pipe shall be reasonably straight and free from injurious imperfections.

ASTM A501 Structural Tubing is for general structural use and not intended for applications involving internal pressure.

Source: *Structural Steel Shapes* (Pittsburgh, Pa.: U.S. Steel Corporation, 1982).

Standard Channels—
Dimensions, Areas, Weights and Section Properties

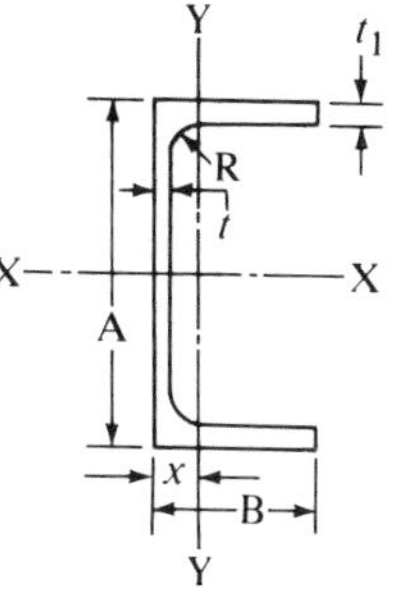

Size							Section Properties③						
				Flange Thick-ness	Web Thick-ness	Fillet Radius	Axis X-X			Axis Y-Y			
Depth A in.	Width B in.	Area① in.²	Weight② lb/ft	t_1 in.	t in.	R in.	I in.⁴	Z in.³	r in.	I in.⁴	Z in.³	r in.	x in.
2.00	1.00	0.491	0.577	0.13	0.13	0.10	0.288	0.288	0.766	0.045	0.064	0.303	0.298
2.00	1.25	0.911	1.071	0.26	0.17	0.15	0.546	0.546	0.774	0.139	0.178	0.391	0.471
3.00	1.50	0.965	1.135	0.20	0.13	0.25	1.41	0.94	1.21	0.22	0.22	0.47	0.49
3.00	1.75	1.358	1.597	0.26	0.17	0.25	1.97	1.31	1.20	0.42	0.37	0.55	0.62
4.00	2.00	1.478	1.738	0.23	0.15	0.25	3.91	1.95	1.63	0.60	0.45	0.64	0.65
4.00	2.25	1.982	2.331	0.29	0.19	0.25	5.21	2.60	1.62	1.02	0.69	0.72	0.78
5.00	2.25	1.881	2.212	0.26	0.15	0.30	7.88	3.15	2.05	0.98	0.64	0.72	0.73
5.00	2.75	2.627	3.089	0.32	0.19	0.30	11.14	4.45	2.06	2.05	1.14	0.88	0.95
6.00	2.50	2.410	2.834	0.29	0.17	0.30	14.35	4.78	2.44	1.53	0.90	0.80	0.79
6.00	3.25	3.427	4.030	0.35	0.21	0.30	21.04	7.01	2.48	3.76	1.76	1.05	1.12
7.00	2.75	2.725	3.205	0.29	0.17	0.30	22.09	6.31	2.85	2.10	1.10	0.88	0.84
7.00	3.50	4.009	4.715	0.38	0.21	0.30	33.79	9.65	2.90	5.13	2.23	1.13	1.20
8.00	3.00	3.526	4.147	0.35	0.19	0.30	37.40	9.35	3.26	3.25	1.57	0.96	0.93
8.00	3.75	4.923	5.789	0.41	0.25	0.35	52.69	13.17	3.27	7.13	2.82	1.20	1.22
9.00	3.25	4.237	4.983	0.35	0.23	0.35	54.41	12.09	3.58	4.40	1.89	1.02	0.93
9.00	4.00	5.927	6.970	0.44	0.29	0.35	78.31	17.40	3.63	9.61	3.49	1.27	1.25
10.00	3.50	5.218	6.136	0.41	0.25	0.35	83.22	16.64	3.99	6.33	2.56	1.10	1.02
10.00	4.25	7.109	8.360	0.50	0.31	0.40	116.15	23.23	4.04	13.02	4.47	1.35	1.34
12.00	4.00	7.036	8.274	0.47	0.29	0.40	159.76	26.63	4.77	11.03	3.86	1.25	1.14
12.00	5.00	10.053	11.822	0.62	0.35	0.45	239.69	39.95	4.88	25.74	7.60	1.60	1.61

Source: *Aluminum Standards and Data,* 5th ed. (New York, N.Y.: The Aluminum Association, © 1976) p. 180.

Standard I-Beams — Dimensions, Areas, Weights and Section Properties

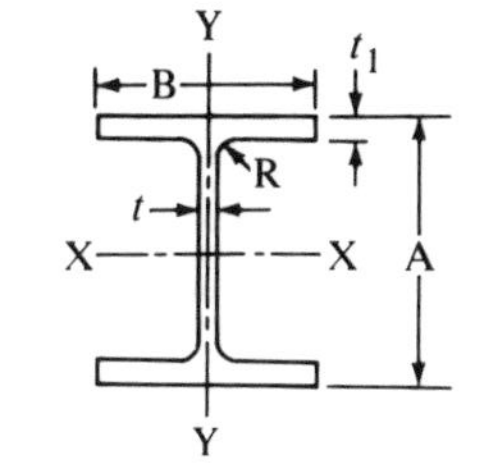

Size							Section Properties③					
							Axis X-X			Axis Y-Y		
Depth A in.	Width B in.	Area① in.²	Weight② lb/ft	Flange Thick-ness t_1 in.	Web Thick-ness t in.	Fillet Radius R in.	I in.⁴	Z in.³	r in.	I in.⁴	Z in.³	r in.
3.00	2.50	1.392	1.637	0.20	0.13	0.25	2.24	1.49	1.27	0.52	0.42	0.61
3.00	2.50	1.726	2.030	0.26	0.15	0.25	2.71	1.81	1.25	0.68	0.54	0.63
4.00	3.00	1.965	2.311	0.23	0.15	0.25	5.62	2.81	1.69	1.04	0.69	0.73
4.00	3.00	2.375	2.793	0.29	0.17	0.25	6.71	3.36	1.68	1.31	0.87	0.74
5.00	3.50	3.146	3.700	0.32	0.19	0.30	13.94	5.58	2.11	2.29	1.31	0.85
6.00	4.00	3.427	4.030	0.29	0.19	0.30	21.99	7.33	2.53	3.10	1.55	0.95
6.00	4.00	3.990	4.692	0.35	0.21	0.30	25.50	8.50	2.53	3.74	1.87	0.97
7.00	4.50	4.932	5.800	0.38	0.23	0.30	42.89	12.25	2.95	5.78	2.57	1.08
8.00	5.00	5.256	6.181	0.35	0.23	0.30	59.69	14.92	3.37	7.30	2.92	1.18
8.00	5.00	5.972	7.023	0.41	0.25	0.30	67.78	16.94	3.37	8.55	3.42	1.20
9.00	5.50	7.110	8.361	0.44	0.27	0.30	102.02	22.67	3.79	12.22	4.44	1.31
10.00	6.00	7.352	8.646	0.41	0.25	0.40	132.09	26.42	4.24	14.78	4.93	1.42
10.00	6.00	8.747	10.286	0.50	0.29	0.40	155.79	31.16	4.22	18.03	6.01	1.44
12.00	7.00	9.925	11.672	0.47	0.29	0.40	255.57	42.60	5.07	26.90	7.69	1.65
12.00	7.00	12.153	14.292	0.62	0.31	0.40	317.33	52.89	5.11	35.48	10.14	1.71

① Areas listed are based on nominal dimensions.
② Weights per foot are based on nominal dimensions and a density of 0.098 pound per cubic inch which is the density of alloy 6061.
③ I=moment of inertia; Z=section modulus; r=radius of gyration.

Answers to Selected Problems

Given here are the answers to problems for which there are unique solutions. Many of the problems for solution in this book are true design problems and individual design decision is required to arrive at the solution. Others are of the review question form for which the answers are in the text of the associated chapter. It should also be noted that some of the problems require the selection of design factors and the use of data from charts and graphs. Because of the judgment and interpolation required, some of the answers may be slightly different from your solution.

CHAPTER 1 Machine Elements in Mechanical Design

15. $D = 44.5$ mm
16. $L = 14.0$ m
17. $T = 1\,418$ N · m
18. $A = 2\,658$ mm^2
19. $Z = 2.43 \times 10^5$ mm^3
20. $I = 3.66 \times 10^7$ mm^4
21. $L2\frac{1}{2} \times 2\frac{1}{2} \times \frac{1}{4}$
22. $P = 5.59$ kW
23. $s_u = 876$ MPa
24. Weight = 48.9 N

CHAPTER 2 Materials in Mechanical Design

9. No. The percent elongation must be greater than 5.0 percent to be ductile.
11. $G = 85.1$ GPa
12. Hardness = 52.8 HRC
13. Tensile strength = 235 Ksi (Approximately)

Questions 14–17 ask what is wrong with the given statements.

14. Annealed steels typically have hardness values in the range from 120 HB to 200 HB. A hardness of 750 HB is extremely hard and characteristic of as-quenched high alloy steels.
15. The HRB scale is normally limited to HRB 100.
16. The HRC hardness is normally no lower than HRC 20.
17. The given relationship between hardness and tensile strength is only valid for steels.
18. Charpy and Izod
19. Iron and carbon. Manganese and other elements are often present.
20. Iron, carbon, manganese, nickel, chromium, molybdenum.
21. Approximately 0.40 percent.
22. Low carbon: Less than 0.30 percent
    Medium carbon: 0.30 to 0.50 percent
    High carbon: 0.50 to 0.95 percent
23. Nominally 1.0 percent.
24. AISI 12L13 steel has lead added to improve machinability.
25. AISI 1045, 4140, 4640, 5150, 6150, 8650.
26. AISI 1045, 4140, 4340, 4640, 5150, 6150, 8650.
27. Wear resistance, strength, ductility. AISI 1080.

28. AISI 5160 OQT 1000 is a chromium steel, having nominally 0.80 percent chromium and 0.60 percent carbon, a high carbon alloy steel. It has fairly high strength and good ductility.
29. Appendixes A-3 and A-4-1 show the properties of oil-quenched AISI 1040 steel. The as-quenched hardness is HB 269, approximately HRC 28. Therefore, it would not harden to HRC 40 and would not be acceptable. But note that more complete references (4) give properties of AISI 1040 when water-quenched showing that it could be hardened to HRC 54. In this form it would be acceptable if tests show it has sufficiently good ductility.

CHAPTER 3 Basic Stress Analysis

1. σ = 31.8 MPa; δ = 0.12 mm
2. For all materials, σ = 34.7 MPa
   Deflection:
   a. δ = 0.277 mm b. δ = 0.277 mm
   c. δ = 0.377 mm d. δ = 0.830 mm
   e. δ = 0.503 mm f. δ = 23.8 mm
   g. δ = 7.56 mm.
   Note: The stress is close to the ultimate strength for (e) and (f).
3. Force = 2 556 lb; σ = 2 506 psi
4. σ = 595 psi
5. Force = 1 061 lb
6. D = 0.274 in
8. τ = 32.6 MPa
9. θ = 0.79°
11. τ = 32 270 psi
14. τ = 70.8 MPa; θ = 1.94°
16. T = 9 624 lb · in; θ = 1.83°
17. Required section modulus = 2.40 in^3. Standard nominal sizes given for each shape:
   a. Each side = 2.50 in
   c. Width = 5.00 in; height = 1.75 in
   e. S4×7.7
   g. TS4×0.188
18. Weights:
   a. 212 lb c. 297 lb
   e. 77.0 lb g. 86.6 lb
19.

	Maximum deflection	Deflection at loads
a.	0.701 in	0.572 in
c.	1.021 in	0.836 in
e.	0.375 in	0.307 in
g.	0.385 in	0.315 in

20. M_A = 330 N · m; M_B = 294 N · m; M_C = −40 N · m
22. a. y_A = 0.238 in; y_B = 0.688 in
   b. y_A = 0.047 in; y_B = 0.042 in
24. σ = 3 480 psi; τ = 172 psi

For Problems 25 through 36, the complete solutions require drawings. Listed below are the maximum bending moments only:

25. 480 lb · in
27. 120 lb · in
29. 93 750 N · mm
31. 8 640 lb · in
33. −11 250 N · mm
35. −1.49 kN · m
37. σ = 86.2 MPa
39. Left: σ = 39 750 psi
   Middle: σ = 29 760 psi
   Right: σ = 31 700 psi
41. σ = 96.4 MPa
43. σ = 32 850 psi

CHAPTER 4 Combined Stresses and Mohr's Circle

1. σ = 62.07 MPa
3. a. σ = 20.94 MPa tension on top of lever
   b. At section B, h = 35.1 mm; at C, h = 18 mm
5. σ = 84.6 MPa tension
7. Sides = 0.50 in
9. Maximum σ = −1.42 MPa compression on the top surface between A and C
11. Required D = 0.643 in. Use D = 11/16 in or 0.800 in.
13. D = 1.800 in
15. τ = 5 345 psi

Answers to Problems 16 to 31:

	Maximum Principal Stress	*Minimum Principal Stress*	*Maximum Shear Stress*
16.	24.14 Ksi	−4.14 Ksi	14.14 Ksi
18.	50.0 Ksi	−50.0 Ksi	50.0 Ksi
20.	124.7 Ksi	0.0 Ksi	62.4 Ksi
22.	20.0 Ksi	−40.0 Ksi	30.0 Ksi
24.	144.3 MPa	−44.3 MPa	94.3 MPa
26.	61.3 MPa	−91.3 MPa	76.3 MPa
28.	168.2 MPa	0.0 MPa	84.1 MPa
30.	250.0 MPa	−80.0 MPa	165.0 MPa

CHAPTER 5 Design for Different Modes of Failure

For problems requiring design decisions, standard or preferred sizes are listed.

1. N = 23.8
3. N = 14.4
5. D = 35 mm; a = 5.85 mm
7. TS3 × 0.156 tube
9. b = 1¾ in
11. N = 9.11
12. σ = 34.7 MPa
   a. N = 5.96 c. N = 8.93
   e. N = 23.8 g. N = 1.30
13. N = 3.37
15. Force = 1 061 lb; D = 5/16 in
17. N = 1.88

19. $N = 5.10$ 21. $D = 2.00$ in 25. $N = 9.03$
27. $N = 3.20$

CHAPTER 6 Columns

1. $P_{cr} = 4\,473$ lb
2. $P_{cr} = 14\,373$ lb
5. $P_{cr} = 4\,473$ lb
6. $P_{cr} = 32.8$ lb
8. a. Pinned ends: $P_{cr} = 7\,498$ lb
   b. Fixed ends: $P_{cr} = 13\,362$ lb
   c. Fixed-pinned ends: $P_{cr} = 11\,484$ lb
   d. Fixed-free ends: $P_{cr} = 1\,874$ lb
10. $D = 1.45$ in required. Use $D = 1.50$ in.
14. $S = 1.248$ in required. Use $S = 1.250$ in.
16. Use $D = 1.50$ in.
23. $P = 1\,189$ lb
25. $P = 1\,877$ lb
27. $\sigma = 212$ MPa; $y = 25.7$ mm
28. $\sigma = 6\,685$ psi; $y = 0.045$ in
30. $P = 48\,000$ psi

CHAPTER 7 Springs

1. $k = 13.3$ lb/in
2. $L_f = 1.497$ in
3. $F_s = 47.8$ lb; $L_f = 1.25$ in
7. $ID = 0.93$ in; $D_m = 1.015$ in; $C = 11.94$; $N = 6.6$ coils
8. $C = 8.49$; $p = 0.241$ in; pitch angle $= 8.70°$; $L_s = 1.12$ in
9. $F_o = 10.25$ lb; stress $= 74\,500$ psi
11. $OD = 0.583$ in when at solid length
12. $F_s = 26.05$ lb; stress $= 189\,300$ psi (High)
28. Bending stress $= 114\,000$ psi; torsion stress $= 62\,600$ psi

CHAPTER 8 Tolerances and Fits

1. RC8: Hole—3.5050/3.5000; shaft—3.4930/3.4895; clearance—0.0070 to 0.0155 in
3. RC8: Hole—0.6313/0.6285; shaft—0.6250/0.6234; clearance—0.0035 to 0.0079 in
5. RC8: Hole—1.2540/1.2500; shaft—1.2450/1.2425; clearance—0.0050 to 0.0115 in
7. RC5: Hole—0.7512/0.7500; shaft—0.7484/0.7476; clearance—0.0016 to 0.0036 in (tighter fit could be used)
9. FN5: Hole—3.2522/3.2500; shaft—3.2584/3.2570; interference—0.0048 to 0.0084 in; pressure $= 13\,175$ psi; stress $= 64\,363$ psi
11. FN5: Interference—0.0042 to 0.0072 in; pressure $= 17\,789$ psi; stress $= 37\,800$ psi at inner surface of aluminum cylinder; stress $= -17\,789$ psi at outer surface of steel rod; stress in aluminum is very high
12. Maximum interference $= 0.001\,77$ in
13. Temperature $= 567°$F
14. Shrinkage $= 0.003\,8$ in; $t = 284°$F
15. Final $ID = 3.497\,1$ in

CHAPTER 9 Shaft Design

All problems in this chapter are design problems for which there are no unique solutions.

CHAPTER 10 Keys and Couplings

1. Use ½ in square key; AISI 1040 cold drawn steel; length $= 3.75$ in.
3. Use ⅜ in square key; AISI 1020 cold drawn steel; required length $= 1.02$ in; use $L = 1.50$ in to match hub length.
5. T = torque; D = shaft diameter; L = hub length. From Table 10-5, $K = T/(D^2L)$.
   a. Data from Problem 1: required $K = 1\,313$; too high for any spline in Table 10-5
   c. Data from Problem 3: required $K = 208$; use 6 splines.
7. Sprocket: ½ in square key; AISI 1020 CD; $L = 1.00$ in
   Wormgear: ⅜ in square key; AISI 1020 CD; $L = 1.75$ in
13. $T = 2\,725$ lb · in
15. $T = 26\,416$ lb · in
19. Data from Problem 16: $T = 313$ lb · in per inch of hub length
    Data from Problem 18: $T = 4\,300$ lb · in per inch of hub length

CHAPTER 11 Spur Gears

1. $N = 44$; $P_d = 12$
   a. $D = 3.667$ in
   b. $P_c = 0.261\,8$ in
   c. $m = 2.117$ mm
   d. $m = 2.00$ mm
   e. $a = 0.083\,3$ in
   f. $b = 0.104\,2$ in
   g. $c = 0.020\,8$ in
   h. $h_t = 0.187\,5$ in
   i. $h_k = 0.166\,7$ in
   j. $t = 0.131$ in
   k. $D_o = 3.833$ in
3. $N = 45$; $P_d = 2$
   a. $D = 22.500$ in
   b. $P_c = 1.571$ in
   c. $m = 12.70$ mm
   d. $m = 12.0$ mm
   e. $a = 0.500\,0$ in
   f. $b = 0.625\,0$ in
   g. $c = 0.125\,0$ in
   h. $h_t = 1.125\,0$ in
   i. $h_k = 1.000\,0$ in
   j. $t = 0.785\,4$ in
   k. $D_o = 23.500$ in
5. $N = 22$; $P_d = 1.75$
   a. $D = 12.571$ in
   b. $P_c = 1.795$ in
   c. $m = 14.514$ mm
   d. $m = 16.0$ mm
   e. $a = 0.571\,4$ in
   f. $b = 0.714\,3$ in
   g. $c = 0.142\,9$ in
   h. $h_t = 1.285\,7$ in

i. $h_k = 1.1429$ in j. $t = 0.8976$ in
k. $D_o = 13.714$ in

7. $N = 180$; $P_d = 80$
a. $D = 2.2500$ in b. $P_c = 0.0393$ in
c. $m = 0.318$ mm d. $m = 0.30$ mm
e. $a = 0.0125$ in f. $b = 0.0170$ in
g. $c = 0.0045$ in h. $h_t = 0.0295$ in
i. $h_k = 0.0250$ in j. $t = 0.0196$ in
k. $D_o = 2.2750$ in

9. $N = 28$; $P_d = 20$
a. $D = 1.4000$ in b. $P_c = 0.1571$ in
c. $m = 1.270$ mm d. $m = 1.25$ mm
e. $a = 0.0500$ in f. $b = 0.0620$ in
g. $c = 0.0120$ in h. $h_t = 0.1120$ in
i. $h_k = 0.1000$ in j. $t = 0.0785$ in
k. $D_o = 1.5000$ in

11. $N = 45$; $m = 1.25$
a. $D = 56.250$ mm b. $P_c = 3.927$ mm
c. $P_d = 20.3$ d. $P_d = 20$
e. $a = 1.25$ mm f. $b = 1.563$ mm
g. $c = 0.313$ mm h. $h_t = 2.813$ mm
i. $h_k = 2.500$ mm j. $t = 1.963$ mm
k. $D_o = 58.750$ mm

13. $N = 22$; $m = 20$
a. $D = 440.00$ mm b. $P_c = 62.83$ mm
c. $P_d = 1.270$ d. $P_d = 1.25$
e. $a = 20.0$ mm f. $b = 25.00$ mm
g. $c = 5.000$ mm h. $h_t = 45.00$ mm
i. $h_k = 40.00$ mm j. $t = 31.42$ mm
k. $D_o = 480.00$ mm

15. $N = 180$; $m = 0.4$
a. $D = 72.00$ mm b. $P_c = 1.26$ mm
c. $P_d = 63.5$ d. $P_d = 64$
e. $a = 0.40$ mm f. $b = 0.500$ mm
g. $c = 0.100$ mm h. $h_t = 0.90$ mm
i. $h_k = 0.80$ mm j. $t = 0.628$ mm
k. $D_o = 72.80$ mm

17. $N = 28$; $m = 0.8$
a. $D = 22.40$ mm b. $P_c = 2.51$ mm
c. $P_d = 31.75$ d. $P_d = 32$
e. $a = 0.80$ mm f. $b = 1.000$ mm
g. $c = 0.200$ mm h. $h_t = 1.800$ mm
i. $h_k = 1.60$ mm j. $t = 1.257$ mm
k. $D_o = 24.00$ mm

19. Problem 6: $P_d = 64$; backlash = 0.0007 to 0.0015 in
Problem 9: $P_d = 20$; backlash = 0.001 to 0.002 in

21. a. $C = 14.000$ in b. $VR = 4.600$
c. $n_G = 48.9$ rpm d. $v_t = 294.5$ ft/min

23. a. $C = 2.266$ in b. $VR = 6.25$
c. $n_G = 552$ rpm d. $v_t = 565$ ft/min

25. a. $C = 90.00$ mm b. $VR = 3.091$
c. $n_G = 566$ rpm d. $v_t = 4.03$ m/s

27. a. $C = 162.0$ mm b. $VR = 1.250$
c. $n_G = 120$ rpm d. $v_t = 1.13$ m/s

For problems 29 through 32, the following are errors in the given statements:

29. The pinion and the gear cannot have different pitches.
30. The actual center distance should be 8.333 in.
31. There are too few teeth in the pinion; interference is to be expected.
32. The actual center distance should be 2.156 in. Apparently, the outside diameters were used instead of the pitch diameters to compute C.
33. $Y = 8.45$ in; $X = 10.70$ in
35. $Y = 44.00$ mm; $X = 58.40$ mm
37. Output speed = 111 rpm CCW
39. Output speed = 144 rpm CW
49. Power capacity = 19.3 hp

CHAPTER 12 Helical Gears, Bevel Gears, and Wormgearing

Helical Gears

1. $P_c = 0.3927$ in $P_{cn} = 0.3401$ in
$P_{dn} = 9.238$ $P_x = 0.680$ in
$d = 5.500$ in $\phi_n = 12.62°$
$F/P_x = 2.94$ axial pitches in the face width
2. $P_d = 8.485$ $P_c = 0.370$ in
$P_{cn} = 0.2618$ in $P_x = 0.370$ in
$\phi_t = 27.2°$ $d = 5.657$ in
$F/P_x = 4.05$ axial pitches in the face width
3. $W_t = 91.6$ lb $W_x = 52.9$ lb
$W_r = 23.7$ lb
4. $W_t = 31.8$ lb $W_x = 31.8$ lb
$W_r = 16.4$ lb
9. $\sigma_t = 2827$ psi
11. $\sigma_t = 20600$ psi
13. $K = 7.71$
15. $K = 48.7$
25. Power capacity = 68.4 hp
$W_x = 647$ lb $W_r = 907$ lb

Bevel Gears

26. $\gamma = 18.43°$ $\Gamma = 71.57°$
$h_t = 0.367$ in $h_k = 0.333$ in
$c = 0.0333$ in $a_G = 0.0985$ in
$a_P = 0.2348$ in $D_o = 7.562$ in
$d_o = 2.9455$ in $OCD = 3.953$ in
27. $W_{tP} = W_{tG} = 599$ lb
$W_{rP} = W_{xG} = 207$ lb
$W_{xP} = W_{rG} = 69$ lb
28. Bending: $\sigma_{tP} = 18300$ psi
Hertz: $\sigma_h = 214500$ psi

Wormgearing

38. $L = 0.3142$ in $\lambda = 4.57°$
$a = 0.100$ in $b = 0.1157$ in

$D_{oW} = 1.450$ in $\quad D_{RW} = 1.0186$ in
$D_G = 4.000$ in $\quad C = 2.625$ in
$VR = 40$

39. $W_{tG} = W_{xW} = 462$ lb $\quad W_{xG} = 54.8$ lb
$W_{rG} = W_{rW} = 120$ lb
Efficiency = 67% $\quad n_W = 1\,200$ rpm
$P_i = 0.65$ hp $\quad \sigma_{tG} = 25\,200$ psi
$W_w = 216$ lb $\quad W_d = 474$ lb

40. Rated output torque based on wear load:
a. $T = 60.5$ lb · in $\quad$ b. $T = 235$ lb · in
c. $T = 884$ lb · in

41.

	a.	b.	c.
Efficiency	69.6%	81.8%	89.3%
Lead angle	4.76°	9.46°	18.44°

45.

	a.	b.
Efficiency	66.4%	75.9%
Forces:		
$W_{tG} = W_{xW}$	480 lb	400 lb
$W_{xG} = W_{tW}$	60 lb	84 lb
$W_{rG} = W_{rW}$	125 lb	106 lb
Power:		
Output	0.38 hp	0.38 hp
Input	0.57 hp	0.50 hp

CHAPTER 13 Belts and Chains

V-belts

1. 3V Belt, 75 inches long
2. $C = 22.00$ in
3. $\theta_1 = 157°$; $\theta_2 = 203°$
10. $v_b = 2405$ ft/min
13. $P = 6.05$ hp

Roller Chain

25. No. 80 chain
28. Design power rating = 12.0 hp; Type II lubrication (bath)
29. Design power rating = 30.0 hp
34. $L = 96$ in, 128 pitches
35. $C = 35.57$ in

CHAPTER 14 Plain Surface Bearings

All problems in this chapter are design problems for which no unique solutions exist.

CHAPTER 15 Rolling Contact Bearings

1. Life = 2.76 million revolutions
2. $C = 12\,745$ lb

CHAPTER 16 Motion Control

1. $T = 495$ lb · in
3. $T = 41$ lb · in
5. Data from Problem 1: $T = 180$ lb · in
   Data from Problem 3: $T = 27.4$ lb · in
7. Clutch: $T = 2\,122$ N · m
   Brake: $T = 531$ N · m
8. $T = 143$ lb · ft
9. $T = 63.6$ lb · ft
11. $T = 223.6$ lb · ft
15. $F_a = 109$ lb
17. $W = 138$ lb
18. $b > 16.0$ in

CHAPTER 17 Electric Motors

13. 480V, 3-Phase because the current would be lower and the motor size smaller
16. $n_s = 1\,800$ rpm in the United States
    $n_s = 1\,500$ rpm in France
17. 2-Pole motor; $n = 3\,600$ rpm at zero load (approximate)
18. $n_s = 12\,000$ rpm
19. 1 725 rpm and 1 140 rpm
20. Variable frequency control
34. a. Single phase, split phase AC motor
    b. $T = 41.1$ lb · in $\quad$ c. $T = 61.7$ lb · in
    d. $T = 144$ lb · in
35. b. $T = 12.6$ N · m $\quad$ c. $T = 18.9$ N · m
    d. $T = 44.1$ N · m
39. Full load speed = synchronous speed = 720 rpm
47. Use a NEMA Type K SCR control to convert 115 VAC to 90 VDC; use a 90 VDC motor.
51. Speed theoretically increases to infinity
52. $T = 20.5$ N · m
54. NEMA 2 starter
55. NEMA 1 starter

CHAPTER 18 Power Screws

5. 2½–3 Acme thread
6. $L > 1.23$ in
7. $T = 6\,974$ lb · in
8. $T = 3\,712$ lb · in
11. Lead angle = 4.72°; self-locking
12. Efficiency = 35%
13. $n = 180$ rpm; $P = 0.866$ hp
17. 24.7 years
21. Grade 2 bolts: 5⁄16–18; $T = 70.3$ lb · in
22. $F = 1\,190$ lb
23. $F = 4.23$ kN
24. Nearest metric thread is M24 × 2. Metric thread is 1.8 mm larger (8% larger).
25. Closest standard thread is M5 × 0.8. (#10–32 is also close.)
26. 3.61 kN
27. a. 1 177 lb $\quad$ c. 2 385 lb
    e. 2 862 lb $\quad$ g. 1 081 lb
    i. 2 067 lb $\quad$ k. 143 lb

CHAPTER 19 Bolted Connections, Welded Joints, and Machine Frames

17.

	Material	Diameter (in)	Weight (lb)
a.	1020 HR steel	0.638	0.0906
c.	Aluminum 2014–T6	0.451	0.0160
e.	Ti–6Al–4V (Aged)	0.285	0.0102

Index

Emmy
the Exaggerating
Elephant
Fenton
the Fearful Frog
Gertie
the Grungy Goat
Herb
the Happy
Hamster
the Impatient
Iguana
Ollie
the Obedient
Ostrich
Perry
the Polite
Porcupine
Queenie
the Quiet Quail
Rupert
the Resourceful
Rhinoceros
Wendy
the Wise
Woodchuck
Xavier
the X-ploring
Xenops
Yori
the Yucky Yak
Ziggy
the Zippy Zebra

NOTE TO PARENTS

Sylvester and the Sand Castle
A story about sportsmanship and fairness

In this story, Sylvester the Stubborn Squirrel misses out on a lot of fun because he insists on having his way. By setting a good example, his AlphaPet friends help him realize the advantages of being adaptable and open-minded.

In addition to enjoying this story with your child, you can use it to teach a gentle lesson about the importance of being a good sport and cooperating with others.

You can also use this story to introduce the letter **S**. As you read about Sylvester the Stubborn Squirrel, ask your child to listen for all the words that start with **S** and point to the objects that begin with **S**. When you've finished reading the story, your child will enjoy doing the activity at the end of the book.

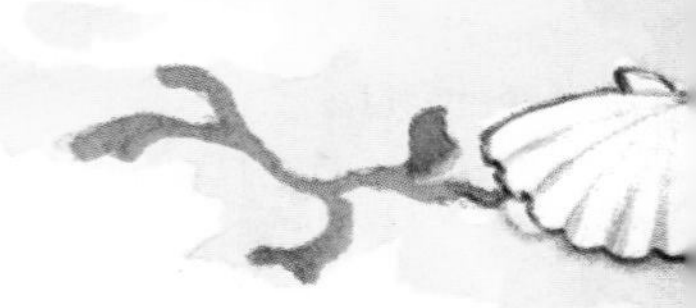

The AlphaPets™ characters were conceived and created by Ruth Lerner Perle.
Characters interpreted and designed by Deborah Colvin Borgo.
Cover/book design and production by Norton & Company.
Logo design by Deborah Colvin Borgo and Nancy S. Norton.
Grolier Books is a Division of Grolier Enterprises, Inc. Printed and Manufactured in the United States of America

Sylvester and the Sand Castle

RUTH LERNER PERLE

Illustrated by Richard Max Kolding

GROLIER
BOOKS

One hot summer Sunday, some of the AlphaPets decided to go to the beach. Sylvester the Stubborn Squirrel was standing on the corner waiting for Vinnie the Vocal Vulture to pick him up.

Before long, Sylvester saw Vinnie driving down the street. Perry the Polite Porcupine, Katy the Kind Koala, Lizzy the Lazy Lamb, and Fenton the Fearful Frog were already in the car with him.

"Good morning, good morning, Sylvester!" Vinnie called. "Hop in the back! There's plenty of room in the middle."

"The middle!" cried Sylvester. "I'm not getting in any middle seat. If I have to sit there, I'm not going."

"Come on, Sylvester," Perry said. "Try to be fair. *Nobody* likes to sit in the middle. That's why we always take turns, and today it's your turn."

"I don't care," Sylvester said, and he started to leave.

"You can take my seat, Sylvester," Katy said. "Get in, and let's have a good time."

So Sylvester took Katy's seat and they drove off.

When they arrived at the beach, Perry asked, "Shall we sit near the water or on the sand dunes?"

"Makes no difference to me," Fenton said. "As long as I'm safe from the sand . . . and sun . . . and sea."

"It's the sandy sand dunes for me!" said Vinnie. "Let's sit in the dunes!"

"Good idea," agreed Lizzy. "It's more relaxing in the sand dunes."

"No sand dunes for me. Oh, no!" Sylvester declared. "When I'm at the beach I like to be near the water and I won't sit anywhere else."

"Well, all right, if it's that important to you, my friend," Vinnie said. "I'm sure we'll all enjoy the waterside just as well." Everyone agreed and they walked down to the waterfront.

The AlphaPets spread out their blankets near the edge of the water. After everybody sat down, Fenton put on his sun hat, socks, and sunglasses. Then he gave everyone a little tube of sunscreen.

"Too much sun is dangerous," he said. "This sunscreen will protect our skin from getting burned."

"Not me! I don't need that sunscreen!" Sylvester yelled.

"But your nose is getting red already," Katy said.

"How can you say that? I *never* burn. Never!" Sylvester insisted. "No sunscreen for me, definitely not."

Katy opened her picnic basket and took out a bag of assorted sandwiches. She offered some to her friends.

"I get to pick a sandwich first!" Sylvester said, pushing his way ahead of Fenton and Perry.

"Please, Sylvester! Wait your turn," Perry said.

Sylvester crossed his arms in front of his chest. "If I don't get to choose a sandwich first, I won't eat at all!" he declared.

So Perry very politely let Sylvester choose first.

Vinnie opened his lunch box and offered everybody slices of ripe avocado.

"What's an avocado?" Sylvester wanted to know. "I never saw one of those before."

"Try it, it's delicious!" Lizzy said.

"Oh, no! Not me! That stuff looks slimy and it's green. I never eat anything that's green," Sylvester announced.

"Why don't you just try a bite?" Perry suggested. "You might like it."

"Uh uh, not me." Sylvester insisted. He unwrapped his sandwich and started to eat.

When they were all finished eating, Vinnie said, "Let's build a sand castle!"

"Good idea!" Fenton agreed. "But first let's put on a little more sunscreen. The sun is really strong today."

So, everyone put on more sunscreen—everyone except Sylvester.

"Your cheeks and forehead are turning bright red," said Fenton, offering him the sunscreen. But Sylvester still refused it. "I never burn," he insisted. "Never."

"Okay," Vinnie said. "Now let's start building our castle. Perry and I will dig a moat and pile up sand for towers. Katy, you and Lizzy can collect seashells and stones for decorations. Sylvester and Fenton can bring pails of water."

"I don't want to bring the water. I want to dig the moat," Sylvester declared.

By now Vinnie was getting tired of Sylvester's stubborn ways.

"Well, you can't always have it your way," Vinnie said. "Sometimes you have to take turns and be considerate of what others want to do."

Sylvester stamped his feet and kicked at the sand. "I want to dig the moat! I want to dig the moat!" he cried.

"Well, I'm afraid you're not going to get your way this time!" Perry said.

"Then I'm not playing with you!" Sylvester screamed. He threw down his pail and stomped off by himself.

The AlphaPets decided to leave Sylvester alone and started to work together happily.

They dug deep holes and trenches . . .

collected shells, seaweed, sticks, and stones . . .

carried buckets
of water . . .

and piled up
a mountain
of sand.

Sylvester could hear the AlphaPets joking and having a good time. He wanted to join them, but instead he dug his heels into the sand and stared at the sea.

"Why won't they let me dig the moat?" he muttered under his breath. "Why can't I have things my way?"

Sylvester was feeling sad and angry. He was also getting hot and thirsty, and his shoulders were starting to burn.

Before long, the sand castle was almost finished. Sylvester watched as Vinnie planted a flag at the top of the castle.

"That looks like fun. I wish I could play," he thought to himself. But he just sat and stared.

"What this castle needs is a big wall all around it," Perry said.

"That's a great idea," Lizzy agreed. "But I think it's too much work for just the five of us. Six people would be just right." Everyone stopped and looked at Sylvester.

Katy called over to Sylvester. "How about helping us?"

At first Sylvester pretended not to hear them.

"Sylvester! Come and help us!" they all called.

"Well, maybe," Sylvester said.

But as he started to get up, he screamed. "*YEOW! OUCH!* My shoulders! My back! My arms!"

"Oh, my word! Looks like you really have a bad sunburn!" Vinnie said, shaking his head.

Fenton ran for his first aid kit and the AlphaPets put ointment on Sylvester's burn. Then Fenton reached into his bag and gave Sylvester an extra hat and shirt that he had in his bag—just in case of an emergency.

The cooling ointment felt good. "Er . . . ah . . . thank you. I guess I should have used the sunscreen after all," Sylvester said. "But now I feel much better, so let's build that wall!"

Everybody worked together, digging, carrying water, and patting the sand.

"Hmmm . . . this is really fun," Sylvester thought. "I should have joined in sooner."

Soon the wall was finished. Everybody admired the beautiful job they had done.

Vinnie poured some sarsaparilla sodas for all the builders. Then he got up to make a speech.

"Ahem, ahem," he started. "I would like to dedicate this most magnificent medieval structure to friendship and cooperation. Without those two ingredients, this castle would not have come to be. No indeed! Not at all. If I said it once, I said it a hundred . . . no, a thousand times. Yes, a thousand times: The secret of getting a big job done is team work. Yes, my friends, team work! Working together to achieve a common goal!"

Vinnie put his arm around Sylvester and offered him a cup of soda. "Here's some nice cool sarsaparilla," he said with a smile. "Let's all drink to friendship and cooperation."

Sylvester started to say that he would never try a drink that has so many sss's in it. But instead, he took the cup and had a little sip.

"Hmmm," Sylvester said with a happy smile. "This stuff doesn't taste too bad. Not bad at all."

Let’s learn these words together.

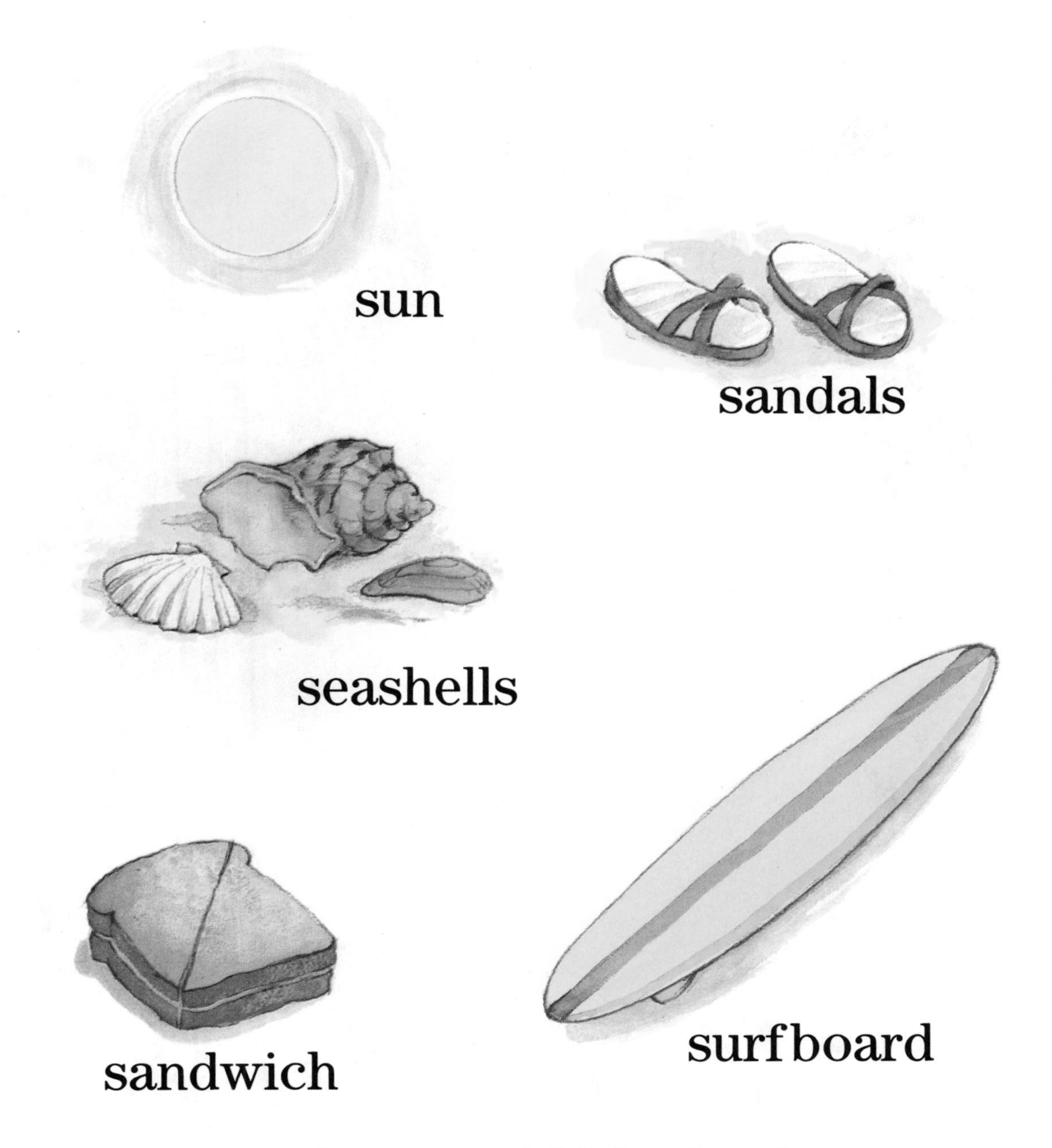

Look back at the pages of this book and try to find these and other words that start with S.

Aa Bb

Gg Hh

Mm Nn Oo Pp

Uu Vv Ww